ANNUAL REVIEW OF GENETICS

ANNUAL REVIEW OF GENETICS

VOLUME 36, 2002

ALLAN CAMPBELL, *Editor*
Stanford University, Stanford

WYATT W. ANDERSON, *Associate Editor*
University of Georgia, Athens

ELIZABETH W. JONES, *Associate Editor*
Carnegie Mellon University, Pittsburgh

www.annualreviews.org   science@annualreviews.org   650-493-4400

ANNUAL REVIEWS
4139 El Camino Way • P.O. Box 10139 • Palo Alto, California 94303-0139

ANNUAL REVIEWS
Palo Alto, California, USA

International Standard Serial Number: 0066-4197
International Standard Book Number: 0-8243-1236-8
Library of Congress Catalog Card Number: 63-8847

TYPESET BY TECHBOOKS, FAIRFAX, VA
PRINTED AND BOUND BY MALLOY INCORPORATED, ANN ARBOR, MI

 *Annual Review of Genetics*
Volume 36, 2002

CONTENTS

ERRATA
   An online log of corrections to *Annual Review of Genetics* chapters
   may be found at http://genet.annualreviews.org/errata.shtml

RELATED ARTICLES

Control and Developmental Timing by MicroRNAs and Their Targets,
Amy E. Pasquinelli and Gary Ruvkun

Remembrance of Things Past: Chromatin Remodeling in Plant Development,
Justin Goodrich and Susan Tweedie

From the ***Annual Review of Ecology and Systematics***, Volume 33 (2002)

Troubleshooting Molecular Phylogenetic Analyses, Michael J. Sanderson
and H. Bradley Shaffer

Reproductive Protein Evolution, Willie J. Swanson and Victor D. Vacquier

Neopolyploidy in Flowering Plants, Douglas W. Schemske and Justin Ramsey

*Estimating Divergence Times from Molecular Data on Phylogenetic
and Population Genetic Timescales,* Brian S. Arbogast, Scott V. Edwards,
John Wakeley, Peter Beerli, and Joseph B. Slowinski

From the ***Annual Review of Microbiology***, Volume 56 (2002)

Bacteriocins: Evolution, Ecology, and Application, Margaret A. Riley
and John E. Wertz

Evolution of Drug Resistance in Candida albicans, Leah E. Cowen,
James B. Anderson, and Linda M. Kohn

Inteins: Structure, Function, and Evolution, J. Peter Gogarten,
Alireza G. Senejani, Olga Zhaxybayeva, Lorraine Olendzenski,
and Elena Hilario

*The Molecular Biology of West Nile Virus: A New Invader of the
Western Hemisphere,* Margo A. Brinton

What Are Bacterial Species? Frederick M. Cohan

Genome Remodeling in Ciliated Protozoa, Carolyn L. Jahn and
Lawrence A. Klobutcher

Bacterial Chromosome Segregation, Geoffrey C. Draper and James W. Gober

Impact of Genomic Technologies on Studies of Bacterial Gene Expression,
Virgil Rhodius, Tina K. Van Dyk, Carol Gross, and Robert A. LaRossa

Control of Chromosome Replication in Caulobacter crescentus,
Gregory T. Marczynski and Lucy Shapiro

Prions as Protein-Based Genetic Elements, Susan M. Uptain and
Susan Lindquist

From the ***Annual Review of Phytopathology***, Volume 40 (2002)

Viral Sequences Integrated into Plant Genomes, Glyn Harper, Roger Hull,
Ben Lockhart, and Neil Olszewski

Richard Lewontin

Annu. Rev. Genet. 2002. 36:1–18
doi: 10.1146/annurev.genet.36.052902.102704

DIRECTIONS IN EVOLUTIONARY BIOLOGY

R. C. Lewontin

*Museum of Comparative Zoology, Harvard University, Cambridge,
Massachusetts 01238; e-mail: lewontin@oeb.harvard.edu*

Key Words phylogeny, natural selection, genomics, ecological genetics, novelties

■ **Abstract** In order to understand both the past and future directions of research in evolutionary biology we need to begin by understanding in what way these programs of research differ from the model of most scientific work. The study of evolutionary processes and, in particular, the genetics of the evolutionary process must confront special difficulties in both the conceptual and the methodological aspects of research. On the conceptual side, unlike for molecular, cellular, and developmental biology, there is no basic mechanism that evolutionists are attempting to elucidate. There is no single cause of the evolutionary change in the properties of members of a species. Natural selection may be involved but so are random events, patterns of migration and interbreeding, mutational events, and horizontal transfer of genes across species boundaries. The change in each character of each species is a consequence of a particular mixture of these causal pathways.

CONTENTS

THE GENERAL SHAPE OF EVOLUTIONARY RESEARCH

In order to understand both the past and future directions of research in evolutionary biology we need to begin by understanding in what way these programs of research differ from the model of most scientific work. The study of evolutionary processes and, in particular, the genetics of the evolutionary process must confront special difficulties in both the conceptual and the methodological aspects of research. On the conceptual side, unlike for molecular, cellular, and developmental biology, there is no basic mechanism that evolutionists are attempting to elucidate. There is no single cause of the evolutionary change in the properties of members of a species. Natural selection may be involved but so are random events, patterns of migration and interbreeding, mutational events, and horizontal transfer of genes across species boundaries. The change in each character of each species is a consequence of a particular mixture of these causal pathways. One of the most detailed and convincing studies of selection in natural populations is the 30-year project of P.R. and B.R. Grant on two species of Darwin's finches, *Geospiza fortis* and *G. scandens*. While they demonstrated changes between generations in bill length, bill shape, and body size that were predictable from their measurements of reproductive fitness, they report that "Natural selection . . . varied from unidirectional to oscillating, episodic to gradual." Moreover, changes over the entire 30-year study were unpredictable from fitness differences and were the consequence of repeated unforeseen hybridizations between the species. Their conclusion after this massive study is that "Continuous, long-term studies are needed to detect and interpret rare but important events and non-uniform evolutionary change" (8). Nor is this only a property of intraspecific change in characters. Different kinds of genetic and environmental events lead to speciation in different cases, and there is no single cause of species' extinction. Not even basic cellular processes are exempt from this contingency. That the genetic code is not universal, but differs between nuclear and mitochondrial genomes and among different mitochondrial genomes is a consequence of multiple independent evolutionary changes in mitochondria from an intracellular symbiont. The very relationship between DNA sequence and protein that is described by the standard dogma for most organisms differs in some protists that edit the transcribed RNA by inserting large numbers of U's into the immature message to make a final set of translatable codons. Of course the DNA code and the machinery of transcription and translation are *nearly* universal so the program of cellular biology can generally be carried out by a judicious choice of an experimental system that will represent the great mass of organisms. But for evolutionary biology both the similarities between organisms that are a consequence of their common descent and the differences between them that have occurred in their evolution are the objects of inquiry. For molecular, cellular, and developmental biology differences between cases are an annoying complication. For the evolutionist, differences are what evolution is all about and are expected, whereas persistent similarities across organisms need to be explained.

Second, evolutionary biology is particularly plagued by the discrepancy between the problematic posed by its agreed-upon subject matter and methods that are available to answer the questions posed. This has not been simply a matter of waiting until an appropriate methodology is developed, although there are indeed such cases, but for many problems the difficulty lies in the relationship between processes that have occurred in the past and the evidence about those processes that can only be recovered from presently existing organisms, with a little help from fossils. In the first place, the fossil record is so sparse that no quasi-continuous process of change and diversification can be followed as a detailed dynamic process. Second, the fossil record provides evidence primarily of morphological change, and nothing can be reconstructed about underlying genetic change and only rarely about physiological and biochemical changes. Nor can living species be used, except under exceptional circumstances, to follow evolutionary changes now occurring. Processes of change and divergence of species are usually extremely slow compared not only to the lifetime of an investigator but of science as an institution. This low speed is, in turn, a consequence of the weakness of most evolutionary forces most of the time. It is now generally agreed that selection differences in nature for most of the alleles segregating at most loci are likely to be of the order of 10^{-3} or less (although the evidence for the intensity of selection, as is discussed below, is indirect). Random changes in allelic composition consequent on extreme fluctuations in population size between generations are not great since, except for very rare alleles, even a population size of 100 is sufficient to keep allelic frequencies within a few percent of their values in the previous generations. Moreover, a small amount of migration between populations, as little as one or two individuals per generation, is sufficient to prevent random divergence among populations. The consequence of the weakness of selective and random forces is that the processes of evolution in living species cannot, except very rarely, be followed as a *dynamic* process in time. Instead, the evolutionary biologist must depend on *static* data, observations of patterns of variation within and between species, to infer the dynamic processes that could not be directly observed.

To infer dynamic processes from static differences within and between species requires a combination of antecedent theoretical developments. In order to estimate the rates and directions of heritable change between species over time, as well as to identify conserved features, it must be possible to reconstruct the chains of common ancestry that connect living species. That is, evolutionary explanation depends upon systematics. The problem is that to reconstruct patterns of ancestry we have only the observed similarities and differences between organisms as data, and we must impose on the data a theory of how character states succeed each other in evolution, a theory that we are trying to construct and verify in the first place by using the inferred relationships. This apparent circularity is not fatal, however, although it leads to ambiguities. The most widely used modern systematic practice depends upon the assumption that a change from character state A in one species to character state B in a descendant species occurs once and once only in the evolutionary process and that this process is irreversible so that B never returns to A. In this scheme, there are no independently derived parallel evolutionary changes, nor convergences from a variety of states to a single one. So, if two species share a character state different from other species, it is because they are more closely related to each other through a recent common ancestor than they are to other species. Given this principle of parsimony, a scheme of common ancestry for all the species is derived that uses all the characters that have been observed. This scheme always contains a number of internal contradictions, so-called homoplasies, inferred character changes that contradict the basic assumptions, but the scheme with the fewest such contradictions is taken as the correct one. Sometimes the choice is easy because one scheme has many fewer homoplasies than its closest competitor, but not infrequently there a several equally or nearly equally parsimonious phylogenies. Given a very well supported and unproblematic phylogeny, the evolutionist can then use the conservations and changes in the characters to study evolutionary phenomena. The homoplasies become particularly revealing because they represent states that have been repeatedly independently derived and so provide evidence of selective or mutational constraints or of hybridization or horizontal genetic transfer between species. For example, Wells (25) studied the evolution of the glycerol-3-phosphate dehydrogenase locus in the genus *Drosophila* by sequencing the gene in a number of species. The phylogeny of *Drosophila* was well supported by the standard parsimony analysis so that he was able to infer the direction of amino acid change for a number of independent evolutionary trajectories. What he discovered was that for four amino acid positions there had been repeated transitions back and forth among several alternatives, providing strong evidence for the existence of a small subspace of amino acid replacements that were presumably roughly functionally equivalent and of high fitness as contrasted with other possible substitutions at the same positions that apparently were selected against.

Within the general field of evolutionary studies, there is a divergence of view about the purpose of research in systematics. For the systematist, the revelation of the true ancestral relationships among organisms is a goal in itself, because a valid part of evolutionary inquiry is the description of what has actually happened in the history of life. The systematist recognizes that inferences about phylogenies are likely to be inexact to some degree and, moreover, that there is no possible independent verification of the details of a phylogeny outside of the methods of systematics itself. So, if there are a number of phylogenies that are more or less equally supported, the systematist accepts the imperfection of knowledge as the closest one can come to the truth. Indeed, this uncertainty is central to one of the methods of phylogenetic reconstruction, the use of maximum likelihood statistics, whose aim is specifically to *estimate* the correct phylogeny, with all the uncertainty that the statistical notion of estimation involves. While not denying that a revelation of the history of species has an independent interest, evolutionary geneticists and those concerned with the dynamical process of evolutionary change see this essentially historical study as only of instrumental value. For the biologist interested in the dynamical processes of evolution, the reconstruction of the correct phyletic relationships among organisms, the bare description of who is related to whom and how closely, is of no interest in itself but only as a prerequisite to making inferences about processes leading to differentiation and stasis. But the uncertainty about the correctness of a phylogeny may have severe consequences when the "correct" phylogeny is needed to make inferences about dynamic processes. If the phylogeny of *Drosophila* species were not so well supported, it would not have been possible to demonstrate the independent substitutions back and forth between a few amino acids in the evolution of the glycerol-3-phosphate dehydrogenase protein. This apparent shifting back and forth might have been taken as strong evidence that the phylogeny was in error. There is a great deal at stake for the reconstruction of evolutionary processes in having a correct phylogeny.

The other methodological apparatus for reconstructing past events from present organisms is meant to estimate the actual magnitude of forces of natural selection and random events that have led to genetic differences between closely related species or to genetic differences between populations of the same species. It involves a mixture of the mathematical theory of population genetics and statistical theory. It begins with a complex mathematical apparatus that is designed to carry the state of a population forward in time from some initial condition. It predicts rates of genetic change from an initial state and possible equilibria that will result from selection, mutation, migration, and recombination. This must be a stochastic, rather than a deterministic, theory to account for random changes that result from genetic drift in finite populations, so that the form of the prediction is not a unique state at future time, but a probability distribution of states. This is the stochastic theory of population genetics originally produced by Wright (26) and further elaborated by Kimura & Ohta (12).

Second, a probabilistic theory is needed that can reverse the deductions of the first theory and infer backwards in time from a particular observed state at present what the most likely dynamical forces were that have led to the actual present situation. But a difficulty arises here. A dynamical theory that predicts the present state generally requires that we know not only the nature and magnitude of the forces that have operated, but also the initial condition and how long the process has been in operation. That means that if we wish to use a backward inference from the present state to estimate the forces that have operated, we would need to know the initial condition and how long the process has been going on as well as assuming that the forces have not changed during the process. But this is precisely what we cannot know. Either we assume that we know the forces, in which case we can make probability statements about the initial conditions, or else we assume that we know the initial conditions, in which case we can make estimates of the forces that have led to the present. We cannot do both. There is one solution to this dilemma. If the evolutionary process has gone on for a sufficiently long time with no changes in the forces, then there is an equilibrium probability distribution of the present states, the so-called steady-state distribution, that is reached irrespective of the original state of the population. So, if we can observe many genetic variations all of which can be assumed to be the result of the same forces, then the distribution of those variations can be used to estimate those forces. The most important development in recent evolutionary theory has been the creation and application of *coalescent theory*, which allows inferences about forces and times since common ancestry from observations of current genetical variation within and between populations on the assumption that the variation is at the stochastic steady state (22).

Third, we require a set of statistical procedures that can test the agreement between the static observations and various hypotheses about the strength of the different forces, especially whether the observations indicate the operation of natural selection as opposed to purely random drift events [for example, (3, 7, 11, 17)]. These mathematical and statistical tools are now the standard methodology for detecting and estimating the selective forces involved in molecular evolution of DNA and protein sequences. As is immediately obvious, this methodology involves many assumptions that we will further examine in the discussion of some specific directions in current research.

The integral importance of formal, mathematical, and statistical structures in the reconstruction of evolutionary relationships and in the testing of hypotheses about evolutionary forces marks out much of the research program of evolutionary biology from most other biological inquiry. Tools and methods are, of course, integral to all research, and a new methodology or machine may take over and remake the problematic of a field, as the mechanization of DNA sequencing has done in molecular genetics. Despite their dramatic effect on research, the development of the methods themselves is not generally a major preoccupation of research in the field. The invention of the PCR machine and automated DNA sequencer had a revolutionizing effect on genetic research, but the engineering of those devices was not a major preoccupation of genetical research. In the case of evolutionary biology and genetics, however, the development of formal conceptual tools has consistently been an important field of research throughout the history of modern evolutionary studies. Only recently, with the use of statistical methods for the localization of quantitative trait loci, has research on mathematical and statistical methods come to play a major role in a biological field outside of evolution.

Both the detailed study of particular natural historical cases of observed dynamical changes and the use of static data to infer unobservable dynamical forces have dealt with a small number of specific examples of general phenomena: How are the changes in bill and body size in Darwin's finches to be explained by the observed reproductive behavior of the finches? Is there evidence that amino acid replacements in alcohol dehydrogenase that occurred in the evolutionary divergence of two species of Drosophila were the result of natural selection (11)?

There are two other directions of evolutionary research that are concerned with other levels of generality. One uses high volume methods of data acquisition to look for some generalities and regularities in the characteristics of organisms, usually their genomes, that are relevant to evolutionary processes. The possibility of evolution is constrained by the amount of genetic variation within species. Evolutionary and population geneticists, conscious of the central position that genetic variation plays in the evolutionary process, struggled for 35 years to assess the amount of genetic variation in natural populations, but the methods available to them were inadequate for the gene-by-gene characterization of genetic variation. Then, with the advent of protein gel electrophoresis in the 1960s and routine DNA sequencing in the 1980s, the description of protein and DNA variation in large numbers of different species from all branches of the living world became possible. The generality that emerged is that a very large amount of protein and DNA variation exists in the genomes of nearly all species. Moreover, the differences in variation among species, among genes, and across different functional regions of DNA have revealed general features of evolution at the gene level. For example, it is now clear that the unequal use of alternative codons for amino acids is

universal and also shows both phylogenetic patterns and amino acid and pro
class specific biases that transcend phylogenetic boundaries. It is also clear th
with the exception of recently evolved pseudogenes, there is no class of DNA
including introns, synonymous codon positions, and downstream flanking DNA
sequences that are not subject to some selective constraint in their variation (20).
Protein gel electrophoresis provides a striking example of how the introduction of
a simple, easily acquired technique can completely dominate and alter the entire
direction of research in a field. Before the mid-1960s experimental evolutionary
genetics was an extremely heterogeneous field of research, encompassing, among
other subjects, studies of the effect of selection on quantitative characters, research
on developmental canalization, on species hybridization, on the sensitivity of fit-
nesses of genotypes to environmental variation, on the evolutionary dynamics of
chromosomal rearrangements, and segregation abnormalities in natural populations. The introduction of protein gel electrophoresis as a tool to investigate the
standing variation within and between species almost totally depauperized evolutionary genetics for 20 years. The immense diversity of research directions was
replaced by a massive program of grinding up every species that lay at hand and
visualizing their proteins by gel electrophoresis. The rest of the program of evolutionary genetic research became marginalized or totally inactive, and it has yet
to recover its diversity. It remains to be seen what generalities will emerge from
the vast amount of DNA sequence that now is being produced by various genome
projects. Although these unquestionably will show some general evolutionary patterns, as did protein gel electrophoresis, it is not clear whether the program of
evolutionary research will be enriched or impoverished.

The other direction of investigation, in many ways the most interesting, is the
discovery and exploration of specific phenomena and relationships that illustrate
the complexity and diversity of evolutionary processes. These include, for example,
the discovery of RNA editing; of the remarkable conservation of the genetic basis
of certain developmental pathways in animals; of the relative ease, in some cases,
with which evolutionary novelties may arise; and in others, of the very constrained
pathways of successive change in mutational space that may allow the passage
from one functional state to another. While none of these are general phenomena of
evolution, their importance is that they are examples of the diversity of evolutionary
processes and products.

SPECIFIC RESEARCH DIRECTIONS

Case Studies in Nature: Ecological Genetics

During the entire history of evolutionary research, there have been research programs that have concentrated on a single polymorphism or observed change in
heritable characters of a species in nature, with the aim of explaining the maintenance of the polymorphism or the rapid evolutionary change that has occurred.
The *locus classicus* of such studies was the attempt to explain the dramatic

increase in the frequency of the melanic form of the peppered moth in Britain during the late nineteenth and early twentieth centuries. The textbook explanation of this evolutionary event is the following story of natural selection involving protection of the moths against bird predators. Light-colored moths resting on tree trunks on which there were patches of grayish lichens were cryptically colored so that bird predators could not see them. The increase in air pollution caused by industrialization resulted in the failure of lichens to grow so that light-colored moths now stood out against the dark lichen-less tree trunks while the dark form of the moth was now cryptically colored and protected. This story was bolstered by field experiments in which noncryptically colored moths were actually observed to be eaten by birds. Unfortunately for the neatness of the story, the actual rate of bird predation appears to be negligible because moths spend rather little time resting on a tree. The field experiments, it is now known, involved tethering the moths to keep them in place. Moreover, the caterpillars of the genetically melanic forms have a higher survival rate although no melanin has yet been formed.

Despite its unsatisfactory result, the moth case has inspired a number of more successful case studies, such as the long-term study of *Geospiza* in the Galapagos in which short-term changes in morphology were successfully explained by differences in reproductive success of individuals of different types, while long-term changes involved occasional hybridizations between species (8). This case, and two other older ones, illustrate the general features of such research and raise the issue of its purpose in the general program of evolutionary studies. Lamotte (14) studied the polymorphism of shell color and banding in a large number of French populations of the snail, *Cepaea nemoralis*. Different local populations differ in the frequency of shells with or without bands. The advantages of this system for a case study are that the presence or absence of bands is determined by a single allelic difference that is easily scored without disturbing the animals, that snails can be captured, marked, and released for migration studies and estimates of population size, that snails have low dispersal rates, that deposits of shells of dead snails leave a record of previous populations, and most important, that such shells can be scored as broken or intact, differentiating those that have been preyed upon by thrushes from those which have not. Lamotte then estimated migration rates, effective population sizes, and searched for a correlation between predation by thrushes and presence of bands in local populations of different vegetational cover as a way of detecting natural selection on banding. He found no such correlation, but this was his only measure of natural selection. He then showed that there was a relation between the physical distance between colonies and the difference in allele frequency between them, a relation that fell off rapidly with distance, as was to be expected from the low measured migration rates of the snails. Finally, he fit the distribution of the allele frequencies among colonies to a stochastic steady-state distribution without natural selection, but with genetic drift and migration. Thus he found negative, although by no means compelling, evidence against selection and positive evidence for the importance of genetic drift and migration. The other illustrative study is that of Christiansen & Frydenberg (4) on an esterase polymorphism in the fish, *Zoarces*. This fish was chosen because it is live-bearing so that offspring broods could be matched to their mothers. By genotyping (using electrophoresis) of mothers and their offspring, the genotype of fathers could also be inferred, and all three pieces of genotypic information could be used to estimate male, female and pair-mating probabilities, fertilities in both sexes, and probabilities of survivorship of genotypes both within broods and between generations in both sexes. This unusually complete set of reproductive information was then used to estimate selection. Despite a sample size of over 1100 females and their broods, no selection was detected.

The examples of the finches, snails, and fish have in common that the model systems were carefully chosen to have biological characteristics that made them particularly suited to the estimation of important parameters of an evolutionary process. Such model systems are very rare. It would not be possible, for example, to carry out such studies in any small mammal or any species of Drosophila. Moreover, they are extremely labor intensive. One of the studies found evidence of selection and two did not. One found evidence of random genetic drift and one of interspecific hybridization. There is no question of making such studies over and over again for many different organisms and many different genes or phenotypes. They neither allow us to make generalizations about forces nor do they uncover new and unexpected evolutionary phenomena. Certainly they do not test any general hypothesis. Rather, they are illustrative of how the forces of evolutionary change may operate in a particular case. There also seems to be some influence of prior theoretical commitment in the choice of the system and the design of the study. While the finch study might have failed to find a correlation between reproductive properties of the birds and the short-term evolution in morphology, its design would not have allowed conclusions about genetic drift affecting the distribution of properties among populations. In contrast, the snail study would not have allowed a complete study of components of natural selection, but was optimally designed to detect genetic drift. The *Zoarces* study, on the other hand, was a purely methodological one with no prior commitment to the size of selective forces, but was rather an exploration of the power of an optimally designed experiment to detect natural selection if it is occurring.

Although well-designed and gratifying in their results, the question then arises as to the role that model natural historical case studies play in evolutionary studies in general. They seem to be entirely illustrative and supportive, to show that a theoretical system of explanation can, in fact, be cashed out in a real case in nature, making the abstract concrete. That is an important goal for a natural science, but the value of yet further examples is unclear. How many finches and snails do we need to convince ourselves that natural selection and genetic drift really do occur?

DETECTING SELECTION FORMALLY

Attempts to measure selection in natural systems by means of natural historical observations of different types is close to Darwin's concrete view of natural selection as the consequence of the differential fit of organisms to their environments leading to differential reproductive success. But one can eliminate the physiological, behavioral, ecological, and morphological underpinnings of selection, i.e., the *biology* of natural selection, and ask simply whether there is evidence that changes in the characters of species must have been biased for or against some types as opposed to others, so that observed differences cannot be accounted for by purely random events. There is no way to answer this question for physiological, morphological, or behavioral characteristics. We do not know the environments of past organisms in any detail and changes in organisms induce changes in their relations to other species and the physical world. Thus, all claims that such changes have been a consequence of natural selection can only be invented stories with no means of verification or falsification. On the other hand, amino acid and nucleotide substitutions in evolution constitute a very large and relatively homogeneous class of changes of known genetic status so that the question can be asked without recourse to natural historical explanations. This question has been a major preoccupation of evolutionary studies since King & Jukes (13) first called attention to the fact that amino acid substitutions in proteins appeared to occur at a constant rate across long periods of evolutionary time. The struggle over a "neutralist" versus "selectionist" view of protein evolution in the broad sense is no longer an active issue in evolutionary studies. If we include in the neutral theory that selection coefficients on amino acid substitutions may be non-zero but of the same order or less than the reciprocal of effective population sizes, then the broad picture of amino acid substitutions over hundreds of millions of years is one of "quasi-neutrality."

The present direction of evolutionary research on this question focuses on much shorter time intervals, on the transitions between closely related species, and the variation within species. The questions are: How much does selection constrain the amino acid and nucleotide variation within species and how important has selection been in driving amino acid substitutions in species divergence? Because it is impossible to observe the dynamics of selection at the amino acid and nucleotide level either within or between species, the techniques of analysis of static data discussed earlier must be used. Attempts to carry out such static data analysis when only amino acid substitution data were available failed because statistical tests were ambiguous in their interpretation (16). Since the availability of complete DNA sequences of genes, however, it has been possible to use the contrast between nucleotide substitutions in different functional parts of the sequence to ask questions about selective constraints and selectively driven substitutions. The data on selective constraints on amino acid variation within species are copious and unambiguous, and no longer an active subject of research. If we assume, as a first approximation, that synonymous subsitutions in codons are selectively unconstrained, then the nucleotide polymorphism within a species will be same for synonymous and replacement substitutions only if replacement substitutions are also selectively neutral. But a large accumulation of sequence data has shown that there is always a large deficiency of amino acid replacement polymorphisms as compared with synonymous variations. In *Drosophila pseudoobscura* this may range from a complete absence of amino acid polymorphisms in a gene with a 5% synonymous polymorphism rate [alcohol dehydrogenase (23)] to the retention of

about 15% of the amino acid variation that is expected from the synonymou.
[xanthine dehydrogenase (21)]. Most mutations leading to amino acid changes
deleterious. An unsolved puzzle is how selection can be so discriminating as
eliminate every amino acid replacement in the alcohol dehydrogenase protein tha
is over 20% leucine, isoleucine, and valine. Can it really be that every leucine-
valine substitution matters significantly to the physiology of this organism in nature
when it can, in fact, survive under general laboratory conditions of culture with
the gene for this alcohol dehydrogenase completely inactivated?

Using the contrast between standing variation in synonymous nucleotide positions and replacement positions, it is also possible to ask whether there is evidence
that amino acid substitutions between species have been driven by selection. These
make use of the statistical tests based on the mathematical apparatus described in
the introductory section, involving data on synonymous and replacement substitutions within and between species. These have been applied to a number of cases
where population samples of sequences are available and the results have varied.
For genes in *Drosophila* a number of cases of selectively driven amino acid replacements between species have been inferred, but different statistical tests have been
used in different cases, and there may be some reporting bias since the failure to
find evidence for selection is less interesting. While each gene is a separate case
study, unlike for natural historical investigation large numbers of different gene
sequences can be determined with modern automated techniques so that an overall picture of how much selection is involved in protein sequence evolution now
becomes possible. Because of the large amount of data accumulating on sequence
differences, it will be possible to make generalizations over genes and species
similar to those made in the past about the extent of genetic diversity based on
gel-electrophoresis and DNA sequencing and to observe patterns of differences in
selective substitution between classes of genes and species with different biology.
The only study so far that uses the same theoretical procedure on a large number of
genes is that of Bustamante et al. (3), which found that of 34 *Drosophila* genes, 32
gave estimates of positive selection of the substitution, of which 10 were deemed
statistically significant, whereas among 12 *Arabidopsis* genes, 10 gave estimates
of some selection against the successful substitutions, of which 6 were statistically
significant. The authors ascribe this difference between organisms to the smaller
effective population size in *Arabadopsis*, which is inbred. The theoretical issue
that remains to be resolved before it is worthwhile pursuing this direction further
is the adequacy of the statistical procedures for turning static data into dynamic
inferences. All the tests so far used make assumptions about the amount of recombination between sites, about the constancy of effective population size during
the evolutionary history of the species, and most important of all, assume that the
populations are in stochastic steady state. That is, they assume that all traces of
initial conditions have disappeared and that no serious disturbances of population
structure have occurred since the divergence of the species. But we know that real
populations are constantly being demographically disturbed and that episodes of
migration, foundation of new populations, and population mixing are a common

feature of population life histories. It is particularly ironic that two species of *Drosophila*, *D. melanogaster* and *D. simulans*, which are human commensals in North America, and a weed, *Arabidopsis*, have been chosen for the application of tests that assume the irrelevance of recent history. Before we can be secure about the results of these studies of selection, we need to know far more about the operating characteristics of the inferential methods when the assumption of steady state is not met.

WHOLESALE SEQUENCING OF GENOMES

The directions of research in any science are a consequence of two interacting forces. First, questions are generated by the phenomena that are the subject matter of the science. What is the chemical nature of the gene? What are the genetic differences that lie at the basis of speciation? How important is natural selection or random drift in driving evolutionary change? But many questions cannot be answered without the development of new experimental techniques, and until those techniques have been introduced the field is filled with reasonable but untestable speculations. The attempts to understand the chemical nature of the gene in the second quarter of the twentieth century relied on the mutational effects of chemicals and radiation, but these observations proved to be inadequate, and the solution had to await the development of crystallographic analysis of macromolecules. Nor could the organization of the genome be understood until methods of DNA sequencing became a routine part of laboratory practice. Once such methods become available, they become a powerful force in directing future directions in the science. Scientists do what they know how to do. As P.B. Medawar put it, science is "the art of the soluble." Any biologist with some money can now sequence vast amounts of DNA, so a major direction of research in evolution will undoubtedly consist in asking, "What will I find out by sequencing the entire genome of . . . ?" There are, however, some problems in evolutionary biology that are already part of the program of study and that will benefit immensely from comparative studies of genomes.

First, comparative genomics is a powerful systematic tool that can resolve problems in phylogenetic reconstruction, which, in turn, will make inferences that use phylogenies more secure. It is already clear that chromosomal and genic rearrangements including inversions, translocations, insertions, deletions, and duplications are an extremely common feature of evolution of the genome. Sequence studies allow the determination of the location of such events down to the nucleotide position, and it is extremely unlikely that the same insertion or rearrangement will recur independently at exactly the same sequence positions in different species or that an insertion, for example, will be excised perfectly at a later time without leaving a trace. For such genomic events, then, the first rule of parsimonious phylogenetic construction holds true, and this provides the possibility of secure phylogenies on which evolutionary influences can be built. Nor is it necessary to sequence the entire genome of species of interest, but only sufficient stretches to include a good sample of syntenies. In the long run this may be the most widespread use of large-scale sequencing in evolutionary studies.

Second, complete genome sequences will be used simply as gene finders. For studies of the evolution of complex phenotypic and physiological traits, the first step in the analysis is to localize regions of the genome that appear to contribute to their variation. This involves the standard techniques of quantitative trait locus identification that associate phenotypic variation with chromosomal regions marked by segregating genetic variants. The next step requires an identification of candidate genes in these regions, genes whose variation can explain the variation of phenotype on the basis of molecular and developmental mechanisms. An important case for evolutionary research at present and in the foreseeable future is the elucidation of the genetic basis of the barriers to gene exchange between closely related species, barriers that include both interference with interspecific mating or fertilization and inviability and infertility of hybrids if they are formed. For example, in *Drosophila* species, gene differences leading to hybrid inviability are located more or less equally on all the chromosomes, but genetic changes causing hybrid sterility are more concentrated on the X chromosome (5). Ultimately, it will be necessary to identify the genes and understand how their reading by the cell in hybrids causes developmental and physiological abnormalities that, in turn, lead to inviability or infertility. The genetic and physiological bases of these mechanisms will vary considerably from one group of organisms to another so that no general mechanism will emerge. Instead, the elucidation of this fundamental feature of evolution will require gene localization and analysis from a variety of forms and the result will be a menu of different explanations.

Third, genome sequencing allows the detection of hidden homologies. The immense diversity of living organisms is based on the divergence and amplification of genomes derived from the earliest life. The mapping between genomic changes and functional and developmental changes is an extremely complex one so that all trace of the common evolutionary origin of diverse phenotypic characters is often lost when we study only the outcome of development. There are parallelisms, convergences, divergences, and novelties at the phenotypic level that make the facile textbook distinction between homology and analogy almost empty. It used to be said that the wings of bats and the wings of birds were homologous, but that the wings of birds and the wings of insects were only analogous because they were based on utterly different developmental processes with different genetic bases. The *Hox* gene complex has changed all that. By matching DNA sequences between species over an immense range of organisms, it is now possible to discover the common origin and trace the evolution of features of organisms irrespective of the degree of their apparent similarity or difference at any phenotypic level. Of course, all trace of common origin may disappear in particular cases as sequence divergence becomes greater, but if enough intermediate forms are sequenced, it will usually be possible to trace their evolution.

The possibility of following DNA and amino acid sequence evolution makes it possible to detect patterns of conservation at other levels. The amino acid differences between yeast and vertebrate lysozymes are sufficient to obscure their homology in the absence of intermediate forms, but the conservation of their three-dimensional structure, despite wholesale amino acid substitution, shows that proper lysozyme function can be achieved by a large array of amino acid compositions, provided that these compositions allow the conservation of a particular three-dimensional structure.

The unification of processes in development of very different organisms that is made possible by the study of sequence similarity is perhaps not so surprising given that the early steps in development of different organisms involve similar serial segmentation patterns. But it is also possible to follow gene evolution through sequence space even when there are radical changes in molecular function. The G-protein–coupled receptor proteins are a group of molecules that include such diverse functions as wing development in *Drosophila*, bone morphogenesis in vertebrates, and animal visual pigment systems. Yet, using sequence information, it has been possible to make a complete phylogeny of this immensely diverse set of molecules (19).

EVOLUTION IN BASIC GENETIC MECHANISMS

The general outline of how genes are transcribed and translated is now well understood, and a great deal of the detailed structural biology of the molecular processes is already known. An examination of the details in a variety of organisms, however, has revealed certain constraints and variations on the basic theme that are evidence of the evolutionary plasticity of this most basic of cellular processes. For example, the genetic code is not universal, but differs between mitochondrial and nuclear DNA and from one mitochondrial genome to another. The difference between nuclear and mitochondrial codes might be expected from the origin of mitochondria from invasion by a symbiont, but even if that symbiont differed in its nuclear code from the current nuclear code the current diversity of mitochondria must be the result of subsequent evolution of the mitochondrial codon set. It is not clear how this ancient set of events can be subject to further experiment or analysis. There are, however, other aspects of variation in the coding system that are part of the project of ongoing research.

Synonymous positions in codons are not unconstrained, as shown by the unequal use of alternative codons (codon usage biases) that are present for all proteins in all species but which differ from protein to protein and species to species. There are some generalities. For example, codon usages of highly transcribed genes are more biased than the average (24) but there are many regularities of codon bias that remain to be understood. Using the direction of species origins from the phylogeny of *Drosophila*, Akashi (1) asked how many codon changes in the divergence of species had occurred from highly used codons to less used ones and the reverse, thus enabling him to estimate the selection intensity in the evolution of codon bias. This is a question that remains part of the active problematic of evolutionary genetics and is accessible to experiment. For example, the Adh protein in *Drosophila* species has 12 or 13 isoleucines but virtually never uses AUA in any species. By in vitro mutagenesis and insertion of a mutated Adh gene into an Adh null line, it would be possible to replace allowed isoleucine codons by the prohibited AUA and measure the fitness decrease and effect on Adh production caused by increasing numbers of such replacements.

It is clear from studies of the variability in polymorphism and species divergence across different regions within a gene that there are patterns of nucleotide conservation within introns and within flanking regions that are without known function. Introns do sometimes contain enhancer elements that need to be conserved, and there are interactions between 5' and 3' ends of RNA sequences that may be important in constraining the evolution of gene sequences. Moreover, the possibility of selective constraints on RNA raises the issue of how much of amino acid conservation in proteins is a consequence of the selective effects at the protein level, as we usually assume, and how much is purely the consequence of structural constraints on messenger RNA. If, say, a certain 6-nucleotide stretch in the message is conserved because the three-dimensional structure of the molecule influences its lifetime or its rate of transcription or translation, then that conservation will appear as a conservation of two amino acids. The question of selection of structural properties of messenger RNA is yet to be investigated.

The most remarkable variation in the coding mechanism is the existence of RNA editing in certain flagellates and ciliates. In one form of editing, the DNA sequence is compressed, leaving out T-A pairs so that the template DNA sequence does not correspond to the final sequence of amino acids. The RNA is then edited to add the missing U's, according to separately coded information in the genome, to producing the mature message (15). In another form of editing, the DNA sequence occurs in segments whose order on the chromosome is scrambled. The production of the message unscrambles this order and assembles it in the correct reading order (6). While protozoa have been the organisms in which these editing mechanisms have so far been found, we cannot assume that they are restricted to that group or that the array of bizarre variations that have evolved has been exhausted.

NOVELTIES AND PHENOTYPE-GENOTYPE MAPS

A number of experiments and observations in evolution call attention to a variety of phenomena that arise from the peculiarities of the relation between phenotypes and genotypes. These phenomena have appeared in a disparate group of experiments held together only by the fact that they are all concerned, in one way or another, with the origin of novel morphologies and physiologies. Precisely because they are novelties they are unpredictable and one cannot design a long-term research program to create them, yet each case has something important to teach us about the nature of the evolutionary process.

Hall (9) set about to select *Escherichia coli* that could use a novel carbon source, lactobionate, for its energy, instead of the usual lactose. For this purpose he used a gene, *Ebg* (extra beta galactosidase) that had a low efficiency for cleaving the galactosidic bond of lactose and could be dispensed with in normal lactose metabolism. Using a mutagen, he finally succeeded in accumulating mutations of *Ebg* that would allow growth on lactobionate, but the evolutionary path to this state was not direct. He was not able to select directly for the new substrate. First, he had to select for increased activity on lactose, followed by a second stage of selection for an intermediate substrate lactulose, and then finally from a strain that could ferment lactulose, he successfully selected strains to grow on lactobionate. Moreover, at each stage of selection there were several strains that acquired the same biochemical phenotype but only some that could be further selected to the next stage. For example, two strains that fermented lactulose did so as a consequence of different mutations, only one of which could be further mutated to a lactobionate fermenter. This result illustrates that the pathway through the space of genotypes from one phenotypic state to another is complex, rather like a maze with many dead ends. Only a restricted subset of all the pathways that lead to the first adaptation are open to the next so that evolution of a novelty may be very difficult to achieve. This suggests one reason for the apparent conservatism of intermediary metabolism.

In contrast, a biochemical novelty may arise by a single very small molecular change. Newcomb et al. (18) found that the acquisition of organo-phosphate herbicide resistance in the blowfly, *Lucilia cuprina*, is a consequence of a single amino acid substitution in the active site of a carboxylesterase that abolished the esterase activity and converted the enzyme to an organophosphatase. Moreover, that qualitative change in specificity was a consequence of a small change in the angle at which the new residue was held in the folded molecule so that a water molecule could now be bound at the active site, a water molecule that could participate in the attack on the phosphate bond of the organophosphate substrate. That this change was not an extraordinary event was shown by the discovery of a second, different amino acid substitution that had the same effect. So, small genetic changes may lead to qualitatively novel adaptive properties.

The last phenomenon, which is likely to be more common, and which can be the object of planned experimental search, concerns the basis for widespread phenotypic conservation of a feature. In *Drosophila*, although there is considerable variation size between species and some variation within species, wing shape is conserved. That is, there are correlations between various wing measurements such that those correlations are characteristic of the variation among individuals within a species, and also between species. Haynes (10) studied a pair of wing measurements that are negatively correlated among individuals within *D. melanogaster* and also among all species of the genus. The assumption originally made was that this represented a developmental constraint that could not be broken. In 15 generations of artificial selection in *D. melanogaster*, she succeeded in reversing the correlation. The same phenomenon was demonstrated for anterior and posterior eyespots on the wings of the butterfly *Bicyclus anynana* by Beldade et al. (2). A strong positive correlation in the size of anterior and posterior eyespot size and other serially repeated features is the rule in butterflies and has been assumed to be a consequence of basic developmental mechanisms of anterio-posterior differentiation. The experiment reversed the correlation within 11 generations of selection. In both cases, despite the universality of the correlations in nature, there was enough genetic variation in growth relations within a population to allow a selective reversal of the pattern within a few generations. It is part of the theoretical commitment of "evo-devo," the study of the evolution of development and the influence of developmental pathways on evolution, that shape is greatly constrained by basic developmental relations resulting from cell-to-cell signaling and gradients in gene transcription that are more or less fixed across a wide range of organisms. That may indeed be true for some features of development, but it is also clear that the observed constancy of some feature is not in itself a demonstration of such genetically determined invariance. At least for wings in flies and moths, we must assume that natural selection is playing a stabilizing role in preventing evolutionary change in these organisms that is already possible with the genetic variability that they possess.

CONCLUSION

What is already known about evolution shows us that there are no universal rules and even what appear to be regularities have many informative exceptions. Evolution is a loose and complex process, the result of a number of interacting, individually weak forces with many alternative outcomes, and at all times contingent on previous history. The best answer to any question about evolution is the lawyer's answer to any general question about the law: "It depends on the jurisdiction." That is why the program of evolutionary investigation never comes to an end—and, so often, never to a conclusion.

The *Annual Review of Genetics* is online at http://genet.annualreviews.org

LITERATURE CITED

1. Akashi H. 1995. Inferring weak selection from patterns of polymorphism and divergence at "silent" sites in *Drosophila* DNA. *Genetics* 139:1067–76
2. Beldade P, Koops K, Brakefield PM. 2002. Developmental constraints versus flexibility in morphological evolution. *Nature* 416:844–47
3. Bustamante CD, Nielsen R, Sawyer SA, Olsen KM, Purugganan MD, et al. 2002. The cost of inbreeding in *Arabidopsis*. *Nature* 416:531–34
4. Christiansen FB, Frydenberg O. 1973. Selection component analysis of natural polymorphisms using population samples including mother-child combinations. *Theor. Pop. Biol.* 4:425–45
5. Coyne JA, Orr HA. 1998. Evolutionary genetics of speciation. *Philos. Trans. R. Soc. London Ser. B* 353:287–305
6. Curtis E, Landweber LF. 1999. The evolution of gene scrambling in ciliate micronuclear genes. *Ann. NY Acad. Sci.* 870:349–50
7. Fu Y-X, Li W-H. 1993. Statistical test of neutrality of mutations. *Genetics* 133:639–709
8. Grant PR, Grant BR. 2002. Unpredictable evolution in a 30-year study of Darwin's finches. *Science* 296:707–11
9. Hall B. 1978. Experimental evolution of a new enzymatic function. II. Evolution of multiple functions for *EBG* enzyme in *E. coli*. *Genetics* 89:453–65

10. Haynes A. 1989. *On developmental constraints in the* Drosophila *wing*. PhD thesis. Harvard Univ., Cambridge. 115 pp.
11. Hudson RR, Kreitman MK, Aguade M. 1987. A test of neutral molecular evolution based on nucleotide data. *Genetics* 116:153–59
12. Kimura M, Ohta T. 1971. *Theoretical Aspects of Population Genetics.* Princeton, NJ: Princeton Univ. Press
13. King JL, Jukes TH. 1969. Non-Darwinian evolution: random fixation of selectively neutral mutations. *Science* 164:788–98
14. Lamotte M. 1951. Recherches sur la structure génétique des populations naturelles de *Cepaea nemoralis* L. *Bull. Biol. Fr. Belg.* 35(Suppl.):1–238
15. Landweber LF. 1992. The evolution of RNA editing in kinetoplastid protozoa. *BioSystems* 28:41–45
16. Lewontin RC. 1974. *The Genetic Basis of Evolutionary Change.* New York: Columbia Univ. Press
17. McDonald JH, Kreitman M. 1991. Adaptive protein evolution at the *Adh* locus in *Drosophila. Nature* 351:652–54
18. Newcomb RD, Campbell PM, Ollis DL, Cheah E, Russell RJ, et al. 1997. A single amino acid substitution converts a carboxylesterase to an organophosporus hydrolase and confers insecticide resistance on a blowfly. *Proc. Natl. Acad. Sci. USA* 94:7464–68
19. Rice K. 1997. *The origin, evolution and classification of G-protein-coupled receptors.* PhD thesis. Harvard Univ., Cambridge. 224 pp.
20. Richter B, Long M, Lewontin RC, Nitasaka E. 1997. Nucleotide variation and conservation at the *Dpp* locus, a gene controlling early development in *Drosophila. Genetics* 145:311–23
21. Riley MA, Hallas ME, Lewontin RC. 1989. Distinguishing the forces controlling genetic variation at the *Xdh* locus in *Drosophila pseudoobscura. Genetics* 123:359–69
22. Sawyer SA, Hartl D. 1992. Population genetics of polymorphism and divergence. *Genetics* 132:1161–76
23. Schaeffer SW, Miller EL. 1992. Molecular population genetics of an electrophoretically monomorphic protein in the alcohol dehydrogenase region of *Drosophila pseudoobscura. Genetics* 132:163–78
24. Sharp PM. 1989. Evolution at "silent" sites in DNA. In *Evolution and Animal Breeding: Reviews in Molecular and Quantitative Approaches in Honour of Alan Robertson*, ed. WG Hill, TFC Mackay. Wallingford, UK: CAB Int.
25. Wells RS. 1996. Excessive homoplasy in an evolutionarily constrained protein. *Proc. R. Soc. London Ser. B* 263:393–400
26. Wright S. 1931. Evolution in Mendelian populations. *Genetics* 16:97–159

Annu. Rev. Genet. 2002. 36:19–45
doi: 10.1146/annurev.genet.36.030602.090831

GENETIC MATING SYSTEMS AND REPRODUCTIVE NATURAL HISTORIES OF FISHES: Lessons for Ecology and Evolution

John C. Avise[1], Adam G. Jones[1,2], DeEtte Walker[1], J. Andrew DeWoody[3], and collaborators[4]

[1]Department of Genetics, University of Georgia, Athens, Georgia, 30602;
e-mail: avise@arches.uga.edu; [2]current address: School of Biology,
Georgia Institute of Technology, 310 Ferst Drive, Atlanta, Georgia, 30332;
e-mail: adam.jones@biology.gatech.edu; [3]Department of Genetics,
Purdue University, West Lafayette, Indiana, 47907-1159;
e-mail: dewoody@fnr.purdue.edu; [4]additional authors (all have been in or associated
with the Avise lab) are Beth Dakin, Anthony Fiumera, Dean Fletcher, Mark Mackiewicz,
Devon Pearse, Brady Porter, and S. David Wilkins.

Key Words   molecular markers, parentage analysis, cuckoldry, sexual selection, alternative reproductive tactics

■ **Abstract**   Fish species have diverse breeding behaviors that make them valuable for testing theories on genetic mating systems and reproductive tactics. Here we review genetic appraisals of paternity and maternity in wild fish populations. Behavioral phenomena quantified by genetic markers in various species include patterns of multiple mating by both sexes; frequent cuckoldry by males and rare cuckoldry by females in nest-tending species; additional routes to surrogate parentage via nest piracy and egg-thievery; egg mimicry by nest-tending males; brood parasitism by helper males in cooperative breeders; clutch mixing in oral brooders; kinship in schooling fry of broadcast spawners; sperm storage by dams in female-pregnant species; and sex-role reversal, polyandry, and strong sexual selection on females in some male-pregnant species. Additional phenomena addressed by genetic parentage analyses in fishes include clustered mutations, filial cannibalism, and local population size. All results are discussed in the context of relevant behavioral and evolutionary theory.

CONTENTS

0066-4197/02/1215-0019$14.00

INTRODUCTION

Molecular markers can unveil and quantify the incidence of organismal reproductive behaviors that otherwise may remain hidden from field naturalists (6, 9, 52). Here we highlight genetic findings on fish mating systems and alternative reproductive tactics in nature as gleaned from DNA-level analyses of maternity and paternity, and interpret the results in conjunction with behavioral observations. Genetic parentage analyses of fish clutches have unearthed several fascinating details of reproductive natural history, and have also yielded fresh insights into broader theories on animal mating systems and sexual selection. This review focuses on conspecific populations of sexually reproducing fishes in the wild; special cases of parentage assessment in interspecific hybrids and in unisexual fishes have been reviewed elsewhere (7, 10).

The Natural History of Fish Reproduction

Fish have remarkably diverse reproductive behaviors (21, 63, 130, 131, 136). A rich natural-history literature documents mating systems ranging from pelagic group spawning to cooperative breeding to social monogamy. Subsequent to spawning, adult care of fertilized eggs and larvae may be nonexistent, confined to one gender, biparental, or communal. When parental care is offered, it may take such varied forms as oral or gill brooding, use of natural or constructed nests, internal gestation by a pregnant mother or by a pregnant father, or open-water guarding of fry.

For reasons that will become apparent, most genetic studies of fish mating behaviors have been conducted on species displaying parental care of offspring (8). In the bony fishes (Osteichthyes), approximately 89 of the 422 taxonomic families (21%) contain at least some species in which adults provide direct postzygotic services, and in nearly 70% of those families, the primary or exclusive custodian is the male (18, 19). Parental care by males alone is otherwise extremely rare in vertebrates other than anuran amphibians (26, 149). Thus, the evolutionary elaboration of paternal devotion makes fishes particularly favorable for testing traditional parental-investment and sexual-selection theories originally motivated by research on mammals and birds, where females typically are the primary caregivers.

Also intriguing for genetic analysis are alternative reproductive tactics *within* a species (61), or sometimes even within an individual during its lif An example of the latter occurs in sequential hermaphroditic species in w an individual fish may switch its gender (and associated mating behavior) fre female to male or vice versa (148). Both sex-changing and nonchanging fish ar present in some populations (146).

Most fish species are gonochoristic (separate sexes), and here too, within a gender, ARTs may be prevalent. In theory, a male fish may maximize the number of eggs he fertilizes by being quicker than rivals in "scramble competition," monopolizing mates or resources such as nests or territories, exploiting the resources of other males via reproductive parasitism, or cooperating or trading with resource holders via mutualism or reciprocity (132). Two or more such tactics are often observed in a population. For example, four types of males co-occur in the ocellated wrasse (*Symphodus ocellatus*): large "bourgeois" males that build nests and tend progeny; small males that parasitize (or cuckold) a bourgeois spawner by sneaking into a nest and "stealing" some of the fertilization events; medium-sized males that defend another male's nest from sneakers, but also court females and occasionally spawn; and extra-large males (pirates) that temporarily usurp the nest of another male (133).

In another example, male salmon spawn either as full-sized anadromous adults after returning from the sea, or as dwarf precocious parr that have remained in fresh water. In marker-based parentage analyses of Atlantic salmon (*Salmo salar*), parr have been shown to fertilize widely varying proportions (5–90%) of the total eggs in various populations (55, 56, 68, 83, 104, 135). They also produce physiologically superior spermatozoa, a feature that partially compensates for their behavioral subordinance to dominant anadromous males (143).

The ARTs of anadromous salmon males and resident parr appear tied to an individual's environmental exposure, but ARTs in some fish species might be genetically hard-wired. Rearing fry under controlled conditions can help in evaluating developmental plasticity, as has been demonstrated with respect to alternative trophic morphs in several fish species (102, 124, 140). Regardless of their mechanistic basis, ARTs are common in fish, their occurrence facilitated by the prevalence in this group of external fertilization, a high incidence of paternal investment, and extensive intrasexual size variation attendant with indeterminate growth (132).

On the female side of the ledger, strategies of mate choice and parental investment can also vary (65). For example, in the peacock wrasse (*Symphodus tinca*), a female normally spawns with territorial males who care for her eggs on the nest, but she also may spawn off-territory with males who provide no parental care (147). Under some ecological circumstances, the cost of nest-searching by a female may outweigh the lower survival of her untended offspring from these latter matings (141). Other ARTs known in female fish include additional variations on patterns of mate choice, parental care, resistance to coercion, and mating mode (1, 65).

Although such ARTs are well described in the fish behavioral literature, their fitness consequences cannot be fully assessed from field observations alone.

enetic markers can shed new light on the realized success of ARTs in nature by disclosing actual biological parentage.

Historical Precedents for Genetic Reassessments

Three broad developments in the 1980s set a stage for refined genetic appraisals of fish reproductive activities. First, highly variable satellite DNA regions (70) were found to be common features of eukaryotic genomes (57), and their utility in parentage assessment was quickly appreciated (24).

Second, revolutionary insights came from analogous genetic studies in other taxonomic groups, notably insects and birds (16). For example, ornithologists supposed that most songbirds were genetically monogamous within a breeding season (92), but the new genetic data often excluded a nest-tending adult as the sire or dam of some nestlings. Such findings led to the realization that extra-pair copulations and other clandestine reproductive activities, including conspecific nest parasitism, are routine phenomena in many avian species (17).

A third development was the realization that patterns of genetic parenthood are important for theoretical models of behavioral evolution (2, 4, 64, 139). For example, surreptitious cuckoldry can yield a "genetic mating system" that departs from the apparent "social mating system" of a population, and this can impact the intensity of sexual selection and the evolution of secondary sexual traits. Furthermore, genetic analyses of birds, insects, and other taxa have shown that realized parentage can reflect postcopulatory processes too, including sperm competition and female sperm choice (47, 129).

BACKGROUND

Microsatellite Markers

Microsatellites are well-suited for genetic parentage analyses because the DNA repeat units are small (2–5 bp each) such that alleles separated through suitable gels usually can be identified cleanly; the assays are applied locus-by-locus so the data can be interpreted in simple Mendelian terms; the polymerase chain reaction (PCR) is employed, so data can be recovered from even small amounts of tissue such as a single fish embryo; and allelic variation in most fish populations is extremely high (35).

The Logic of Molecular Parentage Analysis

A variety of statistical approaches for parentage analysis have been developed for particular biological settings (e.g., 37, 53, 69, 106). However, the basic logic of parentage analysis is generalizable. One common situation in fishes is when the male parent of a brood is known or suspected from genetic or behavioral evidence, and maternity is in question. For each offspring at each locus, the maternal allele can be deduced by subtraction (except when the sire and offspring are identically heterozygous). Then, any female whose genotype is inconsistent with these maternally deduced alleles at multiple loci is excluded as the dam.

An average exclusion probability refers to the mean probability of excluding an unrelated adult as a parent of a randomly chosen juvenile. In nearly all cases considered in this review, genetic markers were sufficiently variable that mean multilocus exclusion probabilities (126) were well above 0.95. Such exclusionary power may earmark the true dam, but this also depends on the number of candidate females in the population, how thoroughly they have been sampled, and their genetic relationships.

Similar logic applies to paternity exclusions when particular offspring display alleles incompatible with those of their male custodian. Cuckoldry (stolen fertilizations by other males), nest piracy, and egg thievery are among the behavioral possibilities that can lead to male foster parentage, and these can often be distinguished by considering details of the particular natural-history setting. Finally, when neither fish parent is available for genetic examination (as is normally true in species lacking parental care), the statistical exclusionary power and, hence, the capacity to draw biologically informative conclusions, usually is reduced considerably.

In most of the genetic appraisals of parentage and reproductive behaviors in fishes published to date, discrete cohorts of embryos within a nest (or inside a gestating parent) were genotyped in conjunction with a custodial adult and other individuals sampled nearby. By straightforward chains of reasoning, these multilocus genotypic data permit powerful deductions about the genetic parentage of particular juveniles, and such information accumulated across hundreds or thousands of fish from multiple nests can reveal the relative success of ARTs as well as the genetic mating system of a natural population.

REPRODUCTIVE PHENOMENA IN VARIOUS FISH GROUPS

What follows are brief synopses of various natural-history phenomena illuminated by genetic parentage analyses in fishes, particularly species with parental care of offspring (Table 1).

North American Sunfishes

In nearly all 30 species of North American sunfish (Centrarchidae), males guard eggs and embryos in shallow depression nests swept in the soft substrate of a lake or stream (20). One species—the bluegill, *Lepomis macrochirus*—has been a model system for the study of ARTs (59–61, 107).

CUCKOLDRY BY MALES   Various routes to paternity are available to bluegills at the study sites in eastern Canada (45, 62). "Parental" or bourgeois males, which mature at seven years of age, construct nests in colonies, attract females, spawn with them on the nest, and vigorously defend the nest and embryos against intruders.

TABLE 1 Summary of salient findings in microsatellite-based studies of genetic parentage in various natural fish populations[a]

Mode of brood care by adult custodian	Species	# Young assayed	# Nests or broods assayed	# Sneaked nests or broods	# Nest takeovers or egg-theft events	% Offspring not parented by custodian	Average minimum (& estimated) # mates[b] per nest or brood	References
Nest-tending by males	*Lepomis auritus* (redbreast sunfish)	996	25	11	2	12	3.6	41
Nest-tending by males	*Lepomis punctatus* (spotted sunfish)	1434	30	13	1	1	4.4 (7.2)	38
Nest-tending by males	*Lepomis macrochirus* (bluegill sunfish)	1677	38	35	0	21	Not estimated	106
Nest-tending by males	*Lepomis marginatus* (dollar sunfish)	1015	23	2	1	4	2.5 (3.7)	Mackiewicz et al, submitted
Nest-tending by males and females	*Micropterus salmoides* (largemouth bass)	1088	26	1	1	7	1.2	42
Nest-tending by males	*Etheostoma olmstedi* (tessellated darter)	610	16	2	3	14	3.2 (9.2)	39
Nest-tending by males	*Spinachia spinachia* (15-spined stickleback)	1307	28	5	4	19	2.6	75
Nest-tending by males	*Pomatoschistus minutus* (sand goby)	981	24	12	1	15	3.4	81, 82[c]
Nest-tending by males	*Etheostoma virgatum* (striped darter)	987	19	0	4	17	4.7 (6.1)	118
Nest-tending by males	*Cottus bairdi* (mottled sculpin)	1259	23	0	1	2	2.8	53a
See footnote d	*Salmo salar* (Atlantic salmon)	250	10	10	0	NA	5.1	97, 103
See footnote d	*Salmo trutta* (brown trout)	85	33	4	0	NA	2.2	56

Internal gestation by females	Gambusia holbrooki (mosquitofish)	823	50	NA	NA	NA	2.0	155
Internal gestation by females	Poecilia reticulata (guppy)	1812	253	NA	NA	NA	1.5	87
Internal gestation by males	Syngnathus scovelli (gulf pipefish)	838	40	0	0	0	1.0	71, 80
Internal gestation by males	Syngnathus floridae (dusky pipefish)	924	22	0	0	0	1.9	72
Internal gestation by males	Syngnathus typhle (Swedish pipefish)	1344	30	0	0	0	3.1	77
External gestation on male's body	Nerophis ophidion (straight-nosed pipefish)	361	15	0	0	0	1.0	101
Internal gestation by males	Hippocampus angustus (W. Australian seahorse)	453	15	0	0	0	1.0	74
Mouthbrooding by females	Pseudotropheus zebra (a Lake Malawi cichlid)	99	7	0	0	0	3.8	110
Mouthbrooding by females	Protomelas spilopterus (a Lake Malawi cichlid)	200	6	4[e]	4[e]	29	1.7	86
Mouthbrooding by females	Seven other species of Lake Malawi cichlids	203	16	0	0	0	2.1	85

[a]Modified and expanded from (36).

[b]Usually, the mean minimum number of female mates of the male parent (but sometimes male mates of the female parent, as in the cases of Salmo and Gambusia) per nest or per brood as deduced directly from counts of relevant gametotypes in the progeny array; estimated numbers (where presented) were statistically corrected by the original authors according to procedures in DeWoody et al. (37, 43).

[c]Data from the Klubban population only (see text for additional information).

[d]Male Salmo do not "tend" nests, so the numbers refer merely to the incidences of multiple parentage.

[e]The genetic data revealed that each of 4 females was mouthbrooding some fry that were not her own biological progeny. Because the field behavior(s) underlying such brood-mixing remain unknown, it is unclear whether these fry should be considered "sneaked" or "stolen".

Precocious cuckolder males, by contrast, attempt to steal fertilizations from nest-holders. These can often be 2- to 3-year-old "sneakers" that dart into a nest and release sperm as the bourgeois male spawns with a female, or older "satellites" that mimic females in color and behavior but release sperm as the primary couple spawns. Cuckolders leave the nest after spawning and show no parental care. They represent an "alternative life history" to that of bourgeois males (62).

What fraction of the reproductive output is attributable to bourgeois versus cuckolder males? Molecular markers provided the answer (29, 106, 116). In the largest genetic study of bluegills, involving 38 nests in one colony, the percentage of offspring per nest sired by the resident male ranged from 26–100 (mean 79%; Figure 1). Cuckoldry by neighboring bourgeois males was rare, so about 20% of the embryos were the result of fertilization thievery by satellites or sneakers (106). The levels of cuckoldry per nest also were evaluated in conjunction with behavioral observations (107). The comparisons showed that as bourgeois males detect paternity lost to cuckolders (by assessing intrusion rates of sneakers and

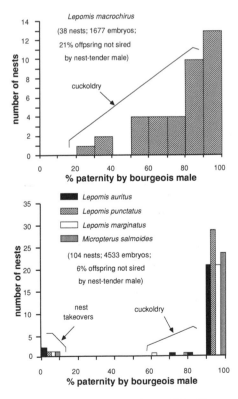

Figure 1 Paternity found for bourgeois male sunfish. Shown are the percentages of progeny per nest sired by the respective attendant males for (*above*), 38 nests of the bluegill, *Lepomis macrochirus*; and (*below*), a total of 104 nests in four other centrarchid species (see Table 1 for literature citations).

perhaps by using olfactory cues on offspring relatedness), they adaptively low their level of parental care.

The overall rate of cuckoldry in this population closely matched the observed proportion (20%) of males at age 2 destined to become cuckolders (62). This raises the possibility that the genetic fitness of an individual may be similar for bourgeois males and cuckolders, a finding consistent with the notion that these two ARTs might be near an equilibrium frequency in an evolutionarily stable system (100) perhaps maintained by frequency-dependent selection (60–62). This conclusion remains tentative, however, because snapshot appraisals of genetic parentage do not yield estimates of lifetime fitness. Furthermore, although not necessarily crucial to the question of evolutionary stability (32, 54), it remains uncertain whether the ARTs in bluegill reflect a genetic polymorphism or a conditional ontogenetic switch regulated by the social or environmental experiences of a male during its development.

Microsatellite assessments of paternity rates for bourgeois males have recently been gathered for four other centrarchid species (Figure 1). Most or all embryos in a majority of nests proved to have been sired by the nest attendant. If we temporarily disregard the nests in which the custodial male fathered none of the young (see below), then mean cuckoldry rates were about 2%, or roughly an order-of-magnitude lower than in *L. macrochirus*. Two factors probably contribute to this difference. First, at the sites studied, the other centrarchid species were either solitary nesters or less colonial than the bluegills assayed, and, all else being equal, lower nesting densities probably reduce the opportunities for cuckoldry (36, 60, 116). Second, specialized cuckolder morphs are not known in *L. auritus*, *L. marginatus*, or *M. salmoides*, and although such morphs have been reported in *L. punctatus* (38), they were rare at the study site.

NEST TAKEOVERS   Among the total of 142 centrarchid nests genetically surveyed, the custodial male in six instances (4.2%) had sired none of the young. One such instance involved a nest tended by a sterile F_1 hybrid between *L. macrochirus* and *L. gibbosus*, so the author concluded that this hybrid male had been 100% cuckolded (106). However, most of the other cases (all intraspecific) probably reflected nest-takeover events, and these nest piracies account for the majority of the documented foster parentage. Perhaps nest takeovers are opportunistic responses to limited nest-site availability, or, perhaps the captured "nest-holder" at the time of sampling was merely a temporary visitor (e.g., was there to cannibalize embryos).

CUCKOLDRY BY FEMALES   Largemouth bass (*Micropterus salmoides*) are unusual among fishes for tendencies toward biparental care of young and for staying with schooling fry for up to a month post-hatching. Most of the 26 offspring cohorts genetically assayed in *M. salmoides* proved to be composed of full-sibs (consistent with the social monogamy suspected for this species), but four cases were documented in which the custodial female was the dam of most but not all of the juveniles that she and her mate tended (42). Most likely, a second female in each case had laid some eggs in another's nest and then left her offspring in the care of their father and stepmother. "Cuckoldry" is normally meant to imply a reproductive behavior by which a breeding individual surreptitiously usurps parental services of another adult of the same gender, so this can qualify as a case of "female cuckoldry" analogous to male cuckoldry discussed above.

MULTIPLE MATING BY BOURGEOIS MALES   In *M. salmoides*, the genetic parentage data show that successful spawning was usually by one female (and one male) per nest (42). However, in three other sunfish species similarly assayed—*L. auritus*, *L. marginatus*, and *L. punctatus*—the genetic data showed that multiple females typically had spawned with each bourgeois male (Table 1). In other words, such nests contained mixtures of full-sib and paternal half-sib embryos. In the spotted sunfish, for example, the mean number of mothers per nest was at least 4.4, and statistical adjustments suggest that the true number occasionally may have been 10 or more (38).

Other Nest-Tending Species

In several other nest-tending fish species from both the freshwater and marine realms, parentage analyses by genetic markers have likewise been used to estimate numbers of dams per nest, rates of non-paternity for bourgeois males, and cases of nest piracy (Table 1). Additional reproductive phenomena have been uncovered as well, as described next.

EGG THIEVERY   Another behavioral route to non-paternity for custodial males is egg-stealing, a nest-raiding phenomenon occasionally observed, for example, in sticklebacks, Family Gasterosteidae (94, 154). A bourgeois male uses kidney-secreted glue to construct a nest in vegetation, into which females lay eggs (14). Then the resident male (and, sometimes, sneakers) swim through the nest, releasing sperm. Occasionally, a bourgeois male is also seen transporting to his own nest a discrete cluster of eggs (clutch) that he stole from a neighbor.

Are these stolen eggs viable, and had they been fertilized by the neighbor? Applying microsatellite markers to the problem, Jones et al. (75) documented probable cases of egg piracy in about 17% of the nests in a population of fifteenspine sticklebacks (*Spinachia spinachia*). Each such instance was adduced when one of two or more clutches of viable embryos in a nest (rather than a few scattered progeny) had been sired by a male other than the bourgeois attendant. Using a slightly different DNA fingerprinting approach, Rico et al. (121) similarly found that about 18% of nests in the threespine stickleback, *Gasterosteus aculeatus*, contained some stolen eggs.

Why would nesting males often pilfer fertilized clutches? Several hypotheses have been advanced for how natural selection on males might have promoted the evolution of egg-stealing tendencies: (*a*) the pump-priming effect—females in several species, including sticklebacks, are known to spawn preferentially in nests that already contain eggs (89, 122, 123); (*b*) the predator-dilution effect—extra eggs in a nest might ameliorate predation on the guardian's own embryos (150);

(c) kin selection—if the larcenist and his victim are close genetic relatives but the thief has much higher prospects for successfully rearing offspring, both individuals might benefit from the theft in terms of inclusive fitness; or (d) the larder-stocking effect—males may steal eggs only to eat them later.

EGG MIMICRY   In *Etheostoma* darters of the eastern United States, males of several species appear to have evolved bodily structures (typically on the tips of fins) that closely resemble eggs of these species and have been interpreted as "egg mimics." In a population of one of these species, *E. virgatum*, the egg mimics are displayed as pigment spots on the pectoral fins. In a genetic maternity analysis of fertilized eggs in the nests of 10 males, Porter et al. (118) found a significant correlation between the number of egg mimic spots on nesting males and their respective numbers of genetically deduced mates. Results are consistent with the hypothesis that egg mimicry by bourgeois males helps to attract gravid females to a nest.

BROOD PARASITISM BY HELPER MALES   Cooperative breeding is fairly common in avian and mammalian species (23), but is known in only eight species of fish (130). In a nest-guarding cichlid fish from Lake Tanganyika, *Neolamprologus pulcher*, a pair of breeders often shares brood-care duties with individuals from previous clutches (134). In general, nest helpers might gain personal benefits such as food, protection, parental experience, or inheritance of a territory or mate (e.g., 23, 49), and/or they might gain in terms of inclusive fitness by rearing kin (64). Might they also profit in the immediate currency of personal fitness by siring some of the offspring within the brood? Yes. Using multilocus DNA fingerprinting assays, Dierkes et al. (44) showed that about 10% of the progeny in seven assayed families of *N. pulcher* were fathered by helpers.

REPRODUCTIVE VARIANCE AND THE OPPORTUNITY FOR SEXUAL SELECTION   One common notion, supported by many studies of avian species, is that extra-pair fertilizations enhance the opportunity for sexual selection by increasing the variance in male reproductive success. By stealing fertilizations from neighbors, some males become bigger winners (and others bigger losers) in the reproductive sweepstakes. Although this view may generally be true in socially monogamous species such as many birds, it may not hold in all situations. Namely, whenever the variance in reproductive success among males is larger in the absence of cuckoldry than in its presence, the opportunity for sexual selection actually may decrease with increased levels of fertilization thievery (82).

Such may well be the case in the sand goby, *Pomatoschistus minutus*, a small European marine species in which males build and defend nests under mussel shells. Nest sites can be at a premium, and males often mate with multiple females. Thus, in total reproductive output, successful bourgeois males might be expected to greatly surpass other males in the population, especially those unable to secure nesting sites. However, as demonstrated genetically by Jones et al. (81, 82), fertilization thievery via sneaking is also extremely common in this species, occurring in about 50% of all nests. By interpreting these empirical findings in the context of models relating the intensity of sexual selection to variances in male reproductive success, the authors conclude that for this species, cuckoldry by sneaker males probably substantially reduces the opportunity for sexual selection.

Oral Brooders

Adults in many fish species protect their offspring by carrying eggs and hatchlings in the mouth or gill cavity. Oral incubation is particularly prevalent, and has evolved many times independently, in fishes of the Family Cichlidae (58, 84, 88). Microsatellite-based paternity analyses (Table 1) in several cichlid species have documented multiple paternity of broods, with up to six males fertilizing a single clutch (85, 110).

INTRASPECIFIC BROOD MIXING   Another otherwise cryptic phenomenon in mouthbrooders, documented by microsatellite markers, is the shuffling of conspecific broods. In four of six orally brooded cohorts of fry examined in a Lake Malawi cichlid, *Protomelas spilopterus*, the proportions of juveniles not dammed by the female who held them ranged from 6% to 65% (86). Several possible explanations for the origin and significance of brood mixing remain highly speculative (86), but based on genetic as well as other evidence (84, 120), this foster behavior in cichlids is remarkably common.

Female-Pregnant Species

In several fish groups, including the Poeciliidae (a large New World Family of live-bearers) and the Embiotocidae (the only Family of marine teleosts that is exclusively viviparous), a female is impregnated by one or more males and carries the resulting embryos internally, giving birth weeks or months later. Internal gestation guarantees that a pregnant female is the biological mother of her brood.

MULTIPLE MATING BY FEMALES   In most nest-tending fishes, there is an inherent gender asymmetry in the genetic power to detect multiple mating. Each clutch typically is associated with a male guardian, so any multiple in situ mating by that male will be apparent in suitable molecular assays of progeny in his focal nest. However, multiple mating by a female can only be revealed if separate nests containing her progeny were included in the field collection, and this may seldom be the case when populations are large or sparsely sampled. Thus, if only for this bias, multiple mating by females has rarely been genetically verified in nest-tending fishes (42, 75), However, this detection bias is reversed in female-pregnant fishes, where multiple mating (if present) by females is normally far easier to document genetically than is multiple mating by males.

In the sailfin molly, *Poecilia latipinna*, allozyme-based paternity analyses revealed that at least 52% of assayed broods were composed of embryos sired by two or more males (137), and that larger females were more likely to have had multiple mates (138). Constantz (31) summarized other allozyme-based estimates of multiple insemination rates in poeciliids. For example, at least 56% of pregnant mosquitofish (*Gambusia affinis*) carried broods of mixed paternity (25). However, marker variability can affect such estimates, as suggested by a later microsatellite analysis of mosquitofish in which multiple paternity was documented in nearly 100% of the surveyed broods (155).

SPERM STORAGE BY FEMALES   Ovarian tissues in poeciliids can store functional sperm for at least 1–2 weeks post-copulation, but storage of viable sperm by female surfperches (Embiotocidae) routinely occurs across several months (145). Using allozyme assays, Darling et al. (33) showed that most broods in the embiotocid shiner perch, *Cymatogaster aggregata*, are sired by multiple males, despite the fact that the matings preceded fertilization by 25 weeks or more (thus evidencing long-term sperm storage by females). This may be the longest known duration in fishes for potential sperm competition and postcopulatory female choice. However, even this pales in comparison to the multiyear utilization of female-stored sperm that has been genetically documented in some turtles (114).

BIOLOGICAL BENEFITS OF FEMALE PROMISCUITY   In female-pregnant fish, pro-miscuous mating tendencies by the males are evident in their vigorous sexual behavior (30) and are easy to understand, but why would females also mate promis-cuously? Multiple mating may expose a female to higher risks from sexually trans-mitted diseases, predation, copulation brutality, or other time or energy expenses associated with the mating process, and these costs might seem to outweigh any benefits in genetic fitness. However, a female also might gain any of several fitness advantages by mating with multiple males, including fertilization insurance against male sterility, access to more or better quality territories, success in "prospecting" better genes for her progeny, production of broods with more diverse and poten-tially adaptive genotypic arrays, and avoidance of inbreeding depression if some of her matings might be with close kin.

Female guppies, *Poecilia reticulata*, often solicit matings from multiple males (67), and many broods have multiple sires, especially in high-predation regimes (87). In microsatellite-based paternity analyses, Evans & Magurran (50) discov-ered that females who had mated with multiple males had shorter gestation times and produced larger broods containing progeny with better-developed schooling behaviors and predator avoidance. These findings provide some of the first ex-perimental evidence in fishes that promiscuity can be genetically rewarding for females as well as males.

Male-Pregnant Species

Two motivations have guided most genetic parentage analyses in fish: intellectual curiosity about a species' natural history and a desire to test broader mating sys-tem theories. Nowhere has the latter objective been more evident than in recent molecular appraisals of the Syngnathidae (73). A universal feature in the more than

200 living species of pipefishes and 30 species of seahorses is male pregnancy. One or more females lay eggs into a male's brood pouch or ventral surface, where they are fertilized by the assured sire and then housed as developing embryos until parturition weeks later. Such high paternal investment in offspring, and a freedom from parental responsibility for females, contrast diametrically with the situation in most mammals and many birds, making the syngnathids ideal subjects for testing, from a mirror-image perspective, traditional notions about gender roles in the context of mating system theories (3, 28, 139, 151).

MATING SYSTEMS, SEXUAL SELECTION, AND SEXUAL DIMORPHISM   In the behavioral literature on syngnathid fishes, "sex-role reversal" is usually defined not as male pregnancy per se, but rather as any situation in which females compete more intensely for access to mates than do males (15, 142). By this definition, some syngnathids are sex-role-reversed and some are not (142). In other words, in some but not all syngnathid species, females potentially produce more eggs during a breeding season than the available brood pouches of males can accommodate, such that males are the limiting resource in reproduction. This situation differs from that in most nest-tending teleosts, where rates of egg care by guardian males usually exceed rates of egg production by females (28).

The reason for defining sex-role reversal in this fashion (whether stemming from male pregnancy, or from any other impacts on the relative reproductive rates of the sexes) is that the phenomenon then ties rather directly to broader theories on mating systems and sexual selection (142). Namely, because sex-role reversal produces a female-biased "operational sex ratio," it presumably is associated with higher intensities of sexual selection on females, a greater potential for the elaboration of secondary sexual traits in that gender, and mating systems tending toward polyandry (Figure 2). All of these predictions fall on the opposite end of a mating-system spectrum from the polygynous behaviors that characterize, for example, many mammal and bird species with traditional gender roles. In these other organisms, males often have the potentially larger variances in fitness, compete actively for females (the limiting resource in reproduction), experience more intense sexual selection, and often display sexually selected behavioral or morphological traits (Figure 2).

In some syngnathid species, sex-role reversal has been evaluated experimentally, for example, as the potential reproductive rates of males versus females (99, 142). In other syngnathid species, the evidence for or against the phenomenon is indirect, involving, for example, the observed degree of dimorphism in secondary sexual characters. In syngnathids, when one gender is more brightly colored or otherwise sexually adorned, it is, indeed, normally the female (34). Given that syngnathid species appear to vary considerably along the sexual-selection continuum, conventional theory suggests that their mating systems may also vary accordingly (Figure 2).

In initial tests of this hypothesis, microsatellite-based appraisals of the genetic mating system have recently been conducted in each of five syngnathid species

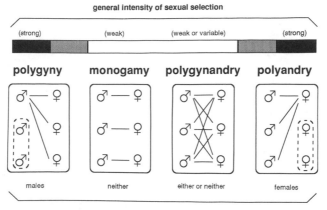

Figure 2   Pictorial definitions of four genetic mating systems possible in fishes. Lines connecting males and females indicate spawning partners that produced offspring. Also shown are the theoretical gradients in sexual-selection intensities and the degrees of gender dimorphism in secondary sexual traits often associated with these mating systems.

(Table 1) that display differing degrees of sexual dimorphism. The results proved to be in general agreement with broader mating system theory in at least two regards (Figure 3). First, the genetic mating systems fell along the monogamy-polyandry end of the mating system continuum, rather than in the monogamy-polygyny range of the spectrum as is normally true in most mammals and birds. Second, the strongly sexually dimorphic pipefish species (*Syngnathus scovelli* and *Nerophis ophidion*) proved to be genetically polyandrous (71, 80, 101), whereas a seahorse species (*Hippocampus subelongatus*, formerly *H. angustus*) in which males and females show no elaboration of secondary sexual traits was genetically monogamous within a breeding episode (74, 91). Furthermore, two pipefish species (*S. typhle* and *S. floridae*) that are intermediate in level of sexual dimorphism displayed a polygynandrous genetic mating system in which many females and males probably had multiple mating partners during the course of a male pregnancy (72, 77, 79).

These initial genetic findings for syngnathids conform to the general expectations of sexual-selection theory and mating-system evolution as applied to taxonomic groups containing role-reversed species. Caution is warranted, however, because many proximate ecological factors (as well as phylogenetic constraints) may also influence mating systems (73). In only a few fish species have molecular parentage analyses been applied to two or more populations, and pronounced geographic variation in the genetic mating system sometimes has (87, 138) but at other times has not (81) been present.

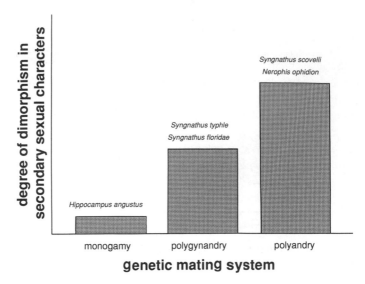

Figure 3 Observed relationships between the degree of sexual dimorphism and the genetic mating system (as deduced from molecular parentage analyses) in each of four pipefish (*Syngnathus* and *Nerophis*) and one seahorse (*Hippocampus*) species (see text).

SEX-ROLE REVERSAL AND BATEMAN'S GRADIENTS   In the literature on animal mating systems, the relative intensities of sexual selection on the two genders have been variously attributed to differences in parental investment (111, 139), operational sex ratio (90), relative variances in reproductive success (112, 144), and potential reproductive rates of the sexes (27, 28), among others. Although such factors certainly impact the nature of sexual selection on males and females, Bateman (12) argued more than 50 years ago that they all do so via one common denominator or first-order factor: the average relationship between the number of mates an individual obtains (its mating success) and the number of offspring it produces (its reproductive success or genetic fitness).

Working with experimental populations of *Drosophila*, Bateman noticed that males' mean genetic fitness tended to increase rapidly with mating success (yielding a steep, linear selection gradient), whereas females gained little in offspring counts by mating with multiple males (a shallow or flat selection gradient). Bateman saw this disparity as the true cause of differential sexual selection; multiple mating afforded to males a higher fitness payoff than it did to females. More recently, "Bateman's gradients" have been touted by some authors as quantitative keys to comparing the intensities of sexual selection across species as well (3, 5).

In sex-role-reversed taxa, steeper slopes in Bateman's gradients are predicted for females than for males (the reverse of the usual situation in *Drosophila*, and in many mammals, birds, and other taxonomic groups). Using microsatellite-based paternity analyses to assay the reproductive success of genetically marked individuals, Jones et al. (76) critically tested this expectation using aquarium populations of a role-reversed pipefish species, *S. typhle*. Consistent with theory, the sexual selection gradients proved to be significantly steeper for females than for males. Results supported the Bateman-gradient approach for characterizing the strength and direction of sexual selection, and its underlying notion that the relationship of mating success to fitness is a cardinal feature in the process of sexual selection.

EXTREME POLYANDRY AND SEX-ROLE REVERSAL   A focused genetic study of the gulf pipefish, *S. scovelli*, uncovered the most compelling data yet available for any vertebrate species that sexual selection in nature can act more strongly on females than on males (80). One small population from a well-demarcated patch of seagrass was sampled exhaustively, thus enabling more complete evaluations than are normally possible on rates of genetic parentage by individuals of both genders. From genetic maternity and paternity analyses of the 21 broods, each pregnant male had mated with only one female, but on average a female had mated with 2.2 males. Furthermore, the standardized variance in female mating success (the variance in the number of embryos dammed by females, divided by the square of the mean—a gauge of the opportunity for sexual selection), proved to be at least seven times greater than the standardized variance in the mating success of males (including those not pregnant). This may represent the highest female-biased asymmetry of reproductive roles yet documented in nature for any vertebrate species, including several of the well-known shorebirds with sex-role reversal such as the phalaropes and jacanas [see citations in (80)].

PHYLOGENETIC CHARACTER MAPPING   A popular approach in recent years is to trace the evolutionary origin and modification of particular morphological or behavioral features through species' phylogenies estimated independently from molecular or other evidence (98). For example, a cladogram for the Syngnathidae, based on mtDNA sequences, was recently generated and used as a phylogenetic backdrop for interpreting the diversification of varied brood pouch morphologies within the Family (152).

This same phylogeny also provided a foundation for interpreting genetic paternity data in the context of the evolutionary rationale for brood pouch elaboration (101). *Nerophis ophidion* is unusual among syngnathid species in that adult males fertilize eggs externally and carry the resulting embryos on the outside of their bodies, rather than in an enclosed brood pouch. This arrangement opens a possibility for fertilization thievery by other males. Nonetheless, paternity analyses based on microsatellites showed that cuckoldry in this species is rare or nonexistent (101). The basal lineage leading to *N. ophidion* branched off early in the syngnathid family tree. Thus, the genetic paternity data suggest that the evolutionary elaboration of enclosed brood pouches in other species of pipefishes and seahorses probably was not in response to strong selection pressures on pregnant males to circumvent cuckoldry, but rather as a means to enhance offspring care and protection (101).

Broadcast Spawners

The majority of fish species provide no parental care to their offspring (18, 19). Typically, the dispersed fry from a spawning event do not remain associated with particular candidate sires or dams, so parentage analyses are far more challenging and problematic. Nonetheless, such focused genetic appraisals have proved fruitful under some circumstances.

Avise & Shapiro (11) used allozyme markers to test the hypothesis (127) that schooling juveniles of an open-water-spawning coral reef fish, *Anthias squamipinnis*, had remained together throughout the pelagic dispersal stage and settled onto a reef as full-sib cohorts. The genetic data proved that juveniles in each school were not close relatives, but instead were a random draw from the local gene pool. In a study of a European minnow, *Phoxinus phoxinus*, Naish et al. (105) likewise showed that discrete schools consisted of unrelated individuals. Conversely, microsatellite markers revealed that discrete fish shoals in the tilapia, *Sarotherodon melanotheron*, often consisted of closely related kin (119).

In experimental populations of free-spawning cod, *Gadus morhua*, microsatellite assessments revealed that although larger males typically achieved a higher reproductive output, multiple males contributed sperm to most of the monitored spawnings (13). Such genetic studies can be used to quantify the variance in reproductive success across males. In cod, these variances were large, and the authors suggest that highly skewed paternity is an important factor contributing to the low ratios of effective population size to census population size often reported in abundant marine organisms.

OTHER PHENOMENA ELUCIDATED

Apart from revealing mating behaviors and reproductive phenomena per se, microsatellite analyses of fish parentage have also found innovative service in some unexpected applications. Three such examples follow.

Clustered Mutations

Population genetic theory traditionally treats each de novo mutation event as producing a variant allele that enters the population as a singleton. However, any mutation arising in a premeiotic germ cell lineage is likely to be copied and distributed to multiple gametes, and, hence, to two or more of that parent's progeny. Such "clustered mutations" have been theoretically neglected and empirically overlooked except in a few organisms such as *Drosophila* (153).

Fish species with large clutches and offspring care by a known parent provide favorable settings to search for clustered mutations. In one microsatellite study of 3195 brooded embryos screened in 110 families of the pipefish species *S. typhle*, a total of 35 de novo mutations were detected (78). These conformed well to a stepwise mutation model, had arisen in both paternal and maternal lines, and yielded standard mutation-rate estimates for microsatellite loci in vertebrates (about 10^{-3} per gamete per generation). Of greater interest, however, was the fact that several of the mutations were clustered—present in multiple progeny within a brood. These findings in a fish extend the documentation of clustered mutations to another major taxonomic group.

Filial Cannibalism

Cannibalism is a widespread phenomenon in fishes: Partially digested remains of conspecific juveniles are often found inside the stomachs of adults (46, 48). Especially intriguing is filial cannibalism, wherein an adult purportedly has eaten some of his own biological offspring (66, 95, 108, 109). This counterintuitive behavior might be rationalized as an adaptive response for removing fungal-invested embryos from a clutch, thereby improving overall offspring survival in the nest (31), or for otherwise enhancing a guardian's genetic fitness if his longer-term net gain in reproductive output (e.g., by avoiding starvation, or circumventing nest desertion) outweighs the immediate fitness cost of reducing offspring numbers via filial cannibalism (66, 123, 125, 128).

Despite extensive theory and field observations on suspected filial cannibalism, the phenomenon itself, by hard criteria, was not definitively confirmed in nature until molecular paternity analyses were applied. Using PCR-based assays, DeWoody et al. (40) genetically proved that the partially digested remains of several dozen embryos recovered from the stomachs of nest-tending tessellated darters (*Etheostoma olmstedi*) and sunfish (*Lepomis punctatus* and *L. auritus*) were indeed the biological progeny of their respective fathers. This outcome was not a foregone conclusion. For reasons described above, fish nests often contain some embryos not sired by the guardian male, and many cold-blooded organisms, including some fish (22), possess refined kin-recognition capabilities (e.g., 51, 93) that might be suspected to play a role in lessening cannibalism rates on close kin (96, 115).

Local Population Sizes

A widespread approach in wildlife biology is to use physical traps in mark-recapture protocols to estimate the contemporary size (n) of a local population. For example, under the commonly employed Lincoln-Peterson statistic to analyze such data, $n = (n_1 + 1)(n_2 + 1)/(m_2 + 1) - 1$, where n_1 is the number of animals captured and physically marked in an initial sample, n_2 is the number of animals caught later, and m_2 is the number of recaptured (marked) animals in the second sample (117).

In a modification of this approach, Jones & Avise (72) used the data from genetic parentage analysis as a novel part of the mark-recapture protocol. In a population of the pipefish *S. floridae*, the initial marks were provided by the deduced genotypes of females (n_1) who had contributed to the broods of pregnant males, the genotypes of assayed adult females were considered the second sample (n_2), and females that matched the maternal genotypes in particular broods were the "recaptures" (m_2).

These observed counts produced an estimate of about 85–192 adult females in the local population.

As discussed by Pearse et al. (113), several modifications of this genetic approach also can be envisioned. For example, in a population that is monitored over multiple breeding seasons, both the marks and recaptures could come from the genetically deduced maternal (or paternal) genotypes in successive clutches of embryos. This method has the distinct advantage for some species in that there is never a need to physically trap the alternate sex because genes provide the marks, and breeding individuals of one sex in effect provide both the captures and the recaptures of the opposite gender (via mating). Also, the resulting estimates of n for a given population refer explicitly to number of successful breeders, a parameter that is of special interest in many circumstances.

CONCLUSIONS

Fish parentage analyses based on microsatellites or other molecular markers have unveiled facets of reproductive natural history and mating systems that would be difficult if not impossible to detect by other means. Furthermore, as we hope to have demonstrated, many of these findings have important ramifications for broader scientific thought about the ecology and evolution of animal reproductive strategies.

ACKNOWLEDGMENTS

Recent work in the Avise laboratory has been supported by funds from the University of Georgia and by a Fellowship in Marine Conservation from the Pew Foundation.

The *Annual Review of Genetics* is online at http://genet.annualreviews.org

LITERATURE CITED

1. Alonzo SH, Warner RR. 2000. Female choice, conflict between the sexes and the evolution of male alternative reproductive behaviours. *Evol. Ecol. Res.* 2:149–70
2. Andersson M. 1994. *Sexual Selection* Princeton, NJ: Princeton Univ. Press
3. Andersson M, Iwasa Y. 1996. Sexual selection. *TREE* 11:53–58
4. Arnold SJ. 1983. Sexual selection: the interface of theory and empiricism. In *Mate Choice*, ed. P Bateson, pp. 67–108. Cambridge, UK: Cambridge Univ. Press

5. Arnold SJ, Duvall D. 1994. Animal mating systems: a synthesis based on selection theory. *Am. Nat.* 143:317–48
6. Avise JC. 1994. *Molecular Markers, Natural History and Evolution.* New York: Chapman & Hall. 511 pp.
7. Avise JC. 2001. Cytonuclear genetic signatures of hybridization phenomena: rationale, utility, and empirical examples from fishes and other aquatic animals. *Rev. Fish Biol. Fish.* 10:461–69
8. Avise JC, ed. 2001. DNA-based profiling

of mating systems and reproductive behaviors in poikilothermic vertebrates. *J.
Hered.* 92:99–211

9. Avise JC. 2002. *Genetics in the Wild.*
Washington, DC: Smithsonian Inst. Press

10. Avise JC, Quattro JM, Vrijenhoek RC.
1992. Molecular clones within organismal
clones: mitochondrial DNA phylogenies
and the evolutionary histories of unisexual vertebrates. *Evol. Biol.* 26:225–46

11. Avise JC, Shapiro DY. 1986. Evaluating
kinship of newly settled juveniles within
social groups of the coral reef fish *Anthias
squamipinnis. Evolution* 40:1051–59

12. Bateman AJ. 1948. Intra-sexual selection
in *Drosophila. Heredity* 2:349–68

13. Bekkevold D, Hansen MM, Loeschcke
V. 2002. Male reproductive competition
in spawning aggregations of cod (*Gadus
morhua*, L.). *Mol. Ecol.* 11:91–102

14. Bell MA, Foster SA, eds. 1994. *The Evolutionary Biology of the Threespine Stickleback.* New York: Oxford Univ. Press

15. Berglund A, Rosenqvist G, Svensson I.
1989. Reproductive success of females
limited by males in two pipefish species.
Am. Nat. 133:506–16

16. Birkhead TR. 2000. *Promiscuity.* Cambridge, MA: Harvard Univ. Press. 272 pp.

17. Birkhead TR, Møller AP. 1992. *Sperm
Competition in Birds: Evolutionary Causes and Consequences.* London: Academic

18. Blumer LS. 1979. Male parental care in
the bony fishes. *Q. Rev. Biol.* 54:149–61

19. Blumer LS. 1982. A bibliography and
categorization of bony fishes exhibiting
parental care. *Zool. J. Linnean Soc.* 76:1–
22

20. Breder CM Jr. 1936. The reproductive
habits of North American sunfishes (family Centrarchidae). *Zoologica* 21:1–47

21. Breder CM, Rosen DE. 1966. *Modes of
Reproduction in Fishes.* Garden City, NY:
Natl. Hist. Press

22. Brown GE, Brown JA. 1996. Kin discrimination in salmonids. *Rev. Fish Biol. Fish.*
6:201–19

23. Brown JL. 1987. *Helping and Communal
Breeding in Birds.* Princeton, NJ: Princeton Univ. Press

24. Burke T, Dolf G, Jeffreys AJ, Wolff R, eds.
1991. *DNA Fingerprinting: Approaches
and Applications.* Basel: Birkhauser
Verlag

25. Chesser RK, Smith MW, Smith MH.
1984. Biochemical genetics of mosquitofish. III. Incidence and significance of
multiple paternity. *Genetica* 74:77–81

26. Clutton-Brock TH. 1991. *The Evolution
of Parental Care.* Princeton, NJ: Princeton Univ. Press

27. Clutton-Brock TH, Parker GA. 1992. Potential reproductive rates and the operation of sexual selection. *Q. Rev. Biol.*
67:437–56

28. Clutton-Brock TH, Vincent ACJ. 1991.
Sexual selection and the potential reproductive rates of males and females. *Nature*
351:58–60

29. Colbourne JK, Neff BD, Wright JM,
Gross MR. 1996. DNA fingerprinting of
bluegill sunfish (*Lepomis macrochirus*)
using $(GT)_n$ microsatelites and its potential for assessment of mating success.
Can. J. Fish. Aquat. Sci. 53:342–49

30. Constantz GD. 1984. Sperm competition
in poeciliid fishes. In *Sperm Competition
and the Evolution of Animal Mating Systems*, ed. RL Smith, pp. 465–86. Orlando,
FL: Academic

31. Constantz GD. 1985. Alloparental care in
the tessellated darter (Pisces: Percidae).
Env. Biol. Fish. 14:175–83

32. Crowley PH. 2000. Hawks, doves, and
mixed-symmetry games. *J. Theor. Biol.*
204:543–63

33. Darling JDS, Noble ML, Shaw E. 1980.
Reproductive strategies in the surfperches. I. Multiple insemination in natural
populations of the shiner perch, *Cymatogaster aggregata. Evolution* 34:271–
77

34. Dawson CE. 1985. *Indo-Pacific Pipefishes.* Ocean Springs, MS: Gulf Coast Res.
Lab.

35. DeWoody JA, Avise JC. 2000. Microsatellite variation in marine, freshwater and anadromous fishes compared with other animals. *J. Fish Biol.* 56:461–73

36. DeWoody JA, Avise JC. 2001. Genetic perspectives on the natural history of fish mating systems. *J. Hered.* 92:167–72

37. DeWoody JA, DeWoody YD, Fiumera AC, Avise JC. 2000. On the number of reproductives contributing to a half-sib progeny array. *Genet. Res. Camb.* 75:95–105

38. DeWoody JA, Fletcher D, Mackiewicz M, Wilkins SD, Avise JC. 2000. The genetic mating system of spotted sunfish (*Lepomis punctatus*): mate numbers and the influence of male reproductive parasites. *Mol. Ecol.* 9:2119–28

39. DeWoody JA, Fletcher DE, Wilkins SD, Avise JC. 2000. Parentage and nest guarding in the tessellated darter (*Etheostoma olmstedi*) assayed by microsatellite markers (Perciformes: Percidae). *Copeia* 2000:740–47

40. DeWoody JA, Fletcher DE, Wilkins SD, Avise JC. 2001. Genetic documentation of filial cannibalism in nature. *Proc. Natl. Acad. Sci. USA* 98:5090–92

41. DeWoody JA, Fletcher DE, Wilkins SD, Nelson WS, Avise JC. 1998. Molecular genetic dissection of spawning, parentage, and reproductive tactics in a population of redbreast sunfish, *Lepomis auritus. Evolution* 52:1802–10

42. DeWoody JA, Fletcher DE, Wilkins SD, Nelson WS, Avise JC. 2000. Genetic monogamy and biparental care in an externally-fertilizing fish, the largemouth bass (*Micropterus salmoides*). *Proc. R. Soc. London Ser. B* 267:2431–37

43. DeWoody JA, Walker D, Avise JC. 2000. Genetic parentage in large half-sib clutches: theoretical estimates and empirical appraisals. *Genetics* 154:1907–12

44. Dierkes P, Taborsky M, Kohler U. 1999. Reproductive parasitism of broodcare helpers in a cooperatively breeding fish. *Behav. Ecol.* 10:510–15

45. Dominey WJ. 1980. Female mimicry in male bluegill sunfish—a genetic polymorphism? *Nature* 284:546–48

46. Dominey WJ, Blumer LS. 1984. Cannibalism of early life history stages in fishes. In *Infanticide: Comparative and Evolutionary Perspectives*, ed. G Hausfater, SB Hrdy, pp. 43–64. New York: Aldine

47. Eberhard WG. 1996. *Female Control: Sexual Selection by Cryptic Female Choice.* Princeton, NJ: Princeton Univ. Press

48. Elgar MA, Crespi BJ. 1992. *Cannibalism: Ecology and Evolution among Diverse Taxa.* Oxford, UK: Oxford Univ. Press

49. Emlen ST. 1991. Evolution of cooperative breeding in birds and mammals. In *Behavioural Ecology*, ed. JR Krebs, NB Davies, pp. 301–37. Oxford, UK: Blackwell

50. Evans JP, Magurran AE. 2000. Multiple benefits of multiple mating in guppies. *Proc. Natl. Acad. Sci. USA* 97:10074–76

51. Evans TA. 1999. Kin recognition in a social spider. *Proc. R. Soc. London Ser. B* 266:287–92

52. Ferraris JD, Palumbi SR. 1996. *Molecular Zoology.* New York: Wiley-Liss. 580 pp.

53. Fiumera AC, DeWoody YD, DeWoody JA, Asmussen MA, Avise JC. 2001. Accuracy and precision of methods to estimate the number of parents contributing to a half-sib progeny array. *J. Hered.* 92:120–26

53a. Fiumera AC, Porter BA, Grossman GD, Avise JC. 2002. Intensive genetic assessment of the mating system and reproductive success in a semi-closed population of the mottled sculpin, *Cottus bairdi. Mol. Ecol.* In press

54. Flaxman SM. 2000. The evolutionary stability of mixed strategies. *TREE* 15:482–84

55. Garant D, Dodson JJ, Bernatchez L. 2001. A genetic evaluation of mating system and determinants of individual reproductive

success in Atlantic salmon (*Salmo salar* L.). *J. Hered.* 92:137–45

56. Garcia-Vazquez E, Morán P, Martinez JL, Perez J, de Gaudemar B, Beall E. 2001. Alternative mating strategies in Atlantic salmon and brown trout. *J. Hered.* 92:146–49

57. Goldstein DB, Schlötterer C, eds. 1999. *Microsatellites: Evolution and Applications.* Oxford, UK: Oxford Univ. Press

58. Goodwin NB, Balshine-Earn S, Reynolds JD. 1998. Evolutionary transitions in parental care in cichlid fish. *Proc. R. Soc London Ser. B* 265:2265–72

59. Gross MR. 1979. Cuckoldry in sunfishes (*Lepomis*: Centrarchidae). *Can. J. Zool.* 57:1507–9

60. Gross MR. 1991. Evolution of alternative reproductive strategies: frequency-dependent sexual selection in male bluegill sunfish. *Philos. Trans. R. Soc. London Ser. B* 332:59–66

61. Gross MR. 1996. Alternative reproductive strategies and tactics: diversity within sexes. *TREE* 11:92–98

62. Gross MR, Charnov EL. 1980. Alternative male life histories in bluegill sunfish. *Proc. Natl. Acad. Sci. USA* 77:6937–40

63. Gross MR, Sargent RC. 1985. The evolution of male and female parental care in fishes. *Am. Zool.* 25:807–22

64. Hamilton WD. 1964. The genetical evolution of social behavior. *J. Theor. Biol.* 7:1–52

65. Henson SA, Warner RR. 1997. Male and female alternative reproductive behaviors in fishes: a new approach using intersexual dynamics. *Annu. Rev. Ecol. Syst.* 28:571–92

66. Hoelzer GA. 1995. Filial cannibalism and male parental care in damselfishes. *Bull. Mar. Sci.* 57:663–71

67. Houde AE. 1997. *Sex, Color, and Mate Choice in Guppies.* Princeton, NJ: Princeton Univ. Press

68. Hutchings JA, Myers RA. 1988. Mating success of alternative maturation phenotypes in male Atlantic salmon, *S. salar. Oecologica* 75:169–74

69. Jamieson A, Taylor SCS. 1997. Comparisons of three probability formulae for parentage exclusion. *Anim. Genet.* 28:397–400

70. Jeffreys AJ, Wilson V, Thein DL. 1985. Hypervariable 'minisatellite' regions in human DNA. *Nature* 314:67–73

71. Jones AG, Avise JC. 1997. Microsatellite analysis of maternity and the mating system in the Gulf pipefish *Syngnathus scovelli*, a species with male pregnancy and sex-role reversal. *Mol. Ecol.* 6:203–13

72. Jones AG, Avise JC. 1997. Polygynandry in the dusky pipefish *Syngnathus floridae* revealed by microsatellite DNA markers. *Evolution* 51:1611–22

73. Jones AG, Avise JC. 2001. Mating systems and sexual selection in male-pregnant pipefishes and seahorses: insights from microsatellite-based studies of maternity. *J. Hered.* 92:150–58

74. Jones AG, Kvarnemo C, Moore GI, Simmons LW, Avise JC. 1998. Microsatellite evidence for monogamy and sex-biased recombination in the Western Australian seahorse *Hippocampus angustus. Mol. Ecol.* 7:1497–505

75. Jones AG, Östlund-Nilsson S, Avise JC. 1998. A microsatellite assessment of sneaked fertilizations and egg thievery in the fifteenspine stickleback. *Evolution* 52:848–58

76. Jones AG, Rosenqvist G, Berglund A, Arnold SJ, Avise JC. 2000. The Bateman gradient and the cause of sexual selection in a sex-role-reversed pipefish. *Proc. R. Soc. London Ser. B* 267:677–80

77. Jones AG, Rosenqvist G, Berglund A, Avise JC. 1999. The genetic mating system of a sex-role-reversed pipefish (*Syngnathus typhle*): a molecular inquiry. *Behav. Ecol. Sociobiol.* 46:357–65

78. Jones AG, Rosenqvist G, Berglund A, Avise JC. 1999. Clustered microsatellite mutations in the pipefish *Syngnathus typhle. Genetics* 152:1057–63

Jones AG, Rosenqvist G, Berglund A, Avise JC. 2000. Mate quality influences multiple maternity in the sex-role-reversed pipefish *Syngnathus typhle*. *Oikos* 90:321–26

80. Jones AG, Walker D, Avise JC. 2001. Genetic evidence for extreme polyandry and extraordinary sex-role reversal in a pipefish. *Proc. R. Soc. London Ser. B* 268:2531–35

81. Jones AG, Walker D, Kvarnemo C, Lindström K, Avise JC. 2001. Surprising similarity of sneaking rates and genetic mating patterns in two populations of the sand goby experiencing disparate sexual selection regimes. *Mol. Ecol.* 10:461–69

82. Jones AG, Walker D, Kvarnemo C, Lindström K, Avise JC. 2001. How cuckoldry can decrease the opportunity for sexual selection: data and theory from a genetic parentage analysis of the sand goby, *Pomatoschistus minutus*. *Proc. Natl. Acad. Sci. USA* 98:9151–56

83. Jordan WC, Youngson AF. 1992. The use of genetic marking to assess the reproductive success of mature male Atlantic salmon parr (*Salmo salar* L.) under natural spawning conditions. *J. Fish Biol.* 41:613–18

84. Keenleyside M, ed. 1991. *Cichlid Fishes: Behavior, Ecology, and Evolution*. London: Chapman & Hall

85. Kellogg KA, Markert JA, Stauffer JR, Kocher TD. 1995. Microsatellite variation demonstrates multiple paternity in lekking cichlid fishes from Lake Malawi, Africa. *Proc. R. Soc. London Ser. B* 260:79–84

86. Kellogg KA, Markert JA, Stauffer JR, Kocher TD. 1998. Intraspecific brood mixing and reduced polyandry in a maternal mouth-brooding cichlid. *Behav. Ecol.* 9:309–12

87. Kelly CD, Godin A-GJ, Wright JM. 1999. Geographical variation in multiple paternity within natural populations of the guppy (*Poecilia reticulata*). *Proc. R. Soc. London Ser. B* 266:2403–8

88. Kornfield I, Smith PF. 2000. African cichlid fishes: model systems for evolutionary biology. *Annu. Rev. Ecol. Syst.* 31:163–96

89. Kraak SBM, Weissing FJ. 1996. Female preference for nests with many eggs: a cost-benefit analysis of female choice in fish with paternal care. *Behav. Ecol.* 7:353–61

90. Kvarnemo C, Ahnesjö I. 1996. The dynamics of operational sex ratios and competition for mates. *TREE* 11:404–12

91. Kvarnemo C, Moore GI, Jones AG, Nelson WS, Avise JC. 2000. Monogamous pair-bonds and mate switching in the Western Australian seahorse *Hippocampus subelongatus*. *J. Evol. Biol.* 13:882–88

92. Lack D. 1968. *Ecological Adaptations for Breeding in Birds*. London: Chapman & Hall

93. Lazzaretto I, Salvato B. 1992. Cannibalistic behaviour in the harpacticoid copepod *Tigriopus fulvus*. *Mar. Biol.* 113:579–82

94. Li SK, Owings DH. 1978. Sexual selection in the three-spined stickleback. II. Nest raiding during the courtship phase. *Behaviour* 64:298–304

95. Lindström K. 2000. The evolution of filial cannibalism and female mate choice strategies as resolutions to sexual conflict in fishes. *Evolution* 54:617–27

96. Loiselle PV. 1983. Filial cannibalism and egg recognition by males of the primitively custodial teleost *Cyprinodon macularius californiensis* Girard (Atherinomorpha: Cyprinodontidae). *Ethol. Sociobiol.* 4:1–9

97. Martinez JL, Morán P, Perez J, DeGaudemar B, Beall E, Garcia-Vazquez E. 2000. Multiple paternity increases effective population size of southern Atlantic salmon populations. *Mol. Ecol.* 9:293–98

98. Martins EP, ed. 1996. *Phylogenies and the Comparative Method in Animal Behavior*. New York: Oxford Univ. Press

99. Masonjones HD, Lewis SM. 2000. Differences in potential reproductive rates

of male and female seahorses related to courtship roles. *Anim. Behav.* 59:11–20

100. Maynard Smith J. 1982. *Evolution and the Theory of Games.* Cambridge, UK: Cambridge Univ. Press

101. McCoy EE, Jones AG, Avise JC. 2001. The genetic mating system and tests for cuckoldry in a pipefish species in which males fertilize eggs and brood offspring externally. *Mol. Ecol.* 10:1793–800

102. Meyer A. 1987. Phenotypic plasticity and heterochrony in *Cichlasoma managuense* (Pisces, Cichlidae) and their implications for speciation in cichlid fishes. *Evolution* 41:1357–69

103. Morán P, Garcia-Vazquez E. 1998. Multiple paternity in Atlantic salmon: a way to maintain genetic variability in relicted populations. *J. Hered.* 89:551–53

104. Morán P, Pendás, Beall E, Garcia-Vazquez E. 1996. Genetic assessment of the reproductive success of Atlantic salmon precocious parr by means of VNTR loci. *Heredity* 77:655–60

105. Naish KA, Carvalho GR, Pitcher TJ. 1993. The genetic structure and microdistribution of shoals of *Phoxinus phoxinus*, the European minnow. *J. Fish Biol.* 43A:75–89

106. Neff BD. 2001. Genetic paternity analysis and breeding success in bluegill sunfish (*Lepomis macrochirus*). *J. Hered.* 92:111–19

107. Neff BD, Gross M. 2001. Dynamic adjustment of parental care in response to perceived paternity. *Proc. R. Soc. London Ser. B* 268:1559–65

108. Nemtzov SC, Clark E. 1994. Intraspecific egg predation by male razorfishes (Labridae) during broadcast spawning: filial cannibalism or intra-pair parasitism? *Bull. Mar. Sci.* 55:133–41

109. Okuda N. 1999. Female mating strategy and male brood cannibalism in a sand-dwelling cardinalfish. *Anim. Behav.* 58:273–79

110. Parker A, Kornfield I. 1996. Polygynandry in *Pseudotropheus zebra*, a cichlid fish from Lake Malawi. *Env. Biol. Fish.* 47:345–52

111. Parker GA, Simmons LW. 1996. Parental investment and the control of sexual selection: predicting the direction of sexual competition. *Proc. R. Soc. London Ser. B* 263:315–21

112. Payne RB. 1979. Sexual selection and intersexual differences in variance of breeding success. *Am. Nat.* 114:447–52

113. Pearse DE, Eckerman CM, Janzen FJ, Avise JC. 2001. A genetic analogue of "mark-recapture" methods for estimating population size: an approach based on molecular parentage assessments. *Mol. Ecol.* 10:2711–18

114. Pearse DE, Janzen FJ, Avise JC. 2001. Genetic markers substantiate long-term storage and utilization of sperm by female painted turtles. *Heredity* 86;378–84

115. Pfennig DW. 1999. Cannibalistic tadpoles that pose the greatest threat to kin are most likely to discriminate kin. *Proc. R. Soc. London Ser. B* 266:57–61

116. Philipp DP, Gross MR. 1994. Genetic evidence for cuckoldry in bluegill *Lepomis macrochirus. Mol. Ecol.* 3:563–69

117. Pollock KH, Nichols JD, Brownie C, Hines JE. 1990. Statistical inference for capture-recapture experiments. *Wildl. Monogr.* 107:1–97

118. Porter BA, Fiumera AC, Avise JC. 2002. Egg mimicry and allopaternal care: two mate attracting tactics by which striped darter (*Etheostoma virgatum*) males enhance reproductive success. *Behav. Ecol. Sociobiol.* 51:350–59

119. Pouyard L, Desmarais E, Chenuil A, Agnese JF, Bonhomme F. 1999. Kin cohesiveness and possible inbreeding in the mouthbrooding tilapia *Sarotherodon melanotheron* (Pisces Cichlidae). *Mol. Ecol.* 8:803–12

120. Ribbink AJ, Marsh AC, Marsh B, Sharp BJ. 1980. Parental behaviour and mixed broods among cichlid fish of Lake Malawi. *S. Afr. J. Zool.* 15:1–6

121. Rico C, Kuhnlein U, Fitzgerrald GJ. 1992. Male reproductive tactics in the threespine stickleback—an evaluation by DNA fingerprinting. *Mol. Ecol.* 1:79–87

122. Ridley M, Rechten C. 1981. Female sticklebacks prefer to spawn with males whose nests contain eggs. *Behaviour* 76:152–61

123. Rohwer S. 1978. Parent cannibalism of offspring and egg raiding as a courtship strategy. *Am. Nat.* 112:429–40

124. Sage RD, Selander RK. 1975. Trophic radiation through polymorphism in cichlid fishes. *Proc. Natl. Acad. Sci. USA* 72: 4669–73

125. Sargent RC. 1992. Ecology of filial cannibalism in fish: theoretical perspectives. In *Cannibalism: Ecology and Evolution among Diverse Taxa*, ed. MA Elgar, BJ Crespi, pp. 38–62. New York: Oxford Univ. Press

126. Selvin S. 1980. Probability of nonpaternity determined by multiple allele codominant systems. *Am. J. Hum. Genet.* 15:997–1008

127. Shapiro DY. 1983. On the possibility of kin groups in coral reef fishes. In *Ecology of Deep and Shallow Reefs*, ed. ML Reaka, pp. 39–45. Washington, DC: NOAA

128. Smith C. 1992. Filial cannibalism as a reproductive strategy in care-giving teleosts? *Neth. J. Zool.* 42:607–13

129. Smith RL, ed. 1984. *Sperm Competition and the Evolution of Animal Mating Systems*. Orlando, FL: Academic

130. Taborsky M. 1994. Sneakers, satellites, and helpers: parasitic and cooperative behavior in fish reproduction. *Adv. Study Behav.* 23:1–100

131. Taborsky M. 1998. Sperm competition in fish: "bourgeois" males and parasitic spawning. *TREE* 13:222–27

132. Taborsky M. 2001. The evolution of bourgeois, parasitic and cooperative reproductive behaviors in fishes. *J. Hered.* 92:100–9

133. Taborsky M, Hudde B, Wirtz P. 1987. Reproductive behaviour and ecology of *Symphodus (Crenilabrus) ocellatus*, a European wrasse with four types of male behaviour. *Behaviour* 102:82–118

134. Taborsky M, Limberger D. 1981. Helpers in fish. *Behav. Ecol. Sociobiol.* 8:143–45

135. Thomaz D, Beall E, Burke T. 1997. Alternative reproductive tactics in Atlantic salmon: factors affecting mature parr success. *Proc. R. Soc. London Ser. B* 264:219–26

136. Thresher RE. 1984. *Reproduction in Reef Fishes*. Neptune City, NJ: TFH

137. Travis J, Trexler JC, Mulvey M. 1990. Multiple paternity and its correlates in female *Poecilia latipinna* (Poeciliidae). *Copeia* 1990:722–29

138. Trexler JC, Travis J, Dinep A. 1997. Variation among populations of the sailfin molly in the rate of concurrent multiple paternity and its implications for mating-system evolution. *Behav. Ecol. Sociobiol.* 40:297–305

139. Trivers RL. 1972. Parental investment and sexual selection. In *Sexual Selection and the Descent of Man*, ed. B Campbell, pp. 136–79. Chicago: Aldine

140. Turner BJ, Grosse DJ. 1980. Trophic differentiation in *Ilyodon*, a genus of stream-dwelling goodeid fishes: speciation versus ecological polymorphism. *Evolution* 34:259–70

141. van den Berghe E. 1990. Variable parental care in a labrid fish: How care might evolve. *Ethology* 84:319–33

142. Vincent A, Ahnesjö I, Berglund A, Rosenqvist G. 1992. Pipefishes and seahorses: Are they all sex-role reversed? *TREE* 7:237–41

143. Vladic TV, Järvi T. 2001. Sperm quality in the alternative reproductive tactics of Atlantic salmon: the importance of the loaded raffle mechanism. *Proc. R. Soc. London Ser. B* 268:2375–81

144. Wade MJ, Arnold SJ. 1980. The intensity of sexual selection in relation to male sexual behavior, female choice and sperm precedence. *Anim. Behav.* 28:446–61

145. Warner RR, Harlan RK. 1982. Sperm competition and sperm storage as determinants of sexual dimorphism in the dwarf surfperch, *Micrometrus minimus*. *Evolution* 36:44–55

146. Warner RR, Hoffman SG. 1980. Local population size as a determinant of mating system and sexual composition in two tropical marine fishes (*Thalassoma* spp.). *Evolution* 34:508–18

147. Warner RR, Lejeune P. 1985. Sex change limited by parental care: a test using four Mediterranean labrid fishes, genus *Symphodus*. *Mar. Biol.* 87:89–99

148. Warner RR, Robertson DR, Leigh EG Jr. 1975. Sex change and sexual selection. *Science* 190:633–38

149. Wells KD. 1977. The social behaviour of anuran amphibians. *Anim. Behav.* 25:666–93

150. Whoriskey FG, FitzGerald GJ. 1994. Ecology of the threespine stickleback on the breeding grounds. See Ref. 14, pp. 188–206

151. Williams GC. 1975. *Sex and Evolution.* Princeton, NJ: Princeton Univ. Press

152. Wilson AB, Vincent A, Ahnesjö I, Meyer A. 2001. Male pregnancy in seahorses and pipefishes (Family Syngnathidae): Rapid diversification of paternal brood pouch morphology inferred from a molecular phylogeny. *J. Hered.* 92:159–66

153. Woodruff RC, Huai H, Thompson JN Jr. 1996. Clusters of identical new mutation in the evolutionary landscape. *Genetica* 98:149–60

154. Wootton RJ. 1984. *A Functional Biology of Sticklebacks.* Berkeley: Univ. Calif. Press

155. Zane L, Nelson WS, Jones AG, Avise JC. 1999. Microsatellite assessment of multiple paternity in natural populations of a live-bearing fish, *Gambusia holbrooki*. *J. Evol. Biol.* 12:61–69

Annu. Rev. Genet. 200.
doi: 10.1146/annurev.genet.36.0416e

GENETICS OF MOTILITY AND CHEMOTAXIS OF A FASCINATING GROUP OF BACTERIA: The Spirochetes

Nyles W. Charon[1] and Stuart F. Goldstein[2]

[1]Department of Microbiology, Immunology, and Cell Biology, Health Sciences Center, Box 9177, West Virginia University, Morgantown, West Virginia 26506-9177; e-mail: ncharon@hsc.wvu.edu

[2]Department of Genetics, Cell Biology and Development, 250 Biological Sciences Center, University of Minnesota, St. Paul, Minnesota 55108; e-mail: golds004@umn.edu

Key Words Lyme disease, syphilis, motility, flagella, sensory transduction, virulence

■ **Abstract** Spirochetes are a medically important and ecologically significant group of motile bacteria with a distinct morphology. Outermost is a membrane sheath, and within this sheath is the protoplasmic cell cylinder and subterminally attached periplasmic flagella. Here we address specific and unique aspects of their motility and chemotaxis. For spirochetes, translational motility requires asymmetrical rotation of the two internally located flagellar bundles. Consequently, they have swimming modalities that are more complex than the well-studied paradigms. In addition, coordinated flagellar rotation likely involves an efficient and novel signaling mechanism. This signal would be transmitted over the length of the cell, which in some cases is over 100-fold greater than the cell diameter. Finally, many spirochetes, including *Treponema*, *Borrelia*, and *Leptospira*, are highly invasive pathogens. Motility is likely to play a major role in the disease process. This review summarizes the progress in the genetics of motility and chemotaxis of spirochetes, and points to new directions for future experimentation.

CONTENTS

0066-4197/02/1215-0047$14.00

INTRODUCTION

These minute filamental organisms dart through the soft medium with great rapidity, first in one direction and then in another, searching for a loose spot which they can pierce through. When encountering an impenetrable obstacle they reverse their progression and start anew. A striking sight is thus presented by these little vermicular organisms darting in all directions (99).

As Hideyo Noguchi noted above in 1918, and as has been witnessed by biologists from Leeuwenhoek in 1681 (33) to the present, watching the movement of spirochetes is indeed fascinating. Noguchi was describing *Leptospira icterohaemorrhagie*, the then newly described bacterial species that causes the highly fatal Weil's disease. The present review focuses on the genetics of motility and chemotaxis of the spirochete group of bacteria. Spirochete motility and chemotaxis were reviewed several years ago (22, 27, 48) and again more recently within the last two years (78, 84). We hope that the reader comes away with an appreciation that this medically important and ecologically significant group of bacteria has evolved some unique and exciting features related to their motility and chemotaxis. In addition, because spirochete genetics is now in the early stages of development, we also hope that this review is timely in formulating important areas of future research.

Unique Morphology of Spirochetes

Spirochetes are a conspicuous group of motile bacteria (58). Most are helically shaped (59), but some species have a flat sinusoidal or meandering waveform (47, 50). In addition to a typical bacterial plasma membrane surrounded by a cell wall containing peptidoglycan, they have an outer lipid bilayer membrane,

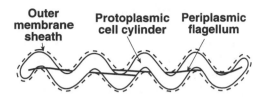

Figure 1 Schematic diagram of spirochete illustrating the outer membrane sheath, protoplasmic cell cylinder, and periplasmic flagella.

also referred to as an outer membrane sheath (Figure 1). The space between the protoplasmic cell cylinder and the outer membrane sheath is referred to as the periplasm. The spirochetes are unique, as their periplasm contains periplasmic flagella that are similar in many respects to the external flagella of rod-shaped bacteria. Each periplasmic flagellum is attached subterminally to only one end of the cell cylinder and extends toward the opposite end. Spirochete species vary with respect to size, number of periplasmic flagella, and whether the periplasmic flagella overlap in the center of the cell. The organisms can be quite large: For example, *Cristispira* are 0.5–3 μm wide, 30–180 μm long, and have over 100 periplasmic flagella attached at each cell end. In contrast, the Leptospiraceae (which include *Leptospira* and *Leptonema* sp.) are approximately 0.1 μm in diameter, 10–20 μm long, and have only one periplasmic flagellum at each end.

Spirochete Diseases and Ecology

Several species of spirochetes cause medically important diseases, some of which are quite prevalent and can have grave consequences. *Treponema pallidum* subspecies *pallidum* causes the dreaded sexually transmitted disease syphilis, and other closely related treponemes cause yaws, bejel, and pinta (2, 100). Worldwide, syphilis is a major disease in developing countries (44). This is of special concern, as syphilis is a recognized cofactor in the acquisition and transmission of human immunodeficiency virus (HIV) (127). *Borrelia burgdorferi* causes Lyme disease, which is the most prevalent vector-borne disease in the United States (23). *Borrelia hermsii* and other *Borrelia* species cause relapsing fever (4). Many *Leptospira* sp. cause leptospirosis (75). This potentially fatal waterborne zoonosis has many possible manifestations and occurs worldwide. *Treponema denticola* and other oral treponemes are associated with periodontal disease (32). *Brachyspira hyodysenteriae* causes swine dysentery, and *Brachyspira pilosicoli* and *Brachyspira aalborgi* are associated with human intestinal infections in developing countries and in immunocompromised individuals (93). Recently, spirochetes have been associated with chronic digital dermatitis in cattle (112).

Spirochetes are a metabolically diverse group of bacteria found in many different habitats, and several of them have yet to be cultured in the laboratory

(19, 21). They vary, for example, with respect to oxygen requirements. *Leptospira* are obligate aerobes; *Spirochaeta* are often facultative; *Borrelia, Brachyspira*, and *T. pallidum* are microaerophilic; and most *Treponema* sp. are obligate anaerobes. They also vary with respect to nitrogen utilization. The free-living *Spirochaeta aurantia* and *Treponema* from termite guts utilize atmospheric nitrogen as a nutrient source. Because the diet of termites is low in nitrogen, these spirochetes may be essential to the termite for nutrient supply (81). Both free-living and commensal spirochetes are widespread in nature. Saprophytic *Leptospira, Spirochaeta*, and other uncultivable species are found in soil, freshwater, and saltwater (57, 91). Other spirochetes are found attached to protozoa sp. in the termite gut. These attached spirochetes enable the protozoa to be motile (22). The large uncultivable *Cristispira* resides in the crystalline style of bivalve mollusks (22, 57). Recently, spirochetes have been found associated with the hydrothermal vent polychaete annelid, *Alvinella pompejana* (20). In sum, besides residing in several different hosts, spirochetes also inhabit soil and saltwater and freshwater.

Selective Advantages of Spirochetes

There are two obvious advantages to being a spirochete. First, these organisms can swim in highly viscous, gel-like media, such as ones containing methylcellulose, that slow down or stop most externally flagellated bacteria (9, 50, 54, 67, 73). Such media include hyaluronic acid and collagen within the mammal (73), and mud or sediments encountered in the environment (22). In fact, the speed of several spirochete species actually accelerates as macroscopic viscosity increases. For example, the speed of *T. denticola* increases from less than 1 μm/sec in liquid media to 19 μm/sec in the presence of 1% methylcellulose (140 centipoise) (108). This attribute may allow *T. denticola* and other spirochetes to penetrate, invade, and adapt to specific ecological niches that exclude other bacterial species (22). Second, the periplasmic flagella reside within the periplasmic space. Because these organelles are not readily accessible to the ambient environment, they do not interact with flagellar-specific antibodies and hence circumvent the immobilization of the intact organism (13, 15, 30). Such antibodies often arise during infection (64, 103). Furthermore, their intracellular location may also offer protection from possibly disruptive extremes in the environment such as pH or salt concentration (36).

Divergent Evolution of Spirochetes

Several decades ago, researchers began investigating the evolutionary origin of spirochetes. Two hypotheses were initially considered (21). First, because these organisms are so markedly diverse with respect to size, metabolism, and ecological niche, and because being a spirochete has selective advantages, their existence could be the result of convergent evolution. Alternatively, the spirochetes could have evolved divergently from a primordial proto-spirochete. To address these hypotheses, the 16S rRNA genes of several pathogenic, parasitic, and free-living species were sequenced. The results indicate that the spirochetes a. an ancient group of Bacteria, and that they evolved from one common ancestra. proto-spirochete. In fact, they are one of the few phyla of Bacteria that can be identified solely from morphology (102). Recently, the entire genomes of T. pallidum, B. burgdorferi, and T. denticola have been or are in the process of being sequenced (39, 40; http://www.hgsc.bcm.tmc.edu/microbial). The results from this analysis also support the conclusion that these spirochetes are evolutionarily closely related.

Spirochetes as Progenitors of Cilia and Flagella

It has been speculated that spirochetes are evolutionary precusors to eukaryotic cilia and flagella (90). Several features, including the similarity of their shapes, their common use of traveling waves for movement, the presence of cytoplasmic filaments that resemble eukaryotic microtubules in some spirochete species (11), and the role of symbiotic spirochetes in the swimming of some protozoans made this an intriguing suggestion. However, several lines of evidence do not support this idea. In contrast to mitochondria and chloroplasts, DNA is not present in flagellar basal bodies (65). An apparently tubulin-like protein identified in some spirochete species has turned out to be a member of the stress family of proteins (10, 97). Further, no tubulin gene sequences were identified in the genomes of the spirochete species that have been sequenced. Instead, these genomes have homologs similar to those encoding the motility genes of other bacteria (39, 40). More importantly, eukaryotic cilia and flagella use ATP-driven dynein arms to produce active sliding between neighboring microtubules (122). In contrast, spirochetes use a proton gradient to power motility (51, 71, 95, 116) and rotate their periplasmic flagella in typical prokaryotic fashion (25, 50). Thus, the suggestion that eukaryotic cilia and flagella arose from spirochetes is less attractive than it once appeared.

Recent Progress in the Genetics of Spirochetes

The genetics of spirochetes is at a very early stage compared to that of other bacterial species. This is due to the complexity of the growth requirements of many spirochete species, their relatively long generation times, and the difficulty of formulating optimal conditions and proper vehicles for targeted mutagenesis. In fact, some medically important spirochete species, such as T. pallidum, have yet to be cultured; consequently, targeted mutagenesis has not been possible for these organisms. As a result, we know very little about spirochete virulence factors.

Two major breakthroughs in spirochete genetics occurred within the past decade. First, in 1992 a successful targeted mutation was obtained by electroporating DNA into B. hyodysenteriae (123). Subsequently, targeted mutagenesis has been achieved with B. burgdorferi (16, 124), T. denticola (80), and Leptospira biflexa (104). In addition, generalized transduction mediated by a bacteriophage-like particle has been shown to readily promote gene exchange in B. hyodysenteriae (120). B. burgdorferi has also been shown to contain a phage that mediates transduction (34). Second, the entire genome sequences of T. pallidum and B. burgdorferi have been determined, and those of *T. denticola* and *L. interrogans* are at different stages of completion (39, 40; http://www.hgsc.bcm.tmc.edu/microbial). Much useful information has been obtained from analyzing genomic sequences and in targeting genes by allelic exchange mutagenesis. The reader is referred to recent reviews describing the advances in spirochete genetics (19, 56, 100, 125).

Why Study Spirochete Motility?

One important question is, "What is the relevance of studying spirochete motility?" We address this question briefly here, with a more detailed discussion later. First, the organelles for motility are attached near each end of the cell within the periplasmic space. As discussed below, translational motility requires the rotation of the two flagellar bundles in opposite directions: This leads to swimming modalities that are quite different and more complex than the well-studied paradigms of *Escherichia coli* and *Salmonella enterica* serovar Typhimurium (referred to as *S. enterica*). Second, the coordinated flagellar rotation, flexion, and direction reversals that occur in spirochetal motility likely involve an efficient and novel signaling mechanism. The coordinating signal would be rapidly transmitted from one end of the cell to the other, which in some cases is over 100-fold greater than the cell diameter. Therefore, the analysis of spirochete motility should lead to a better understanding of the nature of such long traveling signals in both spirochetes and other large bacteria. Finally, many spirochetes, including *Treponema*, *Borrelia*, and *Leptospira* are highly invasive pathogens. Motility is likely to play a major role in the disease process, and there is some recent evidence to support this conjecture (72, 83, 84, 107, 110). Understanding both chemotaxis and motility could lead to the development of new treatments and disease prevention.

Background to Bacterial Motility

Organisms such as *E. coli* and *S. enterica* serve as model systems in helping to decipher spirochete motility and chemotaxis (7, 12, 87, 88). These species swim via the rotation of externally located helically shaped flagellar filaments (Figure 2a; see color insert). A rotary motor at the base of each flagellum powers this rotation, which is driven by a proton gradient. The flagellum structure is complex, with a rotary motor apparatus beginning in the cytoplasm and extending through the cytoplasmic membrane, peptidoglycan cell wall layer, outer membrane, and finally the helical flagellar filament that extends into the ambient medium. In the region below the flagellum, a ring structure (C-ring complex) composed of several different proteins (FliG, FliM, FliN) is attached to another ring (MS ring encoded by *fliF*). This C-ring structure and the MS ring serve as the rotary part of the flagellar motor. MotA and MotB proteins are embedded in the cytoplasmic membrane and cell wall around the MS ring and function as the stator, or nonrotating part of the motor. The P and L rings localize at the peptidoglycan layer and outer-membrane layers, respectively, and serve as bushings. The MS, P, and L rings surround a central rod, which in turn is attached to the hook, an external structure approximately 55 nm long, composed of a polymer of the FlgE protein. The hook serves as a flexible coupling to the flagellar filament, a helically shaped structure that is several μm long. This filament, composed of a polymer of the FliC protein, is a hollow tubular quasi-rigid polymorphic structure that can have either left- or right-handed configurations, depending on the direction it rotates. At the distal tip of the flagellar filament is the FliD protein, which serves as the flagellar cap. This is the site where new monomers that travel through the hollow filament attach to the growing filament. Associated with the flagellar apparatus near the C-ring complex are several different proteins involved in subunit transport and flagellar assembly. External proteins of the hook-basal body complex are excreted by a Type III secretion pathway through the basal body. The only exceptions are some of the proteins that comprise the L and P rings, which are secreted by the SecA-dependent pathway. The assembly of a flagellum is quite remarkable. Indeed, the bacterium constructs the architecturally complex flagellum at a site on its membrane, and it does so by having constant feedback mechanisms governing its assembly at the growing tip (68).

DYNAMICS OF SPIROCHETE MOTILITY

Essential Role of Periplasmic Flagella in Motility

The periplasmic flagella are clearly the spirochetes' organelles of motility. Early on, the analysis of chemically induced and spontaneously occurring mutants and their revertants pointed towards this conclusion (27, 78), but recent targeted mutagenesis studies were conclusive: Mutations that inhibited the synthesis of periplasmic flagella resulted in nonmotility. These mutations include the following for *B. burgdorferi*: *flgE, fliF, flaB* [filament protein, see discussion of *flaB* (below)] (95; M.A. Motaleb, M. Sal & N. W. Charon, unpublished data); *T. denticola: flgE, Tap1* [hook assembly protein (80, 82)]; *B. hyodysenteriae: fliG, flaB1-flaB2* double mutant (C. Li & N. W. Charon, unpublished data, see below); and *L. biflexa* (*flaB*) (104). In some cases, complementation of some mutations restores the wild-type phenotype (28, 111).

Although it has not been directly proven that the periplasmic flagella rotate within the outer membrane sheath, several lines of evidence are strongly suggestive. Protruding periplasmic flagella surrounded by outer membrane sheaths are often seen in stationary phase cells of many spirochete species, and at a high frequency throughout all growth phases in certain mutants of *Treponema phagedenis*. These protrusions have been shown to rotate (25, 50). In addition, as discussed below, many of the motions observed in swimming spirochetes are best explained by rotation of the internally located periplasmic flagella. Finally, both structural and genetic analyses indicate that the periplasmic flagella are very similar to their flagellar counterparts in other bacteria (22, 78). This obviously suggests that they function in a comparable manner.

Skeletal Function of Periplasmic Flagella

One intriguing question is, How can spirochetes swim with their organelles for motility inside the cells? Further, How do the filaments accomplish this feat by rotating within the periplasmic space? The analysis of mutants of several spirochete species is beginning to yield answers. One consistent finding is that the periplasmic flagella of several spirochete species have both motility and skeletal functions, i.e., by skeletal function we mean that the periplasmic flagella influence the shape of the cell in the region where they reside (26, 95, 104, 109). The most remarkable example is *B. burgdorferi*, which is significantly larger than most pathogenic bacteria (10–20 μm long, 0.33 μm wide), and has 7 to 11 periplasmic flagella attached to each end of the cell. These filaments overlap in the center of the cell and form a continuous bundle in the periplasmic space (4, 47). High-voltage electron microscopy and observations of swimming cells indicate that these organisms have a flat-wave morphology (47, 50). Insertion mutations in the gene that encodes the major periplasmic flagellar filament protein FlaB (analogous to FliC of *E. coli* and *S. enterica*) result in cells that lack both periplasmic flagella and the flat-wave morphology. The resulting mutant cells are in fact rod-shaped (Figure 3; see color insert) (95). The *flaB* mutation can be complemented only in *cis*; i.e., only when the plasmid containing wild-type *flaB* recombines with the chromosome does the cell regain wild-type characteristics (111). In addition, mutations in other genes that inhibit flagella synthesis such as *fliF* and *flgE* also result in cells that lack periplasmic flagella and are rod-shaped (M.A. Motaleb, M. Sal & N. W. Charon, unpublished data). Taken together, these results suggest that the periplasmic flagella affect the morphology of *B. burgdorferi* in quite a striking manner.

Model of *Borrelia burgdorferi* Motility

The skeletal function of periplasmic flagella forms the basis of motility models for several spirochete species [see (8, 26, 47, 48, 50, 78) for discussion of these models]. All these models rely on evidence that the periplasmic flagella rotate between the outer membrane sheath and cell cylinder. Recently, we described a model for *B. burgdorferi* motility (47, 50, 78; see video 1, http://www.uic.edu/orgs/blast/videos). This model states that in situ, the periplasmic flagella are juxtaposed with a highly flexible protoplasmic cell cylinder due to containment by the outer membrane sheath. The periplasmic flagella have a defined shape, which is a left-handed helix [a left-handed helix rotates counter-clockwise (CCW) going away from an observer, and a right-handed helix rotates clockwise (CW)]. Because of the close interaction of the more rigid periplasmic flagella with the flexible protoplasmic cell cylinder, the periplasmic flagella influence the shape of the entire cell. High-voltage electron microscopy has shown that the bundles of periplasmic flagella wrap CCW around the cell axis (since *B. burgdorferi* resembles a sine wave, we define the cell axis as the abscissa). In fact, the size of the bundles' left-handed helix pitch approximates that of the cell's flat wave-form. Translational motility occurs as the result of CCW rotation of the periplasmic flagella as viewed from the back of the cell. This rotation causes backward-moving waves to be propagated down the length of the cell. Thus, the action of the periplasmic flagella is likened to that of a worm gear in generating wave propagation. CW rotation of the cell about the body axis (i.e., the center of the protoplasmic cell cylinder—likened to the middle of a sausage) counterbalances the rotation of the periplasmic flagella. Mutants that lack the periplasmic flagella are nonmotile because they have lost the ability to propagate backward-moving waves. In addition, these mutants are rod-shaped as a result of their loss of skeletal function associated with the periplasmic flagella.

Model of *Spirochaeta aurantia* Motility

For some spirochete species, the periplasmic flagella may not have a skeletal function, and an alternative model has been proposed (6). This model is based on observations that some irregularly shaped, rigid cells translate in a low-viscosity medium with the cell rotating completely around the cell axis. For these cells to swim, the model states that the outer membrane sheath rotates or flows around the cell body in one direction, and the protoplasmic cell cylinder rotates in the opposite direction. Forward thrust is obtained by rolling of the cell cylinder in one direction and rotation of sheath in the opposite direction. This model may apply to *S. aurantia* and *Cristispira* sp., but evidence that the sheath rotates or flows in the opposite direction relative to the protoplasmic cell cylinder has not been obtained for any spirochete species.

Unique Structure and Composition of Periplasmic Flagella

The periplasmic flagella of several spirochete genera (*Leptospira*, *Treponema*, *Brachyspira*, and *Spirochaeta*) possess some features that distinguish them from the flagella of other bacteria (Figure 4) (27, 78). Because of a unique protein sheath surrounding the filament core, their flagellar filaments are thicker than those of

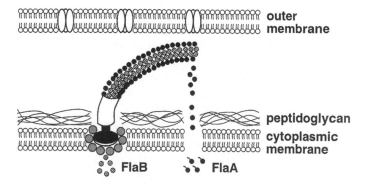

Figure 4 Structure of a spirochete periplasmic flagellum. In several spirochete species, the basal bodies lack the L and P rings found in *E. coli* and *S. enterica* (17, 60, 98). FlaA and FlaB are excreted to the periplasmic space by SecA and type III secretion systems, respectively. This complex periplasmic flagellar structure is found in *Treponema* sp., *S. aurantia*, *Leptospira* sp., and *B. hyodysenteriae* (27, 78).

other bacteria, e.g., 25 nm in diameter for *B. hyodysenteriae* compared to 20 nm for *E. coli* and *S. enterica* (77, 87). This sheath is composed of a protein designated FlaA. Its sequence is well conserved among the spirochetes. The periplasmic flagellar core is composed of a family of FlaB proteins. FlaB proteins show sequence similarity to flagellin proteins of other bacteria, especially at the N- and C-terminal regions. There are 1 to 3 different FlaB proteins (FlaB1, FlaB2, and FlaB3), depending on the species, and a separate gene encodes each one. In addition, there is extensive sequence similarity between the FlaB proteins within a given species, and also between species of different genera. The FlaA protein is excreted into the periplasmic space by a SecA-dependent system, whereas FlaB is excreted by a Type III secretion system. The mechanism of assembly of the sheath around the core is not known, but the confining environment of the periplasmic space likely plays a critical role. The periplasmic flagellar composition of *B. burgdorferi* is somewhat different. It has only one FlaB protein, and there is considerably more FlaB than FlaA. The location of FlaA in *B. burgdorferi* is uncertain, but evidence suggests that it is closely associated with the periplasmic flagella filament (42; M.A. Motaleb & N.W. Charon, unpublished data).

Genetic analysis indicates that the interaction of FlaA with the FlaB subunits affects flagellar filament shape in a manner that has not been observed in any other flagellar system. The composition and genetics of periplasmic flagellar filament synthesis in *B. hyodysenteriae* is best understood, as its genes are more readily manipulated than other spirochetes (106, 120). Because of the extensive homology of its FlaA and FlaB proteins with those of other spirochetes, conclusions drawn from the analysis of its periplasmic flagella are likely to be to relevant to other species such as *T. pallidum* and *T. denticola*. Wild-type cells of *B. hyodysenteriae* have periplasmic flagella that are left-handed with a defined shape, as measured by helix pitch and helix diameter (77). The approximate ratios of flagellin proteins to one another are the following: FlaA : FlaB1 : FlaB2 : FlaB3 of 1 : 0.42 : 0.22 : 0.45. Thus, there is approximately one FlaA per FlaB family subunit. Targeted mutagenesis of each of the genes encoding *flaA*, *flaB1*, *flaB2*, and *flaB3* resulted in cells that were still motile and possessed periplasmic flagella that retained their left-handed configurations. However, comparing mutant and wild-type purified periplasmic flagella shapes yielded surprising results (77). Each of the *flaB* mutants had periplasmic flagella that were slightly different from the wild-type. However, the *flaA* mutant's periplasmic flagella displayed major decreases in both helix pitch and helix diameter compared to the wild type. These results suggest that FlaA impacts the shape of the periplasmic flagella. Because bacterial flagella are often quasi-rigid and undergo helical transformations, perhaps the sheath helps stabilize the FlaB helical core into one of these configurations for optimal thrust as it rotates between the outer membrane sheath and cell cylinder.

The analysis of double mutants of *B. hyodysenteriae* indicates a remarkably complex interaction between the FlaA and FlaB subunits. The location of each FlaB protein along the periplasmic flagellum is unknown, but two different models could account for their arrangement (106). One model proposes that FlaB1,

FlaB2, and FlaB3 reside in specific regions of the flagellum in a manner analogous to the multiple flagellar subunits of *Caulobacter crescentus* (94). For example, perhaps FlaB3 is adjacent to the hook, FlaB2 is in the central region, and FlaB1 is at the distal region. However, an analysis of double mutants supported another model: The different FlaB proteins are distributed along the entire length of the periplasmic flagellum (C. Li & N.W. Charon, unpublished data). In this analysis, the periplasmic flagella from the double mutants *flaA-flaB1*, *flaA-flaB2*, and *flaA-flaB3* were left-handed. Surprisingly, however, their helix pitch and diameter were even less than those of both the wild-type and the single *flaA* mutants. These results suggest that in the absence of FlaA, the inhibition of synthesis of one FlaB protein dramatically influences the shape of the entire periplasmic flagellum. Such a result would not be expected if a given FlaB protein resided in only one domain of the periplasmic flagellum. These results also suggest that FlaA interacts with each FlaB protein to effect periplasmic flagellar shape.

The role of the different flagellin proteins in motility is unclear. Although single mutants of *flaB1*, *flaB2*, *flaB3*, or *flaA* of *B. hyodysenteriae* were motile, observations of swimming cells and swarm-agar-plate assays indicate that all these mutants swam less efficiently than the wild type (72, 77, 106). In addition, the double mutants *flaA-flaB1*, *flaA-flaB2*, and *flaA-flaB3* were also motile, but their swarming deficiency was even more pronounced than the single mutants (72; C. Li & N.W. Charon, unpublished data). The analyses of double *flaB* mutants were intriguing. Whereas the *flaB1–flaB3* mutant was still motile and had periplasmic flagella, the *flaB1-flaB2* mutant was completely nonmotile and lacked periplasmic flagella. These results indicate that the motility and structural functions of FlaB1 and FlaB2 overlap. Clearly, to fully understand the function of the multiple flagellin proteins, we need a more thorough analysis of the dynamics of motility in *B. hyodysenteriae* and the structure of wild-type and mutant periplasmic flagella.

SWIMMING AND CHEMOTAXIS OF SPIROCHETES

Complex Swim Behavior of Spirochetes

Most externally flagellated bacteria have two different swim modalities: runs and tumbles (3, 5, 7, 87). For example, in *E. coli* runs—or translational motility—occur when the flagella rotate CCW (as a frame of reference, the flagella are viewed from the end of the flagella toward its insertion into the cell). During a run, the peritrichous flagella form a bundle behind the cell and act as a propeller. When one or more of the flagella reverse direction and rotate CW, the cell tumbles (3, 87, 126). Some of the filaments leave the flagellar bundle and the cell body reorients. It then runs in a new direction when the flagella revert to the CCW direction. In some species, such as *Rhodobacter sphaeroides* and *Sinorhizobium meliloti*, the flagella stop or slow down instead of changing their direction of rotation.

The swim behavior of spirochetes is more complicated than that of other bacteria as a consequence of their complex geometry. Spirochetes have three motility modes

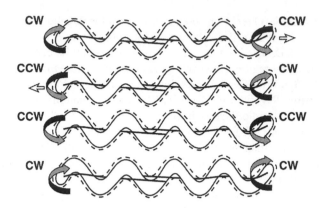

Figure 5 Swimming cells of *B. burgdorferi* as a function of direction of rotation of the periplasmic flagella. Open arrows indicate direction of swimming. Dotted lines represent the outer membrane sheath. Gray arrows indicate direction of rotation of the periplasmic flagella (*thin lines*). For simplification, only one periplasmic flagellum is shown attached at each end of the cell cylinder. In *B. burgdorferi*, there are between 7 and 11. The top two panels show translational forms, and the bottom two nontranslational forms.

(Figure 5): two modes of runs (i.e., with either end leading, top two panels), and one mode of nontranslational motility (bottom two panels) (24, 37, 50, 78). During runs, the anterior bundle of periplasmic flagella rotates in one direction (e.g., CCW), and the posterior bundle rotates in the opposite direction (CW). Thus, the bundles of periplasmic flagella rotate asymmetrically. Reversals occur when both bundles change their direction of rotation. The third mode is a nontranslational interval. During this mode, the bundles rotate in the same direction: both CW or both CCW. In many species of spirochetes during this interval, the cells become quite distorted and may bend in the middle; this is referred to as a flex (37, 50). The two nontranslating forms are morphologically indistinguishable for most spirochete species except for the Leptsopiraceae (8, 24, 37, 48, 49). In the Leptospiraceae, CCW rotation of its periplasmic flagellum results in a spiral-shaped cell end, and CW rotation results in a hook-shaped cell end. Thus, the translating forms are hook-spiral and spiral-hook, and the two nontranslating forms are hook-hook and spiral-spiral. In many ways, the swim behavior of spirochetes is analogous to that of *Spirillum* species (5, 74, 92). These bacteria have a somewhat similar morphology: bundles of flagella attached at each end of a helically shaped cell but with no outer membrane sheath. In addition, there are four different morphological forms of swimming *Spirillum*, which are analogous to those of the Leptospiraceae.

One common question concerns whether both bundles of periplasmic flagella actively rotate during translation as the description above contends. As an alternative model, perhaps one bundle rotates while the other is stopped. In this model, cell reversals occur when motor rotation switches from one bundle to the other.

One prediction of this hypothesis is that the bundles at a given end should show frequent stops. To test this hypothesis, we analyzed spirochete cells tethered to a glass surface. Testing *Treponema phagedenis* gave the most straightforward results. Its periplasmic flagella do not overlap in the center of the cell and are quite short relative to the overall cell length. Periplasmic flagella, as previously mentioned, often have a skeletal function, so they influence the shape of the cell in the domains where they reside. In the case of *T. phagedenis*, these regions are proximal to and include the two cell ends (26). The terminal regions gyrate (i.e., bend in a circular manner without necessarily rotating) as a result of the rotation of relatively short, interior, periplasmic flagella (26). Tethered cell analysis indicated that both of the terminal regions constantly gyrated and often changed direction. In some cells, one end constantly gyrated in a specific direction, and the other end switched direction quite frequently (N.W. Charon, unpublished data). These results indicate that for translation in *T. phagedenis*, both bundles of periplasmic flagella rotate. A similar conclusion based on tethered cell analyses was reached with the Leptsopiraceae and *B. burgdorferi* (24, 76).

Asymmetrical Periplasmic Flagellar Rotation in Spirochetes

Unique to both spirochete and *Spirillum* sp. is asymmetrical rotation of the flagellar bundles during translational motility. Two hypotheses may explain how these bacteria achieve asymmetry. The first hypothesis states that chemotaxis (i.e., directed movement toward an attractant or away from a repellant) plays an essential role in this asymmetry. During chemotaxis, *E. coli* and most bacteria swim by a pattern best described as a biased random walk toward a favorable medium or away from one that is toxic (3, 5, 7, 18, 87). This swim pattern is a function of a combination of runs and tumbles. Initially, an effector molecule binds directly to a membrane-bound chemoreceptor protein referred to as a methyl-accepting chemotaxis protein (MCP), or indirectly to the MCP via a periplasmic binding protein (Figure 2b, see color insert). The MCP regulates autophosphorylation of the sensor histidine kinase CheA. Activated CheA phosphorylates the response regulator CheY forming CheY-P, which then interacts with the switch complex at the flagellar motor to increase the probability of flagellar CW rotation and reduce the probability of CCW rotation. CheA and CheY are essential elements for chemotaxis, as null mutants that encode these proteins are nonchemotactic. These mutants constantly rotate their flagella CCW and continuously run. According to this chemotaxis-based hypothesis, asymmetry in spirochetes depends on localized concentrations of CheY-P, so that during translation one cell end has a higher concentration of CheY-P than the other. This hypothesis predicts that null mutants of *cheA* will continuously rotate their periplasmic flagella in the same direction and will be nontranslating and continuously flex.

The second hypothesis states that the motors at the two ends of the cells are different. This difference results in cells whose bundles rotate asymmetrically independently of CheY-P. Thus, in the default state with no CheY-P present, motors at one end of the cell rotate CCW, and those at the other end rotate CW. Both hypotheses lead to specific predictions. In contrast to the first hypothesis, this hypothesis predicts that null mutants in *cheA* will continually rotate their bundles of periplasmic flagella in opposite directions and constantly run.

Targeted mutagenesis was used to determine the basis for asymmetrical periplasmic flagellar rotation (76). The genomic sequence of *B. burgdorferi* indicated the presence of two *cheA* genes, *cheA1* and *cheA2*. Both *cheA* genes were inactivated to obtain the *cheA1–cheA2* double mutant. This mutant was deficient in chemotaxis toward rabbit serum, which is an attractant for these spirochetes (116). Whereas the wild-type reversed directions quite frequently (19 reversals per minute), the double mutant constantly ran. Thus, in the absence of CheA, i.e., the default state, one bundle of periplasmic flagella rotates CCW, and the other bundle rotates CW. These results support the hypothesis that structural differences exist between the flagellar motors at the opposite ends of the cell. Similar results have been noted in *cheA* mutants of *T. denticola* (85); consequently, it is likely that other spirochetes have asymmetrical rotation of the periplasmic flagella at opposite ends of the cells in the absence of CheA.

Asymmetrical localization of specific proteins and structures has been described for several bacterial species (86, 115). For example, the stalk structure localizes at one end of the cell in *C. crescentus* at a site previously occupied by the flagellum. The ActA protein of *Listeria monocytogenes* and the IcsA protein of *Shigella flexneri* localize at the old cell poles in each of these species. Perhaps in *B. burgdorferi* and other spirochetes there is association of an unknown factor or factors with the flagellar switch complexes at one cell pole. This association could result in the periplasmic flagella at that end rotating CW rather than CCW in the default state.

Model for Asymmetrical Periplasmic Flagellar Rotation

Several candidate effectors could be involved in flagellar asymmetrical rotation. One such possible effector is FliG. Specifically, *B. burgdorferi*, *T. pallidum*, and *T. denticola* have two homologs encoding FliG (39, 40; http://www.hgsc.bcm.tmc.edu/microbial). These genes are the only known duplicate homologs of the flagellar apparatus in *B. burgdorferi* and *T. pallidum*. FliG is part of the switch complex that determines direction of flagellar rotation. Conceivably, one or both of the FliG proteins could differentially localize to motors at one cell end relative to the other. Alternatively, CheX, another well-conserved protein product in spirochetes, could also encode such an effector. Results with *B. burgdorferi* and *T. denticola* suggest that *cheX* mutants are locked in the nontranslational mode (M.A. Motaleb & N.W. Charon, unpublished data; R. Lux & W. Shi, unpublished data). These results are consistent with CheX associating with one cell end leading to periplasmic flagellar asymmetrical rotation. Finally, in *B. burgdorferi* and *T. pallidum*, several open reading frames of unknown function map within putative motility and chemotaxis gene clusters (39, 40, 43). Perhaps one or more of these genes encode proteins involved in asymmetrical flagellar rotation.

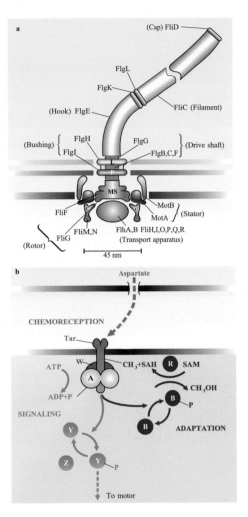

Figure 2 (*a*) Schematic bacterial motor and drive train illustrating the many protein components of the flagellar apparatus of *E. coli* and *S. enterica* serovar Typhimurium, and (*b*) their chemotaxis machinery. The chemotaxis diagram shows some of the components required for chemotaxis toward the amino acid aspartate. Information flows from the outside of the cell (*top*) by way of porins, the periplasmic space, and the cytoplasmic membrane, to the inside of the cell (*bottom*), and then to the flagellar motors (not shown). Dashed arrows indicate physical displacement of chemicals by diffusion. Solid arrows indicate chemical modifications of proteins—phosphorylation or methylation. The cytoplasmic components, all Che proteins (CheW, CheA, CheR, CheB, CheY, CheZ), are identified by their fourth letter only. The receptor complex consists of the MCP Tar, W, and A, with Tar spanning the cytoplasmic membrane. Chemoreception is depicted in orange, signaling in green (for "go"), adaptation in red (for "stop"). SAM is S-adenosylmethionine, the methyl donor. From Motile Behavior of Bacteria, Howard C. Berg, *Physics Today*, 2000, 53:24–29, by permission from the American Institute of Physics.

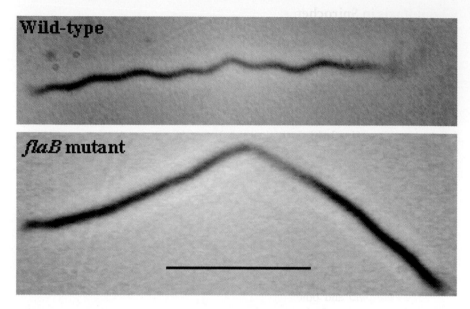

Figure 3 Morphology of *B. burgdorferi* wild-type and *flaB* mutant as observed by phase microscopy. The *flaB* mutant is in the process of division. Bar represents 5 μm.

Chemotaxis in Spirochetes: CheA, CheX, CheY, and Membrane Potential

The fact that spirochetes have three swim modes (two running and one nontranslating interval), whereas most other studied bacteria have two modes (running and tumbling), adds to the challenge of understanding their chemotactic response on a molecular level. Even the well-studied chemotaxis systems of *E. coli*, *S. enterica*, *Bacillus subtilis*, and *R. sphaeroides* seem to grow in complexity as more is known about them (1a, 3, 18, 119)! In spirochetes, the relationship between chemotaxis and reversals in direction is quite intriguing, but very little is known about their regulation.

Swim reversal in spirochetes is quite fast. In the Leptospiraceae, *S. aurantia*, *B. burgdorferi*, and even the very large *Cristispira* it is less than 300 msec (24, 37) (N.W. Charon, E.P. Greenberg, unpublished data). As previously mentioned, a reversal occurs when the sense of rotation of the periplasmic flagella bundles at each end rapidly changes. One hypothesis to explain rapid reversals is that the relative concentration of a signal transducer such as CheY increases at one end of the cell due to contact with an attractant. Along these lines, the MCPs in *S. aurantia* and other bacterial species are polarly localized, and we would expect that those of other spirochete species are similar (45, 86). According to this hypothesis, increased concentration of the signal transducer at one end of the cell causes the periplasmic flagella near that terminus to change direction. The signal transducer also diffuses to the other end of the cell to effect a change of direction at that end. This hypothesis is quite unlikely. The above-mentioned bacteria are relatively large—at least 10 μm for the Leptospiraceae and *B. burgdorferi*, and 30–180 μm for the *Cristispira*. It would take several seconds for a compound the size of CheY to diffuse from one end of the cell to the other—much longer than the time a cell takes to reverse (38, 114). However, analysis of targeted mutants in *cheA* and *cheY* genes of *B. burgdorferi* and *T. denticola* indicate that CheY is indeed involved in chemotaxis (76, 85; R. Lux & W. Shi, unpublished data; M.A. Motaleb & N.W. Charon, unpublished data). Thus, while diffusion of CheY cannot explain rapid cell reversals, this cytoplasmic protein is indeed involved in spirochete chemotaxis.

What are the possible models that could account for rapid cell reversals in spirochetes? In *S. aurantia*, a membrane potential has been found to mediate the chemotactic response (52, 53). Although a membrane potential is not implicated in chemotaxis of *E. coli* and *S. enterica* (18, 89, 117), recent results suggest its involvement in other bacterial species (1). Because spirochetes (and *Spirillum* sp.) are considerably longer than many other bacteria, a change in membrane potential could conceivably allow for communication between cell ends to coordinate periplasmic flagellar rotation (5, 37, 117). However, it is not known if cells with their membrane potential clamped still undergo rapid reversals. Moreover, no one has yet stimulated one end of a spirochete with an attractant, and monitored the motion at the other end, as was done for long filamentous cells of *E. coli* (114).

Such an experiment would clearly resolve whether there is rapid communication between one end of the cell with the other.

The work on *S. aurantia* over a decade ago is significant in analyzing the behavior of individual cells undergoing chemotaxis (37, 38). When these organisms are initially incubated with an attractant, the reversal and flexing frequencies are immediately suppressed, and their run durations are increased. This increase in run duration indicates that spirochetes respond to a temporal gradient. Because flexing and reversals are suppressed, apparently there is a mechanism that coordinates flagellar rotation at each of the cell's ends. These results are consistent with more recent results with *B. burgdorferi* and *T. denticola*, which indicate that *cheA*, *cheX*, *cheW*, and *cheY* not only are involved in chemotaxis, but also influence reversal frequency (76, 85; C. Li, M.A. Motaleb, R. Bakker & N.W. Charon, unpublished data; R. Lux & W. Shi, unpublished data).

With *S. aurantia*, after an initial stimulation with an attractant, the flexing and reversal frequencies eventually approach those of the prestimulated state and indicate that adaptation occurs. In bacteria such as *E. coli*, adaptation is dependent on MCP methylation (Figure 2*b*) (3, 7, 18). This increase in methylation of MCPs has also been observed with *S. aurantia* when cells undergo chemotaxis (70). Recently, MCPs in *T. denticola* and *T. pallidum* have also been shown to undergo methylation, and insertion mutagenesis in *T. denticola* has verified their involvement in chemotaxis (55, 69, 79). Thus, in spirochetes the involvement of MCPs in both the excitation phase and adaptation appears similar to those found in other bacteria.

Models of Spirochete Chemotaxis

In considering what is known about spirochete chemotaxis at this time, two possible models are apparent. Both models are drawn from certain experimental findings from *S. aurantia*, *B. burgdorferi*, and *T. denticola*, and are based on the paradigm of *E. coli* and *S. enterica*. We assume that conclusions drawn from one spirochete species apply to the others. It is also assumed that coordination of the rotation of the periplasmic flagella at both ends of the cell is necessary for chemotaxis, and that this coordination is mediated by attractants binding to polarly located MCP's. We propose CheX, and possibly other proteins such as one of the FliG proteins, localize to motors at one end of the cell. This localization results in asymmetrical rotation of the periplasmic flagella in the default state, i.e., one bundle rotates CCW, the other CW in the absence of CheY-P. As discussed below, this asymmetry imposes real complexity for chemotaxis models of spirochetes.

The first model, the "one-head model," states: Under an isotropic condition, cells run, flex, and reverse at a steady state. These frequencies are a function of the internal CheY-P concentration inside the cell, which is analogous to the frequencies of tumbles and runs found in other bacteria in isotropic conditions. When a chemoattractant binds to a MCP at a cell end (end 1), a signal is transmitted to the CheA bound to that chemoreceptor. CheA autophosphorylation is suppressed at that end of the cell with the concomitant decrease in localized concentration of CheY-P. This decrease in CheY-P suppresses CW rotation of the periplasmic flagella at that cell end with a resultant increase in the frequency of CCW rotation. Concomitantly, either the decrease in CheY-P concentration at that end of the cell, or more likely the bound attractant to the MCP, generates a second signal to the cytoplasmic membrane, which causes a depolarization of the membrane that is rapidly transmitted to the other cell end (end 2). At this end, the MCPs undergo a conformational change due to the depolarization, with a resultant decrease in autophosphorylation of bound CheA almost simultaneously with end 1 and a resultant localized decrease in CheY-P concentration. The decrease in CheY-P at this end leads to suppression of CCW rotation of the periplasmic flagella, an increase in CW rotation, and a concomitant decrease in frequencies of flexes and reversals—the cell runs. Adaptation occurs by mechanisms found in other bacteria, with an increase in flexes and reversals as the MCPs undergo methylation. One consequence of this one-headed model is that the cell swims toward a higher concentration of an attractant with only one end leading, specifically, the end (end 1) that rotates its periplasmic flagella CCW in low CheY-P concentration.

The second is a "two-headed" model that states that both ends of the spirochete are capable of leading a cell up a chemical gradient. How can end 2 lead the cell up a gradient if its default state is CW rotation of the periplasmic flagella? Perhaps there is asymmetry also with the MCPs, or factors that can interact with the MCPs. The model is the same as above, but in this case, when an attractant first binds to end 2, it causes CheA at that end to increase instead of decrease CheY-P concentration at that end. This results in suppression of CW rotation, and an increase in CCW rotation at end 2. The change in membrane potential as described above results in an increase in autophosphorylation at end 1, an increase in CheY-P concentration, and CW rotation of the periplasmic flagella at that end. The cell then runs with end 2 leading the cell. It should be pointed out that in *B. subtilis*, the model of *E. coli* and *S. enterica* is reversed. In *B. subtilis*, CheY-P concentration increases during chemotaxis, CW flagellar rotation is suppressed, and the probability of CCW rotation is increased (1a, 3, 14). Thus, one could speculate that, as a very basic model for spirochetes, one end behaves like *B. subtilis*, and the other end like *E. coli*. The above models are speculative, but it is hoped that they point out the difficulty in applying current knowledge of chemotaxis to spirochetes: Their geometry and asymmetry impose real constraints on possible mechanisms.

REGULATION OF MOTILITY AND CHEMOTAXIS GENE EXPRESSION IN SPIROCHETES

Organization of Motility and Chemotaxis Genes

Complete sequencing of the genomes of *B. burgdorferi* and *T. pallidum* has yielded important information about motility and chemotaxis (39, 40). Sequence homology, transcriptional and Western blot analyses, and targeted mutagenesis are beginning to yield an overall picture. From these analyses, and preliminary results with respect to the *T. denticola* genome (http://www.hgsc.bcm.tmc.edu/microbial), we find that all the genes necessary for flagella-based motility are present in *T. pallidum*, *B. burgdorferi*, and *T. denticola*. Both *T. pallidum* and *B. burgdorferi* have only one copy of the genes that encode the basal body and switch complex. The only exception is *fliG*; both organisms have duplications of this gene. In *B. burgdorferi* and *T. pallidum*, the genes encoding the L and P rings are absent. These results are consistent with electron microscopic observations of the basal body from these spirochetes (59, 61). In addition, the three species share many similarities in their organization of motility and chemotaxis gene clusters.

For a given spirochete, individual chemotaxis and motility genes have the highest homology among other spirochete species compared to other bacteria (78; C. Li & N.W. Charon, unpublished data). Many of the motility and gene clusters in *B. burgdorferi*, and a few in *T. pallidum*, have been shown to function as operons (76, 78). *B. burgdorferi* and *T. pallidum* each have several genes of unknown function that map within the motility and chemotaxis gene clusters. Because of their location, these genes are likely to be involved in either chemotaxis or motility. *T. pallidum* has only one copy of most of the sensory transduction genes (*cheA*, *cheX*, *cheD*, *cheY*, *cheB*, and *cheR*), but it has two copies of *cheW*. In contrast, *B. burgdorferi* has several copies of these genes (2 *cheA*, 3 *cheW*, 3 *cheY*, 2 *cheB*, and 2 *cheR*). Some of the genes were shown to be involved in chemotaxis by targeted mutagenesis in *B. burgdorferi* (*cheA2*, *cheW3*, *cheY1*, *cheY3*, *cheX*) and *T. denticola* (*cheA*, *cheX*, *dmcB*, *dmcA*) (69, 76, 79, 83; C. Li, M.A. Motaleb & N.W. Charon, unpublished data; R. Lux & W. Shi, unpublished data). One of the operons in *B. burgdorferi* (*flaA*, *cheA2*, *cheW3*, *cheX*, *cheY3* operon) is well conserved among the spirochete species (76). Several of the genes in the other chemotaxis operon of *B. burgdorferi* (*cheW2*, *orf566*, *cheA1*, *cheB2*, *orf569*, *cheY2*) are not found in *T. pallidum* and *T. denticola* and are most similar to genes found in other bacteria. This *cheW2* operon is likely the result of recent horizontal gene transfer from a Proteobacteria species (I. Zhulin, unpublished data). Its function is unknown, but the habitats for *B. burgdorferi* are both ticks and mammals, whereas the habitat for *T. pallidum* and *T. denticola* is exclusively mammalian. Perhaps the *cheW2* operon functions primarily in the tick, whereas the other functions in the mammal.

Control of Chemotaxis and Motility Gene Expression

Unique aspects of motility and chemotaxis gene expression in spirochetes are beginning to emerge. In most bacteria studied, there is cascade control of motility gene expression (29, 87). For example, in *E. coli* and *S. enterica* expression of *flhC* and *flhD* (class 1) initiate transcription of several motility genes (class 2). Two of the class 2 genes play a critical role in regulating class 3 gene transcription. Specifically, *fliA* encodes σ^{28}, which binds to RNA polymerase and promotes transcription of class 3 genes such as FliC. The other regulator gene is *flgM*, which encodes the anti-σ^{28} factor FlgM. FlgM is excreted into the medium when the hook basal body is complete. This excretion insures that the initiation of class 3 gene transcription is timed to the assembly of the initial flagellar structure. *T. pallidum, S. aurantia, T. denticola*, and *B. hyodysenteriae* have many motility promoter sequences homologous to σ^{28} recognition sequences (27, 78). Thus, it is likely that flagellar gene regulation is similar in these spirochete species to that found in other bacteria. In contrast, all of the chemotaxis and motility gene promoters in *B. burgdorferi* are initiated by the housekeeping transcription factor σ^{70} (43, 76, 78). No motility-specific sigma factor recognition sequence is evident in genomic analysis. To our knowledge, *B. burgdorferi* is the only bacterial species that lacks transcriptional cascade control of motility gene expression by alternative sigma factors. One hypothesis is that motility and chemotaxis are so vital for the survival of *B. burgdorferi* in both the tick and mammal that many of the genes are constitutively expressed. In support of this hypothesis, *B. burgdorferi* expresses *flaB* message and produces periplasmic flagella in both these hosts (31, 46). Alternatively, perhaps *B. burgdorferi*, and possibly other spirochete species (82), primarily rely on a translational control system to regulate motility and chemotaxis gene expression. Preliminary results with insertion mutants in *flaA, flaB, flgE*, and *fliF* of *B. burgdorferi* support this possibility (M.A. Motaleb, M. Sal & N.W. Charon, unpublished data). For example, mutants in *flaB* of *B. burgdorferi* still synthesize similar amounts of *flaA* message as the wild type, but fail to encode small amounts of FlaA protein.

There are many unknown factors with respect to motility and chemotaxis gene regulation in spirochetes. For example, the periplasmic flagella are inserted into the cell cylinder subterminally at each end. Observations of dividing cells indicate that before the time of division, new periplasmic flagella emerge in both daughter cells proximal to the division plane (59, 63). However, the factors that promote synchronization and localization of basal body synthesis with cell division are unknown. Furthermore, little is known about the factors that control periplasmic flagella length and number, although we know that periplasmic flagella length is tightly controlled in some species (26). Finally, DNA microarray analysis for *B. burgdorferi* indicates that many of the transcripts encoding chemotaxis and motility genes are differentially regulated in the mammal host compared to the tick (105). Genes besides those involved in motility and chemotaxis have been known for some time to be differentially regulated as a function of host (113). Both the environmental factors within the two hosts and the molecular details within the spirochete that promote this differential regulation of gene expression are just beginning to be elucidated (62, 105).

SPIROCHETE MOTILITY AND CHEMOTAXIS AS VIRULENCE FACTORS

Deciphering the roles of motility and chemotaxis for spirochetes in the disease process is at its early stages (4, 84). Many spirochete researchers have long believed that motility and chemotaxis are major virulence factors. This belief is supported by the fact that 5–6% of the genomes of *T. pallidum* and *B. burgdorferi* are putative chemotaxis and motility genes. In addition, several spirochete species are known to be highly invasive to host tissues. This is particularly true for *T. pallidum*, *Leptospira*, the relapsing fever *Borrelia*, *B. burgdorferi*, and the spirochetes associated with digital dermatitis (4, 66, 96, 100, 121). For *B. burgdorferi* in the tick, motility may be essential for the spirochetes to migrate from the intestinal gut to the salivary glands to initiate a new infection. In addition, chemotaxis may play an essential role for the spirochetes to migrate through the skin and localize at the tick bite during feeding. These hypotheses are only speculation, as critical genetic studies have not yet been done. Interestingly, a gene that shows homology to a quorum sensor (*luxS*, gene BB0377 of *B. burgdorferi*) has been identified in the genome of *B. burgdorferi*. Quorum sensing, the ability for bacteria to sense cell density and regulate gene expression, is a key step in the regulation of virulence genes and biofilm formation in other bacterial species (41). In addition, motility plays an important role in biofilm formation in several of these species (101). Spirochetes achieve high densities when attached to host cells and to tissues (4, 59, 118) so it is likely that these organisms are capable of biofilm formation. Thus, the relationship of quorum sensing, motility, and biofilm formation is an intriguing area for future research on spirochetes.

One major concern is that because virulence is often unstable, sorting out whether a given gene is involved in the disease process is complicated. This is particularly true of *B. burgdorferi*, as virulence is associated with plasmids that are often lost on passage (35). The first in vivo studies relating targeted motility mutants to virulence were done with mutants in *flaA*, *flaB1*, and *flaAflaB1* of *B. hyodysenteriae*. The mutants were deficient in colonization and were less virulent than the wild type in the mouse model of swine dysentery (72, 107). Most recently, mutants in *flgE*, *cheA*, and in the MCPs *dmcA* and *dmcB* of *T. denticola* were found to be deficient in the penetration of gingival epithelial cell layers (83, 85). These results suggest that both motility and chemotaxis are involved in tissue penetration. In sum, virulence studies using targeted mutants are at the early stages in spirochetes. Now that the genetic tools have been and are being developed, it should be quite exciting in the next several years to relate specific gene function, including those genes involved with chemotaxis and motility, to virulence.

ACKNOWLEDGMENTS

We thank R. Lux, W. Shi, I. Zhulin, and E.P. Greenberg for sharing unpublished information, and R. Bakker, H. Berg, R. Bourret, J. Campbell, J. Izard, C. Li, S. Lukehart, M. A. Motaleb, S. Norris, M. Sal, and W. Shi for suggestions and critical comments. The research in our laboratories was supported by USPHS grants AI29743, DE012046, USDA Grant 95-37204-2132, and Foundation for Microbiology. We acknowledge the continued interest in our research by our colleagues who attend the "Biology of Spirochetes" and "Sensory Transduction in Microoganisms" Gordon Research Conferences and the "Bacterial Locomotion and Sensory Transduction" meetings.

The *Annual Review of Genetics* is online at http://genet.annualreviews.org

LITERATURE CITED

1. Alexandre G, Zhulin IB. 2001. More than one way to sense chemicals. *J. Bacteriol.* 183:4681–86

1a. Aizawa S-I, IB Zhulin, L Márquez-Magaña, GW Ordal. 2001. Chemotaxis and motility in *Bacillus subtilis*. In Bacillus subtilis *and Its Relatives: From Genes to Cells*, ed. A Sonenshein, R Losick, J Hoch, pp. 437–52. Washington, DC: ASM Press. 629 pp.

2. Antal GM, Lukehart SA, Meheus AZ. 2002. The endemic treponematoses. *Microbes Infect.* 4:83–94

3. Armitage JP. 1999. Bacterial tactic responses. *Adv. Microb. Physiol.* 41:229–89

4. Barbour AG, Hayes SF. 1986. Biology of *Borrelia* species. *Microbiol. Rev.* 50:381–400

5. Berg HC. 1975. Chemotaxis in bacteria. *Annu. Rev. Biophys. Bioeng.* 4:119–36

6. Berg HC. 1976. How spirochetes may swim. *J. Theor. Biol.* 56:269–73

7. Berg HC. 2000. Motile behavior of bacteria. *Physics Today* 53:24–29

8. Berg HC, Bromley DB, Charon NW. 1978. Leptospiral motility. *Symp. Soc. Gen. Microbiol.* 28:285–94

9. Berg HC, Turner L. 1979. Movement of microorganisms in viscous environments. *Nature* 278:349–51

10. Bermudes D, Fracek SP Jr, Laursen RA, Margulis L, Obar R, Tzertzinis G. 1987. Tubulinlike protein from *Spirochaeta bajacaliforniensis*. *Ann. NY Acad. Sci.* 503:515–27

11. Bermudes D, Hinkle G, Margulis L. 1994. Do prokaryotes contain microtubules? *Microbiol. Rev.* 58:387–400

12. Berry RM, Armitage JP. 1999. The bacterial flagellar motor. *Adv. Microb. Physiol.* 41:291–337

13. Bharier MA, Rittenberg SC. 1971. Immobilization effects of anticell and antiaxial filament sera on *Treponema zuelzerae*. *J. Bacteriol.* 105:430–37

14. Bischoff DS, Ordal GW. 1992. *Bacillus subtilis* chemotaxis: a deviation from the *Escherichia coli* paradigm. *Mol. Microbiol.* 6:23–28

15. Blanco DR, Walker EM, Haake DA, Champion CI, Miller JN, Lovett MA. 1990. Complement activation limits the rate of in vitro treponemicidal activity and correlates with antibody-mediated aggregation of *Treponema pallidum* rare outer membrane protein. *J. Immunol.* 144:1914–21

16. Bono JL, Elias AF, Kupko JJ III, Stevenson B, Tilly K, Rosa P. 2000. Efficient targeted mutagenesis in *Borrelia burgdorferi*. *J. Bacteriol.* 182:2445–52

17. Brahamsha B, Greenberg EP. 1988. A biochemical and cytological analysis of the complex periplasmic flagella from *Spirochaeta aurantia*. *J. Bacteriol.* 170:4023–32

18. Bren A, Eisenbach M. 2000. How signals are heard during bacterial chemotaxis: protein-protein interactions in sensory signal propagation. *J. Bacteriol.* 182:6865–73

19. Cabello FC, Sartakova ML, Dobrikova EY. 2001. Genetic manipulation of spirochetes—light at the end of the tunnel. *Trends Microbiol.* 9:245–48

20. Campbell BJ, Cary SC. 2001. Characterization of a novel spirochete associated with the hydrothermal vent polychaete annelid, *Alvinella pompejana*. *Appl. Environ. Microbiol.* 67:110–17

21. Canale-Parola E. 1977. Physiology and evolution of spirochetes. *Bacteriol. Rev.* 41:181–204

22. Canale-Parola E. 1978. Motility and chemotaxis of spirochetes. *Annu. Rev. Microbiol.* 32:69–99

23. Cent. Dis. Control Prev. 2002. Lyme disease—United States. 2000. *Morbid. Mort. Wkly. Rep.* 51:29–31

24. Charon NW, Daughtry GR, McCuskey RS, Franz GN. 1984. Microcinematographic analysis of tethered *Leptospira illini*. *J. Bacteriol.* 160:1067–73

25. Charon NW, Goldstein SF, Block SM, Curci K, Ruby JD, et al. 1992. Morphology and dynamics of protruding spirochete periplasmic flagella. *J. Bacteriol.* 174:832–40

26. Charon NW, Goldstein SF, Curci K, Limberger RJ. 1991. The bent-end morphology of *Treponema phagedenis* is associated with short, left-handed periplasmic flagella. *J. Bacteriol.* 173:4820–26

27. Charon NW, Greenberg EP, Koopman MBH, Limberger RJ. 1992. Spirochete chemotaxis, motility, and the structure of the spirochetal periplasmic flagella. *Res. Microbiol.* 143:597–603

28. Chi B, Limberger RJ, Kuramitsu HK. 2002. Complementation of a *Treponema denticola flgE* mutant with a novel coumermycin A1-resistant *T. denticola* shuttle vector system. *Infect. Immun.* 70: 2233–37

29. Chilcott GS, Hughes KT. 2000. Coupling of flagellar gene expression to flagellar assembly in *Salmonella enterica* serovar Typhimurium and *Escherichia coli*. *Microbiol. Mol. Biol. Rev.* 64:694–708

30. Coleman JL, Benach JL. 1989. Identification and characterization of an endoflagellar antigen of *Borrelia burgdorferi*. *J. Clin. Invest.* 84:322–30

31. Das S, Barthold SW, Giles SS, Montgomery RR, Telford SR III, Fikrig E. 1997. Temporal pattern of *Borrelia burgdorferi p21* expression in ticks and the mammalian host. *J. Clin. Invest.* 99:987–95

32. Dewhirst FE, Tamer MA, Ericson RE, Lau CN, Levanos VA, et al. 2000. The diversity of periodontal spirochetes by 16S

rRNA analysis. *Oral Microbiol. Immunol.* 15:196–202

33. Dobell C. 1960. *Antony van Leeuwenhoek and His Little Animals*, p. 225. New York: Dover. 435 pp.

34. Eggers CH, Kimmel BJ, Bono JL, Elias AF, Rosa P, Samuels DS. 2001. Transduction by phiBB-1, a bacteriophage of *Borrelia burgdorferi*. *J. Bacteriol.* 183:4771–78

35. Elias AF, Stewart PE, Grimm D, Caimano MJ, Eggers CH, et al. 2002. Clonal polymorphism of *Borrelia burgdorferi* strain B31 MI: implications for mutagenesis in an infectious strain background. *Infect. Immun.* 70:2139–50

36. Fenno JC, McBride BC. 2002. Virulence factors of oral treponemes. *Anaerobe* 4:1–17

37. Fosnaugh K, Greenberg EP. 1988. Motility and chemotaxis of *Spirochaeta aurantia*: computer-assisted motion analysis. *J. Bacteriol.* 170:1678–774

38. Fosnaugh K, Greenberg EP. 1989. Chemotaxis mutants of *Spirochaeta aurantia*. *J. Bacteriol.* 171:606–11

39. Fraser CM, Casjens S, Huang WM, Sutton GG, Clayton R, et al. 1997. Genomic sequence of a Lyme disease spirochaete, *Borrelia burgdorferi*. *Nature* 390:580–86

40. Fraser CM, Norris SJ, Weinstock CM, White O, Sutton GG, et al. 1998. Complete genome sequence of *Treponema pallidum*, the syphilis spirochete. *Science* 281:375–88

41. Fuqua C, Parsek MR, Greenberg EP. 2001. Regulation of gene expression by cell-to-cell communication: acyl-homoserine lactone quorum sensing. *Annu. Rev. Genet.* 35:439–68

42. Ge Y, Li C, Corum L, Slaughter CA, Charon NW. 1998. Structure and expression of the FlaA periplasmic flagellar protein of *Borrelia burgdorferi*. *J. Bacteriol.* 180:2418–25

43. Ge Y, Old IG, Saint Girons I, Charon NW. 1997. Molecular characterization

of a large *Borrelia burgdorferi* motility operon which is initiated by a consensus σ^{70} promoter. *J. Bacteriol.* 179:2289–99

44. Gerbase AC, Rowley JT, Heymann DH, Berkley SF, Piot P. 1998. Global prevalence and incidence estimates of selected curable STDs. *Sex Transm. Infect.* 74(Suppl. 1):S12–16

45. Gestwicki JE, Lamanna AC, Harshey RM, McCarter LL, Kiessling LL, Adler J. 2000. Evolutionary conservation of methyl-accepting chemotaxis protein location in Bacteria and Archaea. *J. Bacteriol.* 182:6499–502

46. Gilmore RD Jr, Mbow ML, Stevenson B. 2001. Analysis of *Borrelia burgdorferi* gene expression during life cycle phases of the tick vector *Ixodes scapularis*. *Microbes Infect.* 3:799–808

47. Goldstein SF, Buttle KF, Charon NW. 1996. Structural analysis of *Leptospiraceae* and *Borrelia burgdorferi* by high-voltage electron microscopy. *J. Bacteriol.* 178:6539–45

48. Goldstein SF, Charon NW. 1988. The motility of the spirochete *Leptospira*. *Cell Motil. Cytoskel.* 9:101–10

49. Goldstein SF, Charon NW. 1990. Multiple exposure photographic analysis of a motile spirochete. *Proc. Natl. Acad. Sci. USA* 87:4895–99

50. Goldstein SF, Charon NW, Kreiling JA. 1994. *Borrelia burgdorferi* swims with a planar waveform similar to that of eukaryotic flagella. *Proc. Natl. Acad. Sci. USA* 91:3433–37

51. Goulbourne EA Jr, Greenberg EP. 1980. Relationship between proton motive force and motility in *Spirochaeta aurantia*. *J. Bacteriol.* 143:1450–57

52. Goulbourne EA Jr, Greenberg EP. 1981. Chemotaxis of *Spirochaeta aurantia*: involvement of membrane potential in chemosensory signal transduction. *J. Bacteriol.* 148:837–44

53. Goulbourne EA Jr, Greenberg EP. 1983. A voltage clamp inhibits chemotaxis of *Spirochaeta aurantia*. *J. Bacteriol.* 153:916–20

54. Greenberg EP, Canale-Parola E. 1977. Relationship between cell coiling and motility of spirochetes in viscous environments. *J. Bacteriol.* 131:960–69

55. Hagman KE, Porcella SF, Popova TG, Norgard MV. 1997. Evidence for a methyl-accepting chemotaxis protein gene (*mcp1*) that encodes a putative sensory transducer in virulent *Treponema pallidum*. *Infect. Immun.* 65:1701–9

56. Hardham JM, Rosey EL. 2000. Antibiotic selective markers and spirochete genetics. *J. Mol. Microbiol. Biotechnol.* 2:425–32

57. Harwood CS, Canale-Parola E. 1984. Ecology of spirochetes. *Annu. Rev. Microbiol.* 38:161–92

58. Holt JG, Krieg NR, Sneath PHA, Staley JT, Williams ST. 1994. The Spirochetes. In *Bergey's Manual of Determinative Bacteriology*, pp. 27–31. Baltimore: Williams & Wilkins. 787 pp. 9th ed.

59. Holt SC. 1978. Anatomy and chemistry of spirochetes. *Microbiol. Rev.* 42:114–60

60. Hovind-Hougen K. 1976. Determination by means of electron microscopy of morphological criteria of value for classification of some spirochetes, in particular treponemes. *Acta Pathol. Microbiol. Scand. Sect. B* (Suppl.) 255:1–41

61. Hovind Hougen K. 1984. Ultrastructure of spirochetes isolated from *Ixodes ricinus* and *Ixodes dammini*. *Yale J. Biol. Med.* 57:543–48

62. Hubner A, Yang X, Nolen DM, Popova TG, Cabello FC, Norgard MV. 2001. Expression of *Borrelia burgdorferi* OspC and DbpA is controlled by a RpoN-RpoS regulatory pathway. *Proc. Natl. Acad. Sci. USA* 98:12724–29

63. Izard J, Samsonoff WA, Kinoshita MB, Limberger RJ. 1999. Genetic and structural analyses of cytoplasmic filaments of wild-type *Treponema phagedenis* and a flagellar filament-deficient mutant. *J. Bacteriol.* 181:6739–46

64. Johnson BJB, Robbins KE, Bailey RE, Cao BL, Sviat SL, et al. 1996. Serodiagnosis of Lyme disease: accuracy of a two-step approach using a flagella-based ELISA and immunoblotting. *J. Infect. Dis.* 174:346–53

65. Johnson KA, Rosenbaum JL. 1990. The basal bodies of *Chlamydomonas reinhardtii* do not contain immunologically detecable DNA. *Cell* 62:615–19

66. Johnson RC, Marek N, Kodner C. 1984. Infection of syrian hamsters with Lyme disease spirochetes. *J. Clin. Microbiol.* 20:1099–101

67. Kaiser GE, Doetsch RN. 1975. Enhanced translational motion of *Leptospira* in viscous environments. *Nature* 255:656–57

68. Karlinsey JE, Tanaka S, Bettenworth V, Yamaguchi S, Boos W, et al. 2000. Completion of the hook-basal body complex of the *Salmonella typhimurium* flagellum is coupled to FlgM secretion and *fliC* transcription. *Mol. Microbiol.* 37:1220–31

69. Kataoka M, Li H, Arakawa S, Kuramitsu H. 1997. Characterization of a methyl-accepting chemotaxis protein gene, *dmcA*, from the oral spirochete *Treponema denticola. Infect. Immun* 65:4011–16

70. Kathariou S, Greenberg EP. 1983. Chemoattractants elicit methylation of specific polypeptides in *Spirochaeta aurantia. J. Bacteriol.* 156:95–100

71. Kefford B, Marshall KC. 1984. Adhesion of *Leptospira* at a solid-liquid interface: a model. *Arch. Microbiol.* 138:84–88

72. Kennedy MJ, Rosey EL, Yancey RJ Jr. 1997. Characterization of *flaA⁻* and *flaB⁻* mutants of *Serpulina hyodysenteriae*: both flagellin subunits, FlaA and FlaB, are necessary for full motility and intestinal colonization. *FEMS Mirobiol. Let.* 153:119–28

73. Kimsey RB, Spielman A. 1990. Motility of Lyme disease spirochetes in fluids as viscous as the extracellular matrix. *J. Infect. Dis.* 162:1205–8

74. Krieg NR, Tomelty JP, Wells JS Jr. 1967. Inhibition of flagellar coordination in *Spirillum volutans. J. Bacteriol.* 94:1431–36

75. Levett PN. 2001. Leptospirosis. *Clin. Microbiol. Rev.* 14:296–326

76. Li C, Bakker RG, Motaleb MA, Sartakova ML, Cabello FC, Charon NW. 2002. Asymmetrical flagellar rotation in *Borrelia burgdorferi* non-chemotactic mutants. *Proc. Natl. Acad. Sci. USA* 99:6169–74

77. Li C, Corum L, Morgan D, Rosey EL, Stanton TB, Charon NW. 2000. The spirochete FlaA periplasmic flagellar sheath protein impacts flagellar helicity. *J. Bacteriol.* 182:6698–706

78. Li C, Motaleb MA, Sal M, Goldstein SF, Charon NW. 2000. Spirochete periplasmic flagella and motility. *J. Mol. Microbiol. Biotechnol.* 2:345–54

79. Li H, Arakawa S, Deng QD, Kuramitsu H. 1999. Characterization of a novel methyl-accepting chemotaxis gene, *dmcB*, from the oral spirochete *Treponema denticola. Infect. Immun.* 67:694–99

80. Li H, Ruby J, Charon N, Kuramitsu H. 1996. Gene inactivation in the oral spirochete *Treponema denticola*: construction of a *flgE* mutant. *J. Bacteriol.* 178:3664–67

81. Lilburn TG, Kim KS, Ostrom NE, Byzek KR, Leadbetter JR, Breznak JA. 2001. Nitrogen fixation by symbiotic and free-living spirochetes. *Science* 292:2495–98

82. Limberger RJ, Slivienski LL, Izard J, Samsonoff WA. 1999. Insertional inactivation of *Treponema denticola tap1* results in a nonmotile mutant with elongated flagellar hooks. *J. Bacteriol.* 181:3743–50

83. Lux R, Miller JN, Park NH, Shi W. 2001. Motility and chemotaxis in tissue penetration of oral epithelial cell layers by *Treponema denticola. Infect. Immun.* 69:6276–83

84. Lux R, Moter A, Shi W. 2000. Chemotaxis

in pathogenic spirochetes: directed movement toward targeting tissues? *J. Mol. Microbiol. Biotechnol.* 2:355–64

85. Lux R, Sim J, Tsai JP, Shi W. 2002. Construction and characterization of a *cheA* mutant of *Treponema denticola. J. Bacteriol.* 184:3130–34

86. Lybarger SR, Maddock JR. 2001. Polarity in action: asymmetric protein localization in bacteria. *J. Bacteriol.* 183:3261–67

87. Macnab, RM. 1996. Flagella and motility. In Escherichia coli *and* Salmonella. *Cellular and Molecular Biology*, ed. FC Neidhardt, R Curtiss III, JL Ingraham, ECC Lin, KB Low, B Magasanik, WS Reznikoff, M Riley, M Schaechter, HE Umbarger. 1:123–45. Washington, DC: ASM Press. 1459 pp. 2nd ed.

88. Macnab RM. 1999. The bacterial flagellum: reversible rotary propeller and type III export apparatus. *J. Bacteriol.* 181: 7149–53

89. Margolin Y, Eisenbach M. 1984. Voltage clamp effects on bacterial chemotaxis. *J. Bacteriol.* 159:605–10

90. Margulis L. 1970. *Origin of Eukaryotic Cells*, pp. 208–45. New Haven: Yale Univ. Press. 349 pp.

91. Margulis L, Ashen JB, Sole M, Guerrero R. 1993. Composite, large spirochetes from microbial mats: spirochete structure review. *Proc. Natl. Acad. Sci. USA* 90:6966–80

92. Metzner P. 1920. Die Bewegung und Reizbeantwortung der bipolar gegeisselten Spirillen. *Jahrb. Wiss. Bot.* 59:325–412

93. Mikosza AS, Hampson DJ. 2001. Human intestinal spirochetosis: *Brachyspira aalborgi* and/or *Brachyspira pilosicoli? Anim. Health Res. Rev.* 2:101–10

94. Minnich SA, Ohta N, Taylor N, Newton A. 1988. Role of the 25-, 27-, and 29-kilodalton flagellins in *Caulobacter crescentus* cell motility; method for constuction of deletion and Tn5 insertion mutants by gene replacement. *J. Bacteriol.* 170:3953–60

95. Motaleb MA, Corum L, Bono JL, Elias AF, Rosa P, et al. 2000. *Borrelia burgdorferi* periplasmic flagella have both skeletal and motility functions. *Proc. Natl. Acad. Sci. USA* 97:10899–904

96. Moter A, Leist G, Rudolph R, Schrank K, Choi BK, et al. 1998. Fluorescence in situ hybridization shows spatial distribution of as yet uncultured treponemes in biopsies from digital dermatitis lesions. *Microbiology* 144 (Pt. 9):2459–67

97. Munson D, Obar R, Tzertzinis G, Margulis L. 1993. The 'tubulin-like' S1 protein of *Spirochaeta* is a member of the Hsp65 stress protein family. *BioSystems* 31:161–67

98. Nauman RK, Holt SC, Cox CD. 1969. Purification, ultrastructure, and composition of axial filaments from *Leptospira. J. Bacteriol.* 98:264–80

99. Noguchi H. 1918. Morophological characteristics and nomenclature of *Leptospira (Spirochaeta) icterohaemorrhagiae* (Inada and Ido). *J. Exp. Med.* 27:575–92

100. Norris SJ, Cox DL, Weinstock GM. 2001. Biology of *Treponema pallidum*: correlation of functional activities with genome sequence data. *J. Mol. Microbiol. Biotechnol.* 3:37–62

101. O'Toole G, Kaplan HB, Kolter R. 2000. Biofilm formation as microbial development. *Annu. Rev. Microbiol.* 54:49–79

102. Paster BJ, Dewhirst FE. 2000. Phylogenetic foundation of spirochetes. *J. Mol. Microbiol. Biotechnol.* 2:341–44

103. Pedersen NS, Petersen CS, Vejtorp M, Axelsen NH. 1982. Serodiagnosis of syphilis by an enzyme-linked immunosorbent assay for IgG antibodies against the Reiter treponeme flagellum. *Scand. J. Immunol.* 15:341–48

104. Picardeau M, Brenot A, Saint Girons I. 2001. First evidence for gene replacement in *Leptospira* spp. Inactivation of *L. biflexa flaB* results in non-motile mutants deficient in endoflagella. *Mol. Microbiol.* 40:189–99

105. Revel AT, Talaat AM, Norgard MV. 2002. DNA microarray analysis of differential gene expression in *Borrelia burgdorferi*, the Lyme disease spirochete. *Proc. Natl. Acad. Sci. USA* 99:1562–67

106. Rosey EL, Kennedy MJ, Petrella DK, Ulrich RG, Yancey RJ Jr. 1995. Inactivation of *Serpulina hyodysenteriae flaA1* and *flaB1* periplasmic flagellar genes by electroporation-mediated allelic exchange. *J. Bacteriol.* 177:5959–70

107. Rosey EL, Kennedy MJ, Yancey RJ Jr. 1996. Dual *flaA1 flaB1* mutant of *Serpulina hyodysenteriae* expressing periplasmic flagella is severely attenuated in a murine model of swine dysentery. *Infect. Immun.* 64:4154–62

108. Ruby JD, Charon NW. 1998. Effect of temperature and viscosity on the motility of the spirochete *Treponema denticola*. *FEMS Microbiol. Lett.* 169:251–54

109. Ruby JD, Li H, Kuramitsu H, Norris SJ, Goldstein SF, et al. 1997. Relationship of *Treponema denticola* periplasmic flagella to irregular cell morphology. *J. Bacteriol.* 179:1628–35

110. Sadziene A, Thomas DD, Bundoc VG, Holt SC, Barbour AG. 1991. A flagella-less mutant of *Borrelia burgdorferi*. Structural, molecular, and in vitro functional characterization. *J. Clin. Invest.* 88:82–92

111. Sartakova ML, Dobrikova EY, Motaleb MA, Godfrey HP, Charon NW, Cabello FC. 2001. Complementation of a non-motile *flaB* mutant of *Borrelia burgdorferi* by chromosomal integration of a plasmid containing a wild-type *flaB* allele. *J. Bacteriol.* 183:6558–64

112. Schrank K, Choi BK, Grund S, Moter A, Heuner K, et al. 1999. *Treponema brennaborense* sp. nov., a novel spirochaete isolated from a dairy cow suffering from digital dermatitis. *Int. J. Syst. Bacteriol.* 49:43–50

113. Schwan TG, Piesman J, Golde WT, Dolan MC, Rosa PA. 1995. Induction of an outer surface protein of *Borrelia burgdorferi* during tick feeding. *Proc. Natl. Acad. Sci. USA* 2909–13

114. Segall JE, Ishihara A, Berg HC. 1985. Chemotactic signaling in filamentous cells of *Escherichia coli*. *J. Bacteriol.* 161:51–59

115. Shapiro L, Losick R. 2000. Dynamic spatial regulation in the bacterial cell. *Cell* 100:89–98

116. Shi W, Yang ZM, Geng Y, Wolinsky LE, Lovett MA. 1998. Chemotaxis in *Borrelia burgdorferi*. *J. Bacteriol.* 180:231–35

117. Snyder MA, Stock JB, Koshland DE Jr. 1981. Role of membrane potential and calcium in chemotactic sensing by bacteria. *J. Mol. Biol.* 149:241–57

118. Soames JV, Davies RM. 1975. The structure of subgingival plaque in a beagle dog. *J. Periodont.* 9:333–41

119. Sourjik V, Berg HC. 2002. Receptor sensitivity in bacterial chemotaxis. *Proc. Natl. Acad. Sci. USA* 99:123–27

120. Stanton TB, Matson EG, Humphrey SB. 2001. *Brachyspira* (*Serpulina*) *hyodysenteriae gyrB* mutants and interstrain transfer of coumermycin A(1) resistance. *Appl. Environ. Microbiol.* 67:2037–43

121. Stavitsky AB. 2002. Studies on the pathogenesis of leptospirosis. *J. Infect. Dis.* 76:179–92

122. Summers KE, Gibbons IR. 1971. Adensoine-triphosphate-induced sliding of tubules in trypsin-treated flagella of sea urchin sperm. *Proc. Natl. Acad. Sci. USA* 68:3092–96

123. ter Huurne AAHM, van Houten M, Muir S, Kusters JG, Van der Zeijst BAM, Gaastra W. 1992. Inactivation of a *Serpula* (*Treponema*) *hyodysenteriae* hemolysin gene by homologous recombination: importance of this hemolysin inpathogenesis of *S. hyodysenteriae* in mice. *FEMS Microbiol. Lett.* 92:109–14

124. Tilly K, Casjens S, Stevenson B, Bono JL, Samuels DS, et al. 1997. The *Borrelia*

burgdorferi circular plasmid cp26: conservation of plasmid structure and targeted inactivation of the *ospC* gene. *Mol. Microbiol.* 25:361–73

125. Tilly K, Elias AF, Bono JL, Stewart P, Rosa P. 2000. DNA exchange and insertional inactivation in spirochetes. *J. Mol. Microbiol. Biotechnol.* 2:433–42

126. Turner L, Ryu WS, Berg HC. 2000. Real-time imaging of fluorescent flagellar filaments. *J. Bacteriol.* 182:2793–801

127. Wasserheit JN. 1992. Epidemiological synergy. Interrelationships between human immunodeficiency virus infection and other sexually transmitted diseases. *Sex Transm. Dis.* 19:61–77

Annu. Rev. Genet. 2002. 36:75–97
doi: 10.1146/annurev.genet.36.040202.111115

RECOMBINATION IN EVOLUTIONARY GENOMICS

David Posada[1,2], Keith A. Crandall[3,4], and Edward C. Holmes[5]

[1]Variagenics Inc. Cambridge, Massachusetts 02139, [2]Center for Cancer Research,
Massachusetts Institute of Technology, Cambridge, Massachusetts 02139, [3]Department
of Integrative Biology, [4]Department of Microbiology and Molecular Biology, Brigham
Young University, Provo, Utah 84602, and [5]Department of Zoology, University of Oxford,
Oxford OX1 3PS, United Kingdom; e-mail: dposada@variagenics.com

Key Words   phylogeny, incongruence, maximum likelihood, bioinformatics,
linkage disequilibrium

■ **Abstract**   Recombination can be a dominant force in shaping genomes and associated phenotypes. To better understand the impact of recombination on genomic
evolution, we need to be able to identify recombination in aligned sequences. We review
bioinformatic approaches for detecting recombination and measuring recombination
rates. We also examine the impact of recombination on the reconstruction of evolutionary histories and the estimation of population genetic parameters. Finally, we review
the role of recombination in the evolutionary history of bacteria, viruses, and human
mitochondria. We conclude by highlighting a number of areas for future development
of tools to help quantify the role of recombination in genomic evolution.

CONTENTS

INTRODUCTION

The comparative analysis of genome sequence data is transforming evolutionary biology. Not only does genomic analysis allow us to reconstruct phylogenetic patterns and processes with more accuracy than ever before, but it also provides new insights to the fundamental mechanisms of evolutionary change. One such mechanism is recombination. Already, the bioinformatic analysis of genome sequence data has revolutionized our understanding of this central evolutionary process, including its impact on genome structure (104) and on phenotypic variation (146), and its relationship to the study of genetic disease (12). Further, there is now a greater understanding of how recombination confounds our attempts to infer phylogenetic history and other key evolutionary parameters, and that lateral gene transfer has been a common occurrence in the evolutionary history of many species, so that taxa cannot always be related by single phylogenetic tree (73).

Given the central importance of recombination in evolutionary biology, it is crucial that we have bioinformatic tools that are able to accurately detect its occurrence and understand how it affects the inference of phylogenetic relationships. Our review covers current tools available for detecting recombination and discusses the impact of recombination on phylogeny estimation. For this purpose, it is important to distinguish between homologous recombination, which affects related gene sequences, from nonhomologous recombination, which does not. Although both conform to a broad definition of recombination—an evolutionary event that has as a consequence the horizontal exchange of genetic material—our discussion of the phylogenetic impact and detection of recombination implicitly assumes that we are dealing with homologous sequences. We also consider recombination in both prokaryotes and eukaryotes, where traditionally the process of recombination is thought to act differently. This distinction is significant because the concept of recombination prevalent in evolutionary genetics is based on meiosis in eukaryotic organisms, where recombination is a complex molecular process by which a fragment of DNA is reciprocally exchanged between homologous chromosomes. On the other hand, prokaryotes provide several possible pathways of recombination—conjugation, transformation, and transduction—that are more accurately denominated lateral gene transfer or gene conversion, as they involve the nonreciprocal replacement or addition of sequences rather than their exchange, involving either homologous or nonhomologous sequences (although gene conversion is also a frequent process in eukaryotic multigene families). It is also important to distinguish between recombinational events that occur between different genes (intergenic recombination) or between alleles of the same gene (intragenic recombination). Hence, whereas there are many different mechanisms to generate recombinant genomes (in our broadly defined sense), the evolutionary outcomes of recombination are largely the same in whichever system is analyzed. It is the impact of these outcomes that we address here.

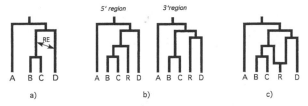

Figure 1   Recombination may generate different phylogenies for different regions of a gene or alignment. (*a*) A recombinational event between the ancestor of B&C and D generates a recombinant R that is present in the sample. (*b*) This recombinant will cluster with the ancestor of B&C in the region 3′ to the recombination breakpoint, whereas in the 5′ region it will cluster with D. (*c*) The fact that there is more than one history underlying the data is often represented as loops or reticulations (and therefore the term "reticulate evolution" is frequently used).

A central theme of our review is the impact of recombination on phylogenetic inference. The reconstruction of phylogenies has been the subject of considerable and often intemperate debate for many years, and more recently, the accumulation of molecular data has added a new level of interest and analytical power (47). Although there are many examples of the myriad uses of molecular phylogenies (40), most of these applications rely on an accurate estimation of the phylogenies themselves. Traditional methods of phylogeny estimation, such as maximum parsimony (MP), minimum evolution (ME), or maximum likelihood (ML) [see (125)], assume that only one evolutionary history underlies the sample under study. However, this assumption is violated by the occurrence of recombination, which can lead to samples with several underlying phylogenies, in which case it is more accurate to describe relationships in terms of reticulate evolution (Figure 1). Indeed, it is important to remember that a bifurcating tree is a hypothesis about how taxa are related, not a truism. In those studies that have explored the possibility of recombination, it has had a significant impact in our understanding of the history of gene genealogies and arguments based on these phylogenies (14, 16, 38, 48, 49, 97, 108, 110, 143, 147). Technological advances have allowed for even larger regions of DNA to be sequenced, thereby increasing the chances for recombination to have occurred in the sample under study. A clear understanding of how we can detect and estimate the rate at which recombination occurs is therefore essential.

THE DETECTION OF RECOMBINATION

Given the importance of recombination in the evolutionary analysis of sequence data and as a potentially dominant force in the rearrangement of genetic variation, it is essential to be able to identify whether a given set of sequences has been affected by recombination, to identify the boundaries of the recombinational units, and to evaluate the impact of recombination on our ability to reconstruct evolutionary histories and estimate population genetic parameters. In the following sections we summarize different methods for detecting the presence of recombination and their relative performance. By detecting recombination we mean just to answer the question of whether recombination has occurred or not. How to measure the amount of recombination is discussed in the next section.

Statistical Methods for Detecting Recombination

During the past 15 years numerous methods have been developed to test for the occurrence of recombination, to identify the parental and recombinant individuals, and to determine the location of the recombinational break-points. These techniques differ greatly in approach and applicability, but may be (tentatively) classified into five nonexclusive general categories: similarity, distance, phylogenetic, compatibility, and nucleotide substitution distribution methods. Here we provide a brief overview of current methods within each of these categories. For a more detailed review of these methods see Crandall & Templeton (11), or the supplementary material in Posada & Crandall (101). David Robertson (Department of Zoology, University of Oxford) also offers a web site with links to the implementations of these methods at http://grinch.zoo.ox.ac.uk/RAP_links.html.

(a) *Similarity Methods*   These methods infer gene conversion when synonymous substitutions at variable regions exceed those at conserved regions (85, 94). However, they have not been used extensively, in part because they are most useful for detecting gene conversion in multigene families and can be applied only to coding regions.

(b) *Distance Methods*   Several methods look for inversions of distance patterns among the sequences (138). In general, they use a sliding window approach and the estimation of some statistic based on the genetic distances among the sequences. Because the phylogeny does not need to be known, these methods are highly computationally efficient.

(c) *Phylogenetic Methods*   Other methods infer recombination when phylogenies from different parts of the genome result in discordant topologies or when orthologous genes from different species are clustered. When comparisons of adjacent sequences yield topological incongruence, there is good reason to suspect the involvement of recombination (3, 29, 35, 41, 42, 49, 55, 56, 64, 75, 79–82, 107–109, 116, 117). Such phylogeny-based methods are currently the most common in use to detect recombination.

(d) *Compatibility Methods*   Compatibility methods test for partition phylogenetic incongruence on a site-by-site basis. These methods do not require a phylogeny of the sequences under study (14, 17, 59, 60, 120).

(e) *Substitution Distribution*   This family of methods include strategies that examine sequences for a significant clustering of substitutions or fit to an expected statistical distribution (5, 11, 16, 32, 57, 77, 83, 111, 112, 118, 119, 121, 127, 131, 142).

Performance of Recombination Detection Methods

The performance of several methods for detecting recombination has been evaluated through the analysis of simulated (7, 76, 101, 141) and empirical data (15, 99). These studies have focused on the detection of the presence of recombination rather than on the identification of parentals and recombinant individuals, or on the location of the specific break-points, and hence give an incomplete picture of our ability to accurately detect every aspect of recombination.

Recombination detection methods differ in performance depending on the amount of recombination, the genetic diversity of the data, and the degree of rate variation among sites. Most methods are efficient, showing more power with increasing recombination rates, although some methods are more efficient than others. Most methods also show better performance at higher levels of divergence, most likely because of an increase in the amount of signal for recombination present in the data. For the majority of methods, a minimum nucleotide diversity of 5% seems necessary to obtain substantial power, and several recombination events are needed to infer the presence of recombination. Recombination is also difficult to detect when the phylogeny has long terminal, and short internal, branches (141). Rate variation among sites (145) can also be confounded with recombination, and in some cases it leads to false positives (99, 101, 114, 142). Perhaps the most interesting consensus result from these studies is that methods that use the substitution patterns or incompatibility among sites seem to be more powerful than methods based on phylogenetic incongruence. This might be partially explained by the fact that, in general, phylogenetic methods can only detect recombination events that change the topology of the tree, and at high recombination rates there should be many such events.

Note also that there are two different contexts in which we may wish to detect recombination: rare, sporadic recombination or frequent, repeated recombination (76). Not surprisingly, most methods have trouble detecting rare recombinational events, especially when sequence divergence is low. Indeed, recent events should be more easily identifiable than older events, as the latter may be obscured by subsequent mutation. On the other hand, when recombination rates are extremely high, leading to situations close to linkage equilibrium, we would expect substitution methods to have difficulty in identifying site patterns (76), although this is not what is observed with real data sets (99). Indeed, we are interested in maximizing the chances of detecting recombination while minimizing the chances of false positives. In order to do so, we need to take into account levels of variation. For example, for data sets with very low divergence (1%), the homoplasy test (77) appears to be a reasonable method, as long as there is little among-site rate variation. For higher levels of divergence the homoplasy test is not adequate, and methods like the modified maximum chi-square (101, 141), GENECONV (113) or RDP (75) are more powerful. However, perhaps the key conclusion from simulation and analytical studies is that one should not rely on a single method to detect recombination (101, 141).

The Proportion of Undetectable Recombination

In general, our ability to detect recombination depends on the amount of genetic variation in the population. If recombination occurs between two identical strands of DNA, then this event is undetectable. Therefore, estimates of recombination will always be underestimates due to our inability to detect recombination between identical or nearly identical sequences. Hudson & Kaplan (54) studied the theoretical sampling distribution of the number of recombination events that have occurred during the history of a sample of DNA sequences. Through computer simulation they compared the known number of recombination events with the number inferred by a detection technique based on parsimony (four-gamete test), and found that only a small fraction of known recombination events were detected. In this context, recombination events can be divided in two categories (122): those that do not result in any observable effect on the DNA sequences and hence are undetectable with any analytical method, and those that do affect the DNA sequence and here are potentially detectable. The first category includes recombination events between identical sequences (Figure 2a), between sequences that differ at a single site (Figure 2b,c) and between sequences that differ at several sites with the crossover point flanking the segregating sites (Figure 2d,e). The number of nucleotide differences, d, between a random pair of DNA sequences is related to the quantity $\theta = 4N_e\mu$, where N_e is the effective population size and μ the mutation rate. Stephens (122) has shown that even for relatively high values of θ, a substantial fraction of the recombination events cannot be detected, even with $d \geq 2$. Undetected recombinant events occur mainly because of inefficiency in the detection, and less commonly because of redundant recombination events. Quite clearly, attempts to estimate recombination rates should take into account the fraction of undetectable events (122). Similarly, and from a phylogenetic point of view, recombination events can be classified with respect to their effect on tree topology, as (a) events that do not change the branch lengths, (b) events that do change the branch lengths, but do not change the tree topology, and (c) events that change the tree topology (Figure 3). Wiuf et al. (141) give theoretical expectations for the tree types of events.

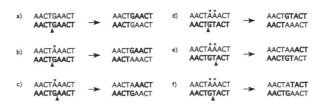

Figure 2   Different types of recombination events depicted in substitution patterns. Events *a–e* belong to the category of undetectable events. Event *f* is a detectable event and so belongs to the category of detectable events. Sites with the * symbol are variable sites. A triangle indicates the location of the recombination break-point.

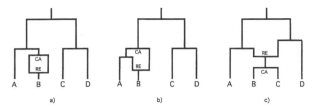

Figure 3 Different types of recombination events and how they affect tree topology and branch lengths. CA indicates a coalescent event, while RE indicates a recombination event. (*a*) No change in topology nor in branch lengths. (*b*) No change in topology but a change in branch lengths. (*c*) A change both in topology and branch lengths. Events of type (*a*) and (*b*) can occur in samples of any size, but events of type (*c*) can only occur in samples of size ≥ 4. Most events in samples of large size are of type (*c*) (141).

Recombination in Polymerase Chain Reaction

Once recombination is detected, questions of the validity of the recombination event in vivo remain because recombination can also be produced in vitro during the polymerase chain reaction (PCR) used to amplify the desired region of DNA (6, 87, 93). Recombination is a particular concern when attempting to amplify long products by PCR, as has been demonstrated with HIV-1 sequences. Phylogenetic analyses offer a way to distinguish between PCR-induced recombination events and actual in vivo recombination events. If mutational events are mapped along the branches of the reconstructed histories, recent and historical recombination events can be differentiated by the number of accumulated mutations after the recombination event. In PCR recombination, we do not expect to accumulate additional substitutions, so that if such additional substitutions are present along the recombinant branch, then the inferred event most likely occurred in vivo.

ESTIMATING RECOMBINATION RATES

Recombination can play a dominant role in the generation of novel genetic variants through the rearrangement of existing genetic variation generated through mutation. Recombination also plays a role in the dissipation of linkage disequilibrium. Hence, when coupled with selection, recombination can be a key evolutionary force (26, 46). To understand the role of recombination in the generation of genetic diversity relative to the role of mutation we need to be able to accurately estimate recombination rates. Whereas in the previous section we focused on detection of the presence of recombination, here we focus on the rate of recombination. Indeed, recombination rate estimators can be used to build tests for the presence of recombination (7).

The population recombination parameter is defined as $\rho = 4N_e r$, where N_e is the effective population size and r is the per-locus (or per-site) recombination rate per generation. The population mutation parameter (genetic diversity) can be similarly defined as $\theta = 4N_e\mu$, where μ is the per-locus (or per-site) mutation rate per generation. If we can accurately estimate both ρ and θ, we can then define the relative rate of recombination compared to mutation as

$$\varepsilon = \frac{\rho}{\theta} = \frac{4N_e r}{4N_e \mu} = \frac{r}{\mu}. \qquad\qquad 1.$$

On the rare occasion that this quantity has been estimated from nucleotide sequence data, it has provided keen insights into the population dynamics of the organism under study (e.g., 23, 103).

Recombination Rate Estimators

As there are a variety of methods to estimate θ (19, 28, 31, 66, 126, 132, 136), there are also a number of methods to estimate recombination rates in populations. Like recent methods for estimating θ, most approaches for estimating ρ are based on neutral coalescent theory (62) [reviewed in (52)] with recombination (37, 50). The estimators of recombination generally take one of two approaches, either quantifying recombination as a summary statistic or estimating recombination rates by considering all the data. Hudson (51) took the former approach to derive an estimator of ρ based on the observed variance of the number of pairwise differences. The expected variance in pairwise differences decreases with decreasing amounts of linkage disequilibrium between segregating sites as a result of increasing recombination. Therefore, this observed variance in pairwise differences is a measure of the amount of linkage disequilibrium and hence also of the recombination rate (134). Wakeley (133) improved this estimator by considering only nonidentical pairs of sequences, which reduces the bias and standard error on this estimator of recombination rate.

The second approach is to use a maximum likelihood framework to provide a joint estimate of mutation and recombination rate that uses the maximal information in the sample, rather than a summary statistic. The first such estimator was developed by Griffiths & Marjoram (36); they used a coalescent process with recombination resulting in a genealogy with reticulations that they termed an "ancestral recombination graph." Kuhner et al. (67) also used a "recombinant genealogy" to co-estimate the recombination rate and mutation rate (our ε; Equation 1) using a Metropolis-Hastings sampling strategy across genealogies. Fearnhead & Donnelly (24) similarly present a full-likelihood–based approach to the joint estimate of recombination and mutation rates. Their method develops an improved importance sampling scheme, which should result in more accurate estimates. The advantage of these methods is that they use all the data to estimate recombination rate instead of a summary statistic. However, they accomplish this at the expense of computational efficiency. A compromise solution was proposed by Hey & Wakeley (45) whose approach averages likelihood estimates of recombination rate for subsets of sequences. An alternative approach proposed by Wall (134) uses the number of distinct haplotypes to estimate ρ by using maximum likelihood on summary statistics. A Bayesian approach to estimating recombination rate was recently proposed by Falush et al. (23), but the statistical properties of this method are unexplored. An alternative Bayesian approach was suggested by Nielsen (91), and the statistical foundation of this approach is much better laid out. This method is similar to that of Kuhner et al. except that a Bayesian approach is used in parameter estimation instead of importance sampling. Hudson (53) recently proposed an alternative approach that considers polymorphic sites in pairs and then utilizes likelihood methods appropriate for analyzing a pair of polymorphic sites. This composite-likelihood estimator has the advantage of being more computationally efficient relative to the full-likelihood methods but without summarizing the data in a single statistic. McVean et al. (83) extended this to accommodate different models of evolution (including, importantly, rate variation) and to relax the infinite-sites assumption (typically violated by many empirical data sets). All of these approaches assume constant population sizes, independence of sites, neutral evolution, and an infinite-sites model of evolution (with the exception of McVean et al.'s method). However, they differ considerably in terms of the required population sampling, level of nucleotide polymorphism, and number and type of nucleotide positions surveyed.

Performance of Estimators of Recombination Rate

At least two studies have extensively compared different estimators of recombination rate. The first compared ten estimators and found that their relative performance depended greatly on the amount of genetic diversity (θ), with most methods performing poorly at low levels of genetic diversity (134). The best performing estimator in these simulations was that of Kuhner et al., which had the smallest mean squared error, the greatest proportion of estimates within a factor of two of the actual value, and the second smallest bias (134). The second study compared the relative performance of the full-maximum–likelihood methods. In this simulation study, the authors distinguish between two possible comparisons. One can either compare how accurately the methods approximate the likelihood surface or the properties of the methods' ability to estimate ρ and θ (24). While the former study based comparisons on the second criterion, these authors argue that the first is more fundamental and therefore report results from this approach. They show that their new sampling method is up to four orders of magnitude more efficient than the previous method of Griffiths & Marjoram (36). In addition, they showed their approach outperformed Kuhner et al.'s method and also that this method often gave misleading results. The discrepancy between these results and those obtained by Wall and Kuhner et al. are presumably due to the different criterion of assessment (likelihood surface instead of the parameter estimates themselves) and the difference in the relative amount of recombination to mutation (Kuhner et al. simulated data with mutation rates much higher than recombination rates, whereas Fearnhead & Donnelly simulated under the opposite conditions) (24). Clearly, the comparisons of these methods have just begun. As new methods emerge from our better understanding of existing methods, further research is needed to discern how robust such estimators are to violations of the standard coalescent assumptions (135).

RECOMBINATION AND PHYLOGENETIC INFERENCE

Phylogenetic studies typically ignore the potential occurrence of recombination, which may produce sequence regions with different evolutionary histories. An accurate history of such mosaic sequences cannot be estimated by traditional phylogenetic methods that assume a single nonreticulate tree. If recombination is present and we have ignored it, can we expect the inferred phylogeny to represent any of the underlying evolutionary histories? Furthermore, what happens if we then use these trees to estimate relevant evolutionary parameters? Partial answers to these questions have been only recently investigated, and in this section we outline our current understanding of the impact of recombination in phylogenetic studies.

The Effect of Recombination on Phylogeny Estimation

Recombination has long been recognized as a serious confounding factor for phylogeny estimation. However, only a few studies have explicitly addressed this question. Wiens (140) carried out a simulation study to explore the effect of combining data sets with different phylogenetic histories. This problem is identical to the problem of recombination, only that when we combine data sets we have already defined the potential partitions of the data. Wiens explicitly investigated the effect of combining genes generated under different genealogies on the estimation of the true "species tree." The main conclusion was that a combined analysis provides a poor estimate of the species tree in areas where the gene genealogies are very different, but an improved estimate in regions where the gene genealogies agree. Wiens also provides a simple strategy to deal with such situations, consisting of (*a*) defining the data partitions (e.g., by gene), (*b*) performing a separate analysis on each partition, and (*c*) undertaking a combined analysis, with caution directed toward nodes not supported in the analyses of separate partitions (i.e., step *b*).

However, in many cases we only have a single data set with no obvious partitions and the question then becomes what happens when recombination has occurred, but is ignored? In such a case can we expect the inferred phylogeny to represent any of the underlying evolutionary histories? Posada & Crandall (102) examined this question by applying traditional phylogenetic reconstruction methods to mosaic sequence alignments. Their results suggest that the effect of recombination on phylogeny estimation is dependent upon the relatedness of the sequences involved in the recombination event and on the relative size of the regions with different phylogenetic histories. When recombination occurred between closely related taxa, or when recombination was ancient, one of the histories underlying the data was inferred. In these cases, the phylogeny under which the majority of sites were evolved was generally recovered. On the other hand, when recombination occurred recently among divergent taxa and the recombinational break-point divided the alignment in two regions of similar length, a phylogeny that was very different from any of the true phylogenies underlying the data was inferred. Hence, recombination can be very misleading, resulting in the inference of wrong topologies, but only in some circumstances. More extensive simulations are needed to determine the generality of these conclusions.

Estimating Parameters from Recombinant Trees

While recombination can have a major impact on phylogenetic trees, the tree is seldom the endpoint of a phylogenetic analysis. Indeed, trees are now used to infer many relevant evolutionary parameters and to test different evolutionary hypotheses. Schierup & Hein (114) characterized some of the consequences of ignoring recombination when using phylogenies to make demographical, chronological, or substitutional inferences. Long terminal branches appear in a more star-shaped phylogeny, which suggests apparent exponential growth when the population size is actually constant. Further, parallel mutations are postulated to fit the data to a single tree and the extent of rate heterogeneity among-sites is wrongly inferred. Crucially, however, recombination affects different phylogenetic methods in different ways. While distance methods underestimate the time to the most common ancestor, maximum likelihood leads to an overestimate of the total number of mutations. The amount of recombination needed for these effects to be evident is not high, and such effects were found with just 100 bp in *Drosophila*, or 2000 bp in humans, although obviously the recombination rate varies extensively over the genome.

Recombination and the Molecular Clock

Ignoring recombination may lead to the false rejection of the molecular clock if phylogenetic methods like the likelihood ratio test (27) are used (115). To appropriately test the clock hypothesis in the presence of recombination, we need to use a test that is independent of tree topology. Muse & Weir (89) proposed a triplet likelihood ratio test to test for equality of evolutionary rates for two species at a time using a third species as an outgroup. Posada (98) has shown that this can be used as a conservative test for recombinant sequences if an outgroup is selected that did not recombine with the ingroup.

REPRESENTING RETICULATE EVOLUTION

The presence of recombination in an evolutionary history presents a significant problem for the representation of that history. A typical representation of an evolutionary history consists of a bifurcating evolutionary tree. However, with recombination, the true underlying history is reticulate in nature. Therefore, a bifurcating tree is, at best, only a partial representation of the actual evolutionary history. To better represent the actual reticulate evolutionary history, researchers have developed network approaches for estimating phylogenetic trees. Similar to standard tree estimation approaches, network approaches employ a variety of optimality criteria, including parsimony, distance, and likelihood approaches. Many combine parsimony, distance, and/or likelihood approaches into a single method. One of the first methods developed for comparing sequences in a network fashion was statistical geometry (18). This approach considers quartet combinations of nucleotide sequences and then develops a geometric configuration to represent the combination of quartets. Statistical parsimony (129), implemented in the software package TCS (10), makes minimum pairwise connections among sequence variants up to a point determined by the calculation of a probability of parsimonious connections. Netting (30) is an approach that represents all the most parsimonious trees in a single network by presenting homoplasies as networked connections in different dimensions. Molecular variance parsimony (20) takes into account haplotype frequencies and their geographic distributions to estimate network relationships. Split decomposition (4) takes sequence characters and divides them into partitions of mutually exclusive sets and then compares these splits across characters. When the splits are incompatible, loops are formed in the graphical representation of genealogical relationships. Finally, a likelihood network procedure (123, 124) allows for a directed graphical model (where nodes are stochastic variables and branches indicate correlation between these variables) to represent the evolutionary history of sequences along a network. These methods and their theoretical advantages over standard bifurcating approaches for the representation of gene genealogies have been recently reviewed in detail (100). However, much work remains in terms of testing the accuracy of these methods in reconstructing evolutionary histories and their relative performances.

THE IMPACT OF RECOMBINATION: EMPIRICAL EXAMPLES

The detection and estimation of recombination has led to major biological insights in a variety of cases. These studies have been particularly important in microbiology where the application of the bioinformatic tools described in this review to the growing data base of gene sequences has radically changed our perspective on how frequently recombination occurs in both bacteria and viruses. Such findings have wide-ranging implications, from the successful reconstruction of evolutionary and epidemiology history, to preventing the development of drug resistance and the evolution of virulence. Moreover, the high levels of genetic variation in many microbial species, particularly RNA viruses, also mean that they constitute ideal model organisms to assess the reliability of different estimators of the presence and rate of recombination.

Recombination in Bacteria

Because bacteria reproduce by binary fission, it was generally assumed that they evolved entirely by asexual mechanisms. Such a view was initially confirmed by studies of *Escherichia coli* using multilocus enzyme electrophoresis (MLEE), which indicated that populations were characterized by high levels of linkage disequilibrium and phylogenetic congruence [reviewed in (78)]. However, the growing availability of genetic data gradually shifted opinion toward the view that recombination could occur, sometimes frequently, among bacterial species other than

E. coli (78). A more radical overhaul came with the availability of large-scale nucleotide sequence data, either in the guise of multilocus sequence typing [MLST (74)] or large regions of bacterial genomes. The bioinformatic analyses of these data indicated that recombination rates in bacterial species can be both extremely high and extremely variable, and that lateral gene transfer can occur among very distantly related species, even between bacteria and eukaryotes (43). This last observation has important implications for reconstructing the evolutionary history of cellular life forms (W. F. Doolittle, personal communication).

MLST data provide a genome-wide subsample of housekeeping genes, but from a very large number of isolates. This makes it possible to measure a range of evolutionary parameters, including recombination frequency. This can be done indirectly by assessing the degree of incongruence between phylogenies of different genes within MLST data sets. Using a maximum likelihood method, in which the differences in log likelihood between the trees estimated for each gene are compared to a null distribution generated using random tree topologies, very high rates of recombination were inferred in *Neisseria meningitidis*, *Staphylococcus aureus*, *Streptococcus pneumoniae*, and *Streptococcus pyogenes*, as tree topologies from different genes were no more similar than random (25, 48). As expected, less incongruence was found in *E. coli* (25, 105), indicating that the rate of recombination is far lower in this species. In some cases, it has also been possible to estimate ε (see above) from MLST data, $\sim$5 for *N. meningitidis* (25), and $\sim$1 for both *N. gonorrheae* (103) and *Helicobacter pylori* (23), that are broadly similar to the recombination rates estimated for human genes. In contrast, the evidence that lateral gene transfer can occur between very distantly related bacterial species has more often been obtained through studies of aberrant $G + C$ content than incongruence (71, 92). Such studies have revealed lateral gene transfer to be a recurrent evolutionary event; in *E. coli*, for example, foreign DNA is estimated to have been imported at a frequency of up to16 Kb per million years (71).

Despite the growing evidence that recombination is a fundamental process in bacterial evolution, the precise mechanisms by which it occurs, and why rates vary so extensively, are less clear. In species that are competent for the uptake of naked DNA from the environment, such as *Neisseria* sp., transformation clearly plays a major role. In many other cases, conjugation, usually involving the transfer of plasmids between bacteria that have come into physical contact, has been described. A role for bacteriophage-mediated transduction is also a frequent suggestion (92). At present, the best evidence for this latter process is that known attachment sites of bacteriophage integrases are frequently found next to imported regions of bacterial DNA, such as the LEE pathogenicity island of *E. coli* (39). However, as bacteriophages from natural environments have received little study, their overall role in bacterial recombination remains uncertain.

Recombination in Viruses

Recombination in DNA and RNA viruses occurs by very different processes. In DNA viruses recombination is likely to take place in the same manner as in other

DNA genomes, i.e., involving an enzyme-mediated breakage-reunion mechanism. This appears to be a relatively common process and can also result in the capture of host genes, which may allow the virus to mimic or block host proteins, thereby assisting in the development of persistent infection (9). Recombination in RNA viruses can occur through either reassortment or "copy-choice" replication. Reassortment describes the process by which viruses with segmented genomes shuffle those segments during mixed infection. This has been described in detail in influenza A virus, where it is associated with the production of novel strains that can evade pre-existing immunity through an "antigenic shift" (137). In copy-choice replication, the viral RNA-dependent RNA polymerase switches from one RNA molecule to another during replication, generating mosaic genomes (90). This process is now thought to occur in a wide variety of positive-sense RNA viruses and retroviruses (143).

As with bacteria, much of the evidence for homologous RNA virus recombination involves the detection of phylogenetic incongruence (143). Such topological mismatching has been documented at a variety of phylogenetic levels, from within single species, to different viral families, in one case between a RNA and a DNA virus (33). However, it is equally clear that RNA viruses vary greatly in their ability to undergo recombination, although, to date, rates of recombination relative to that of mutation have not been estimated through sequence comparisons. For example, hepatitis C virus (HCV) and GBV-C are members of the same viral family (the *Flaviviridae*), yet recombination in HCV is rare, whereas GBV-C appears to recombine at high frequency (144).

It is possible that recombination rate in RNA viruses is a selectively determined trait. Indeed, recombination has been shown to result in direct fitness increases, for example by bringing together different genomes carrying individual drug-resistance mutations in HIV (88), and there is also evidence that reassortment can allow viruses to escape from deleterious mutation accumulation (8). Conversely, it is also possible that recombination rate is simply an outcome of mechanistic constraints set by genome structure and is not a selected entity at all. For example, the highest rates of recombination are found in retroviruses, which carry two copies of their genome within each mature virion, so making recombination easier, and in viruses that can frequently reassort their segmented genomes. Far lower rates of recombination are found in positive-sense, single-strand RNA viruses, where it is easily detected as mosaic sequences, and more so in negative-strand RNA viruses, which have genomes packaged into filamentous ribonucleoprotein (RNP) structures, greatly limiting their ability to recombine. Furthermore, most recombinants, unless they are very similar in sequence, will be deleterious and hence removed by purifying selection. This further emphasizes how estimates of recombination rate based on sequence comparisons are likely to be underestimates of the actual number of recombination events. Determining the basis for the variation in viral recombination rates, and whether they correlate with other biological features such as virulence, is clearly an important area for future study.

Recombination in Human Mitochondrial DNA

One of the most controversial claims in evolutionary genetics in recent years is that, contrary to mainstream opinion, the human mitochondrial genome may undergo recombination. Although signals of genetic exchange have been found in other animal mitochondria (68, 69), the claim that it can occur in human mtDNA has provoked intense debate (21, 44), not least because it has serious implications for our attempts to infer the origin and migration of modern humans.

Two pieces of evidence have been cited in support of mtDNA recombination: that there is excessive homoplasy at polymorphic sites, as revealed in the homoplasy test (22), and that there is a decrease in linkage equilibrium (LD) with physical distance in the mtDNA genome (2). Because recombination creates convergent/parallel evolutionary change, the occurrence of widespread homoplasy at face value represents strong evidence for this process. However, homoplasy can also occur through excessive multiple substitution at single sites, such as those that are known to be hypervariable in mtDNA. Indeed, if the complex pattern of among-site rate heterogeneity is taken into account through the use of the gamma distribution, the evidence for recombination in mtDNA seemingly disappears (142).

The evidence for some degree of linkage equilibrium in the mitochondrial genome has also been questioned. In particular, LD values appear to be highly dependent on the analytical method used, with different estimates obtained with r^2 and $|D'|$ (86), and also the particular data in question. Most notably, a recent survey of 53 complete mitochondrial genome sequences from a variety of geographical regions provided no evidence for any decline in LD with physical distance (58). What causes these conflicting signals is unknown but clearly requires explanation (84). More fundamentally, there is as yet no evidence for incongruence in mtDNA phylogenies. In sum, the evidence for recombination in human mtDNA appears to be weak on current data.

CONCLUSIONS

Several important messages regarding the detection of recombination stem from our review. First, the fact that many recombinational events cannot be detected implies that current methods detect less recombination than is possible. As a consequence, we are consistently underestimating the number of recombination events that have occurred among sequences, and therefore also the overall recombination rate. In addition, we should keep in mind that the power to detect recombination decreases with the degree of genetic variation. Significantly, no single recombination detection strategy seems to perform optimally under all scenarios, so that using a combination of methods currently appears to be the best strategy.

It is also important to distinguish between frequent recombination happening within a population and rare recombination generating mosaic sequences. In the first case, nonrecombinant regions may be very difficult to identify, and therefore genealogies or phylogenies will be difficult to reconstruct. In such cases, network approaches may offer a general idea of the (reticulate) evolutionary history. Indeed, population genetic estimates should be interpreted with care in the light of assumptions made regarding the presence of recombination. When recombination is rare, mosaic and parental sequences can be identified, as well as recombination break-points. In this case, independent phylogenies can be reconstructed for the nonrecombinant regions and then compared to decipher the recombinant history. Alternatively, recombinants can be "peeled off" the tree to reveal the underlying phylogenetic structure (1, 128, 130).

Still outstanding are issues regarding our ability to measure and depict the action of recombination. First, it is not known how robust the estimators for recombination rate are to violations in the key assumptions, particularly that real populations are subdivided and not panmictic, and that natural selection, as well as genetic drift, may have shaped patterns of genetic diversity. The impact of natural selection may be particularly important because most estimators of ε available at present make use of θ, a measure of genetic diversity that assumes exclusively neutral evolution. Whether natural selection can seriously bias estimates of recombination rate is clearly an area that needs urgent attention. Second, although there is a growing appreciation that network methods are a more appropriate representation of evolutionary relationships when recombination is relatively frequent, the accuracy and power of the network methods proposed to date has yet to be tested. Studies using both simulated and real data, such as those used to determine the accuracy of recombination detection models, are clearly a goal for the immediate future.

Recombination clearly plays a significant role in shaping the genetic architecture of organisms. As a case in point, recombination within introns allows for the shuffling of exons and domains (63, 72, 96), and provides a powerful mechanism for the evolution and adaptation of genomes. Over the next few years, we will see an increasing application of "genome shuffling" techniques to rapidly generate "improved" organisms (146). Another of the most exciting promises of the genomics era is the mapping of genes for common human diseases. Here, again, studies of recombination have a key role to play. Whole-genome association studies have been proposed as an indirect strategy to find genes for disease (70, 106) and within these strategies, whole-genome linkage disequilibrium (LD) scans seem to be the most feasible approach (65). Understandably, this has raised considerable interest in revealing the patterns of LD in human populations. Recently, it has been suggested that human haplotypes are structured into discreet blocks of high LD and low diversity, separated by hot spots of recombination (12, 34, 61, 95), although whether this block structure is a general property of human genomes across populations is still to be demonstrated. In any event, the usefulness and completion of a haplotype map (or maps) [see (139)] of the human genome will be dependent upon a good understanding and description of recombination at every level.

ACKNOWLEDGMENTS

This work was supported by the National Science Foundation (DP, KAC), the National Institutes of Health (DP, KAC), the US-UK Fulbright Commission (KAC), the Wellcome Trust (KAC, ECH), the Royal Society (ECH), and the Brigham Young University International Studies Office (KAC).

The *Annual Review of Genetics* is online at http://genet.annualreviews.org

LITERATURE CITED

1. Antunes A, Templeton AR, Guyomard R, Alexandrino P. 2002. The role of nuclear genes in intraspecific evolutionary inference: genealogy of the *transferrin* gene in the brown trout. *Mol. Biol. Evol.* In press

2. Awadalla P, Eyre-Walker A, Maynard Smith J. 1999. Linkage disequilibrium and recombination in hominid mitochondrial DNA. *Science* 286:2524–25

3. Balding DJ, Nichols RA, Hunt DM. 1992. Detecting gene conversion: primate visual pigment genes. *Proc. R. Soc. London Ser. B* 1992:275–80

4. Bandelt H-J, Dress AWM. 1992. Split decomposition: a new and useful approach to phylogenetic analysis of distance data. *Mol. Phylogenet. Evol.* 1:242–52

5. Betrán E, Rozas J, Navarro A, Barbadilla A. 1997. The estimation of the number and the length distribution of gene conversion tracts from population DNA sequence data. *Genetics* 146:89–99

6. Bradley RD, Hillis DM. 1997. Recombinant DNA sequences generated by PCR amplification. *Mol. Biol. Evol.* 14:592–93

7. Brown CJ, Garner EC, Dunker KA, Joyce P. 2001. The power to detect recombination using the coalescent. *Mol. Biol. Evol.* 18:1421–24

8. Chao L, Tran TT. 1997. The advantage of sex in the RNA virus phi6. *Genetics* 147:953–59

9. Chaston TB, Lidbury BA. 2001. Genetic 'budget' of viruses and the cost to the infected host: a theory on the relationship between the genetic capacity of viruses, immune evasion, persistence and disease. *Immunol. Cell Biol.* 79:62–66

10. Clement M, Posada D, Crandall KA. 2000. TCS: a computer program to esti-

mate gene genealogies. *Mol. Ecol.* 9:1657–60

11. Crandall KA, Templeton AR. 1999. Statistical methods for detecting recombination. In *The Evolution of HIV*, ed. KA Crandall, pp. 153–76. Baltimore, MD: Johns Hopkins Univ. Press

12. Daly MJ, Rioux JD, Schaffner SF, Hudson TJ, Lander ES. 2001. High-resolution haplotype structure in the human genome. *Nat. Genet.* 29:229–32

13. Deleted in proof

14. Drouin G, Dover GA. 1990. Independent gene evolution in the potato actin gene family demonstrated by phylogenetic procedures for resolving gene conversions and the phylogeny of angisperm actin genes. *J. Mol. Evol.* 31:132–50

15. Drouin G, Prat F, Ell M, Clarke GDP. 1999. Detecting and characterizing gene conversions between multigene family members. *Mol. Biol. Evol.* 16:1639–90

16. DuBose RF, Dykhuizen DE, Hartl DL. 1988. Genetic exchange among natural isolates of bacteria: recombination within the *phoA* gene of *Escherichia coli. Proc. Natl. Acad. Sci. USA* 85:7036–40

17. Eastbrook G. 1978. Some concepts for the estimation of evolutionary relationships in systematic biology. *Syst. Bot.* 3:146–58

18. Eigen M, Winkler-Oswatitsch R, Dress A. 1988. Statistical geometry in sequence space: a method of quantitative sequence analysis. *Proc. Natl. Acad. Sci. USA* 85: 5917

19. Ewens WJ. 1979. *Mathematical Population Genetics*. Berlin: Springer-Verlag. 325 pp.

20. Excoffier L, Smouse PE. 1994. Using allele frequencies and geographic subdivision to reconstruct gene trees within a

species: molecular variance parsimony. *Genetics* 136:343–59

21. Eyre-Walker A, Awadalla P. 2001. Does human mtDNA recombine? *J. Mol. Evol.* 53:430–35

22. Eyre-Walker A, Smith NH, Maynard Smith J. 1999. How clonal are human mitochondria? *Proc. R. Soc. London Ser. B* 266:477–83

23. Falush D, Kraft C, Taylor NS, Correa P, Fox JG, et al. 2001. Recombination and mutation during long-term gastric colonization by Helicobacter pylori: estimates of clock rates, recombination size, and minimal age. *Proc. Natl. Acad. Sci. USA* 98:15056–61

24. Fearnhead P, Donnelly P. 2001. Estimating recombination rates from population genetic data. *Genetics* 159:1299–318

25. Feil EJ, Holmes EC, Bessen DE, Chan M-S, Day NPJ, et al. 2001. Recombination within natural populations of pathogenic bacteria: short-term empirical estimates and long-term phylogenetic consequences. *Proc. Natl. Acad. Sci. USA* 98:182–87

26. Felsenstein J. 1974. The evolutionary advantage of recombination. *Genetics* 78:737–56

27. Felsenstein J. 1981. Evolutionary trees from DNA sequences: a maximum likelihood approach. *J. Mol. Evol.* 17:368–76

28. Felsenstein J. 1992. Estimating effective population size from samples of sequences: inefficiency of pairwise and segregating sites as compared to phylogenetic estimates. *Genet. Res. Cambridge* 59:139–47

29. Fitch DHA, Goodman M. 1991. Phylogenetic scanning: a computer assisted algorithm for mapping gene conversions and other recombinational events. *CABIOS* 7:207–15

30. Fitch WM. 1997. Networks and viral evolution. *J. Mol. Evol.* 44:S65–S75

31. Fu Y-X. 1994. A phylogenetic estimator of effective population size or mutation rate. *Genetics* 136:685–92

32. Gibbs MJ, Armstrong JS, Gibbs AJ. 2000. Sister-Scanning: a Monte Carlo procedure for assessing signals in recombinant sequences. *Bioinformatics* 16:573–82

33. Gibbs MJ, Weiller GF. 1999. Evidence that a plant virus switched hosts to infect a vertebrate and then recombined with a vertebrate-infecting virus. *Proc. Natl. Acad. Sci. USA* 96:8022–27

34. Goldstein DB. 2001. Islands of linkage disequilibrium. *Nat. Genet.* 29:109–11

35. Grassly NC, Holmes EC. 1997. A likelihood method for the detection of selection and recombination using nucleotide sequences. *Mol. Biol. Evol.* 14:239–47

36. Griffiths RC, Marjoram P. 1996. Ancestral inference from samples of DNA sequences with recombination. *J. Comput. Biol.* 3:479–502

37. Griffiths RC, Tavare S. 1994. Ancestral inference in population genetics. *Stat. Sci.* 9:307–19

38. Guttman DS, Dykhuizen DE. 1994. Clonal divergence in *Escherichia coli* as a result of recombination, not mutation. *Science* 266:1380–83

39. Hacker J, Blum-Oehler G, Muhldorfer I, Tschape H. 1997. Pathogenicity islands of virulent bacteria: structure, function and impact on microbial evolution. *Mol. Microbiol.* 23:1089–97

40. Harvey PH, Holmes EC, Mooers AO, Nee S. 1994. Inferring evolutionary processes from molecular phylogenies. In *Models in Phylogeny Reconstruction*, ed. RW Scotland, DJ Siebert, DM Williams, pp. 313–33. Oxford: Clarendon

41. Hein J. 1990. Reconstructing evolution of sequences subject to recombination using parsimony. *Math. Biosci.* 98:185–200

42. Hein J. 1993. A heuristic method to reconstruct the history of sequences subject to recombination. *J. Mol. Evol.* 36:396–405

43. Heinemann JA, Sprague GFJ. 1989. Bacterial conjugative plasmids mobilize DNA transfer between bacteria and yeast. *Nature* 340:205–9

44. Hey J. 2000. Human mitochondrial DNA

recombination: Can it be true? *Trends Ecol. Evol.* 15:181–82

45. Hey J, Wakeley J. 1997. A coalescent estimator of the population recombination rate. *Genetics* 145:833–46

46. Hill WG, Robertson A. 1966. The effect of linkage on limits to artificial selection. *Genet. Res.* 8:269–94

47. Hillis DM, Huelsenbeck JP, Cunningham CW. 1994. Application and accuracy of molecular phylogenies. *Science* 264:671–77

48. Holmes EC, Urwin R, Maiden MCJ. 1999. The influence of recombination on the population structure and evolution of the human pathogen *Neisseria meningitidis. Mol. Biol. Evol.* 16:741–49

49. Holmes EC, Worobey M, Rambaut A. 1999. Phylogenetic evidence for recombination in dengue virus. *Mol. Biol. Evol.* 16:405–9

50. Hudson RR. 1983. Properties of a neutral allele model with intragenic recombination. *Theor. Popul. Biol.* 23:183–201

51. Hudson RR. 1987. Estimating the recombination parameter of a finite population model without selection. *Genet. Res. Cambridge* 50:245–50

52. Hudson RR. 1990. Gene genealogies and the coalescent process. *Oxford Surveys Evol. Biol.* 7:1–44

53. Hudson RR. 2001. Two-locus sampling distributions and their application. *Genetics* 159:1805–17

54. Hudson RR, Kaplan NL. 1985. Statistical properties of the number of recombination events in the history of a sample of DNA sequences. *Genetics* 111:147–64

55. Husmeier D, Wright F. 2001. Detection of recombination in DNA alignments with hidden Markov models. *J. Comput. Biol.* 8:401–27

56. Husmeier D, Wright F. 2001. Probabilistic divergence measure for detecting interspecies recombination. *Bioinformatics* 17:S123–S31

57. Imanishi T. 1996. DNA polymorphisms shared among different loci of the major histocompatibility complex genes. In *Current Issues in Molecular Evolution*, ed. M Nei, N Takahata, pp. 89–96. Hayama, Jpn.: Inst. Mol. Evol. Genet., Penn. State Univ. and Grad. Sch. Adv. Stud.

58. Ingman M, Kaessmann H, Paabo S, Gyllensten U. 2000. Mitochondrial genome variation and the origin of modern humans. *Nature* 408:708–13

59. Jakobsen IB, Easteal S. 1996. A program for calculating and displaying compatibility matrices as an aid to determining reticulate evolution in molecular sequences. *Comput. Appl. Biosci.* 12:291–95

60. Jakobsen IB, Wilson SE, Easteal S. 1997. The partition matrix: exploring variable phylogenetic signals along nucleotide sequences alignments. *Mol. Biol. Evol.* 14:474–84

61. Jeffreys AJ, Kauppi L, Neummann R. 2001. Intensely punctate meiotic recombination in the class II region of the major histocompatibilty complex. *Nat. Genet.* 29:217–22

62. Kingman JFC. 1982. The coalescent. *Stoch. Process. Appl.* 13:235–48

63. Kolkman JA, Stemmer WP. 2001. Directed evolution of proteins by exon shuffling. *Nat. Biotechnol.* 19:423–28

64. Koop BF, Siemieniak D, Slightom JL, Goodman M, Dunbar J, et al. 1989. Tarsius delta- and beta-globin genes: conversions, evolution and systematic implications. *J. Biol. Chem.* 264:68–79

65. Kruglyak L. 1999. Prospects for whole-genome linkage disequilibrium mapping of common disease genes. *Nat. Genet.* 22:139–44

66. Kuhner MK, Yamato J, Felsenstein J. 1998. Maximum likelihood estimation of population growth rates based on the coalescent. *Genetics* 149:429–34

67. Kuhner MK, Yamato J, Felsenstein J. 2000. Maximum likelihood estimation of recombination rates from population data. *Genetics* 156:1393–401

68. Ladoukakis ED, Zouros E. 2001.

Direct evidence for homologous recombination in mussel (*Mytilus galloprovincialis*) mitochondrial DNA. *Mol. Biol. Evol.* 18:1168–75

69. Ladoukakis ED, Zouros E. 2001. Recombination in animal mitochondrial DNA: evidence from published sequences. *Mol. Biol. Evol.* 18:2127–31

70. Lander ES. 1996. The new genomics: global views of biology. *Science* 274:536–39

71. Lawrence JG, Ochman H. 1998. Molecular archaeology of the *Escherichia coli* genome. *Proc. Natl. Acad. Sci. USA* 95: 9413–17

72. Li WH, Gu Z, Wang H, Nekrutenko A. 2001. Evolutionary analyses of the human genome. *Nature* 409:847–49

73. Maddison W. 1997. Gene trees in species trees. *Syst. Biol.* 46:523–36

74. Maiden MC, Bygraves JA, Feil E, Morelli G, Russell JE, et al. 1998. Multilocus sequence typing: a portable approach to the identification of clones within populations of pathogenic microorganisms. *Proc. Natl. Acad. Sci. USA* 95:3140–45

75. Martin D, Rybicki E. 2000. RDP: detection of recombination amongst aligned sequences. *Bioinformatics* 16:562–63

76. Maynard Smith J. 1999. The detection and measurement of recombination from sequence data. *Genetics* 153:1021–27

77. Maynard Smith J, Smith NH. 1998. Detecting recombination from gene trees. *Mol. Biol. Evol.* 15:590–99

78. Maynard Smith J, Smith NH, O'Rourke M, Spratt BG. 1993. How clonal are bacteria? *Proc. Natl. Acad. Sci. USA* 90:4384–88

79. McGuire G, Wright F. 1998. TOPAL: recombination detection in DNA and protein sequences. *Bioinformatics* 14:219–20

80. McGuire G, Wright F. 2000. TOPAL 2.0: improved detection of mosaic sequences within multiple alignments. *Bioinformatics* 16:130–34

81. McGuire G, Wright F, Prentice MJ. 1997. A graphical method for detecting recombination in phylogenetic data sets. *Mol. Biol. Evol.* 14:1125–31

82. McGuire G, Wright F, Prentice MJ. 2000. A Bayesian model for detecting past recombination events in DNA multiple alignments. *J. Comput. Biol.* 7:159–70

83. McVean G, Awadalla P, Fearnhead P. 2002. A coalescent-based method for detecting and estimating recombination from gene sequences. *Genetics* 160:1231–41

84. McVean GAT. 2001. What do patterns of genetics variability reveal about mitochondrial recombination? *Heredity* 87:613–20

85. Menotti-Raymond M, Starmer WT, Sullivan DT. 1991. Characterization of the structure and evolution of the Adh region of *Drosophila hydei*. *Genetics* 127:355–66

86. Meunier J, Eyre-Walker A. 2001. The correlation between linkage disequilibrium and distance: implications for recombination in hominid mitochondria. *Mol. Biol. Evol.* 18:2132–35

87. Meyerhans A, Vartanian J-P, Wain-Hobson S. 1990. DNA recombination during PCR. *Nucleic Acids Res.* 18:1687–91

88. Moutouh L, Corbeil J, Richman DD. 1996. Recombination leads to the rapid emergence of HIV-1 dually resistant mutants under selective drug pressure. *Proc. Natl. Acad. Sci. USA* 93:6106–11

89. Muse SV, Weir BS. 1992. Testing for equality of evolutionary rates. *Genetics* 132:269–76

90. Nagy PD, Simon AE. 1997. New insights into the mechanisms of RNA recombination. *Virology* 235:1–9

91. Nielsen R. 2000. Estimation of population parameters and recombination rates from single nucleotide polymorphisms. *Genetics* 154:931–42

92. Ochman H, Lawrence JG, Groisman EA. 2000. Lateral gene transfer and the nature of bacterial innovation. *Nature* 405:299–304

93. Odelberg SJ, Weiss RB, Hata A, White

R. 1995. Template-switching during DNA synthesis by *Thermus aquaticus* DNA polymerase I. *Nucleic Acids Res.* 23:2049–57

94. Ohta T, Basten CJ. 1992. Gene conversion generates hypervariability at the variable regions of kallikreins and their inhibitors. *Mol. Phylogenet. Evol.* 1:87–90

95. Patil N, Berno AJ, Hinds DA, Barrett WA, Doshi JM, et al. 2001. Blocks of limited haplotype diversity revealed by high-resolution scanning of human chromosome 21. *Science* 294:1719–23

96. Patthy L. 1999. Genome evolution and the evolution of exon-shuffling—a review. *Gene* 238:103–14

97. Popadic A, Anderson WW. 1995. Evidence for gene conversion in the amylase multigene family of *Drosophila pseudobscura*. *Mol. Biol. Evol.* 12:564–72

98. Posada D. 2001. Unveiling the molecular clock in the presence of recombination. *Mol. Biol. Evol.* 18:1976–78

99. Posada D. 2002. Evaluation of methods for detecting recombination from DNA sequences: empirical data. *Mol. Biol. Evol.* 19:708–17

100. Posada D, Crandall KA. 2001. Intraspecific gene genealogies: trees grafting into networks. *Trends Ecol. Evol.* 16:37–45

101. Posada D, Crandall KA. 2001. Performance of methods for detecting recombination from DNA sequences: computer simulations. *Proc. Natl. Acad. Sci. USA* 98:13757–62

102. Posada D, Crandall KA. 2002. The effect of recombination in phylogeny reconstruction. *J. Mol. Evol.* 54:396–402

103. Posada D, Crandall KA, Nguyen M, Demma JC, Viscidi JC. 2000. Population genetics of the *porB* gene of *Neisseria gonorrheae*: different dynamics in different homology groups. *Mol. Biol. Evol.* 17:423–36

104. Reich DE, Cargill M, Bolk S, Ireland J, Sabeti PC, et al. 2001. Linkage disequilibrium in the human genome. *Nature* 411:199–204

105. Reid SD, Herbelin CJ, Bumbaugh AC, Selander RK, Whittam TS. 2000. Parallel evolution of virulence in pathogenic *Escherichia coli*. *Nature* 406:64–67

106. Risch N, Merikangas K. 1996. The future of genetic studies of complex human diseases. *Science* 273:1516–17

107. Robertson DL, Hahn BH, Sharp PM. 1995. Recombination in AIDS viruses. *J. Mol. Evol.* 40:249–59

108. Robertson DL, Sharp PM, McCutchan FE, Hahn BH. 1995. Recombination in HIV-1. *Nature* 374:124–26

109. Salminen MO, Carr JK, Burke DS, McCutchan FE. 1996. Identification of breakpoints in intergenotypic recombinants of HIV-1 by bootscanning. *AIDS Res. Hum. Retrovir.* 11:1423–25

110. Sanderson MJ, Doyle JJ. 1992. Reconstruction of organismal and gene phylogenies from data on multigene families: concerted evolution, homoplasy, and confidence. *Syst. Biol.* 41:4–17

111. Satta Y. 1992. Balancing selection at *HLA* loci. In *Population Paleo-Genetics*, ed. N Takahata, pp. 129–49. Tokyo: Jpn. Sci. Soc. Press

112. Sawyer S. 1989. Statistical tests for detecting gene conversion. *Mol. Biol. Evol.* 6:526–38

113. Sawyer SA. 1999. GENECONV: a computer package for the statistical detection of gene conversion. *Distributed by the author, Dep. Math., Wash. Univ. St. Louis, available at* http://www.math.wustl.edu/~sawyer

114. Schierup MH, Hein J. 2000. Consequences of recombination on traditional phylogenetic analysis. *Genetics* 156:879–91

115. Schierup MH, Hein J. 2000. Recombination and the molecular clock. *Mol. Biol. Evol.* 17:1578–79

116. Siepel AC, Halpern AL, Macken C, Korber BTM. 1995. A computer program designed to screen rapidly for HIV type 1 intersubtype recombinant sequences. *AIDS Res. Hum. Retrovir.* 11:1413–16

117. Siepel AC, Korber BK. 1995. Scanning the data base for recombinant HIV-1 genomes. In *Human Retroviruses and AIDS 1995: A Compilation and Analysis of Nucleic Acid and Amino Acid Sequences*, ed. G Myers, B Korber, B Hahn, K-T Jeang, J Mellors, et al. Los Alamos, NM: Theor. Biol. Biophys. Group, Los Alamos Natl. Lab.

118. Sneath PHA. 1995. The distribution of the random division of a molecular sequence. *Binary* 7:148–52

119. Sneath PHA. 1998. The effect of evenly spaced constant sites on the distribution of the random division of a molecular sequence. *Bioinformatics* 14:608–16

120. Sneath PHA, Sackin MJ, Ambler RP. 1975. Detecting evolutionary incompatibilities from protein sequences. *Syst. Zool.* 24:311–22

121. Stephens JC. 1985. Statistical methods of DNA sequence analysis: detection of intragenic recombination or gene conversion. *Mol. Biol. Evol.* 2:539–56

122. Stephens JC. 1986. Of the frequency of undetectable recombination events. *Genetics* 112:923–26

123. Strimmer K, Moulton V. 2000. Likelihood analysis of phylogenetic networks using directed graphical models. *Mol. Biol. Evol.* 17:875–81

124. Strimmer K, Wiuf C, Moulton V. 2001. Recombination analysis using directed graphical models. *Mol. Biol. Evol.* 18:97–99

125. Swofford DL, Olsen GJ, Waddell PJ, Hillis DM. 1996. Phylogenetic inference. In *Molecular Systematics*, ed. DM Hillis, C Moritz, BK Mable, pp. 407–514. Sunderland, MA: Sinauer

126. Tajima F. 1983. Evolutionary relationships of DNA sequences in finite populations. *Genetics* 105:437–60

127. Takahata N. 1994. Comments on the detection of reciprocal recombination or gene conversion. *Immunogenetics* 39:146–49

128. Templeton AR, Clark AG, Weiss KM, Nickerson DA, Boerwinkle E, Sing CF. 2000. Recombinational and mutational hotspots within the human lipoprotein lipase gene. *Am. J. Hum. Genet.* 66:69–83

129. Templeton AR, Crandall KA, Sing CF. 1992. A cladistic analysis of phenotypic associations with haplotypes inferred from restriction endonuclease mapping and DNA sequence data. III. Cladogram estimation. *Genetics* 132:619–33

130. Templeton AR, Weiss KM, Nickerson DA, Boerwinkle E, Sing CF. 2000. Cladistic structure within the human *Lipoprotein Lipase* gene and its implications for phenotyopic association studies. *Genetics* 156:1259–75

131. Valdés AM, Piñero D. 1992. Phylogenetic estimation of plasmid exchange in bacteria. *Evolution* 46:641–56

132. Vasco D, Crandall KA, Fu Y-X. 2000. Molecular population genetics: coalescent methods based on summary statistics. In *Computational and Evolutionary Analysis of HIV Molecular Sequences*, ed. AG Rodrigo, GH Learn, pp. 173–216. Dordrecht: Kluwer

133. Wakeley J. 1997. Using the variance of pairwise differences to estimate the recombination rate. *Genet. Res.* :45–458

134. Wall JD. 2000. A comparison of estimators of the population recombination rate. *Mol. Biol. Evol.* 17:156–63

135. Wall JD. 2001. Insights from linked single nucleotide polymorphisms: What we can learn from linkage disequilibrium. *Curr. Opin. Genet. Dev.* 11:647–51

136. Watterson GA. 1975. On the number of segregating sites in genetical models without recombination. *Theor. Popul. Biol.* 7:256–76

137. Webster RG, Laver WG, Air GM, Schild GC. 1982. Molecular mechanisms of variation in influenza viruses. *Nature* 296:115–21

138. Weiller GF. 1998. Phylogenetic profiles: a graphical method for detecting genetic recombination in homologous sequences. *Mol. Biol. Evol.* 15:326–35

139. Weiss KM, Clark AG. 2002. Linkage disequilibrium and the mapping of complex human traits. *Trends Genet.* 18:19–24

140. Wiens JJ. 1998. Combining data sets with different phylogenetic histories. *Syst. Biol.* 47:568–81

141. Wiuf C, Christensen T, Hein J. 2001. A simulation study of the reliability of recombination detection methods. *Mol. Biol. Evol.* 18:1929–39

142. Worobey M. 2001. A novel approach to detecting and measuring recombination: new insights into evolution in viruses, bacteria, and mitochondria. *Mol. Biol. Evol.* 18:1425–34

143. Worobey M, Holmes EC. 1999. Evolutionary aspects of recombination in RNA viruses. *J. Gen. Virol.* 80:2535–43

144. Worobey M, Holmes EC. 2001. Homologous recombination in GB virus C/hepatitis G virus. *Mol. Biol. Evol.* 18:254–61

145. Yang Z. 1996. Among-site rate variation and its impact on phylogenetic analysis. *Trends Ecol. Evol.* 11:367–72

146. Zhang YX, Perry K, Vinci VA, Powell K, Stemmer WP, del Cardayre SB. 2002. Genome shuffling leads to rapid phenotypic improvement in bacteria. *Nature* 415:644–46

147. Zhou J, Bowler LD, Spratt BG. 1997. Interspecies recombination, and phylogenetic distortions, within the glutamine synthetase and skikimate dehydrogenase genes of *Neisseria meningitidis* and commensal *Neisseria* species. *Mol. Microbiol.* 23:799–812

Annu. Rev. Genet. 2002. 36:99–124
doi: 10.1146/annurev.genet.36.040102.131941

DEVELOPMENT AND FUNCTION OF THE ANGIOSPERM FEMALE GAMETOPHYTE

Gary N. Drews[1] and Ramin Yadegari[2]

[1]Department of Biology, University of Utah, Salt Lake City, Utah 84112;
e-mail: drews@biosci.biology.utah.edu; [2]Department of Plant Sciences, University
of Arizona, Tucson, Arizona 85721; e-mail: yadegari@ag.arizona.edu

Key Words   female gametophyte, embryo sac, plant reproduction, Arabidopsis

■ **Abstract**   The plant life cycle alternates between a diploid sporophyte generation and a haploid gametophyte generation. The angiosperm female gametophyte is critical to the reproductive process. It is the structure within which egg cell production and fertilization take place. In addition, the female gametophyte plays a role in pollen tube guidance, the induction of seed development, and the maternal control of seed development. Genetic analysis in Arabidopsis has uncovered mutations that affect female gametophyte development and function. Mutants defective in almost all stages of development have been identified, and analysis of these mutants is beginning to reveal features of the female gametophyte developmental program. Other mutations that affect female gametophyte function have uncovered regulatory genes required for the induction of endosperm development. From these studies, we are beginning to understand the regulatory networks involved in female gametophyte development and function. Further investigation of the female gametophyte will require complementary approaches including expression-based approaches to obtain a complete profile of the genes functioning within this critical structure.

CONTENTS

0066-4197/02/1215-0099$14.00

INTRODUCTION

Relative to animals, a unique feature of the plant life cycle is the existence of a multicellular haploid generation referred to as the gametophyte. Gametophytes arise in plants because meiosis gives rise to haploid spores, which undergo cell proliferation and develop into gametophytes. A major function of the gametophyte generation is to produce haploid gametes. As in animals, egg and sperm fuse to form diploid zygotes, which then undergo cell proliferation and differentiate into multicellular diploid individuals called sporophytes. A major function of the diploid sporophyte generation in plants is to produce the haploid spores that are the beginning of the gametophyte generation. This type of life cycle is referred to as alternation of generations (91).

In lower plants (bryophytes), the gametophyte is the conspicuous and dominant generation, whereas sporophytes are attached to and physiologically dependent upon gametophytes. Plant evolution is associated with several changes in the gametophyte. First, gametophytes became increasingly reduced in size and complexity relative to sporophytes. Second, gametophytes became embedded within and physiologically dependent upon sporophytes. Third, the advent of heterospory, or two spore types (megaspores and microspores), resulted in the production of two types of unisexual gametophytes (female gametophytes and male gametophytes). Finally, once the haploid gametophyte generation became embedded within the sporophyte, it acquired several important functions associated with the reproductive process (discussed below) (35).

In angiosperms, the gametophytes are comprised of very few cells and are embedded within the sexual organs of the flower (Figure 1A). The male gametophyte, also referred to as the pollen grain or microgametophyte, develops within the anther and is comprised of two sperm cells encased within a vegetative cell (75, 89). The female gametophyte, also referred to as the embryo sac or megagametophyte, develops within the ovule, which, in turn, is embedded within the ovary of the carpel. Among angiosperms, the female gametophyte has a variety of forms. The most common form, depicted in Figure 1B,C, consists of seven cells and four different cell types: three antipodal cells, two synergid cells, one egg cell, and one central cell (27, 46, 119).

The gametophytes play a central role in the angiosperm reproductive process. Sexual reproduction is initiated when the male gametophyte is transferred from the anther to the carpel's stigma. Shortly thereafter, the male gametophyte forms a pollen tube that grows great distances through the carpel's internal tissues to deliver its two sperm cells to the female gametophyte. One sperm cell fertilizes

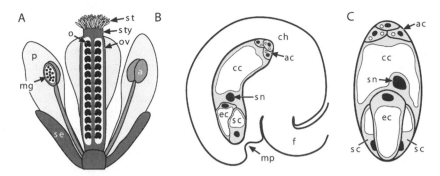

Figure 1   Depictions of the Arabidopsis reproductive structures. (*A*) Flower. (*B*) Ovule. (*C*) Female gametophyte. View in (*C*) is perpendicular to that in (*B*). The mature female gametophyte in Arabidopsis is approximately 105 μm in length and approximately 25 μm in width (16). In (*B*) and (*C*), the gray areas represent cytoplasm, the white areas represent vacuoles, and the black areas represent nuclei. Abbreviations: a, anther; ac, antipodal cells; cc, central cell; ch, chalazal region of the ovule; ec, egg cell; f, funiculus; mg, male gametophyte; mp, micropyle; o, ovule; ov, ovary; p, petal; sc, synergid cell; se, sepal; sn, secondary nucleus; st, stigma; sty, style.

the egg cell, and the second sperm cell fuses with the central cell. Following fertilization, the egg cell gives rise to the seed's embryo, which is the beginning of the sporophyte generation, and the central cell's polar nuclei give rise to the seed's endosperm, which surrounds and provides nutrients to the developing embryo. Embryo and endosperm comprise the major portion of the seed (27, 97, 99, 110).

Analysis of the female gametophyte is important for two reasons. Biologically, the female gametophyte is an integral aspect of the plant life cycle and is essential for seed formation. In addition, it plays a critical role in essentially every step of the reproductive process. During the fertilization process, the female gametophyte participates in directing the pollen tube to the ovule (43, 49, 94, 102) and mediates fertilization of egg and polar nuclei (97, 99). Upon fertilization, female gametophyte-expressed genes participate in inducing seed development (15, 39, 82). Finally, during seed development, female gametophyte-expressed gene products may play a role in controlling embryo and/or endosperm development (14, 27).

Furthermore, female gametophyte mutations exhibit reduced or no transmission through the female gametophyte and appear at reduced frequency in subsequent generations. As a consequence, female gametophyte mutations often cannot become homozygous during the diploid sporophyte generation, thus preventing genetic analysis during this phase. Many important cellular and developmental processes occur during female (and male) gametophyte development including mitosis, cell wall formation, nuclear fusion, and cell death (discussed below). Thus, genetic analysis of many important cellular and developmental processes is likely to require gametophytic screens and analysis of the gametophyte generation (27, 37).

Although extensively studied at the cytological level, very little is known about the molecular and genetic processes controlling female gametophyte development and function. However, genetic approaches are beginning to provide some understanding. In this review, we describe female gametophyte structure and development, summarize the female gametophyte's reproductive functions, and discuss the molecular and genetic approaches used to understand these processes at the molecular level. Comprehensive reviews of the female gametophyte have been published previously (40, 46, 70, 112, 119); this review emphasizes recent progress made in Arabidopsis, in which genetic analysis of female gametophyte development and function is well under way.

GENE ACTIVITY DURING FEMALE GAMETOPHYTE DEVELOPMENT

Gene expression within the mature female gametophyte has not been well characterized. However, several lines of evidence suggest that the female gametophyte is extremely active in gene expression. First, the number of female gametophyte-expressed genes might be expected to be comparable to the number of genes expressed in the male gametophyte. Although not determined in Arabidopsis, the number of male gametophyte-expressed genes has been estimated to be approximately 20,000 in other species, including maize (120, 121). Thus, by comparison, the female gametophyte probably expresses many thousands of genes. Second, genetic studies have shown that chromosomal deficiencies, in general, exhibit reduced transmission through the female gametophyte (9, 84, 85, 113, 118), which implies that essential female gametophyte-expressed genes are scattered throughout the genome. Third, as discussed below, female gametophyte mutations occur at high frequency. Taken together, these data suggest very strongly that plant genomes contain a large number of female gametophyte-expressed genes required for megagametogenesis and/or female gametophyte function. As discussed below, genetic approaches have begun to identify and assign functions to these genes.

FEMALE GAMETOPHYTE MUTATIONS

Gametophytic Mutations Exhibit Non-Mendelian Segregation Patterns

In contrast to animals, plants possess two broad classes of mutations that exhibit fundamentally different segregation patterns. Sporophytic mutations affect the diploid sporophyte phase of the plant life cycle and exhibit Mendelian 1:2:1 segregation patterns. Gametophytic mutations, by contrast, affect the gametophyte phase of the plant life cycle and, as a consequence, exhibit altered segregation patterns. The existence of gametophytic mutations exhibiting aberrant segregation patterns was first reported over a century ago (8, 18, 21, 52).

Gametophytic mutations can affect the female gametophyte and/or the male gametophyte. We refer to mutations affecting the female gametophyte but not the male gametophyte as female gametophyte-specific mutations, to mutations affecting the male gametophyte but not the female gametophyte as male gametophyte-specific mutations, and to mutations affecting both gametophytes as general gametophytic mutations. The segregation patterns of sporophytic mutations and the three classes of gametophytic mutations are summarized in Table 1.

Gametophytic mutations exhibit no or reduced transmission through the female and/or male gametophytes. For example, fully penetrant female gametophyte-specific mutations transmit through the male gametophyte but not through the female gametophyte and, as a consequence, cannot become homozygous in the sporophyte generation (Table 1). Thus, gametophytic mutations exhibit non-Mendelian segregation patterns and transmit from generation to generation as heterozygotes. If a mutation affects both gametophytes (general gametophytic mutation) and is fully penetrant, it cannot transmit to subsequent generations (Table 1). However,

TABLE 1   Segregation of sporophytic and gametophytic mutations in a self cross of a heterozygous individual (A/a × A/a)

| Mutation class | Functional gametes | | Progeny genotypes[c] | R/S[d] |
	MG[a]	FG[b]		
Sporophytic				
Typical (not lethal)[e]	A & a (1:1)	A & a (1:1)	A/A A/a a/a (1:2:1)	3.0
Embryo defective[f]	A & a (1:1)	A & a (1:1)	A/A A/a a/a* (1:2:1)	2.0
Haplo-insufficient endosperm defective[g]	A & a (1:1)	A/A & a/a (1:1)	A/A/A A/a/a* A/A/a a/a/a* (1:1:1:1)	1.0
Paternally imprinted embryo/ endosperm defective[h]	A̲ & a̲ (1:1)	A & a (1:1)	A/A A/a* a̲/A a̲/a* (1:1:1:1)	1.0
Gametophytic				
FG specific	A & a (1:1)	A	A/A A/a (1:1)	1.0
MG specific	A	A & a (1:1)	A/A A/a (1:1)	1.0
General gametophytic	A	A	A/A	0

[a]MG, male gametophyte.

[b]FG, female gametophyte. Genotypes are for the egg cell except for haplo-insufficient endosperm defective mutations, in which the genotype is for the central cell.

[c]Genotypes are for the embryo except for haplo-insufficient endosperm defective mutations, in which the genotype is for the endosperm. Genotypes with asterisks are lethal.

[d]Seedling resistance ratios if the mutation is caused by a T-DNA or transposon that carries an antibiotic/herbicide resistance gene.

[e]Mutations that do not cause seed/seedling lethality (e.g., floral morphology mutations).

[f]Mutations that cause lethality during embryo development.

[g]Mutations in genes in which at least two wild-type copies are required for endosperm development. The central cell is homodiploid and therefore has two alleles. The endosperm is triploid and therefore has three alleles.

[h]Mutations in genes that are paternally imprinted and that are required during embryo or endosperm development. Underlined alleles are imprinted and, thus, are inactive in the progeny genotypes.

as discussed below, many general gametophytic mutations are partially penetrant and thus can be transmitted to subsequent generations.

Several classes of sporophytic mutations segregate in a pattern identical or similar to that of gametophytic mutations. These sporophytic-mutation classes include embryo-defective mutations, haplo-insufficient endosperm-defective mutations, and mutations affecting paternally imprinted genes required for embryo and/or endosperm development. As summarized in Table 1, these mutation classes exhibit 1:2:1 segregation; however, because they cause seed lethality, their segregation patterns appear to deviate from 1:2:1 when seedlings or mature plants are scored. Because of these similarities, one must be cautious in concluding that a given mutation affects the gametophytic phase. However, as discussed below, specific genetic and molecular tests can be applied to distinguish among the various mutant classes.

Female Gametophyte Mutants are Identified Using Segregation and Seed Set Screens

Female gametophyte mutants typically are identified using two criteria: reduced seed set and segregation distortion (27, 77). Reduced seed set results because in a plant heterozygous for a female gametophyte mutation, approximately half of the female gametophytes are mutant and nonfunctional. Ovules harboring nonfunctional female gametophytes fail to undergo seed development and eventually desiccate. Thus, heterozygous female gametophyte mutants have fruits (a silique in Arabidopsis) containing 50% normal seeds and 50% desiccated ovules.

Segregation distortion results because, as described above, female gametophyte mutations are transmitted to subsequent generations at reduced frequency (Table 1). To facilitate segregation analysis, lines can be mutagenized with a T-DNA or transposon carrying a gene conferring antibiotic/herbicide resistance. Resistant:sensitive (R:S) ratios can be used to identify lines containing an insert disrupting a female gametophyte gene. For example, in the progeny of a heterozygous plant, R:S is 1:1 for lines with female gametophyte-specific mutations, as compared to 3:1 for most lines with sporophytic mutations (Table 1).

Several factors complicate the identification of female gametophyte mutants. First, reduced seed set also is caused by a variety of environmental (e.g., high growth temperature or water stress) or genetic (e.g., chromosomal rearrangements or female-sterile mutations with 50% penetrance) conditions (7, 77, 100, 113). Second, segregation distortion also is caused by male gametophyte mutations (Table 1). Third, several classes of sporophytic mutations appear to exhibit segregation distortion (Table 1).

Most of these ambiguities can be resolved by carrying out reciprocal crosses with heterozygous mutants. For example, in a cross of a heterozygous female (genotype A/a) with a wild-type male (genotype A/A), the progeny are all wild type (genotype A/A) if the mutation affects the female gametophyte (Table 2).

TABLE 2   Segregation of sporophytic and gametophytic mutations in a cross of a heterozygous female (A/a) with a wild-type male (A/A)

Mutation class	Functional gametes		Progeny genotypes[c]	R/S[d]
	MG[a]	FG[b]		
Sporophytic				
Typical (not lethal)[e]	A	A & a (1:1)	A/A A/a (1:1)	1.0
Embryo defective[f]	A	A & a (1:1)	A/A A/a (1:1)	1.0
Haplo-insufficient endosperm defective[g]	A	A/A & a/a (1:1)	A/A/A A/a/a* (1:1)	0
Paternally imprinted embryo/ endosperm defective[h]	<u>A</u>	A & a (1:1)	<u>A</u>/A <u>A</u>/a* (1:1)	0
Gametophytic				
FG specific	A	A	A/A	0
MG specific	A	A & a (1:1)	A/A A/a (1:1)	1.0

[a]MG, male gametophyte.

[b]FG, female gametophyte. Genotypes are for the egg cell except for haplo-insufficient endosperm-defective mutations, in which the genotype is for the central cell.

[c]Genotypes are for the embryo except for haplo-insufficient endosperm defective mutations, in which the genotype is for the endosperm. Genotypes with asterisks are lethal.

[d]Seedling resistance ratios if the mutation is caused by a T-DNA or transposon that carries an antibiotic/herbicide resistance gene.

[e]Mutations that do not cause seed/seedling lethality (e.g., floral morphology mutations).

[f]Mutations that cause lethality during embryo development.

[g]Mutations in genes in which at least two wild-type copies are required for endosperm development. The central cell is homodiploid and therefore has two alleles. The endosperm is triploid and therefore has three alleles.

[h]Mutations in genes that are paternally imprinted and that are required during embryo or endosperm development. Underlined alleles are imprinted and thus are inactive in the progeny genotypes.

However, using this criterion, both haplo-insufficient endosperm-defective mutations and mutations affecting paternally imprinted genes required for embryo and/or endosperm development appear to exhibit reduced transmission through the female gametophyte (Table 2). Thus, in practice, additional criteria must be applied to determine whether a mutation truly affects the female gametophyte. For example, microscopic analysis showing abnormal embryo sac structure clearly demonstrates an effect on the female gametophyte. This criterion becomes ambiguous only where female gametophyte structure appears to be unaffected. In these cases, further tests must be carried out.

Female Gametophyte Mutants in Arabidopsis

Because female gametophyte development and function require many fundamental cellular and developmental processes, female gametophyte mutants are expected to occur at high frequency (27). This appears to be the case. In several screens in Arabidopsis, the frequency of female gametophyte mutants was approximately

1% among T-DNA/transposon lines (6, 45, 77) (C.A. Christensen, R.H. Brown, J. Brown & G.N. Drews, unpublished data). In a saturation screen (53, 114), approximately 180,000 T-DNA insertions must be screened to achieve a 95% probability of mutating any gene of median size (66). These data suggest that approximately 1800 [(180,000)(0.01)] female gametophyte mutants could be identified in a saturation screen. At 95% saturation, approximately three alleles per gene are isolated on average, suggesting that approximately 600 [(1800)(.33)] independent female gametophyte loci could be identified via mutation.

The Arabidopsis female gametophyte mutations identified to date are listed in Table 3. The vast majority of these are partially penetrant and affect both the female gametophyte and the male gametophyte. For example, from an analysis of 39 mutants, approximately 85% of female gametophyte mutations affect both gametophytes and essentially all are partially penetrant in the female gametophyte and/or male gametophyte (C.A. Christensen, R.H. Brown, J. Brown & G.N. Drews, unpublished data). These data suggest that most of the genes expressed in the female gametophyte also are expressed in the male gametophyte and that mutations affecting both gametophytes (i.e., general gametophytic mutations) can be efficiently isolated. As discussed below, many of these mutants have been analyzed at the phenotypic and molecular levels, and specific functions in female gametophyte development and function have been assigned.

TABLE 3   Female gametophyte mutants in Arabidopsis

Mutant	Reference
ada, line2, line3, line4, line6, line8, lpa, tya	(45)
bod1–bod3	(37)
ctr1	(56)
fem1–fem38	(17; C.A. Christensen, R.H. Brown, J. Brown and G.N. Drews, unpublished data)
fie/fis3	(15, 82)
fis2	(15)
gf	(16, 95)
gfa1–gfa7	(17, 31)
hdd	(77)
maa1–maa4	(102)
mea/fis1/emb173	(10, 15, 39)
prl	(107)
trp1; trp4	(80)
ttd1–ttd38	(6)

GENETIC ANALYSIS OF MEGAGAMETOGENESIS

Structure and Development of the Arabidopsis Female Gametophyte

Over 15 different patterns of female gametophyte development have been described (40, 70, 119). The developmental pattern exhibited by Arabidopsis is referred to as the Polygonum type because it was first described in *Polygonum divaricatum* (76, 108). The Polygonum-type female gametophyte is found in about 70% of the species examined (70) and is thought to be the ancestral type (46). Thus, Arabidopsis is an experimental system in which the most common form of female gametophyte development can be studied.

Many descriptive studies of Arabidopsis female gametophyte structure and development have been carried out (16, 73, 76, 78, 88, 101, 116, 117). Arabidopsis female gametophyte development can be divided into two phases: megasporogenesis and megagametogenesis. During megasporogenesis, a diploid megaspore mother cell undergoes meiosis to produce four haploid megaspores. The chalazal-most megaspore survives and the other three undergo cell death (16, 101, 116).

Megagametogenesis in Arabidopsis is depicted in Figure 2. The functional megaspore undergoes three rounds of mitosis, producing an eight-nucleate cell. During the third mitosis, phragmoplasts and cell plates form between sister and nonsister nuclei (117). Shortly thereafter, the female gametophyte cells become completely surrounded by cell walls. During cellularization, two nuclei (the polar nuclei), one from each pole, migrate toward the center of the developing female gametophyte and eventually fuse (16, 88, 101, 117). Fusion of the polar nuclei occurs after cellularization is completed (16). In the final stage of megagametogenesis, the three antipodal cells undergo cell death (16, 78, 88, 101). Thus, the mature female gametophyte in Arabidopsis consists of one egg cell, one central cell, and two synergid cells (16, 73, 78, 101), which together comprise the female

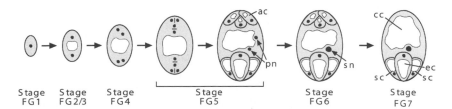

| Stage FG1 | Stage FG2/3 | Stage FG4 | Stage FG5 | Stage FG6 | Stage FG7 |

Figure 2 Depiction of megagametogenesis in Arabidopsis. The steps are described in the text and in Reference (16). The gray areas represent cytoplasm, the white areas represent vacuoles, and the black areas represent nuclei. In the depiction of early stage FG5, the black lines between the nuclei represent partial cell walls. Abbreviations: ac, antipodal cells; cc, central cell; ec, egg cell; pn, polar nuclei; sc, synergid cell; sn, secondary nucleus (fused polar nuclei).

germ unit (29). Megagametogenesis occurs over a 3- to 4-day period in Arabidopsis (C.A. Christensen & G.N. Drews, unpublished results).

The molecular processes controlling cell specification during megagametogenesis are not understood. Microscopic analysis in many species including Arabidopsis suggests that the female gametophyte nuclei become specified at or before the four-nucleate stage (16, 33, 48, 98). At the four-nucleate stage, the micropylar nuclei initially lie along a line orthogonal to the chalazal-micropylar axis. They then migrate so that they lie along a line parallel to the chalazal-micropylar axis. The micropylar-most nucleus forms a transverse spindle and gives rise to the two synergid nuclei, and the chalazal-most nucleus forms a longitudinal spindle and gives rise to the egg nucleus and the micropylar polar nucleus (Figure 2). Similarly, the chalazal nuclei migrate to occupy a position parallel to the chalazal-micropylar axis. Here, the micropylar-most nucleus divides to form one polar nucleus and one antipodal nucleus, and the chalazal-most nucleus gives rise to two antipodal nuclei (Figure 2) (16, 33, 48, 98). Two models for cell specification during megagametogenesis are consistent with these data. First, the female gametophyte nuclei may become specified after migration via positional information present in the four-nucleate embryo sac. Alternatively, the female gametophyte nuclei may become specified before migration via the asymmetric segregation of nuclear determinants during mitosis (36). Whether the developing female gametophyte contains positional information and/or nuclear determinants is currently unknown.

The ovule and female gametophyte have several structural features that facilitate the fertilization process. First, the ovule and female gametophyte are polarized structures. The ovule's chalazal pole is the end that joins the funiculus, and the micropylar pole is the end at which the integuments form a pore. Within the female gametophyte, the egg and synergid cells occupy the micropylar pole, and the antipodal cells lie at the chalazal pole (Figure 1). This polarity is important for the fertilization process because the pollen tube reaches the female gametophyte by growing through the micropyle and entering the embryo sac via the synergid cells (see below). Second, the egg cell and central cells are polarized such that the egg cell nucleus and secondary nucleus (nucleus of the central cell) lie very close to each other (Figure 1). This feature is important for the fertilization process because the egg cell nucleus and secondary nucleus are the targets of the two sperm nuclei. Finally, in the regions where the egg, synergid, and central cells meet, the cell walls are absent or discontinuous, and the plasma membranes of these cells are in direct contact with each other. This eliminates the cell wall barrier that would otherwise obstruct the sperm cell's access to the egg and central cells (16, 46, 73, 78, 119). The molecular processes establishing these structural features during female gametophyte development are unknown.

Megagametogenesis Mutants

As discussed above, many thousands of genes are likely to be expressed during megagametogenesis. These genes presumably regulate and mediate the developmental and cellular processes occurring during female gametophyte development.

Just several years ago, only a few mutants affected in megagametogenesis were known (27). However, over the past few years, many screens for Arabidopsis female gametophyte mutants have been carried out, and over 100 Arabidopsis female gametophyte mutants have been identified (6, 17, 31, 37, 45, 77, 102; C.A. Christensen, R.H. Brown, J. Brown & G.N. Drews, unpublished data). Many of these mutants are affected in megagametogenesis, and analysis of these mutants is beginning to reveal features of the molecular processes governing female gametophyte development.

Over the past few years, the phenotypes of over 50 Arabidopsis female gametophyte mutants have been characterized (Table 4). These mutants fall into five phenotypic categories. Category 1 mutants are affected at the earliest possible step and fail to progress beyond the one-nucleate stage (stage FG1). The existence of this class shows that expression of the haploid genome is required very early in megagametogenesis. Category 2 mutants have defects during the nuclear division phase of megagametogenesis (stages FG2 to early FG5) and fail to cellularize. Category 2 mutants have defects in nuclear number and/or positions, and developmental arrest at stages FG2 to early FG5. Category 3 mutants become cellularized but have defects in cellular morphology, including abnormal nuclear positions within cells, misshapen cells, and unusual cell features. Category 3 mutations, thus, appear to affect the cellularization process. Category 4 mutants have defects in fusion of the polar nuclei. With these mutants, the polar nuclei migrate properly, come to lie side by side, but fail to fuse. Category 5 mutants have phenotypically wild-type female gametophytes at the terminal developmental stage (stage FG7),

TABLE 4 Phenotypes of Arabidopsis female gametophyte mutants

Phenotypic category	Mutants in category	References
1	*ada, bod1, bod2, bod3, gf, gfa4, gfa5, fem2, fem3, fem9, fem12, fem16, fem18, fem26, fem29, fem31, fem35, fem37, fem38, tya*	(16, 17, 37, 45; C.A. Christensen, R.H. Brown, J. Brown & G.N. Drews, unpublished data)
2	*fem5, fem10, fem11, fem13, fem14, fem15, fem19, fem20, fem21, fem22, fem23, fem24, fem25, fem30, fem33, fem34, fem36, hdd, prl*	(17, 77, 107; C.A. Christensen, R.H. Brown, J. Brown & G.N. Drews, unpublished data)
3	*fem4, fem6, fem7, fem8*	(17) (C.A. Christensen, R.H. Brown, J. Brown & G.N. Drews, unpublished data)
4	*gfa2, gfa3, maa1, maa3*	(17, 102)
5	*ctr1, fem17, fie, fis2, mea*	(15, 39, 82; C.A. Christensen, R.H. Brown, J. Brown & G.N. Drews, unpublished data)

which suggests that megagametogenesis is not affected. Category 5 mutants are discussed in more detail below. The one developmental step for which no mutants were recovered is degeneration of the antipodal cells. This suggests that such mutants occur at low frequency or that antipodal degeneration does not require female gametophyte-expressed genes.

Of the mutants analyzed, the vast majority ($\geq$90%) fall into categories 1 to 4, which indicates that megagametogenesis mutants occur at high frequency. These data suggest that megagametogenesis is governed extensively by female gametophyte-expressed genes and that the vast majority of female gametophyte-expressed genes are required during the steps of megagametogenesis. However, caution must be applied in making this extrapolation because most of the mutants were generated using T-DNA mutagenesis, the primary target of which is the embryo sac (3, 23, 123), using the floral dip method (4); thus, mutations affecting the later steps of female gametophyte development and function may be eliminated immediately upon mutagenesis.

This survey analysis also illustrates several features of mutant phenotypes in megagametogenesis. First, mutants with defects throughout megagametogenesis, including the late steps, can be identified. Second, mutations affecting specific cellular processes can be identified. For example, the *gfa2* mutation is required for nuclear fusion (17). Third, based on phenotype, the affected genes can be ordered, to some extent, within the female gametophyte developmental program. For example, category 1 and category 2 genes function before category 3 and category 4 genes. Finally, this analysis indicates that a genetic approach is an effective method to identify genes important for megagametogenesis.

The affected genes for most of these mutants are not yet known. However, many of these mutants were generated using insertional mutagens such as T-DNAs or transposons, which should facilitate rapid gene identification in these lines. A molecular understanding of megagametogenesis is likely to progress rapidly over the next few years.

GENETIC ANALYSIS OF FEMALE GAMETOPHYTE FUNCTION

Perhaps the most exciting prospect of female gametophyte mutant screens is the potential of identifying genes required for the female gametophyte's reproductive functions, including pollen tube guidance, fertilization, induction of seed development, and maternal control of seed development. A number of female-gametophyte-function mutants have been identified, and the molecular processes regulating female gametophyte function are beginning to be understood.

Pollen Tube Guidance

Shortly after pollen is transferred from anther to stigma, the male gametophyte forms a pollen tube that grows via a tip growth process through the carpel's sporophytic tissue to the female gametophyte. The pollen tube initially grows into the cells of the stigma and then down through the cells of the transmitting tract. Within the ovary, the pollen tube changes its growth direction to emerge from the transmitting tract and then grows along the placental surface toward an ovule. Upon reaching an ovule, the pollen tube grows up the ovule's funiculus, through the ovule's micropyle, and into the female gametophyte (49, 54, 89, 90).

How does the pollen tube find its way to the female gametophyte and does the female gametophyte play a role in this process? To address these questions, several groups have assayed pollen tube growth to ovules containing abnormal female gametophytes. The abnormal female gametophytes were generated in Arabidopsis using a variety of genetic methods including sporophytic mutations, gametophytic mutations, and chromosomal translocations. Such studies consistently have shown that ovules harboring defective female gametophytes fail to receive pollen tubes (22, 49, 94, 102). These data suggest very strongly that the female gametophyte is the source of a guidance cue that attracts pollen tubes to the ovule. More recently, laser ablation studies in Torenia have shown that the synergid cells are the source of the female gametophyte guidance cue (43).

The synergid cell guidance cue has not yet been identified. One approach to identify the attractant is biochemical isolation. The recent development of an in vitro pollen-tube-guidance assay in Torenia (41, 43) may facilitate biochemical isolation, but this approach is likely to be technically difficult because of the small amounts of tissue available for biochemical analysis. An alternative approach is identification of pollen-tube-guidance mutants. Because the synergid cells produce the attractant, mutations affecting attractant production will exhibit the behavior of female gametophyte mutations. Thus far, female gametophyte mutants defective in pollen tube guidance have not been reported. The chemical nature of the synergid cell guidance cue still remains to be determined.

Fertilization

The pollen tube enters the female gametophyte by growing through the filiform apparatus, an elaboration of the micropylar synergid cell wall, and then into one of the two synergid cells. The synergid cell penetrated by the pollen tube undergoes cell death before or upon arrival of the pollen tube. Shortly after arrival, growth of the pollen tube ceases. An opening then forms at or near the pollen tube tip and the pollen tube contents, including the two sperm cells, are released into the degenerating synergid cytoplasm. The two sperm cells then migrate to the egg cell and central cell. Finally, the plasma membranes of the sperm cells fuse with those of the egg and central cells, and the sperm nuclei are transmitted into these cells for karyogamy (97, 99, 110).

Molecular understanding of the angiosperm fertilization pathway has been hampered by the inaccessibility of the fertilization targets to harvesting and manipulation. To circumvent this problem, in vitro fertilization systems have been developed (63). Such systems should allow manipulation and molecular dissection of the angiosperm fertilization process. An alternative approach is identification of mutants defective in the fertilization process. Because fertilization takes place within the embryo sac, many female gametophyte-expressed gene products should be required for this process. Thus far, female gametophyte mutants affected in fertilization have not been reported. Thus, the molecular events mediating and regulating the angiosperm fertilization process remain to be determined.

Synergid Cell Death

As discussed above, one and in some cases both of the synergid cells undergoes cell death during reproduction in angiosperms (97, 99, 110, 119). The degenerating synergid, most likely, provides lowered resistance to both pollen tube penetration and sperm cell migration during fertilization (27, 46, 119). In addition, synergid degeneration is accompanied by a cytoskeletal reorganization that may facilitate male gamete migration (34, 98). Thus, for several reasons, synergid cell death is likely to be a prerequisite for fertilization.

The timing and mechanism of synergid cell death may depend upon developmental signals and/or pollination. In some species including Arabidopsis, synergid cell death requires pollination, i.e., it does not occur if pollination is prevented (16, 47, 51), which indicates that synergid degeneration is not a feature of the megagametogenesis developmental program in these species. By contrast, in other species, synergid cell death occurs in the absence of pollination (97, 110, 119), which suggests that synergid degeneration is an integral part of the megagametogenesis developmental program in these species. In species such as Arabidopsis, in which synergid cell death requires pollination, synergid cell death could occur via two general mechanisms that are not mutually exclusive. First, synergid cell death could occur via a mechanical process: The release of pollen tube contents into the synergid cell may increase the volume and pressure within this cell and result in rupture of the synergid membrane (42, 97, 110). Second, synergid cell death could be a physiological response induced by pollination or pollen tubes growing within the carpel tissue. The induction could entail a long-range signal or may require pollen tube-synergid cell contact. In many species, synergid degeneration appears to be initiated prior to arrival of the pollen tube at the female gametophyte (97, 110, 119), which could suggest a long-range signal. The timing of synergid cell death relative to pollen tube growth has not yet been determined in Arabidopsis.

The induction of synergid cell death by the male gametophyte may, in turn, trigger responses in the pollen. Synergid cells may contain a pollen tube guidance signal that is released upon cell death. For example, in some species, synergid cells contain high concentrations of calcium (11–13, 50), which can act as a directional signal for pollen tube guidance (74, 96). The male gametophyte may therefore send a signal to the female gametophyte to induce synergid cell death, which, in turn, releases a signal for pollen tube guidance.

Molecular and genetic analysis of synergid cell death is still in its infancy. Thus far, only one mutation, *gfa2*, affecting synergid cell death has been reported. Following pollination, *gfa2* female gametophytes attract pollen tubes but fail to undergo synergid cell death. The *gfa2* mutation also affects nuclear fusion during megagametogenesis (17). The *gfa2* mutation does not affect megaspore or antipodal cell death, which suggests that synergid cell death has unique features or that the other two cell death events are rescued by the surrounding sporophytic tissue. The *GFA2* gene encodes a chaperone that functions in the mitochondrial matrix, and the yeast ortholog is required for mitochondrial function. These data suggest that synergid cell death requires functional mitochondria, which also are required for cell death in animals. Analysis of *gfa2* mutants also suggests that synergid cell death does not occur via a mechanical process (because the *gfa2* defect does not affect pollen tube attraction and should not prevent mechanical rupture of the synergid cell) and that synergid cell death is not required for pollen tube attraction (because *gfa2* embryo sacs fail to undergo synergid cell death and yet attract pollen tubes) (C.A. Christensen, S.W. Gorsich, J.M. Shaw & G.N. Drews, unpublished data). Molecular aspects of the synergid cell death process remain to be determined.

Induction of Seed Development

Double fertilization initiates seed development in angiosperms. During seed development, the fertilized egg cell develops into the embryo, the fertilized polar nuclei give rise to the endosperm, the sporophytic ovule integuments expand to form the seed coat, and the sporophytic ovary develops into the fruit (91). In sexually reproducing plants, development of seed and fruit occurs only following fertilization, i.e., under normal circumstances, fertilization induces development of embryo, endosperm, seed coat, and fruit in angiosperms. The mechanism by which fertilization induces seed and fruit development remains to be determined.

In some asexually reproducing species, development of seed and fruit is not dependent upon fertilization. Many angiosperms undergo apomictic growth in which one or more of the processes associated with seed/fruit development occur in the absence of fertilization. For example, in diplosporic gametophytic apomixis, an unreduced megaspore mother cell gives rise to a diploid female gametophyte that then produces an embryo without fertilization. In these and other apomictic species, the resulting embryos are genotypically identical to the maternal parent (70, 81). Genetic analysis indicates that the apomixis trait is genetically inherited and controlled by one or a few dominant loci in the species studied thus far (38, 60, 61). These and other data suggest that genetic lesions can cause development of embryo and endosperm to occur in the absence of fertilization.

The existence of apomictic strains led a number of groups to screen for Arabidopsis mutants that undergo seed development in the absence of fertilization (15, 82). Such screens led to the identification of three loss-of-function mutations: *fertilization-independent endosperm* (*fie*), *medea* (*mea*), and *fertilization-independent seed2* (*fis2*) (15, 39, 82). In all three mutants, diploid endosperm, seed coat, and fruit development take place in the absence of fertilization. However, embryo development does not occur when fertilization is prevented in these mutants. Genetic analysis indicates that the *fie*, *mea*, and *fis2* mutations segregate as female gametophyte-specific mutations (Table 1). Taken together, these data suggest that the female gametophyte expresses a series of proteins, including FIE, MEA, and FIS2, that function to suppress endosperm development until fertilization.

The *FIE, MEA, and FIS2* genes have recently been isolated. *FIE* encodes a WD-repeat polycomb protein related to Drosophila Esc, mammalian Eed, and Caenorhabditis Mes6 (83); *MEA* encodes a SET-domain polycomb protein related to Drosophila E(z) and Caenorhabditis Mes2 (39, 59, 69); and *FIS2* encodes a zinc-finger protein related to the Drosophila Su(z)12 Polycomb-group protein (5, 69). In Drosophila, complexes of related Polycomb-group (PcG) proteins regulate gene expression during early embryogenesis by conferring heritable silencing of the downstream homeotic genes that can be maintained through cell divisions as long as PcG proteins are present (87, 103).

By analogy to the Drosophila PcG proteins, it has been proposed that the FIE, MEA, and FIS2 proteins form a complex that functions to suppress endosperm gene expression within an unfertilized central cell (69, 83). Consistent with this model, all three genes are expressed before fertilization in the central cell (68, 105, 111, 122). Also, in support of this model, the FIE and MEA proteins have been shown to interact physically using two-hybrid analysis and in vitro pull-down assays (68, 105, 122). On the other hand, no interaction has been observed between FIE and FIS2 or MEA and FIS2 in yeast two-hybrid experiments (68, 122). The lack of detectable interaction between FIS2-MEA and FIS2-FIE suggests that these proteins may not interact. Alternatively, the interactions may be weak and undetectable or could be mediated by additional proteins. A more detailed understanding of this complex will require biochemical analysis and/or isolation of additional mutants affecting the induction of endosperm development.

It is still unclear how fertilization induces endosperm development. If, as is likely, the FIE, FIS2, and MEA proteins form a complex that represses endosperm proliferation, fertilization must somehow inactivate this complex. One possibility is that sperm cells deliver factors that inactivate or alter the FIE/FIS2/MEA complex (69, 83). In Drosophila and mammals, repressive PcG complexes are counteracted by activating complexes containing trithorax-group (trxG) protein(s) (55, 104). By analogy, plant trxG proteins may inactivate the FIE/FIS/MEA complex upon fertilization. Consistent with this possibility, trxG-like coding sequences have been identified within the Arabidopsis genome (2).

Analysis of the *fie, fis2,* and *mea* mutants suggests that sexually reproducing plants contain inhibitors of endosperm development that are inactivated upon fertilization. Likewise, the existence of apomictic strains suggests that sexually reproducing plants also contain inhibitors of embryo development. Mutations in genes encoding such factors should undergo embryo development in the absence of fertilization. However, apomictic mutants have yet to be identified in Arabidopsis. Thus, nothing is currently known about the molecular processes controlling the induction of embryo development by fertilization.

Maternal Control of Seed Development

In many animal systems, embryo pattern (e.g., anterior-posterior polarity) is established in part by maternal-effect genes, which are expressed prior to fertilization and/or in the maternal tissue surrounding the egg and developing embryo (36). As discussed above, angiosperms undergo double fertilization; thus, embryo and endosperm development could both be influenced by maternal-effect genes. Because the embryo/endosperm precursor cells (i.e., the egg cell and central cell) are within the haploid female gametophyte, maternal-effect genes may be female gametophyte expressed. We refer to such genes as gametophytic maternal-effect genes. However, as with animal systems, maternal-effect genes may also be expressed in the maternal tissue surrounding the egg and developing embryo, which, in plants, is the diploid sporophytic tissue of the ovule. We refer to this latter gene class as sporophytic maternal-effect genes (14, 27, 92). Several sporophytic maternal mutations have been reported, including the Arabidopsis *sin* (93) and barley *seg* (32) mutations, as well as the Petunia *FBP7/FBP11* co-suppression lines (19). However, unambiguous examples of female gametophyte-expressed genes influencing embryo or endosperm development have not yet been reported.

Despite the absence of data supporting the existence of gametophytic maternal-effect genes, several lines of evidence suggest that the egg cell and central cell are preprogrammed to undergo embryo and endosperm development, respectively. First, shortly after fertilization (within hours in Arabidopsis), the fertilized egg cell and central cell initiate dramatically different developmental programs (71, 72). This behavior apparently is not dependent on the sporophytic maternal tissue or the choice of which sperm nucleus fertilizes the egg or the central cell because in in vitro fertilization systems (e.g., in maize), invariably fertilized eggs undergo embryo development (62, 64) and fertilized central cells undergo endosperm development (65). Second, the egg cell of almost all angiosperms examined has a cytoplasm that is highly polarized along its apical-basal axis. This polarity has caused botanists for over 50 years to speculate that the asymmetric distribution of cytoplasm within the egg may play a role in the specification of apical and basal cell fates following zygotic division (70). Third, in many species, mutant strains have been identified that uncouple development of embryo and/or endosperm from fertilization. For example, in the Arabidopsis *fie*, *fis2*, and *mea* mutants discussed above, endosperm development occurs in the absence of fertilization (15, 82), and in many apomictic species, development of embryo and endosperm occurs in the absence of fertilization (38, 60, 81). Taken together, these data suggest very strongly that the embryo and endosperm developmental programs are specified, at least in part, within the unfertilized egg cell and central cell.

One approach to identify gametophytic maternal-effect genes is via the identification of gametophytic maternal mutations. Heterozygous mutants should exhibit reduced transmission of the mutant allele through the female gametophyte, have phenotypically wild-type female gametophytes, and have siliques containing 50%

abnormal seeds. The abnormal seeds should be white seeds if the mutation affects embryo development or brown shrunken seeds if the mutation affects endosperm (27).

One complication associated with the identification of gametophytic maternal mutations is that at least two classes of sporophytic mutations may behave in a similar manner. These sporophytic mutation classes include haplo-insufficient endosperm mutations and mutations affecting paternally imprinted genes required for embryo and/or endosperm development (Tables 1 and 2). As summarized in Table 2, both of these mutation classes produce 50% abnormal seeds when heterozygous mutants are pollinated with wild-type pollen. Therefore definitive demonstration that a mutant falls into gametophytic maternal class requires additional analysis. Crosses with triploid pollen donors can be used to determine whether a mutant falls into the haplo-insufficient endosperm class (39). Molecular analysis is required to distinguish between gametophytic maternal mutations and mutations in paternally imprinted genes. For example, expression of a gene prefertilization but not postfertilization would suggest very strongly that the gene falls into the gametophytic maternal effect class.

Several mutants exhibiting the phenotype expected of gametophytic maternal mutations have recently been reported; these include the Arabidopsis *constitutive triple response 1* (*ctr1*) (57), *prolifera* (*prl*) (107), *fie* (82), *fis2* (15), and *mea* (39) mutants, as well as the maize *maternal effect lethal1* (*mel1*) mutation (30). The *FIE*, *FIS2*, and *MEA* gene products are discussed above, the *PRL* gene encodes a homologue of the DNA replication licensing factor Mcm7 (107), the *CTR1* gene encodes a Raf serine/threonine protein kinase involved in ethylene signal transduction (57), and the *MEL1* gene product is unknown. In all of these mutants, embryo and endosperm development are abnormal following pollination of mutant female gametophytes with wild-type pollen. However, it remains ambiguous as to whether the postfertilization defects exhibited by these mutants are solely due to maternal effects. With the *CTR1* and *MEL1* genes, expression pre- and postfertilization have not been analyzed, and with the *FIE*, *FIS2*, *MEA*, and *PRL* genes, expression is detected both pre- and postfertilization (68, 105, 107, 111, 122). With the *MEA* gene, haplo-insufficiency has been ruled out through the use of crosses with triploid pollen donors (39), and this gene is paternally imprinted in the endosperm (58, 111). With the *FIS2* gene, triploid crosses have not been carried out and no direct evidence exists to indicate that this gene is paternally imprinted. However, *fis2* female gametophytes can generate viable seeds if hypomethylated pollen is used, which suggests that the *FIS2* gene is paternally imprinted (68). These data suggest that the *MEA* and *FIS2* genes might in fact be involved in regulation of seed development after fertilization and not represent strict maternal-effect genes. By contrast, with the *PRL* and *FIE* genes, paternal imprinting has been shown not to occur (106, 122). Thus, these gene products may produce a gametophytic maternal effect; however, a clear connection between the postfertilization defects and pre-fertilization expression has yet to be established. Thus, clear examples of gametophytic maternal-effect genes still remain to be uncovered.

EXPRESSION-BASED APPROACHES

Understanding female gametophyte development and function ultimately will require functional analysis of all female gametophyte-expressed genes. However, as discussed above, the female gametophyte most likely expresses many thousands of genes, and only approximately 600 of these could be identified using genetic screens. This discrepancy suggests that the majority of female gametophyte-expressed genes cannot be identified using traditional genetic approaches, possibly because of functional redundancy. Therefore, alternative expression-based approaches must be pursued to identify the majority of the genes expressed in the female gametophyte.

Using expression-based approaches, a relatively small number of female gametophyte-expressed genes have been identified, including the maize *ZmES/ defensins* (20), *eIF-5A* (25), and ribosomal protein (24) genes; the Arabidopsis *PRL* (44, 106, 107), *AGL15* (86), and *AGL18* (1) genes; a Brassica *SKP1*-like gene (28); the orchid O108 gene (79); and several wheat genes (67). However, these genes clearly represent only a small fraction of those expressed in the female gametophyte.

Large-scale identification of female gametophyte-expressed genes is problematic because the embryo sac is small and inaccessible. Two approaches have been attempted to circumvent these problems. First, in vitro methods have been developed to isolate large numbers of female gametophytes (62, 115), and this has led to the construction of cDNA libraries of embryo sac mRNAs in maize (26) and wheat (67). Second, gene-trap and enhancer-trap lines have been generated and screened for female gametophyte expression patterns (109). However, together, these approaches have led to the identification of only a few genes (20, 24, 25, 67).

Although very little progress has been made in isolating female gametophyte genes, this is nevertheless an important goal for several reasons. First, this information will enable a functional analysis of the gene products required for normal female gametophyte development and function. Second, such genes will serve as a source of useful markers for different female gametophyte cell types. Third, the promoters of such genes can be used to drive expression of other genes in the female gametophyte to analyze or modify female gametophyte or seed development. Finally, these genes could be used in promoter dissection experiments and thereby serve as entry points into the regulatory networks underlying female gametophyte development. Therefore, the application of expression-based approaches to identify the genes expressed during female gametophyte development and in the mature female gametophyte is an important goal to emphasize in future studies.

SUMMARY

The female gametophyte is a biologically important structure that mediates several reproductive processes, including pollen tube guidance, fertilization, induction of seed development, and maternal control of seed development. The female gametophyte is extremely active genetically and probably expresses many thousands of genes. Over the past several years, genetic approaches in Arabidopsis have yielded much information about the molecular and genetic processes governing megagametogenesis and the induction of seed development. By contrast, little is known at the molecular level about pollen tube guidance, fertilization, and the maternal control of seed development. Also, little progress has been made in the identification of genes expressed in the female gametophyte. Thus, the goals over the next several years must emphasize obtaining a complete profile of all genes expressed during female gametophyte development. In combination with genetic approaches, this information will lead to a functional understanding of how the haploid female gametophyte develops and carries out its important reproductive functions.

ACKNOWLEDGMENTS

Our work on female gametophyte development was supported by grants from the National Science Foundation (Number IBN-9630371) and Ceres, Inc. to G.N.D.

The *Annual Review of Genetics* is online at http://genet.annualreviews.org

LITERATURE CITED

1. Alvarez-Buylla ER, Liljegren SJ, Pelaz S, Gold SE, Burgeff C, et al. 2000. MADS-box gene evolution beyond flowers: expression in pollen, endosperm, guard cells, roots and trichomes. *Plant J.* 24:457–66
2. Baumbusch LO, Thorstensen T, Krauss V, Fischer A, Naumann K, et al. 2001. The *Arabidopsis thaliana* genome contains at least 29 active genes encoding SET domain proteins that can be assigned to four evolutionarily conserved classes. *Nucleic Acids Res.* 29:4319–33
3. Bechtold N, Jaudeau B, Jolivet S, Maba B, Vezon D, et al. 2000. The maternal chromosome set is the target of the T-DNA in the in planta transformation of *Arabidopsis thaliana*. *Genetics* 155:1875–87
4. Bechtold N, Pelletier G. 1998. In planta Agrobacterium-mediated transformation of adult *Arabidopsis thaliana* plants by vacuum infiltration. *Methods Mol. Biol.* 82:259–66
5. Birve A, Sengupta AK, Beuchle D, Larson J, Kennison JA, et al. 2001. Su(z)12, a novel Drosophila Polycomb group gene that is conserved in vertebrates and plants. *Development* 128:3371–79
6. Bonhomme S, Horlow C, Vezon D, de Laissardiere S, Guyon A, et al. 1998. T-DNA mediated disruption of essential gametophytic genes in Arabidopsis is unexpectedly rare and cannot be inferred from segregation distortion alone. *Mol. Gen. Genet.* 260:444–52
7. Brink RA, Burnham CR. 1929. Inheritance of semi-sterility in maize. *Am. Nat.* 63:301–16
8. Brink RA, MacGillivray JH. 1924. Segregation for the waxy character in maize pollen and differential development of the male gametophyte. *Am. J. Bot.* 11:465–69
9. Buckner B, Reeves SL. 1994. Viability of female gametophytes that possess deficiencies for the region of chromosome 6 containing the *Y1* gene. *Maydica* 39:247–54
10. Castle LA, Errampalli D, Atherton TL,

Franzmann LH, Yoon ES, Meinke DW. 1993. Genetic and molecular characterization of embryonic mutants identified following seed transformation in *Arabidopsis*. *Mol. Gen. Genet.* 241:504–14

11. Chaubal R, Reger BJ. 1990. Relatively high calcium is localized in synergid cells of wheat ovaries. *Sex. Plant Reprod.* 3:98–102

12. Chaubal R, Reger BJ. 1992. Calcium in the synergid cells and other regions of pearl millet ovaries. *Sex. Plant Reprod.* 5:34–46

13. Chaubal R, Reger BJ. 1992. The dynamics of calcium distribution in the synergid cells of wheat after pollination. *Sex. Plant Reprod.* 5:206–13

14. Chaudhury AM, Berger F. 2001. Maternal control of seed development. *Semin. Cell Dev. Biol.* 12:381–86

15. Chaudhury AM, Ming L, Miller C, Craig S, Dennis ES, Peacock WJ. 1997. Fertilization-independent seed development in *Arabidopsis thaliana*. *Proc. Natl. Acad. Sci. USA* 94:4223–28

16. Christensen CA, King EJ, Jordan JR, Drews GN. 1997. Megagametogenesis in *Arabidopsis* wild type and the *Gf* mutant. *Sex. Plant Reprod.* 10:49–64

17. Christensen CA, Subramanian S, Drews GN. 1998. Identification of gametophytic mutations affecting female gametophyte development in *Arabidopsis*. *Dev. Biol.* 202:136–51

18. Collins GN, Kempton JH. 1911. *Inheritance of waxy endosperm in hybrids of Chinese maize*. Presented at Fourth Int. Conf. Genet., Paris, France

19. Colombo L, Franken J, Van der Krol AR, Wittich PE, Dons HJM, Angenent GC. 1997. Downregulation of ovule-specific MADS box genes from petunia results in maternally controlled defects in seed development. *Plant Cell* 9:703–15

20. Cordts S, Bantin J, Wittich PE, Kranz E, Lörz H, Dresselhaus T. 2001. ZmES genes encode peptides with structural homology to defensins and are specifically expressed in the female gametophyte of maize. *Plant J.* 25:103–14

21. Correns C. 1902. Scheinbare Ausnahmen von der Mendel'shen Spaltungsregel fur Bastarde. *Ber. Dtsch. Bot. Ges.* 20:159–72

22. Couteau F, Belzile F, Horlow C, Grandjean O, Vezon D, Doutriaux MP. 1999. Random chromosome segregation without meiotic arrest in both male and female meiocytes of a dmc1 mutant of Arabidopsis. *Plant Cell* 11:1623–34

23. Desfeux C, Clough SJ, Bent AF. 2000. Female reproductive tissues are the primary target of Agrobacterium-mediated transformation by the Arabidopsis floral-dip method. *Plant Physiol.* 123:895–904

24. Dresselhaus T, Cordts S, Heuer S, Sauter M, Lörz H, Kranz E. 1999. Novel ribosomal genes from maize are differentially expressed in the zygotic and somatic cell cycles. *Mol. Gen. Genet.* 261:416–27

25. Dresselhaus T, Cordts S, Lörz H. 1999. A transcript encoding translation initiation factor eIF-5A is stored in unfertilized egg cells of maize. *Plant Mol. Biol.* 39:1063–71

26. Dresselhaus T, Lörz H, Kranz E. 1994. Representative cDNA libraries from few plant cells. *Plant J.* 5:605–10

27. Drews GN, Lee D, Christensen CA. 1998. Genetic control of female gametophye development and function. *Plant Cell* 10:1–15

28. Drouaud J, Marrocco K, Ridel C, Pelletier G, Guerche P. 2000. A Brassica napus skp1–like gene promoter drives GUS expression in *Arabidopsis thaliana* male and female gametophytes. *Sex. Plant Reprod.* 13:29–35

29. Dumas C, Knox RB, McConchie CA, Russell SD. 1984. Emerging physiological concepts in fertilization. *What's New Plant Physiol.* 15:17–20

30. Evans MM, Kermicle JL. 2001. Interaction between maternal effect and zygotic effect mutations during maize seed development. *Genetics* 159:303–15

31. Feldmann KA, Coury DA, Christianson ML. 1997. Exceptional segregation of a selectable marker (KanR) in *Arabidopsis* identifies genes important for gametophytic growth and development. *Genetics* 147:1411–22

32. Felker FC, Peterson DM, Nelson OE. 1985. Anatomy of immature grains of eight maternal effect shrunken endosperm barley mutants. *Am. J. Bot.* 72:248–56

33. Folsom MW, Cass DD. 1990. Embryo sac development in soybean: cellularization and egg apparatus expansion. *Can. J. Bot.* 68:2135–47

34. Fu Y, Yuan M, Huang B-Q, Yang HY, Zee SY, O'Brien TP. 2000. Changes in actin organization in the living egg apparatus of *Torenia fournieri* during fetilization. *Sex. Plant Reprod.* 12:315–22

35. Gifford EM, Foster AS. 1988. *Morphology and Evolution of Vascular Plants.* New York: Freeman. 626 pp.

36. Gilbert SF. 2000. *Developmental Biology.* Sunderland, MA: Sinauer. 749 pp.

37. Grini PE, Schnittger A, Schwarz H, Zimmermann I, Schwab B, et al. 1999. Isolation of ethyl methanesulfonate-induced gametophytic mutants in *Arabidopsis thaliana* by a segregation distortion assay using the multimarker chromosome 1. *Genetics* 151:849–63

38. Grossniklaus U, Nogler GA, van Dijk PJ. 2001. How to avoid sex: the genetic control of gametophytic apomixis. *Plant Cell* 13:1491–97

39. Grossniklaus U, Vielle-Calzada J-P, Hoeppner MA, Gagliano WB. 1998. Maternal control of embryogenesis by *MEDEA*, a polycomb group gene in *Arabidopsis. Science* 280:446–50

40. Haig D. 1990. New perspectives on the angiosperm female gametophyte. *Bot. Rev.* 56:236–75

41. Higashiyama T, Kuroiwa H, Kawano S, Kuroiwa T. 1998. Guidance *in vitro* of the pollen tube to the naked embryo sac of torenia fournieri. *Plant Cell* 10:2019–32

42. Higashiyama T, Kuroiwa H, Kawano S, Kuroiwa T. 2000. Explosive discharge of pollen tube contents in *Torenia fournieri. Plant Physiol.* 122:11–14

43. Higashiyama T, Yabe S, Sasaki N, Nishimura Y, Miyagishima S, et al. 2001. Pollen tube attraction by the synergid cell. *Science* 293:1480–83

44. Holding DR, Springer PS. 2002. The Arabidopsis gene PROLIFERA is required for proper cytokinesis during seed development. *Planta* 214:373–82

45. Howden R, Park SK, Moore JM, Orme J, Grossniklaus U, Twell D. 1998. Selection of T-DNA-tagged male and female gametophytic mutants by segregation distortion in *Arabidopsis. Genetics* 149:621–31

46. Huang B-Q, Russell SD. 1992. Female germ unit: organization, isolation, and function. *Int. Rev. Cytol.* 140:233–92

47. Huang B-Q, Russell SD. 1992. Synergid degeneration in *Nicotiana*: a quantitative, fluorochromatic and chlorotetracycline study. *Sex. Plant Reprod.* 5:151–55

48. Huang B-Q, Sheridan WF. 1994. Female gametophyte development in maize: microtubular organization and embryo sac polarity. *Plant Cell* 6:845–61

49. Hülskamp M, Schneitz K, Pruitt RE. 1995. Genetic evidence for a long-range activity that directs pollen tube guidance in *Arabidopsis. Plant Cell* 7:57–64

50. Jensen WA. 1965. The ultrastructure and histochemsitry of the synergids of cotton. *Am. J. Bot.* 52:238–56

51. Jensen WA, Ashton ME, Beasley CA. 1983. Pollen tube-embryo sac interaction in cotton. In *Pollen: Biology and Implications for Plant Breeding*, ed. DL Mulcahy, E Ottaviano, pp. 67–72. New York: Elsevier

51a. Johri BM, ed. 1984. *Embryology of Angiosperms.* Berlin: Springer-Verlag

52. Jones DF. 1924. Selective fertilization among the gametes from the same individuals. *Proc. Natl. Acad. Sci. USA* 10:218–21

53. Jürgens G, Mayer U, Torres-Ruiz RA, Berleth T, Misera S. 1991. Genetic

analysis of pattern formation in the *Arabidopsis* embryo. *Dev. Suppl.* 1:27–38

54. Kandasamy MK, Nasrallah JB, Nasrallah ME. 1994. Pollen-pistil interactions and developmental regulation of pollen tube growth in Arabidopsis. *Development* 120:3405–18

55. Kennison JA. 1995. The polycomb and trithorax group proteins of Drosophila: trans-regulators of homeotic gene function. *Annu. Rev. Genet.* 29:289–303

56. Kieber JJ, Ecker JR. 1994. Molecular and genetic analysis of the constitutive ethylene response mutation *ctr1*. In *Plant Molecular Biology; Molecular Genetic Analysis of Plant Development and Metabolism*, ed. G Coruzzi, P Puigdomenech, pp. 193–201. Berlin: Springer-Verlag

57. Kieber JJ, Rothenberg M, Roman G, Feldman KA, Ecker JR. 1993. *CTR1*, a negative regulator of the ethylene response pathway in *Arabidopsis*, encodes a member of the Raf family of protein kinases. *Cell* 72:427–41

58. Kinoshita T, Yadegari R, Harada JJ, Goldberg RB, Fischer RL. 1999. Imprinting of the MEDEA polycomb gene in the Arabidopsis endosperm. *Plant Cell* 11:1945–52

59. Kiyosue T, Ohad N, Yadegari R, Hannon M, Dinneny J, et al. 1999. Control of fertilization-independent endosperm development by the MEDEA polycomb gene in Arabidopsis. *Proc. Natl. Acad. Sci. USA* 96:4186–91

60. Koltunow AM. 1993. Apomixis: embryo sacs and embryos formed without meiosis or fertilization in ovules. *Plant Cell* 5:1425–37

61. Koltunow AM, Bicknell RA, Chaudhury AM. 1995. Apomixis: molecular strategies for the generation of genetically identical seeds without fertilization. *Plant Physiol.* 108:1345–52

62. Kranz E, Bautor J, Lörz H. 1991. *In vitro* fertilisation of single, isolated gametes of maize (*Zea mays* L.). *Sex. Plant Reprod.* 4:12–16

63. Kranz E, Dresselhaus T. 1996. *In vitro* fertilization with isolated higher plant gametes. *Trends Plant Sci.* 1:82–89

64. Kranz E, Lörz H. 1993. In-vitro fertilization with isolated, single gametes results in zygotic embryogenesis and fertile maize plants. *Plant Cell* 5:739–46

65. Kranz E, von Wiegen P, Quader H, Lörz H. 1998. Endosperm development after fusion of isolated, single maize sperm and central cells *in vitro*. *Plant Cell* 10:511–24

66. Krysan PJ, Young JC, Sussman MR. 1999. T-DNA as an insertional mutagen in Arabidopsis. *Plant Cell* 11:2283–90

67. Kumlehn J, Kirik V, Czihal A, Altschmied L, Matzk F, et al. 2001. Parthenogenetic egg cells of wheat: cellular and molecular studies. *Sex. Plant Reprod.* 14:239–43

68. Luo M, Bilodeau P, Dennis ES, Peacock WJ, Chaudhury A. 2000. Expression and parent-of-origin effects for FIS2, MEA, and FIE in the endosperm and embryo of developing Arabidopsis seeds. *Proc. Natl. Acad. Sci. USA* 97:10637–42

69. Luo M, Bilodeau P, Koltunow A, Dennis ES, Peacock WJ, Chaudhury AM. 1999. Genes controlling fertilization-independent seed development in *Arabidopsis thaliana*. *Proc. Natl. Acad. Sci. USA* 96:296–301

70. Maheshwari P. 1950. *An Introduction to the Embryology of Angiosperms*. New York: McGraw-Hill. 453 pp.

71. Mansfield SG, Briarty LG. 1990. Development of the free-nuclear endosperm in *Arabidopsis thaliana*. *Arabidopsis Inf. Serv.* 27:53–64

72. Mansfield SG, Briarty LG. 1991. Early embryogenesis in *Arabidopsis thaliana*. II. The developing embryo. *Can. J. Bot.* 69:461–76

73. Mansfield SG, Briarty LG, Erni S. 1991. Early embryogenesis in *Arabidopsis thaliana*. I. The mature embryo sac. *Can. J. Bot.* 69:447–60

74. Mascarenhas JP. 1962. The hormonal control of the directional growth of

pollen tubes. *Vitam. Hormone* 20:347–72

75. McCormick S. 1993. Male gametophyte development. *Plant Cell* 5:1265–75

76. Misra RC. 1962. Contribution to the embryology of *Arabidopsis thaliana* (Gay and Monn.). *Agra Univ. J. Res. Sci.* 11:191–99

77. Moore JM, Vielle-Calzada JP, Gagliano W, Grossniklaus U. 1997. Genetic characterization of *hadad*, a mutant disrupting female gametogenesis in *Arabidopsis thaliana*. *Cold Spring Harbor Symp. Quant. Biol.* 62:35–47

78. Murgia M, Huang B-Q, Tucker SC, Musgrave ME. 1993. Embryo sac lacking antipodal cells in *Arabidopsis thaliana* (Brassicaceae). *Am. J. Bot.* 80:824–38

79. Nadeau JA, Zhang XS, Li J, O'Neill SD. 1996. Ovule development: identification of stage-specific and tissue-specific cDNAs. *Plant Cell* 8:213–39

80. Niyogi KK, Last RL, Fink GR, Keith B. 1993. Suppressors of *trp1* fluorescence identify a new *Arabidopsis* gene, *TRP4*, encoding the anthranilate synthase beta subunit. *Plant Cell* 5:1011–27

81. Nogler GA. 1984. Gametophytic apomixis. See Ref. 51a, pp. 159–96

82. Ohad N, Margossian L, Hsu Y-C, Williams C, Repetti P, Fischer RL. 1996. A mutation that allows endosperm development without fertilization. *Proc. Natl. Acad. Sci. USA* 93:5319–24

83. Ohad N, Yadegari R, Margossian L, Hannon M, Michaeli D, et al. 1999. Mutations in FIE, a WD polycomb group gene, allow endosperm development without fertilization. *Plant Cell* 11:407–16

84. Patterson EB. 1978. Properties and uses of duplicate deficient chromosome compliments in maize. In *Maize Breeding and Genetics*, ed. D Walden, pp. 693–710. New York: Wiley

85. Patterson EB. 1994. Translocations as genetic markers. In *The Maize Handbook*, ed. M Freeling, V Walbot, pp. 361–63. New York: Springer

86. Perry S, Nichols KW, Fernandez DE. 1996. The MADS domain protein AGL15 localizes to the nucleus during early stages of seed development. *Plant Cell* 8:1977–89

87. Pirrotta V. 1997. PcG complexes and chromatin silencing. *Curr. Opin. Genet. Dev.* 7:249–58

88. Poliakova TF. 1964. Development of the male and female gametophytes of *Arabidopsis thaliana* (L) Heynh. *Issled. Genet. USSR* 2:125–33

89. Preuss D. 1995. Being fruitful: genetics of reproduction in *Arabidopsis*. *Trends Genet.* 11:147–53

90. Pruitt RE, Hülskamp M, Kopczak SD, Plownse SE, Schneitz K. 1993. Molecular genetics of cell interactions in *Arabidopsis*. *Dev. Suppl.* 77–84

91. Raven PH, Evert RF, Eichhorn SE. 1999. *Biology of Plants*. New York: Freeman. 944 pp.

92. Ray A. 1997. Three's company: regulatory cross-talk during seed development. *Plant Cell* 9:665–67

93. Ray S, Golden T, Ray A. 1996. Maternal effects of the *short integument* mutation on embryo development in *Arabidopsis*. *Dev. Biol.* 180:365–69

94. Ray S, Park S-S, Ray A. 1997. Pollen tube guidance by the female gametophyte. *Development* 124:2489–98

95. Redei GP. 1965. Non-Mendelian megagametogenesis in *Arabidopsis*. *Genetics* 51:857–72

96. Reger BJ, Chaubal R, Pressey R. 1992. Chemotropic responses by pearl millet pollen tubes. *Sex. Plant Reprod.* 5:47–56

97. Russell SD. 1992. Double fertilization. *Int. Rev. Cytol.* 140:357–88

98. Russell SD. 1993. The egg cell: development and role in fertilization and early embryogenesis. *Plant Cell* 5:1349–59

99. Russell SD. 1996. Attraction and transport of male gametes for fertilization. *Sex. Plant Reprod.* 9:337–42

100. Schneitz K, Hülskamp M, Kopczak SD, Pruitt RE. 1997. Dissection of sexual

organ ontogenesis: a genetic analysis of ovule development in *Arabidopsis thaliana. Development* 124:1367–76

101. Schneitz K, Hülskamp M, Pruitt RE. 1995. Wild-type ovule development in *Arabidopsis thaliana*: a light microscope study of cleared whole-mount tissue. *Plant J.* 7:731–49

102. Shimizu KK, Okada K. 2000. Attractive and repulsive interactions between female and male gametophytes in Arabidopsis pollen tube guidance. *Development* 127:4511–18

103. Simon J. 1995. Locking in stable states of gene expression: transcriptional control during Drosophila development. *Curr. Opin. Cell Biol.* 7:376–85

104. Simon JA, Tamkun JW. 2002. Programming off and on states in chromatin: mechanisms of Polycomb and trithorax group complexes. *Curr. Opin. Genet. Dev.* 12:210–18

105. Spillane C, MacDougall C, Stock C, Köhler C, Vielle-Calzada JP, et al. 2000. Interaction of the Arabidopsis polycomb group proteins FIE and MEA mediates their common phenotypes. *Curr. Biol.* 10:1535–38

106. Springer PS, Holding DR, Groover A, Yordan C, Martienssen RA. 2000. The essential Mcm7 protein PROLIFERA is localized to the nucleus of dividing cells during the G(1) phase and is required maternally for early Arabidopsis development. *Development* 127:1815–22

107. Springer PS, McCombie WR, Sundaresan V, Martienssen RA. 1995. Gene trap tagging of *PROLIFERA*, an essential *MCM2-3-5*-like gene in *Arabidopsis. Science* 268:877–80

108. Strasburger E. 1879. *Die Angiospermen und die Gymnospermen.* Jena

109. Sundaresan V, Springer P, Volpe T, Haward S, Jones JDG, et al. 1995. Patterns of gene action in plant development revealed by enhancer trap and gene trap transposable elements. *Genes Dev.* 9:1797–810

110. van Went JL, Willemse MTM. 1984. Fertilization. See Ref. 51a, pp. 273–317

111. Vielle-Calzada JP, Thomas J, Spillane C, Coluccio A, Hoeppner MA, Grossniklaus U. 1999. Maintenance of genomic imprinting at the *Arabidopsis medea* locus requires zygotic *DDM1* activity. *Genes Dev.* 13:2971–82

112. Vijayaraghavan MR, Seth N, Jain A. 1988. Ultrastructure and histochemistry of angiosperm embryo sac—an overview. *Proc. Indian Natl. Sci. Acad.* 54:93–110

113. Vollbrecht E, Hake S. 1995. Deficiency analysis of female gametogenesis in maize. *Dev. Genet.* 16:44–63

114. Waddington CH. 1940. The genetic control of wing development in Drosophila. *J. Genet.* 41:75–139

115. Wagner VT, Song Y, Matthys-Rochon E, Dumas C. 1988. The isolated embryo sac of *Zea mays*: structural and ultrastructural observations. In *Sexual Reproduction in Higher Plants*, ed. M Cresti, P Gori, E Pacini, pp. 125–30 Berlin: Springer-Verlag

116. Webb MC, Gunning BES. 1990. Embryo sac development in *Arabidopsis thaliana*. I. Megasporogenesis, including the microtubular cytoskeleton. *Sex. Plant Reprod.* 3:244–56

117. Webb MC, Gunning BES. 1994. Embryo sac development in *Arabidopsis thaliana*. II. The cytoskeleton during megagametogenesis. *Sex. Plant Reprod.* 7:153–63

118. Weber DF. 1983. Monosomic analysis in diploid crop plants. In *Cytogenetics of Crop Plants*, ed. PK Gupta, U Sinha, pp. 352–78 New Delhi: Macmillan India

119. Willemse MTM, van Went JL. 1984. The female gametophyte. See Ref. 51a, pp. 159–96

120. Willing RP, Bashe D, Mascarenhas JP. 1988. An analysis of the quantity and diversity of messenger RNAs from pollen and shoots of *Zea mays. Theor. Appl. Genet.* 75:751–53

121. Willing RP, Mascarenhas JP. 1984. Analysis of the complexity and diversity of

mRNAs from pollen and shoots of *Trades-cantia. Plant Physiol.* 75:865–68

122. Yadegari R, Kinoshita T, Lotan O, Cohen G, Katz A, et al. 2000. Mutations in the *FIE* and *MEA* genes that encode interacting polycomb proteins cause parent-of-origin effects on seed development by distinct mechanisms. *Plant Cell* 12:2367–82

123. Ye G-N, Stone D, Pang S-Z, Creely W, Gonzalez K, Hinchee M. 1999. *Arabidopsis* ovule is the target for *Agrobacterium in planta* vacuum infiltration transformation. *Plant J.* 19:249–57

Annu. Rev. Genet. 2002. 36:125–51
doi: 10.1146/annurev.genet.36.031902.105056
Copyright © 2002 by Annual Reviews. All rights reserved

PRIMORDIAL GENETICS:
Phenotype of the Ribocyte

Michael Yarus

*Department of Molecular, Cellular and Developmental Biology, University of Colorado,
Boulder, Colorado 80309-0347; e-mail: yarus@stripe.colorado.edu*

Key Words SELEX, selection-amplification, RNA world, replication, translation,
RNA cell

■ **Abstract** The idea that the ancestors of modern cells were RNA cells (ribocytes)
can be investigated by asking whether all essential cellular functions might be performed by RNAs. This requires isolating suitable molecules by selection-amplification
when the predicted molecules are presently extinct. In fact, RNAs with many properties required during a period in which RNA was the major macromolecular agent
in cells (an RNA world) have been selected in modern experiments. There is, accordingly, reason to inquire how such a ribocyte might appear, based on the properties of
the RNAs that composed it. Combining the intrinsic qualities of RNA with the fundamental characteristics of selection from randomized sequence pools, one predicts
ribocytes with a cell cycle measured (roughly) in weeks. Such cells likely had a rapidly
varying genome, composed of many small genetic and catalytic elements made of
tens of ribonucleotides. There are substantial arguments that, at the mid-RNA era, a
subset of these nucleotides are reproducibly available and resemble the modern four.
Such cells are predicted to evolve rapidly. Instead of modifying preexisting genes,
ribocytes frequently draw new functions from an internal pool containing zeptomoles
(<1 attomole) of predominantly inactive random sequences.

CONTENTS

0066-4197/02/1215-0125$14.00

INTRODUCTION

The RNA world hypothesis predicts a biota antecedent to our own that used an RNA-like molecule for a variety of tasks today performed by RNA, DNA, and proteins together. Is RNA sufficiently versatile to have once been the major genetic, catalytic, and structural molecule of cells? Is it possible there once was an RNA cell (a ribocyte)?

Such questions are a major epistemological challenge. The posited events probably occurred near the origin of life, so early in the history of the planet that contemporary rocks are not expected to survive (49). Accordingly, there may never be the usual kind of fossil evidence of the ribocyte, though isotopic abundance data (67) and "molecular fossils" (72) may give information about the RNA era. In addition, RNA cells were subject to the constraints that apply to all cells (77). However, I argue here that evidence of an independent kind, from SELEX or selection-amplification, supports and defines the existence of an RNA cell. Further, it seems that the ribocyte was sufficiently circumscribed by the laws of chemistry and physics that we can likely say something about it and about the processes that gave rise to it.

RNA Versatility

RNA organisms are old in concept (76, 110), as is some of the best evidence for them (104, 109). More recently, the RNA world has been conceptually unified under its present name (28). Nevertheless, for all this time, a basic implausibility confounded thought about any hypothetical RNA world. Because RNA has a much-diminished present role, the existence of a ribocyte necessarily implied a large group of extinct RNA functions. Proof that RNA is capable of many unknown reactions, an idea explored extensively over the past two decades, is therefore an essential general support for any RNA world hypothesis. Because it is easy to forget the more limited picture of RNA capabilities once prevalent, I begin by summarizing some notable additions to the RNA repertoire.

Any such summary begins with the discovery of RNA enzymes (ribozymes[1]); that is, with the accurate self-splicing of the *Tetrahymena* large ribosomal RNA (58)

---

[1]In what follows I do not distinguish between acceleration of a reaction (a directly and frequently selected RNA capability) and ribozyme activity (rarer; when strictly defined, requires identical turnovers). This facilitates exposition, and although strictly incorrect, does not seem to introduce crucial errors.

and the 5' nucleolytic processing of tRNA precursors by ribonuclease P (33). These phosphoester transfer reactions were joined by the self-splicing group II introns (66) and by other natural hydrolytic catalysts: the hammerhead (102), the hairpin ribozyme (13), the hepatitis delta motif (5), and the *Neurospora* VS ribozyme (80). In fact, ribonucleotide backbone cleavage can be accelerated by metallo-RNA structures as small as seven nucleotides (51). While these examples might have given the impression that RNA accelerated a limited set of phosphodiester cleavages and transfers, many other kinds of reactions have now been found. Selection-amplification (24, 81, 101) for new RNA activities has established that, for example, carbonyl chemistry (38), carbon-carbon bond formation (44), carbon-nitrogen (glycosidic) bond creation in nucleosides (103), and the Michael reaction (89) were also within the repertoire of RNA folds containing only the standard four nucleotides. In fact, the acceleration of nucleophilic attack at a carbonyl by RNAs correctly predicted (38) that the ribosomal peptidyl transferase, performing similar chemistry, could be a ribozyme (72). The stabilization of transition states by RNA is not well understood and could mostly rely on accurate approximation of the reactants (entropic catalysis). But RNA catalysis also extends at least to general acid/base action to shuttle protons to or from transition states (71, 91). In addition, though not all ribozymes use bound metal ions near the active site (84), many mobilize divalent metals (90) and monovalent metals (2) for varied catalytic roles.

Though selection methods usually cannot identify RNA-mediated reactions between multiple small substrates, such multisubstrate RNA reactions have been designed and constructed (36). Such juxtaposition of several relatively small reactants on RNA surfaces is required for an anabolic metabolism. Further, multidomain ribozymes (94) can respond to the presence of unrelated small-molecule effectors by changing their activity (81). Regulation might also occur via oligonucleotide effectors (106). Thus metabolic networks regulated by small metabolites or other regulators can be rationalized for RNAs, as for protein enzyme networks.

Notably, RNA-catalyzed ribonucleotide polymerization using nucleoside triphosphate substrates (47) can be selected (see below). Furthermore, amino acid activation, aminoacyl-RNA synthesis, peptide synthesis, and coding (116) appear to be within the RNA repertoire. Whereas RNA may suffer as a general catalyst by comparison to proteins, its apparent ability to both replicate and facilitate chemical reactions is unique [compare (85)]. The demonstration of these capabilities enables us to conceptualize a plausible ancestral biota, related to us but simpler by omission of a major molecular component (proteins). Furthermore, for the first time the succession from such a simpler situation to the present nucleoprotein biology can be envisioned. Thus the demonstration that RNA might replicate on its own and also devise the templated synthesis of peptides argues for an RNA world, and our descent from it. RNAs also can synthesize (35) and exploit ribonucleotide cofactors, such as CoA, that also supply proteins with the chemical versatility required for metabolism (43).

To these catalytic feats, we must add RNA's ability to fold into binding sites for unusual non-nucleotidyl ligands. Examples include free amino acids (114), lipids (28), hydrocarbons (63), heme (100), vitamin B12 (95), and even phospholipid membranes (55). In fact, only a few RNA nucleotides are apparently required for some such activities. RNA seems to have virtually unlimited ability to make ribonucleotide surfaces complementary to peptide surfaces found in nucleic-acid binding and nonbinding proteins alike (30). Accordingly, the range of specific ligands for RNA enzymes is much broader than once thought.

Thus the implication of cryptic RNA capacities is sound. Taken all together, these findings make RNA an unusually versatile molecule, one whose reactions perhaps span a range that might be appropriate to a ribocyte. Perhaps it will help to imagine for a moment that carbohydrates had been shown to possess the above list of activities. The role accorded to carbohydrates in the history of life on Earth would surely be under radical revision. As for this hypothetical carbohydrate, so for the real oligoribonucleotide.

However, all the above capacities should not by any means obscure what RNA is still not known to do. Many known RNA reactions require rates and specificities better than those presently measured in order to be biologically useful. The genesis of the RNA-like ancestral molecule itself is a continuing problem, since chemically impure precursors (inevitable early on) inhibit the synthesis and replication of RNA itself (50). Nucleosides and nucleotides are difficult to synthesize by plausible prebiotic means, and are, like RNA itself, unstable on the time scales that are likely to be involved (60). These difficulties are sometimes taken as evidence of the existence of a pre-RNA world (50).

Because not all new capacities are equally important, it is necessary to emphasize those particular reactions "appropriate to a ribocyte." The rest of this essay is devoted to that consideration.

The Method

What do modern RNA isolation (selection-amplification, SELEX) experiments (24, 81, 101) mean and how do they apply to the question of interest, which is about the history of biology? To answer, we must ask what selection-amplifications do. They are cyclical, repetitive purifications from randomized sequences, where the ability of nucleic acids to replicate after each selection allows almost any degree of purification, via any number of purification cycles. We are not directly studying history when we show which RNA activities are isolable by a modern selection; that is, latent in a particular pool of randomized oligoribonucleotides. Instead, significant experiments (positive ones) establish only that some number of RNA (or DNA) active sites for a given task exist.

We know after successful selection of a new RNA that a particular reaction is available to RNA. In fact, practice suggests that we know a bit more. The overlapping of proven reactions is also likely. For example, the isolation of a self-aminoacylating RNA (38) makes more likely an amino-acid specific self-aminoacylating RNA. This is especially so in light of the isolation of other RNA folds that specifically bind different amino acids (114). The amino acid-specific self-acylator was later confirmed by isolation (41). Proof can be provided, then, that a group of related reactions would have been accessible to RNA. This may seem modest progress, but compare it with the situation before the experiment. Instead of speculation ranging over an unlimited range of outcomes, we have particular results and particular measured characteristics, known to be plausible. The advance from having no data to having some data is dramatic, and such results will be ignored at the usual peril to theorists who ignore relevant data.

Furthermore, establishing that an activity is "possible" has a special meaning within these experiments. Selection amplifications isolate new folds from randomized sequences. Therefore the idea that cellular activities could arise from groups of arbitrary inactive RNAs is strengthened by successful selections. In fact, below I argue that we can see the identifiable impact of this process on the course taken by life in an RNA world.

There are also quantitative consequences of selection from arbitrary sequences. Calculations suggest (118) that active sites derived by this process cannot contain more than 15–30 required nucleotides, or the size of the implied initial pool of sequences would be unattainable. In fact, the smallest active site(s) capable of meeting the selection will be predominant, because the number of RNA folds with a given length increases rapidly as the fold becomes shorter, about tenfold more abundant when a desired motif is shortened by 1.6 nucleotides.

To make an RNA world plausible, RNA activities that could initiate, sustain, and end such an era must exist. In the following sections, I summarize work that bears on the plausibility of each of these events by showing that selection produces simple RNAs that perform related tasks.

Beginning the Period of RNA Self-Sufficiency

Though it is possible to debate what particular event would be sufficient to establish that an RNA world had begun, many geneticists would accept the beginning of Darwinian evolution as the crucial watershed. Once one has replication with variation, selection will follow. Prior events adhere to a chemical program, bearing downhill on a shifting free energy surface. Subsequent events accelerate, taking a new and specifically biological direction responsive to the requirements of life for the first time. Therefore there has been a long-term effort to show that replication of RNAs by an RNA is possible.

Templated Synthesis of RNA

One might think that RNA synthesis by RNA could be conceded, as the classical RNA-catalyzed reactions are all nucleophilic attacks at phosphorus, paralleling the ribose-OH attack that adds a nucleotide to a growing RNA chain. However, not until the group II intron was shown to be capable of releasing pyrophosphate from a nucleoside triphosphate (68) was it clear that RNA catalysis could use the same reactants and leaving group as does a modern RNA replicase. Even then, the path to templated activity has required several successive, cleverly planned selective steps. Early experiments utilized group I self-splicing RNA activity in various ways, including ligation of complementary oligonucleotides paired to a continuous template strand (4, 17, 32). However, the closest approach to an RNA RNA replicase currently results from elaboration of a ligase activity (1) selected directly from randomized sequences. The original selection demanded that an oligonucleotide be joined to the 5' triphosphate of the ribozyme, releasing pyrophosphate, to create the template for further amplification (Figure 1a). From 1.4×10^{15} RNA molecules (moored to beads to prevent aggregation), each having

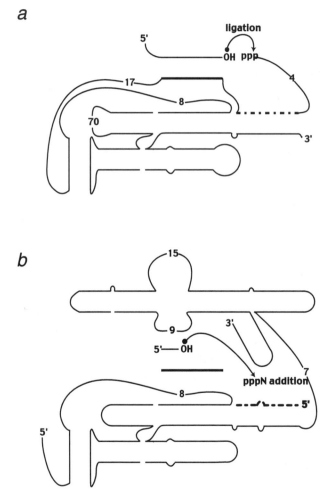

Figure 1 Related ligase and RNA replicase activities. (a) 185-mer class I ligase activity (23) joins the 5' α phosphate to the 3' hydroxyl oxygen, as shown, releasing pyrophosphate. Tracts related to template sequences are shown as solid bold, and a sequence ultimately required as a separate oligonucleotide cofactor is shown as a dotted bold line. Intercalated numbers are the size of that curved segment, in nucleotides. (b) 196-mer RNA replicase (47) (189-mer plus 7-mer cofactor). Notation as in panel (a).

three sections totaling 220 randomized nucleotides, more than 65 sequences were detected that ligated the 5′ tag. These were of three structural families, primarily 2′,5′ ligators, though a unique molecule with a large active center was isolated that made 3′,5′ phosphodiester bonds as its only activity (23). An outline of the secondary structure of a 186-mer 3′,5′ ligator is shown in Figure 1a. 3′ and 5′ terminal deletion analysis, and elimination of internal sequences that were not conserved, reduced this unique RNA (called the class I sequence) from the initial 220 to 112 nucleotides. A definite step toward replicase activity was that the template strand/substrate oligonucleotide could be cut free in a minimized and optimized construct to produce a ribozyme core/free template-junction system that formed phosphodiesters at k_{cat} of 100 min^{-1} [22° (23)].

The next step was to detach the pppG to be ligated from the primer template complex (22). The resulting ribozyme would now bind and add the free nucleotide pppG opposite C in the template, extending the primer at its 3′ end by the templated pG (releasing the pyrophosphate) at 0.3 min^{-1}. In fact, by allowing the template strand to slide on a repetitive pairing sequence linking it to the core, 6 nucleotides could be added in 144 hours [22° (22)]. Most interestingly, primer extension was substantially complementary to the template strand; C opposite G and so on. At this point the ribozyme is, remarkably, replicating the template of a noncovalently bound, small double-stranded primer-template, using nucleoside triphosphates. Replication of an initial GC pair is about 20-fold faster than AU pairs ($\approx$50 vs 2.2 M^{-1} min^{-1}), and this tendency can be seen to carry over into longer products. Accuracy of addition of the first nucleotide varies from 0.71 (U template) to a very impressive 0.996 (C template). Salient errors include the expected UG/GU wobbles, but more subtly, also the insertion of G > A = C opposite template A (in addition to U).

In the latest and most complex selections of all (47), the core sequence above was asked to add several nucleotides (to select better extension) to primer-templates of different sequence in each round (to avoid any sequence preference that would prejudice the replication of RNAs of arbitrary sequence). For this selection, a short primer was leashed to the core by 5′,5′ linkage to a deoxynucleotide tail on the previously selected catalyst. These newly introduced, unusual linkages could be deleted in the end, leaving extension of a noncovalently bound primer-template by the separate catalyst, as had been hoped. A 76-nucleotide 3′ domain was also added to help supply general affinity for the primer template. Remutagenesis and reselection for extension of longer templates gave the current most elaborate (round 18) ribozyme (Figure 1b); it is a 189-mer that requires a 7-mer cofactor (dashed line in Figure 1) to complete an essential paired region in the catalytic core. The 196-nucleotide catalytic unit will add up to 14 templated nucleotides to varied free primer-template complexes in 24 h at 22° (template is solid bold in Figure 1). The linkages are 3′,5′ and therefore identical to biological RNAs. Even more interestingly, the round 18 ribozyme has different qualities than earlier, simpler activities. Now addition of A opposite U (41 M^{-1} min^{-1}) and C opposite G (87 M^{-1} min^{-1}) are the fastest cognate addition reactions, by 8- to 17-fold.

Accuracy (measured again as fraction correct at the first added nucleotide) has increased substantially, varying from 0.92 at template U, to 0.9996 at template C. Only GU and UG wobble pairs now cause salient errors. Comparison with the earlier RNA shows strikingly that differential polymerization rates and polymerization accuracies respond to selection, even when changes are only indirectly favored. Thus the round 18 RNA not only fills in varied primer-template helices with new precision, but its properties suggest that further directed selection can yield more accurate replication, and of longer RNAs. These experiments therefore argue that an RNA replicase composed of RNA is possible, and that by extension, RNA cells could have multiplied using RNA-catalyzed replication.

Exit from an RNA World

The RNA era ends after the invention of translation (templated peptide synthesis), which provides specific peptide competitors that ultimately replace RNAs as major biological agents. If the RNA world immediately preceded our present one, then RNA-like molecules must have devised the first translational system. Accordingly, it must be determined whether RNAs alone are capable of reactions required for translation (116). We enumerate these as follows: RNAs must be capable of coding (perhaps thereby generating some of the current genetic code) in order to act as the templates for specific peptides. To utilize these hypothetical templates, RNAs must activate amino acids in order to drive peptide bond formation thermodynamically, and probably link the activated amino acids to other RNAs (tRNA progenitors) to read the templates. Finally, the hypothesis provides only one agent to actually speed the formation of peptide bonds, so an all-RNA peptidyl transferase must be possible.

Coding

The prediction above that RNA-like molecules invented part of the code is a variant of the "stereochemical" hypothesis for the origin of the genetic code (110). A counterhypothesis is often called the "frozen accident" (15), the idea that the code, although it cannot be changed, has no basis in chemistry.

A recent review of the interaction of mono- to trinucleotides with amino acids finds no consistent or strong pattern in the interactions of these small molecules (56). Instead, such interactions apparently require larger RNAs. This argument again places the origin of the code in the RNA world, when larger functional RNAs first appeared.

The experimental evidence bearing on this possibility originates in the discovery that the group I self-splicing RNA's active center specifically binds the amino acid arginine within its splicing cofactor (G) site (113). Strikingly, bound arginine is in contact with conserved sequences that are arginine codons and only arginine codons (117). This natural RNA could therefore be a molecular fossil bearing the imprint of the code for arginine. Given the means for selecting RNA-amino acid binding sites systematically (24, 81, 101), we can test the more general proposition that other coding triplets (codons and anticodons) were initially parts of primordial binding sites for amino acids. The testable implication is that, given a modern selection for new amino acid binding sites, some of them will have an improbable concentration of cognate triplets within their binding sites. The experimental design is to compare sequences inside and outside the site in the same molecule (57). Cognate sequences outside the binding site are used as controls for the background frequency of triplets. The nucleotides near the bound amino acid can be enumerated by using nucleotide protections, modification interference, and conservation to define the active sequences. Where available, as for arginine, a high-resolution structure is most helpful (112).

My laboratory is proceeding through the standard 20 amino acids, incorporating the work of others (65) where possible, to answer the question about binding site sequences for most amino acids. As a result, there is a substantial published discussion; the most recent as of writing is for phenylalanine binding RNAs [see references in (42)]. I will not repeat this discussion in detail, but instead summarize the present state of the argument.

RNA is proficient at binding varied free amino acids, as would be required for the stereochemical hypothesis. At present, there are defined, newly selected RNA binding sites selected by affinity chromatography for six different amino acids: arginine, glutamine, isoleucine, leucine, phenylalanine, and tyrosine (natural arginine binding sites are not considered). Thus hydrophobic (aliphatic and aromatic) as well as polar amino acids are included. Figure 2 shows one example, which is evidently the most probable RNA site for isoleucine, isolated five times independently in a selection for ile-agarose affinity followed by ile elution (64). The site is a simple internal loop, with a conserved, essential isoleucine codon within the amino acid site (open font in Figure 2).

However, to test the hypothesis, the pooled data on all amino acids should be more definitive than any one example. Taken together, these presently include 26 independently isolated amino acid binding folds. The 26 sites reside in molecules containing 1657 initially randomized nucleotides in all, where 364 of these nucleotides are classified as being "within the site" by the molecular criteria above. Codon triplets within sites are 26.4% of nucleotides (96 of 364 nucleotides);

Figure 2 Example of a selected RNA-amino acid site, specific for L-isoleucine binding. All nucleotides whose identity is important are shown, though the site is usually stabilized by nonspecific base pairs in dotted regions. Outline font marks an AUU isoleucine codon conserved within the binding site in five independently isolated examples of the site (64).

outside the sites in the 26 molecules codons are 10.6% of nucleotides (137 of 1293 total). This concentration of codons in sites would occur with a probability of 3.5 $\times$ 10^{-13} [G test (92)], were the codons and the sites unrelated. The improbable tendency of codons to be within sites is very significant overall, despite the fact that it is not observable for glutamine, leucine, or phenylalanine. That is, arginine, tyrosine, and isoleucine (Figure 2) seem to strongly concentrate codons within their sites, suggesting that their codons are remnants of a primordial period of codon assignment in the RNA world. It therefore seems that a substantial fraction of the code may be a stereochemical consequence of the likely structure of amino acid binding sites. This impression is not dependent on any one amino acid. Eliminating the most codon-biased set of sites (for arginine) still leaves a very highly significant argument from the remainder.

Further, with a sample of this size now in hand, the argument is not unique to the codons. Specific anticodons are 25.1% of total binding site nucleotides and 11.3% of non-site nucleotides. The G test suggests the probability is 2.6 $\times$ 10^{-11} that anticodons and binding site nucleotides are unrelated. Therefore there is an only slightly less significant argument from the total data that some anticodons originated as likely site sequences, as for codons. Again, the overall probability results from the fact that only half the amino acids (isoleucine, leucine, and tyrosine) are potentially significant as individuals. The extension to anticodons as well as codons certainly raises the question: How many sets of triplets would give such correlations? To answer this question, coding assignments were scrambled to make 10^7 new codes using each one of five distinct reassignment rules [randomize the amino acids assigned to modern blocks, randomize the amino acids to arbitrary individual codons, and three others (56)]. In every case, codons and anticodons from new codes tended to appear in observed binding sites much less frequently than from the real genetic code. Thus the tendency for coding sequences to be recovered in amino acid binding sites is a very unusual property of the real genetic code.

To summarize: on the basis of newly selected site structures, there presently appear to be three classes of amino acids. Arginine behaves straightforwardly; it follows the natural example of the group I active center and strongly overrepresents its codons (but not anticodons) within its binding sites. A second anticipated type includes glutamine and phenylalanine; with due allowance for the fact that these are negative results, these amino acids have no detectable relation to their coding sequences. The comment about negative results is not trivial; the glutamine codon CAG turns up in an RNA recognition domain for gln-cyanomethyl ester (61), adjacent to a metal site required for a 5′ self-aminoacylation reaction. This association is a reminder that negative results can always be superceded. However, these amino acids might have been assigned their codons on another basis (such as preservation of protein structures) during the later evolutionary history of the code (27). Finally, the data suggest an unanticipated class of amino acids—isoleucine and tyrosine are bound in sites containing disproportionate numbers of both codon and anticodon sequences (though the potentially complementary triplets are not paired). The genetic code therefore seems, in part, stereochemical. It is based on primordial RNA-like amino acid binding sites whose parts seem to have been adapted to become the mRNAs and tRNAs of a later, more evolved translation apparatus. To return to the argument about RNA activities, the need for primitive RNA-based coding could be met with ordered amino acid binding sites as peptide templates (114), as well as in other ways (97).

Amino Acid Activation

The activation of the carboxyl of an amino acid for translation presently occurs via a universal reaction with ATP to give a mixed anhydride, aminoacyl-AMP:

$$R\text{-}CH(NH_3^+)C(O)O^- + pppA \rightarrow R\text{-}CH(NH_3^+)C(O)O\text{-}pA + PP_i$$

amino acid   ATP   adenylate   pyrophosphate

The reaction is thermodynamically uphill, and the product is kinetically unstable to hydrolysis, even at pH 7 and 0°C. Perhaps for these reasons, an RNA that performed the activation reaction was elusive. However, by using a carboxylic acid instead of an amino acid (to give a more stable product) and a lowered selection pH to preserve product RNAs, at least 6 different RNAs among 2×10^{14} initial sequences were found (59) to activate 3-mercaptopropionic acid using their 5′ triphosphates (the sulfhydryl served for capture of reactive RNAs):

$$HS\text{-}H_2C\text{-}H_2C\text{-}C(O)O^- + pppRNA \rightarrow HS\text{-}H_2C\text{-}H_2C\text{-}C(O)O\text{-}pRNA + PP_i$$

3-mercaptopropionic   RNA   activated carboxylic   pyrophosphate

     acid   transcript   acid

One characterized 114-mer (80 initially randomized positions; an outline of the structure is shown in Figure 3) was shown to also activate phenylalanine and

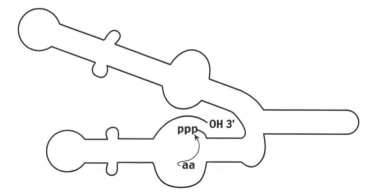

Figure 3 An RNA that activates leucine and phenylalanine (59). The carboxyl of the amino acid reacts at the RNA's 5′ α-phosphate, releasing pyrophosphate and forming an activated amino acid similar to that formed by a protein aminoacyl-tRNA synthetase.

leucine, requiring only Ca^{2+} and pH 4.5. Reaction with leucine was in the average range for selected ribozymes; $k_{cat}/K_m = 23$ M^{-1} min^{-1}. Thus activation of carboxyl groups for general biochemistry, and for translation in particular, is within the RNA catalytic repertoire. However, smaller catalysts with higher pH optima would be welcome.

Aminoacyl-RNA Synthesis

Self-aminoacylating 95-mers (50 initially randomized nucleotides) were first selected (38) using the unique α-amino group of the amino acid as a tag. A hydrophobic label attached there allowed resolution of self-aminoacylating RNA catalysts by reverse phase HPLC. At least 5 independent types (37) of self-aminoacylating RNAs were present in 1.7×10^{14} initial randomized sequences. Rates were about 120 M^{-1} min^{-1} for phenylalanylation from phe-AMP (at $0°$):

$$\text{Phe-AMP} + \text{RNA} \rightarrow \text{RNA2}'(3')\text{-phe} + \text{AMP}$$

phe-adenylate   aminoacyl-RNA

The 5' triphosphate (shown in Figure 4), although not consumed in the reaction, was required for maximum rate and seems to be a part of the active site (37). One self-aminoacylating RNA that met the initial selection is shown in Figure 4a.

Self-aminoacylation by this class of RNAs does not require a large active site, perhaps because an adenylate offers the readily bound AMP as a handle (39). An AMP-phe-dependent aminoacylator can be reduced to only 29 nucleotides in total without loss of reaction velocity (40) (structure in Figure 4b). Only 13 of these 29 nucleotides are nucleotides that were originally randomized. However, deletion has enhanced a secondary reaction in which a second phe-AMP is used to make a phe-phe peptide bond (40). Thus a simple RNA structure of 29 nucleotides can accelerate two of the three chemical steps required for translation.

For the above RNAs, characterized aminoacylation activity did not discriminate between amino acid adenylates. However, selected RNA binding sites for the amino acid phenylalanine can be very specific (42). Thus it is not surprising that a 90-mer self-aminoacylating RNA selected in the same way was highly specific for phenylanine (41) (outline structure in Figure 4c). The chemically related amino acid tyrosine was the only other amino acid used, and rejection of isoleucine, alanine, glutamine, and serine was better than 10,000-fold, a specificity that exceeds even modern protein PheRS. The cognate reaction was also very fast, $k_{cat}/K_m = 60,000$ M^{-1} min^{-1} with k_{cat} at 430 min^{-1} ($0°$, pH 7.25), too fast for manual measurement. This phe-specific RNA therefore accelerates its reaction at least 6×10^7-fold over background, and performs aminoacyl transfer faster than similar proteins. Thus specific aminoacyl-RNA synthesis is a readily accessible RNA function. Interestingly, a subsequent selection for self-aminoacylation [at 36 M^{-1} min^{-1} (87)] with cyanomethyl-phe-biotin gave similar phe sidechain specificity, suggesting that RNA sites for activated phenylalanine with these molecular preferences are easily selected.

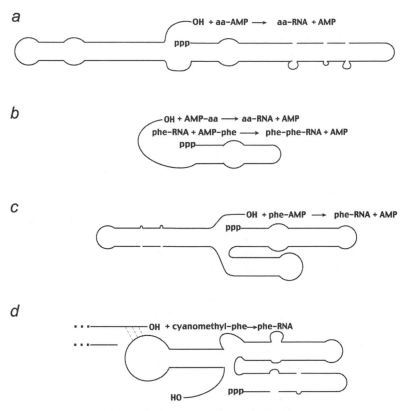

Figure 4   RNAs that accelerate the synthesis of aminoacyl-RNA from activated amino acids. (*a*) Initially selected parental 95-mer (38), deleted to give the RNA in panel (*b*). The aminoacylation reaction, using amino acid adenylate as substrate is shown. (*b*) The 29-mer RNA that aminoacylates itself, then forms a peptide at an accelerated rate, using a second adenylate (40) as shown. (*c*) A highly phenylalanine-specific, rapidly self-aminoacylating 90-mer (41). (*d*) A catalytic RNA (*right*) that aminoacylates a separate acceptor RNA (*left*). The acceptor RNA binds to the catalyst using three base pairs (*dashed lines*), and can be a minihelix, a modified tRNA or a canonical tRNA molecule (87).

Independently isolated RNAs can use other kinds of activated amino acids, for example, a thioester (86). Moreover, a ribozyme can aminoacylate a tRNA-like molecule in *trans* (86). This suggests that the present division of functions into a protein aminoacyl-tRNA synthetase and a tRNA could have been preceded by a translation system in which both molecules were RNAs. This pair of molecules is outlined in Figure 4*d*; the acceptor helix of the acylated RNA (at the left), which can be a tRNA, a deleted tRNA, or a mini-acceptor helix is base-paired to the catalytic portion (on the right). Taken together, the above findings suggest that

RNA-mediated aa-RNA synthesis could have appeared early and had the qualities required for a successful translation system.

The Peptidyl Transferase

The literature of peptide synthesis and related reactions associated with RNAs must be taken cautiously. Claims that in vitro transcripts of ribosomal RNA domains formed peptide bonds (73) were apparently erroneous because a byproduct was mistaken for the peptide (54, 74). A reaction related to peptidyl transfer, amide hydrolysis (16) within a group I RNA active site, was later reduced in magnitude due to mistaken product identification (48). A report of selected ribosome-like peptide formation (121) is apparently true peptide synthesis, but from aa-AMP precursors (31), rather than from the aminoacyl ester precursors used by the ribosome; RNA-catalyzed peptide formation from adenylates had been identified in the interim (40).

However, RNAs can accelerate nucleophilic reactions at carbonyl carbon, the class of reaction (107) performed by the peptidyl transferase. The group I active site accelerates hydrolysis of the ester bond of formyl-met-ACCAAC 5-fold (78). RNAs have been selected that move acyl groups (46) and that aminoacylate themselves (38). Thus it seemed probable that the peptidyl transferase, an arguably related reaction, might be RNA-mediated (38).

It now seems that the peptidyl transferase is an RNA enzyme, as was foreshadowed by biochemical data. Triply-deproteinized (detergent, protease, phenol) large subunit ribosomal RNA carries out a model peptide synthesis reaction (the "fragment reaction") virtually as well as does the ribosome itself (75). Surprisingly, the rigorously treated RNA still contains virtually unaltered ribosomal proteins (53). Removal of the proteins unfolds the RNA and prevents the fragment reaction, probably because the conformation of the active site cannot be maintained in protein-free RNA. Strong evidence that this is the correct interpretation came from the crystallographic structure of the archaeal 50S ribosome (72). Diffusion of a synthetic transition state analogue directed to the peptidyl transferase active site [CCdApPuro (107)] into 50S ribosome crystals revealed one resolved bound molecule, thereby locating the enzymatic center. There was no protein residue close to the site of reaction; the nearest peptide was 18.4 Å away, almost a protein's diameter (72). Thus the peptidyl transferase reaction takes place in an RNA milieu. The ribosome is a ribozyme (12). In prior, independent experiments, selection of RNAs that bind CCdApPuro strongly (108) produced, from randomized sequences, a binding site that had the same nucleotide octamer as was found in contact with CCdApPuro within the highly conserved core of the ribosomal peptidyl transferase itself (72). Thus it is likely that the modern peptidyl transferase was first formed by selection from relatively short randomized RNA sequences by favoring those that could host reacting amino acids, as also occurred in this modern selection (119). While the peptidyl transferase reaction is still the subject of intense discussion (3, 69, 79), the controversy is now at the level of the role of particular protons adjacent to the reacting groups (70, 111). In other words, there is not only a persuasive argument that peptidyl transferase is an RNA reaction, but in addition that the universal modern enzyme is an RNA selected from varied RNA sequences in an RNA world.

Other Essential Matters

I have reckoned (115) that the simplest satisfactory account of an RNA world would justify the appearance of cellularity, replication, translation, and metabolism. The text above deals with replication and translation, leaving cells and metabolism. Membrane-bounded RNA cells or ribocytes are widely regarded as essential for Darwinian evolution (98), because otherwise selection cannot bear on the products of genetic activity. However, whereas membranes made of amphipathic lipids are plausible prebiotic structures (19), the broad and specific modulation of permeability required for cellular life seems to require activities that parallel modern protein pores, channels, and porters. Thus it may be significant that RNAs can be selected to bind directly to phospholipid bilayers, thereby altering their ion permeabilities (55). Even human plasma membranes can be affected. Under different conditions, complexes of RNAs have affinity for such phosphatidylcholine bilayers, and can assemble as large supramolecular assemblies that alter the permeability of liposomes, and ultimately disrupt the integrity of a planar membrane (105). Some mutants of *Saccharomyces* suggest that similar RNAs could exist in modern cells (62). Thus while there is much to be done, there are hints that a ribocyte could exist within a bilayer membrane by modulating its permeabilities using membrane-bound RNAs.

One cannot envision a replicating RNA genome, commanding a translation system within a specifically modified membrane boundary, without also supposing an energy metabolism and a substantial system of anabolic pathways. Cells would otherwise quickly run down and exhaust their supply of precursors. Analysis of modern metabolism suggests that some modern reactions do indeed descend from an RNA ancestor (7, 45). There are also suggestions of long standing that the ribonucleotide cofactors, which supply chemical versatility to modern proteins, are actually a legacy from a pre-existing generation of RNA enzymes (104, 109). In fact, the reactive functional groups of cofactors like CoA and NAD can be synthesized by plausible prebiotic chemistry (14, 18, 52). In addition, ribozyme action can assemble related products into the full-fledged cofactors CoA, NAD, and FAD (35). Finally, for CoA, *trans*-acting ribozymes will place the cofactor covalently at the 5′ terminus of any RNA with a specific terminal sequence (43). Alternatively, group I-like activities might covalently link cofactors to RNAs (9), although this is slower and appears to require a more complex catalyst containing hundreds of nucleotides. Cofactor-RNAs can subsequently use CoA-SH to attack activated acyl groups, generating acetyl-CoA and butyryl-CoAs (43). Thus a method of shuttling carbon fragments, as well as all other reactions of acyl-CoAs, may have been accessible in the RNA world. Because these latter experiments potentially generalize to all phosphorylated cofactors, these data suggest that a substantial metabolism might be built on the four standard nucleotides plus the likely ancient ribonucleotide cofactors. The appearance of redox properties, already available for

CoA-RNA and prospectively available from NAD- and FAD-RNAs, is particularly welcome. These suggest that the absence of redox properties in modern RNA is not a bar to an oxidative ribocyte energy metabolism.

Probability of the Ribocyte

The argument that a ribocyte existed in some form is substantial. In particular, an RNA translation apparatus is a prediction based on the idea that ribocytes were our immediate progenitors. Ribocytes were not necessarily present at life's origin; instead, the experimental evidence only suggests that they devised the modern translation system, giving rise to the present nucleoprotein world. Thus everything said below is consistent with the existence of a pre-RNA world in which the problem (50) of synthesis of an RNA-like molecule itself was solved.

This view now rests on many independent items of evidence. I extend a previous argument (116) using a relevant form of Bayes' theorem. Bayes' is one algebraic step from the definition of conditional probability, and is thus a fundamental theorem of the probability calculus. For this use, it can be written to specify a rational way to update beliefs about an RNA world in view of experiments whose outcome is predicted by the RNA world hypothesis.

$$\frac{P(RNA\ world\ |\ Experiments)}{P(RNA\ world)} = \frac{P(Experiments\ |\ RNA\ world)}{P(Experiments)}.$$

The P are probabilities and the vertical bars mean "given." The left-hand side is the guidance we seek, the ratio of the probability of the RNA world given certain results, to the probability we would have assigned before knowing the results. The right-hand term is the probability of the experimental results given that an RNA world did exist, divided by the *a priori* probability that the experiments would have given the predicted results. Supposing that an RNA world did exist predicts an RNA RNA replicase, the four translational reactions, membrane RNAs, and RNA-based metabolism, and other required ribocyte functions. Accordingly, had there been an RNA world the probability of existence of these RNA reactions is 1 (numerator on the right). If n such experimental predictions are borne out,

$$\frac{P(RNA\ world\ |\ Experiments)}{P(RNA\ world)} = \frac{1}{P^n}.$$

Thus the factor by which the improbability of an RNA world should be increased depends on the *a priori* probability we could independently find n arbitrary RNA activities (all are assigned the same probability P for simplicity). Bayes' abstraction is useful because it converts an almost unthinkable question into a more tangible one. The less likely it appears that RNA would have previously unknown replicative, membrane, translational, and biochemical reactions, the more our respect for the existence of the RNA world should be elevated by their discovery. For example, if we conservatively take $n = 7$ for the list of major reactions confirmed in this review, but think one of three arbitrarily selected reactions could be performed by

RNA in any case, then studies thus far have increased the likelihood of the RNA world $1/0.333^7 \approx 2000$-fold. Moreover, the exponent n will increase if other reactions required for a ribocyte can be confirmed, and the likelihood of succession from an RNA world to the present one would accordingly rapidly increase.

This analysis also circumvents a recurrent gibe: "RNA appears unique only because an experimental method for examining it exists. If we had the means to do the experiments, carbohydrates or lipids would be candidates as impressive as RNA for a primordial biological role." The first and best counter-argument is that it is not accidental that RNAs are the focus of our arguments. Their ability to replicate is essential for their ancient biological role, as well as being required for modern selection-amplification that uses them. The absence of replication mechanisms for lipids is therefore not a defect to be taken lightly. However, elevation of the probability of the ribocyte via Bayes' theorem operates independently of a skeptical predilection for other molecules. Eventually, the Bayesian argument can become so persuasive that hypothetical alternatives are substantially reduced in potential significance, though of course not ruled out.

In sum, the argument from selection can be made more pointed than described above (in the section on **RNA Versatility**): RNA performs, at least in rudimentary form, the very reactions most essential for cellular life. This seems very unlikely to be entirely accidental, especially when all that is known is considered together. Most important, these results are necessarily relevant to the probability of a ribocyte.

Phenotype of the Ribocyte

Supposing that a ribocyte probably existed, I now summarize data gathered from selection-amplification and bearing on some cellular properties. The reader will appreciate the positives and negatives of such an effort. On one hand, such a Frankenstein's monster has the unprecedented virtue (among speculations about the early history of life on Earth) of being composed of ingredients that are known to exist. As is shown below, there is ancillary evidence that such a composite may be more than casually accurate. On the other hand, the reader will realize that only diffusely illuminated spots on a vast dark plain of possibilities are known. Errors will arise from generalizing the properties of the few RNAs that are well characterized. Therefore at this early stage, only properties already multiply confirmed should be considered; unique RNA activities should be avoided.

The Ribocyte Emerged by Selection from Random Sequences

There is a remarkable property of the data on translation, evident since early studies on the peptidyl transferase (119). Modern selections sometimes reproduce the route that biology apparently took. The most striking example is the recurrence of the core octamer of the peptidyl transferase within the sites from RNAs selected to bind CCdApPuro (108). CCdApPuro models the peptidyl transferase reaction; thus the need to embrace reacting aminoacyl-nucleotides appears to have shaped the peptidyl transferase active site. However, the tendency for coding sequences (codons and anticodons) to appear within oligoribonucleotide sites selected for amino acid affinity (42) is another example of the same kind. These tendencies have been explained as exiguity, i.e., sparseness or fewness (119); that is, as an expression of the fact that there are only a few molecular solutions to the problem presented. Therefore, biological selection and modern selection can sometimes find the same solutions. But how is it possible that only a few structures are eligible for selection?

It is nearly axiomatic that a molecular property chosen for selection-amplification must be as simple as possible. Some of the art of designing selections lies in simplifying the task for which the randomized sequences are to be screened. Only one function and therefore one active site must be required. This imperative comes from the limit on the size of the sequence that may be selected in one experiment (118). Cryptic double simultaneous selections can demand more selected nucleotide sequences and therefore more initial molecules than could possibly be available in a practical initial pool. This problem is even more pressing than it first appears. Two active sites will not be independent, but each will likely interfere with the other's fold, which will make it yet more difficult to simultaneously select more than one function or site per molecule. Selected RNAs containing multiple sites therefore demand specialized approaches (11).

Thus the rationale for exiguity is likely that the simplest sites, having the fewest specified nucleotides, will be selected because they are the most numerous. Because these smallest solutions are few, they can be repetitively found. In modern selections, the hammerhead ribozyme is apparently the simplest motif for cleavage at specified sequences (88), and therefore may have emerged repetitively when RNA-catalyzed RNA cleavage was advantageous. However, the broad conservation of the translation apparatus makes it clear that its origin predates the separation of the three domains of life. Thus with regard to present translation, there was likely one effective origin.

The advantage of being the simplest molecular solution to a problem cannot have persisted indefinitely. As the ribocyte became more competent, it would probably have become able to afford the most functional solutions rather than the simplest ones. Nevertheless, at the time of the appearance of the translation apparatus, major components such as the genetic code and the peptidyl transferase were still those that are the most readily selected. Therefore the process of selection from pools of arbitrary sequences must have still been a major route to evolutionary novelty in the mid-RNA world. Accordingly, the ribocyte contained or had access to suitably sized RNA pools of arbitrary sequence. To put this point differently, paralogous derivation of new genes, as in a modern organism, was frequently not possible because the precursor genes did not yet exist. Therefore the principal route to new gene functions had a different logic; to wit, selection from largely functionless reservoirs.

This may intuitively seem a cumbersome route for evolving new functions. For one thing, given that a cell membrane was mandatory for Darwinian evolution, the RNA pool from which new functions would come would likely have been internal to the ribocyte. However, calculations suggest (118) that the many folds available to individual RNA molecules unexpectedly reduce the number of RNA molecules required for the evolution of early ribozymic functions. For purposes of estimation of the smallest pool that could support an RNA biota, we seek the smallest group of RNA molecules that could give rise to useful RNA catalysts. If the small RNA (40) that self-aminoacylates and makes peptide bonds is tentatively taken as a model catalyst, this suggests that RNAs should have at least 11 to 19 specific nucleotides in their active sites. This in turn implies the presence of an RNA pool having the equivalent of zeptomoles (602 molecules; ≈33 attograms or more) of randomized 80-mers (118). If only zeptomoles (that is, less than one attomole, 602,000 molecules; ≈33 femtograms) of randomized RNAs are required to begin evolving catalysts, the RNA world and its ribocytes could have emerged quickly. Early ribocytes potentially require less RNA than found in a modern bacterium. Experimental confirmation of this zeptomole world for RNA would therefore provide strong support for the hypothesis of a ribocyte that evolves by drawing new activities from its portable pool.

The Principal Agents of the Ribocyte are Comparatively Small Molecules

An impressive convergence of different kinds of evidence limits emergent ribocytes to RNA active sites containing required 15 to 30-mers (but perhaps embedded in somewhat longer molecules with unspecified sequences). The error rates of RNA-mediated replication, especially early on, limit the size of the molecule that can be accurately reproduced (49). Observed accuracies (47) make sequences whose activity depends on the identity of several tens of nucleotides probable (as long as the effects of mutation are moderate). Second, although the origins of RNA are still obscure (50), the untemplated polymerization of activated nucleosides seems a likely source of early RNAs. Model processes of this type produce sequences in this same size range, a few tens of nucleotides (25). Third, the number of molecules submitted to a selection limits the size of the active site that is likely to occur (118). Again, the amount of RNA available to select the functions of a ribocyte is not likely to have been enough to fix more than a few tens of nucleotides in the active sequences of RNA molecules.

We can even say something about the form of these active sites. They will likely be constructed by folding together separated shorter specified sequences embedded in longer molecules. Thus they will contain intercalated non-specific sequences as material for further evolution or oligomerization of the initially evolved molecules. This prediction seems relatively sure because of the statistics of folding. It should be much easier to find an active site with any number of specified nucleotides if it is composed of separated sequence pieces that may be found at many places and folded together (118), instead of requiring contiguous nucleotides in the fold, found in few places. When selection itself is better understood, its fundamental properties will likely further constrain the nature of the RNAs available to the ribocyte.

While catalysts might be combined in a large genomic fragment, it is more likely at first that the most primitive arrangement was adopted. Thus the RNAs carrying the active sites probably existed separately, though there might have been more stable archival double-stranded RNA (93), as well as exposed and active single-stranded forms of each gene. The assumption of simplicity is most congenial with large numbers of small RNAs apportioned statistically during cell division. This uses the inevitable small size of the first catalysts advantageously, to avoid the need for a specific partitioning machine.

The Ribocyte Lived Slowly

Modern ribozymes can be sophisticated, deploying hundreds of nucleotides to construct active sites that accelerate their reactions as much as do some protein enzymes, or to carry out a succession of steps, each with its own demanding stereochemistry. Both are well exemplified by the group I RNA self-splicing active center (34), which performs two sequence-specific reactions, transesterification chemistry at 350 min^{-1}, and attacks an oligonucleotide substrate at $k_{cat}/K_m = 9 \times 10^7$ M^{-1} min^{-1}. However, this performance is probably irrelevant to the emergence of the ribocyte. Laboratory selection-amplification uses larger pools that should yield more capable active sites than those that would be plausible in an early RNA world (118). Nevertheless, much experience suggests that newly selected ribozymes initially react with small substrates at $k_{cat}/K_m = 10^2 M^{-1} min^{-1}$ or less, as reviewed above. However, selection for faster rates can be successful (1). This is not necessarily a comment on the (in)competence of RNA, but perhaps a more general property of newly evolved catalysts. This difference of about a million-fold between evolved and selected RNAs (and more if modern proteins are compared) seems likely to be reflected in all cellular properties. For example, consider cell division: To divide rapidly, rapid catalysis would be required of every RNA on the critical path through the cell cycle. Therefore a nascent ribocyte probably grew and divided much more slowly than most contemporary laboratory organisms.

Taking cell division as an example, the cell cycle cannot be faster than the period required to replicate all elements of the genome. A state-of-the-art RNA replicating activity adds up to 14 nucleotides to a primer in 24 h at 22° (47), though the majority of extended products are shorter. Because selected RNAs of this kind have universally slowed polymerization as they proceeded, and because strand separation is not attempted in these experiments, it seems fair to take the average time of replication of one of our 15 to 30 essential-nucleotide catalysts as "days". This is notably close to chemical limits. At the pH of the RNA replicase reactions (pH 8.5), the hydrolytic rate for ribonucleotide linkages in normal RNA molecules ranges about an order of magnitude, up to 3.6×10^{-5} min^{-1} [data of Soukup & Breaker (93) at 23°]. If there are 20 linkages to crucial ribonucleotides where hydrolytic breakage inactivates, RNAs waiting to be replicated will only survive for an average of 1 to 12 days. Present 24-h replicase assays are said to be limited by decay of the RNA (47), so these calculated times appear realistic. Correcting to pH 7 at 23° (120), replication will have to be finished in ≈2 to 24 weeks to stay ahead of RNA decay. Three conclusions are worth emphasis. First, the present RNA RNA replicase activity is close to the biological break-even point for overall velocity. Second, this is a fourth argument that the ribocyte's RNA agents were short sequences, with no more than tens of essential nucleotides in their active sites (see section on principal agents). Lastly, the early ribocyte is likely to have divided in a time appropriately measured in weeks. Even this rate will require, on simplest assumptions, that more than half the RNA in cells be replicase and replicase complements.

The Ribocyte Used a Constant Set of Nucleotides that Resemble Modern Ones

The ribocyte's basic activities probably relied on nucleotides resembling the present four nucleotides. The properties of polynucleotides can likely vary within only narrow limits to produce the favorable structural and solubility properties required for catalysis (8). Because many active folds can be selected containing deoxynucleotides (10), we cannot be sure that the primordial nucleotide set was all ribonucleotides. However, usually the general substitution of deoxynucleotides for ribonucleotides inactivates a selected motif (30), though a few exceptions for simple G-quartet binding sites are known (100). In addition, structures selected using nonstandard bases usually do not tolerate their replacement with normal ones (99). Even mixed structures with similar bases are unlikely. Were there more than one nucleotide that paired like, say, adenine, this would probably be tolerable only if the variant nucleotide (A') is very rare, because equal amounts of A and A' give rise to a myriad of sequences of which only a tiny fraction would have any given structure. For a molecule that has 20 positions required to be adenine-like, the equal mixture reproduces the A/A' sequence only about once in a million molecules. Taking all this together, the need to reproducibly place a complex set of chemical features in space (to reliably construct active sites) probably means that only a few positions near an RNA active site will tolerate changes in the bases or backbone. The ribocyte therefore must make its ribopolynucleotides of a constant set of building blocks, preserving both backbone and bases. In addition, omission of even one of the modern four nucleotides decisively impairs the activity of folds that can be selected using only the other three nucleotides (83). Thus a nucleotide set as complex as the present four is probably a necessity, and it has been also suggested that the optimum number is four nucleotides (96). However, up to six base pairs (12 nucleotides) may be possible (6).

Finally, the peptidyl transferase and the genetic code, executed with modern ribonucleotides, appear to resemble the most probable structures in the mid-RNA world, when translation appeared. This suggests the use of the standard four nucleotides. To assemble these separate observations: the ribocyte at the point we are considering it was sufficiently evolved to provide substantially pure supplies of nucleic acid monomers, each having a unique sugar-phosphate and base-pairing pattern, and four of its monomers must have resembled the modern four. There is a parallel argument [see (116)] that the ribocyte's translation apparatus itself also resembled the modern one.

The Ribocyte was a Complex Cell

A ribocyte likely decorated its membranes with RNAs that controlled the shape and permeability of the cellular boundary. It might have varied these properties as needed, for example, to serve a cell cycle. It had a metabolism that necessarily provided everything needed for Darwinian evolution, at a minimum activated nucleotides for replication, and ultimately activated amino acids for translation. To serve this end, RNA cells performed a metabolism that at least extended to the limits set by catalytic RNA folds with four nucleotides, which can now be seen to provide sites for highly varied small molecules. In addition, it could probably deploy a set of reactive, specially functionalized ribonucleotide cofactors to perform reactions between such varied bound molecules. In particular, there is promise of reactions like those required for energy metabolism, difficult to imagine using only the four standard modern ribonucleotides.

The ribocyte was probably complex in another sense. Because of the substantial error rate of RNA copying (47) and potentially inexact segregation, ribocytes will not have a unique genome. Instead they comprise something like a quasispecies (21), though a quasispecies is a mutation-selection steady-state, and ribocytes may not have always existed in such a steady-state. In any case, the ribocyte will have a distribution of quite varied genotypes whose centroid we can call the genotype. In addition, newly selected catalysts often perform several related reactions and are not highly specific (43). This tendency should be particularly pronounced among the earliest, simplest RNAs. For example, a self-aminoacylating RNA acquired enhanced peptide synthesis as it was progressively simplified by deletion of structural elements (40). With its variable genome and gene products with multiple activities, the ribocyte presented genetic variance greater than any contemporary organism. Because the rate of evolution is proportional to the genetic variance of fitness (20, 26), the ribocyte evolved very quickly compared to our intuition based on familiar creatures. Even successive generations might change noticeably during the early history of the RNA cell. In so saying, I have mixed feelings. We can presently know the ribocyte only while it remains simple, and it can quickly leave simplicity behind.

The *Annual Review of Genetics* is online at http://genet.annualreviews.org

LITERATURE CITED

1. Bartel DP, Szostak JW. 1993. Isolation of new ribozymes from a large pool of random sequences. *Science* 261:1411–18
2. Basu S, Rambo RP, Strauss-Soukup J, Cate JH, Ferre-D'Amare AR, et al. 1998. A specific monovalent metal ion integral to the AA platform of the RNA tetraloop receptor. *Nat. Struct. Biol.* 5:986–92
3. Batfield MA, Dahlberg AF, Schulmeister

U, Dorner S, Barta A. 2001. A conformational change in the ribosomal peptidyl transferase center upon active/inactive transition. *Proc. Natl. Acad. Sci. USA* 98: 10096–101

4. Been MD, Cech TR. 1988. RNA as an RNA polymerase: net elongation of an RNA primer catalyzed by the *Tetrahymena* ribozyme. *Science* 239:1412–16

5. Been MD, Wickham GS. 1997. Self-cleaving ribozymes of hepatitis delta virus RNA. *Eur. J. Biochem.* 247:741–53

6. Benner SA, Battersby TR, Eschgfäller B, Hutter D, Kodra JT, et al. 1998. Redesigning nucleic acids. *Pure Appl. Chem.* 70:263–66

7. Benner SA, Ellington AD, Tauer A. 1989. Modern metabolism as a palimpsest of the RNA world. *Proc. Natl. Acad. Sci. USA* 86:7054–58

8. Benner SA. 1999. Did the RNA world exploit an expanded genetic alphabet? See Ref. 27a, pp. 163–81

9. Breaker RR, Joyce GF. 1995. Self-incorporation of coenzymes by ribozymes. *J. Mol. Evol.* 40:551–58

10. Breaker RR. 2000. Molecular biology color: making catalytic DNAs. *Science* 290:2095–96

11. Burke DH, Willis JH. 1998. Recombination, RNA evolution, and bifunctional RNA molecules isolated through chimeric SELEX. *RNA* 4:1165–75

12. Cech TR. 2000. Structural biology. The ribosome is a ribozyme. *Science* 289:878–79

13. Chowrira BM, Burke JM. 1992. Extensive phosphorothioate substitution yields highly active and nuclease-resistant hairpin ribozymes. *Nucleic Acids Res.* 20: 2835–40

14. Cleaves HJ, Miller SJ. 2001. The nicotinamide biosynthetic pathway is a byproduct of the RNA world. *J. Mol. Evol.* 52:73–77

15. Crick FHC. 1968. The origin of the genetic code. *J. Mol. Biol.* 38:367–79

16. Dai X, De Mesmaeker A, Joyce GF. 1995.

Cleavage of an amide bond by a ribozyme. *Science* 267:237–40

17. Doudna JA, Couture S, Szostak JW. 1991. A multisubunit ribozyme that is a catalyst of and template for complementary strand RNA synthesis. *Science* 251:1605–9

18. Dowler MJ, Fuller WD, Orgel LE, Sanchez RA. 1970. Prebiotic synthesis of propiolaldehyde and nicotinamide. *Science* 169:1320–21

19. Dworkin JP, Deamer DW, Sandford AS, Allamandola LJ. 2001. Self-assembling amphiphilic molecules: synthesis in simulated interstellar/precometary ices. *Proc. Natl. Acad. Sci. USA* 98:815–19

20. Edwards AWF. 1994. The fundamental theorem of natural selection. *Biol. Rev.* 69:443–74

21. Eigen M, Schuster P. 1977. The hypercycle. A principle of natural self-organization. Part A: Emergence of the hypercycle. *Naturwissenschaften* 64:541–65

22. Ekland EH, Bartel DP. 1996. RNA-catalysed RNA polymerization using nucleoside triphosphates. *Nature* 382:373–76

23. Ekland EH, Szostak JW, Bartel DP. 1995. Structurally complex and highly active RNA ligases derived from random RNA sequences. *Science* 269:364–70

24. Ellington A, Szostak JW. 1990. In vitro selection of RNA molecules that bind specific ligands. *Nature* 346:818–22

25. Ferris JP, Hill AR, Liu R, Orgel LE. 1996. Synthesis of long prebiotic oligomers on mineral surfaces. *Nature* 381:59–61

26. Fisher RA. 1958. The fundamental theories of natural selection. In *The Genetical Theory of Natural Selection,* ch. II. New York: Dover Publ. 2nd ed.

27. Freeland SJ, Hurst LD. 1998. The genetic code is one in a million. *J. Mol. Evol.* 47:238–48

27a. Gesteland RF, Cech TR, Atkins JF, eds. 1999. *The RNA World.* Cold Spring Harbor, NY: Cold Spring Harbor Lab. Press. 2nd ed.

28. Gilbert BA, Sha M, Wathen ST, Rando

RR. 1997. RNA aptamers that specifically bind to a K *Ras*-derived farnesylated peptide. *Bioorg. Med. Chem.* 5:1115–22

29. Gilbert W. 1986. The RNA world. *Nature* 319:618

30. Gold L, Polisky B, Uhlenbeck O, Yarus M. 1995. Diversity of oligonucleotide functions. *Annu. Rev. Biochem.* 64:763–97

31. Gottlieb RL, Cech TR, Cui Z, Sun L, Zhang B. 2001. *Adenylated amino-acids reprise their central role as intermediates in multiple natural and selected peptidyl transferase systems.* In Abstr. Meet. Cent. Extreme Early Evol. Ind. Univ., RNA-based life, Nov. 15–17, p. 49

32. Green R, Szostak JW. 1992. Selection of a ribozyme that functions as a superior template in a self-copying reaction. *Science* 258:1910–15

33. Guerrier-Takada C, Gardiner K, Marsh T, Pace N, Altman S. 1983. The RNA moiety of ribonuclease P is the catalytic subunit of the enzyme. *Cell* 35:849–57

34. Herschlag D, Cech TR. 1990. Catalysis of RNA cleavage by the *Tetrahymena thermophila* ribozyme. 1. Kinetic description of the reaction of an RNA substrate complementary to the active site. *Biochemistry* 29:10159–71

35. Huang F, Bugg CW, Yarus M. 2000. RNA-catalyzed CoA, NAD, and FAD synthesis from phosphopantetheine, NMN, and FMN. *Biochemistry* 39:15548–55

36. Huang F, Yang Z, Yarus M. 1998. RNA enzymes with two small-molecule substrates. *Chem. Biol.* 5:669–78

37. Illangasekare M, Kovalchuke O, Yarus M. 1997. Essential structures of a self-aminoacylating RNA. *J. Mol. Biol.* 274:519–29

38. Illangasekare M, Sanchez G, Nickles T, Yarus M. 1995. Aminoacyl-RNA synthesis catalyzed by an RNA. *Science* 267:643–47

39. Illangasekare M, Yarus M. 1997. Small-molecule-substrate interactions with a self-aminoacylating ribozyme. *J. Mol. Biol.* 268:631–39

40. Illangasekare M, Yarus M. 1999. A tiny RNA that catalyzes both aminoacyl-RNA and peptidyl-RNA synthesis. *RNA* 5:1482–89

41. Illangasekare M, Yarus M. 1999. Specific, rapid synthesis of Phe-RNA by RNA. *Proc. Natl. Acad. Sci. USA* 96:5470–75

42. Illangasekare M, Yarus M. 2002. Phenylalanine-binding RNAs and genetic code evolution. *J. Mol. Evol.* 54:298–311

43. Jadhav VR, Yarus M. 2002. Acyl-CoAs from coenzyme ribozymes. *Biochemistry* 41:723–29

44. Jaschke A. 2001. RNA-catalyzed carbon-carbon bond formation. *Biol. Chem.* 382:1321–25

45. Jeffares DC, Poole AM, Penny D. 1998. Relics from the RNA world. *J. Mol. Evol.* 46:18–36

46. Jenne A, Famulok M. 1998. A novel ribozyme with ester transferase activity. *Chem. Biol.* 5:23–34

47. Johnston WK, Unrau PJ, Lawrence MS, Glasner ME, Bartel DP. 2001. RNA-catalyzed RNA polymerization: accurate and general RNA-templated primer extension. *Science* 292:1319–25

48. Joyce GF, Dai X, De Mesmaeker A. 1996. Amide cleavage by a ribozyme: correction. *Science* 272:18–19

49. Joyce GF, Orgel LE. 1999. Prospects for understanding the origin of the RNA world. See Ref. 27a, pp. 49–77

50. Joyce GF. 1989. RNA evolution and the origins of life. *Nature* 338:217–24

51. Kazakov S, Altman S. 1992. A trinucleotide can promote metal ion-dependent specific cleavage of RNA. *Proc. Natl. Acad. Sci. USA* 89:7939–43

52. Keefe AD, Newton GL, Miller SL. 1995. A possible prebiotic synthesis of pantetheine, a precursor to coenzyme A. *Nature* 373:683–85

53. Khaitovich P, Mankin AS, Green R, Lancaster L, Noller HF. 1999. Characterization of functionally active subribosomal particles from *Thermus aquaticus*. *Proc. Natl. Acad. Sci. USA* 96:85–90

54. Khaitovich P, Tenson T, Mankin AS, Green R. 1999. Peptidyl transferase activity catalyzed by protein-free 23S ribosomal RNA remains elusive. *RNA* 5:605–8

55. Khvorova A, Kwak TG, Tamkun M, Majerfeld I, Yarus M. 1999. RNAs that bind and change the permeability of phospholipid membranes. *Proc. Natl. Acad. Sci. USA* 96:10649–54

56. Knight RD, Landweber L, Yarus M. 2002. Tests of a stereochemical genetic code. In *Translation Mechanisms*, ed. J Lapointe, L Brakier-Gingras. Houston: Landes Biosci. In press

57. Knight RD, Landweber LF. Rhyme or reason: RNA-arginine interactions and the genetic code. *Chem. Biol.* 5:R215–20

58. Kruger K, Grabowski PJ, Zaug AJ, Sands J, Gottschling DE, Cech TR. 1982. Self-splicing RNA: autocxcision and autocyclization of the ribosomal RNA intervening sequence of Tetrahymena. *Cell* 31:147–57

59. Kumar RK, Yarus M. 2001. RNA-catalyzed amino acid activation. *Biochemistry* 40:6998–7004

60. Lazcano A, Miller SL. 1996. The origin and early evolution of life: prebiotic chemistry, the pre-RNA world, and time. *Cell* 85:793–98

61. Lee N, Suga H. 2001. Essential roles of innersphere metal ions for the formation of the glutamine binding site in a bifunctional ribozyme. *Biochemistry* 40:13633–43

62. MacIntosh GC, Bariola PA, Newbigin E, Green P. 2001. Characterization of Rny1, the *Saccharomyces cerevisiae* member of the T2 RNase family of RNases: unexpected functions for ancient enzymes? *Proc. Natl. Acad. Sci. USA* 98:1018–23

63. Majerfeld I, Yarus M. 1994. An RNA pocket for an aliphatic hydrophobe. *Nat. Struct. Biol.* 1:287–92

64. Majerfeld I, Yarus M. 1998. Isoleucine: RNA sites with associated coding sequences. *RNA* 4:471–78

65. Mannironi C, Scherch C, Fruscoloni P, Tocchini-Valentini GP. 2000. Molecular recognition of amino acids by RNA aptamers: the evolution into an L-tyrosine binder of a dopamine-binding RNA motif. *RNA* 6:520–27

66. Michel F, Ferat JL. 1995. Structure and activities of group II introns. *Annu. Rev. Biochem.* 64:435–61

67. Mojzsis SJ, Arrhenius G, McKeegan KD, Harrison TM, Nutman AP, Friend CRL. 1996. Evidence for life on Earth before 3800 million years ago. *Nature* 384:55–59

68. Mörl M, Niemer I, Schmelzer C. 1992. New reactions catalyzed by a group II intron ribozyme with RNA and DNA substrates. *Cell* 70:803–10

69. Muth GW, Chen L, Kosek AB, Strobel SA. 2001. pH-dependent conformational flexibility within the ribosomal peptidyl transferase center. *RNA* 7:1403–15

70. Muth GW, Ortoleva-Donnelly L, Strobel SA. 2000. A single adenosine with a neutral pKa in the ribosomal peptidyl transferase center. *Science* 289:947–50

71. Nakano S, Chadalavada DM, Bevilaqua PC. 2000. General acid-base catalysis in the mechanism of a hepatitis delta virus ribozyme. *Science* 287:1493–97

72. Nissen P, Hansen J, Ban N, Moore PB, Steitz TA. 2000. The structural basis of ribosome activity in peptide bond synthesis. *Science* 289:920–30

73. Nitta I, Ueda T, Watanabe K. 1998. Possible involvement of *Escherichia coli* 23S ribosomal RNA in peptide bond formation. *RNA* 4:257–67

74. Nitta I, Ueda T, Watanabe K. 1999. Retraction. *RNA* 5:707

75. Noller HF, Hoffarth V, Zimniak L. 1992. Unusual resistance of peptidyl transferase to protein extraction procedures. *Science* 256:1416–19

76. Orgel L. 1968. Evolution of the genetic apparatus. *J. Mol. Biol.* 38:381–93

77. Pace NR. 2001. The universal nature of biochemistry. *Proc. Natl. Acad. Sci. USA* 98:805–8

78. Piccirilli JA, McConnell TS, Zaug AJ, Noller HF, Cech TR. 1992. Aminoacyl esterase activity of the Tetrahymena ribozyme. *Science* 256:1420–24

79. Polacek N, Gaynor M, Yassin A, Mankin AS. 2001. Ribosomal peptidyl transferase can withstand mutations at the putative catalytic nucleotide. *Nature* 411:498–501

80. Rastogi T, Collins RA. 1998. Smaller, faster ribozymes reveal the catalytic core of Neurospora VS RNA. *J. Mol. Biol.* 227:215–24

81. Roberson D, Joyce J. 1990. Selection in vitro of an RNA enzyme that specifically cleaves single-stranded DNA. *Nature* 344:467–68

82. Robertson MP, Ellington AD. 2000. Design and optimization of effector-activated ribozyme ligases. *Nucleic Acids Res.* 28:1751–59

83. Rogers J, Joyce GF. 2001. The effect of cytidine on the structure and function of an RNA ligase ribozyme. *RNA* 7:395–404

84. Rupert PB, Ferre-D'Amare AR. 2001. Crystal structure of a hairpin ribozyme-inhibitor complex with implications for catalysis. *Nature* 410:780–86

85. Saghatelian A, Kokobayashi Y, Soltani K, Ghadiri MR. 2001. A chiroselective peptide replicator. *Nature* 409:797–801

86. Saito H, Kourouklis D, Suga H. 2001. An in vitro evolved precursor tRNA with aminoacylation activity. *EMBO J.* 20:1797–806

87. Saito H, Watanabe K, Suga H. 2001. Concurrent molecular recognition of the amino acid and tRNA by a ribozyme. *RNA* 7:1867–78

88. Salehi-Ashtiani K, Szostak JW. 2001. In vitro evolution suggests multiple origins for the hammerhead ribozyme. *Nature* 414:82–84

89. Sengle G, Eisenfuhr A, Arora PS, Nowick JS, Famulok M. 2001. Novel RNA catalysts for the Michael reaction. *Chem. Biol.* 8:459–73

90. Shan S, Kravchuk AV, Piccirilli JA, Herschlag D. 2001. Defining the catalytic metal ion interactions in the Tetrahymena ribozyme reaction. *Biochemistry* 40:5161–71

91. Shih IH, Been MD. 2001. Involvement of a cytosine side chain in proton transfer in the rate-determining step of ribozyme self-cleavage. *Proc. Natl. Acad. Sci. USA* 98:1489–94

92. Sokal RR, Rohlf FJ. 1995. *Biometry: The Principles and Practice of Statistics in Biological Research.* New York: Freeman

93. Soukup GA, Breaker RR. 1999. Relationship between internucleotide linkage geometry and the stability of RNA. *RNA* 5:1308–25

94. Soukup GA, DeRose EC, Koizumi M, Breaker RR. 2001. Generating new ligand-binding RNAs by affinity maturation and disintegration of allosteric ribozymes. *RNA* 7:524–36

95. Sussman D, Nix JC, Wilson C. 2000. The structural basis for molecular recognition by the vitamin B12 RNA aptamer. *Nat. Struct. Biol.* 7:53–57

96. Szathmary E. 1992. What is the optimum size for the genetic alphabet? *Proc. Natl. Acad. Sci. USA* 89:2614–18

97. Szathmary E. 1999. The origin of the genetic code: amino acids as cofactors in an RNA world. *Trends Genet.* 15:223–29

98. Szostak JW, Bartel DP, Luisi PL. 2001. Synthesizing life. *Nature* 409:387–90

99. Tarasow TM, Tarasow BL, Eaton SE. 1997. RNA-catalysed carbon-carbon bond formation. *Nature* 389:54–57

100. Travascio P, Bennet AJ, Wang DY, Sen D. 1999. A ribozyme and a catalytic DNA with peroxidase activity: active sites versus cofactor-binding sites. *Chem. Biol.* 6:779–87

101. Tuerk C, Gold L. 1990. Systematic evolution of ligands by exponential enrichment: RNA ligands to bacteriophage T4 DNA polymerase. *Science* 249:505–10

102. Uhlenbeck OC. 1987. A small catalytic oligoribonucleotide. *Nature* 328:596–600

103. Unrau PJ, Bartel DP. 1998. RNA-catalysed nucleotide synthesis. *Nature* 395:260–63

104. Visser CM, Kellogg RM. 1978. Bioorganic chemistry and the origin of life. *J. Mol. Evol.* 11:163–68

105. Vlassov A, Khvorova A, Yarus M. 2001. Binding and disruption of phospholipid bilayers by supramolecular RNA complexes. *Proc. Natl. Acad. Sci. USA* 98:7706–11

106. Wang DY, Sen D. 2001. A novel mode of regulation of an RNA-cleaving DNAzyme by effectors that bind to both enzyme and substrate. *J. Mol. Biol.* 310:723–34

107. Welch M, Chastang J, Yarus M. 1995. An inhibitor of ribosomal peptidyl transferase using transition-state analogy. *Biochemistry* 34:385–90

108. Welch M, Majerfeld I, Yarus M. 1997. 23S rRNA similarity from selection for peptidyl transferase mimicry. *Biochemistry* 36:6614–23

109. White HB. 1976. Coenzymes as fossils of an earlier metabolic state. *J. Mol. Evol.* 7:101–4

110. Woese CR. 1967. The basic nature of the genetic code. In *The Genetic Code*, pp. 150–78. New York: Harper & Row

111. Xiong L, Polacek N. Sander P, Bottger EC, Mankin AS. 2001. pKa of adenine 2451 in the ribosomal peptidyl transferase center remains elusive. *RNA* 7:1365–69

112. Yang Y, Kochoyan M, Burgstaller P, Westhof E, Famulok M. 1996. Structural basis of ligand discrimination by two related RNA aptamers resolved by NMR spectroscopy. *Science* 272:1343–47

113. Yarus M. 1988. A specific amino acid binding site composed of RNA. *Science* 240:1751–58

114. Yarus M. 1998. Amino acids as RNA ligands: a direct-RNA-template theory for the code's origin. *J. Mol. Evol.* 47:109–17

115. Yarus M. 1999. Boundaries for an RNA world. *Curr. Opin. Chem. Biol.* 3:260–67

116. Yarus M. 2001. On translation by RNAs alone. *Cold Spring Harbor Symp. Quant. Biol.* 66:207–15

117. Yarus M, Christian EL. 1989. Genetic code origins. *Nature* 342:349–50

118. Yarus M, Knight, RD. 2002. The scope of selection. In *The Genetic Code and the Origin of Life,* ed. LR Pouplana. Houston: Landes Biosci. In press

119. Yarus M, Welch M. 2000. Peptidyl transferase: ancient and exiguous. *Chem. Biol.* 7:R187–90

120. Yi Y, Breaker RR. 1999. Kinetics of RNA degradation by specific base catalysis of transesterification involving the 2′-OH group. *J. Am. Chem. Soc.* 121:5364–72

121. Zhang B, Cech TR. 1997. Peptide bond formation by in vitro selected ribozymes. *Nature* 390:96–100

Annu. Rev. Genet. 2002. 36:153–73
doi: 10.1146/annurev.genet.36.041002.120114
Copyright © 2002 by Annual Reviews. All rights reserved

STUDYING GENE FUNCTION IN EUKARYOTES BY CONDITIONAL GENE INACTIVATION

Manfred Gossen[1] and Hermann Bujard[2]

[1]Max Delbrück Centrum, Robert-Rössle-Strasse 10, D-13125 Berlin, Germany;
e-mail: mgossen@mdc-berlin.de; [2]ZMBH, Universität Heidelberg, Im Neuenheimer
Feld 282, D-69120 Heidelberg, Germany; e-mail: h.bujard@zmbh.uni-heidelberg.de

Key Words tetracycline-controlled expression system, site-specific recombinases, transgenic animals, conditional overexpression, knock-in mice

■ **Abstract** The prospect of specifically controlling gene activities in vivo has become a defining hallmark of many model organisms of biological research. Where once the aim was to gain control over gene activities using endogenous control elements, new technologies have emerged that owe their remarkable specificity to heterologous components derived from evolutionarily distant species. This review highlights inducible transcriptional systems and site-specific recombination. Their quantitative and qualitative characteristics are discussed, with examples of how recent developments have expanded the spectrum of cells and organisms that are now accessible to genetic dissection of unprecedented precision. Transgenesis has already converted the mouse into a prime model for mammalian genetics. Combined with the new approaches of conditional activation or inactivation of genes, this model has opened up new horizons for the analysis of gene function in mammals.

CONTENTS

INTRODUCTION

Experimental approaches that have been developed and refined over the past decade now allow us to conditionally modulate individual gene activities in eukaryotes in a highly specific, temporally defined, and cell-type restricted manner. Analysis of the accompanying phenotypic changes has provided new insights into numerous biological mechanisms and processes hitherto not amenable to genetic dissection. There have been two basic strategies: The first targets genes directly by site-specific recombination, resulting in activation, inactivation, or alteration of the gene of interest; the second aims at quantitatively and reversibly controlling a gene's function, generally leaving the endogenous gene itself untouched within its genetic context. In this chapter, we emphasize concepts, with specific examples to illustrate methodological progress in the field, and summarize recent reports that often provide impressive new insights.

The most important single precondition for the sensible application of gene inactivation strategies is specificity: Interference should be strictly limited to the gene under study. This demanding goal has been reached, to differing extents, by use of heterologous components or by specific alterations of endogenous proteins. Systems derived from prokaryotes, evolutionarily the most distant, meet this criterion and are therefore the most broadly applied. The Cre/Lox recombination system of phage P1 of *Escherichia coli* has been used very successfully in eukaryotes, and the conditional knockout approach in transgenic mice has yielded a wealth of information. In the second strategy, elements from prokaryotes seem to provide the highest specificity when utilized for controlling gene expression in the eukaryotic cell, as exemplified in the transcription control system based on elements of the tetracycline (Tc) resistance operon (*tet* operon) of *E. coli*. When the two basic strategies for modulating gene activities—recombination and interference with gene expression—are compared, genetic analysis via temporally and spatially restricted alteration of a gene by site-specific recombination has proven to be very powerful. Nevertheless, the potential of reversibly interfering with a gene's activity, for example at the level of transcription, adds another level of sophistication to the study of gene function in vivo. As is shown below, the ability to reversibly perturb a system permits not only comparison between the normal and the perturbed state, but also analysis of the system's reactions after interference has ceased and during repeated cycles of gene inactivation/reactivation.

In the discussion that follows, we briefly point out the principles underlying different approaches, generally referring to other reviews covering the various areas of research. We touch only briefly on the straight recombination approach via the Cre/Lox and the yeast-derived Flp/FRT system because this topic has already been covered by several excellent recent reviews (36, 39, 56). In our view, an essential feature of a "truly conditional" interference with a gene's activity is reversibility of the induced perturbation. Hence we focus on controlled gene expression and discuss, in addition, the emerging synergies between this approach and site-specific recombination.

Despite the wealth of valuable information gained by controlling gene activities in cultured cells of various origins, we restrict ourselves to applications in model systems where genomic loci can be targeted in a defined manner and which, therefore, hold particularly great promise for the study of gene function in vivo: *Saccharomyces cerevisiae*, the chicken B cell line DT40, and the mouse.

EXPERIMENTAL PRINCIPLES

Site-Specific Recombination

At present, the site-specific recombinases Cre (58) and Flp (51) are most frequently used for modulating gene activities via recombination. They target their specific 34-bp recognition site, loxP and FRT, respectively, and catalyze a recombination event within these sites (Figure 1). As no cofactors are required for this recombination, both systems function in a heterologous environment, and after the temperature optimum of Flp has been adjusted to 37°C (8), both recombinases function efficiently in mice, for example. Both systems allow DNA segments to be inverted or deleted in *cis* or to be connected in *trans*. Thus, besides conditional gene (in)activation, there are several other applications such as chromosome engineering (84), recycling of selectable markers (1), etc. Here, we sketch in brief the most generally applied approach whereby genes are activated or silenced by deleting DNA fragments located between directly repeated recombination sites. In most cases, loxP/FRT sites are placed in introns. Therefore, the respective gene will generally exhibit wild-type properties until the gene or gene parts are deleted. Thus, when a mouse containing a "floxed" or a "flrted" gene is crossed with an animal expressing the respective recombinase under the control of a tissue specific promoter, recombination will occur only when the promoter driving the *cre* or *flp* gene becomes active within the developmental program of the mouse. Obviously, in many cases this approach allows potential embryonic lethality to be circumvented and leads to a temporally and spatially well-defined genetic alteration. Both recombination systems show an exquisite specificity that has been documented in a large number of experiments. Nevertheless, an increasing number of reports have described degenerate loxP sites in mammalian genomes that can be targeted by the recombinases in vivo and in vitro, albeit with low efficiency (40, 59, 64, 70), leading to undesirable effects. Therefore, high constitutive levels of Cre or Flp should be avoided. Controlling the activity of the recombinases, as is discussed below, is an alternative that expands the applicability of the recombination principle, even though it is more demanding experimentally.

Tools for the Transcriptional Control of Gene Activity

The conditional transcription systems briefly outlined below either repress or activate promoters via "synthetic," i.e., specifically designed or heterologous, transcription factors. The first repression system modeled according to the prokaryotic

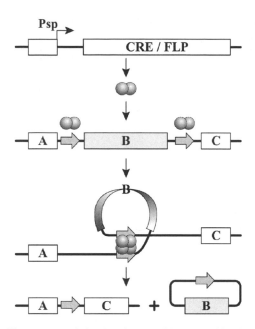

Figure 1 Controlling gene activity by site-specific recombination. Site-specific re-combinases, usually Cre or Flp, are produced under the control of a tissue-specific promoter, P_{sp}. Dimeric recombinases interact with their specific binding sites (Lox or FRT), consisting of a 13-bp inverted repeat and an 8-bp asymmetric core sequence, which provides directionality. Association of two DNA-bound recombinase dimers catalyzes a homologous recombination event within the 8-bp core sequence. Depend-ing on the orientation of the respective sites, this recombination event will either delete or invert a "floxed" or "flrted" DNA fragment. Typically, a gene is knocked out by dele-tion of B provided it encodes an essential gene function. If B contains a stop sequence separating a regulatory region A from the functional gene C, deletion of B will place C under the control of A and activate the C. In the usual binary approach, animals of a Cre mouse line, for instance, are crossed with individuals of a floxed line. Whenever P_{sp} becomes active in a double transgenic animal, recombination will occur in cells specified by P_{sp}.

paradigm made use of repressor and operator of the *lac* operon in *E. coli*, which were shown to function in mammalian cell cultures (7, 27) and recently also in transgenic mice (13). Another prokaryotic repressor/operator system derived from the *tet* operon has been used in transgenic plants (21) and in mammalian tissue cultures (83). In these experimental designs where operator sequences are placed within an RNA polymerase II promoter, the operator-bound repressor appears to sterically hinder the formation of transcription-competent RNA polymerase II complexes. The action of the respective repressors is controlled at the level of DNA binding by administering IPTG or Tc, respectively. To date, use of these repres-sion approaches has been limited. Why it appears more demanding to establish repression systems than activation systems (described below) has been discussed elsewhere (22).

The various transcription activation systems have one requirement in common: "minimal" RNA polymerase II promoters. Ideally, such promoters are totally inactive by themselves whereas they are highly activated upon binding of artificial transcription factors in their immediate vicinity. Depending upon the integration site within a genome, such conditions can indeed be established [discussed in (2, 17)]. The function of the respective transcription factors is subject to external control, which, in most cases, acts at the level of DNA binding. Examples for such artificial transcription factors are fusion proteins where the DNA binding specificity is provided by, for example, the yeast Gal4 protein (79), the *Drosophila* ecdysone receptor EcR (as heterodimer with the retenoid X receptor, RXR) (57), the repressors of the *lac* (37) or *tet* operon in *E. coli*. These DNA binding proteins are fused to a transcription activation domain capable of stimulating RNA polymerase II. Lac and Tet repressor fusions can be controlled by IPTG and Tc's, respectively, and EcR/RXR heterodimers are responsive to muristerone or ponasterone A. By contrast, as external control via carbohydrates is not feasible in mammalian cells, Gal4 fusions require the addition of a hormone binding domain, e.g., from the progesterone receptor, that mediates regulation via mifepristone.

In a fundamentally different approach, split transcription factors were developed, with one moiety constituting the DNA binding domain, the other a transcriptional transactivation domain (55). Each of these domains is fused to a portion of the human FK506 binding protein that interacts with the drug. In the presence of ligands like FK1012, a dimer of the immunosuppressant FK506, a functional transcription factor is constituted by dimerization of the two fusion proteins.

In principle, all these systems can be used to modulate gene activities, but we focus on transcription activation systems that are controlled by tetracyclines, particularly by doxycycline (Dox), simply because these systems are by far the most widely used, not only at the level of cultured cells of plant, amphibian, insect, avian, and mammalian origin but also in whole organisms including *S. cerevisiae* (see below), *Dictyostelium* (5), *Drosophila* (4), plants, e.g., *Arabidopsis* (41), and mammals such as mice (see below). Despite this broad spectrum of applications and the exciting new insights into a variety of biological phenomena that many have yielded, we restrict this review primarily to the experimental model systems outlined in the introduction.

Controlling Transcription via Tetracyclines

The favorable properties of the Tet regulatory system as outlined in Figure 2 are based on two crucial parameters: (*a*) the unusual specificity of interaction between the Tet repressor (TetR) and its specific DNA binding site, the tet operator (*tet*O) and between Tet repressor and its inducer, particularly Dox, and (*b*) the well-studied chemical and physiological properties of the inducing agents, i.e., various tetracyclines. Some of these, particularly Dox, have been widely used in human and animal medicine for several decades. As outlined in Figure 2, there

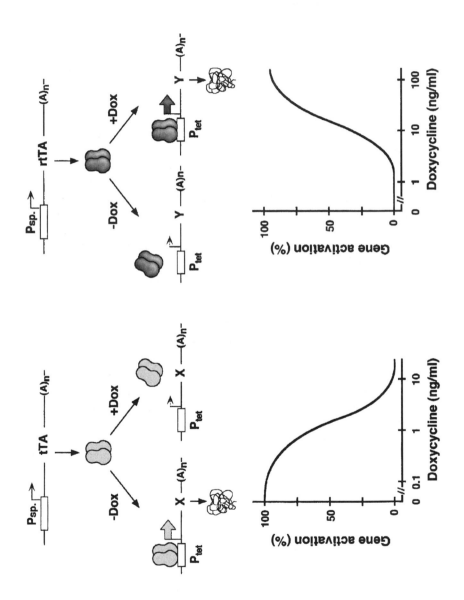

are two complementary Tet control systems. In the tTA (tetracycline controlled transactivator) or Tet-Off system, Dox prevents binding of tTA to P_{tet}, and thus abolishes transcription. By contrast, the rtTA (reverse tetracycline controlled transactivator) or Tet-On system requires Dox for binding to and activation of P_{tet} (25). Although the various modifications and improvements of the Tet systems have been described (2, 23), a few properties of the Tet systems should be noted. Both systems can be set up in a way that allows extremely tight control and, at the same time, regulation over a wide range, although to attain this quality of control is sometimes experimentally demanding (2, 17). To quantitatively control expression in cultured cells with either system, Dox concentrations are required (1 to 5 ng/ml for tTA, 50 to 100 ng/ml for rtTA-M2, see below) that are far below any toxicity threshold (5 to 10 μg/ml in HeLa cells). Due to the excellent cell and tissue penetration properties of Dox, these concentrations can be readily achieved, even in different compartments of the mouse (34) including the placenta and the milk of lactating mothers (see below). Thus, Tet regulation can be imposed on the developing embryo as well as on the offspring before and during weaning. On the other hand, Dox penetrates the blood/brain barrier less efficiently. Thus, the use of the rtTA system in the brain was problematic before the recent advent of the M2 modifications (72). Finally, as Dox penetrates cells by diffusion, a ridgeless regulation at the single cell level can be achieved wherever Dox concentration can be reliably titrated [see (35) and work described below].

Figure 2   Outline of the Tet regulatory principle. Left upper part shows the mode of action of the Tc-controlled transactivator (tTA). In the absence of the effector molecule Dox, tTA binds to the *tet*O sequence within P_{tet} and activates transcription of gene x. Addition of Dox prevents tTA from binding and, thus, abolishes the initiation of transcription. Left lower part depicts the dose response curve for the effects of Dox on tTA-dependent gene expression. Gene activity is maximal in the absence of the antibiotic but as effector concentrations increase, transcription gradually decreases to background levels at Dox concentration ≥ 5 ng/ml. Right upper part illustrates the mechanism of action of the reverse Tc-controlled transactivator (rtTA). rtTA is identical to tTA except for 4 amino acid substitutions in the TetR moiety. rtTA requires Dox for binding to *tet*O sequences within P_{tet} in order to activate transcription of gene y. Right lower part outlines the dose response curve for the effects of Dox on the rtTA-dependent transcription activation. By increasing the effector concentration beyond 20 ng/ml of Dox, rtTA-dependent gene expression is gradually stimulated. P_{tet} is a minimal promoter fused downstream of an array of 7 *tet* operators (24). It interacts with tTA as well as with rtTA. The new rtTA2^S-M2 (72) is a highly improved version of the original rtTA. It exhibits an enhanced sensitivity towards Dox and a negligible residual affinity towards *tet*O.

CONDITIONAL GENE ACTIVATION IN *S. CEREVISIAE*

The yeast *Saccharomyces cerevisiae* is an excellent model organism for the study of gene function in eukaryotes, particularly as efficient homologous recombination and reliable transcriptional induction systems can be exploited. Since the commonly used transcription control systems such as the endogenous GAL promoter in combination with the inducer galactose cause pleiotropic effects, the Tet system was transferred to *S. cerevisiae* early on. Gari et al. (20) and Nagahashi et al. (47) were the first to successfully make this transfer.

Belli and colleagues (3) established a strategy whereby cellular promoters are substituted by P_{tet}, placing the respective genes under Dox control. This versatile principle (63, 65) is now applied systematically as part of the EUROFAN projects (http://mips.gsf.de/proj/eurofan/; J. Hegemann, personal communication), which analyze a large number of essential yeast genes of unknown function. By systematically substituting authentic promoters for P_{tet}, it is anticipated that insights into the function of genes will be gained where null alleles or genomic sequences do not reveal direct clues to their physiological roles. The applicability of Tet regulation in fungal systems is underlined by the work of Nakayama and colleagues (48, 49), who transferred the essential regulatory elements into two *Candida* species where, in tTA-expressing strains, endogenous promoters were again replaced by P_{tet}, subjecting the respective genes to Tc control. Rather than using *Candida* as an eukaryotic model system, the focus of this work is to elucidate gene functions involved in pathogenicity.

TARGETED CONDITIONAL MUTANTS OF THE CHICKEN CELL LINE DT40

The chicken B cell line DT40 is distinct from other vertebrate cell lines in its ability to promote extraordinary high rates of homologous chromosomal integration upon introduction of suitable vector DNA (10). The frequency of recombination by far exceeds even that observed in murine embryonic stem (ES) cells (14) and allows for the efficient sequential inactivation of all alleles (mostly two) of any given DT40 gene. Specific homologous recombination is carried out with cloned genomic copies of the gene of interest, which is interrupted by an expression unit for a selection marker. Upon transfer of the targeting vectors into DT40 cells, appropriate selection conditions yield a high percentage of stable clones, which will have one copy of the gene of interest inactivated by homologous recombination. If this procedure is repeated with a different or, after recycling, even with the same selection marker, the second genomic copy can be knocked out. This strategy yields highly defined cell lines amenable to precise analysis. The experimental design is, however, limited to genes that are not essential for cell viability; it also lacks the advantage of reversible gene inactivation. This limitation has been overcome by a conditionally active allele of the gene under investigation using the

Tet system (11, 18, 19, 45, 46, 66–69, 75–78, 85). In most reports, when the first targeting round was performed as described, it was noted that the second round did not yield any homozygous knockout cells. To bypass this experimental bottleneck, expression vectors containing the tTA gene under control of either Pcmv or a chicken beta-actin promoter were cotransfected with the gene of interest under control of P_{tet} and tested for tetracycline-dependent expression. With the transgene unit "on," the second endogenous gene copy could be targeted, thereby creating a conditional knockout line. Two examples illustrate how such approaches aid the phenotypic analysis of vertebrate gene function. Cleavage stimulation factor CstF is an evolutionarily conserved heterotrimeric protein factor involved in the processing of 3′ ends of pre-mRNAs. In particular, in B-cells the expression of IgM heavy chain genes switches via alternative splicing of the immunoglobin pre-mRNA from the membrane-bound form, as found in pre-B cells, to the secreted form, as found in plasma cells, presumably owing to increased concentrations of CstF. Takagaki & Manley (68) directly addressed the function of CstF in this process by placing the endogenous DT40 CstF-64 gene under control of the Tet system, according to the strategy described above. Not only did they find that this gene is essential for viability, as depletion of its protein product causes apoptosis, but reduced concentrations of the proteins resulted in a reversible G_0/G_1 block of the cell cycle. More specifically, when CstF-64 concentrations were adjusted to about 10% of the endogenous levels by varying the tetracycline concentration in the medium, IgM expression, while greatly reduced on the overall mRNA level, was shifted toward the membrane-bound form of the immunoglobulin. This indicated a direct role of CstF concentration in the expression patterns of IgM heavy chain in a developmentally regulated way. Another study aimed at unraveling the role of individual centromere proteins in chromosome segregation. Whereas the centromere proteins CENP-A and CENP-C had been analyzed in some detail, a function for the more recently identified CENP-H remained to be established. Fukagawa and colleagues (18) generated a conditional loss-of-function mutant of DT40 CENP-H by introducing a Tet controlled transgene into these cells and subsequently inactivating the endogenous gene copy. As expected for a gene essential for centromere function, cells depleted of CENP-H arrested in metaphase. Concomitant immunocyto-chemistry to localize CENPs with the transgene in either "on" or "off" state revealed that in the assembly of the centromere, CENP-A localization is an early event, independent of CENP-H. By contrast, CENP-C could not be recruited to the centromere in CENP-H-depleted cells. Thus, these experiments established, in part, the order in which protein components assemble in the formation of a functional centromere.

Other studies were extended to reintroduce additional, modified copies of the gene under investigation. While the conditional transgene was in the "off" state, this allowed, for example, for mutation and domain analyses (46, 69, 76, 78) as well as interspecies complementation experiments (75). Evidence from phenotypic examination in some studies provided direct support for Tc-regulated expression to occur homogeneously at the single cell level in most if not all cells of a clonal population [see, for example, Fukagawa et al. (19)]. The possibility of precisely titrating intermediate rates of gene expression by varying the tetracycline concentration in the tissue culture medium was demonstrated by Wang & Dreyfuss [(76), see also below]. Furthermore, conditional overexpression studies demonstrated that in the DT40 cell line the Tet system permits substantial (>100-fold) overexpression of transgenes when compared to rates of the endogenous genes (19).

Despite the exciting opportunities that this chicken cell line offers, so far only a relatively small (but growing) community of researchers have taken advantage of it. However, DT40 cells will likely find more proponents in the future as the potential of combining knockout technology with a reliable conditional gene expression system becomes more obvious (9) and as the chicken genome is sequenced (http://www.nih.gov/science/models/nmm/appd.html and references therein).

CONDITIONAL GENE ACTIVATION IN THE MOUSE

Transgenesis and ES cell technologies have opened up a unique access to the mouse genome and converted this rodent into the most advanced model organism for mammalian genetics. Predictably, approaches allowing for conditional mutagenesis and for reversible gene activation/inactivation were introduced in this model system early on, achieving an unprecedented degree of precision in the study of in vivo gene function. Placing the *cre* gene under control of defined tissue-specific promoters limits the recombination event to defined stages of the animal's development and to specific tissues, frequently preventing embryonic lethality or developmental adaptation. The impact of this approach is reflected by the "zoo" of mouse lines that either express the *cre* gene under the control of various tissue-specific promoters or contain a variety of genes equipped with Lox sites for gene inactivation, activation, or alteration (for information on such mouse lines, see http://www.mshri.on.ca/nagy/cre.htm). The synergism generated by the number of mouse lines available is rapidly increasing the impact of this technology. On the other hand, Cre-induced recombination results in irreversible genetic alterations that reflect the temporal and spatial activity spectrum, i.e., the activity "history" of the promoter driving the *cre* gene. Thus, unless Cre activity is controlled from outside, this approach is limited not only by the irreversibility but also by the rigidity of the developmental and differentiation program of the organism, which, for example, prevents recombination in a fully differentiated tissue at a later period in the animal's life. In principle, these limitations can be overcome by systems that allow gene expression to be controlled from outside at will. As noted, there are excellent reviews on the achievements of the Cre/Lox and the Flp/FRT system in the mouse. We therefore focus here on applications of the Tet regulatory systems that particularly demonstrate the power of reversible gene inactivation. We also discuss approaches for controlling recombinase activities from outside and point out the beginning synergism generated by combining the Cre/Lox with the

Tet system by making use of the collection of transgenic mouse lines from both worlds.

Controlling Dominant Negative Gene Products

The most straightforward approach for studying gene functions in a transgenic mouse uses a dominant negative mutation version of the gene of interest placed under the control of the Tet system. When activated at the desired time point within the life span of the animal, an altered phenotype is expected that may be abrogated when the expression of the mutant gene is switched off again. Tissue- or cell type–specificity is achieved by placing the tTA/rtTA gene under control of the appropriate promoter. Impressive first examples of this approach were described by Kandel's group (42, 43); they generated mouse lines expressing tTA or rtTA, respectively, under the control of the αCamKII promoter that restricts the presence of tTA/rtTA to defined regions of the forebrain, particularly the hippocampus. Crossing these mice with animals containing dominant negative versions of αCamKII or calcineurin, respectively, yielded animals suitable for the study of synaptic plasticity in the context of learning and memory formation. Thus, animals could not master a spatial learning task when, for example, the dominant negative version of the αCamKII gene was active (43). The full potential of conditional expression was revealed when the mice were put through the learning procedure before the dominant negative genes were activated, for activation resulted in loss of memory. Most interestingly, however, when switched back to the wild-type situation by supplying Dox in the drinking water, spatial memory returned. This indicates that αCamKII is required not only for learning and memory formation but also for the retrieval of information from memory.

Today, more than 40 mouse lines have been published where tTA or rtTA is controlled by a variety of promoters, and about the same number of lines containing various genes and mutant versions of genes under P_{tet} control have been described [for review see (60)]. Among the many exciting findings are some where reversibility of gene activation was a prerequisite for gaining new insights into tumor pathology. Several oncogenes, including H-Ras^{V}12^G (12), K-Ras4b^{G}12^D (16), c-myc (15), ErbB2 (81), and BCR-ABL-1 (28), have been placed under Tet control. In all cases, malignant tumors could be induced in specific tissues in the adult animal. Unexpectedly, however, in all systems the vast majority of tumors remained fully dependent on the continued expression of the tumor-initiating oncogene, as demonstrated by the complete regression of tumors upon inactivation of the respective oncogene. In another exciting result, Yamamoto et al. (82) reported that activation of a mutant huntingtin gene in the brain of adult mice induced massive anatomical and neurological symptoms in the animals. Upon inactivation of the mutant huntingtin gene by Dox, the deposits in the animal's brain dissolved and the animals appeared to fully recover. These are the first results to indicate that Huntington disease may be reversible even after the manifestation of massive symptoms, provided that the activity of the huntingtin gene can be abrogated.

Surviving Development by Conditional Transgene Expression

There are numerous well-documented examples for embryonic or neonatal phenotypes in mice after gene knockouts or after transgene over- or misexpression during early development. In gene knockouts, the time of death only reflects the first stage of development where a particular gene is indispensable when no redundant gene activity is available for compensation. Even though this might be informative, embryonic death is rarely the result hoped for when analysis of the targeted gene was envisaged. In biomedical research, to name just one example, most questions addressed to an animal model require adulthood. Obviously, conditionality in transgene expression could overcome these limitations, provided that the systems warrant exogenous experimental control in pre- and perinatal development. In the case of the Tet system, this would require maternal delivery of Dox in utero and via breast milk (38, 53). Several groups (53) have demonstrated that, as expected, Dox fulfills these conditions. By regulating expression rates of IL-11 in the embryonic, neonatal, and adult lung, Ray and colleagues could discriminate between development-dependent and -independent phenotypic consequences of IL-11 overexpression on lung growth and morphology. A second example is the conditional expression of the receptor tyrosine kinase ErbB2 in transgenic mice (81). Expression of this oncogene in epithelial cells via the keratin 14 (K14) promoter resulted in hyperplasia and abnormal hair follicle morphogenesis but death of the animals around birth prevented the analysis of ErbB2 in adult mice. However, when the K14 promoter was used to drive expression of rtTA and ErbB2 was placed under the control of P_{tet}, adult transgenic animals were obtained and controlled expression via Dox administration established a direct link between overexpression of ErbB2 and skin hyperplasia, which was reversible upon withdrawal of the inducing substance. Meanwhile, there are many examples of transgenic mouse lines where embryonic Tet control over transgenes prevented an otherwise lethal outcome of the experiments [see (28, 31, 54)].

Inducible Knock-ins

Homologous recombination as used for the generation of knockout mice proved to be an indispensable tool for the analysis of gene function in mammals. In its "simplest" version, this technology is used to inactivate one copy of an endogenous gene by recombination with a suitable targeting construct at the level of ES cells. Subsequent breeding of ES cell-derived mice to homozygosity with respect to the targeted gene then leads to the desired genotype. Using this basic approach, it is also possible to modify, rather than to inactivate, an endogenous gene by substituting parts or the entire gene with sequences of differing function. Such "knock-ins" provide unparalleled possibilities in dissecting the function of an individual gene in vivo, but again potential embryonic lethality or compensatory development limits this approach. Early studies now show that these limitations can be overcome by combining conditional gene expression with knock-in approaches. Thus, Tilghman and coworkers (62) analyzed the temporal requirements for expression of the

Ednrb, the gene for the endothelin receptor B in murine development. Homozygous null mutant mice for this gene die as juveniles. Shin et al. (62) created two different knock-in lines, the first by substituting Ednrb for either tTA or rtTA. This strategy ensures that the spatial and temporal expression patterns of the transactivators most likely follow that of the targeted gene while inactivating one copy of that gene. In the second line, the promoter of Ednrb was substituted for P_{tet}-1, which also resulted in the inactivation of that gene copy. Crosses between these two lines made expression of Ednrb conditionally dependent on Dox. Reverting the functional null mutation during defined time windows revealed that Ednrb is only required between days 10 and 12.5 of embryonic development. In another study, Adelman and coworkers (6) analyzed a gene for one of the subunits of calcium-activated potassium channels, SK3. Here, only one knock-in targeting construct was used to bring tTA under control of the endogenous SK3 promoter and to simultaneously place P_{tet}-1 in front of the endogenous gene. When the respective mice were bred to homozygosity, the functional null mutation (induced by administration of Dox) did not cause a distinct phenotype. By contrast, overexpression of SK3 in the absence of Dox clearly demonstrated a role of SK3 in respiratory patterns under hypoxic conditions. Not only did that study exemplify the tight control that the Tet system can exert over endogenous genes, it also demonstrated that expression rates for endogenous genes can be achieved well in excess of the natural situation.

Yet another strategy was followed in an attempt to model Charcot-Marie-Tooth disease in transgenic mice, a demyelinating neuropathy believed to be a result of overexpression of peripheral myelin protein 22 (PMP22). Previous studies had shown that a yeast artificial chromosome (YAC) encompassing the region of that gene faithfully reflects the expression patterns of PMP22 in mice. Perea et al. (52) therefore chose such a YAC and placed tTA under control of the PMP22 promoter. Upon transgenesis, tTA was expressed as expected, and when respective animals were crossed with transgenic mice harboring the PMP22 gene under control of P_{tet}, the resulting double transgenic mice allowed the time course of demyelination and the development of characteristic pathology upon overexpression of PMP22 to be followed. Again, by downmodulating PMP22 expression via administration of Dox, demyelination and pathology were corrected.

Inducible Knockouts

As indicated above, controlling the activity of recombinases in vivo would add precision and another degree of freedom to the conditional knockout approaches via site-specific recombination. Besides controlling *cre* or *flp* gene expression, e.g., by the Tet system (see below), modified recombinases were developed whereby activity can be modulated directly by respective effector molecules. By fusing Cre to mutated ligand binding domains (LBD) of the progesterone (32) or estrogen (44) receptor, transport of the fusion protein and thus of the enzymatic activity in the cell nucleus can be controlled by the synthetic steroid Ru486 or tamoxifen, respectively. High recombination efficiencies were obtained in the mouse skin (74) and in adipocytes (29), for example, by using a Cre-ER fusion and tamoxifen.

Two experimental parameters initially limited the Cre-ER-tamoxifen approach: the high, almost toxic concentrations of inducer required and the residual recombinase activity in the absence of the inducer. The first problem has apparently been solved by generating novel fusions between Cre and LBDs of the progesterone (80) and estrogen receptors (30), with enhanced sensitivities towards their respective ligands. However, the tightness of regulation still poses a problem, particularly when targeting cell types with a long lifetime, such as most neuronal cells. Thus, in the respective tissues, cells will accumulate over time that have undergone Cre-dependent deletions, and even at low recombination rates a phenotype may eventually be seen in the uninduced state. On the other hand, in many tissues with a high turnover of cells such as skin, hormone-controlled recombinases appear to be suitable tools.

An obvious alternative approach is to control the expression of recombinases. Even though this strategy may be experimentally demanding, there are examples where the Tet system, for instance, was used to efficiently control *cre* gene expression. Thus, Lee and coworkers (73) generated mouse lines in which the rtTA gene, controlled by the retinoblastoma or the whey acidic promoter and the P_{tet}-Cre expression unit, were transferred as a single construct. By breeding animals of these lines with mice containing floxed genes, they demonstrated accurate control of recombination via Dox. In an interesting though rather complex variation of that theme, Tsien and coworkers (61) generated a mouse line where the tTA gene is placed under control of an ubiquitous β-actin promoter, separated, however, by a floxed stop-sequence, which rendered tTA expression sensitive to Cre action. Thus, whenever Cre is produced under the control of a tissue-specific promoter, the block of tTA expression is removed in the respective cell type and expression of the P_{tet}-controlled transgene becomes Dox dependent. As in these experiments floxed copies of an endogenous gene were deleted simultaneously with the stop-sequence of the tTA gene, this particular experimental design seems highly, if not prohibitively, demanding with respect to the breeding efforts required. However, simplified versions of this approach can be envisaged, as already indicated by the work of Utomo et al. (73).

As Tet control can be set up in mice such that there is very tight regulation of Cre (K. Schönig, F. Schwenk, K. Rajewsky & H. Bujard, in preparation), respective mouse lines would allow two sizable collections of transgenics to be exploited: mouse lines that express tTA or rtTA tissue specifically, and mouse lines carrying various floxed genes. Even though this approach will require significant breeding efforts, it allows many questions to be addressed without the need for generating an additional transgenic mouse line.

CONCLUDING REMARKS

In recent years, application of the methodological principles discussed here has added considerably to our understanding of gene function in vivo and is beginning to provide fundamental insights into such complex biological phenomena as development, behavior, and disease. The impact of conditional gene activation/

inactivation approaches is likely to increase as the various experimental systems become more sophisticated and are adapted to further model organisms. At the same time, consideration of the intricate action of genes in vivo reveals both the strength and the limitations of current experimental capabilities. For example, many genes are active within distinct time windows and in different cell types or tissues during development and adulthood of an organism. Their proper function also depends on the rate of expression, which in turn may be influenced by environmental signals. Accordingly, there are many examples showing that misexpression of a gene, i.e., over- or underexpression or expression outside the genuine developmental or differentiation program, results in pathologies. Mimicking such misexpression in transgenic mice has led to numerous disease models that are of particular interest whenever the disease-causing gene activity can be controlled from outside. Important questions concerning the onset, the progression, and particularly the potential reversibility of a disease can thus be addressed. As discussed above, several mouse models were generated whereby the expression of oncogenes could be controlled from outside. In all model systems, malignant tumors were induced that, considering the lag time between oncogene induction and invasive tumor growth, appear to have accumulated additional mutations. This is in agreement with the prevailing hypothesis that multiple genetic changes are required for persistent tumor growth (33). Surprisingly, upon inactivation of the tumor-initiating gene, the overwhelming number of tumors regressed, demonstrating that the respective oncogenes are also required for tumor maintenance. Moreover, mice that recovered from such tumors and were kept under the respective conditions had a normal life expectancy. As tumor regression in these systems is largely due to apoptosis induced by the shutoff of the initiating oncogene, it appears that the tumor cells had adapted to a relatively rigid oncogene-dependent expression program that cannot be reverted quickly enough to prevent activation of apoptosis (16). These findings not only point to and validate promising drug targets, they also challenge some generally accepted views on tumor development and persistence. There are further reports where disease models based on the above strategy uncovered unexpected mechanisms of disease development and regression [e.g. (71, 82)]. Together with the exciting work of Kandel's group on the mechanisms of brain plasticity and memory (see above), these reports exemplify the power of genetic analysis by reversible perturbations that may even be induced and released repeatedly during the life span of an animal.

A major experimental challenge is the controlled perturbation of expression rates to examine genes involved in regulatory networks, for example. Generally, the products of such genes interact with multiple partners at different affinities and, thus, minor variations in intracellular equilibria are likely to invoke distinct phenotypes. A. Smith's group (50), in an interesting example, showed in mouse ES cells that by titrating the expression of transcription factor Oct3/4 within a small margin, three distinct cell fates can be induced. Thus, it appears likely that by varying intracellular concentrations of gene products, a large number of phenotypes obtained by knockout strategies will separate into different phenotypes, indicating that the respective gene product participates in multiple equilibria and possibly reveals novel interaction partners.

Although expression rates can be readily fine-tuned in cultured cells or unicellular organisms, such fine-tuning in transgenic mice poses an experimental challenge. Nevertheless, it may well be feasible to some extent, provided there is favorable information on tissue penetration and distribution, biological half-life time, etc., of the inducing agent. Indeed, by coregulating a marker gene whose activity can be monitored noninvasively, the activity of a gene of interest can be followed over long periods of time and adjusted to desired rates (26).

Despite the great potential of the various experimental approaches available today, many problems remain. Thus, as long as we know little about the influence of chromatin structure on gene expression and its sensitivity to perturbation, particularly when modifying regulatory regions, reliable experimental designs where cell type specificity is maintained and position effect variegation prevented will remain difficult, if not impossible. Precise and minimal disturbance gene targeting and the application of the BAC technology will certainly be most valuable in future. On the other hand, approaches that leave the gene of interest fully untouched in its genomic context may be available soon. For example, controlling custom-made, zinc finger–based transcription factors recognizing specific sequences within the promoter of the gene of interest or regulated synthesis of RNAs or polypeptides (including modified antibodies) that attack specific gene products may develop into systems of choice. Finally, despite the rapid kinetics by which, for example, Tet-controlled transcription can be induced (<5 min in cultured cells, <1 h in the liver of transgenic mice), many biological processes are governed by vastly faster mechanisms that can hardly be addressed by methods currently available. Obviously, in our strivings toward a better understanding of in vivo gene function, there are no limits for new ideas.

ACKNOWLEDGMENTS

The authors thank Drs. Johannes Hegemann and Günther Schütz for helpful scientific discussions. Ms Sibylle Reinig is recognized for help in the preparation of this manuscript. Work in the authors' laboratories is supported by the VolkswagenStiftung (program: "Conditional Mutagenesis," grants to M.G. and H.B.), an EC-grant (H.B.), and the Fonds der Chemischen Industrie Deutschlands (H.B.).

The *Annual Review of Genetics* is online at http://genet.annualreviews.org

LITERATURE CITED

1. Abuin A, Bradley A. 1996. Recycling selectable markers in mouse embryonic stem cells. *Mol. Cell Biol.* 16:1851–56
2. Baron U, Bujard H. 2000. Tet repressor-based system for regulated gene expression in eukaryotic cells: principles and advances. *Methods Enzymol.* 327:401–21

3. Belli G, Gari E, Aldea M, Herrero E. 1998. Functional analysis of yeast essential genes using a promoter-substitution cassette and the tetracycline-regulatable dual expression system. *Yeast* 14:1127–38
4. Bello B, Resendez-Perez D, Gehring WJ.

1998. Spatial and temporal targeting of gene expression in Drosophila by means of a tetracycline-dependent transactivator system. *Development* 125:2193–202

5. Blaauw M, Linskens MH, van Haastert PJ. 2000. Efficient control of gene expression by a tetracycline-dependent transactivator in single *Dictyostelium discoideum* cells. *Gene* 252:71–82

6. Bond CT, Sprengel R, Bissonnette JM, Kaufmann WA, Pribnow D, et al. 2000. Respiration and parturition affected by conditional overexpression of the Ca^{2+}-activated K^{+} channel subunit, SK3. *Science* 289:1942–2222

7. Brown M, Figge J, Hansen U, Wright C, Jeang KT, et al. 1987. lac repressor can regulate expression from a hybrid SV40 early promoter containing a lac operator in animal cells. *Cell* 49:603–12

8. Buchholz F, Angrand PO, Stewart AF. 1998. Improved properties of FLP recombinase evolved by cycling mutagenesis. *Nat. Biotechnol.* 16:657–62

9. Buerstedde JM, Arakawa H, Watahiki A, Carninci PP, Hayashizaki YY, et al. 2002. The DT40 web site: sampling and connecting the genes of a B cell line. *Nucleic Acids Res.* 30:230–31

10. Buerstedde JM, Takeda S. 1991. Increased ratio of targeted to random integration after transfection of chicken B cell lines. *Cell* 67:179–88

11. Chen Z, Manley JL. 2000. Robust mRNA transcription in chicken DT40 cells depleted of TAF(II)31 suggests both functional degeneracy and evolutionary divergence. *Mol. Cell Biol.* 20:5064–76

12. Chin L, Tam A, Pomerantz J, Wong M, Holash J, et al. 1999. Essential role for oncogenic Ras in tumour maintenance. *Nature* 400:468–72

13. Cronin CA, Gluba W, Scrable H. 2001. The lac operator-repressor system is functional in the mouse. *Genes Dev.* 15:1506–17

14. Dhar PK, Sonoda E, Fujimori A, Yamashita YM, Takeda S. 2001. DNA repair studies: experimental evidence in support of chicken DT40 cell line as a unique model. *J. Environ. Pathol. Toxicol. Oncol.* 20:273–83

15. Felsher DW, Bishop JM. 1999. Reversible tumorigenesis by MYC in hematopoietic lineages. *Mol. Cell* 4:199–207

16. Fisher GH, Wellen SL, Klimstra D, Lenczowski JM, Tichelaar JW, et al. 2001. Induction and apoptotic regression of lung adenocarcinomas by regulation of a K-Ras transgene in the presence and absence of tumor suppressor genes. *Genes Dev.* 15: 3249–62

17. Freundlieb S, Schirra-Muller C, Bujard H. 1999. A tetracycline controlled activation/repression system with increased potential for gene transfer into mammalian cells. *J. Gene Med.* 1:4–12

18. Fukagawa T, Mikami Y, Nishihashi A, Regnier V, Haraguchi T, et al. 2001. CENP-H, a constitutive centromere component, is required for centromere targeting of CENP-C in vertebrate cells. *EMBO J.* 20:4603–17

19. Fukagawa T, Pendon C, Morris J, Brown W. 1999. CENP-C is necessary but not sufficient to induce formation of a functional centromere. *EMBO J.* 18:4196–209

20. Gari E, Piedrafita L, Aldea M, Herrero E. 1997. A set of vectors with a tetracycline-regulatable promoter system for modulated gene expression in *Saccharomyces cerevisiae*. *Yeast* 13:837–48

21. Gatz C, Frohberg C, Wendenburg R. 1992. Stringent repression and homogeneous derepression by tetracycline of a modified CaMV 35S promoter in intact transgenic tobacco plants. *Plant J.* 2:397–404

22. Gossen M, Bonin AL, Bujard H. 1993. Control of gene activity in higher eukaryotic cells by prokaryotic regulatory elements. *Trends Biochem. Sci.* 18:471–75

23. Gossen M, Bujard H. 2001. Tetracyclines in the control of gene expression in eukaryotes. In *Tetracyclines in Biology, Chemistry and Medicine*, ed. M Nelson,

W Hillen, RA Greenwald, pp. 139–57. Basel: Birkhäuser Verlag

24. Gossen M, Bujard H. 1992. Tight control of gene expression in mammalian cells by tetracycline-responsive promoters. *Proc. Natl. Acad. Sci. USA* 89:5547–51

25. Gossen M, Freundlieb S, Bender G, Muller G, Hillen W, et al. 1995. Transcriptional activation by tetracyclines in mammalian cells. *Science* 268:1766–69

26. Hasan MT, Schonig K, Berger S, Graewe W, Bujard H. 2001. Long-term, noninvasive imaging of regulated gene expression in living mice. *Genesis* 29:116–22

27. Hu MC, Davidson N. 1987. The inducible lac operator-repressor system is functional in mammalian cells. *Cell* 48:555–66

28. Huettner CS, Zhang P, Van Etten RA, Tenen DG. 2000. Reversibility of acute B-cell leukaemia induced by BCR-ABL1. *Nat. Genet.* 24:57–60

29. Imai T, Jiang M, Chambon P, Metzger D. 2001. Impaired adipogenesis and lipolysis in the mouse upon selective ablation of the retinoid X receptor alpha mediated by a tamoxifen-inducible chimeric Cre recombinase (Cre-ERT2) in adipocytes. *Proc. Natl. Acad. Sci. USA* 98:224–28

30. Indra AK, Warot X, Brocard J, Bornert JM, Xiao JH, et al. 1999. Temporally-controlled site-specific mutagenesis in the basal layer of the epidermis: comparison of the recombinase activity of the tamoxifen-inducible Cre-ER(T) and Cre-ER(T2) recombinases. *Nucleic Acids Res.* 27:4324–27

31. Jerecic J, Schulze CH, Jonas P, Sprengel R, Seeburg PH, et al. 2001. Impaired NMDA receptor function in mouse olfactory bulb neurons by tetracycline-sensitive NR1 (N598R) expression. *Brain Res. Mol. Brain Res.* 94:96–104

32. Kellendonk C, Tronche F, Casanova E, Anlag K, Opherk C, et al. 1999. Inducible site-specific recombination in the brain. *J. Mol. Biol.* 285:175–82

33. Kinzler KW, Vogelstein B. 1996. Lessons from hereditary colorectal cancer. *Cell* 87:159–70

34. Kistner A, Gossen M, Zimmermann F, Jerecic J, Ullmer C, et al. 1996. Doxycycline-mediated quantitative and tissue-specific control of gene expression in transgenic mice. *Proc. Natl. Acad. Sci. USA* 93:10933–38

35. Kringstein AM, Rossi FMV, Hofmann A, Blau HM. 1998. Graded transcriptional response to different concentrations of a single transactivator. *Proc. Natl. Acad. Sci. USA* 95:13670–75

36. Kühn R, Schwenk F. 2002. Conditional knockout mice. See Ref. 73a, pp. 159–86

37. Labow MA. 1995. Use of lac activator proteins for regulated expression of oncogenes. *Methods Enzymol.* 254:375–87

38. Lee P, Morley G, Huang Q, Fischer A, Seiler S, et al. 1998. Conditional lineage ablation to model human diseases. *Proc. Natl. Acad. Sci. USA* 95:11371–76

39. Lewandoski M. 2001. Conditional control of gene expression in the mouse. *Nat. Rev. Genet.* 2:743–55

40. Loonstra A, Vooijs M, Beverloo HB, Allak BA, van Drunen E, et al. 2001. Growth inhibition and DNA damage induced by Cre recombinase in mammalian cells. *Proc. Natl. Acad. Sci. USA* 98:9209–14

41. Love J, Scott AC, Thompson WF. 2000. Technical advance: stringent control of transgene expression in *Arabidopsis thaliana* using the Top10 promoter system. *Plant J.* 21:579–88

42. Mansuy IM, Winder DG, Moallem TM, Osman M, Mayford M, et al. 1998. Inducible and reversible gene expression with the rtTA system for the study of memory. *Neuron* 21:257–65

43. Mayford M, Bach ME, Huang YY, Wang L, Hawkins RD, et al. 1996. Control of memory formation through regulated expression of a CaMKII transgene. *Science* 274:1678–83

44. Metzger D, Clifford J, Chiba H, Chambon P. 1995. Conditional site-specific recombination in mammalian cells using a

ligand-dependent chimeric Cre recombinase. *Proc. Natl. Acad. Sci. USA* 92:6991–95

45. Morrison C, Shinohara A, Sonoda E, Yamaguchi-Iwai Y, Takata M, et al. 1999. The essential functions of human Rad51 are independent of ATP hydrolysis. *Mol. Cell Biol.* 19:6891–97

46. Morrison C, Sonoda E, Takao N, Shinohara A, Yamamoto K, et al. 2000. The controlling role of ATM in homologous recombinational repair of DNA damage. *EMBO J.* 19:463–71. Erratum. 2000. *EMBO J.* 19(4):786

47. Nagahashi S, Nakayama H, Hamada K, Yang H, Arisawa M, et al. 1997. Regulation by tetracycline of gene expression in *Saccharomyces cerevisiae*. *Mol. Gen. Genet.* 255:372–75

48. Nakayama H, Izuta M, Nagahashi S, Sihta EY, Sato Y, et al. 1998. A controllable gene-expression system for the pathogenic fungus *Candida glabrata*. *Microbiology* 144:2407–15

49. Nakayama H, Mio T, Nagahashi S, Kokado M, Arisawa M, et al. 2000. Tetracycline-regulatable system to tightly control gene expression in the pathogenic fungus *Candida albicans*. *Infect. Immun.* 68:6712–19

50. Niwa H, Miyazaki J, Smith AG. 2000. Quantitative expression of Oct-3/4 defines differentiation, dedifferentiation or self-renewal of ES cells. *Nat. Genet.* 24:372–76

51. O'Gorman S, Fox DT, Wahl GM. 1991. Recombinase-mediated gene activation and site-specific integration in mammalian cells. *Science* 251:1351–55

52. Perea J, Robertson A, Tolmachova T, Muddle J, King RH, et al. 2001. Induced myelination and demyelination in a conditional mouse model of Charcot-Marie-Tooth disease type 1A. *Hum. Mol. Genet.* 10:1007–18

53. Ray P, Tang W, Wang P, Homer R, Kuhn C, et al. 1997. Regulated overexpression of Interleukin 11 in the lung. Use to dissociate development-dependent and -independent phenotypes. *J. Clin. Invest.* 100:2501–11

54. Rhoades KL, Hetherington CJ, Harakawa N, Yergeau DA, Zhou L, et al. 2000. Analysis of the role of AML1-ETO in leukemogenesis, using an inducible transgenic mouse model. *Blood* 96:2108–15

55. Rivera VM. 1998. Controlling gene expression using synthetic ligands. *Methods* 14:421–29

56. Ryding AD, Sharp MG, Mullins JJ. 2001. Conditional transgenic technologies. *J. Endocrinol.* 171:1–14

57. Saez E, No D, West A, Evans RM. 1997. Inducible gene expression in mammalian cells and transgenic mice. *Curr. Opin. Biotechnol.* 8:608–16

58. Sauer B, Henderson N. 1989. Cre-stimulated recombination at loxP-containing DNA sequences placed into the mammalian genome. *Nucleic Acids Res.* 17:147–61

59. Schmidt EE, Taylor DS, Prigge JR, Barnett S, Capecchi MR. 2000. Illegitimate Cre-dependent chromosome rearrangements in transgenic mouse spermatids. *Proc. Natl. Acad. Sci. USA* 97:13702–7

60. Schönig K, Bujard H. 2002. Generating conditional mouse mutants via tetracycline controlled gene expression. See Ref. 73a, pp. 69–104

61. Shimizu E, Tang Y-P, Rampon C, Tsien JZ. 2000. NMDA receptor-dependent synaptic reinforcement as a crucial process for memory consolidation. *Science* 290:1170–2222

62. Shin MK, Levorse JM, Ingram RS, Tilghman SM. 1999. The temporal requirement for endothelin receptor-B signalling during neural crest development. *Nature* 402:496–501

63. Shirra MK, Patton-Vogt J, Ulrich A, Liuta-Tehlivets O, Kohlwein SD, et al. 2001. Inhibition of acetyl coenzyme A carboxylase activity restores expression of the INO1 gene in a snf1 mutant strain

of *Saccharomyces cerevisiae. Mol. Cell. Biol.* 21:5710–90

64. Silver DP, Livingston DM. 2001. Self-excising retroviral vectors encoding the Cre recombinase overcome Cre-mediated cellular toxicity. *Mol. Cell* 8:233–43

65. Simon E, Clotet J, Calero F, Ramos J, Arino J. 2001. A screening for high copy suppressors of the sit4 hal3 synthetically lethal phenotype reveals a role for the yeast Nha1 antiporter in cell cycle regulation. *J. Biol. Chem.* 276:29740–32190

66. Sonoda E, Sasaki MS, Buerstedde JM, Bezzubova O, Shinohara A, et al. 1998. Rad51-deficient vertebrate cells accumulate chromosomal breaks prior to cell death. *EMBO J.* 17:598–608

67. Sonoda E, Sasaki MS, Morrison C, Yamaguchi-Iwai Y, Takata M, et al. 1999. Sister chromatid exchanges are mediated by homologous recombination in vertebrate cells. *Mol. Cell Biol.* 19:5166–69

68. Takagaki Y, Manley JL. 1998. Levels of polyadenylation factor CstF-64 control IgM heavy chain mRNA accumulation and other events associated with B cell differentiation. *Mol. Cell* 2:761–71

69. Takami Y, Nakayama T. 2000. N-terminal region, C-terminal region, nuclear export signal, and deacetylation activity of histone deacetylase-3 are essential for the viability of the DT40 chicken B cell line. *J. Biol. Chem.* 275:16191–201

70. Thyagarajan B, Guimaraes MJ, Groth AC, Calos MP. 2000. Mammalian genomes contain active recombinase recognition sites. *Gene* 244:47–54

71. Tremblay P, Meiner Z, Galou M, Heinrich C, Petromilli C, et al. 1998. Doxycycline control of prion protein transgene expression modulates prion disease in mice. *Proc. Natl. Acad. Sci. USA* 95:12580–85

72. Urlinger S, Baron U, Thellmann M, Hasan MT, Bujard H, et al. 2000. Exploring the sequence space for tetracycline-dependent transcriptional activators: novel mutations yield expanded range and sensitivity. *Proc. Natl. Acad. Sci. USA* 97:7963–68

73. Utomo AR, Nikitin AY, Lee WH. 1999. Temporal, spatial, and cell type-specific control of Cre-mediated DNA recombination in transgenic mice. *Nat. Biotechnol.* 17:1091–96

73a. van Deursen J, Hofker M, eds. 2002. *The Transgenic Mouse—Methods and Protocols.* Totowa, NJ: Humana

74. Vasioukhin V, Degenstein L, Wise B, Fuchs E. 1999. The magical touch: genome targeting in epidermal stem cells induced by tamoxifen application to mouse skin. *Proc. Natl. Acad. Sci. USA* 96:8551–56

75. Wang J, Dreyfuss G. 2001. A cell system with targeted disruption of the SMN gene. Functional conservation of the SMN protein and dependence of gemin2 on SMN. *J. Biol. Chem.* 276:9599–605

76. Wang J, Dreyfuss G. 2001. Characterization of functional domains of the SMN protein in vivo. *J. Biol. Chem.* 276:45387–449

77. Wang J, Takagaki Y, Manley JL. 1996. Targeted disruption of an essential vertebrate gene: ASF/SF2 is required for cell viability. *Genes Dev.* 10:2588–99

78. Wang J, Xiao SH, Manley JL. 1998. Genetic analysis of the SR protein ASF/SF2: interchangeability of RS domains and negative control of splicing. *Genes Dev.* 12:2222–33

79. Wang Y, Tsai SY, O'Malley BW. 1999. Antiprogestin regulable gene switch for induction of gene expression in vivo. *Methods Enzymol.* 306:281–94

80. Wunderlich FT, Wildner H, Rajewsky K, Edenhofer F. 2001. New variants of inducible Cre recombinase: a novel mutant of Cre-PR fusion protein exhibits enhanced sensitivity and an expanded range of inducibility. *Nucleic Acids Res.* 29:E47

81. Xie W, Chow LT, Paterson AJ, Chin E, Kudlow JE. 1999. Conditional expression

of the ErbB2 oncogene elicits reversible hyperplasia in stratified epithelia and up-regulation of TGFalpha expression in transgenic mice. *Oncogene* 18:3593–607

82. Yamamoto A, Lucas JJ, Hen R. 2000. Reversal of neuropathology and motor dysfunction in a conditional model of Huntington's disease. *Cell* 101:57–66

83. Yao F, Svensjo T, Winkler T, Lu M, Eriksson C, et al. 1998. Tetracycline repressor, tetR, rather than the tetR-mammalian cell transcription factor fusion derivatives, regulates inducible gene expression in mammalian cells. *Hum. Gene Ther.* 9: 1939–50

84. Yu Y, Bradley A. 2001. Engineering chromosomal rearrangements in mice. *Nat. Rev. Genet.* 2:780–90

85. Zhao W, Manley JL. 1998. Deregulation of poly(A) polymerase interferes with cell growth. *Mol. Cell Biol.* 18:5010–20

Annu. Rev. Genet. 2002. 36:175–203
doi: 10.1146/annurev.genet.36.032902.111815

DNA TOPOLOGY-MEDIATED CONTROL OF GLOBAL GENE EXPRESSION IN *ESCHERICHIA COLI*

G. Wesley Hatfield
Department of Microbiology and Molecular Genetics, College of Medicine, University of California, Irvine, California 92697; e-mail: gwhatfie@uci.edu

Craig J. Benham
Department of Mathematics, University of California, Davis, California 95616; cjbenham@ucdavis.edu

Key Words DNA supercoiling, SIDD, transcriptional coupling, energy charge, metabolic regulation

■ **Abstract** Because the level of DNA superhelicity varies with the cellular energy charge, it can change rapidly in response to a wide variety of altered nutritional and environmental conditions. This is a global alteration, affecting the entire chromosome and the expression levels of all operons whose promoters are sensitive to superhelicity. In this way, the global pattern of gene expression may be dynamically tuned to changing needs of the cell under a wide variety of circumstances. In this article, we propose a model in which chromosomal superhelicity serves as a global regulator of gene expression in *Escherichia coli*, tuning expression patterns across multiple operons, regulons, and stimulons to suit the growth state of the cell. This model is illustrated by the DNA supercoiling-dependent mechanisms that coordinate basal expression levels of operons of the *ilv* regulon both with one another and with cellular growth conditions.

CONTENTS

INTRODUCTION

Gene expression patterns must be stringently regulated according to the nutritional needs of the cell and environmental conditions. For bacteria, these circumstances can change rapidly and drastically. For example, an enteric organism like *Escherichia coli* that has been growing under relatively nutrient-rich, stable conditions can suddenly find itself relocated to a hostile environment that is essentially devoid of nutrients. To survive such a transition, this organism possesses a wide variety of metabolic and genetic regulatory networks that enable it to rapidly adjust to new conditions.

Our current understanding of the regulatory systems that coordinate gene expression in *E. coli* involves several hierarchical levels—local control of individual operons, regional control of multiple operons within a regulon, and of multiple regulons within a stimulon or modulon (59), and global control of overall expression patterns (Figure 1). The most basic and best-understood level is the regulation of individual operons. Many different types of operon-specific control mechanisms have been described, each of which responds to regulatory signals that are closely related to its function. For example, the expression of an operon encoding genes for a biosynthetic pathway commonly is repressed by pathway-specific end-products, while expression of an operon encoding genes for a catabolic pathway often is activated by pathway-specific substrates. The DNA binding proteins that mediate operon-specific regulation commonly are present in small numbers

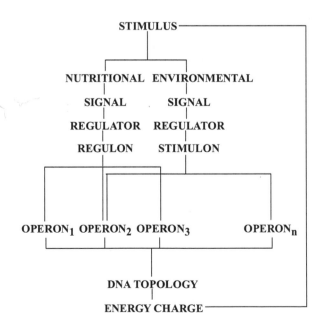

Figure 1   Hierarchical levels of genetic regulatory circuits in bacteria. See text for discussion.

in the cell and bind in a highly site-specific manner to only a few DNA target sites.

At the next hierarchical level are regulons, which are groups of operons that are coordinately regulated by nutrient or environmental conditions. The operons within a regulon usually participate in a common function, such as nitrogen or carbon utilization, and share a common regulator, usually a protein repressor or activator, that recognizes a DNA target sequence common to all members. The regulatory proteins involved in regulon control are more abundant in the cell, and bind to multiple DNA target sites on the chromosome.

At an even higher hierarchical level, certain environmental changes, such as in the osmolarity or oxygen content of the growth medium, may generate a signal that induces operons contained in multiple regulons. These overlapping networks are referred to as stimulons. The DNA binding proteins that regulate stimulons commonly are even more abundant, and bind to many, often quite degenerate, DNA target sites.

These levels of regulation target specific operons or sets of operons, adjusting their relative expression levels to optimize their activities to the current conditions of the cell. They do not address the highest level of regulation, the global coordination of the expression levels of all the genes in the genome that enables the cell to efficiently accommodate to changing conditions. This global level of gene regulation requires the integration of a variety of nutritional and environmental signals, and the generation of a coordinated response that adjusts the basal levels of expression of all genes so as to optimize cell growth and survival under a broad range of possibly rapidly changing conditions. This arrangement does not override operon-, regulon-, or stimulon-specific controls, but rather tunes the global gene expression patterns to the demands of the cell under the prevailing conditions. For example, a lower basal level of amino acid biosynthesis is needed during stationary phase than during log phase growth. So the expression levels of genes encoding enzymes required for amino acid biosynthesis are lowered in stationary phase, but must rapidly increase during transition to log phase. Throughout this transition, all operon-, regulon-, and stimulon-specific controls on these genes must remain operative, fine-tuning their expression to specific circumstances.

We propose that this highest level of hierarchical gene regulation is mediated by DNA supercoiling whose level is regulated by the energy charge of the cell, which in turn is modulated by nutrient and environmental signals. Here we review the evidence supporting this model.

DNA SUPERCOILING, ENERGY CHARGE, AND TRANSCRIPTIONAL REGULATION

Control of DNA Supercoiling

With very few exceptions, DNA extracted from prokaryotic sources has a linking number deficiency, $\Delta Lk < 0$, and is therefore negatively supercoiled [for reviews see (27) and (16)]. This in vivo level of negative supercoiling is tightly controlled by the combined influences of many factors, including DNA binding proteins, transcription, replication, and the activities of topoisomerase enzymes (21, 27, 51).

Nucleoids that have been carefully removed from *E. coli* are bound by a variety of proteins. Some of these DNA binding proteins stabilize supercoils by wrapping the DNA into stably wound toroidal loops. These constrained supercoils are not lost when the DNA is nicked (66). However, when these binding proteins are removed the constraint on these supercoils is released and the measured superhelix density changes, even though the actual linking number is invariant (75, 76, 91). The portion of the overall supercoiling that is lost when the DNA is nicked is called unconstrained supercoiling, or superhelical tension. About 50% of the supercoiling in *E. coli* is constrained by protein binding (10, 11, 66, 69, 82). A major source of this restraint comes from binding proteins that separate DNA strands, including proteins of the replication apparatus and RNA polymerase (RNAP). Architectural proteins such as HNS, IHF, and the histone-like HU proteins also restrain negative supercoils. It has been observed that HU-deficient mutant strains exhibit levels of supercoiling approximately 15% lower than those of wild-type strains (39). Further reductions are seen in strains that also contain deletions of genes encoding other chromatin organizing proteins such as HNS and IHF (38, 65).

The level of unconstrained superhelical tension within cells is determined by the activities of the enzymes DNA gyrase, and the DNA topoisomerases Topo I and Topo IV (27, 28, 34, 84). DNA gyrase introduces negative supercoils into DNA by a reaction requiring the hydrolysis of ATP. Topo I removes negative supercoils, and thereby relaxes negative superhelical tension, in an ATP-independent reaction (99). Although Topo IV also removes negative supercoils, its primary function appears to be resolving DNA knots and catenanes by a process that requires ATP (24).

Early genetic studies revealed that secondary lesions in the genes encoding DNA gyrase (*gyrA* or *gyrB*) are always observed in strains in which the entire functional *top*A gene that encodes Topo I has been deleted. Null mutations in *gyrB* are lethal in these strains (25, 52, 63, 68, 70). These results suggest that tight control of the interplay between Topo I and gyrase is necessary for optimal growth.

A variety of counterbalancing controls are maintained on gyrase and Topo I, involving regulation both of the activities of these enzymes and of the level of expression of their encoding genes. These controls collectively act as a homeostatic feedback system, maintaining the negative superhelical tension on the *E. coli* chromosome within narrow limits around an optimal level whose value depends on the growth phase of the cell and on environmental conditions (20, 28, 53, 87).

Cellular Energy Charge and Global DNA Supercoiling Levels

The elegant and extensive studies by Atkinson and co-workers have shown that the energy charge of the adenylate pool, defined as $([ATP] + \frac{1}{2}[ADP])/([ATP] + [ADP] + [AMP])$, is the parameter that correctly describes the amount of metabolically available energy for the cell (1, 2, 18, 19). They also demonstrated that, during states of metabolic adjustment when the energy charge transiently decreases, such as in the transition from aerobic to anaerobic growth, enzymes involved in ATP-regenerating reactions are activated and enzymes involved in ATP-utilizing reactions are inhibited. In general, decreases in the energy charge induce increases in the rates of enzymes that produce ATP and decreases in the rates of enzymes that consume ATP, while increases in the energy charge have the opposite effect. Together these changes maintain the energy charge in homeostatic balance. These findings explain why, although the absolute levels of adenylate pools differ under different growth conditions, the energy charge of the cell remains constant at a value of ∼0.85 during balanced growth under all growth conditions. In stationary phase, however, the energy charge is maintained at a lower level (41). To facilitate the following discussion, we refer to energy charge as the cellular [ATP/ADP] ratio, realizing that this is a simplification of Atkinson's definition.

Many studies have shown that the level of global negative supercoiling is controlled by the cellular energy charge. Because the enzymatic activity of gyrase is controlled by the intracellular [ATP]/[ADP] ratio, not by the free ATP concentration (38, 39), high negative superhelical densities occur when cellular energy charge is high, and low negative superhelical densities occur when it is low (27, 29, 39, 95, 96). It is well known that energy charge and DNA supercoiling play coordinated roles in cellular adaptation and survival, both under suboptimal growth conditions and during growth state transitions. In nongrowing *E. coli* cells in stationary phase where the [ATP/ADP] ratio is low, the superhelical density of a reporter plasmid is $\sigma \approx -0.03$. As cells recover and enter into log phase, the [ATP/ADP] ratio increases and the global negative superhelical density moves into the midphysiological level of $\sigma \approx -0.05$, a value typical during balanced growth (45). Physical stresses alter both cellular energy charge and DNA supercoiling levels. For example, osmotic stress (salt shock) causes the cellular [ATP/ADP] ratio to transiently increase four-fold, and the negative superhelical density of the bacterial chromosome to increase to a value as high as $\sigma = -0.09$ (38). During transitions from aerobic to anaerobic growth the cellular [ATP/ADP] ratio decreases, and the global negative superhelical density transiently falls from $\sigma = -0.05$ to $\sigma = -0.038$ (20).

Effects of DNA Supercoiling on Gene Expression

Steck et al. used O'Farrell 2-D protein gel electrophoresis techniques to quantify the relative abundances of proteins expressed in *E. coli* strains containing non-lethal mutations in either *gyrB* or *topA* that alter the global superhelical density of the chromosome (84). Of the 88 proteins whose abundances were quantified, 39% showed changes of abundance, and inferentially of cognate gene expression levels, during steady-state growth in oversupercoiling *topA* versus undersupercoiling *gyrB* mutants. Maximal abundances of some proteins occurred at supercoiling levels below that of the wild type, while others were most abundant at elevated negative superhelix densities. A third class exhibited optimum expression at a normal physiological supercoiling level.

There are many ways in which DNA template topology (i.e., an imposed linking difference ΔLk) can influence gene expression. These could involve changes of helicity, ΔTw, and/or changes of tertiary structure, ΔWr. The former include alterations of the helical twist of the B-form, and/or local transitions to alternative secondary structures having different helicities from that of B-DNA, while the latter require bending deformations. Supercoils that are stabilized by interactions with other molecules can be toroidal, but unstabilized supercoils commonly are plectonemically interwound. [For a recent review of the ways in which DNA topology can influence transcriptional activity see (23)].

REGULATION BY SUPERHELICAL MODULATION OF HELICAL TWIST   Drlica and co-workers showed that the level of supercoiling which gives optimal expression for promoters correlates with the length of the spacer region between their -35 and the -10 regions (84). Promoters whose maximum expression occurs at low superhelical densities tend to have spacer regions that are shorter than 17 base pairs. Promoters whose maximal activity occur at high levels of supercoiling commonly have spacers that are longer than 17 base pairs. Promoters with 17 base pair spacers were preferentially optimized for normal physiological levels of supercoiling and showed less sensitivity to changes in supercoiling than did the others.

The relative orientation between the -35 and the -10 regions can strongly affect the ability of the σ^{70} subunit of RNA polymerase (RNAP) to locate and bind to a promoter (92). As negative supercoiling untwists DNA (i.e., $\Delta Lk < 0$, so $\Delta Tw < 0$), its effect on the helical twist of the spacer can explain the correlation between promoter spacer length, superhelicity, and activity. A long spacer region, with a larger intrinsic twist than would provide optimal alignment, will in this model become more active at higher negative superhelical densities because this deformation decreases twist. Conversely, a short spacer whose twist is less than optimal would have its activity decreased by negative supercoiling, and hence would be more active at smaller superhelix densities. Optimal recognition occurs with 17-base pair spacing containing the negative helical twist characteristic of a midphysiological superhelical density. Cell growth conditions that alter the global superhelical density of the chromosome, by changing the helical alignment of recognition elements within σ^{70} promoters, would affect their activities accordingly. Examples of such a supercoiling-induced realignment mechanism have been demonstrated in promoters involved in processes as diverse as the cold-shock and osmotic-shock responses, amino acid biosynthesis, and carbon utilization (14, 42, 43, 45, 90, 92).

REGULATION BY SUPERHELICAL MODULATION OF TERTIARY STRUCTURE   The imposition of a negative linking difference on a topological domain of DNA may alter the tertiary structure through its effect on Wr. In short regions, this can cause looping, while in longer regions it can induce plectonemic interwinding. Looped structures can form microdomains, which act as small independent topological domains. This process has been proposed to be involved in prokaryotic transcriptional activation (56). The formation of an interwound structure brings regions that are remote along the sequence into close physical proximity. If the DNA reptates through such a structure, eventually any site will find itself close in space to any other site. In this way, plectonemic interwinding can greatly enhance the opportunities for sites remote along the duplex, or molecules bound thereto, to interact. This effect is the basis for the activity of the NtrC-dependent enhancer in *E. coli*, which can act in a supercoil-dependent manner over large distances, on the order of 2000 bp (50).

REGULATION BY SUPERHELICALLY DRIVEN TRANSITIONS TO ALTERNATIVE SECONDARY STRUCTURES   DNA supercoiling is known to drive transitions to a wide variety of alternative secondary structures, including local denaturation (44), transitions to Z-form (5) and to H-form (40), and cruciform extrusion (48). The formation of alternative DNA structures can serve regulatory functions, either by forming or modifying a regulatory binding site, or by altering the level of unconstrained supercoiling in the balance of the domain (3, 4, 7, 8). Dai & Rothman-Denes have shown that the bacteriophage N4 virion RNA polymerase (vRNAP) promoters contain short inverted repeat sequences centered at position −12 (22). These sites extrude cruciforms that are required for vRNAP recognition at physiological levels of supercoiling.

Negative superhelicity is known to destabilize the DNA duplex at specific locations (44). This effect, known as stress-induced duplex destabilization (SIDD), has been implicated in a variety of regulatory processes, including many that are involved in transcriptional regulation. SIDD can decrease the energy required for open complex formation and thereby increase transcriptional activity. Travers and coworkers used in vitro transcription and S1 nuclease to probe the structure of the *tyr*T promoter. They found that negative supercoiling increased the rate of initiation of transcription from this promoter by inducing unwinding that assisted open complex formation (26). SIDD also can affect the binding affinities of single strand-specific DNA binding proteins. Levens and coworkers have shown that the FUSE element located 1500 bp upstream from the promoter region of the human *c-myc* gene regulates the initiation of transcription from that gene (54). The mechanism for this activation is formation of a SIDD site at the FUSE element, which then reacts with the single strand-specific regulatory binding protein FPB. Although this eukaryotic system has a large set of other regulatory transcription factors and elements, in the absence of this specific system endogenous *c-myc* expression cannot be sustained.

Local denaturation is the most extreme form of duplex destabilization. By altering the helicity of the region involved, local denaturation diminishes the level of unconstrained superhelicity experienced by the rest of the topological domain. This process can affect any regulatory activity that is attuned to supercoiling levels. The converse process, whereby the binding of a DNA duplex binding protein to a destabilized site can force it back to B-form, also can be important. For example, this binding-induced reassociation can transmit the destabilization from the original SIDD site to the next most susceptible location, which can be a substantial distance away. We describe below cases where this process regulates promoter activity.

SIDD ANALYSIS

Local transitions to conformations that are less twisted in the right-handed sense than the B-form can be driven by negative superhelicity (i.e., $\Delta Lk < 0$). By accommodating some of the imposed linking difference as a net decrease of the local twist at the transition site(s), they diminish by a corresponding amount the superhelical deformation imposed on the balance of the molecule. Although transition to an alternative conformation requires energy, it also releases energy by this partial relaxation. In a domain containing a single susceptible site, transition will occur when the energy it releases exceeds its energy costs.

Experimentally observed superhelical transitions include cruciform extrusions (48), transitions to Z-form (83), and to H-form (40, 44). While regulatory roles have been suggested for each of these types of transitions, to date only strand separation has been shown to be widely involved in regulation.

The constraint imposed by DNA superhelicity is the constancy of the linking number within a topological domain. Suppose this domain contains N base pairs, and that it is in a state in which n base pairs are denatured. If the helical twist of B-DNA is A bp/turn, then n/A helical turns are untwisted when these n base pairs change from B-form to the unstressed denatured state. If $\Delta Lk = -n/A$, then this transition completely relaxes the imposed superhelicity. However, if n has any other value, complete relaxation does not occur and a residual deformation remains. Because single-stranded DNA is highly flexible, this residual deformation will cause the two single strands within an unpaired region to helically rotate around each other with a twist rate of τ radians per bp. The residual linking difference ΔLk_{res} is the amount of the imposed linking difference that remains after these two twist effects of the transition are accounted for. In this way, the superhelical constraint is expressed by the equation

$$\Delta Lk = \Delta Lk_{res} - (n/A) + (n\tau/2\pi).$$

Although thermal denaturation in linear or nicked DNA involves only near-neighbor interactions, when the transition is driven by superhelicity, this constraint globally couples together the conformational states of all the base pairs within a topological domain. This coupling occurs because transition of any base pair alters its helicity, which by the above equation changes the partitioning of ΔLk throughout the domain. So whether transition occurs at a given site depends not just on its local sequence properties, but also on how that transition competes with all others to which the domain is susceptible. For this reason, strictly local methods are not appropriate for analyzing superhelical transitions. Instead, these must be analyzed as global events, including competitions among all local transitions that the base sequence of the domain permits.

The behavior of a superhelical domain that contains multiple sites susceptible to transitions will be determined by a possibly complex competition among them. Whether or not transition occurs at a specific site will depend not just on its local properties, but also on how well that transition competes with others elsewhere in the domain. For example, Lilley and colleagues have shown that local denaturation of an A + T-rich region contained within a supercoiled DNA plasmid requires a significantly higher level of negative supercoiling when a $(TG)_{12}$ region is inserted in the plasmid (12). This insertion creates a site that can form left-handed Z-DNA under superhelical tension, and this transition competes effectively with denaturation. This illustrates how superhelically induced structural transitions compete with one another for the negative superhelical energy required for their formation, with the presence of a more susceptible site inhibiting less favorable transitions at other positions. However, this competition is not determined exclusively by the energy costs of the competing transitions themselves, but rather by the net energy difference for each between its cost and the relaxation it affords. Thus, even though the B-Z transition may be energetically more expensive than denaturation, it still can be favored because it produces significantly more relaxation per transformed base pair.

In general, a given linking difference imposed on a DNA domain can be accommodated by many combinations of torsional and flexural deformations, and by those conformational transitions to which its base sequence renders it susceptible. All of these responses to superhelicity require energy, and they all are topologically coupled together by the superhelical constraint arising from the requirement that the linking number must remain constant. In consequence, there are very many conformational states accessible to such a molecule. The competition among these alternative conformational states determines a thermodynamic equilibrium distribution. That is, a population of identical molecules at equilibrium distributes itself among all accessible states according to their energies, with low-energy states being exponentially more populated than high-energy states.

Benham and coworkers have developed theoretical techniques to analyze the superhelical stress-induced destabilization of the DNA double helix (5, 32, 86). Although these methods focus explicitly on denaturation, the same formalism can be applied to analyze other types of transitions. The values of the energy parameters used in these calculations have all been taken from experimental measurements. In particular, the free energy required for denaturation is known to depend significantly on base sequence and environmental conditions (15, 85). This is why destabilization is not uniform along a sequence, but instead is concentrated at specific locations.

These methods calculate two SIDD properties at single base-pair resolution in domains of specified base sequence and superhelicity. The transition profile is the graph of the probability of denaturation of each base pair along the DNA sequence. A more sensitive measure of destabilization may be calculated as the incremental free energy $G(x)$ needed to force the base pair at position x to always be separated (7, 25). A value of $G(x)$ near or below zero indicates an essentially completely destabilized base pair, which is predicted to denature with high probability at equilibrium. Positive values of $G(x)$ occur for base pairs where incremental free energy is needed to assure separation. Regions of partial destabilization are indicated by intermediate $G(x)$ values. Stress-induced duplex destabilization (SIDD) profiles are plots of $G(x)$ versus x. Figure 2 shows the SIDD profile (A) and the

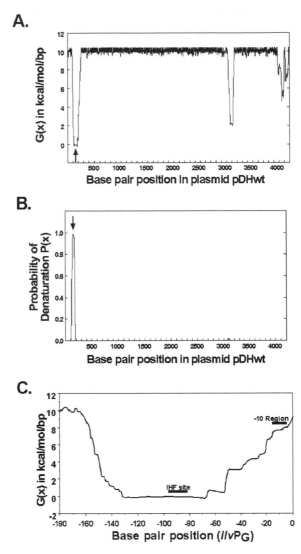

Figure 2   Stress-induced duplex destabilization (SIDD) profile of plasmid pDHΔwt at $\Delta Lk = \sigma - 0.05$. (*A*) SIDD profile of pDHΔwt indicating the predicted free energy G(x) required for DNA duplex destabilization as a function of base pair location x. The arrow indicates the location of the IHF binding site in the UAS region of the *ilv*P$_G$ promoter insert. (*B*) P(x) is the probability of DNA duplex destabilization as a function of base pair location x. The arrow indicates the location of the IHF binding site in the UAS1 region. (*C*) Closeup SIDD profile [G(x)] of pDHΔwt from bp positions −180 to +1 relative to the *ilv*P$_G$ transcriptional start site (bp positions 76 to 256 in the plasmid).

denaturation probability profile (*B*) for the experimental pDHΔwt plasmid containing the *ilv*P$_G$ promoter that was used in experiments described in the following section.

SIDD profiles are more informative than transition profiles because they also depict sites where the amount of free energy needed to induce denaturation is fractionally decreased, but not enough to denature with a significant probability. This will be important when duplex opening occurs by processes that can provide sufficient free energy to cause local denaturation only if the DNA site involved already is marginally destabilized by stresses. Such partially destabilized regions could be biologically important as facilitators of strand separation by enzymatic or other processes.

Calculations have been performed of the predicted stress-induced destabilization properties of numerous genomic DNA sequences. These calculations commonly assume a superhelix density of −0.055, which corresponds to a midphysiological value; however, calculations at any superhelical density are possible. The deformation and transition energy parameters are given their experimentally measured values, so there are no free parameters in the analysis. Yet the predictions of these calculations are in precise quantitative agreement with experimental results in all cases for which experimental data on the locations and extents of superhelical denaturation are available. The sites that denature are predicted exactly, and the calculated extents of transition at each site agree precisely with the experimental measurements. In every case, the predicted superhelicity required to drive a specific amount of separation is within one turn of the observed value over the whole range where experiments were performed (6, 9, 32). This reflects the limit of accuracy with which extents of transition can be experimentally measured. And most importantly, the major changes in the locations of destabilized regions that result from minor sequence alterations are precisely predicted.

GLOBAL REGULATION OF GENE EXPRESSION BY DNA TOPOLOGY-DEPENDENT MECHANISMS

To enable an organism to be both metabolically efficient and rapidly adaptive, mechanisms must exist to coordinate its global patterns of gene expression to its growth and environmental conditions. We propose that global levels of gene expression from specific promoters are coupled to the growth and nutritional states and environmental conditions of the cell through the regulation of transcriptional initiation by mechanisms that are sensitive to DNA superhelicity. This implies, for example, that the basal level expression of operons encoding structural genes for the biosynthesis of intermediary metabolites should be coordinated above the operon-specific level, so their basal expression levels are low when the chromosomal superhelical density is low, and are high when it is high. Work in our laboratories has shown this to be the case for operons of the *ilv* regulon, which encode the structural genes for the enzymes required for the biosynthesis of the branched chain amino acids, L-isoleucine, L-valine, and L-leucine (Figure 3). We also have

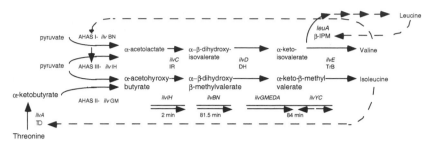

Figure 3   The biosynthetic pathways for the synthesis of the branched chain amino acids L-isoleucine, L-valine, and L-leucine in *E. coli*. The enzymes involved in the common pathway for branched chain amino acid biosynthesis are abbreviated as follows: AHAS, acetohydroxyacid synthase; IR, acetohydroxyacid isomeroreductase; DH, dihydroxyacid dehydrase; TrB, transaminase B; TD, threonine deaminase; β−IPM, β−isopropylmalate synthase. Genes encoding each of these enzymes are indicated in italics. Feedback inhibition patterns are indicated by dashed lines. The genomic organization of the operons of the *ilv* regulon is shown below the metabolic pathway.

shown that genes encoding tRNA^{Leu}, which are not in the *ilv* regulon, are also expressed in a manner that is tuned to biosynthetic demand through chromosomal supercoiling.

Unlike genes for catabolic systems that are transcriptionally inactive in the absence of a catabolite-inducer, genes required for the biosynthesis of intermediary metabolites such as amino acids must be continuously expressed at levels tuned to the amounts of their pathway end-products. For example, operons regulated by attenuation, such as the *ilvGMEDA*, *ilvBN*, and *leu* operons of the *ilv* regulon, continuously transcribe a leader RNA whose translation into a leader polypeptide monitors the intracellular levels of their pathway end-products (46). The *ilvY* gene of this regulon also must be continuously expressed to maintain an IlvY protein-DNA complex that continuously monitors cellular levels of the α-acetohydroxyacid isomeroreductase substrate, and adjusts the expression of the *ilvC* gene accordingly (71, 72). Because these monitoring activities that are typical of biosynthetic systems present a high energy cost to the cell, one expects global mechanisms to exist that coordinate them both with each other and with cellular demand. These global mechanisms would be expected to respond to the energy charge of the cell in a manner independent of operon-specific controls.

In this section we describe the *ilv* regulon, a well-understood system involving hierarchical levels of global and operon-specific regulation that together coordinate the biosynthesis of branched chain amino acids with the metabolic demands of the cell and its nutritional and environmental growth conditions. We begin with a brief description of the metabolic and operon-specific genetic regulatory mechanisms of this regulon that coordinate rates of carbon flow through these branched chain amino acid biosynthetic pathways with the expression levels of the genes encoding the enzymes of these pathways. Then we describe the energy charge-coupled, DNA supercoiling-dependent, mechanisms that coordinate the basal levels of expression of operons of the *ilv* regulon with one another, and with the nutritional and environmental growth conditions of the cell, in ways that are independent of the operon-specific controls.

Metabolic and Operon-Specific Regulatory Mechanisms of the *ilv* Regulon

METABOLIC REGULATION   Carbon flow through biosynthetic pathways is regulated by end-product inhibition of the first enzyme specific for each pathway (89), so threonine deaminase is end-product inhibited by L-isoleucine, and β-isopropylmalate synthase is end-product inhibited by L-leucine (Figure 3). However, because the parallel pathways for L-valine and L-isoleucine biosynthesis are catalyzed by a single set of bifunctional enzymes, L-valine inhibition of the first enzyme specific for its synthesis would compromise the cell for L-isoleucine biosynthesis. This type of a regulatory problem is often solved by using multiple isozymes that are differentially regulated by multiple end-products. In this case, there are three α-acetohydoxy acid synthase (AHAS) isozymes that catalyze the first step of the L-valine pathway, which is also the second step of the L-isoleucine pathway. AHAS I and AHAS III have substrate preferences for condensation of the two pyruvate molecules required for L-valine and L-leucine biosynthesis, and are both end-product inhibited by L-valine. AHAS III is also end-product inhibited by L-leucine. The third isozyme, AHAS II, which has a substrate preference for the condensation of pyruvate and α-ketobutyrate required for L-isoleucine biosynthesis, is not inhibited by any of the branched chain amino acids. These intricate metabolic circuits respond to substrate inputs and end-product outputs to insure a balanced flow of carbon substrates through these pathways under all growth conditions.

GENETIC REGULATION   The ilv regulon contains 15 structural genes organized into five operons, *ilvGMEDA*, *ilvBN*, *ilvIH*, *ilvYC*, and *leuABCD* (Figure 3). A variety of genetic regulatory mechanisms are involved in regulating these operons. The *ilvGMEDA*, *ilvBN*, and *leuABCD* operons are controlled by transcriptional attenuation mechanisms that respond to the levels of aminoacylated-, leucyl-, valyl-, and isoleucyl tRNA in the cell (31, 33, 36, 47, 57). The two remaining operons, *ilvIH* and *ilvYC*, are each regulated by other operon-specific mechanisms that are uniquely suited to the biosynthetic roles of their gene products.

The *ilvIH* structural genes encode the heterodimeric subunits of AHAS III. At the metabolic level, this isozyme is end-product inhibited by L-valine and L-leucine. At the genetic level, *ilvIH* operon expression is repressed by free L-leucine, as mediated by the global regulatory L-leucine-responsive protein, Lrp. In the absence of L-leucine, Lrp cooperatively binds to six highly degenerate sites within a 250-base pair upstream region, which activates transcription from the downstream promoter of the *ilvIH* operon. In the presence of L-leucine, this higher-order Lrp-DNA complex dissociates, and transcription from the promoter decreases (17). The supercoiling dependence of this operon has not yet been investigated.

Each of the four operons of the *ilv* regulon described to this point is regulated at the genetic level by mechanisms that respond to the intracellular levels of their pathway-specific end-products, either a free branched chain amino acid (*ilvIH*) or the cognate branched chain aminoacyl-tRNAs (*ilvGMEDA, ilvBN,* and *leu*). In contrast, the *ilvC* gene of the *ilvYC* operon is regulated by its substrates, not by pathway end-products. To understand this unusual situation, consider the biochemical role of the *ilvC* gene product, α-acetohydroxyacid isomeroreductase. This enzyme catalyzes the rate-limiting step in parallel pathways, and hence must be responsive to changing concentrations of substrates produced by the three differentially regulated AHAS isozymes. By tuning its gene expression level to the concentration of substrate, it can keep carbon efficiently flowing through any or all of these parallel pathways.

The *ilvY* and *ilvC* genes comprising the *ilvYC* operon are oppositely oriented and divergently transcribed from overlapping promoter sites (Figure 4). Operon-specific regulation is mediated by the IlvY protein, the product of the *ilvY* gene. IlvY is a member of the LysR family of regulatory proteins (78). It is a homodimer that cooperatively binds to adjacent operator sites O_1 and O_2 in the divergent promoter region in a manner independent of substrate-inducer concentration. The O_1 site covers the region between positions -10 and $+1$ of the *ilvY* promoter. By modulating the expression of its own gene, IlvY effectively autoregulates its own synthesis. The O_2 site is located on the opposite face of the helix from O_1, and overlaps the -35 region of the *ilvC* promoter (93, 78). In the absence of substrate-inducers, the IlvY protein causes a $60°$ bend centered at the -35 region of the inactive *ilvC* promoter. When substrate-inducers bind to this preformed IlvY protein-DNA complex, the bend is relaxed and the affinity for RNA polymerase is increased 100-fold (72). The IlvY protein is autoregulated at a level that keeps these operators nearly saturated at all times. In this way, the IlvY protein-DNA complex acts as a sensor of the intracellular abundance of the α-acetohydroxyacid isomeroreductase substrates, thereby continuously adjusting expression of the *ilvC* gene to the abundance of its substrates, synthesized by the three AHAS isozymes.

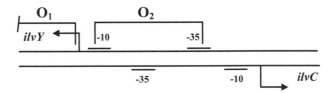

Figure 4   The divergent promoter region of the *ilvYC* operon. Arrows identify the transcription start sites for the *ilvY* and *ilvC* genes. The -10 and -35 hexanucleotide regions of the *ilvY* and *ilvC* promoters are identified with horizontal lines. The O_1 (truncated) and O_2 operator sites are denoted by brackets.

Global Control of Basal Level Expression of the Operons of the *ilv* Regulon

At this time, three DNA supercoiling-dependent transcriptional regulatory mechanisms related to branched chain amino acid biosynthesis and utilization have been documented (62, 61, 79, 80; M.L. Opel & G.W. Hatfield, unpublished data). These act independently of the operon-specific controls described above to coordinate the expression levels of the operons of the *ilv* regulon with each another, and with the nutritional needs and growth state of the cell in its physical environment. Two of these mechanisms (those governing the *ilvGMEDA* and *LeuV* operons) modulate basal level transcription into their structural genes by a protein-mediated (IHF or Fis) transmission of local superhelical energy from an upstream SIDD site to a downstream promoter site. This influences the rate of RNAP-promoter open complex formation and/or the rate of RNAP binding (79, 80; M. L. Opel & G.W. Hatfield, unpublished data). In the third case, the basal level expression of the *ilvC* gene is enhanced by additional local superhelical energy contributed to the *ilvC* promoter region by divergent transcription of the *ilvY* gene (61, 62, 71). In each case, local superhelical energy is provided to the promoter regions to amplify promoter activity over the entire range of global physiological superhelical densities in a manner that coordinates the basal level expression of these operons within multiple regulons, both with one another and with the energy charge of the cell. Each of these mechanisms is described separately below.

REGULATION OF TRANSCRIPTIONAL INITIATION BY PROTEIN-MEDIATED TRANSLOCATION OF SUPERHELICAL ENERGY   As described above, negative superhelicity imposed on a DNA domain can drive local transitions to alternative, nonB-DNA conformations. This transition behavior can be influenced by proteins that bind at or near susceptible sites. For example, if a DNA region that would be favored to form a superhelix-induced alternative structure becomes trapped in the B-form by the binding of a protein, transition may instead occur at the next most favored site, which could be remote along the sequence. In this way, protein binding events can cause the translocation of destabilization to other sites. If this second site is in the -10 region of a promoter, this binding-induced translocation of destabilization can activate transcriptional initiation by facilitating open complex formation. This is the basic mechanism employed by the *ilvGMEDA* and *leuV* operons.

The ilvGMEDA operon   Nested 5'-deletions extending into the *ilv*P_G promoter of the *ilvGMEDA* operon identified an upstream activating region (UAS). This UAS contains a high-affinity Integration Host Factor (IHF) target binding site located 92 base pairs upstream from the transcriptional start site. Biochemical and genetic experiments have shown that IHF binding to this site, both in vivo and in vitro on a supercoiled DNA template, causes a fivefold activation of transcription from the downstream *ilv*P_G promoter (97). Several experimental approaches have established that this activation occurs in the absence of specific protein interactions

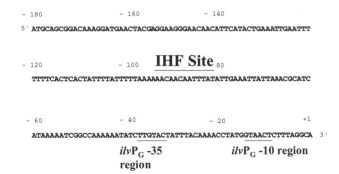

Figure 5

Figure 5   DNA sequence of the *ilv*P$_G$ UAS-promoter region. See text for discussion.

between IHF and RNA polymerase; it is not the consequence of a DNA looping mechanism; and it requires a superhelical DNA template (64, 65, 79–81).

The possible presence of a SIDD site in the UAS was first suggested by the observation that the base-pair composition of the *ilv*P$_G$ promoter-regulatory region is exceptionally A + T rich, the 80-bp segment from bp positions −67 to −153 being approximately 88% (A + T) (Figure 5). In order to determine if this region does indeed contain a SIDD site, SIDD profiles were calculated for the pBR322-based plasmid pDHΔwt (79). This plasmid contains the *ilv*P$_G$ promoter region from positions −248 to +6, together with transcriptional terminators located downstream from the *ilv*P$_G$ start site. The results of these calculations are presented as destabilization profiles in Figure 2. Subsequent chemical probing experiments confirmed that, in the absence of IHF, the UAS region was indeed destabilized at the superhelix densities where activation occurs. Moreover, IHF binding in the UAS region of a superhelical DNA template resulted in the transmission of this duplex destabilization into the −10 region of the downstream *ilv*P$_G$ promoter site. Abortive transcription assays showed that this DNA structural change at the downstream promoter site is correlated both with an increase in the rate of open complex formation, and with a concomitant increase in the rate of transcriptional initiation (79).

These results suggested that a novel, protein-mediated, DNA supercoiling-dependent, DNA structural transmission mechanism regulates basal level transcription from the *ilv*P$_G$ promoter (Figure 6). In this mechanism, the binding of IHF prevents superhelical destabilization at the SIDD site in the A + T-rich UAS region, which transfers that destabilization to the −10 region of the nearby *ilv*P$_G$ promoter. This explanation accounts for the DNA supercoiling-dependence of the IHF-mediated activation of transcription from this promoter, and for the fact that this activation occurs in the absence of specific interactions with RNAP.

According to this mechanism, the primary determinant for IHF-mediated activation is predicted to be superhelically induced DNA destabilization. Neither specific DNA sequences nor specific IHF-RNAP interactions are required. To

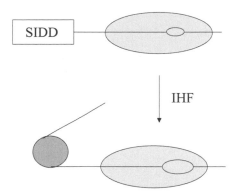

Figure 6 DNA structural transmission mechanism for protein-mediated DNA superhelical-dependent transcriptional activation of the *ilv*PG promoter. In this model, IHF binding prevents the superhelical destabilization of the SIDD region of the UAS, which transfers the transition to the −10 region of the downstream *ilv*PG promoter, thereby facilitating open complex formation during transcription initiation. Open complex formation is represented by a bubble (*denoted by an oval*) in the duplex DNA (*designated by a single line*). Increased open complex formation is indicated by the increased size of the bubble.

directly test this prediction, the sequence (CG)13AATT(CG)22 was inserted into plasmid pDHΔwt approximately 500 base pairs upstream from the UAS-*ilv*P$_G$ promoter to yield plasmid pSSΔZ (80). This sequence is susceptible to a superhelically induced transition to Z-form DNA (69). Because the B-Z transition at this remote site competes effectively with the SIDD site in the UAS, this insertion can alter the destabilization properties of the UAS without changing the DNA sequence in any part of the *ilv*P$_G$ regulatory-promoter region. This transition was shown to absorb 13 negative superhelical turns, thereby relaxing the global superhelix density of the remainder of the supercoiled DNA template by a corresponding amount. Since this B-Z transition occurs at a lower threshold superhelical density ($\sigma = -0.025$) than does the destabilization of UAS ($\sigma = -0.038$), it inhibits UAS destabilization until approximately 13 additional negative superhelical turns have been added to the DNA template. If the energy required for IHF-mediated transcriptional activation is indeed derived by IHF binding-induced transfer from the destabilized UAS region, then the superhelicity required for IHF activation in the pSSΔZ plasmid should be offset by approximately 13 turns. The results of transcription assays on DNA templates of defined superhelix densities showed this to be the case. The superhelicities required both for half-maximal basal level and for IHF-activated transcription were indeed offset by 13 turns (79). This experiment clearly demonstrated that IHF-mediated transcriptional activation of the *ilv*P$_G$ promoter is solely DNA supercoiling-dependent and confirmed the predictions of the protein-mediated, DNA structural transmission mechanism of transcriptional activation proposed above.

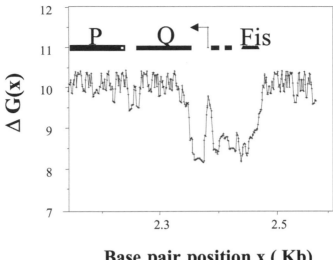

Base pair position x (Kb)

Figure 7 SIDD profile of the *leuV* promoter region. Predicted free energies G(x) for duplex destabilization at a superhelical density of $\sigma = -0.055$ at base position x are expressed in kcal/mol/bp. The positions of relevant features are indicated above the SIDD plot. The transcriptional start site is indicated by an arrow. The locations of the Fis binding site, the -10 and -35 hexanucleotide regions, and the promoter-proximal structural genes are identified by thick horizontal lines. The discriminator is the peak of duplex stability in the SIDD region near the transcriptional start site. The three identical structural genes for the tRNA^{Leu1} isoacceptor of the *leuV* operon are *leuQ*, *leuP*, and *leuV* (not shown).

The leuV operon   The *leuV* operon of *E. coli* encodes three of the five genes for tRNA^{Leu1} isoacceptor. Like other stable RNA-encoding genes, it has a strong promoter, with near-consensus RNAP recognition sequences and a G + C-rich discriminator region located between base pair positions -8 and $+1$ (Figure 7). Transcription from the *leuV* promoter is enhanced by a third RNAP recognition element located between base pairs -39 to -47. This A + T-rich UP sequence makes contacts with the α-subunits of RNAP, stabilizes closed complex formation, and activates *leuV* expression more than tenfold (30, 67, 73). The upstream activating sequence (UAS) of this promoter contains a Fis protein binding site centered at base pair position -71. Fis binding to this site enhances *leuV* expression an additional threefold (74).

A large body of evidence demonstrates that Fis is a class-I activator, enhancing transcription by increasing RNAP binding affinity through direct contacts with the C-terminal domain of its α-subunits (12, 13, 15, 60, 98). However, other mechanisms also are involved in Fis activation of stable RNA promoters. For example, Fis binding has been shown to increase the rate of promoter clearance at the *rrnD* promoter (77). Muskhelishvili & Travers have shown that Fis activates transcription from the *tyrT* promoter by enhancing the rate of open complex formation and promoter clearance, as well as RNA polymerase binding affinity (56). Most recently, Opel et al. have obtained evidence that basal level expression of the *leuV* promoter is also activated by a Fis-mediated translocation of superhelical energy mechanism similar to the IHF-mediated, DNA supercoiling-dependent mechanism previously described for the *ilv*P_G operon (M.L. Opel & G.W. Hatfield, unpublished data).

A SIDD profile of the promoter-regulatory region of the *leuV* operon is shown in Figure 8. At a midphysiological superhelical density of $\sigma = -0.05$, this region is predicted to be destabilized from base pair positions +43 to −94. The Fis protein binding site is located in the upstream region of this SIDD site centered at base-pair position −72. An interesting feature of the SIDD profile is the sharp peak of duplex stability at the G + C-rich discriminator region between positions +1 and −8. The presence of this region of high duplex stability at a site where strand separation must occur predicts a high energy of activation for open complex formation. Indeed, *leuV* transcription is exceptionally sensitive to negative DNA supercoiling, increasing over 100-fold from its lowest level on a relaxed DNA template to its highest level on a more supercoiled DNA template. The SIDD profile further predicts that the upstream region of the SIDD site should be stabilized by Fis binding to its target site, and that this binding should destabilize the downstream portion of this SIDD region containing the *leuV* promoter sequences. This prediction was confirmed with in vitro transcription assays and KMnO$_4$ structural probing experiments performed with supercoiled DNA template topoisomers in the presence or absence of Fis and/or RNAP. These experiments showed that Fis binding enhances the rate of open complex formation in a DNA supercoiling-dependent manner. At subsaturating concentrations of RNAP, Fis activation is facilitated both by protein-protein interactions between Fis and RNAP, and by DNA supercoiling-dependent enhancement of open complex formation. At saturating RNAP concentrations, only the enhancement of open complex is seen. Mutant Fis proteins that do not form contacts with the α-subunits of RNAP but bind to the target site with wild-type affinities were used to demonstrate that this activation does not require Fis-RNAP interactions (M.L. Opel & G.W. Hatfield, unpublished data). These mutant proteins maintain their ability to facilitate DNA supercoiling-dependent enhancement of open complex formation. Thus, Fis activation of basal level transcription from the *leuV* promoter involves at least two mechanisms: stabilization of the closed complex by protein interactions with the α-subunit of RNAP, and increasing open complex formation by translocation of superhelical energy from the upstream portion of the SIDD region containing the Fis target site to the downstream portion of this region containing the *leuV* promoter sequences.

Unlike the *ilv*P_G promoter that reaches its peak transcriptional activity at a high physiological superhelical density near $\sigma = -0.10$, transcriptional activity of the *leuV* promoter peaks at a superhelical density near $\sigma = -0.07$ and decreases thereafter to the level observed on a relaxed DNA template. This decrease in

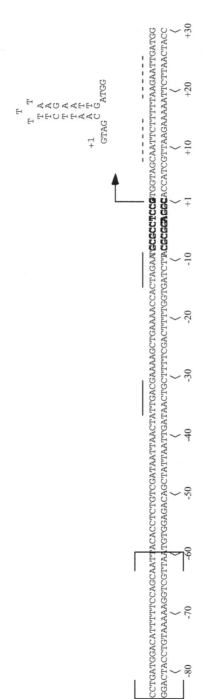

Figure 8   Nucleotide sequence of the *leuV* promoter region from −84 to +30. The transcriptional start site for *leuV* is identified by an arrow. The −10 and −35 hexanucleotide regions are identified by horizontal lines. The Fis binding site is denoted by brackets. The discriminator is identified by bold-face type. An inverted repeat is identified by a dashed line. Cruciform extrusion at the inverted repeat is depicted above the inverted repeat.

transcriptional activity at high superhelix densities is accompanied by the formation of a cruciform structure located between base-pair positions +8 to +26 relative
to the transcriptional start site. Structural probing experiments with DNA topoisomers show that the threshold superhelical density required for extrusion of this
cruciform structure is $\sigma = -0.069$, the same superhelical density beyond which
transcription from the *leuV* promoter decreases. This suggests that, at superhelical
densities beyond this threshold, the global and Fis-transferred local superhelical
energy in the promoter region is absorbed by the cruciform. This transition has two
effects on *leuV* gene expression: It usurps the local superhelical energy that would
otherwise have been transferred to the promoter region by Fis for open complex
formation, and it physically blocks RNAP binding.

A schematic diagram illustrating how Fis binding and DNA supercoiling-
induced structural transitions regulate transcription from the *leuV* promoter is
presented in Figure 9. At the low physiological superhelical densities typical of
stationary phase growth, transcription from the *leuV* promoter is very low. This is
due to the energy barrier for open complex formation caused by the G + C-rich discriminator near the transcription start site. Under these conditions, Fis can activate
transcription about threefold by increasing RNAP binding through interactions
with its α-subunits. Since this low level of global superhelicity is insufficient for
SIDD site formation, no additional activation by Fis-mediated translocation of superhelical energy to the promoter site is possible. As the global superhelicity of the
DNA template is increased to the midphysiological range, the energy barrier for
open complex formation caused by the discriminator is decreased and transcription

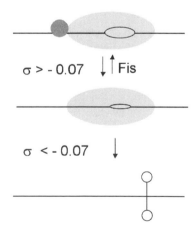

Figure 9   DNA supercoiling and Fis-mediated regulation of the *leuV* promoter. RNA
polymerase is represented by a large oval. Fis is represented by a small circle. Open
complex formation is represented by a bubble in the duplex DNA (*designated by a
single line*). The strength of open complex formation is indicated by the size of the
bubble. A cruciform is indicated by a stem-loop structure. See text for details.

increases up to 100-fold. As the SIDD site now is present, Fis binding can activate transcription both by enhancing RNAP binding and by protein-mediated translocation of superhelical energy to the promoter region. As the global superhelical density passes beyond the midphysiological range, the superhelical energy at the SIDD site is usurped by formation of the cruciform structure near the transcription start site, and RNAP binding is inhibited.

REGULATION OF TRANSCRIPTIONAL INITIATION BY TRANSCRIPTIONAL COUPLING
LysR-type regulated operons are the largest class of positively regulated operons, and are found in many prokaryotic species (37, 78). The prevalence of a divergent gene arrangement among the LysR-type regulated operons suggests an evolutionary conservation of potential regulatory significance. According to the twin-domain model (49), a local domain having a high level of DNA supercoiling can be generated between, and influence the activities of, divergently transcribed promoters. Mojica & Higgins have used in vivo psoralen cross-linking techniques to demonstrate that localized domains of increased negative DNA supercoiling are indeed generated upstream from an actively transcribed promoter (55). They demonstrated DNA supercoiling-mediated transcriptional coupling between the divergently oriented *tetA* and mutant *leu-500* promoters.

The *ilvYC* operon of *E. coli* K-12 is a prototypic LysR-type regulated system (78, 37). Rhee et al. used double-reporter gene constructs to provide the first in vivo evidence for transcriptional coupling in a naturally occurring system, the *ilvYC* operon of *E. coli* (71). They showed that each of these promoters is intrinsically sensitive to global DNA supercoiling, and that a 13-fold decrease in transcriptional activity from the *ilvY* promoter results in an 11-fold decrease in transcription from the divergent *ilvC* promoter. This transcriptional coupling was shown to be the consequence of transcription-induced negative DNA supercoiling. In this situation, a highly stressed local topological domain is created in the promoter region by divergent transcription, in which the total supercoiling is the sum of the basal, global superhelicity plus the supercoiling arising from divergent transcription. This suggested a strategy to document and characterize transcriptional coupling in a purified in vitro transcription system. When a set of DNA topoisomer templates containing the wild-type, divergently oriented *ilvY* and *ilvC* promoters were transcribed in a purified system, optimal transcriptional activity was observed to occur at superhelical density $\sigma = -0.065$ for the *ilvY* promoter, and $\sigma = -0.11$ for the *ilvC* promoter (71). If the level of negative DNA supercoiling in the divergently transcribed promoter region were in fact proportional to the sum of transcription-induced (local) DNA supercoiling and the global superhelical density of the DNA template, then a decrease in transcription (hence in transcription-induced supercoiling) from either promoter should require a compensating increase in global DNA supercoiling to maintain maximal transcription from its divergently transcribed partner. Conversely, an increase in transcription from either promoter should require a corresponding decrease in global supercoiling to maintain maximal transcription from the divergently oriented other promoter.

Further, the twin-domain model of Liu & Wang (49) also predicts that the levels of transcription-induced negative DNA supercoiling in the divergent promoter domain region should be proportional to the lengths of the transcripts. Opel & Hatfield have performed in vitro experiments using a purified transcription system and DNA topoisomer templates containing the genes of the *ilvYC* operon that confirm both of these predictions (62).

SUMMARY AND PERSPECTIVES

We have proposed a model in which chromosomal superhelicity serves as a global regulator of gene expression in *E. coli*, tuning expression patterns across multiple operons, regulons, and stimulons to suit the current growth state, nutritional requirements, and environmental conditions of the cell. Because the level of DNA superhelicity varies with the cellular energy charge, it can change rapidly—often in less than a minute—in response to a wide variety of altered cellular and environmental conditions. This is a global alteration, affecting the entire chromosome and the expression levels of all operons whose promoters are sensitive to superhelicity. In this way, the global pattern of gene expression may be dynamically tuned to changing needs of the cell under a wide variety of circumstances.

This model is illustrated by the DNA supercoiling-dependent mechanisms that control basal level expression of the *ilvYC* and *ilvGMEDA* operons. The basal expression levels of these operons are coordinated by distinct, DNA supercoiling-dependent mechanisms. However, both mechanisms have the same effect—to provide additional local superhelical energy to the promoter regions to lower the energy of activation for open complex formation and to increase the rate of transcription initiation from the promoter. This additional energy is important for biosynthetic promoters that must coordinate their transcriptional activity over the entire superhelical density range. Such promoters must possess a high superhelical density-activity optimum. This makes them, by definition, intrinsically weak promoters because of the high energy of activation required for open complex formation. However, by supplementing the global superhelicity of the chromosome with locally generated superhelicity, their promoter activities can be amplified to a level sufficient for their basal level functions, while at the same time maintaining their ability to increase their activity over the entire range of physiological superhelical densities. In the case of the *ilvGMEDA* operon, this is accomplished by IHF-mediated translocation of superhelical energy from an upstream SIDD site to the downstream promoter region. In the case of the *ilvYC* operon, this is accomplished by the additional local superhelical energy contributed to the *ilvC* promoter region by the divergently transcribed *ilvY* promoter. In both cases, promoter activities increase tenfold over the physiological superhelical densities that the chromosome encounters. Therefore, these different mechanisms serve to coordinate the basal level of expression of these two operons of a common biosynthetic pathway with one another and with the nutritional and physical signals that determine energy charge of the cell.

Since free amino acids are penultimate end-products for protein synthesis, it is perhaps not surprising that the operon encoding genes for the major leucyl-tRNA isoacceptors is also regulated by a similar DNA supercoiling-dependent mechanism. However, in this case the global regulatory, abundant, DNA architectural protein that mediates translocation of superhelical energy from the upstream portion of the SIDD site to the downstream promoter is Fis rather than IHF. Since the *leuV* operon is concerned with consumption rather than production, it has evolved a supercoiling-dependent mechanism to augment L-leucine conservation during periods of stress where the superhelical density of the chromosome is high. Under these conditions, *leuV* expression is controlled by a complex competition between an activating duplex destabilization, which is favored at midphysiological superhelical densities, and an inhibiting cruciform extrusion that occurs at more extreme superhelicity levels.

Note that these DNA supercoiling-dependent global regulatory mechanisms described here affect operon expression in a manner independent of the specific controls whereby these operons respond to the in vivo levels of metabolically important, small-molecule coregulators and other cellular signals. Instead, these mechanisms serve to adjust the basal level, or capacity, for operon expression according to changes in DNA supercoiling that reflect the energy charge of the cell.

There are many possible approaches for coupling superhelicity in a physiologically meaningful way. We have documented two distinct types of mechanisms— transcriptional coupling at divergent promoters, and a novel, binding-induced transmission of destabilization. The prevalence of divergently oriented ORFs in *E. coli* suggests that some form of transcriptional coupling could represent a widespread regulatory mechanism. Although to date the transmission-of-destabilization mechanism has only been demonstrated to regulate two operons, other evidence suggests that this too could be a widely used regulatory strategy. Perhaps the most intriguing clue is the fact that both operons are regulated by highly abundant DNA architectural proteins, IHF and Fis, that bind to hundreds of sites on the *E. coli* chromosome. Another intriguing observation is that recently completed calculations of the SIDD profiles of the entire *E. coli* chromosome suggest the presence of approximately 1100 SIDD sites. Furthermore, the great majority of these sites are found in intergenic regions close to promoters (C. J. Benham, unpublished data).

Certainly, much work remains to assess the generality of our proposed model. Careful experimental evaluation of the structure and sequence motifs of IHF and Fis binding sites will enable us to search the *E. coli* genome for binding sites, and cross-linking immunoprecipitation experiments with DNA microarrays will allow an estimation of in vivo occupancy at these sites. These results together with the results of gene expression profiling experiments will allow us identify genes in the vicinity of IHF or Fis binding sites that overlap SIDD sites. This approach will illuminate which genes are regulated by supercoiling-dependent mechanisms, which genes are regulated by IHF or Fis, and which among these are also regulated in a supercoiling-dependent manner.

The *Annual Review of Genetics* is online at http://genet.annualreviews.org

LITERATURE CITED

1. Atkinson DE. 1965. Biological feedback control at the molecular level. *Science* 150:851–57
2. Atkinson DE. 1969. Regulation of enzyme function. *Annu. Rev. Microbiol.* 23:47–68
3. Benham CJ. 1979. Torsional stress and local denaturation in supercoiled DNA. *Proc. Natl. Acad. Sci. USA* 76:3870–74
4. Benham CJ. 1980. Theoretical analysis of transitions between B- and Z-conformations in torsionally stressed DNA. *Nature* 286:637–38
5. Benham CJ. 1990. Transitions in superhelical DNA molecules of specified sequence. *J. Chem. Phys.* 92:6294–305
6. Benham CJ. 1992. Energetics of the strand separation transition in superhelical DNA. *J. Mol. Biol.* 225:835–47
7. Benham CJ. 1993. Sites of predicted stress-induced DNA duplex destabilization occur preferentially at regulatory loci. *Proc. Natl. Acad. Sci. USA* 90:2999–3003
8. Benham CJ. 1996. Duplex destabilization in superhelical DNA is predicted to occur at specific transcriptional regulatory regions. *J. Mol. Biol.* 255:425–34
9. Benham C, Kohwi-Shigematsu T, Bode J. 1997. Stress-induced duplex DNA destabilization in scaffold/matrix attachment regions. *J. Mol. Biol.* 274:181–96
10. Bensaid A, Almeida A, Drlica K, Rouviere J. 1996. Cross-talk between topoisomerase I and HU in *Escherichia coli. J. Mol. Biol.* 256:292–300
11. Bliska JB, Cozzarelli NR. 1987. Use of site-specific recombination as a probe of DNA structure and metabolism in vivo. *J. Mol. Biol.* 194:205–18
12. Bokal AJ, Ross W, Gaal T, Johnson RC, Gourse RL. 1997. Molecular anatomy of a transcription activation patch: FIS-RNA polymerase interactions at the *Escherichia coli* rrnB P1 promoter. *EMBO J.* 16:154–62
13. Bokal AJ, Ross W, Gourse RL. 1995. The transcriptional activator protein FIS: DNA interactions and cooperative interactions with RNA polymerase at the *Escherichia coli* rrnB P1 promoter. *J. Mol. Biol.* 245:197–207
14. Borowiec JA, Gralla JD. 1985. Supercoiling response of the lac ps promoter in vitro. *J. Mol. Biol.* 184:587–98
15. Breslauer KJ, Frank R, Blocker H, Marky LA. 1986. Predicting DNA duplex stability from the base sequence. *Proc. Natl. Acad. Sci. USA* 83:3746–50
16. Calladine CR. 1997. *Understanding DNA: The Molecule and How it Works.* New York: Academic
17. Calvo JM, Matthews RG. 1994. The leucine-responsive regulatory protein, a global regulator of metabolism in *Escherichia coli. Microbiol. Rev.* 58:466–90
18. Chapman AG, Atkinson DE. 1977. Adenine nucleotide concentrations and turnover rates. Their correlation with biological activity in bacteria and yeast. *Adv. Microb. Physiol.* 15:253–306
19. Chapman AG, Fall L, Atkinson DE. 1971. Adenylate energy charge in *Escherichia coli* during growth and starvation. *J. Bacteriol.* 108:1072–86
20. Cortassa S, Aon MA. 1993. Altered topoisomerase activities may be involved in the regulation of DNA supercoiling in aerobic-anaerobic transitions in *Escherichia coli. Mol. Cell Biochem.* 126:115–24
21. Cozzarelli NR. 1980. DNA gyrase and the supercoiling of DNA. *Science* 207:953–60
22. Dai X, Greizerstein MB, Nadas Chinni K, Rothman Denes LB. 1997. Supercoil-induced extrusion of a regulatory DNA hairpin. *Proc. Natl. Acad. Sci. USA* 94:2174–79
23. Dai X, Rothman-Denes LB. 1999. DNA

structure and transcription. *Curr. Opin. Microbiol.* 2:126–30

24. Deibler RW, Rahmati S, Zechiedrich EL, 2001. Topoisomerase IV, alone, unknots DNA in *E. coli. Genes Dev.* 15:748–61

25. DiNardo S, Voelkel KA, Sternglass R, Reynolds AE, Wright A. 1983. *Escherichia coli* DNA topoisomerase I mutants have compensatory mutations at or near DNA gyrase genes. *Cold Spring Harbor Symp. Quant. Biol.* 47:779–84

26. Drew HR, Weeks JR, Travers AA. 1985. Negative supercoiling induces spontaneous unwinding of a bacterial promoter. *EMBO J.* 4:1025–32

27. Drlica K. 1992. Control of bacterial DNA supercoiling. *Mol. Microbiol.* 6:425–33

28. Drlica K, Franco RJ. 1988. Inhibitors of DNA topoisomerases. *Biochemistry* 27:2253–59

29. Drlica K, Zhao X. 1997. DNA gyrase, topoisomerase IV, and the 4-quinolones. *Microbiol. Mol. Biol. Rev.* 61:377–92

30. Estrem ST, Gaal T, Ross W, Gourse RL. 1998. Identification of an UP element consensus sequence for bacterial promoters. *Proc. Natl. Acad. Sci. USA* 95:9761–66

31. Friden P, Newman T, Freundlich M. 1982. Nucleotide sequence of the ilvB promoter-regulatory region: a biosynthetic operon controlled by attenuation and cyclic AMP. *Proc. Natl. Acad. Sci. USA* 79:6156–60

32. Fye RM, Benham CJ. 1999. Exact method for numerically analyzing a model of local denaturation in superhelically stressed DNA. *Phys. Rev.* 59:3408–26

33. Gemmill RM, Wessler SR, Keller EB, Calvo JM. 1979. leu operon of *Salmonella typhimurium* is controlled by an attenuation mechanism. *Proc. Natl. Acad. Sci. USA* 76:4941–45

34. Giaever GN, Snyder L, Wang JC. 1988. DNA supercoiling in vivo. *Biophys. Chem.* 29:7–15

35. Gosink KK, Ross W, Leirmo S, Osuna R, Finkel SE, et al. 1993. DNA binding and bending are necessary but not sufficient for

Fis-dependent activation of rrnB P1. *J. Bacteriol.* 175:1580–89

36. Hauser CA, Hatfield GW. 1983. Nucleotide sequence of the ilvB multivalent attenuator region of *Escherichia coli* K12. *Nucleic Acids Res.* 11:127–39

37. Henikoff S, Haughn GW, Calvo JM, Wallace JC. 1988. A large family of bacterial activator proteins. *Proc. Natl. Acad. Sci. USA* 85:6602–6

38. Higgins CF, Dorman CJ, Stirling DA, Waddell L, Booth IR, et al. 1988. A physiological role for DNA supercoiling in the osmotic regulation of gene expression in *S. typhimurium* and *E. coli. Cell* 52:569–84

39. Hsieh LS, Rouviere-Yaniv J, Drlica K. 1991. Bacterial DNA supercoiling and [ATP]/[ADP] ratio: changes associated with salt shock. *J. Bacteriol.* 173:3914–17

40. Htun H, Dahlberg JE. 1989. Topology and formation of triple-stranded H-DNA. *Science* 243:1571–76

41. Jensen PR, Loman L, Petra B, Vanderwe C, Westerhouse HV. 1995. Energy buffering of DNA structure fails when *Escherichia coli* runs out of substrate. *J. Bacteriol.* 177:3420–26

42. Jones PG, Inouye M. 1994. The cold-shock response—a hot topic. *Mol. Microbiol.* 11:811–18

43. Jones PG, VanBogelen RA, Neidhardt FC. 1987. Induction of proteins in response to low temperature in *Escherichia coli. J. Bacteriol.* 169:2092–95

44. Kowalski D, Natale DA, Eddy MJ. 1988. Stable DNA unwinding, not "breathing," accounts for single-strand-specific nuclease hypersensitivity of specific A+T-rich sequences. *Proc. Natl. Acad. Sci. USA* 85:9464–68

45. Kusano S, Ding QQ, Fujita N, Ishihama A. 1996. Promoter selectivity of *Escherichia coli* RNA polymerase E sigma 70 and E sigma 38 holoenzymes. Effect of DNA supercoiling. *J. Biol. Chem.* 271:1998–2004

46. Landick R, Yanofsky C. 1987. Transcription attenuation. See Ref. 58, 1:1276–301

47. Lawther RP, Hatfield GW. 1980. Multivalent translational control of transcription termination at attenuator of ilvGEDA operon of *Escherichia coli* K-12. *Proc. Natl. Acad. Sci. USA* 77:1862–66

48. Lilley DM. 1980. The inverted repeat as a recognizable structural feature in supercoiled DNA molecules. *Proc. Natl. Acad. Sci. USA* 77:6468–72

49. Liu LF, Wang JC. 1987. Supercoiling of the DNA template during transcription. *Proc. Natl. Acad. Sci. USA* 84:7024–27

50. Liu Y, et al. 2001. DNA supercoiling allows enhancer action over a large distance. *Proc. Natl. Acad. Sci. USA* 98:14883–88

51. Luttinger A. 1995. The twisted 'life' of DNA in the cell: bacterial topoisomerases. *Mol. Microbiol.* 15:601–6

52. McEachern F, Fisher LM. 1989. Regulation of DNA supercoiling in *Escherichia coli*: genetic basis of a compensatory mutation in DNA gyrase. *FEBS Lett.* 253:67–70

53. Menzel R, Gellert M. 1983. Regulation of the genes for *E. coli* DNA gyrase: homeostatic control of DNA supercoiling. *Cell* 34:105–13

54. Michelotti GA, Michelotti EF, Pullner A, Duncan RC, Eick D. 1996. Multiple single-stranded cis elements are associated with activated chromatin of the human c-myc gene in vivo. *Mol. Cell Biol.* 16:2656–69

55. Mojica FJ, Higgins CF. 1996. Localized domains of DNA supercoiling: topological coupling between promoters. *Mol. Microbiol.* 22:919–28

56. Muskhelishvili G, Buckle M, Heumann H, Kahmann R, Travers AA. 1997. FIS activates sequential steps during transcription initiation at a stable RNA promoter. *EMBO J.* 16:3655–65

57. Nargang FE, Subrahmanyam CS, Umbarger HE. 1980. Nucleotide sequence of ilvGEDA operon attenuator region of *Escherichia coli*. *Proc. Natl. Acad. Sci. USA* 77:1823–27

58. Neidhardt FC, ed. 1987. *Escherichia coli and Salmonella typhimurium: Cellular and Molecular Biology*. Washington, DC: ASM

59. Neidhardt FC, Savageau MA. 1996. Regulation beyond the operon. See Ref. 58, 2:1310–24

60. Newlands JT, Josaitis CA, Ross W, Gourse RL. 1992. Both fis-dependent and factor-independent upstream activation of the rrnB P1 promoter are face of the helix dependent. *Nucleic Acids Res.* 20:719–26

61. Opel ML, Arfin SM, Hatfield GW. 2001. The effects of DNA supercoiling on the expression of operons of the ilv regulon of *Escherichia coli* suggest a physiological rationale for divergently transcribed operons. *Mol. Microbiol.* 39:1109–15

62. Opel ML, Hatfield GW. 2001. DNA supercoiling-dependent transcriptional coupling between the divergently transcribed promoters of the ilvYC operon of *Escherichia coli* is proportional to promoter strengths and transcript lengths. *Mol. Microbiol.* 39:191–98

63. Oram M, Fisher LM. 1992. An *Escherichia coli* DNA topoisomerase I mutant has a compensatory mutation that alters two residues between functional domains of the DNA gyrase A protein. *J. Bacteriol.* 174:4175–78

64. Parekh BS, Hatfield GW. 1996. Transcriptional activation by protein-induced DNA bending: evidence for a DNA structural transmission model. *Proc. Natl. Acad. Sci. USA* 93:1173–77

65. Parekh BS, Sheridan SD, Hatfield GW. 1996. Effects of integration host factor and DNA supercoiling on transcription from the ilvPG promoter of *Escherichia coli*. *J. Biol. Chem.* 271:20258–64

66. Pettijohn DE, Pfenninger O. 1980. Supercoils in prokaryotic DNA restrained in vivo. *Proc. Natl. Acad. Sci. USA* 77:1331–35

67. Pokholok DK, Redlak M, Turnboug CL, Dylla S, Holmes WM. 1999. Multiple mechanisms are used for growth rate and stringent control of leuV transcriptional initiation in *Escherichia coli*. *J. Bacteriol.* 181:5771–82

68. Pruss GJ, Manes SH, Drlica K. 1982. *Escherichia coli* DNA topoisomerase I

mutants: increased supercoiling is corrected by mutations near gyrase genes. *Cell* 31:35–42

69. Rahmouni AR, Wells RD. 1989. Stabilization of Z DNA in vivo by localized supercoiling. *Science* 246:358–63

70. Raji A, Zabel DJ, Laufer CS, Depwe RE. 1985. Genetic analysis of mutations that compensate for loss of *Escherichia coli* DNA topoisomerase I. *J. Bacteriol.* 162:1173–79

71. Rhee KY, Opel M, Ito E, Hung SP, Arfin SM, Hatfield GW. 1999. Transcriptional coupling between the divergent promoters of a prototypic LysR-type regulatory system, the ilvYC operon of *Escherichia coli*. *Proc. Natl. Acad. Sci. USA* 96:14294–99

72. Rhee KY, Senear DF, Hatfield GW. 1998. Activation of gene expression by a ligand-induced conformational change of a protein-DNA complex. *J. Biol. Chem.* 273:11257–66

73. Ross W, Gosink KK, Salomon J, Igarashi K, Zou C, et al. 1993. A third recognition element in bacterial promoters: DNA binding by the alpha subunit of RNA polymerase. *Science* 262:1407–13

74. Ross W, Salomon J, Holmes WM, Gourse RL. 1999. Activation of *Escherichia coli* leuV transcription by FIS. *J. Bacteriol.* 181:3864–68

75. Rouviere-Yaniv J. 1978. Localization of the HU protein on the *Escherichia coli* nucleoid. *Cold Spring Harbor Symp. Quant. Biol.* 42 (Pt. 1):439–47

76. Rouviere-Yaniv J, Yaniv M, Germond JE. 1979. *E. coli* DNA binding protein HU forms nucleosomelike structure with circular double-stranded DNA. *Cell* 17:265–74

77. Sander P, Langert W, Mueller K. 1993. Mechanisms of upstream activation of the rrnD promoter P1 of *Escherichia coli*. *J. Biol. Chem* 268:16907–16

78. Schell MA. 1993. Molecular biology of the LysR family of transcriptional regulators. *Annu. Rev. Microbiol.* 47:597–626

79. Sheridan SD, Benham CJ, Hatfield GW. 1998. Activation of gene expression by a novel DNA structural transmission mechanism that requires supercoiling-induced DNA duplex destabilization in an upstream activating sequence. *J. Biol. Chem.* 273:21298–308

80. Sheridan SD, Benham CJ, Hatfield GW. 1999. Inhibition of DNA supercoiling-dependent transcriptional activation by a distant B-DNA to Z-DNA transition. *J. Biol. Chem.* 274:8169–74

81. Sheridan SD, Opel ML, Hatfield GW. 2001. Activation and repression of transcription initiation by a distant DNA structural transition. *Mol. Microbiol.* 40:684–90

82. Sinden RR, Kochel TJ. 1987. Reduced 4,5′,8-trimethylpsoralen cross-linking of left-handed Z-DNA stabilized by DNA supercoiling. *Biochemistry* 26:1343–50

83. Singleton CK, Klysik J, Stirdiva SM, Wells RD. 1982. Left-handed Z-DNA is induced by supercoiling in physiological ionic conditions. *Nature* 299:312–16

84. Steck TR, Franco RJ, Wang JY, Drlica K. 1993. Topoisomerase mutations affect the relative abundance of many *Escherichia coli* proteins. *Mol. Microbiol.* 10:473–81

85. Steger G. 1994. Thermal denaturation of double-stranded nucleic acids: prediction of temperatures critical for gradient gel electrophoresis and polymerase chain reaction. *Nucleic Acids Res.* 22:2760–68

86. Sun HZ, Mezei M, Fye R, Benham CJ. 1995. Monte Carlo analysis of conformational transitions in superhelical DNA. *J. Chem. Phys.* 103:8653–65

87. Tse-Dinh YC. 1985. Regulation of the *Escherichia coli* DNA topoisomerase I gene by DNA supercoiling. *Nucleic Acids Res.* 13:4751–63

88. Tse-Dinh YC, Beran RK. 1988. Multiple promoters for transcription of the *Escherichia coli* DNA topoisomerase I gene and their regulation by DNA supercoiling. *J. Mol. Biol.* 202:735–42

89. Umbarger HE. 1996. Biosynthesis of the branched-chain amino acids. See Ref. 58, 2:442–57

90. Urios A, Herrera G, Aleixand V, Blanco

M. 1990. Expression of the recA gene is reduced in *Escherichia coli* topoisomerase I mutants. *Mutat. Res.* 243:267–72

91. Varshavsky AJ, Nedospa SA, Bakayev VV, Bakayeva TG, Georgiev GP. 1977. Histone-like proteins in the purified *Escherichia coli* deoxyribonucleoprotein. *Nucleic Acids Res.* 4:2725–45

92. Wang JY, Syvanen M. 1992. DNA twist as a transcriptional sensor for environmental changes. *Mol. Microbiol.* 6:1861–66

93. Wek RC, Hatfield GW. 1986. Nucleotide sequence and in vivo expression of the ilvY and ilvC genes in *Escherichia coli* K12. Transcription from divergent overlapping promoters. *J. Biol. Chem.* 261:2441–50

94. Wek RC, Hatfield GW. 1988. Transcriptional activation at adjacent operators in the divergent-overlapping ilvY and ilvC promoters of *Escherichia coli*. *J. Mol. Biol.* 203:643–63

95. Westerhoff HV, van Workum M. 1990. Control of DNA structure and gene expression. *Biomed. Biochim. Acta* 49:839–53

96. Westerhoff HV, Odea MH, Maxwell A, Gellert M. 1988. DNA supercoiling by DNA gyrase. A static head analysis. *Cell Biophys.* 12:157–81

97. Winkelman JW, Hatfield GW. 1990. Characterization of the integration host factor binding site in the ilvPG1 promoter region of the ilvGMEDA operon of *Escherichia coli*. *J. Biol. Chem.* 265:10055–60

98. Zacharias M, Goringer HU, Wagner R. 1992. Analysis of the Fis-dependent and Fis-independent transcription activation mechanisms of the *Escherichia coli* ribosomal RNA P1 promoter. *Biochemistry* 31:2621–28

99. Zechiedrich EL, Khodursk AB, Bachelli S, Schneider R, Chem DR, et al. 2000. Roles of topoisomerases in maintaining steady-state DNA supercoiling in *Escherichia coli*. *J. Biol. Chem* 275:8103–13

Annu. Rev. Genet. 2002. 36:205–32
doi: 10.1146/annurev.genet.36.041102.113929

MEIOTIC RECOMBINATION AND CHROMOSOME SEGREGATION IN DROSOPHILA FEMALES

Kim S. McKim, Janet K. Jang, and Elizabeth A. Manheim

*Waksman Institute and Department of Genetics, Rutgers, The State University of
New Jersey, Piscataway, New Jersey 08854-8020; e-mail: mckim@rci.rutgers.edu*

Key Words   double-strand break, meiotic recombination, Drosophila, crossing
over, oogenesis, synaptonemal complex

■ **Abstract**   In this review, we describe the pathway for generating meiotic crossovers in *Drosophila melanogaster* females and how these events ensure the segregation
of homologous chromosomes. As appears to be common to meiosis in most organisms, recombination is initiated with a double-strand break (DSB). The interesting
differences between organisms appear to be associated with what chromosomal events
are required for DSBs to form. In Drosophila females, the synaptonemal complex is
required for most DSB formation. The repair of these breaks requires several DSB
repair genes, some of which are meiosis-specific, and defects at this stage can have
effects downstream on oocyte development. This has been suggested to result from
a checkpoint-like signaling between the oocyte nucleus and gene products regulating oogenesis. Crossovers result from genetically controlled modifications to the DSB
repair pathway. Finally, segregation of chromosomes joined by a chiasma requires a
bipolar spindle. At least two kinesin motor proteins are required for the assembly of
this bipolar spindle, and while the meiotic spindle lacks traditional centrosomes, some
centrosome components are found at the spindle poles.

CONTENTS

INTRODUCTION

Drosophila females, like the majority of sexually reproducing organisms, generate crossovers between homologous chromosomes to direct segregation at the first meiotic division. This important function of crossovers is likely related to the three other important features of meiosis. First, the number and distribution of crossovers along a chromosome is nonrandom and under genetic controls. Second, the meiotic chromosomes have a unique structure, including the synaptonemal complex (SC) that forms between homologs during meiotic prophase, and an assembly of cohesion proteins that regulate separation of sister chromatids at the two meiotic divisions. Third, the segregation of homologous chromosomes requires the formation of a bipolar spindle.

The SC is a proteinaceous structure with remarkably similar structure in a variety of organisms (Figure 1). Although the structure of the SC has been conserved, its relationship to recombination has diverged significantly. This review concentrates on recent studies that provide insight into the relationship between the SC and recombination in *Drosophila melanogaster* females, and the mechanisms by which chiasmata function to segregate the homologs. Since Drosophila males lack meiotic recombination, including crossing over (93) and gene conversion (30), this review focuses on meiosis in Drosophila females. We also give only scant coverage to the extensive and rich background of classical genetic studies in Drosophila, because they are elegantly described in three previous reviews (6, 73, 111).

Our goal is to review some of the recent findings that are explaining many of the classical genetic observations on the molecular and cytological level. Since the publication of a similar review on meiosis almost 10 years ago (55), several major advances have been made.

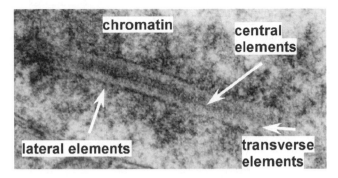

Figure 1   Homolog synapsis during meiosis in Drosophila. Homologs pair via the SC and engage in recombination initiated with a DSB. Recombination is thought to occur at sites marked by a recombination nodule (not shown). This is an electron micrograph from a *mei-P22* mutant, showing that normal SC can form in the absence of DSBs (85).

1. The SC has two roles: It is required for initiating meiotic recombination events and for the repair of DSBs into crossovers.
2. Identification and cloning of genes required for initiating meiotic recombination have shown that double-strand break formation (DSB) is a conserved mechanism.
3. Defects in DSB repair cause reductions in crossing over and abnormal oogenesis.
4. Crossing over may be regulated by controlling which mechanism of DSB repair is utilized. This could involve the generation of an intermediate that only forms in the crossing over pathway.
5. Segregation occurs on an acentrosomal spindle, although this spindle may have a defined structure or complex of proteins at the spindle poles.

THE GENETICS OF MEIOTIC RECOMBINATION AND HOMOLOG DISJUNCTION IN DROSOPHILA FEMALES

Most of the meiotic mutants in Drosophila were isolated because they increased the frequency of X-chromosome nondisjunction. Fortunately, with only four major chromosomes, *D. melanogaster* females are fertile and produce abundant euploid progeny even if there is random segregation of all chromosomes. The first genetic screens to isolate meiotic mutants were performed in Larry Sandler's laboratory by testing for X-chromosome nondisjunction (5, 112). More recently, genetic screens on a larger scale have been performed using P-element mutagenesis (117) or EMS mutagenesis (K. McKim, unpublished results; J. Sekelsky, unpublished results). Some of the mutants identified in these and other screens are listed in Tables 1 to 3.

TABLE 1   Early meiotic gene mutant phenotypes

	Common name (homologs)	SC formation	DSBs/Gene conversion	Crossing over (% of WT)	Crossover distribution	References[b]
c(3)G	Zip1/SCP1	No	Reduced	.01	Abnormal[a]	(47, 98)
c(2)M	NC	Abnormal	Normal	10	Abnormal	E. Manheim & K. McKim, unpublished
mei-W68	Spo11	Yes	No	.01	Abnormal[a]	(87)
mei-P22	NC	Yes	No	.01	Abnormal[a]	(76, 85)
ord	NC	ND	ND	10	Abnormal	(9, 10)

ND = not determined, NC = not conserved, WT = wild-type.
[a]Crossover distribution measured in hypomorphs when null alleles eliminate recombination.
[b]Representitive references and not meant to be a comprehensive list.

There Are Multiple Pathways for Segregating Homologs

CHIASMATE SEGREGATION   The canonical pathway for chromosome segregation is based on the ability of chiasmata to hold and orient homologs on the meiotic spindle (Figure 2). A single medially placed crossover appears ideal for this process. Not only is that where the majority of crossovers are located (73), but there is evidence that distally or proximally located crossovers, or the

TABLE 2   Middle meiotic gene mutant phenotypes: these genes have no effect on SC formation

	Common name (homologs)	DSB repair or gene conversion[a]	Crossing over (% of WT)[b]	Crossover distribution	References
spnB	Rad51/Dmc1	Defective	30	Abnormal	(38)
spnD	ND	Defective	30	Abnormal	(38)
okr	rad54	Defective	30[c]	Abnormal	(38)
mei-41	ATM/mec1	ND	50[c]	Abnormal	(5, 48)
mei-218	NC	Normal	10	Abnormal	(28, 84)
mei-217	NC	Normal	10	Abnormal	(75)
mei-9	rad1/XPF	PMS	10	Normal	(28, 116)
mus312	NC	ND	10	Normal	(44), J. Sekelksy, per. comm.
mei-S282	ND	ND	50	Abnormal	(101)
rec	ND	ND	5	Normal	(45)

ND = not determined, NC = not conserved, WT = wild-type.
[a]Any defects in DSB repair or gene conversion, if known.
[b]Approximate.
[c]Null alleles are sterile, crossover date comes from hypomorphs.

TABLE 3   Late meiotic gene mutant phenotypes: no effect on SC or crossing over

	Common name (homolog)	MI nondisjunction	MII nondisjunction	Chiasmate nondisjunction	Spindle defect	References
ncd	Kinesin	Yes	No	Yes	Yes	(50, 83)
sub	Kinesin	Yes	No	Yes	Yes	(41)
nod	Kinesin	Yes	No	No	No	(134, 135)
msps	XMAP215	ND	ND	ND	Yes	(32)
d-tacc	D-TACC	ND	ND	ND	Yes	(32)
mei-38	ND	Yes	No	Yes	No	(5)
mei-S332	NC	No	Yes	Yes	No	(63, 64)
ald	ND	Yes	No	Yes	No	(97)
Axs	NC	Yes	No	No	Yes	(130)

NC = not conserved, ND = not determined.

presence of multiple crossovers, may fail to ensure or even hinder segregation (67, 90).

ACHIASMATE SEGREGATION   Several observations reveal the presence of additional mechanisms that influence the segregation of chromosomes. For instance, the Drosophila fourth chromosomes segregate correctly, even though they never form a chiasma. Also, heterozygosity for a single balancer, which suppresses crossing over on one chromosome, does not increase the frequency of nondisjunction. These observations are explained by the presence of systems that are responsible for the segregation of achiasmate chromosomes. Further analysis has led to a model in

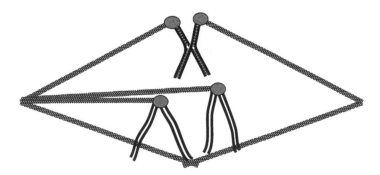

Figure 2   Segregation of homologs requires chiasmata and sister chromatid cohesion. One bivalent is joined by a chiasma and the centromeres are being pulled in opposite directions. Sister chromatid cohesion distal to the crossover site prevents the bivalent from slipping apart. At anaphase, the distal cohesion is released but centromere cohesion must be maintained in order to keep sister chromatids together until meiosis II. Two other chromosomes are not joined by a chiasmata. This could have occurred because no crossover formed or because there was no arm cohesion. The result is that the two univalents do not orient on the spindle and are free to attach randomly to a spindle pole.

which there are two achiasmate segregation systems (54, 56). One achiasmate system is based on heterochromatic homology (the homologous achiasmate system). These pairings persist until the meiosis I spindle assembles and may be instrumental in the orientation of achiasmate homologs (34). In contrast, the nonhomologous achiasmate system segregates chromosomes without centromeric alignment and is based on parameters such as shape and size. Genetic assays for defects in the achiasmate system may actually facilitate the identification of spindle-associated proteins. Mutations in several genes have phenotypes that are specific to the achiasmate system (Table 3). Many of these genes may also be important for chiasmate segregation, but the achiasmate system is more sensitive to mild spindle defects.

If there is a system to segregate achiasmate homologs, why do crossover-defective mutants cause nondisjunction? An important insight into this question comes from the observation that the relationship between the frequencies of X-chromosome nondisjunction and nonexchange tetrads is not linear. In fact, X-chromosome nondisjunction is proportional to the frequency of nonexchange tetrads cubed (6). The reason either achiasmate system fails in crossover-defective mutants is that it can accurately segregate only one pair of large chromosomes. When an achiasmate pair of X-chromosomes is present in the same oocyte as a pair of achiasmate autosomes, equal numbers of chromosomes are sent to each pole without respect for homology. If there are simultaneously two X and two 2nd chromosomes in the achiasmate pool, XX $\longleftrightarrow$ 22 segregation is just as likely as X2 $\longleftrightarrow$ X2, but XX2 $\longleftrightarrow$ 2 is less likely. Thus, when it comes to large chromosomes, the preference of the achiasmate system is for the equal distribution of chromosome numbers. A detailed analysis of possible mechanisms for achiasmate segregation has been published by Hawley & Theurkauf (56).

FEMALE MEIOSIS IN THE DEVELOPMENTAL CONTEXT

The Oocyte Develops Alongside 15 Nurse Cells

Meiosis is one of the first programmed events in the development of the oocyte. Drosophila females have two ovaries, each comprised of several ovarioles containing chains of developing oocytes. At the anterior end of each ovariole is the germarium, where four rounds of incomplete mitotic divisions produce a 16-cell cyst with intercellular junctions termed ring canals. The germarium is divided into four regions based on the morphology of the 16-cell cysts (Figure 3, see color insert). The cysts move down the germarium as they mature such that a cyst in a more posterior position is in a later stage of meiotic prophase than a cyst in a more anterior position. Soon after the last premeiotic division (region 2a), meiotic prophase begins and recombination is initiated (21). Note that the position of a cyst within the germarium is only a rough guide to its meiotic stage (18). While the cysts are arrayed in order of developmental age, their absolute position in the germarium does not equate to a specific stage in meiotic prophase. This organization, however, makes it possible to compare oocytes at different stages of meiosis within a single germarium.

Cytological Descriptions of Meiosis in Drosophila

A series of electron microscopy studies by Carpenter laid the foundation for much of the current work on meiosis in Drosophila (18–22, 26, 27). Several cells enter meiosis in each 16-cell cyst, but only one will become the oocyte while the rest become nurse cells. Specifically, the SC first develops in the two cells with four ring canals, one of which will become the oocyte (18, 21). Later, SC develops to variable extents in the other cells that have fewer ring canals. Recombination is probably initiated in these pro-nurse cells as well (27, 76). Eventually, all cells except the oocyte exit the meiotic program and SC is maintained in only the oocyte. These studies also revealed that the SC is a dynamic structure; early in pachytene the SC progressively shortens and thickens and then towards the end of pachytene, lengthens (18, 21).

It was from Carpenter's studies that the significance of recombination nodules (RN) was first discerned (19, 21). Like most organisms, Drosophila females develop two temporally and morphologically distinct types of RN. Ellipsoidal RNs appear first and have a random distribution. Spherical RNs appear later and have a nonrandom distribution that resembles crossovers. In addition, the number of spherical RNs is reduced and their morphology is abnormal in *mei-218* mutants, in which there is a specific failure in crossover formation (20). Based on these observations, it is reasonable to conclude that ellipsoidal RNs are the sites of early DSB repair while spherical RNs are those sites where a crossover will occur. Note that in Drosophila, both types of RN do not appear until after the homologs are fully synapsed (pachytene). This is later than some other organisms, such as the plants *Lycopersicon esculentum*, *Allium cepa*, and *A. fistulosum*, where early RNs first appear during zygotene (2, 121).

MOLECULES IN THE PATHWAY FOR EXCHANGE

Evidence that Double-Strand Breaks Initiate Meiotic Recombination

Double-strand breaks (DSB) were initially shown to be efficient initiators of recombination in meiotic cells of *Saccharomyces cerevisiae* (72). Extending this observation to other organisms was not trivial because of the inability to physically detect DSBs. Recent evidence from several experiments now strongly indicates that DSBs initiate meiotic recombination in Drosophila females.

mei-W68 ENCODES A spo11 HOMOLOG   One of the genes required for all meiotic recombination in Drosophila females, *mei-W68* (85), encodes a homolog of *spo11* of *S. cerevisiae* (87). The *spo11* proteins, which are homologous to a novel type II topoisomerase from archaebacteria (topo6A), are proposed to be the enzymatic factor for DSB formation (62). Similarly, Spo11 homologs have now been identified in many organisms, including *Caenorhabditis elegans* (33) and mammals (61, 109), lending support to the idea that most or all organisms initiate meiotic recombination with a DSB.

X-IRRADIATION INDUCES MEIOTIC CROSSING OVER   When *mei-W68* mutants were exposed to 4000R of radiation, crossing over was efficiently induced to almost 50% of the wild-type frequency, a 200-fold increase over unirradiated controls (R. Bhagat & K. McKim, unpublished). These results suggest that *mei-W68* mutants lack meiotic recombination due to an absence of DSBs. Similarly, artificial creation of DSBs by X-irradiation partially rescues the meiotic recombination defects in mutants unable to generate DSBs, including the *spo11* homolog mutants of *S. cerevisiae* (126), *C. cinereus* (29), and *C. elegans* (33).

ELIMINATING RECOMBINATION SUPPRESSES THE STERILITY OF DSB REPAIR-DEFECTIVE MUTANTS   Mutations in DSB repair-defective genes such as *spnB* and *okr* cause sterility because defects in meiotic DSB repair cause the oocyte to develop abnormally (38, 39). Mutants that eliminate DSBs, such as *mei-W68*, suppress the oogenesis defects and partially restore fertility to *spnB* and *okr* mutants (35; R. Patel & K. McKim, unpublished results).

DSB SITES CAN BE DETECTED WITH ANTIBODIES   The previous three genetic results provide convincing, albeit indirect, evidence that DSBs initiate meiotic recombination in Drosophila. More direct evidence has come from experiments using a human antibody to the phosphorylated form of histone H2AX (γ-H2AX), which detects a chromatin response to a DSB in mammalian mitotic and meiotic cells (79, 108). Using this antibody, nuclear foci have been detected during Drosophila female meiotic prophase (Figure 3). These γ-H2AX foci are probably the sites of DSB repair in Drosophila since they are absent in *mei-W68* mutant females (D. Sherizen & K. McKim, unpublished). Although it has not been conclusively shown, this antibody probably detects the phosphorylated form of H2AvD in Drosophila.

The Nonrandom Frequency and Distribution of Crossing Over

Meiotic crossovers and gene conversions are products of double-strand break (DSB) repair. Crossing over is a tightly regulated process as shown by the nonrandom frequency and distribution of events along each chromosome. In an example cited by Carpenter & Sandler (25), the average number of crossovers per meiosis on the X-chromosome is approximately 1.35 (67 m.u.). With an average of 1.35 crossovers, the Poisson distribution predicts that 26% of the chromosomes would lack a crossover. Interestingly, experimental data show that only 5% of the homologous chromosomes lack a crossover (129). Also, there are fewer chromosomes with multiple crossovers than would be expected. This abundance of single-crossover bivalents is logical because a single chiasma is usually sufficient to direct segregation (53).

When multiple crossovers do occur, however, they rarely happen close to each other (e.g., within 10 cM) owing to the phenomenon of interference. In addition to interference, there are other factors that influence the distribution of crossovers, the most important being proximity to the centromere or telomere (Figure 4). This was shown experimentally by Beadle (8), who used a translocation that placed a fourth chromosome centromere in the proximal third of chromosome 3R. In its new position, the fourth chromosome centromere repressed crossing over in the

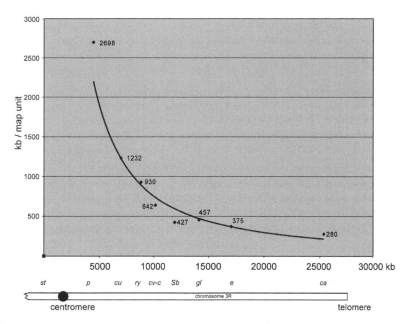

Figure 4 Nonrandom distribution of crossovers on chromosome 3R. This graph indicates that crossing over occurs much less frequently within intervals close to the centromere. The spacing of the genes is based on the physical map, except for the *st–p* interval, which includes the centric heterochromatin. However, the centric heterochromatin was not included in the bp/m.u. calculation. The graph was generated by dividing the published map distances (74) by the physical distance between the genes (94).

adjacent third chromosome intervals. More recent but similar experiments have demonstrated the same phenomenon in budding yeast (68). The results from intragenic recombination studies suggest that differences in crossover rates reflect differences in the probability that DSB repair generates a crossover. At the *rosy* locus the frequency of gene conversion is approximately 1×10^{-5} (57). Gene conversion at *maroon-like* (*mal*) occurs at a similar frequency as *rosy*; however, the intragenic events at *mal* are not associated with crossovers (120). This is not surprising, given that *mal* is located in a very proximal region of the X-chromosome euchromatin (19D1) where crossing over is exceedingly rare; rather it suggests that the initiation of recombination is random whereas crossovers occur at a nonrandomly selected subset of these sites.

The Order of Events: What Regulates DSBs and Crossovers?

Lindsley & Sandler (73) used three terms: prepachytene, pachytene, and postpachytene functions, to denote the important phases leading up to the segregation of homologous chromosomes in Drosophila. These terms are useful because they highlight the important finding, summarized below, that the initiation of meiotic recombination in Drosophila is dependent on homolog synapsis. Mutants with abnormal SC, crossing over, and disjunction affect prepachytene functions.

Mutants with abnormal crossing over and disjunction but normal SC affect pachytene functions. Mutants with abnormal disjunction and normal crossing over affect postpachytene functions.

Prepachytene: The SC Regulates DSB Formation

Drosophila mutants that lack meiotic DSBs, such as *mei-W68*, form SC early in region 2a just like wild type (85). Therefore, there are no delays in SC formation when DSBs are not created. This observation is identical to one in *C. elegans* (33), but differs from experiments in budding yeast and mouse that show SC does not form correctly in the absence of DSBs (7, 107, 110). One interpretation of the Drosophila results is that SC formation is required for DSBs, but an alternative possibility is that the two events are independent.

One approach to investigating the relationship between DSB formation and the SC is to simultaneously monitor the development of SC and the activity of proteins involved in DSB formation or repair. Three recent developments have made these types of experiments possible. First, a component of Drosophila SC, C(3)G, has been identified. Second, another gene required for DSB formation, *mei-P22*, has been identified, and the protein can be detected at putative DSB sites. Third, DSBs can be detected with an antibody that recognizes a phosphorylated histone.

C(3)G IS A TRANSVERSE ELEMENT PROTEIN   Since its discovery, *c(3)G* has been the focus of considerable interest because mutants have a severe meiotic phenotype (43). In *c(3)G* mutants, SC does not develop (102, 119) and meiotic recombination is eliminated (17, 47). Based on these phenotypes, there were two possible functions for C(3)G: (*a*) C(3)G could be required for the initiation of recombination with the synapsis defects being secondary, or (*b*) C(3)G could be required for SC formation with recombination initiation being dependent on synapsis or an SC component. The cloning of *c(3)G* has shown the latter to be the case (98). *c(3)G* encodes a coiled-coil protein, and immunohistochemical studies have shown that it is probably a component of the SC. In wild-type pachytene oocytes, C(3)G is found along the length of each meiotic chromosome, giving it a ribbon-like appearance. Coiled-coil proteins that are known components of the transverse filaments (TF) of the SC have been identified in yeast (ZIP1) and mammals (SCP1). Thus, C(3)G may also be a TF protein. Remarkably, none of these proteins shows primary sequence conservation.

mei-P22 ENCODES A CHROMOSOME-ASSOCIATED PROTEIN REQUIRED FOR DSB FORMATION   Strong mutations in *mei-P22* have a phenotype identical to *mei-W68*: All meiotic recombination is eliminated, γ-H2AX foci do not accumulate on chromosomes, and the sterility of DSB repair-defective mutants is suppressed (76). In addition, crossing over in *mei-P22* mutants can be restored to almost 50% of wild type by X-irradiation. These data strongly support the hypothesis that *mei-P22* is required for DSB formation.

Using an epitope-tagged *mei-P22* transgene, Liu et al. have shown that MEI-P22 is nuclear and colocalizes with meiotic chromosomes (76). Localization of

MEI-P22 does not depend on DSB formation (it is present in a *mei-W68* mutant), suggesting that MEI-P22 binds to the sites that will become DSBs. By observing the ordered array of oocytes at different stages of meiosis within a single germarium, it has been possible to conclude that MEI-P22 appears on chromosomes after SC formation begins, suggesting that the SC has a role in facilitating binding of the protein to the chromosomes. In confirmation of this hypothesis, no MEI-P22 chromosomal localization was observed in *c(3)G* mutants, indicating that SC formation alters the chromosomes in some way to allow binding of DSB-inducing proteins. MEI-P22 foci are short lived, appearing on the chromosomes briefly in early meiotic prophase (within region 2a), even though the SC is present for a much longer time (region 2a–3 and early vitellarium).

γ-H2AX FOCI ARE DEPENDENT ON AND APPEAR AFTER THE SC   Similar to MEI-P22, γ-H2AX foci do not appear until after SC formation is completed, a finding that contrasts with the studies in male mouse meiosis using this same antibody (79). However, unlike MEI-P22, the γ-H2AX foci are observed in *c(3)G* mutant females but in significantly fewer numbers than in wild type (R. Bhagat & K. McKim, unpublished results). Thus, SC formation, or C(3)G accumulation in particular, appears to have a large positive effect on, but is not absolutely required for, DSB formation. This effect of the SC could occur through the recruitment of DSB-inducing proteins. In the absence of *c(3)G*, very low levels of proteins such as MEI-P22 are recruited to the chromosomes (see above), resulting in inefficient DSB induction.

REGULATING DSB FORMATION BY LIMITING THE TIME WHEN MEI-P22 CAN BIND TO CHROMOSOMES   Regulating DSB formation is an important part of the meiotic program. Once a sufficient number of DSBs have been induced, the activity of generating breaks must then be attenuated, otherwise, excessive chromosomal breakage could lead to cell lethality. In Drosophila, DSB formation appears to be regulated by restricting when proteins like MEI-P22 have access to the chromosomes. The restriction of MEI-P22 foci to early pachytene suggests that DSB formation is regulated by limiting the time during which proteins like MEI-P22 can bind to potential DSB sites. Interestingly, the ultimate inhibition or removal of MEI-P22 foci occurs in the absence of DSBs and therefore is unrelated to DSB repair (76). Thus, the factors that regulate MEI-P22 and DSB formation may be dependent on the structure of meiotic chromosomes. We suggest that the SC is required for the formation of MEI-P22 foci, but its maturation may contribute to their disappearance. The disappearance of MEI-P22 foci early in pachytene may correspond to changes in SC morphology previously noted by Carpenter (18). As the oocyte progresses through pachytene, the SC progressively shortens and thickens. The corresponding structural changes of SC components could modify the chromosomes such that MEI-P22 can no longer bind and, therefore, prevent further DSB formation.

The SC Is Required for Meiotic Crossing Over

The SC promotes DSB formation but also has a second, independent function in the generation of crossovers. This conclusion is based on data from experiments using exogenous sources of DSBs to induce recombination. As described above, when *mei-W68* and *mei-P22* mutants were exposed to 4000R of X-irradiation, crossing over was efficiently induced. When *c(3)G* mutant females were irradiated, crossing over was not induced (105; R. Bhagat & K. McKim, unpublished results), suggesting that the SC is required for the DSB repair pathway that yields crossovers. Therefore, the SC has two separable functions: the initiation of meiotic recombination and promoting the specialized DSB repair pathway that leads to crossing over.

A recently identified gene, *c(2)M*, encodes a novel SC component that may be responsible for this crossover-specific function. Strong alleles of *c(2)M* reduce crossing over by greater than 90%, whereas staining with the γ-H2AX antibody indicates that DSBs are induced (E. Manheim & K. McKim, unpublished). Therefore, *c(2)M* mutants have a crossover-specific defect. The predicted *c(2)M* protein has no homologs in the databases, but using a polyclonal antibody, C(2)M was found to colocalize with C(3)G, suggesting that *c(2)M* encodes a component of the synaptonemal complex (Figure 5, see color insert).

The localization of C(2)M to meiotic chromosomes suggests it could have a role in SC formation. In fact, C(3)G staining is severely aberrant in *c(2)M* mutants. In *c(2)M* mutant oocytes that would normally be in pachytene, C(3)G staining fails to develop into complete strands along the lengths of each chromosome, but instead appears as a variable number of small patches (E. Manheim & K. McKim, unpublished). The SC-like localization of C(2)M is surprising considering that *c(3)G* has a more severe effect on crossing over. While c(3)G is requried for all recombination, c(2)M may be more critical for crossover formation. It is possible that the DSBs in *c(2)M* mutants occur at the sites where SC (C(3)G) formation can initiate, but *c(2)M* is required for these DSBs to be repaired as crossovers. *c(2)M* is an SC component with the specific function of providing the chromosomal context required for crossing over to occur.

In summary, these studies have identified two functions of the SC: It initiates meiotic recombination and facilitates the repair of DSBs into crossovers. In contrast, in mutants of the *c(3)G* homolog in budding yeast, *zip1*, DSBs occur at a normal frequency and crossing over is reduced by only 50% (123). Thus, *zip1* does not exhibit a DSB-initiation function like *c(3)G* but does have the crossover function. The relatively high incidence of crossing over in *zip1* mutants indicates that there is an SC-independent mechanism of crossing over in budding yeast. We suggest that all crossovers in Drosophila female meiosis occur by a single SC-dependent DSB repair pathway. A restriction of crossover formation by a single pathway also appears to be the situation in *C. elegans* (133).

Pachytene Functions: Regulating the Outcome of DSB Repair

Genes required after SC and DSB formation for crossing over can be divided into two groups: the DSB repair proteins and the crossover proteins. In other organisms, several genes have been implicated in meiotic DSB repair, most notably members of the Rad52 epistasis group (46). This group includes strand-exchange proteins (e.g., Rad51) and other factors that stimulate and regulate the repair process (e.g., Rad52 and Rad54). In some cases, there are meiosis-specific DSB repair proteins, such as the Rad51 homolog Dmc1 in yeast (12). The second group of proteins are only required for crossing over. DSB repair occurs in the absence of these proteins, but crossovers are reduced in frequency. This group includes the mismatch repair proteins in budding yeast, *C. elegans*, and mammals (49, 127).

Defects in DSB Repair Lead to Sterility

The identification of DSB repair-defective mutants in Drosophila came from an unexpected source, the *spindle* genes, which were originally identified because the mutants are sterile due to defects in establishing dorsal-ventral polarity of the oocyte (42). Oocytes from a *spindle* mutant mother are ventralized because of reduced *gurken* (*grk*) levels, a signaling protein required for specification of the dorsal fate (42). Two of the *spindle* genes, *spnB* and *okr*, encode homologs of the DSB repair proteins Rad51 and Rad54, respectively (38). Consistent with an important function in DSB repair, the oocyte patterning defects in *spnB* and *okr* mutants are suppressed by *mei-W68* mutations that eliminate DSBs (39). Conversely, mutants, such as *mei-218*, that reduce crossing over but not DSB formation do not suppress *spnB* sterility (75). These results suggest that meiotic DSB repair must occur properly before oocyte development can proceed. One proposal suggests that a checkpoint response to unrepaired DSBs and the subsequent cell-cycle arrest are responsible for the failure to translate *grk* (39). The involvement of a checkpoint pathway is supported by the observation that the patterning defects in *spnB* and *okr* mutants are suppressed by mutations in the *mei-41* gene. *mei-41* encodes the Drosophila homolog of *rad3/ATR*, which is a DNA-dependent protein kinase required for a checkpoint response to DSBs. Indeed, the budding yeast homolog of *mei-41*, *mec1*, is also required for the pachytene arrest that occurs in DSB repair-defective mutants such as *dmc1* (78).

Direct evidence of a DSB repair defect has come from the observation that *okr*^{WS} and *spnB*^{BU} mutants exhibit prolonged staining of γ-H2AX foci compared to wild type (D. Sherizen & K. McKim, unpublished). Whereas the γ-H2AX foci in wild-type are limited to early pachytene cells (region 2a of the germarium), in *okr*^{WS} and *spnB*^{BU} mutants the foci always persist until the end of pachytene (beyond region 3, Figure 3). By showing that meiotic DSBs are not repaired until late in prophase, these data are consistent with the checkpoint hypothesis.

Several important questions concerning the checkpoint response remain. The signal that links DSB repair defects to *grk* signaling is not known. Most importantly, whereas the presence of a system that responds to DSB repair defects is not in doubt, the function of this checkpoint and its role in meiotic crossing over is not known. Thus, there should be some caution in concluding that there is a cell-cycle arrest. Currently, there is no evidence that any cell cycle events are delayed in the DSB repair-defective mutants. In budding yeast meiosis, for example, the standard measure for a pachytene arrest due to a DSB repair defect is the amount of time until metaphase (e.g., 78). Two important events appear to occur without delay in the Drosophila DSB repair mutants: The SC dissolves by vitellarium stage 6 and a metaphase I spindle is built in stage 14 oocytes as in wild-type (88; D. Sherizen & K. McKim, unpublished). Thus, by these measures, *spnB* and *okr* mutants appear to have no cell-cycle arrest. Given the biology of the ovariole, there is actually little rationale behind an arrest. Each 16-cell cyst is propelled down the ovariole by external forces that are beyond its control. Even if a cell is under a "cell-cycle arrest," it has no choice but to continue developing. Thus, in strong *okr* and *spnB* mutants, there appears to be no benefit for the oocyte to arrest in response to unrepaired DSBs.

The Role of Crossover-Specific Genes

In *mei-218* mutants, crossing over (28, 84), but not gene conversion (23, 24), is reduced by more than 90%. The interpretation of this phenotype is that these mutants are able to initiate recombination but are severely impaired in an aspect of DSB repair that leads to crossovers but not to gene conversion. In addition, a reduction in the crossover frequency is accompanied by a change in distribution; the frequency of residual crossovers is reduced less drastically in regions of the chromosome closer to the centromere. Changes in crossover distribution are one of the defining characteristics of the "precondition" mutants (112). One interpretation of this phenotype is that these genes establish the preconditions for exchange by determining the frequency and distribution of events (73, 112).

While precondition genes such as *mei-218* may be required to determine which DSB sites will be repaired as crossovers, genes such as *mei-9* may be directly required for the resolution reaction (such as resolving the Holliday junction intermediate) (5, 23, 24, 116). Consistent with this idea, several lines of evidence suggest that *mei-218* is required before *mei-9* in the recombination pathway. *mei-218* mutants reduce the frequency and morphology of spherical (or late) RNs while these structures appear normal in *mei-9* mutants (20). *mei-9* mutants exhibit high rates of post-meiotic segregation (PMS), whereas *mei-218* mutants do not (23). A PMS phenotype indicates a failure to repair mismatches that form when strands exchange during DSB repair. The *mei-9 mei-218* double mutant lacks PMS, indicating that this requirement for *mei-9* is dependent on an earlier function by *mei-218* (D. Sherizen & K. McKim, unpublished). *mei-9* encodes an endonuclease homologous to XPF/Rad1, which is also consistent with a direct role in the resolution of DSB repair to generate crossovers (116). In *mei-9* mutants, the crossover sites are probably determined correctly but cannot be resolved as crossovers. Thus, *mei-218* appears to be required solely for specifying the sites of crossovers while *mei-9* seems to participate directly in the resolution event.

While several genes have precondition mutant phenotypes (see below), *mus312* is the only other Drosophila gene with a crossover phenotype similar to *mei-9*. Also like *mei-9*, this gene has a role in somatic DNA repair. Interestingly, *mus312* was identified in a two-hybrid screen for proteins that interact with MEI-9 (J. Sekelsky, unpublished). A protein complex, including MEI-9 and MUS312, most likely functions in the exchange process, perhaps in the actual resolution event itself. The finding that a complex in *S. pombe* containing a protein related to the Rad1/MEI-9 family (Mus81) binds and cleaves Holliday junctions supports this hypothesis (13). It is impressive that the crossover distribution phenotype is an accurate predictor of gene function (see below for model).

Do Precondition Genes Have a Role in Directing the Outcome of DSB Repair?

The mutant phenotype of *mei-9* and *mei-218* mutants demonstrates that crossing over can be eliminated while leaving DSB repair intact. One hypothesis to explain this phenotype is that crossing over and gene conversion result from different pathways of DSB repair and that *mei-9* and *mei-218* have roles in the crossover-specific aspect of the DSB repair pathway. We suggest that the mechanism by which *mei-218* specifies crossover sites is by promoting the formation of a crossover-specific recombination intermediate (possibly the Holliday junction). This conclusion is based on the finding that mutations in *mei-218* are epistatic to *mei-9*. The reduction of PMS observed in the *mei-9 mei-218* double mutant suggests that *mei-218* is required to promote the formation of a recombination intermediate containing mismatches that *mei-9* normally repairs. Therefore, these results support the hypothesis that *mei-218* is required during the repair of DSBs for the production of a recombination intermediate that leads to crossover formation. A divergence of crossover and noncrossover pathways has also been suggested from physical studies of meiotic DSB repair in *S. cerevisiae* (3, 58). DSB repair by a noncrossover pathway may not involve a Holliday junction (100) and conversely, resolution of Holliday junctions may be biased towards crossovers (31).

How Is the Frequency and Distribution of Crossing Over Regulated in Drosophila Females

In the precondition class of Drosophila recombination-defective mutants, a reduction in the frequency of crossing over is accompanied by a change in distribution of the residual events. One interpretation of this phenotype is that these genes determine the frequency and location of crossovers (73, 112). The problem with this model is that defects at several stages of the meiotic recombination pathway, both before and after DSB formation, can alter the distribution of crossovers. For example, hypomorphic alleles of genes required for SC formation, such as *c(3)G* (98), and the initiation of recombination, such as *mei-W68* (cited in 73) and *mei-P22* (76), reduce crossing over by approximately 50% and alter the distribution. DSB repair-defective mutants also cause changes in the distribution of crossovers (D. Sherizen & K. McKim, unpublished). Thus, defects at all stages of meiotic recombination prior to the resolution step can have an impact on the distribution of crossovers.

Above we proposed that generating crossovers involves the formation of an intermediate that does not form in the noncrossover pathway. A model in which the frequency of crossing over is regulated by monitoring a single intermediate late in the DSB repair pathway would explain why defects in several stages of the meiotic recombination pathway alter the distribution of crossovers. We suggest that all of the precondition mutants alter the distribution of crossovers because they fail to produce the crossover-specific intermediate. For example, in precondition mutants, low numbers of the crossover intermediate may trigger a feedback response involving either the creation of additional DSBs or converting more existing

DSBs into crossovers. Such an effect could increase the frequency of crossovers in regions that normally experience few of these events, thus altering the overall distribution. Using the same reasoning, mutants that do not affect the formation of this intermediate, like *mei-9*, do not alter the distribution of crossovers because they affect a step (resolution) after the crossover-specific intermediate has formed. Thus, we are proposing that this intermediate has a critical role in regulating the distribution and frequency of crossovers.

Carpenter & Sandler (28) proposed a similar model as an alternative to the idea that the function of genes like *mei-218* is to determine where the crossovers will occur. They suggested that some crossover-defective mutants trigger the interchromosomal effect (IE). The IE is the phenomenon where suppression of crossing over on one chromosome results in an increased frequency and altered distribution of crossing over on the other chromosomes (77). It seems reasonable that the effects of precondition mutants, the IE, and interference on crossover distribution are mechanistically related: All three phenomena may be responding to the presence or absence of the same recombination intermediate that only forms when a crossover is going to be produced. Interestingly, most models for interference involve interactions between crossover sites on the same chromosome (e.g., 36), whereas the IE appears to involve a *trans*-acting component.

Molecular Analysis of *mei-218*: Cytoplasmic Regulators of Crossing Over

More than ten alleles of *mei-218* have been characterized and the gene encodes a novel protein (84). Expression of the *mei-218* RNA is restricted to the region of the ovary where the oocyte is determined and enters meiotic prophase (the germarium) (80). Similarly, the *mei-218* protein first appears when the oocyte enters meiotic prophase but, surprisingly, is cytoplasmic. The cytoplasmic localization of MEI-218 suggests it regulates other genes involved in meiotic crossing over. Such indirect regulators of meiotic recombination are not without precedent in Drosophila. Hypomorphic alleles of *Sxl* (K. Cook, personal communication; 14) and *mei-P26* (99), two genes required for oocyte differentiation, have a precondition phenotype. MEI-218 could be part of a pathway linking germline differentiation and the regulation of meiotic recombination events.

POST-PACHYTENE: SEGREGATING THE CHROMOSOMES

For chiasmata to direct segregation of the homologs, two other systems must operate correctly (Figure 2). First, the mechanical process of segregating homologous chromosomes requires a bipolar meiotic spindle and associated factors such as motor proteins. Second, the segregation of chiasmate homologs requires sister chromatid cohesion. Mutations in genes required for either of these processes have a distinct phenotype in Drosophila; homologous chromosomes do not segregate properly during meiosis even though chiasmata form between them. The specific processes these mutants affect can be easily determined through cytological analysis. Mutants that affect the segregation machinery typically have defects in spindle

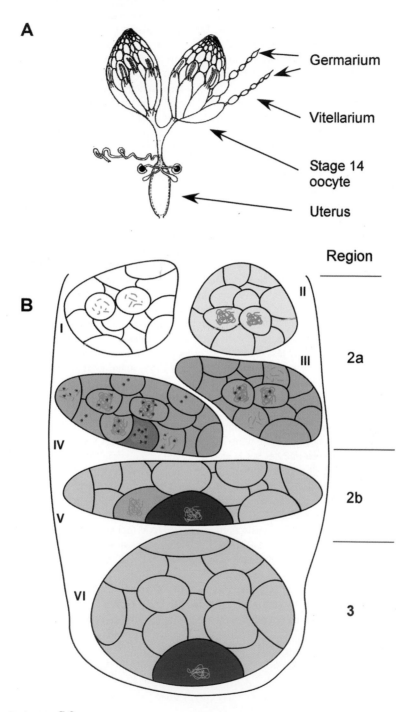

See text page C-2

Figure 3 (page C-1) Schematic of the Drosophila ovary. (*A*) Each ovary is composed of strings of developing oocytes called ovariols. At the tip of each ocariole is the germarium, where each oocyte is born as part of a 16-cell cyst. Approximately 4 days elapse between the completion of recombination (in the germarium) and oocyte maturation (stage 14) when the metaphase spindle forms (65) [figure adapted from King (65)]. (*B*) Schematic diagram of the Drosophila germarium and a model for meiotic progression. Not shown at the anterior (*top*) end is region 1, where four rounds of mitotic divisions create the 16-cell cysts. Following the last mitosis, the 16-cell cyst is in region 2a and undergoes premeiotic S-phase (also not shown). Each 16-cell cyst contains two cells with four intercellular connections or ring canals (the pro-oocytes), one of which will become the oocyte. The order of events depicted here is based on the observations of SC and RNs at the EM level (21) and ORB (blue) (69), C(3)G (green) (98), MEI-P22 (not shown) (76), and γ-H2AX (red) (D. Sherizen & K. McKim, unpublished results) antibody stainings. Early and persistent SC and C(3)G staining mark the two pro-oocytes, but ORB accumulates to high levels in the oocyte before the restriction of SC occurs. Whereas the cysts are usually [but not always (21)] arranged in temporal order, the relationship between the stage of pachytene and either absolute cyst position or ORB staining is partially independent. (I) Zygotene: The pro-oocytes have incomplete SC formation and no γ-H2AX foci. (II) Early pachytene: The pro-oocytes have complete C(3)G staining but have few or no γ-H2AX foci. In some cases, these cysts will not have detectable ORB staining. Most germaria have only one of the two early cyst types (I and II), suggesting they are relatively brief stages in prophase. For example, relative to pachytene, zygotene is a comparatively brief stage in Drosophila, possibly less than 12 h (21). In contrast, for each of the following region 2 cysts (III—V), there may be multiple examples in a given germarium. (III) Pachytene: There are increasing numbers of γ-H2AX foci. Ellipsoidal RNs show a distribution similar to the γ-H2AX staining and are usually present one cyst before spherical RNs, at which point they may coexist. MEI-P22 foci may appear in larger numbers at this stage than γ-H2AX. SC begins to form in 3-ring cells. (IV) Pachytene: Cysts have C(3)G staining in additional cells, such as those with 3- or 2-ring canals. γ-H2AX staining reaches its highest levels in the pro-oocytes but is also observed in the nurse cells, suggesting that DSBs form in any cell where C(3)G assembles on the chromosomes. There is a gradient in each 16-cell cysts, whereas SC forms to the greatest degree in the two pro-oocytes. RNs, MEI-P22, and γ-H2AX foci appear in pro-nurse cells, indicating that in addition to SC, DSBs are generated in these cells (27, 76; D. Sherizen & K. McKim, unpublished results). (V) Region 2b pachytene: The cysts flatten out, γ-H2AX foci and ellipsoidal RNs are either infrequent or absent, and MEI-P22 foci are never observed. Spherical RNs may persist into region 3. The observation that there are fewer RNs or γ-H2AX foci in the earliest and latest staining oocytes than in the middle ones suggests that the appearance and disappearance of these structures is asynchronous. (VI) Region 3 pachytene: By the end of the gernarium SC is usually restricted to the oocyte, and γ-H2AX foci and ellipsoidal RNs are absent. SC will continue to be present until stage 6 of the vitellarium.

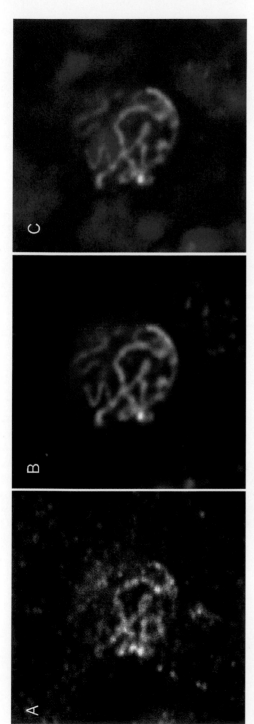

Figure 5 C(2)M colocalizes with C(3)G. (*A*) Pachytene stage oocyte stained for C(2)M. (*B*) Same oocyte stained for C(3)G to show the SC. C(3)G is probably a component of the transverse elements. (*C*) Color merge, with C(2)M (red), C(3)G (green) and DNA (blue).

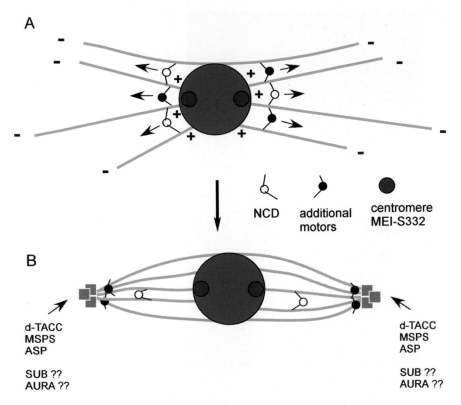

Figure 6 Motor proteins and centrosome components on the Drosophila oocyte meiosis I spindle [(adapted from (83, 125)]. (*A*) The chromosomes capture microtubules that are not organized into a bipolar array. (*B*) Beginning at the karyosome, motor proteins may bind parallel microtubules and taper them into defined poles. This process may involve several motor proteins, such as dynein, NCD and SUB, spindle pole associated proteins such as MSPS and D-TACC, and include processes like microtubule bundling and sliding between parallel and/or antiparallel microtubules. There may also be a "polar ejection force" exerted by motors associated with the chromosomes. In addition, some (MSPS), but not all (D-TACC) proteins require NCD motor activity for spindle pole association.

formation while mutants that affect sister chromatid cohesion are able to form bipolar spindles.

Segregation of Homologs on an Acentrosomal Spindle

A common feature of female meiotic spindles is the absence of centrosomes and centrioles. In such diverse organisms as humans, nematodes, insects, and plants, acentriolar spindles have a barrel shape due to the absence of defined poles [reviewed in (86)]. In Drosophila oocytes some centrosome components, such as γ-tubulin, do not concentrate at the spindle poles (83). However, γ-tubulin is required for Drosophila female meiosis (124). In the absence of a spindle pole that acts as a microtubule-organizing center, a bipolar spindle must organize via an alternative mechanism. There is substantial evidence that chromosomes have a significant role in organizing the microtubules of a meiotic spindle (86). In mature Drosophila oocytes (stage 14, Figure 3), there is no congression of bivalents to the metaphase plate. Instead, meiotic spindle formation begins with a mass of microtubules emanating from the chromosomes, suggesting that the chromosomes somehow nucleate or capture microtubules that are later shaped into a bipolar spindle. In fact, individual chromosomes form bipolar minispindles in Drosophila oocytes (125) and other organisms (128).

Prior to fertilization, meiosis in Drosophila females normally arrests at metaphase I. In the absence of chiasmata between homologs, the oocytes precociously enter anaphase I (89), suggesting that the metaphase I arrest is dependent upon tension. Further experiments showed that tension, created by the interaction of the chiasmate bivalents with the spindle, is responsible for the metaphase arrest rather than the crossovers themselves (60). Another important finding from these studies is that despite the absence of centrosomes and chiasmata, bipolar spindles form.

The first gene to be identified with a function in spindle pole formation was *ncd*. Mutations in *ncd* cause defects in meiotic chromosome segregation during meiosis I. Cytological studies of *ncd* mutants revealed a variety of microtubule organization defects, including the development of multipolar or apolar spindles (51, 83). A study of *ncd* in living oocytes led Matthies et al. (83) to conclude that NCD has a role in spindle pole formation. They proposed that, in an acentrosomal spindle, NCD tapers and bundles microtubules into a pole as its minus-end directed motor moves away from the chromosomes (Figure 6, see color insert).

Recently, the situation has become more complex as more genes have been identified with mutant phenotypes similar to *ncd*. For example, *subito (sub)*, which also encodes a kinesin-like motor protein, has a critical role in meiotic spindle pole formation (41). Like *ncd, sub* mutants have defects in spindle pole formation. *sub^1794* is a fertile recessive mutant that is female specific and affects sex and autosomal chromosome segregation during meiosis I. Genetic tests showed that in *sub^1794* mutant females, nondisjunction involves homologs at the first division of meiosis while sister chromatid segregation and male meiosis are unaffected. These experiments also showed that meiotic crossing over is not affected in *sub^1794* females and, in fact, chiasmata fail to ensure segregation of the homologs in *sub* mutants.

Matthies et al. (83) proposed that multiple motors are required for tapering the spindles into defined poles (Figure 6). One interesting possibility is that SUB and

NCD have similar and partially redundant roles in tapering the meiotic spindle. This model predicts that a *sub; ncd* double mutant would have a more severe defect in spindle pole formation than either single mutant. However, a double mutant was able to make spindle poles with a similar array of defects as the single mutants. Therefore, it appears likely that both *ncd* and *sub* are involved in the same process of spindle formation.

Consistent with this conclusion, mutations in genes that do not encode motor proteins have been identified with similar defects in spindle pole formation. Cytological experiments have shown that mutations in *mini spindles* (*msps*) and *transforming acidic coiled-coil protein* (*d-tacc*) have meiotic phenotypes that are similar to *sub* and *ncd* mutants (32). The common mutant phenotype of these genes reveals a group of potentially interacting proteins required for spindle bipolarity. Indeed, the interpretation that NCD activity is directly involved in tapering the spindle poles must be reassessed since several different genes, including some encoding centrosome-associated non-motor proteins (see below), have *ncd*-like phenotypes. All of these genes are required for the formation of a bipolar spindle, but the functions of specific motor proteins like NCD and SUB remain to be elucidated.

The analysis of *msps* and *d-tacc* demonstrates that some centrosome components are present at the poles of oocyte meiosis I spindles. Even though the Drosophila female meiotic spindle lacks critical centrosomal components, MSPS and D-TACC form a complex of proteins at the poles (32). Furthermore, the product of a third gene, *abnormal spindle* (*asp*), is also found localized to female meiotic spindle poles (104). Immunolocalization of these proteins in different mutant backgrounds has shown that NCD is required for the localization of MSPS to the spindle poles (32). These results suggest a new role for the motor proteins in that they may be required to facilitate the localization of centrosomal proteins to the spindle poles. The mechanism by which D-TACC gets to the poles is not known and does not require NCD (32) or SUB (J.K. Jang & K.S. McKim, unpublished results).

Many of the genes with roles in female meiotic spindle pole formation are required for centrosome function in mitotic cells. Mutations in genes such as *Klp3A* (131), *αTubulin67C* (81) *wispy* (16), and *sub* (41, 113) cause female sterility due to a failure in pronuclear migration following fertilization. As proposed for *d-tacc* and *asp* mutants, a centrosome defect in the early embryo may result in the failure to grow microtubules from the sperm aster (37). All of these genes may have a role in centrosome function that is also critical during female meiosis where centrosomes are lacking.

A second kinesin gene implicated in female meiosis is *nod*. Unlike *ncd*, this gene is not required for spindle formation and pole organization. Instead, NOD is an example of a chromokinesin and facilitates interactions between the spindle and the chromosomes (1). Interestingly, NOD may not have motor activity, but instead may act as a break to prevent chromosomes from moving prematurely to the poles (82). Although this function appears most important for achiasmate chromosomes, *nod* also shows dominant genetic interactions with the spindle-forming kinesins such as *ncd* (66) and *sub* (41). In addition, double mutants involving *nod* and a crossover-defective gene like *mei-218* cannot form a mature spindle (86). Thus, *nod* may have a role in spindle formation that is redundant with the chromosomes or chiasmata.

Sister Chromatid Cohesion at Centromeres and Chromosome Arms

The orientation of homologs by chiasmata at meiosis I requires sister chromatid cohesion (Figure 2). The Drosophila genome contains all of the basic components known to function in sister chromatid cohesion [reviewed in (71, 96)]. As such, some of the basic concepts of sister chromatid cohesion that have been described in other systems probably also apply to Drosophila. However, the analysis of two mutants in Drosophila has yielded important insights into the differential regulation of arm and centromere sister chromatid cohesion.

One of the specific requirements of meiosis I is that cohesion distal to the chiasma must be released for anaphase to occur. In contrast, cohesion must be maintained between the centromeres until meiosis II. The importance of these requirements can be seen in the mutant phenotypes of two genes, *mei-S332* and *ord*. Mutations of *mei-S332* cause nondisjunction of sister chromatids through premature separation prior to meiosis II (63). This phenotype is consistent with the finding that MEI-S332 is located in the centromeric regions of the meiotic chromosomes (64, 92). MEI-S332 localizes to centromeres in mitotic cells as well, but the meiotic defects in mutants are by far the most dramatic because centromeric cohesion is most important when arm cohesion has been released (70). Interestingly, MEI-S332 dissociates from the centromeres at anaphase II, although protein levels stay the same, suggesting that it is not degraded (92).

Chiasmata are required for the homologous chromosomes to remain attached and resist the pulling force of the spindle microtubules towards the poles until anaphase begins. Chiasmata are not sufficient for this process, however, as additional binding forces must act to keep the chiasma at the crossover site and prevent the homologs from slipping apart. The importance of cohesion distal to the chiasmata has been shown with mutations in *orientation disrupter* (*ord*), which cause a dramatic increase in reductional and equational nondisjunction. Genetic studies have shown that nondisjunction occurs at both meiotic divisions in *ord* mutants (11, 91), and cytological studies have confirmed that in *ord* mutants, there is premature separation of sister chromatids prior to the first division and chiasmata are unable to hold homologs together at metaphase I (9). The result is that chiasma cannot direct the segregation of homologs. The phenotype of *ord* mutants is what would be expected for general sister chromatid cohesion proteins such as SMC1 and SMC3. ORD is a novel protein, although it might be predicted that it would be a component of the sister chromatid cohesion complex (10). The lack of a mitotic phenotype in *ord* mutants suggests that ORD represents a specialized component of meiotic sister chromatid cohesion (11). It is also interesting, considering that sister chromatid cohesion proteins are assembled during S-phase, that *ord* has a function much earlier in meiotic prophase, it is required for normal levels of meiotic recombination (91).

SUMMARY AND FUTURE DIRECTIONS

When Are Crossovers Sites Chosen?

DSB formation, and thus the activity of MEI-W68 and MEI-P22, are dependent on C(3)G and SC formation. While these events are certainly some of the earliest in meiotic recombination pathway, caution is sometimes needed when trying to infer the order of other functions later in the pathway. For example, *mei-218* is required for meiotic crossing over, and has no known effects on DSB repair. But it is not known when during the DSB initiation and repair process *mei-218* is required. C(2)M is required for crossing over and, unlike *mei-218*, the first defects in mutants are observed prior to or concomitant with DSB formation. However, DSB repair-defective mutants alter the distribution of crossovers, suggesting that events after DSB formation can influence crossover site determination. Further studies are required to determine at what points in the DSB repair pathway crossovers can be determined.

The Genome Sequence and What Is Dispensable

With the complete sequence of the genome, one can determine what gene products are missing from Drosophila meiosis (115). This is a difficult task for components of the synaptonemal complex because these proteins are often not conserved. With more certainty, we can state that Drosophila females generate crossovers without the Msh4 and Msh5 homologs (assuming they are not buried in heterochromatin). DSB repair in Drosophila also utilizes a smaller set of factors than some other organisms. For instance, there is no Rad52 homolog and only one Rad54. Many organisms have at least one Rad52 homolog and two Rad54 homologs, one of which has a role in interhomolog recombination (118). Drosophila does have two Rad51 homologs, and one is meiosis-specific like Dmc1. While there is poor sequence conservation (38), the phenotype of *spnB* mutants makes it likely that this gene encodes a functional homolog Dmc1.

The Synapsis of Homologs and Making Double-Strand Breaks

The data summarized here establish the dependence of DSB formation on SC formation in Drosophila. This raises two significant questions. First, what component(s) of the SC is responsible for the stimulation of DSBs and second, what is the mechanism for homolog alignment and synapsis? Some of these questions have become more amenable now that there are antibodies to observe the SC (C(3)G, C(2)M), and DSB sites (MEI-P22, γ-H2AX).

The dependence of DSB formation on an SC component is not unique to Drosophila. DSBs are reduced to 10% of wild-type levels by mutations in the *hop1* gene of *S. cerevisiae*, which encodes a component of the axial elements, a precursor of SC lateral elements (95, 114). However, mutants in the *zip1* gene, a homolog of *c(3)G*, do not form SC but have normal levels of DSBs (122). Therefore, in *S. cerevisiae* and Drosophila, DSBs are dependent on different components of the synaptonemal complex; in Drosophila, the transverse elements stimulate DSBs, whereas in *S. cerevisiae*, the axial elements do. In addition, DNA replication and chromatin structure have a role in DSB formation in budding yeast (15, 132). Whether these factors play a role in Drosophila has yet to be determined. Interestingly, the analysis of MEI-P22 suggests that SC-mediated changes in chromatin structure may have a significant role in regulating DSB formation in Drosophila. Critical to understanding DSB formation, therefore, will be to understand what changes in chromosome structure predipose a region to a break and how these changes are regulated.

The mechanism for homolog synapsis, which occurs normally in the absence of DSBs, remains elusive. It is an old observation that translocation heterozygotes suppress crossing over in Drosophila females (e.g., 35). Because translocations can suppress crossing over for considerable distances from their breakpoints, it was concluded that a defect in synapsis leads to the reductions in crossing over (106). A careful comparison of translocation breakpoints and their respective crossover-suppressed regions has led to the mapping of discrete "pairing sites" on the X and third chromosomes (44; D. Sherizen & K. McKim, unpublished). It is possible that these sites, in the absence of DSBs, align large chromosomal domains and promote SC formation. However, while the pairing site hypothesis is attractive, most of its predictions have yet to be tested.

Female Meiotic Spindle Assembly Uses a Subset of Centrosome Components

The identification and understanding of genes required for meiotic spindle formation in oocytes has only just begun. Most likely, additional proteins will be identified in the near future. For example, it might be expected that kinases regulate the kinesin-like motors or centrosome-associated proteins. *Aurora* A kinase phosphorylates D-TACC and is required for its localization in mitotic cells (40) while POLO kinase phosphorylates ASP to activate its microtubule organizing properties (4). It is not known if either of these kinases has a role in the first meiotic division. In the case of *polo* and *asp*, hypomorphic mutations do not disrupt the structure of the meiosis I spindle poles (103, 104), and the effects of *aur* mutations are not known. These proteins are, however, required for the formation of the meiosis II spindle. This occurs through the reorganization of the central spindle microtubules at telophase I and the development of a single MTOC between the two meiosis II spindles.

Regulation of Meiotic Entry

Perhaps least is known about regulation of meiotic entry and the related question of how crossing over occurs in females but not males. The observation that hypomorphic alleles of two genes required for oocyte differentiation, *Sxl* and *mei-P26* (14, 99), have reduced levels of crossing over suggests there is an important link between the genes that regulate oocyte differentiation and meiotic recombination.

Immediately following the last mitotic division and premeiotic S-phase, the two prooocytes begin to build SC and induce DSBs. In *BicD* and *egl* mutants, SC forms in all 16 cells of each cyst (27, 59, 98). Thus, it appears as if the default state is for all cells to enter meiosis and build SC. So far, however, no regulatory gene has been identified with the specific function of promoting SC formation. Such an avenue of research may also provide insights into the chromosomal changes that facilitate SC formation and meiotic recombination.

ACKNOWLEDGMENTS

We are grateful to Scott Hawley and Jeff Sekelsky for insighful comments on the manuscript. Research in the laboratory of KMcK has been supported the American Cancer Society and the National Science Foundation.

The *Annual Review of Genetics* is online at http://genet.annualreviews.org

LITERATURE CITED

1. Afshar K, Barton NR, Hawley RS, Goldstein LSB. 1995. DNA binding and meiotic chromosomal localization of the Drosophila Nod kinesin-like protein. *Cell* 81:129–38
2. Albini SM, Jones GH. 1987. Synaptonemal complex speading in *Allium cepa* and *A. fistulosum*. I. The initiation and sequence of pairing. *Chromosoma* 95:324–38
2a. Ashburner M, Novitski E, eds. 1976. *The Genetics and Biology of Drosophila*. New York: Academic
3. Allers T, Lichten M. 2001. Differential timing and control of noncrossover and crossover recombination during meiosis. *Cell* 106:47–57
4. Avides MC, Tavares A, Glover DM. 2001. Polo kinase and Asp are needed to promote the mitotic organizing activity of centrosomes. *Nat. Cell. Biol.* 3:421–24
5. Baker BS, Carpenter ATC. 1972. Genetic analysis of sex chromosomal meiotic mutants in *Drosophila melanogaster*. *Genetics* 71:255–86
6. Baker BS, Hall JC. 1976. Meiotic mutants: genetic control of meiotic recombination and chromosome segregation. See Ref. 2a, pp. 351–434
7. Baudat F, Manova K, Yuen JP, Jasin M, Keeney S. 2000. Chromosome synapsis defects and sexually dimorphic meiotic progression in mice lacking Spo11. *Mol. Cell* 6:989–98
8. Beadle G. 1932. A possible influence of the spindle fiber on crossing over in *Drosophila melanogaster*. *Proc. Natl. Acad. Sci. USA* 18:160–65
9. Bickel SE, Orr-Weaver TL, Balicky EM. 2002. The sister-chromatid cohesion protein ORD is required for chiasma maintenance in Drosophila oocytes. *Curr. Biol.* In Press
10. Bickel SE, Wyman DW, Miyazaki WY, Moore DP, Orr-Weaver TL. 1996. Identification of ORD, a *Drosophila* protein essential for sister chromatid cohesion. *EMBO J.* 15:1451–59
11. Bickel SE, Wyman DW, Orr-Weaver TL. 1997. Mutational analysis of the Drosophila sister-chromatid cohesion protein ORD and its role in the maintenance of centromeric cohesion. *Genetics* 146: 1319–31
12. Bishop DK, Park D, Xu L, Kleckner N.

1992. *DMC1*: a meiosis-specific yeast homolog of *E. coli recA* required for recombination, synaptonemal complex formation, and cell cycle progression. *Cell* 69:439–56

13. Boddy MN, Gaillard PH, McDonald WH, Shanahan P, Yates JR 3rd, Russell P. 2001. Mus81-Eme1 are essential components of a Holliday junction resolvase. *Cell* 107: 537–48

14. Bopp D, Schutt C, Puro J, Huang H, Nothiger R. 1999. Recombination and disjunction in female germ cells of Drosophila depend on the germline activity of the gene *sex-lethal*. *Development* 126:5785–94

15. Borde V, Goldman AS, Lichten M. 2000. Direct coupling between meiotic DNA replication and recombination initiation. *Science* 290:806–9

16. Brent AE, MacQueen A, Hazelrigg T. 2000. The Drosophila *wispy* gene is required for RNA localization and other microtubule-based events of meiosis and early embryogenesis. *Genetics* 154:1649–62

17. Carlson PS. 1972. The effects of inversions and the *c(3)G* mutation on intragenic recombination in Drosophila. *Genet. Res. Camb.* 19:129–32

18. Carpenter ATC. 1975. Electron microscopy of meiosis in *Drosophila melanogaster* females. I. Structure, arrangement, and temporal change of the synaptonemal complex in wild-type. *Chromosoma* 51:157–82

19. Carpenter ATC. 1975. Electron microscopy of meiosis in *Drosophila melanogaster* females. II. The recombination nodule—a recombination-associated structure at pachytene? *Proc. Natl. Acad. Sci. USA* 72:3186–89

20. Carpenter ATC. 1979. Recombination nodules and synaptonemal complex in recombination-defective females of *Drosophila melanogaster*. *Chromosoma* 75: 259–92

21. Carpenter ATC. 1979. Synaptonemal complex and recombination nodules in wild-type *Drosophila melanogaster* females. *Genetics* 92:511–41

22. Carpenter ATC. 1981. EM Autoradiographic evidence that DNA synthesis occurs at recombination nodules during meiosis in *Drosophila melanogaster* females. *Chromosoma* 83:59–80

23. Carpenter ATC. 1982. Mismatch repair, gene conversion, and crossing over in two recombination-defective mutants of *Drosophila melanogaster*. *Proc. Natl. Acad. Sci. USA* 79:5961–65

24. Carpenter ATC. 1984. Meiotic roles of crossing over and of gene conversion. *Cold Spring Harbor Symp. Quant. Biol.* 49:23–29

25. Carpenter ATC. 1988. Thoughts on recombination nodules, meiotic recombination, and chiasmata. See Ref. 67a, pp. 529–48

26. Carpenter ATC. 1989. Are there morphologically abnormal early recombination nodules in the *Drosophila melanogaster* meiotic mutant *mei-218*? *Genome* 31:74–80

27. Carpenter ATC. 1994. *egalitarian* and the choice of cell fates in *Drosophila melanogaster* oogenesis. *Ciba Found. Symp.* 182:223–54

28. Carpenter ATC, Sandler L. 1974. On recombination-defective meiotic mutants in *Drosophila melanogaster*. *Genetics* 76: 453–75

29. Celerin M, Merino ST, Stone JE, Menzie AM, Zolan ME. 2000. Multiple roles of Spo11 in meiotic chromosome behavior. *EMBO J.* 19:2739–50

30. Chovnick A, Ballantyne GH, Baillie DL, Holm DG. 1970. Gene conversion in higher organisms: half-tetrad analysis of recombination within the rosy cistron of *Drosophila melanogaster*. *Genetics* 66: 315–29

31. Cromie GA, Leach DRF. 2000. Control of crossing over. *Mol. Cell* 6:815–26

32. Cullen CF, Ohkura H. 2001. Msps protein is localized to acentrosomal poles to

ensure bipolarity of Drosophila meiotic spindles. *Nat. Cell. Biol.* 3:637–42

33. Dernburg AF, McDonald K, Moulder G, Barstead R, Dresser M, Villeneuve AM. 1998. Meiotic recombination in *C. elegans* initiates by a conserved mechanism and is dispensable for homologous chromosome synapsis. *Cell* 94:387–98

34. Dernburg AF, Sedat JW, Hawley RS. 1996. Direct evidence of a role for heterochromatin in meiotic chromosome segregation. *Cell* 85:135–46

35. Dobzhansky T. 1931. The decrease of crossing-over observed in translocations, and its probable explanation. *Am. Nat.* 65:214–32

36. Foss E, Lande R, Stahl FW, Steinberg CM. 1993. Chiasma interference as a function of genetic distance. *Genetics* 133:681–91

37. Gergely F, Kidd D, Jeffers K, Wakefield JG, Raff JW. 2000. D-TACC: a novel centrosomal protein required for normal spindle function in early *Drosophila* embryo. *EMBO J.* 19:241–52

38. Ghabrial A, Ray RP, Schupbach T. 1998. *okra* and *spindle-B* encode components of the RAD52 DNA repair pathway and affect meiosis and patterning in Drosophila oogenesis. *Genes Dev* 12:2711–23

39. Ghabrial A, Schupbach T. 1999. Activation of a meiotic checkpoint regulates translation of Gurken during Drosophila oogenesis. *Nat. Cell. Biol.* 1:354–57

40. Giet R, McLean D, Descamps S, Lee MJ, Raff JW, et al. 2002. Drosophila Aurora A kinase is required to localize D-TACC to centrosomes and to regulate astral microtubules. *J. Cell Biol.* 156:437–51

41. Giunta KL, Jang JK, Manheim EM, Subramanian G, McKim KS. 2002. *subito* encodes a kinesin-like protein required for meiotic spindle pole formation in *Drosophila melanogaster*. *Genetics* 160:1489–501

42. Gonzalez-Reyes A, Elliot H, St. Johnston D. 1997. Oocyte determination and the origin of polarity in *Drosophila*: the role of the *spindle* genes. *Development* 124:4927–37

43. Gowen MS, Gowen JW. 1922. Complete linkage in *Drosophila melanogaster*. *Am. Nat.* 61:286–88

44. Green MM. 1981. *mus(3)312^{D1}*, a mutagen sensitive mutant with profound effects on female meiosis in *Drosophila melanogaster*. *Chromosoma* 82:259–66

45. Grell RF, Generoso EE. 1980. Time of recombination in the *Drosophila melanogaster* oocyte. *Chromosoma* 81:339–48

46. Haber JE. 2000. Partners and pathways repairing a double-strand break. *Trends Genet.* 16:259–64

47. Hall JC. 1972. Chromosome segregation influenced by two alleles of the meiotic mutant *c(3)G* in *Drosphila melanogaster*. *Genetics* 71:367–400

48. Hari KL, Santerre A, Sekelsky JJ, McKim KS, Boyd JB, Hawley RS. 1995. The *mei-41* gene of *D. melanogaster* is a structural and function homolog of the human ataxia telangiectasia gene. *Cell* 82:815–21

49. Hassold TJ. 1996. Mismatch repair goes meiotic. *Nat. Genet.* 13:261–62

50. Hatsumi M, Endow SA. 1992. The *Drosophila* ncd microtubule motor protein is spindle-associated in meiotic and mitotic cells. *J. Cell Sci.* 103:1013–20

51. Hatsumi M, Endow SA. 1992. Mutants of the microtubule motor protein, nonclaret disjunctional, affect spindle structure and chromosome movement in meiosis and mitosis. *J. Cell Sci.* 101:547–59

52. Hawley RS. 1980. Chromosomal sites necessary for normal levels of meiotic recombination in *Drosophila melanogaster*. I. Evidence for and mapping of the sites. *Genetics* 94:625–46

53. Hawley RS. 1988. Exchange and chromosomal segregation in eucaryotes. In *Genetic recombination*, ed. R Kucherlapati, G Smith, pp. 497–527. Washington, DC: ASM

54. Hawley RS, Irick HA, Zitron AE, Haddox DA, Lohe AR, et al. 1992. There are

two mechanisms of achiasmate segregation in *Drosophila* females, one of which requires heterochromatic homology. *Dev. Genet.* 13:440–67

55. Hawley RS, McKim KS, Arbel T. 1993. Meiotic segregation in *Drosophila melanogaster* females: molecules, mechanisms and myths. *Annu. Rev. Genet.* 27: 281–317

56. Hawley RS, Theurkauf WE. 1993. Requiem for distributive segregation: achiasmate segregation in *Drosophila* females. *Trends Genet.* 9:310–17

57. Hilliker AJ, Chovnick A. 1981. Further observations on intragenic recombination in *Drosophila melanogaster*. *Genet. Res.* 38:281–96

58. Hunter N, Kleckner N. 2001. The single-end invasion: an asymmteric intermediate at the double-strand break to double-holliday junction transition of meiotic recombination. *Cell* 106:59–70

59. Huynh J, St. Johnston D. 2000. The role of BicD, Egl, Orb and the microtubules in the restriction of meiosis to the Drosophila oocyte. *Development* 127:2785–94

60. Jang JK, Messina L, Erdman MB, Arbel T, Hawley RS. 1995. Induction of metaphase arrest in Drosophila oocytes by chiasma-based kinetochore tension. *Science* 268:1917–19

61. Keeney S, Baudat F, Angeles M, Zhou ZH, Copeland NG, et al. 1999. A mouse homolog of the *Saccharomyces cerevisiae* meiotic recombination DNA transesterase Spo11p. *Genomics* 61:170–82

62. Keeney S, Giroux CN, Kleckner N. 1997. Meiosis-specific DNA double-strand breaks are catalyzed by Spo11, a member of a widely conserved protein family. *Cell* 88:375–84

63. Kerrebrock AW, Miyazaki WY, Birnby D, Orr-Weaver TL. 1992. The Drosophila *mei-S332* gene promotes sister-chromatid cohesion in meiosis following kinetochore differentiation. *Genetics* 130:827–41

64. Kerrebrock AW, Moore DP, Wu JS, Orr-

Weaver TL. 1995. Mei-S332, a Drosophila protein required for sister-chromatid cohesion, can localize to meiotic centromere regions. *Cell* 83:247–56

65. King RC. 1970. *Ovarian Development in Drosophila melanogaster*. New York: Academic

66. Knowles BA, Hawley RS. 1991. Genetic analysis of microtubule motor proteins in *Drosophila*: A mutation at the *ncd* locus is a dominant enhancer of *nod*. *Proc. Natl. Acad. Sci. USA* 88:7165–69

67. Koehler KE, Boulton CL, Collins HE, French RL, Herman KC, et al. 1996. Spontaneous X chromosome MI and MII nondisjunction events in *Drosophila melanogaster* oocytes have different recombinational histories. *Nat. Genet.* 14: 406–14

67a. Kucherlapati R, Smith G, eds. 1988. *Genetic Recombination*. Washington, DC: Am. Soc. Microbiol.

68. Lambie EJ, Roeder GS. 1988. A yeast centromere acts in *cis* to inhibit meiotic gene conversion of adjacent sequences. *Cell* 52:863–73

69. Lantz V, Chang JS, Horabin JI, Bopp D, Schedl P. 1994. The Drosophila ORB RNA-binding protein is required for the formation of the egg chamber and establishment of polarity. *Genes Dev.* 8:598–613

70. LeBlanc HN, Tang TT, Wu JS, Orr-Weaver TL. 1999. The mitotic centromeric protein MEI-S332 and its role in sister-chromatid cohesion. *Chromosoma* 108: 401–11

71. Lee JY, Orr-Weaver TL. 2001. The molecular basis of sister-chromatid cohesion. *Annu. Rev. Cell Dev. Biol.* 17:753–77

72. Lichten M, Goldman A. 1995. Meiotic recombination hotspots. *Annu. Rev. Genet.* 29:423–44

73. Lindsley DL, Sandler L. 1977. The genetic analysis of meiosis in female *Drosophila melanogaster*. *Philos. Trans. R. Soc. London Ser. B* 277:295–312

74. Lindsley DL, Zimm GG. 1992. *The*

Genome of Drosophila melanogaster. San Diego: Academic

75. Liu H, Jang JK, Graham J, Nycz K, McKim KS. 2000. Two genes required for meiotic recombination in Drosophila are expressed from a dicistronic message. *Genetics* 154:1735–46

76. Liu H, Jang JK, Kato N, McKim KS. 2002. *mei-P22* encodes a chromosome-associated protein required for the initiation of meiotic recombination in *Drosophila melanogaster. Genetics.* In press

77. Lucchesi JC, Suzuki DT. 1968. The interchromosomal control of recombination. *Annu. Rev. Genet.* 2:53–86

78. Lydall D, Nikolsky Y, Bishop DK, Weinert T. 1996. A meiotic recombination checkpoint controlled by mitotic checkpoint genes. *Nature* 382:840–43

79. Mahadevaiah SK, Turner JM, Baudat F, Rogakou EP, de Boer P, et al. 2001. Recombinational DNA double-strand breaks in mice precede synapsis. *Nat. Genet.* 27:271–76

80. Manheim EA, Jang JK, Dominic D, McKim KS. 2002. Cytoplasmic localization and evolutionary conservation of MEI-218, a protein required for meiotic crossing over in Drosophila. *Mol. Biol. Cell* 13:84–95

81. Matthews KA, Rees D, Kaufman TC. 1993. A functionally specialized α-tubulin is required for oocyte meiosis and cleavage mitoses in Drosophila. *Development* 117:977–91

82. Matthies HJ, Baskin RJ, Hawley RS. 2001. Orphan kinesin NOD lacks motile properties but does possess a microtubule-stimulated ATPase activity. *Mol. Biol. Cell* 12:4000–12

83. Matthies HJ, McDonald HB, Goldstein LS, Theurkauf WE. 1996. Anastral meiotic spindle morphogenesis: role of the *non-claret disjunctional* kinesin-like protein. *J. Cell Biol.* 134:455–64

84. McKim KS, Dahmus JB, Hawley RS.

1996. Cloning of the *Drosophila melanogaster* meiotic recombination gene *mei-218*: a genetic and molecular analysis of interval 15E. *Genetics* 144:215–28

85. McKim KS, Green-Marroquin BL, Sekelsky JJ, Chin G, Steinberg C, et al. 1998. Meiotic synapsis in the absence of recombination. *Science* 279:876–78

86. McKim KS, Hawley RS. 1995. Chromosomal control of meiotic cell division. *Science* 270:1595–601

87. McKim KS, Hayashi-Hagihara A. 1998. *mei-W68* in *Drosophila melanogaster* encodes a Spo11 homolog: evidence that the mechanism for initiating meiotic recombination is conserved. *Genes Dev.* 12: 2932–42

88. McKim KS, Jang JK, Sekelsky JJ, Laurencon A, Hawley RS. 2000. *mei-41* is required for precocious anaphase in Drosophila females. *Chromosoma* 109:44–49

89. McKim KS, Jang JK, Theurkauf WE, Hawley RS. 1993. Mechanical basis of meiotic metaphase arrest. *Nature* 362: 364–66

90. Merriam JR, Frost JN. 1964. Exchange and nondisjunction of the *X* chromosomes in female *Drosophila melanogaster. Genetics* 49:109–22

91. Miyazaki WY, Orr-Weaver TL. 1992. Sister-chromatid misbehavior in Drosophila *ord* mutants. *Genetics* 132:1047–61

92. Moore DP, Page AW, Tang TT, Kerrebrock AW, Orr-Weaver TL. 1998. The cohesion protein MEI-S332 localizes to condensed meiotic and mitotic centromeres until sister chromatids separate. *J. Cell Biol.* 140:1003–12

93. Morgan TH. 1914. No crossing over in the male of Drosophila of genes in the second and third pairs of chromosomes. *Biol. Bull. Wood's Hole* 26:195—204

94. Myers EW, Sutton GG, Delcher AL, Dew IM, Fasulo DP, et al. 2000. A whole-genome assembly of Drosophila. *Science* 287:2196–204

95. Nag DK, Scherthan H, Rockmill B, Bhargava J, Roeder GS. 1995. Heteroduplex DNA formation and homolog pairing in yeast meiotic mutants. *Genetics* 141:75–86

96. Nasmyth K. 2001. Disseminating the genome: joining, resolving, and separating sister chromatids during mitosis and meiosis. *Annu. Rev. Genet.* 35:673–745

97. O'Tousa J. 1982. Meiotic chromosome behavior influenced by mutation-altered disjunction in *Drosophila melanogaster* females. *Genetics* 102:503–24

98. Page SL, Hawley RS. 2001. *c(3)G* encodes a *Drosophila* synaptonemal complex protein. *Genes Dev.* 15:3130–43

99. Page SL, McKim KS, Deneen B, Van Hook TL, Hawley RS. 2000. Genetic studies of *mei-P26* reveal a link between the processes that control germ cell proliferation in both sexes and those that control meiotic exchange in Drosophila. *Genetics* 155:1757–72

100. Paques F, Haber JE. 1999. Multiple pathways of recombination induced by double-strand breaks in *Saccharomyces cerevisiae. Micro. Mol. Biol. Rev.* 63:349–404

101. Parry DM. 1973. A meiotic mutant affecting recombination in female *Drosophila melanogaster. Genetics* 73:465–86

102. Rasmussen SW. 1975. Ultrastructural studies of meiosis in males and females of the *c(3)G* mutant in *Drosophila melanogaster* meigen. *C.R. Trav. Lab. Carlsberg* 40:163–73

103. Riparbelli MG, Callaini G, Glover DM. 2000. Failure of pronuclear migration and repeated divisions of polar body nuclei associated with MTOC defects in polo eggs of Drosophila. *J. Cell Sci.* 113:3341–50

104. Riparbelli MG, Callaini G, Glover DM, Avides Md Mdo C. 2002. A requirement for the Abnormal Spindle protein to organise microtubules of the central spindle for cytokinesis in Drosophila. *J. Cell Sci.* 115:913–22

105. Roberts PA. 1969. Some components of

X ray-induced crossing over in females of *Drosophila melanogaster. Genetics* 63:387–404

106. Roberts PA. 1976. The genetics of chromosome aberration. See Ref. 2a, pp. 68–184

107. Roeder GS. 1997. Meiotic chromosomes: It takes two to tango. *Genes Dev.* 11:2600–21

108. Rogakou EP, Boon C, Redon C, Bonner WM. 1999. Megabase chromatin domains involved in DNA double-strand breaks in vivo. *J. Cell Biol.* 146:905–16

109. Romanienko PJ, Camerini-Otero RD. 1999. Cloning, characterization, and localization of mouse and human SPO11. *Genomics* 61:156–69

110. Romanienko PJ, Camerini-Otero RD. 2000. The mouse *spo11* gene is required for meiotic chromosome synapsis. *Mol. Cell* 6:975–87

111. Sandler L, Lindsley DL. 1974. Some observations on the study of the genetic control of meiosis in *Drosophila melanogaster. Genetics* 78:289–97

112. Sandler L, Lindsley DL, Nicoletti B, Trippa G. 1968. Mutants affecting meiosis in natural populations of *Drosophila melanogaster. Genetics* 60:525–58

113. Schupbach T, Wieschaus E. 1989. Female sterile mutations on the second chromosome of *Drosophila melanogaster.* I. Maternal effect mutations. *Genetics* 121:101–17

114. Schwacha A, Kleckner N. 1994. Identification of joint molecules that form frequently between homologs but rarely between sister chromatids during yeast meiosis. *Cell* 76:51–63

115. Sekelsky JJ, Brodsky MH, Burtis KC. 2000. DNA repair in Drosophila: insights from the Drosophila genome sequence. *J. Cell Biol.* 150:F31–36

116. Sekelsky JJ, McKim KS, Chin GM, Hawley RS. 1995. The Drosophila meiotic recombination gene *mei-9* encodes a homologue of the yeast excision repair protein Rad1, *Genetics* 141:619–27

117. Sekelsky JJ, McKim KS, Messina L, French RL, Hurley WD, et al. 1999. Identification of novel Drosophila meiotic genes recovered in a P-element screen. *Genetics* 152:529–42

118. Shinohara M, Shita-Yamaguchi E, Buerstedde JM, Shinagawa H, Ogawa H, Shinohara A. 1997. Characterization of the roles of the *Saccharomyces cerevisiae* RAD54 gene and a homologue of RAD54, RDH54/TID1, in mitosis and meiosis. *Genetics* 147:1545–56

119. Smith PA, King RC. 1968. Genetic control of synaptonemal complexes in *Drosophila melanogaster*. *Genetics* 60: 335–51

120. Smith PD, Finnerty VG, Chovnick A. 1970. Gene conversion in *Drosophila*: Non-reciprocal events at the maroon-like cistron. *Nature* 228:442–44

121. Stack SM, Anderson L. 1986. Two-dimensional spreads of synaptonemal complexes from solanaceous plants. II. Synapsis in *Lycopersicon esculentum* (tomato). *Am. J. Bot.* 73:264–81

122. Sym M, Engebrecht J, Roeder GS. 1993. ZIP1 is a synaptonemal complex protein required for meiotic chromosome synapsis. *Cell* 72:365–78

123. Sym M, Roeder GS. 1994. Crossover interference is abolished in the absence of a synaptonemal complex protein. *Cell* 79:283–92

124. Tavosanis G, Llamazares S, Goulielmos G, Gonzalez C. 1997. Essential role for gamma-tubulin in the acentriolar female meiotic spindle of Drosophila. *EMBO J.* 16:1809–19

125. Theurkauf WE, Hawley RS. 1992. Meiotic spindle assembly in *Drosophila* females: behavior of nonexchange chromosomes and the effects of mutations in the *nod* kinesin-like protein. *J. Cell Biol.* 116:1167–80

126. Thorne LW, Byers B. 1993. Stage-specific effects of X-irradiation on yeast meiosis. *Genetics* 134:29–42

127. Villeneuve AM, Hillers KJ. 2001. Whence meiosis. *Cell* 106:647–50

128. Waters JC, Salmon ED. 1995. Chromosomes take an active role in spindle assembly. *BioEssays* 17:911–14

129. Weinstein A. 1936. The theory of multiple-strand crossing over. *Genetics* 21:155–99

130. Whyte WL, Irick HA, Arbel T, Yasuda G, French RL, et al. 1993. The genetic analysis of achiasmate segregation in *Drosophila melanogaster*. III. The wild-type product of the *Axs* gene is required for the meiotic segeregation of achiasmate homologs. *Genetics* 134:825–35

131. Williams BC, Dernburg AF, Puro J, Nokkala S, Goldberg ML. 1997. The *Drosophila* kinesin-like protein KLP3A is required for proper behavior of male and female pronucleii at fertilization. *Development* 124:2365–76

132. Wu T-C, Lichten M. 1994. Meiosis-induced double-strand break sites determined by yeast chromatin structure. *Science* 263:515–18

133. Zalevsky J, MacQueen AJ, Duffy JB, Kemphues KJ, Villeneuve AM. 1999. Crossing over during *Caenorhabditis elegans* meiosis requires a conserved MutS-based pathway that is partially dispensable in budding yeast. *Genetics* 153: 1271–83

134. Zhang P, Hawley RS. 1990. The genetic analysis of distributive segregation in *Drososphila melanogaster*. II. Further genetic analysis of the *nod* locus. *Genetics* 125:115–27

135. Zhang P, Knowles BA, Goldstein LSB, Hawley RS. 1990. A kinesin-like protein required for distributive chromosome segregation in *Drosophila*. *Cell* 63:1053–62

Annu. Rev. Genet. 2002. 36:233–78
doi: 10.1146/annurev.genet.36.042902.092433

Xist RNA AND THE Mechanism of X Chromosome Inactivation

Kathrin Plath, Susanna Mlynarczyk-Evans, Dmitri A. Nusinow, and Barbara Panning

Department of Biochemistry & Biophysics, University of California San Francisco, San Francisco, California 94143; e-mail: kathrin@itsa.ucsf.edu, smlynar@itsa.ucsf.edu, meter@itsa.ucsf.edu, bpanning@biochem.ucsf.edu

Key Words dosage compensation, imprinting, *Tsix* RNA, facultative heterochromatin, differentiation

■ **Abstract** Dosage compensation in mammals is achieved by the transcriptional inactivation of one X chromosome in female cells. From the time X chromosome inactivation was initially described, it was clear that several mechanisms must be precisely integrated to achieve correct regulation of this complex process. X-inactivation appears to be triggered upon differentiation, suggesting its regulation by developmental cues. Whereas any number of X chromosomes greater than one is silenced, only one X chromosome remains active. Silencing on the inactive X chromosome coincides with the acquisition of a multitude of chromatin modifications, resulting in the formation of extraordinarily stable facultative heterochromatin that is faithfully propagated through subsequent cell divisions. The integration of all these processes requires a region of the X chromosome known as the *X-inactivation center*, which contains the *Xist* gene and its *cis*-regulatory elements. *Xist* encodes an RNA molecule that plays critical roles in the choice of which X chromosome remains active, and in the initial spread and establishment of silencing on the inactive X chromosome. We are now on the threshold of discovering the factors that regulate and interact with *Xist* to control X-inactivation, and closer to an understanding of the molecular mechanisms that underlie this complex process.

CONTENTS

INTRODUCTION

Over 40 years ago, Mary Lyon postulated that equalization of X-linked gene dosage between male and female mammals occurs by the transcriptional silencing of one X chromosome in female cells (86). Early observations established that this process occurs at random, such that the X chromosome inherited from either parent is silenced in 50% of cells. The silencing of one X chromosome occurs early in development and roughly coincides with differentiation of pluripotent cells to restricted lineages (105). Analysis of X chromosome aneuploidies showed that one X chromosome remains active in a diploid cell, while all additional X chromosomes are silenced (53). Genetic studies indicated that the silencing of the inactive X chromosome (Xi) is initiated at one location on the X chromosome, termed the *X-inactivation center* (*Xic* in mouse and *XIC* in human). Silencing spreads in *cis* from the *Xic/XIC* (137, 155), indicating that this *cis*-regulatory element contains sequences that initiate a chromosome-wide alteration in chromatin structure. Cytological analysis revealed that, once established, the Xi is clonally propagated, such that females are functionally mosaic for X-linked traits (86, 136). Thus, random X-inactivation has often been described as a multistep process involving choice of the active X chromosome (Xa), initiation and spread of silencing on the Xi, and subsequent maintenance of the Xi's silent state. Mouse embryonic stem (ES) cells have become a valuable tool in the study of X-inactivation because they undergo random X-inactivation upon induction to differentiate in culture (115, 150). ES cells can be genetically manipulated and subsequently transmitted through the germline, allowing study of X-inactivation both in culture and in the embryo. As a result, the molecular mechanisms involved in regulation of X-inactivation are best understood in the mouse.

Xist/XIST RNA IS CRITICAL FOR X-INACTIVATION

The *Xist/XIST* Gene Is Located in the *Xic* /XIC

The location of the *Xic/XIC* was determined by analyzing X chromosome translocations that resulted in inactivation of the autosome to which the X chromosome fragment had fused (56). The minimal *Xic/XIC* regions are syntenic between mouse and human, although multiple inversions have resulted in shuffled gene order between the species, as depicted in Figure 1 (42, 56). The discovery in 1991 of one gene located within the *Xic/XIC*, the *Xi-specific transcript* (*Xist* in mouse, and

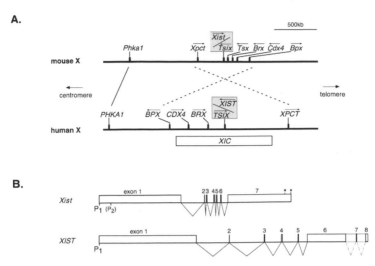

Figure 1 Mouse and human *Xic/XIC* and *Xist/XIST*. *A*. Schematic diagram of the area surrounding the *XIST/Xist* gene on the human and mouse X-chromosomes. The *XIC/Xic* and the surrounding regions of the human and mouse X-chromosomes are depicted; both include the *XIST/Xist* gene. In human, the candidate region for the *XIC* encompassing 700–1200 kb is located within band Xq13 on the proximal long arm of the X chromosome (56). In mouse, the *Xic* region identified by X chromosome rearrangements is considerably larger than that in human (56). However, the regions required to recapitulate all functions of the *Xic*, as defined by transgenic analysis in mouse, contain only the *Xist* gene and minimal surrounding sequences (57, 62, 79, 81, 82, 92). The segment of the X chromosome containing the *Xic/XIC* locus is syntenic but inverted between mouse and human (*dashed lines*) (42, 56). Small black boxes depict the location of genes in the *Xic/XIC* area, and the arrow above the gene name gives the orientation of transcription. Linkage of the two orthologous regions shows some discontinuity as local inversions of transcription units have occurred (42). The existence of a human TSX homolog is unclear (100, 108). *B*. Comparison of the human and mouse *Xist/XIST* genes. P1 indicates the major transcriptional start site for *Xist/XIST*. In mouse, two RNA isoforms of 17 and 17.9 kb may be produced by usage of alternative polyadenylation sites denoted by stars. In mouse, a second promoter, P2, has been described (69) but is not conserved in other *Xist/XIST* genes analyzed to date (108). Extensive alternative splicing of the human gene has been described, yielding *XIST* RNA isoforms lacking, for example, exon 4, half of exon 6, or exon 7 (18, 20). Only the linear splicing events are shown here. As indicated by the broken lines, the two 3′ introns may not always be excised from the processed RNA, yielding an RNA molecule of up to 19.3 kb in length (64). Short, colinear stretches of high homology have been identified throughout the genes including the very 3′ ends (64, 108).

XIST in human), has revolutionized our understanding of X-inactivation (13–15, 18, 20). *Xist/XIST* is the only gene transcribed from the Xi and not from Xa in somatic cells. No significant open reading frame has been identified, suggesting that *Xist/XIST* does not encode a protein. *Xist* has subsequently been shown to be the pivotal player in choice of which X chromosome remains active (89), and in the spread of silencing on the future Xi (90, 117). In simplest terms, the *Xic* can be thought of as the *Xist* gene and *cis*-elements that ensure its correct developmental regulation on both the Xa and Xi.

Xist/XIST Gene Structure Is Conserved, But Sequence Varies

The genomic lengths of mouse *Xist* and human *XIST* are approximately 23 kb and 35 kb, respectively. Transcripts of both genes are spliced and polyadenylated. Mouse *Xist* is comprised of unusually large initial and terminal exons flanking five smaller exons (15, 63, 95). Transcription initiating from the promoter designated P1 may give rise to *Xist* transcripts of up to 17.9 kb depending on polyadenylation site usage (63, 95). The overall structure of the gene, including the transcription start site P1 and exon/intron boundaries, is similar in the human counterpart (17, 18). The longest human *XIST* transcript may be 19.3 kb in length (64). In human, many different *XIST* RNA isoforms are produced by extensive alternative splicing occurring within the 3' half of the gene (17, 18, 64). So far, the significance of the heterogeneity of *Xist/XIST* transcripts is unknown.

On the sequence level, all *Xist/XIST* genes sequenced to date display a relatively low degree of conservation (60, 61, 108). Mouse and human *Xist/XIST* show 49% sequence identity overall, with exons being slightly more conserved than introns (60, 108). The homology between rodent and human *Xist/XIST* exons is considerably lower than the average identity for protein-coding regions (85%) and even lower than identity between 5' or 3' untranslated regions of rodent and human mRNA (~65–80%) (108). Although several short stretches of unique sequence show relatively high homology between the species (64, 108), the most salient feature of the known *Xist/XIST* sequences is the conservation of six repeated elements, designated A through F (15, 18, 108).

Xist/XIST RNA Is Localized to the Xi in Female Somatic Cells

From initial sequence analysis of the *Xist/XIST* gene, it was unknown whether its function in X-inactivation, if any, would be mediated by the DNA sequence of the gene or its untranslated RNA product. Analysis of *Xist/XIST* RNA distribution quickly led investigators to search for a role for the RNA. *Xist/XIST* transcripts are retained in the nucleus of female somatic cells (15, 18). When analyzed by fluorescence in situ hybridization (FISH), *Xist/XIST* RNA appears highly particulate and is located to a large nuclear domain that corresponds to the space occupied by the Xi (18, 31, 114). *Xist/XIST* RNA particles are released from the Xi during mitosis (31, 44). A three-dimensional analysis showed that *XIST* RNA is not just bound to the surface of the Xi, but seems to reside within the entire space delineated by the Xi (31). Such an extensive and intimate association of an RNA with an entire chromosome was unprecedented at the time, but now appears to be shared by the *roX1* and *roX2* RNAs that coat the dosage-compensated X chromosome in male *Drosophila* (71). The close association of *Xist/XIST* RNA with the Xi supported a model in which the RNA acts as a functional molecule that regulates X-inactivation.

Cis-Spread of *Xist* RNA Correlates with X-Inactivation During Mouse Development

In male and female mouse embryos, *Xist* is transcribed in the pluripotent epiblast lineage prior to its differentiation into the embryonic tissues. By FISH, *Xist* RNA is detected as a pinpoint signal at the site of its transcription on all X chromosomes, which are active at this point (113, 143). Mouse ES cells, which are derived from pluripotent cells of the early embryo, show this same pinpoint *Xist* expression (114). Consistent with a role in X-inactivation, *Xist* levels increase dramatically in female but not male cells in the developmental window in which silencing of the X chromosome occurs (70). When epiblast cells differentiate into the embryonic germ layers during gastrulation, and when ES cells are differentiated in vitro, both cell types undergo X-inactivation (115, 150, 153) and exhibit similar dynamic *Xist* expression patterns (113, 143). *Xist* RNA coats the presumptive Xi, while the Xa retains a low-level pinpoint of *Xist* RNA that is subsequently extinguished (113, 143). FISH studies on ES cells indicated that genes are silenced soon after *Xist* transcripts coat the Xi (114, 143). That initial coating of the X chromosome by *Xist* RNA coincides with the onset of silencing suggested a causative role for *Xist* in X-inactivation.

Xist RNA Is Necessary and Sufficient for Chromosome-Wide Gene Silencing

To investigate *Xist* function, two groups engineered large deletions, removing the promoter and first exon of the *Xist* gene (117) or the majority of exon 1 through exon 5 (90). The X chromosome bearing the loss-of-function allele could never be inactivated in ES cells or in the embryonic tissues of mice heterozygous for the deletion. Instead, only the wild-type X chromosome ever became the Xi (89, 90, 117), indicating that *Xist* is required in *cis* for silencing of the X chromosome. In addition, use of antisense oligonucleotide analogs to block *Xist* RNA's ability to coat the presumptive Xi during differentiation prevented the formation of a silenced X chromosome (7). These results demonstrate that expression of *Xist* RNA and coating of the chromosome are required to silence the Xi during X-inactivation.

An elegant series of studies from Wutz and colleagues has demonstrated unequivocally that expression of *Xist* RNA is sufficient to induce chromosome-wide silencing (164, 165). A transgene was generated containing a 15-kb *Xist* cDNA, lacking approximately 3 kb at the end of the terminal exon, under control of a tetracycline-inducible promoter. Induction of *Xist* transcription from this transgene in male ES cells results in *cis*-spread of *Xist* RNA into linked autosomal sequences and silencing of genes (164). This result indicates that high-level expression of *Xist* RNA is sufficient for chromosome coating and silencing, even in male cells. It has been suggested that *Xist* RNA mediates these functions as part of a ribonucleoprotein complex (162, 165). Although *Xist* RNA can induce chromosome-wide silencing in ES cells (164), activation of *XIST/Xist* expression and subsequent coating of the Xa in differentiated cells is insufficient to cause silencing (30, 49, 156, 164). Thus, *Xist* transcripts can always coat the chromosome in *cis*, but the RNA can only mediate silencing in a narrow window during differentiation (164). Proper spatial and developmental regulation of *Xist* expression is therefore critical, so that a female cell can reliably keep one X chromosome active, and inactivate the other.

CHOOSING THE ACTIVE X CHROMOSOME

Xa Choice Is Regulated by Several Classes of *cis*-Elements

In contrast to the dosage compensation mechanisms of *Drosophila* and *Caenorhabditis elegans*, where all X chromosomes in the dosage-compensated sex are treated equivalently (33), the two X chromosomes in the cells of female mammals take on radically different fates. In random X-inactivation, mammalian cells designate a single Xa, and then carry out X-inactivation on any remaining X chromosome(s). Cells must choose between two equivalent X chromosomes and randomly make a differentiating mark on only one of them, the future Xa. The simplest model to explain the marking of one X chromosome as Xa-elect hypothesizes that a blocking factor, so named since it functions to block X-inactivation, is responsible for making this mark (Figure 2) (128). Since Xa number increases with the number of autosome complements (21, 46, 161), the amount of blocking factor activity must be determined by autosomal ploidy such that each diploid set of autosomes produces sufficient blocking factor to choose only one Xa. The region of the *Xic* required to mediate blocking factor activity is referred to as the counting element. Modification of the counting element by the blocking factor represents the first molecular difference between the Xa-elect and the presumptive Xi. The consequence of blocking factor's interaction with the counting element on the Xa-elect is the interference with *Xist* RNA's silencing function on this chromosome. The counting element can therefore be seen as the fundamental *cis*-element required for choosing the Xa.

When two X chromosomes bearing identical *Xic* regions are present in a female cell, Xa choice occurs randomly because of the equal probability that blocking factor will interact with either counting element (99). Genetic studies of the *Xic* region have shown the existence of additional *cis*-elements, termed choice elements, that can skew the Xa choice event such that one X chromosome becomes Xa more frequently than the other (103). Skewing of random X-inactivation in female cells heterozygous for a choice element is thought to be the outcome of better or worse competition for the blocking factor interaction compared to the wild-type X. A

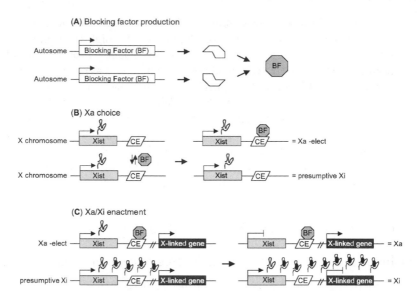

Figure 2   Xa choice by blocking factor. (*A*) An autosomally encoded blocking factor (BF) is produced in sufficient quantity to choose one Xa. (*B*) During the Xa choice process, both X chromosomes compete for the limiting blocking factor at the *cis*-acting counting element (CE). Choice of the Xa is achieved when blocking factor interacts with the counting element on one X chromosome. In this time period, *Xist* RNA molecules (indicated by *black squiggles*) are produced but restricted to the site of transcription. (*C*) During the enactment period, the activity of *Xist* RNA is repressed on the Xa-elect as a consequence of its modification by blocking factor, and *Xist* expression is eventually shut off, resulting in commitment to the Xa fate. On the presumptive Xi, *Xist* RNA, most likely in a ribonucleoprotein complex (indicated by the addition of the *black oval*), coats the chromosome in *cis*, leading to inactivation of X-linked genes, and committing the chromosome to the Xi fate. The location of the counting element and the implication of the drawing that blocking factor is bound at the counting element are speculative.

choice element must therefore affect the affinity of the *cis*-linked counting element for blocking factor, such that there is an increase or decrease in the likelihood that the blocking factor will interact with it. Thus, a combination of *cis*-elements within the *Xic* are employed to allow cells to differentiate an Xa from an Xi during random X-inactivation.

The Minimal Counting Element Is Closely Linked to *Xist*

The definitive test of counting element function is the ability of a sequence to induce inactivation of the single X chromosome in a male cell when integrated as an autosomal transgene. The presence of the counting element in the transgene titers blocking factor away from the endogenous *Xic* such that the single

X becomes the presumptive Xi in a fraction of cells. Several *Xist*-containing transgenes have been reported to induce silencing of the endogenous X chromosome
in male cells (57, 62, 79, 81, 82, 92). The minimal region sufficient for counting
element function is defined by a 35-kb transgene containing genomic *Xist* flanked
by 9 kb of upstream and 3 kb of downstream sequences [(62) based on the recent
3′ end mapping by Hong and colleagues (63)], demonstrating that the counting
element is very closely linked to *Xist* (Figure 3). The counting element is unlikely
to reside within the transcribed region of the *Xist* gene sufficient for silencing as the
Xist cDNA transgene is unable to induce ectopic X-inactivation in male cells (164).
Like all functional *Xic* transgenes examined so far, the 35-kb transgene containing
counting element activity is present in multiple tandem copies (62). No single-
copy transgene has been shown to faithfully recapitulate *Xic* activity (57). These

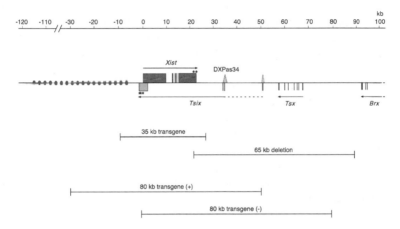

Figure 3   *Xist* and *Tsix* region on the mouse X chromosome. The figure displays
the 200-kb region surrounding the mouse *Xist* gene in detail with the zero point of
the scale set at *Xist* promoter P1. Transcripts derived from the sense and antisense
strands are drawn above and below the line, respectively. Exons are depicted as solid
blocks. Major and minor *Tsix* transcription start sites are each associated with CpG-
rich region (indicated by the *triangles*). The CpG-rich region at the major promoter
is known as DXPas34 (37). Both *Tsix* transcription start sites are the 5′ ends of short
∼200-bp exons that are spliced to an acceptor site located in the antisense region
complementary to the 5′ end of *Xist* exon 1. The final exon of *Tsix* is from 2 to 4 kb
in length, depending on polyadenylation site usage (indicated by *stars*) (139). The
chromatin domain immediately upstream of the *Xist* transcriptional start site P1 is
characterized by methylation of histone H3 on lysine 9 (indicated by *ovals*) (59). A
35-kb transgene containing minimal *Xic* activity and sufficient for counting element
function (62), and a 65-kb deletion thought to remove counting element function (32)
are shown. Two 80-kb *Xic* transgenes, one able (+) and one unable (−) to enact the Xi
fate upon differentiation (81), are shown.

results suggest either that position effects, to which single-copy transgenes are of-ten subject, interfere with counting element function, or that the counting element is multipartite, perhaps requiring a certain threshold number of minor elements to function. Even in the cases when multicopy transgenes induce X-inactivation of the endogenous X chromosome, they do not confer truly random choice, because the endogenous X chromosome is chosen as Xa more frequently than expected (81, 82). In contrast, X: A translocations containing the *Xic* compete equally with a normal X chromosome for choice as Xa (128a). Thus, although the transgene contains the counting element, additional sequences regulating the choice process are missing from these transgenes.

Given its pivotal role in receiving the blocking factor-mediated Xa choice sig-nal, the counting element should be absolutely required for an X chromosome to be chosen as Xa. Therefore, deletion of the counting element is predicted to result in ectopic X-inactivation of the single X chromosome in male cells. A 65-kb deletion implicates the region 3' of *Xist*, beginning in *Xist* exon 7 and continuing downstream, as containing the counting element (Figure 3) (32). Nearly all cells carrying this deletion on their only X chromosome died upon differentiation due to ectopic *Xist* RNA coating and silencing. Although this deletion may have removed the counting element such that the single X chromosome could not be chosen as the Xa, it is also possible that the counting element is unaffected and that a second critical element required for downregulation of *Xist* activity on the Xa, acting down-stream of Xa choice, may have been perturbed. Three smaller regions contained within the 65-kb region have been deleted independently in male ES cells without resulting in ectopic X-inactivation upon differentiation (80, 130, 139). Identifica-tion of the counting element may provide important clues about the mechanism of Xa choice and may allow isolation of the blocking factor hypothesized to interact with this sequence.

Xist Acts in *cis* to Influence Xa Choice

Like a counting element deletion, deletion of a choice element causes skewed X-inactivation in female cells. In contrast to a counting element deletion, however, deletion of a choice element does not result in aberrant X-inactivation in males, since choice elements affect the relative affinity of the *cis*-linked counting element for blocking factor. Thus, mutations that skew X-inactivation in females but do not cause ectopic X-inactivation in males define choice elements.

Xist has been implicated as a choice element. Upon differentiation of female cells heterozygous for *Xist* deletions, only the wild-type X chromosome was ever inactivated (90, 117). There were two possible explanations for this result. Choice could have occurred normally, but the 50% of cells choosing the wild-type X chromosome as the Xa may have died owing to the inability to silence the *Xist* deletion-bearing presumptive Xi. Alternatively, choice may have been completely skewed such that the deletion-bearing X chromosome was always chosen as Xa and thus the wild-type X chromosome would always be inactivated. By distinguishing between the wild type and deleted *Xist* alleles using FISH probes, it was determined that the wild-type X chromosome was never selected as the Xa in differentiating embryonic cells (89). The effect of this deletion implicates the transcribed region of *Xist* or *Xist* RNA itself as a choice element that reduces the affinity of the *cis*-linked counting element for blocking factor.

Although loss of *Xist* function increased the likelihood of choosing the mutant X chromosome as the Xa, a chromosome with greater *Xist* expression showed a decreased chance of becoming the Xa. In male ES cells, replacement of 2.5 kb of *Xist* upstream sequences with a selectable marker driven by a constitutive *Pgk-1* promoter resulted in an increase in Xist RNA levels (109). The somatic cells of female mice heterozygous for this replacement exhibited skewed X-inactivation with the targeted X chromosome becoming the Xa in only 10–20% of cells. When the *Pgk-1* promoter and selectable marker were removed, random X-inactivation was restored. These results suggest that increased transcription of *Xist* has a negative influence on Xa choice, consistent with the hypothesis that *Xist* RNA decreases the affinity of a *cis*-linked counting element for blocking factor. Since all the factors required for *Xist* RNA-mediated coating and silencing are present in ES cells (164), one mechanism by which *Xist* RNA could negatively influence Xa choice is by causing local coating and silencing, and regulating accessibility of the closely linked counting element for the blocking factor.

Tsix, an Antisense *Xist* Transcript, Positively Influences Xa Choice

In 1999, several groups simultaneously reported the identification of antisense transcription through the *Xist* locus in ES cells (41, 78, 102). The transcript, named *Tsix* in recognition of it being antisense to *Xist*, was found to initiate at a major transcription start site 13 kb downstream of the *Xist* 3′ end, and to extend across *Xist* into its promoter region (Figure 3) (78). Subsequently, a minor *Tsix* promoter has been identified, and mature *Tsix* transcripts of up to 4 kb have been shown to be produced by splicing (139). Like *Xist*, *Tsix* has no significant open reading frame and is not thought to encode a protein.

The expression pattern of *Tsix* yielded some exciting clues about its function. *Tsix* RNA is detected in ES cells and differentiating cells, and not in somatic cells (41, 78, 102), suggesting that its role might be in regulating initiation of X-inactivation. Whenever low-level *Xist* pinpoint expression is observed in male and female ES cells, antisense *Tsix* RNA is detected as an overlapping pinpoint signal (78). Upon differentiation, *Tsix* expression is extinguished from the presumptive Xi concomitant with *Xist* spreading to coat the chromosome (41, 78). Pinpoint expression from the Xa continues for several days and persists longer than pinpoint *Xist* expression (78, 139). *Tsix* shows the same dynamic expression patterns during random X-inactivation in developing embryonic tissues (77, 139). The correlation between *Tsix* expression and unstable *Xist* expression in *cis* led to speculation that *Tsix* might be a negative regulator of *Xist*.

Deletion of the major *Tsix* promoter in female cells dramatically reduced antisense transcription and skewed Xa choice away from the targeted *Tsix* allele (80, 139). In female cells, the X chromosome bearing the deletion was chosen as Xa in at most 4% of cells, but male cells correctly chose a deletion-bearing X chromosome as Xa, demonstrating that *Tsix* is not the counting element. Insertion of a transcriptional stop signal between the *Tsix* promoter and the 3' end of *Xist*, with (139) or without (85) deleting the major *Tsix* promoter, abolished antisense transcription and caused a similar dramatic skewing phenotype in female cells. Ectopic X-inactivation occurred in a small proportion of male cells (85, 139). Thus, when in competition with a wild-type *Xic*, the chromosome carrying the *Tsix* loss-of-function allele became the Xi in nearly every cell (41, 80, 85, 139). These results suggest that the normal role of *Tsix* RNA, or the process of transcription antisense to *Xist*, is to promote Xa choice by increasing the affinity of the *cis*-linked counting element for blocking factor. Surprisingly, insertion of the constitutive EF-1α promoter increased *Tsix* RNA levels in undifferentiated cells (146) but did not skew X-inactivation, indicating that higher than normal *Tsix* levels cannot further promote Xa choice.

In addition to skewing Xa choice, *Tsix* transcription affects the abundance of *Xist* RNA in *cis*. In cells with the strongest *Tsix* loss-of-function alleles, *Xist* RNA levels were dramatically higher than in wild-type cells (85, 139). In the presence of the allele that retained transcription from the *Tsix* minor promoter, *Xist* RNA levels rose threefold compared to wild type (78). These results are consistent with *Tsix* transcription normally acting to reduce *Xist* RNA steady-state levels. In contrast, increased *Tsix* transcription from the EF-1α promoter insertion did not affect *Xist* RNA levels (146).

Tsix May Act Through *Xist* to Influence Xa Choice

Both mutations that increase *Xist* RNA levels, such as the introduction of the *Pgk-1* promoter (109), and mutations that decrease *Xist* RNA levels, such as an internal *Xist* deletion (89), skew Xa choice. Interestingly, only those *Tsix* alleles that affect *Xist* RNA levels in *cis* also skew Xa choice, suggesting that *Tsix* could act through *Xist* to regulate Xa choice. Indeed, the levels of *Xist* RNA in cells with different *Tsix* alleles correlate with the degree of skewed Xa choice observed upon differentiation (Figure 4) (80, 85, 139, 146). These results are consistent with a model in which *Tsix* transcription acts to reduce the amount of *Xist* RNA in *cis*, thus reducing *Xist*'s negative influence on the likelihood of blocking factor interacting with the counting element.

How could the process of *Tsix* antisense transcription or the *Tsix* antisense transcript itself affect *Xist* RNA levels? If the process of antisense transcription through the *Xist* locus is important, then this process might prevent efficient transcription of *Xist*, lowering the amount of functional *Xist* molecules that are produced. Alternatively, if the antisense RNA is functional, then *Tsix* RNA may actively destabilize *Xist* transcripts or prevent *Xist* RNA from becoming fully active. One possibility

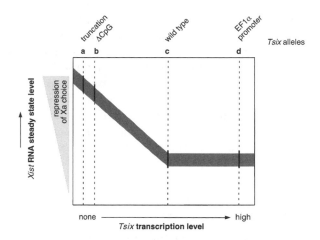

Figure 4   *Tsix* functions through modulation of the *Xist* steady-state level in Xa choice. This figure depicts the effects of *Tsix* transcription on the abundance of *Xist* RNA transcribed in *cis*. The gray band indicates a range of *Xist* RNA steady-state levels achieved by a population of cells bearing the corresponding *Tsix* allele. Four *Tsix* alleles are depicted in order of increasing *Tsix* transcription. *Tsix* alleles are described in the text [allele a (85, 139); allele b (80); allele c, wild type; allele d (146)]. Hypothetically, in a female cell homozygous for a certain *Tsix* allele, each chromosome produces identical steady-state levels of *Xist* RNA in *cis*, and choice occurs randomly. In cells heterozygous for *Tsix* alleles, the allele with lower *Xist* RNA levels will more likely be chosen as Xa; for example, in a cell heterozygous for a wild-type *Tsix* allele (*c*) and a strong *Tsix* loss-of-function allele (*a*), the wild-type *Tsix* allele out-competes the other allele for Xa choice, since the steady-state level of *Xist* is lower on the wild-type allele. This model could be tested by generating female cells heterozygous for these *Tsix* alleles and determining the skewing phenotype.

is that a *Tsix/Xist* duplex stimulates a double-stranded RNA-mediated turnover of both transcripts (121). Another possibility is that *Tsix* transcripts interfere with *Xist* RNA folding, *Xist* ribonucleoprotein complex formation, or *cis*-spread along the Xi. Interestingly, the spliced form of *Tsix* RNA contains only 2 kb of overlap with the mature *Xist* transcript (139). This overlap occurs within a domain of *Xist* that is critical for silencing activity (165), suggesting that *Tsix* modulation of Xa choice could involve regulation of *Xist* RNA's silencing activity by affecting its interactions with *trans*-acting factors. ES cells carrying a 65-kb deletion removing both *Tsix* promoters, and possibly the counting element, produced higher levels of *Xist* transcripts that appeared to freely diffuse from the site of transcription (106). Reinsertion of the *Tsix* promoter restored normal transcript levels and re-established pinpoint localization of *Xist* (106). These results are consistent with the possibility that *Tsix* can affect some aspect of *Xist* RNA metabolism that regulates *cis*-spread.

Tsix was the first example of a presumably nonfunctional RNA whose transcription is critical to negative regulation of its functional gene partner. Recently, transcription of another antisense RNA, *Air*, has been shown to be required for regulating imprinted expression of the coding sense gene, *Igf2r*, with which it overlaps (145). As both *Air* and *Tsix* are antisense to the genes they regulate, this may indicate a common antisense transcription or antisense RNA-mediated mechanism for gene regulation in mammals. Antisense transcription appears to be a common phenomenon associated with imprinted genes, and has been detected at the *UBE3A* gene in the Prader-Willi/Angelman syndrome locus (135) and at the *Gnas* locus (83).

Xce Effects May Be Correlated with *Xist* RNA Steady-State Levels

When some divergent mouse strains are crossed, F1 mice exhibit skewing of Xa identity away from the normal 50/50 ratio (56). This skewing is mediated by the X choosing element (*Xce*) locus, which lies downstream of *Xist*, beyond *Tsix* (144). X chromosomes bearing a stronger *Xce* allele are thought to be chosen more frequently as Xa. Interestingly, female somatic tissues from a strain with a strong *Xce* allele showed markedly lower levels of *Xist* RNA than those from a strain with a weak *Xce* allele (14). Indeed, even before differentiation, a female ES cell line heterozygous for *Xce* alleles shows lower levels of *Xist* RNA and somewhat higher levels of *Tsix* transcripts from the stronger *Xce* allele (146). This observation raises the possibility that, like *Tsix*, the *Xce* acts by modulating *Xist* RNA levels in *cis*, either directly or perhaps indirectly through *Tsix*. One study correlated *Xce* strength with methylation levels at *DXPas34* (37), the CpG island located at the major *Tsix* promoter (Figure 3), though differential methylation appeared to follow X-inactivation (122).

ENACTMENT OF THE ACTIVE AND INACTIVE X FATES

After choosing the number of active X chromosomes appropriate to autosome dosage, the next step in establishing proper dosage compensation is to carry out, or enact, the fate assigned to each X chromosome. In addition to the events occurring on the presumptive Xi to achieve inactivation, a distinctly different series of events occurs on the Xa-elect. To protect the Xa-elect from *Xist*-mediated silencing, this chromosome must be shielded from the factors that are required to enact the Xi fate.

An Increase in *Xist* RNA Levels Correlates with Coating and Silencing

The silencing of the presumptive Xi is mediated by *cis*-spread of *Xist* RNA to coat the chromosome (164). Quantitative RT-PCR and slot blot experiments demonstrated a 10- to 20-fold difference in the amount of *Xist* RNA in ES cells versus female somatic cells (80, 106, 143, 146). These results suggest that an increase in steady-state levels of *Xist* RNA coincides with spreading of *Xist* transcripts in *cis* to coat the X chromosome and mediate gene silencing. Indeed, increased levels of *Xist* RNA in undifferentiated cells can result in premature enactment of the Xi fate (85, 109, 139, 164, 165). The mature *Xist* transcript in somatic cells has a 10- to 20-fold longer half-life than in embryonic cells (113, 143). Thus, an attractive hypothesis is that an increase in *Xist* RNA half-life on the presumptive Xi is a prerequisite for its inactivation.

Multiple Mechanisms Regulate Increases in *Xist* RNA Abundance

The increase in *Xist* RNA half-life that occurs upon differentiation could be achieved by changes in the *Xist* transcript itself or changes in factors that regulate *Xist* RNA stability. Several groups examined whether the increased amount of *Xist* RNA that correlates with onset of X chromosome silencing is caused by production of an alternative transcript possessing higher stability. When analyzed by RT-PCR, the major splicing pattern appeared to be the same in ES cells and female somatic cells (113, 143), suggesting that there are no major changes in processing occurring during Xi-enactment. With an RNA of this enormous size, however, alternative processing may be difficult to detect. An attractive model proposed that *Xist* transcripts might be produced from different promoters in undifferentiated and differentiated cells (69). Transcripts from the somatic promoter P1 might be inherently stable, whereas an alternative embryonic promoter P0, located 6.6 kb further upstream, was suggested to produce unstable transcripts (69). Upon differentiation, a switch from P0 to P1 promoter usage on the presumptive Xi could trigger production of stable *Xist* transcripts, enacting the Xi fate. However, subsequent studies have been unable to detect *Xist* transcription from P0, and transcripts from P1 have been shown to be unstable in ES cells, effectively ruling out the P0 hypothesis (160). The presence of *Tsix* antisense transcription and a highly conserved ribosomal protein pseudogene immediately downstream of the P0 promoter was found to have complicated transcript analysis in this region (133, 160).

Since *Tsix* expression reduces *Xist* RNA levels in *cis* during Xa choice (80, 85, 139), developmentally regulated shutoff of *Tsix* on the presumptive Xi is an excellent candidate for the mechanism of *Xist* RNA stabilization. *Tsix* is shut off as soon as *Xist* transcripts can be seen coating the presumptive Xi (41, 78). Furthermore, when *Tsix* cannot be shut off on the presumptive Xi due to constitutive high-level expression from the EF-1α promoter, *Xist* cannot coat and silence the presumptive Xi (146). Taken in combination, these results suggest that silencing of *Tsix* on the Xi could result in *Xist* RNA stabilization, facilitating its *cis*-spread and X-inactivation. However, lack of *Tsix* transcription does not result in ectopic X-inactivation in the majority of ES cells prior to differentiation (80, 85, 139). Thus, additional developmentally regulated mechanisms contribute to the increase in *Xist* RNA coating and silencing activity on the presumptive Xi upon differentiation.

The increase in *Xist* RNA half-life could be controlled by the developmentally regulated production of *trans* factors that stabilize the RNA. High-level *Xist* RNA produced from the cDNA transgene in ES cells has a half-life equivalent to that of *Xist* RNA in somatic cells and can coat and silence the chromosome in *cis* (164). These results indicate that *Xist* RNA can recruit the factors required for its stabilization in ES cells. However, since *Xist* was expressed at an extremely high level in these experiments, it is still possible that a developmentally regulated stabilizing factor could increase *Xist* activity upon differentiation.

Developmentally regulated activation of *Xist* transcription is also an attractive explanation for the increase in *Xist* activity seen upon differentiation of female cells, since increased *Xist* transcription can enact the Xi fate (109, 165). *Xist* transcription rates in male ES cells and female somatic cells are roughly comparable (113, 143), suggesting that any transcriptional activation at the time of *Xist* RNA *cis*-spread and coating may be minor or transient. Such a minor or transient boost might be sufficient to account for the observed increase in *Xist* RNA half-life on the presumptive Xi. There may be a threshold level of *Xist* RNA that must reached in order to form stable ribonucleoprotein complexes. In theory, *Xist* transcription rates could be tuned precisely relative to the *Tsix*-mediated destabilization rate to limit the amounts of these complexes that assemble in undifferentiated ES cells or even prevent their formation altogether. A slight increase in *Xist* transcription upon entry into the enactment phase could result in assembly of or increase in abundance of functional *Xist* ribonucleoprotein complexes. Since *Xist* RNA can silence very closely linked genes (42), it is reasonable to propose that it could also shut off *Tsix* expression. *Tsix* shutoff by the *Xist* ribonucleoprotein complex would positively reinforce *Xist* activity, since *Tsix*-mediated destabilization of *Xist* transcripts would be eliminated.

Regions Flanking *Xist* Are Implicated in Xi-Enactment

Though many *Xic* transgenes are capable of pinpoint *Xist* expression in ES cells as analyzed by FISH, no single-copy transgene has been found to produce *Xist* RNA that spreads to coat the chromosome in *cis* and mediates silencing upon differentiation (57). Instead, a few days into differentiation, some cells exhibit a faint, dispersed *Xist* signal emanating from the transgene, and all cells ultimately extinguish *Xist* expression. This result suggests that *cis*-elements required to direct increases in *Xist* RNA levels or that facilitate *Xist* RNA spread are missing from these transgenes, or that single-copy transgenes cannot overcome position effects. In contrast, multicopy transgenes as small as 35 kb in size can recapitulate *Xic* function (62), suggesting that duplicate elements present in neighboring copies of the transgene can substitute for endogenous elements missing in a single-copy transgene, or that multicopy transgenes are not as sensitive to position effects (57).

One 80-kb *Xic* transgene containing less than 1 kb of sequence upstream of the *Xist* promoter P1 is notable in that it does not support X chromosome coating and silencing by *Xist* RNA even when present in more than 10 copies (81). The functioning of other 80-kb transgenes containing 30 kb of *Xist* upstream sequence argues that a critical Xi-enactment element lies in the approximately 30 kb of additional upstream sequence contained in the functional transgene (Figure 3) (81). The minimal region containing this element may be defined by a functional multicopy 35-kb transgene containing 9 kb of upstream sequence (62). A 2.5-kb region containing DNase I hypersensitive sites lies within this candidate region, but its deletion had no effect on X-inactivation, further delineating the location of the *cis*-element required for enactment of the Xi fate (109).

Xa Enactment Employs Several Mechanisms to Repress *Xist* in *cis*

During the period in which stable *Xist* RNA initiates silencing on the presumptive Xi, the Xa-elect continues to express *Tsix* and unstable *Xist* (78, 113, 143). Although *Tsix* destabilizes *Xist* in *cis*, it is not strictly required to block *Xist* RNA-mediated silencing during differentiation since the vast majority of male cells lacking *Tsix* function still enact the Xa fate on their single X chromosome (80, 85, 139). One possibility is that blocking factor functions to prevent the increase in steady-state levels of *Xist* RNA on the Xa-elect, perhaps by interfering with the hypothesized transcriptional activation of *Xist*.

A role for DNA methylation in controlling *Xist* expression during Xa-enactment was suggested by differential methylation of the *Xist* promoter on the Xa and the Xi (111). The *Xist* promoter is methylated on the silent *Xist* allele on the Xa in somatic cells, and less methylated on the *Xist*-expressing Xi in somatic cells and on both X chromosomes in undifferentiated ES cells (6). The phenotype of the *DNA methyltransferase 1* (*Dnmt1*) mutant demonstrated that proper regulation of DNA methylation is involved in enactment of the Xa fate. The *Xist* promoter showed extremely low levels of DNA methylation in undifferentiated male *Dnmt1* mutant ES cells, and did not acquire methylation upon differentiation (6). Mutant male ES cells exhibited normal pinpoint expression of *Xist* RNA. Upon differentiation, these cells showed *Xist* RNA coating and silencing of X-linked genes on the presumed Xa-elect (114). A low but significant percentage of male and female somatic cells in *Dnmt1* mutant embryos also exhibited ectopic X-inactivation (114). These results indicate that DNA methylation is employed in the Xa-enactment mechanism, most likely in repressing *Xist* transcription from the Xa-elect.

An Integrated View of Choice and Enactment

When the data relating to Xa choice and enactment of the Xa and Xi fates are taken together, a coordinated view of these processes emerges. Prior to differentiation, an embryonic cell is uncommitted in its choice of the Xa. Upon differentiation the choice of the Xa is made. During the choice process, the two X chromosomes in a female cell compete for the limiting quantity of blocking factor. Choice elements act in *cis* to influence the counting element's affinity for blocking factor interaction. *Xist* RNA is a choice element, negatively influencing the likelihood of blocking factor interaction with the *cis*-linked counting element. This negative choice activity may be a local manifestation of the *Xist* ribonucleoprotein complex's silencing activity, affecting the chromatin structure or accessibility of the counting element. *Tsix* transcription, on the other hand, promotes Xa choice, most likely by destabilizing *Xist* RNA and lowering the abundance of functional *Xist* complexes acting on the counting element in *cis*. The blocking factor interaction with the counting element of the Xa-elect represents the first differentiation between the two X chromosomes, allowing the cell to distinguish the Xa-elect from the presumptive Xi.

The Xa and Xi fates are carried out in the enactment phase. To enact the fate of the presumptive Xi, *Xist* activity increases in *cis*. A *trans*-acting factor modulating *Xist* transcription or stability could be regulated by entry into the enactment phase. Even a small boost in *Xist* steady-state levels may lead to an increase in the abundance of functional *Xist* complexes. An increase in *Xist* activity could lead to local silencing of *Tsix*, thus shutting off *Tsix*-mediated destabilization of *Xist* and further increasing *Xist* activity. *Xist* would then be free to spread along the presumptive Xi, silencing genes. Choice as the Xa protects the Xa-elect from this developmentally regulated boost in *Xist* activity; perhaps modification of the counting element by blocking factor does so directly. *Tsix* continues to be expressed on the Xa, contributing to low *Xist* steady-state levels, and DNA methylation may be responsible for turning off *Xist* expression. When the cell exits the enactment window, the active and silent states of the Xa and Xi become irrevocable.

DESIGNATION OF THE ACTIVE AND INACTIVE
X CHROMOSOMES IN IMPRINTED X-INACTIVATION

In contrast to random X-inactivation in embryonic tissues, X-inactivation occurs in an imprinted manner in the extraembryonic tissues that support the developing mouse embryo (152). In imprinted X-inactivation, the maternally inherited X chromosome (Xm) always remains active and the paternal X chromosome (Xp) is silenced. Since normal male embryos lack Xp and normal females have a single Xp, this imprinting system results in appropriate dosage compensation. *Xist* and *Tsix* expression patterns during early embryogenesis suggest that differential regulation of both transcripts may be critical to setting the imprint or carrying out the information contained in it. The dynamics of *Xist* and *Tsix* expression are depicted in Figure 5. *Xist* expression begins from the Xp at the onset of zygotic transcription (166) and transcript levels rise quickly during the early cleavage stages (55). By FISH, *Xist* RNA appears to partially coat the Xp (41, 143), which exhibits partial silencing during the cleavage stages (76, 154), but has not yet committed to the Xi fate. On the Xm, a weak pinpoint of *Xist* RNA can be detected (41). While *Tsix* is not expressed from the Xp, it is strongly expressed as a pinpoint signal from the Xm (41, 77). By the late-blastocyst stage of development, when the pluripotent epiblast has reversed *Xist* RNA coating, the extraembryonic lineages have differentiated

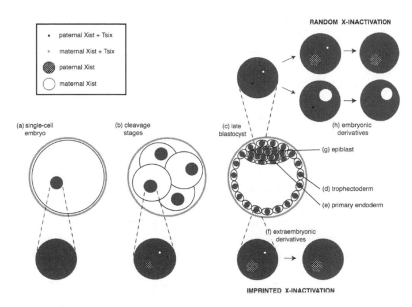

Figure 5 *Xist* and *Tsix* expression during early female mouse development. At the single-cell stage of female mouse embryogenesis (*a*), *Xist* expression is undetectable by RT-PCR or FISH. *Xist* expression commences at the 2-cell stage at the onset of zygotic transcription (166). By FISH, cleavage stage embryos (*b*) exhibit differential biallelic *Xist* expression starting at the 2-cell stage, with *Xist* RNA appearing to coat the Xp at least partially (41, 91, 107, 143), and a weak *Xist* pinpoint signal at the Xm (41). The late blastocyst (*c*) consists of the differentiated extraembryonic lineages, the trophectoderm (*d*) and primary endoderm (*e*), and the pluripotent embryonic lineage precursor, the epiblast (*g*). The trophectoderm and primary endoderm have undergone imprinted X-inactivation by the mid and late blastocyst stages, respectively (151). The Xp (which is now the Xi) has become fully coated by *Xist* RNA (113, 143). After the early embryo implants into the uterus, the extraembryonic tissues derived from the trophectoderm and primary endoderm (*f*) shut off low-level *Xist* expression on the Xm and continue to exhibit *Xist* RNA coating of the Xp throughout subsequent cell divisions. At the late blastocyst stage, the cells of the epiblast have reversed the partial *Xist* RNA coating of the Xp and now exhibit low-level *Xist* RNA pinpoint signals from both the Xm and the Xp (113, 143). Between implantation and completion of gastrulation, epiblast cells differentiate into the embryonic germ layers and undergo random X-inactivation (105, 153, 154). During this time period, *Xist* transcription once again displays differential biallelic expression in the embryonic derivatives (*h*) (113, 143). After completion of gastrulation, embryonic cells cease to express the *Xist* pinpoint signal from the Xa (113, 143). *Tsix* antisense RNA is coexpressed whenever low-level pinpoint *Xist* expression is found (41, 78), and persists from the Xa for a limited period of time after *Xist* RNA shutoff (139).

and the Xp has become completely coated and silenced by *Xist* RNA (113, 143). The Xp in the extraembryonic lineages has now become a committed Xi (151). Thus, high-level *Xist* expression from the Xp and *Tsix* transcription from the Xm correlate with their fates as the Xi and Xa, respectively. *Xist* and *Tsix* expression programmed in the gametes may allow imprinted tissues to bypass the Xa choice mechanism of random X-inactivation.

Xist and *Tsix* Have Opposite, Parent-of-Origin–Specific Effects

Xist deletions are lethal when inherited from the father, but not from the mother. The lethality arising from paternal inheritance of an *Xist* deletion is due to the total lack of X-inactivation in the extraembryonic tissues, indicating that *Xist* is required for imprinted silencing of the Xp (90). As the Xm, which bears a functional *Xist* allele, does not upregulate *Xist* to compensate for the lack of paternal X-inactivation, it appears that the imprint cannot be overridden. *Tsix* promoter deletions have the opposite parent-of-origin–specific effects. *Tsix* deletions are lethal only when inherited from the mother, owing to ectopic *Xist* expression and X-inactivation of the Xm in most extraembryonic cells (77, 139). Therefore, the *Tsix* promoter region, *Tsix* antisense transcription through the *Xist* locus, or the *Tsix* transcript itself is implicated in repression of *Xist* activity on the Xm during imprinted X-inactivation. As expected for genes with opposite effects on the same process, a paternally inherited *Xist* deletion can be rescued by maternal inheritance of a *Tsix* loss-of-function allele (139). The parent-of-origin–specific effects of *Xist* and *Tsix* deletions are limited to the extraembryonic tissues, as X-inactivation occurs normally in the embryonic tissues due to the random choice mechanism. Lethality is due to extraembryonic defects.

A small number of pups inheriting a maternal *Tsix* deletion survive to term, indicating that proper dosage compensation can be achieved in some mutant extraembryonic cells (77, 139). Imprinted X-inactivation is not absolute in extraembryonic tissues, suggesting that the random choice mechanism can function to some extent to establish proper dosage compensation in extraembryonic cells (65). The imprint may normally direct binding of the blocking factor to the Xm in a constitutive manner during imprinted X-inactivation (103). One explanation for occasional maternal X-inactivation in the extraembryonic tissues is that the imprint may be lost at low frequency, allowing the Xm and Xp to compete equally for blocking factor and resulting in random X-inactivation in a subset of cells (65). The ultimate test of the hypothesis that blocking factor is involved in imprinted X-inactivation will require deleting the counting element or knocking out blocking factor activity and testing the effects on imprinted X-inactivation in the mouse.

The Xm Epigenotype Is More Rigid than that of the Xp

Embryos inheriting exclusively paternal X chromosomes or extra maternal X chromosomes have defects in dosage compensation and aberrantly express other imprinted genes, resulting in embryonic death midgestation. Embryos with exclusively paternal X chromosomes exhibit *Xist* RNA coating of all X chromosomes during the cleavage stages. However, XpXp androgenetic embryos (75, 112) and XpO embryos (91) are progressively able to repress *Xist* expression from the Xp over time and ultimately achieve nearly correct dosage compensation. Correction of dosage compensation in XmXm parthenogenotes by upregulation of *Xist* on at least one X chromosome was highly variable between embryos (75, 91, 107). Whereas XpO embryos had shut off *Xist* expression from the single Xp in nearly 50% of cells by the blastocyst stage, XmXmY embryos activated *Xist* expression from one Xm in at most 10% of cells at this stage (91). The greater lability of the Xp epigenotype compared with the Xm may explain why some embryos survive maternal inheritance of *Tsix* deletions (77, 139), while paternal inheritance of an *Xist* deletion is always lethal (90). *Xist* may be downregulated on the Xp in the former more easily than it may be upregulated from the Xm in the latter. This differential ability to reprogram *Xist* expression patterns on the Xp and the Xm suggests that the imprint controlling X-inactivation is located on the Xm, designating the Xa fate. Thus, as in the choice phase of random X-inactivation, the imprinting mechanism may specifically designate the Xa rather than the Xi. Nuclear transplantation experiments with cytologically marked X chromosomes suggest that the maternal X chromosome's resistance to inactivation is acquired during oocyte growth between prophase of meiosis I and meiosis II, a time when other imprints are placed on the maternal genome (149).

The Xa and Xi Epigenotypes Are Functionally Equivalent to the Xm and Xp

Eggan and colleagues examined X-inactivation patterns in mice cloned by nuclear transfer of female somatic nuclei. Resulting mouse clones exhibited complete non-random X-inactivation of the former Xi in the imprinted extraembryonic tissues, whereas X-inactivation was random in embryonic lineages (45). This result indicates that the epigenetic states of the *Xist* alleles on the Xa and Xi in somatic cells are functionally equivalent to the epigenetic states of the Xm and the Xp, respectively. It would be an elegant solution to the imprinting problem if the imprints controlling *Xist* expression in the early embryo were the same as the somatic epigenotypes controlling *Xist* expression from the Xi and the Xa.

Changes in *Xist* Expression During Gametogenesis Correlate with Parental Imprints

Like female somatic cells, the primordial germ cells (PGCs) that will give rise to the germline undergo X-inactivation during gastrulation. However, prior to entry into meiosis midgestation, female PGCs reactivate their Xi (154). This reactivation is accompanied by a loss of *Xist* expression without an apparent restoration of

Tsix expression (107). *Xist* cannot be detected in mature oocytes by RT-PCR, suggesting that transcripts are absent or extremely rare (3, 70). This repression of *Xist* expression during female gametogenesis correlates with extremely low *Xist* expression from the Xm upon fertilization (41, 77).

Male PGCs do not contain an Xi or express *Xist* RNA. During the process of spermatogenesis, however, *Xist* becomes expressed at low levels and is present in mature spermatids (93, 111, 132, 140). By in situ RT-PCR, *Xist* RNA appears to be associated with the heterochromatic sex chromatin body, which consists of the X and Y chromosomes that have been sequestered away from the autosomes during meiosis I of spermatogenesis (4). Even though it is expressed in the male germline, *Xist* is dispensable for spermatogenesis (90), suggesting that *Xist* expression in the male germline may be the result of epigenetic modifications that facilitate *Xist* expression from the Xp early in development.

DNA Methylation Is a Candidate for Establishing or Maintaining the Imprint

Many imprints in mammals require DNA methylation, acquired in the male or female germline, for establishment or maintenance (66). Several groups have searched for differential methylation of the *Xist* gene. In the male germline, CpG dinucleotides at the *Xist* promoter are methylated prior to meiosis, but this methylation is erased during spermatogenesis and is absent in mature spermatids (2, 167) and on the Xp in preimplantation embryos (2, 94, 167). In the female germline, methyl-sensitive restriction analysis suggested that CpGs at the *Xist* promoter and in the first exon become methylated during oogenesis and retain this methylation upon fertilization (2, 167). Using bisulfite sequencing to detect CpG methylation, another group failed to confirm methylation of *Xist* on the Xm in oocytes and early embryos (94). It is therefore unclear whether the *Xist* gene on the Xm is methylated in the early embryo, which would correlate with lack of *Xist* transcription from the Xm. Both male-specific demethylation and female-specific methylation of *Xist* would require reprogramming in germ cells, and suggest that both parental genomes have the potential to acquire epigenetic marks.

Another attractive target for the X-inactivation imprint is the *Tsix* gene. A maternal imprint would direct high maternal *Tsix* expression, whereas a paternal imprint would direct low or absent expression. It has been suggested that the CpG island at the *Tsix* promoter, *DXPas34*, is an imprinting center involved in keeping the *Xist* allele on the Xm silent during early embryogenesis (77). Although *DXPas34* is differentially methylated on the Xa and Xi (37), bisulfite sequencing in oocytes and spermatocytes failed to find any differences in the methylation patterns between the two (122). Binding sites for <u>C</u>CC<u>TC</u> binding <u>f</u>actor (CTCF) have been identified in *DXPas34*, the CpG island associated with the major *Tsix* promoter, and CTCF binds to this sequence in vitro (27). Differential methylation of CTCF binding sites has been implicating in establishing the imprint required for expression of *H19/Igf2* gene pair (8, 52, 148). While CpG methylation has relatively little effect on CTCF binding to *DXPas34* in vitro, non-CpG methylation abolishes CTCF binding in this assay. If CTCF binding occurs in vivo during Xa choice, then searching for an imprint in the form of differential non-CpG methylation at *DXPas34* would be a logical next step.

ACTIVE X CHROMOSOME CHOICE IN HUMANS

Humans and Mice Differ in Developmental Regulation of *XIST/Xist*

In comparison to mouse, little is known about the developmental regulation of X-inactivation in humans. Human *XIST* RNA can be detected by RT-PCR in oocytes and in both male and female preimplantation embryos, increasing in abundance until the blastocyst stage (40, 129). This early *XIST* expression is not exclusively paternal in humans, consistent with a lack of strict paternal X-inactivation in human extraembryonic tissues (54, 134). In addition, both maternally and paternally derived supernumerary X chromosomes are tolerated in humans (88, 123), in contrast to the mouse. These data suggest that X-inactivation does not occur in an imprinted fashion in humans. Thus, the skewing toward paternal X-inactivation detected in some extraembryonic tissue samples (48) could either be a primary effect due to nonrandom Xa choice or a secondary effect due to differential cell survival (99). Given that secondary effects have been demonstrated to be significant in humans (141), this issue can only be addressed by analyzing tissues immediately after X-inactivation.

The Xa Choice Machinery Exhibits Limited Species Conservation

The molecular machinery involved in random choice of a single Xa must be largely conserved between mouse and human. Male mouse ES cells containing 480-kb human *XIC* transgenes exhibited rare cases of *Xist* expression and coating of the single mouse X chromosome upon differentiation (58, 101), suggesting that the human and mouse counting elements can both compete for mouse blocking factor. Surprisingly, human *XIST* is expressed at high levels prior to differentiation of the transgenic ES cells, suggesting that *XIST* is not unstable in undifferentiated mouse cells. Human *TSIX* transcription has been identified in transgenic mouse ES cells as well as in human embryocarcinoma cells and cell lines derived from human PGCs (100). However, the mouse and human *Tsix/TSIX* promoter regions are poorly conserved (78, 100), and the *TSIX* transcript does not appear to traverse the entire *XIST* gene (100). These differences between human *TSIX* and its mouse counterpart could render *TSIX* RNA unable to destabilize *XIST*. RNA FISH for *XIST* and *TSIX* in human preimplantation embryos and ES cells would allow determination of the distribution of these two transcripts in early development and provide insights into their role in Xa choice in humans.

SILENCING AND ALTERATIONS
IN CHROMATIN STRUCTURE

Silencing by *Xist* Can Be Divided Into Three Phases

The enactment of the Xi fate requires an increase in the steady-state level of *Xist* RNA on the presumptive Xi. At this time, *Xist* transcripts spread to coat the chromosome in *cis*, and induce gene silencing within 24 hours (164), leading to the formation of extraordinarily stable heterochromatin. Studies employing the inducible *Xist* cDNA in male ES cells suggest that the silencing process can be divided into three steps: initiation, establishment, and maintenance (Figure 6) (164). In the initiation phase, high-level *Xist* expression can cause de novo silencing. Initiation of gene silencing can occur as long as the cells remain undifferentiated and up to 1.5 days after induction of differentiation. In this phase, silencing is reversible, such that *cis*-linked genes can be reactivated if *Xist* expression is extinguished. Extending this result to female development, endogenous *Xist* expression is normally upregulated within one to two days of differentiation in female ES cells (143). The period in which enactment of the Xi fate occurs corresponds to the initiation phase of silencing. A differentiation milestone triggers the establishment phase, which is characterized by the requirement for continued coating by

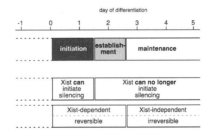

Figure 6   The three steps of the silencing process. The *Xist* RNA-mediated silencing process can be divided into initiation, establishment, and maintenance phases (164). Only during the initiation phase (up to the first 1.5 days of ES cell differentiation), can induction of *Xist* RNA initiate gene silencing. Both the initiation phase and the subsequent establishment phase are characterized by their dependence on coating by *Xist* RNA and the reversibility of the silent state such that continued *Xist* expression is required to prevent loss of transcriptional repression. The maintenance phase is distinguished by the irreversible and *Xist* RNA-independent propagation of silencing, but shares the resistance to *Xist* RNA-mediated initiation of silencing with the establishment phase. The brief 24-hour establishment phase can be envisioned as a window in which the silenced state is locked in. This locking in is tightly linked to ES cell differentiation since it does not occur when silencing by *Xist* RNA takes place in undifferentiated cells. The fact that silencing by *Xist* RNA can be initiated in undifferentiated ES cells is indicated by the extension of the initiation phase to a time before differentiation (−1 days) with dashed lines.

Xist RNA for transcriptional repression and the resistance to de novo silencing by *Xist* RNA. After a defined period of time, which is approximately 24 hours in length, silencing becomes irreversible and *Xist*-independent, demonstrating entry into the maintenance phase. The establishment step may therefore be seen as the process of locking in the silent state, which is then stably maintained.

Silencing Is Accompanied by a Cascade of Chromatin Modifications

The Xi heterochromatin is characterized by a multitude of chromatin modifications that distinguish it from the Xa. The Xi replicates late in S-phase (151), is methylated at promoter CpG islands (163), and hypoacetylated at histones H4, H2A, and H3 (10, 68). In addition, increased histone H3 lysine-9 (H3 Lys-9) methylation and decreased histone H3 lysine-4 (H3 Lys-4) methylation were recently identified as components of the Xi chromatin (12, 59, 97, 119). Each of the above modifications is generally associated with regions of heterochromatin (131), suggesting that X-inactivation employs general mechanisms that are used to regulate gene expression in other contexts. The remodeled Xi chromatin is further characterized by the accumulation of variant histone H2A isoforms termed macroH2A (34). In addition, in human cells, the exclusion of the Barr Body–deficient histone H2A variant (H2ABBD) from the Xi has been reported (26). The function of these H2A variants is unclear.

The appearance of chromatin modifications on the Xi has been catalogued relative to the induction of differentiation, *Xist* RNA coating and gene silencing in female ES cells (Figure 7). *Xist* RNA spread occurs between the first and second days of differentiation and is followed closely by methylation of histone H3 Lys-9,

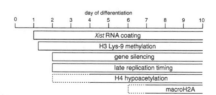

Figure 7   Chromatin modifications on the Xi occur sequentially during the differentiation of female ES cells. The scheme shows the timing of appearance of chromatin modifications that characterize the heterochromatin on the Xi. Solid lines indicate cytological observations (by FISH or immunofluorescence) made when cells were differentiated in embryoid body cultures; the dotted parts of the bars depict results obtained from retinoic acid–induced ES cell differentiation, which appear to place the appearance of modifications at earlier times (59, 73, 96, 127). Once set up, all features of the Xi are stably maintained throughout all somatic cell divisions, as indicated by the open ends of the bars. Although methylation of promoter CpG islands has been observed on the Xi in somatic cells, DNA methylation is not depicted in this figure due to insufficient information on the timing of acquisition of this modification during differentiation.

such that the region of H3 Lys-9 methylation always appears either smaller than or equal in size to the *Xist* RNA domain (59, 97). Concomitant with the appearance of H3 Lys-9 methylation, hypoacetylation of H3 Lys-9 and hypomethylation of H3 Lys-4 on the Xi were observed, indicating that these modifications may be co-ordinated. These data are consistent with the fact that deacetylation of lysine 9 is a prerequisite for methylation of this residue. The spread of *Xist* RNA and H3 Lys-9 methylation precedes gene silencing and late replication timing, both of which begin on day 2 and are essentially complete around day 4 and day 6, respectively (73). The appearance of H4 hypoacetylation is currently placed either at day 2 or day 4, depending on the differentiation method used; therefore, this modification may occur coincident with or subsequent to gene silencing (59, 73). The accumulation of macroH2A has been detected around day 6 or 7 (96, 127), although one study placed the appearance earlier (126). The timing of the first appearance of promoter CpG methylation has not been analyzed, but has been detected at day 21 post-differentiation (73, 84). When these data are taken together, a sequential progression of chromatin changes appears to underlie the X-inactivation process in female ES cells.

If any one of these modifications is crucial for setting up the silent state of the Xi, then it should be present in the initiation phase of silencing. Using the *Xist* cDNA transgene, the initiation stage can be separated from the differentiation-dependent establishment phase by expressing the *Xist* RNA in undifferentiated ES cells (Figure 6) (164). *Xist* cDNA-induced gene silencing was achieved in undifferentiated ES cells without a shift in the affected chromosomes' replication timing or H4 acetylation status. The relatively early appearance of both modifications during the differentiation of female ES cells suggests, however, that they play a role in the establishment of silencing (73, 164). The co-localization of macroH2A with the Xi takes place only after the transition to the irreversible maintenance phase (127, 164), suggesting that macroH2A does not play a role in establishment. Although it has yet to be shown that modifications on histone H3 occur during the *Xist* RNA-induced initiation phase in undifferentiated ES cells, H3 Lys-9 methylation is the only alteration that appears nearly coincident with *Xist* RNA spread in differentiating female ES cells, implying that modifications of histone H3 are involved in the initiation of silencing (59, 97). Studies to determine whether H3 Lys-9 methylation is necessary for the initation of silencing on the Xi would be of great interest.

A Histone H3 Methylation Hotspot May Be Required for the Initiation of Silencing

A hotspot of H3 Lys-9 methylation, spanning a region of roughly 100 kb 5' of the *Xist* P1 promoter, was identified by both immunostaining and chromatin immuno-precipitation in undifferentiated female and male ES cells (Figure 3) (59). As ES cells are induced to differentiate, this hotspot gives way to chromosome-wide H3 Lys-9 methylation on the future Xi, while it is retained for up to 5 days on the

Xa-elect. Heard and colleagues proposed that this hotspot of H3 Lys-9 methylation may serve as a nucleation site for the spread of *Xist* RNA along the X chromosome (59). The H3 Lys-9 hotspot is also present on single- and multicopy 480-kb transgenes in undifferentiated male ES cells (59). Upon differentiation, the hotspot disappears from single-copy transgenes more rapidly than associated *Xist/Tsix* expression, suggesting that the lack of *Xist* RNA spread from single-copy transgenes may be due to lack of the H3 Lys-9 methylation nucleation center. The absence of a crucial *cis*-element, such as a boundary element that blocks spread of chromatin modifications from sequences flanking the site of transgene insertion, may subject single-copy transgenes to position effects that prevent proper maintenance of the H3 Lys-9 methylation hotspot. Presumably, multicopy transgenes can be shielded from such position effects. Whether or not pinpoint *Xist* expression plays a role in the formation of this hotspot has yet to be tested.

As H3 Lys-9 methylation is a self-propagating modification in other systems (5, 74), it is possible that spread of H3 Lys-9 methylation on the Xi is also self-propagating and does not require *Xist* RNA. Alternatively, *Xist* RNA could simultaneously coat the Xi and propagate H3 Lys-9 methylation along the chromosome by binding nucleosomes that contain H3 Lys-9 methylation, and then inducing this modification on adjacent nucleosomes in a self-reinforcing mechanism. However, H3 Lys-9 methylation per se is not sufficient to mediate *Xist* RNA spread, as *Xist* RNA is absent from regions of constitutive heterochromatin such as centromeres (31, 44), which are characterized by H3 Lys-9 methylation (120). Lys-9 of H3 in constitutive heterochromatin of centromeres is methylated by the histone methyltransferases Suv39h1 and Suv39h2 (120), neither of which is required for H3 methylation on the Xi (119). In addition, the heterochromatin protein HP1, which interacts with H3 methylated on Lys-9 (74), is enriched on centromeric heterochromatin but not on the Xi (119). Thus, facultative heterochromatin of the Xi may employ different histone methyltransferases and methyl-histone binding proteins than constitutive heterochromatin.

MacroH2A Is Recruited to the Xi by *Xist*

The enrichment of macroH2A histone variants on the Xi in female somatic cells revealed the first evidence of differential protein composition between the Xa and Xi (34, 36). MacroH2A1 contains an amino-terminal domain with 64% homology to core histone H2A, which mediates macroH2A's incorporation into nucleosomes as well as its enrichment on the Xi, and a large carboxy-terminal domain of unknown function (24, 116, 118). Two subtypes of macroH2A1 (macroH2A1.1 and 1.2), differing in only a putative leucine zipper within the non-histone domain, are generated by differential splicing of the same gene (116, 125). The third subtype, macroH2A2, is produced from a separate gene and has a histone domain highly similar to that of macroH2A1, but differs slightly throughout the entire non-histone domain (25, 35). Most studies have concentrated on the macroH2A1 subtypes. MacroH2A1 may not only be involved in X-inactivation since it is expressed at similar levels in male and female ES and somatic cells, is found throughout the entire nucleus, and is located on centrosomes (34, 96, 98, 125, 126). When *Xist* is conditionally deleted from the Xi in female somatic cells (39), or when *Xist* RNA is prevented from coating the Xi (7), macroH2A1 disappears from the Xi. These results suggest that this histone variant is recruited to the Xi in an *Xist* RNA-dependent fashion. A tight association between *Xist* RNA and the Xi chromatin is suggested by the fact that *Xist* RNA co-immunoprecipitates with macroH2A1 and also with core histones (47).

Multiple Mechanisms Contribute to the Maintenance of the Xi

The continued expression and association of *Xist* RNA with the Xi in somatic cells hinted that this RNA plays a role in the maintenance of the Xi (31). However, X-inactivation appears to be stable without continued *Xist/XIST* expression in differentiated cells, suggesting that the RNA is not strictly required to maintain the silent state (19, 38, 39, 124, 164). To investigate a possible role for *Xist* RNA in maintenance, Csankovszki and colleagues generated somatic cells containing an Xi that allowed for the conditional loss of the *Xist* gene (38). The stability of the Xi was analyzed by monitoring the reactivation of two *cis*-linked genes, *Hprt* and a *GFP* transgene. The loss of *Xist* expression on the Xi allowed reactivation of the formerly silent *GFP* and *Hprt* genes in a significant number of cells. Cells generally reactivated only one of the two genes, suggesting that silencing is maintained on a gene-by-gene basis. The approximately 50-fold greater effect of the conditional *Xist* deletion on the endogenous *Hprt* gene compared to the *GFP* transgene suggested that bona fide X-linked genes are more dependent on *Xist*-mediated maintenance of silencing. These data are consistent with the idea that *Xist* RNA coating contributes to the stability of the Xi. When the loss of *Xist* was coupled with treatment by DNA demethylating agents, DNA methyltransferase mutants, and/or deacetylase inhibitors, synergistic reactivation of the *GFP* and *Hprt* genes was observed (38), indicating that multiple redundant mechanisms maintain the silent state of the Xi.

DNA methylation appears to be extremely important for the stability and maintenance of gene silencing on the X chromosome. The use of DNA demethylating agents on human Xi-containing somatic cell hybrids resulted in reactivation of several genes in vitro (104). When similar studies were performed on mouse somatic cell lines, there was a 10- to 20-fold increase in reactivation of X-linked genes (38). A greater effect was seen upon loss of the *Dnmt1* gene from a conditional allele, which resulted in genome-wide hypomethylation. A 1500-fold increase in reactivation of the X-linked *GFP* transgene occurred upon loss of *Dnmt1* activity (38). Similarly, mouse embryos with a mutation in *Dnmt1* cannot maintain the Xi in embryonic tissues, leading to the reactivation of a formerly silenced X-linked *LacZ* transgene in many cells (138).

Inactivation of autosomal genes in cells with X:autosome translocations can be stably maintained without the spread of late replication from the Xi into the translocated autosomal region (142). Therefore, late replication timing is not absolutely necessary for continued silencing, most likely due to the redundancy in Xi maintenance mechanisms. The importance of late replication timing is only apparent in the absence of CpG island DNA methylation. Humans with ICF syndrome (Immunodeficiency, Centromeric instability, and Facial anomalies syndrome), which is caused by mutations in the de novo DNA methyltransferase *DNMT3b*, can properly establish an Xi even with global hypomethylation at promoter regions of X-linked genes (50, 51). Reactivation of formerly silenced genes in these patients' cells was observed only when regions on the Xi had shifted to an earlier replication timing.

Although delayed appearance of H4 hypoacetylation during differentiation suggests a role in establishment and maintenance of X-inactivation, the use of histone deacetylase inhibitors on female cells did not result in reactivation of the Xi (38, 73). However, when deacetylase inhibitors were combined with treatments that reduce DNA methylation, a twofold increase in reactivation was seen over demethylation treatment alone, indicating that histone hypoacetylation plays some role in the stable maintenance of silencing (38). Considering the importance of proper dosage compensation during development, it is not surprising that multiple mechanisms ensure the stability of the Xi.

Genes that Escape X-Inactivation Lack Chromatin Modifications Characteristic of the Xi

A subset of genes escapes X-inactivation in humans and mice (43). The mechanisms by which X-linked genes can escape *Xist* RNA-mediated silencing remain mysterious. FISH provides insufficient resolution to determine whether escapees are coated by *Xist* RNA. However, it appears that *Xist/XIST* RNA does not uniformly coat the X chromosome in mouse and human cells as the RNA is absent from constitutive heterochromatin (31, 44). Chromatin isolated using antibodies directed against histone H3 methylated on Lys-9 contained promoter regions of silenced genes on the human Xi, but not of two X-linked genes escaping silencing (12). Instead, the promoter regions of these escapees could be precipitated using antibodies directed against H3 methylated on Lys-4 (12), consistent with the association of H3 Lys-4 methylation with active regions of transcription and H3 Lys-9 methylation with inactive regions (67). Extending the correlation between escaping genes and modifications associated with actively transcribing regions, the pseudoautosomal region (PAR), which contains a number of genes that escape X-inactivation (43), is methylated on H3 Lys-4 in human cells (12), and acetylated on H4 in both mouse and human cells (68). The human X chromosome region Xp11.2–3, known to harbor escaping genes, and the homologous region A2 in mouse show the same pattern of histone methylation and acetylation as the PAR (11, 68). Furthermore, escapees do not show late replication timing or methylation of CpG islands in their promoter regions (22, 157). Thus, active genes on the Xi are not subject to the same chromatin modifications as silent genes, indicating that there may be *cis*-elements that determine whether *Xist* RNA can mediate silencing on particular regions of the X chromosome.

The genes that escape silencing in humans tend to be found in clusters, suggesting that their expression may be regulated at the level of chromatin domains (23). DNA boundary elements have been demonstrated to separate regions of differentially regulated chromatin (9, 29, 72, 110). Boundary elements could play a similar role in insulating genes that escape X-inactivation from the surrounding heterochromatic environment of the Xi. Another hypothesis is that escaping genes are deficient in sequences promoting *Xist*-mediated silencing. It has recently been suggested that *long interspersed nuclear elements* (*LINEs*) enhance *Xist* RNA spread, since these repetitive elements are enriched on the X chromosome relative to autosomes (87). However, *Xist* cDNA transgenes can coat and silence *LINE*-poor chromosomes (101, 164), indicating that high *LINE* density is not absolutely required for chromosome coating and silencing.

Xi Chromatin Modifications Play Different Roles in the Extraembryonic Tissues

The appearance as well as the importance of Xi chromatin modifications differ in mouse extraembryonic and embryonic tissues, which undergo imprinted and random X-inactivation, respectively. Whereas macroH2A association with the Xi appears to be a late event in random X-inactivation, this histone variant becomes enriched on one X chromosome between the 12-cell stage and blastocyst formation (36). In blastocysts, macroH2A accumulation is mostly restricted to the differentiated extraembryonic cells, suggesting that macroH2A might be involved in the silencing of the Xp in this lineage (36). Aside from *Xist* RNA spread, macroH2A recruitment is the earliest known chromatin modification in extraembryonic tissues. The presence of H3 Lys-9 hypermethylation and hypoacetylation, H3 Lys-4 hypomethylation, and H4 hypoacetylation, chromatin modifications that are associated with the Xi in embryonic lineages, have yet to be examined in preimplantation embryos and the extraembryonic tissues.

From the blastocyst stage until gastrulation, the Xi in the extraembryonic lineages replicates very early in S-phase, before autosomes and the Xa (151). The Xi then shifts to the more familiar late replication timing within a single cell cycle (147, 153). It is unclear why the Xi in the extraembryonic tissues initially replicates very early and subsequently shifts to late replication, although the shift does follow the first appearance of a late replicating Xi in the embryo proper (153).

Whereas DNA methylation is critical for X-inactivation in embryonic tissues, it does not appear to play a major role in extraembryonic tissues. There is less CpG methylation in extraembryonic cells than in those of the embryo proper (28), with genes such as *Hprt* seemingly devoid of CpG methylation compared to their counterparts in the embryonic tissue (84). Consistent with a minor role for DNA methylation in extraembryonic tissues, Xi maintenance in these lineages is unaffected by mutations in *Dnmt1* (138).

Since there is little DNA methylation in extraembryonic tissues, one might expect the maintenance of the extraembryonic Xi to rely more heavily on other mechanisms. Wang and colleagues showed that mutations in the Polycomb group protein *extraembryonic deficient* (*eed*) cause defects in the development of certain extraembryonic tissues, with females showing a more severe phenotype (159). Further analysis using a paternally inherited X-linked *GFP* transgene showed that *eed* mutants are deficient in maintaining the Xi in the extraembryonic, but not the embryonic tissues (159). Following imprinted X-inactivation, extraembryonic cells were negative for GFP fluorescence. Subsequently, GFP expression reappeared, supporting the idea that these tissues initially underwent X-inactivation, but then failed to maintain it. Since the *eed* protein has been shown to interact with histone deacetylases, and their activity is crucial to *eed*-mediated repression (158), maintenance of the Xi could be more reliant on histone hypoacetylation in extraembryonic than in embryonic lineages. It remains to be determined whether the extraembryonic Xi is hypoacetylated, and if so, whether deacetylation occurs in an *eed*-dependent manner. Even though X-inactivation in the embryo proper was shown to be stable, it would be interesting to challenge embryonic cells from *eed* mutants by compromising an Xi maintenance mechanism to see if *eed* may also play a role in maintaining the Xi in embryonic tissues.

Xist RNA CONSISTS OF FUNCTIONAL MODULES

Xist/XIST RNA Displays a Conserved Repeat Structure

Mouse *Xist* and human *XIST* RNA contain 6 direct repeats, designated *A* through *F*, whose order and sequence are highly conserved (Figure 8). The delineation of these repeats was reinforced and extended by comparison to *Xist* sequences of four common vole species (108). The *Xist/XIST* RNA repeats *A*, *B*, and *F* are highly conserved in copy number, whereas the more complex *C*, *D*, and *E* repeats have been differentially amplified in mouse and human (15, 18, 108). In total, almost half of the *Xist/XIST* transcript is composed of tandem repeats. The strong conservation of repeats contrasts with the overall low homology between mouse *Xist* and human *XIST* RNA, suggesting that the *Xist/XIST* repeats play an important role in the function of the RNA. The differential amplification of *C*, *D*, and *E* repeats in human and mouse suggests that they may be functionally redundant.

Xist Contains Multiple Redundant Domains that Mediate X Chromosome Coating

The inducible *Xist* cDNA transgene approach allowed the first thorough analysis of the requirement of *Xist* RNA sequences for chromosome coating and silencing. In an impressive study, Wutz and colleagues generated 43 *Xist* cDNA constructs containing different deletions, which were individually integrated into the same site on the X chromosome in male ES cells (165). *Xist* expression levels, RNA spread along the X chromosome, and transcriptional inactivation of genes located on the single X chromosome were examined.

A.

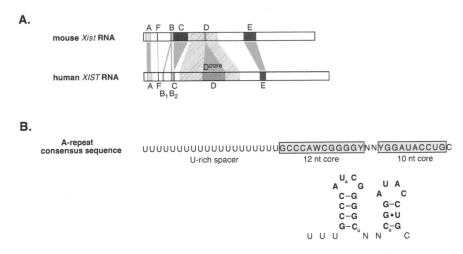

B.

Figure 8 The repeat structure of *Xist/XIST* RNA. *A*. Repeats are conserved between human and mouse *Xist/XIST* RNA. *XIST/Xist* RNA contains six types of repeats, labeled A–F. The E-repeat is located at the 5′ end of exon 6 of *XIST* and at the 5′ end of exon 7 of *Xist*, respectively; all other repeats are contained within the large first exon of *XIST/Xist*. The *A*-repeat, located within the first 1000 nt of *XIST/Xist* RNA, consists of 8.5 copies of a 43–59 nt unit in the human RNA and of 7.5 copies of a 42–74 nt unit in mouse counterpart [for a more detailed description, see Figure 8*B* (15, 18)]. The short *F*-repeat is located approximately 750 nt downstream of the *A*-repeat (108). It contains two copies of a G/C-rich 10-nt sequence motif (UGGCGGGCUU) separated by 8 nt in mouse *Xist* and 16 nt in human *XIST*. This repeat was identified recently based on its extension to 5 copies in vole *Xist* and may actually function as DNA element (108). Besides the *A*-repeat, the *B*-repeat is the most conserved repeat in *Xist/XIST* RNA (15, 18, 108). In the mouse RNA, the low-complexity *B*-repeat is composed of 32 C-rich 4–8-nt units, with 21 of those containing the motif (A/U)GCCCC. In human *XIST* RNA, the B-repeat is split. Twelve C-rich 6–9-nt units, eight of which contain the motif ACCCCCC, are separated by approximately 700 nt of unrelated sequence from 17 C-rich 4–11-nt units, ten of which contain the motif PuPuCCC.

In contrast to the *A*, *B*, and *F* repeats, the three most 3′ *Xist/XIST* repeats have been differentially amplified in mouse and human (15, 18). In mouse, the C-repeat contains 14 copies of 120-nt units, which show more than 90% similarity to each other. Only one weakly homologous copy of the *C*-repeat unit can be found in human *XIST* RNA, which, however, is located in the equivalent part of the RNA. In contrast to the *C*-repeat, the *D*-repeat is emphasized in the human RNA (15, 18). The *D*-repeat appears to have the greatest sequence complexity. Recently, the *D*-repeat region of human *XIST* RNA has been extended. The originally described *D*-core region, comprised of seven 300-nt units with more than 75% homology, was shown to be flanked on both sides by truncated copies of the *D*-repeat unit for a total of 19 truncated copies, such that the entire *D*-repeat region encompasses 6000 nt (108). In mouse *Xist* RNA, only 10 partial copies of the *D*-repeat unit were identified within a region of 3000 nt (108). The *E*-repeat encompasses 1350 nt in mouse and 700 nt in human *Xist/XIST* RNA, and contains highly variable copies of 20–25-nt units (15, 18, 108).

This analysis allowed designation of functions to certain domains of *Xist* RNA (Figure 9). Large regions, covering roughly the first two thirds of *Xist* RNA, mediate the association of *Xist* transcripts with the Xi, and are functionally redundant such that not all of them have to be present for coating to occur (165). The regions implicated in *Xist* RNA localization contain all of the *Xist* repeats (*A*, *B*, *D*, *E*, *F*, and at least parts of the *C* repeat; Figure 9). Therefore, it seems likely that all 6 *Xist* repeats are involved in localizing *Xist* transcripts to the Xi, but no single one of them is essential. These observations could be explained if the repeats and surrounding sequences contain low-affinity binding sites for *trans*-acting factors whose occupancy is increased by cooperative interactions, conferring full *Xist* RNA coating function (165). In a different experimental approach, oligonucleotide analogs antisense to *Xist* RNA were used to physically disrupt *Xist* RNA function. By this approach, a highly conserved sequence within the *C*-repeat was found to be important for localizing *Xist* RNA to the Xi, since antisense oligonucleotides to this region induced dissociation of *Xist* transcripts from the Xi in female somatic cells (7) (Figure 9). As this manipulation may have caused a general disruption of *Xist* RNA secondary structure and/or ribonucleoprotein complex formation, these results do not necessarily implicate the *C*-repeat directly in *Xist* RNA localization.

Trans-Acting Factors Required for X Chromosome Coating Are Species-Specific

Interspecies experiments support the idea that *trans*-acting factors are required for the association of *Xist* RNA with the Xi and may give some insights into their nature. In rodent somatic cell hybrids containing a human Xi, *XIST* is highly

←——————————————————————————

Figure 8 (*Continued*) *B*. Each *A*-repeat unit forms two conserved stem loops. The consensus sequence of the *A*-repeat unit for mouse and human *XIST/Xist* RNA is shown (W = U or A, Y = C or U, (18)). Each *A*-repeat unit contains, within a 25-nt core region, two highly conserved GC-rich motifs of 10 nt and 12 nt, respectively. A variable, primarily U-rich stretch, here depicted as a stretch of 20 uracil residues, serves as spacer between the GC-rich regions of neighboring *A*-repeat units. The length of the spacer (17–49 nt) and its composition vary between individual repeat units, and can be decreased down to 8 residues in every unit without interfering with the function of the *A*-repeat (165). Both the 12-nt and 10-nt GC-rich regions fold into stem loop structures as supported by secondary structure predication and mutational analysis (165). In the secondary structure model of the two GC-rich regions depicted here, nucleotides shown in small font size occur less frequently at the corresponding positions. Mouse *Xist* RNA contains 7.5 copies of the *A*-repeat unit, comprising 8 of the first, larger, and 7 of the second, smaller stem loop; human *XIST* RNA contains 8.5 copies with 9 large stem loops and 8 small ones.

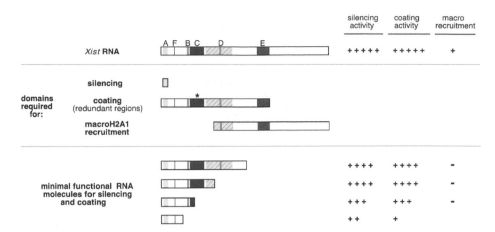

Figure 9 Functional domains of *Xist* RNA. *Xist* RNA is depicted as in Figure 8*A*. Domains required for silencing, chromosome coating, and macroH2A1 recruitment as determined by deletion analysis are shown (165). Deletion of the *A*-repeat region results in an RNA molecule that can coat but not silence the chromosome. The coating function requires multiple partly redundant domains contained within a large sequence block at the 5′ end. A short antisense molecule directed against a highly conserved sequence of the *C*-repeat (indicated by a *star*) completely disrupts localization of *Xist* RNA to the Xi (7).

The deletion study by Wutz et al. (165) defined various large 3′ truncations that for the most part do not interfere with the silencing function of *Xist* RNA. The silencing activity correlates tightly with the coating activity such that on a scale from high (+++++) to no (−) activity, the silencing activity decreases with a reduction of the coating activity. The smaller the minimal *Xist* RNA molecule becomes, the lower the coating activity due to the requirement of multiple domains for coating. MacroH2A1 colocalization is not essential for the initiation of the silencing process, since minimal functional RNA molecules do not recruit macroH2A1 (indicated by + or −).

expressed but not localized tightly to the X chromosome (30). In these cells, some *XIST* RNA is located within the region delineated by the human X chromosome, but it is mostly distributed throughout the rodent nucleus, having a particulate appearance (30, 49). Similar results were obtained when human *XIC* transgenes were analyzed in mouse ES cells. Human *XIST* RNA was found to incompletely coat the mouse autosome carrying the transgene showing a less compact accumulation around the transgenic chromosome (58, 101). Since mouse ES cells contain all the *trans*-acting factors required for chromosome coating by mouse *Xist* RNA (164, 165), it seems that these factors are unable to confer the ability of human *XIST* RNA to coat the chromosome. Thus, factors that mediate localization of mouse *Xist* and human *XIST* transcripts may be species-specific (30). The low degree of primary sequence conservation between mouse and human *Xist/XIST* RNA within regions required for chromosome association, and the differential amplification of the C, D, and E repeats, could explain the requirement for species-specific factors in *Xist*/*XIST* RNA localization.

Despite the fact that *XIST* RNA is not properly associated with the Xi in rodent cell hybrids or transgenic mouse ES cells, the half-life, and presumably the steady-state level, of *XIST* RNA is similar to that of *XIST* transcripts in human female somatic cells (30, 49, 58, 101). Therefore, in contrast to localization factors, *trans*-acting factors that stabilize the RNA could be conserved between mouse and human. The hybrid cell data suggest *Xist*/*XIST* RNA can be stable without chromosomal association. This idea is supported by the fact that *Xist* RNA accumulates at a steady-state level similar to that normally seen in female somatic cells when association with the chromosome is blocked by addition of antisense oligonucleotides (7). While all short *Xist* RNA molecules encoded by cDNA fragments that are able to coat the chromosome are also stable, no fragment that is defective in coating has been shown to be stable (165). Thus, a minimal length of *Xist* RNA may be required to achieve stability.

The A-Repeat Mediates *Xist* RNA's Silencing Function

Wutz and coauthors clearly demonstrated that the silencing function of *Xist* RNA can be attributed to the *A*-repeat region (Figure 9) (165). Deletion of the *A*-repeats in inducible *Xist* cDNA transgenes does not disrupt Xi localization of *Xist* RNA, but abolishes silencing activity. Thus, chromosome coating alone is not sufficient for initiation of X-inactivation (165). When the *A*-repeat region is present, however, coating of the Xi by *Xist* RNA is a prerequisite for silencing (7, 165). Since the *A*-repeat region functions properly even when moved to the 3' end of *Xist*, it can be viewed as a position-independent silencing module (165). It seems most likely that the *A*-repeat functions by recruiting proteins that mediate silencing.

Wutz and colleagues provide data that support a model in which the highly conserved 25-nucleotide core region of each *A*-repeat RNA unit forms 2 stem loops that are necessary for silencing (Figure 8*B*) (165). Mutations that should disrupt a base pair in the first stem results in the loss of silencing activity. Mutational analysis further showed that the first stem loop, but not the sequence of either the stem or the loop, is essential (165). Detailed analysis of the second stem loop has not been reported. Inversion of the *A*-repeat region in the *Xist* cDNA transgene resulted in production of an RNA containing antisense *A*-repeats that did not confer silencing function (165), indicating some requirement for sequence-specificity of the two stem loops for the silencing function. The length and sequence of the spacer separating the core regions is less important (165). At least 5.5 copies of the *A*-repeat unit, replacing the endogenous *A*-repeat region, are necessary to confer silencing function (165). Higher *A*-repeat copy number improved the efficiency of silencing only slightly. Since no silencing at all is observed with 4 copies, it is very likely that proteins mediating the silencing activity of *Xist*

RNA interact with the stem loops of different A-repeat units in a cooperative fashion.

Based on the high degree of conservation of the A-repeat sequence between $Xist$ and $XIST$ RNA, proteins interacting with the A-repeat may be conserved between mouse and human. The ability of mouse A-repeat-interacting proteins to bind the human A-repeat and mediate silencing is suggested by the fact that $XIST$ RNA expressed from human XIC transgenes in mouse cells can mediate silencing of linked autosomal genes (58, 101). Two groups have identified proteins that interact with the A-repeat in vitro. Using an entire A-repeat region, hnRNP C1/2 were isolated from nuclear extracts (16). It is perhaps not surprising that two proteins with affinity for polyU stretches, which are highly enriched in the A-repeat spacer regions, were isolated. In a second approach, a synthetic RNA oligonucleotide comprising just the first stem loop bound a 120-kD protein of unknown identity (1). Given that the secondary structure rather than the primary sequence of first stem loop is crucial for A-repeat function, it is unclear whether specificity will lie within this region of the A-repeat. Clearly, identification of the hypothesized A-repeat-interacting proteins would provide insight into the mechanism of A-repeat-mediated silencing.

Developmental regulation of A-repeat-interacting proteins could explain how $Xist$ RNA can induce silencing only during the initiation phase of X-inactivation (164). Alternatively, hypothetical A-repeat-interacting proteins may be present throughout development, but changes in chromatin structure that occur upon differentiation could interfere with their silencing activity. Of the $Xist$ RNA regions required for Xi association, the A-repeat region could contribute indirectly to localization, since A-repeat-interacting proteins may bind the Xi-chromatin to mediate their silencing function. Thus, the $Xist$ RNA domains required for coating and silencing could be non-overlapping. It will also be of interest to determine whether the A-repeat is required for $Xist$ RNA's role in Xa choice or maintenance of the Xi, which would implicate the silencing activity of $Xist$ RNA as crucial in these processes.

Wutz and colleagues found two large deletions in the $3'$ half of $Xist$ RNA that still allow spreading along the chromosome but are deficient in macroH2A recruitment (165) (Figure 9). The deletion of the A-repeat region has no effect on macroH2A localization, indicating that the $Xist$ RNA-mediated accumulation of macroH2A on the Xi occurs independently of $Xist$ RNA's silencing function. This finding further supported the idea that macroH2A is not involved in the initiation or establishment of silencing but could play a role in the maintenance of the silent state of the Xi (96, 126, 127). It remains possible that the macroH2A histone variants could mediate $Xist$ RNA's role in the maintenance of the Xi heterochromatin. $Xist$ cDNA transgenes lacking domains required for macroH2A colocalization should be tested for their ability to confer $Xist$ RNA's Xi maintenance function.

Taken together, initial analysis of the sequence requirements for $Xist$ RNA function indicates that the RNA can be divided into separate domains conferring the activities of chromosome coating in cis, gene silencing, and macroH2A recruitment. The modular nature of *Xist* RNA suggests that the RNA may act as a scaffold to coordinate multiple functions.

CONCLUSION

The different fates taken on by the two X chromosomes in female mammals have long been attributed to the action of a single chromosomal locus, the *X-inactivation center*, which responds to developmental cues orchestrating the X-inactivation process. *Xist* RNA, encoded within the *Xic*, has turned out to be the pivotal player in X-inactivation, as it is both necessary and sufficient for initiation and spread of silencing on the Xi. *Cis*-regulatory elements of the *Xist* gene direct its expression and control the RNA's activity during the period in which the Xa chromosome is chosen. The existence of a critical sequence required to choose an Xa, the counting element, has been inferred, but this sequence remains to be identified. Antisense transcription of *Xist* has recently been identified, generating *Tsix* RNA. Both *Xist* RNA and *Tsix* transcription have been shown to regulate the random choice of the Xa. *Cis*-limited control of *Xist* activity ensures that the Xa will be protected from *Xist* RNA-mediated silencing during the period in which *Xist* activity becomes upregulated on the future Xa. In imprinted X-inactivation, gametic imprints control *Tsix* and *Xist* expression patterns, which direct the maternal Xa fate and the paternal Xi fate. Most advances have been made through the study of X-inactivation in the mouse. Early evidence suggests that the process of Xa choice during human X-inactivation may exhibit some differences.

Silencing of the X chromosome by *Xist* RNA is a multistep process that can only be initiated during a brief developmental window. *Xist* RNA coordinates multiple chromosome-wide chromatin modifications, possibly by recruiting modifiers of chromatin structure. These modifications, characteristic of heterochromatin, stably maintain the silent state of the Xi. Out of all known modifications, methylation on histone H3 Lys-9 spreads on the Xi concomitant with coating by *Xist* RNA, immediately before gene silencing. The spread of methylation on H3 Lys-9, which is involved in transcriptional repression in other contexts, suggests that *Xist* RNA may initiate silencing through recruitment of a histone methyltransferase. Determination of the composition and function of the presumed *Xist* ribonucleoprotein complex will ultimately allow an understanding of how the mammalian dosage compensation system initiates, establishes, and maintains silencing on the Xi.

No *Xic* transgene analyzed to date directs truly random X-inactivation, suggesting that researchers must look beyond the known *Xic* elements, and outside the largest transgene, for sequences that control random choice of the active X and ensure enactment of the inactive X fate. Despite significant advances in the genetics and molecular biology of X-inactivation, many challenges remain before a comprehensive view of the molecular mechanisms directing the two diverge fates of the mammalian X chromosome will emerge.

ACKNOWLEDGMENTS

The authors would like to thank Gail Martin, Angela Andersen, Hannah Cohen, Cecile de la Cruz, Faraz Farzin, Oliver Liu, Dale Talbot, Dennis Wykoff, and Gyorgyi Csankovszki for critical reading of the manuscript. The authors extend special thanks to D. S. T. for his patience during the manuscript preparation process. B. P. is funded by HHMI Research Resources Grant 76296-549901, NIH grant GM 63671-01, and a grant from the Sandler Family Supporting Foundation. K. P. is an O'Donnell Foundation Fellow of the Life Sciences Research Foundation. S. M.-E. is supported by a NSF graduate fellowship.

The *Annual Review of Genetics* is online at http://genet.annualreviews.org

LITERATURE CITED

1. Allaman-Pillet N, Djemai A, Bonny C, Schorderet DF. 2000. The 5′ repeat elements of the mouse *Xist* gene inhibit the transcription of X-linked genes. *Gene Expr.* 9:93–101

2. Ariel M, Robinson E, McCarrey JR, Cedar H. 1995. Gamete-specific methylation correlates with imprinting of the murine *Xist* gene. *Nat. Genet.* 9:312–15

3. Avner R, Wahrman J, Richler C, Ayoub N, Friedmann A, et al. 2000. X inactivation-specific transcript expression in mouse oocytes and zygotes. *Mol. Hum. Reprod.* 6:591–94

4. Ayoub N, Richler C, Wahrman J. 1997. *Xist* RNA is associated with the transcriptionally inactive XY body in mammalian male meiosis. *Chromosoma* 106: 1–10

5. Bannister AJ, Zegerman P, Partridge JF, Miska EA, Thomas JO, et al. 2001. Selective recognition of methylated lysine 9 on histone H3 by the HP1 chromo domain. *Nature* 410:120–24

6. Beard C, Li E, Jaenisch R. 1995. Loss of methylation activates *Xist* in somatic but not in embryonic cells. *Genes Dev.* 9:2325–34

7. Beletskii A, Hong YK, Pehrson J, Egholm M, Strauss WM. 2001. PNA interference mapping demonstrates functional domains in the noncoding RNA

Xist. Proc. Natl. Acad. Sci. USA 98: 9215–20

8. Bell AC, Felsenfeld G. 2000. Methylation of a CTCF-dependent boundary controls imprinted expression of the Igf2 gene. *Nature* 405:482–85

9. Bell AC, West AG, Felsenfeld G. 2001. Insulators and boundaries: versatile regulatory elements in the eukaryotic. *Science* 291:447–50

10. Belyaev N, Keohane AM, Turner BM. 1996. Differential underacetylation of histones H2A, H3 and H4 on the inactive X chromosome in human female cells. *Hum. Genet.* 97:573–78

11. Boggs BA, Allis CD, Chinault AC. 2000. Immunofluorescent studies of human chromosomes with antibodies against phosphorylated H1 histone. *Chromosoma* 108:485–90

12. Boggs BA, Cheung P, Heard E, Spector DL, Chinault AC, Allis CD. 2002. Differentially methylated forms of histone H3 show unique association patterns with inactive human X chromosomes. *Nat. Genet.* 30:73–76

13. Borsani G, Tonlorenzi R, Simmler MC, Dandolo L, Arnaud D, et al. 1991. Characterization of a murine gene expressed from the inactive X chromosome. *Nature* 351:325–29

14. Brockdorff N, Ashworth A, Kay GF,

Cooper P, Smith S, et al. 1991. Conservation of position and exclusive expression of mouse *Xist* from the inactive X chromosome. *Nature* 351:329–31

15. Brockdorff N, Ashworth A, Kay GF, McCabe VM, Norris DP, et al. 1992. The product of the mouse *Xist* gene is a 15 kb inactive X-specific transcript containing no conserved ORF and located in the nucleus. *Cell* 71:515–26

16. Brown CJ, Baldry SE. 1996. Evidence that heteronuclear proteins interact with *XIST* RNA in vitro. *Somat. Cell Mol. Genet.* 22:403–17

17. Brown CJ, Ballabio A, Rupert JL, Lafreniere RG, Grompe M, et al. 1991. A gene from the region of the human X inactivation centre is expressed exclusively from the inactive X chromosome. *Nature* 349:38–44

18. Brown CJ, Hendrich BD, Rupert JL, Lafreniere RG, Xing Y, et al. 1992. The human *XIST* gene: analysis of a 17 kb inactive X-specific RNA that contains conserved repeats and is highly localized within the nucleus. *Cell* 71:527–42

19. Brown CJ, Willard HF. 1994. The human X-inactivation centre is not required for maintenance of X-chromosome inactivation. *Nature* 368:154–56

20. Brown SD. 1991. *XIST* and the mapping of the X chromosome inactivation centre. *BioEssays* 13:607–12

21. Carr DH. 1971. Chromosome studies in selected spontaneous abortions. Polyploidy in man. *J. Med. Genet.* 8:164–74

22. Carrel L, Clemson CM, Dunn JM, Miller AP, Hunt PA, et al. 1996. X inactivation analysis and DNA methylation studies of the ubiquitin activating enzyme E1 and PCTAIRE-1 genes in human and mouse. *Hum. Mol. Genet.* 5:391–401

23. Carrel L, Cottle AA, Goglin KC, Willard HF. 1999. A first-generation X-inactivation profile of the human X chromosome. *Proc. Natl. Acad. Sci. USA* 96: 14440–44

24. Chadwick BP, Valley CM, Willard HF.

2001. Histone variant macroH2A contains two distinct macrochromatin domains capable of directing macroH2A to the inactive X chromosome. *Nucleic Acids Res.* 29:2699–705

25. Chadwick BP, Willard HF. 2001. Histone H2A variants and the inactive X chromosome: identification of a second macroH2A variant. *Hum. Mol. Genet.* 10:1101–13

26. Chadwick BP, Willard HF. 2001. A novel chromatin protein, distantly related to histone H2A, is largely excluded from the inactive X chromosome. *J. Cell Biol.* 152:375–84

27. Chao W, Huynh KD, Spencer RJ, Davidow LS, Lee JT. 2002. CTCF, a candidate trans-acting factor for X-inactivation choice. *Science* 295:345–47

28. Chapman V, Forrester L, Sanford J, Hastie N, Rossant J. 1984. Cell lineage-specific undermethylation of mouse repetitive DNA. *Nature* 307:284–86

29. Chung JH, Whiteley M, Felsenfeld G. 1993. A 5' element of the chicken beta-globin domain serves as an insulator in human erythroid cells and protects against position effect in Drosophila. *Cell* 74:505–14

30. Clemson CM, Chow JC, Brown CJ, Lawrence JB. 1998. Stabilization and localization of *Xist* RNA are controlled by separate mechanisms and are not sufficient for X inactivation. *J. Cell Biol.* 142:13–23

31. Clemson CM, McNeil JA, Willard HF, Lawrence JB. 1996. *XIST* RNA paints the inactive X chromosome at interphase: evidence for a novel RNA involved in nuclear/chromosome structure. *J. Cell Biol.* 132:259–75

32. Clerc P, Avner P. 1998. Role of the region 3' to *Xist* exon 6 in the counting process of X-chromosome inactivation [see comments]. *Nat. Genet.* 19:249–53

33. Cline TW, Meyer BJ. 1996. Vive la différence: males vs females in flies vs worms. *Annu. Rev. Genet.* 30:637–702

34. Costanzi C, Pehrson JR. 1998. Histone macroH2A1 is concentrated in the inactive X chromosome of female mammals. *Nature* 393:599–601

35. Costanzi C, Pehrson JR. 2001. MACRO-H2A2, a new member of the MAR-COH2A core histone family. *J. Biol. Chem.* 276:21776–84

36. Costanzi C, Stein P, Worrad DM, Schultz RM, Pehrson JR. 2000. Histone macroH2A1 is concentrated in the inactive X chromosome of female preimplantation mouse embryos. *Development* 127:2283–89

37. Courtier B, Heard E, Avner P. 1995. Xce haplotypes show modified methylation in a region of the active X chromosome lying 3′ to *Xist*. *Proc. Natl. Acad. Sci. USA* 92:3531–35

38. Csankovszki G, Nagy A, Jaenisch R. 2001. Synergism of *Xist* RNA, DNA methylation, and histone hypoacetylation in maintaining X chromosome inactivation. *J. Cell Biol.* 153:773–84

39. Csankovszki G, Panning B, Bates B, Pehrson JR, Jaenisch R. 1999. Conditional deletion of *Xist* disrupts histone macroH2A localization but not maintenance of X inactivation [letter]. *Nat. Genet.* 22:323–24

40. Daniels R, Zuccotti M, Kinis T, Serhal P, Monk M. 1997. *XIST* expression in human oocytes and preimplantation embryos. *Am. J. Hum. Genet.* 61:33–39

41. Debrand E, Chureau C, Arnaud D, Avner P, Heard E. 1999. Functional analysis of the DXPas34 locus, a 3′ regulator of *Xist* expression. *Mol. Cell. Biol.* 19:8513–25

42. Debrand E, Heard E, Avner P. 1998. Cloning and localization of the murine Xpct gene: evidence for complex rearrangements during the evolution of the region around the *Xist* gene. *Genomics* 48:296–303

43. Disteche CM. 1995. Escape from X inactivation in human and mouse. *Trends Genet.* 11:17–22

44. Duthie SM, Nesterova TB, Formstone EJ, Keohane AM, Turner BM, et al. 1999. *Xist* RNA exhibits a banded localization on the inactive X chromosome and is excluded from autosomal material in cis. *Hum. Mol. Genet.* 8:195–204

45. Eggan K, Akutsu H, Hochedlinger K, Rideout W 3rd, Yanagimachi R, Jaenisch R. 2000. X-Chromosome inactivation in cloned mouse embryos. *Science* 290:1578–81

46. Endo S, Takagi N, Sasaki M. 1982. The late-replicating X chromosome in digynous mouse triploid embryos. *Dev. Genet.* 3:165–76

47. Gilbert SL, Pehrson JR, Sharp PA. 2000. *XIST* RNA associates with specific regions of the inactive X chromatin. *J. Biol. Chem.* 275:36491–94

48. Goto T, Wright E, Monk M. 1997. Paternal X-chromosome inactivation in human trophoblastic cells. *Mol. Hum. Reprod.* 3:77–80

49. Hansen RS, Canfield TK, Stanek AM, Keitges EA, Gartler SM. 1998. Reactivation of *XIST* in normal fibroblasts and a somatic cell hybrid: abnormal localization of *XIST* RNA in hybrid cells. *Proc. Natl. Acad. Sci. USA* 95:5133–38

50. Hansen RS, Stoger R, Wijmenga C, Stanek AM, Canfield TK, et al. 2000. Escape from gene silencing in ICF syndrome: evidence for advanced replication time as a major determinant. *Hum. Mol. Genet.* 9:2575–87

51. Hansen RS, Wijmenga C, Luo P, Stanek AM, Canfield TK, et al. 1999. The DNMT3B DNA methyltransferase gene is mutated in the ICF immunodeficiency syndrome. *Proc. Natl. Acad. Sci. USA* 96:14412–17

52. Hark AT, Schoenherr CJ, Katz DJ, Ingram RS, Levorse JM, Tilghman SM. 2000. CTCF mediates methylation-sensitive enhancer-blocking activity at the H19/Igf2 locus. *Nature* 405:486–89

53. Harnden DG. 1961. Nuclear sex in triploid XXY human cells. *Lancet* ii:488

54. Harrison KB. 1989. X-Chromosome inactivation in the human cytotrophoblast. *Cytogenet. Cell Genet.* 52:37–41

55. Hartshorn C, Rice JE, Wangh LJ. 2002. Developmentally-regulated changes of *Xist* RNA levels in single preimplantation mouse embryos, as revealed by quantitative real-time PCR. *Mol. Reprod. Dev.* 61:425–36

56. Heard E, Clerc P, Avner P. 1997. X-chromosome inactivation in mammals. *Annu. Rev. Genet.* 31:571–610

57. Heard E, Mongelard F, Arnaud D, Avner P. 1999. *Xist* yeast artificial chromosome transgenes function as X-inactivation centers only in multicopy arrays and not as single copies. *Mol. Cell. Biol.* 19: 3156–66

58. Heard E, Mongelard F, Arnaud D, Chureau C, Vourc'h C, Avner P. 1999. Human *XIST* yeast artificial chromosome transgenes show partial X inactivation center function in mouse embryonic stem cells. *Proc. Natl. Acad. Sci. USA* 96:6841–46

59. Heard E, Rougeulle C, Arnaud D, Avner P, Allis CD, Spector DL. 2001. Methylation of histone H3 at Lys-9 is an early mark on the X chromosome during X inactivation. *Cell* 107:727–38

60. Hendrich BD, Brown CJ, Willard HF. 1993. Evolutionary conservation of possible functional domains of the human and murine *XIST* genes. *Hum. Mol. Genet.* 2:663–72

61. Hendrich BD, Plenge RM, Willard HF. 1997. Identification and characterization of the human *XIST* gene promoter: implications for models of X chromosome inactivation. *Nucleic Acids Res.* 25:2661–71

62. Herzing LB, Romer JT, Horn JM, Ashworth A. 1997. *Xist* has properties of the X-chromosome inactivation centre [see comments]. *Nature* 386:272–75

63. Hong YK, Ontiveros SD, Chen C, Strauss WM. 1999. A new structure for the murine *Xist* gene and its relationship to chromosome choice/counting during X-chromosome inactivation. *Proc. Natl. Acad. Sci. USA* 96:6829–34

64. Hong YK, Ontiveros SD, Strauss WM. 2000. A revision of the human *XIST* gene organization and structural comparison with mouse *Xist*. *Mamm. Genome* 11:220–24

65. Huynh KD, Lee JT. 2001. Imprinted X inactivation in eutherians: a model of gametic execution and zygotic relaxation. *Curr. Opin. Cell Biol.* 13:690–97

66. Jaenisch R. 1997. DNA methylation and imprinting: why bother? [see comments]. *Trends Genet.* 13:323–29

67. Jenuwein T, Allis CD. 2001. Translating the histone code. *Science* 293:1074–80

68. Jeppesen P, Turner BM. 1993. The inactive X chromosome in female mammals is distinguished by a lack of histone H4 acetylation, a cytogenetic marker for gene expression. *Cell* 74:281–89

69. Johnston CM, Nesterova TB, Formstone EJ, Newall AE, Duthie SM, et al. 1998. Developmentally regulated *Xist* promoter switch mediates initiation of X inactivation. *Cell* 94:809–17

70. Kay GF, Penny GD, Patel D, Ashworth A, Brockdorff N, Rastan S. 1993. Expression of *Xist* during mouse development suggests a role in the initiation of X chromosome inactivation. *Cell* 72:171–82

71. Kelley RL, Kuroda MI. 2000. Noncoding RNA genes in dosage compensation and imprinting. *Cell* 103:9–12

72. Kellum R, Schedl P. 1992. A group of scs elements function as domain boundaries in an enhancer-blocking assay. *Mol. Cell. Biol.* 12:2424–31

73. Keohane AM, O'Neill LP, Belyaev ND, Lavender JS, Turner BM. 1996. X-Inactivation and histone H4 acetylation in embryonic stem cells. *Dev. Biol.* 180:618–30

74. Lachner M, O'Carroll D, Rea S, Mechtler K, Jenuwein T. 2001. Methylation of histone H3 lysine 9 creates

binding site for HP1 proteins. *Nature* 410:116–20

75. Latham KE, Patel B, Bautista FD, Hawes SM. 2000. Effects of X chromosome number and parental origin on X-linked gene expression in preimplantation mouse embryos. *Biol. Reprod.* 63: 64–73

76. Latham KE, Rambhatla L. 1995. Expression of X-linked genes in androgenetic, gynogenetic, and normal mouse preimplantation embryos. *Dev. Genet.* 17:212–22

77. Lee JT. 2000. Disruption of imprinted X inactivation by parent-of-origin effects at *Tsix. Cell* 103:17–27

78. Lee JT, Davidow LS, Warshawsky D. 1999. *Tsix*, a gene antisense to *Xist* at the X-inactivation centre [see comments]. *Nat. Genet.* 21:400–4

79. Lee JT, Jaenisch R. 1997. Long-range cis effects of ectopic X-inactivation centres on a mouse autosome [see comments]. *Nature* 386:275–79

80. Lee JT, Lu N. 1999. Targeted mutagenesis of *Tsix* leads to nonrandom X inactivation. *Cell* 99:47–57

81. Lee JT, Lu N, Han Y. 1999. Genetic analysis of the mouse X inactivation center defines an 80–kb multifunction domain. *Proc. Natl. Acad. Sci. USA* 96:3836–41

82. Lee JT, Strauss WM, Dausman JA, Jaenisch R. 1996. A 450 kb transgene displays properties of the mammalian X-inactivation center. *Cell* 86:83–94

83. Li T, Vu TH, Zeng ZL, Nguyen BT, Hayward BE, et al. 2000. Tissue-specific expression of antisense and sense transcripts at the imprinted Gnas locus. *Genomics* 69:295–304

84. Lock LF, Takagi N, Martin GR. 1987. Methylation of the Hprt gene on the inactive X occurs after chromosome inactivation. *Cell* 48:39–46

85. Luikenhuis S, Wutz A, Jaenisch R. 2001. Antisense transcription through the *Xist* locus mediates *Tsix* function in embryonic stem cells. *Mol. Cell. Biol.* 21:8512–20

86. Lyon MF. 1961. Gene action in the X-chromosome of the mouse (*Mus musculus* L). *Nature* 190:372–73

87. Lyon MF. 1998. X-chromosome inactivation: a repeat hypothesis. *Cytogenet. Cell Genet.* 80:133–37

88. MacDonald M, Hassold T, Harvey J, Wang LH, Morton NE, Jacobs P. 1994. The origin of 47,XXY and 47,XXX aneuploidy: heterogeneous mechanisms and role of aberrant recombination. *Hum. Mol. Genet.* 3:1365–71

89. Marahrens Y, Loring J, Jaenisch R. 1998. Role of the *Xist* gene in X chromosome choosing. *Cell* 92:657–64

90. Marahrens Y, Panning B, Dausman J, Strauss W, Jaenisch R. 1997. *Xist*-deficient mice are defective in dosage compensation but not spermatogenesis [see comments]. *Genes Dev.* 11:156–66

91. Matsui J, Goto Y, Takagi N. 2001. Control of *Xist* expression for imprinted and random X chromosome inactivation in mice. *Hum. Mol. Genet.* 10:1393–401

92. Matsuura S, Episkopou V, Hamvas R, Brown SD. 1996. *Xist* expression from an *Xist* YAC transgene carried on the mouse Y chromosome. *Hum. Mol. Genet.* 5:451–59

93. McCarrey JR, Dilworth DD. 1992. Expression of *Xist* in mouse germ cells correlates with X-chromosome inactivation. *Nat. Genet.* 2:200–3

94. McDonald LE, Paterson CA, Kay GF. 1998. Bisulfite genomic sequencing-derived methylation profile of the *Xist* gene throughout early mouse development. *Genomics* 54:379–86

95. Memili E, Hong YK, Kim DH, Ontiveros SD, Strauss WM. 2001. Murine *Xist* RNA isoforms are different at their 3′ ends: a role for differential polyadenylation. *Gene* 266:131–37

96. Mermoud JE, Costanzi C, Pehrson JR, Brockdorff N. 1999. Histone macro-H2A1.2 relocates to the inactive X

chromosome after initiation and propagation of X-inactivation. *J. Cell Biol.* 147:1399–408

97. Mermoud JE, Popova B, Peters AH, Jenuwein T, Brockdorff N. 2002. Histone h3 lysine 9 methylation occurs rapidly at the onset of random X chromosome inactivation. *Curr. Biol.* 12:247–51

98. Mermoud JE, Tassin AM, Pehrson JR, Brockdorff N. 2001. Centrosomal association of histone macroH2A1.2 in embryonic stem cells and somatic cells. *Exp. Cell Res.* 268:245–51

99. Migeon BR. 1998. Non-random X chromosome inactivation in mammalian cells. *Cytogenet Cell Genet.* 80:142–48

100. Migeon BR, Chowdhury AK, Dunston JA, McIntosh I. 2001. Identification of TSIX, encoding an RNA antisense to human *XIST*, reveals differences from its murine counterpart: implications for X inactivation. *Am. J. Hum. Genet.* 69:951–60

101. Migeon BR, Winter H, Kazi E, Chowdhury AK, Hughes A, et al. 2001. Low-copy-number human transgene is recognized as an X inactivation center in mouse ES cells, but fails to induce cis-inactivation in chimeric mice. *Genomics* 71:156–62

102. Mise N, Goto Y, Nakajima N, Takagi N. 1999. Molecular cloning of antisense transcripts of the mouse *Xist* gene. *Biochem. Biophys. Res. Commun.* 258:537–41

103. Mlynarczyk SK, Panning B. 2000. X inactivation: *Tsix* and *Xist* as yin and yang. *Curr. Biol.* 10:R899–903

104. Mohandas T, Sparkes RS, Shapiro LJ. 1981. Reactivation of an inactive human X chromosome: evidence for X inactivation by DNA methylation. *Science* 211:393–96

105. Monk M, Harper MI. 1979. Sequential X chromosome inactivation coupled with cellular differentiation in early mouse embryos. *Nature* 281:311–13

106. Morey C, Arnaud D, Avner P, Clerc P.

2001. *Tsix*-mediated repression of *Xist* accumulation is not sufficient for normal random X inactivation. *Hum. Mol. Genet.* 10:1403–11

107. Nesterova TB, Barton SC, Surani MA, Brockdorff N. 2001. Loss of *Xist* imprinting in diploid parthenogenetic preimplantation embryos. *Dev. Biol.* 235:343–50

108. Nesterova TB, Slobodyanyuk SY, Elisaphenko EA, Shevchenko AI, Johnston C, et al. 2001. Characterization of the genomic *Xist* locus in rodents reveals conservation of overall gene structure and tandem repeats but rapid evolution of unique sequence. *Genome Res.* 11:833–49

109. Newall AE, Duthie S, Formstone E, Nesterova T, Alexiou M, et al. 2001. Primary non-random X inactivation associated with disruption of *Xist* promoter regulation. *Hum. Mol. Genet.* 10:581–89

110. Noma K, Allis CD, Grewal SI. 2001. Transitions in distinct histone H3 methylation patterns at the heterochromatin domain boundaries. *Science* 293:1150–55

111. Norris DP, Patel D, Kay GF, Penny GD, Brockdorff N, et al. 1994. Evidence that random and imprinted *Xist* expression is controlled by preemptive methylation. *Cell* 77:41–51

112. Okamoto I, Tan S, Takagi N. 2000. X-chromosome inactivation in XX androgenetic mouse embryos surviving implantation. *Development* 127:4137–45

113. Panning B, Dausman J, Jaenisch R. 1997. X chromosome inactivation is mediated by *Xist* RNA stabilization. *Cell* 90:907–16

114. Panning B, Jaenisch R. 1996. DNA hypomethylation can activate *Xist* expression and silence X-linked genes. *Genes Dev.* 10:1991–2002

115. Paterno GD, McBurney MW. 1985. X chromosome inactivation during induced differentiation of a female mouse embryonal carcinoma cell line. *J. Cell Sci.* 75:149–63

116. Pehrson JR, Fried VA. 1992. MacroH2A, a core histone containing a large nonhistone region. *Science* 257:1398–400

117. Penny GD, Kay GF, Sheardown SA, Rastan S, Brockdorff N. 1996. Requirement for *Xist* in X chromosome inactivation [see comments]. *Nature* 379:131–37

118. Perche PY, Vourc'h C, Konecny L, Souchier C, Robert-Nicoud M, et al. 2000. Higher concentrations of histone macroH2A in the Barr body are correlated with higher nucleosome density. *Curr. Biol.* 10:1531–34

119. Peters AH, Mermoud JE, O'Carroll D, Pagani M, Schweizer D, et al. 2002. Histone H3 lysine 9 methylation is an epigenetic imprint of facultative heterochromatin. *Nat. Genet.* 30:77–80

120. Peters AH, O'Carroll D, Scherthan H, Mechtler K, Sauer S, et al. 2001. Loss of the Suv39h histone methyltransferases impairs mammalian heterochromatin and genome stability. *Cell* 107:323–37

121. Plasterk RH, Ketting RF. 2000. The silence of the genes. *Curr. Opin. Genet. Dev.* 10:562–67

122. Prissette M, El-Maarri O, Arnaud D, Walter J, Avner P. 2001. Methylation profiles of DXPas34 during the onset of X-inactivation. *Hum. Mol. Genet.* 10:31–38

123. Quan F, Janas J, Toth-Fejel S, Johnson DB, Wolford JK, Popovich BW. 1997. Uniparental disomy of the entire X chromosome in a female with Duchenne muscular dystrophy. *Am. J. Hum. Genet.* 60:160–65

124. Rack KA, Chelly J, Gibbons RJ, Rider S, Benjamin D, et al. 1994. Absence of the *XIST* gene from late-replicating isodicentric X chromosomes in leukaemia. *Hum. Mol. Genet.* 3:1053–59

125. Rasmussen TP, Huang T, Mastrangelo MA, Loring J, Panning B, Jaenisch R. 1999. Messenger RNAs encoding mouse histone macroH2A1 isoforms are expressed at similar levels in male and female cells and result from alternative splicing. *Nucleic Acids Res.* 27:3685–89

126. Rasmussen TP, Mastrangelo MA, Eden A, Pehrson JR, Jaenisch R. 2000. Dynamic relocalization of histone MacroH2A1 from centrosomes to inactive X chromosomes during X inactivation. *J. Cell Biol.* 150:1189–98

127. Rasmussen TP, Wutz AP, Pehrson JR, Jaenisch RR. 2001. Expression of *Xist* RNA is sufficient to initiate macrochromatin body formation. *Chromosoma* 110:411–20

128. Rastan S. 1983. Non-random X-chromosome inactivation in mouse X-autosome translocation embryos—location of the inactivation centre. *J. Embryol. Exp. Morphol.* 78:1–22

128a. Rastan S, Robertson EJ. 1985. X-chromosome deletions in embryo-derived (EK) cell lines associated with lack of X-chromosome inactivation. *J. Embryol. Exp. Morphol.* 90:378–88

129. Ray PF, Winston RM, Handyside AH. 1997. *XIST* expression from the maternal X chromosome in human male preimplantation embryos at the blastocyst stage. *Hum. Mol. Genet.* 6:1323–27

130. Reizis B, Lee JT, Leder P. 2000. Homologous genomic fragments in the mouse pre-T cell receptor alpha (pTa) and *Xist* loci. *Genomics* 63:149–52

131. Richards EJ, Elgin SC. 2002. Epigenetic codes for heterochromatin formation and silencing: rounding up the usual suspects. *Cell* 108:489–500

132. Richler C, Soreq H, Wahrman J. 1992. X inactivation in mammalian testis is correlated with inactive X-specific transcription. *Nat. Genet.* 2:192–95

133. Romer JT, Ashworth A. 2000. The upstream region of the mouse *Xist* gene contains two ribosomal protein pseudogenes. *Mamm. Genome* 11:461–63

134. Ropers HH, Wolff G, Hitzeroth HW. 1978. Preferential X inactivation in human placenta membranes: Is the paternal

X inactive in early embryonic development of female mammals? *Hum. Genet.* 43:265–73

135. Rougeulle C, Cardoso C, Fontes M, Colleaux L, Lalande M. 1998. An imprinted antisense RNA overlaps UBE3A and a second maternally expressed transcript. *Nat. Genet.* 19:15–16

136. Russell LB. 1961. Genetics of mammalian sex chromosomes. *Science* 133:1795–803

137. Russell LB. 1963. Mammalian X-chromosome action: inactivation limited in spread and region of origin. *Science* 140:976–78

138. Sado T, Fenner MH, Tan SS, Tam P, Shioda T, Li E. 2000. X inactivation in the mouse embryo deficient for Dnmt1: distinct effect of hypomethylation on imprinted and random X inactivation. *Dev. Biol.* 225:294–303

139. Sado T, Wang Z, Sasaki H, Li E. 2001. Regulation of imprinted X-chromosome inactivation in mice by *Tsix. Development* 128:1275–86

140. Salido EC, Yen PH, Mohandas TK, Shapiro LJ. 1992. Expression of the X-inactivation-associated gene *XIST* during spermatogenesis. *Nat. Genet.* 2:196–99

141. Sharp A, Robinson D, Jacobs P. 2000. Age- and tissue-specific variation of X chromosome inactivation ratios in normal women. *Hum. Genet.* 107:343–99

142. Sharp A, Robinson DO, Jacobs P. 2001. Absence of correlation between late-replication and spreading of X inactivation in an X;autosome translocation. *Hum. Genet.* 109:295–302

143. Sheardown SA, Duthie SM, Johnston CM, Newall AE, Formstone EJ, et al. 1997. Stabilization of *Xist* RNA mediates initiation of X chromosome inactivation [see comments]. *Cell* 91:99–107

144. Simmler MC, Cattanach BM, Rasberry C, Rougeulle C, Avner P. 1993. Mapping the murine Xce locus with (CA)n repeats. *Mamm. Genome* 4:523–30

145. Sleutels F, Zwart R, Barlow DP. 2002. The non-coding Air RNA is required for silencing autosomal imprinted genes. *Nature* 415:810–13

146. Stavropoulos N, Lu N, Lee JT. 2001. A functional role for *Tsix* transcription in blocking *Xist* RNA accumulation but not in X-chromosome choice. *Proc. Natl. Acad. Sci. USA* 98:10232–37

147. Sugawara O, Takagi N, Sasaki M. 1983. Allocyclic early replicating X chromosome in mice: genetic inactivity and shift into a late replicator in early embrogenesis. *Chromosoma* 88:133–38

148. Szabo P, Tang SH, Rentsendorj A, Pfeifer GP, Mann JR. 2000. Maternal-specific footprints at putative CTCF sites in the H19 imprinting control region give evidence for insulator function. *Curr. Biol.* 10:607–10

149. Tada T, Obata Y, Tada M, Goto Y, Nakatsuji N, et al. 2000. Imprint switching for non-random X-chromosome inactivation during mouse oocyte growth. *Development* 127:3101–5

150. Tada T, Tada M, Takagi N. 1993. X chromosome retains the memory of its parental origin in murine embryonic stem cells. *Development* 119:813–21

151. Takagi N. 1974. Differentiation of X chromosomes in early female mouse embryos. *Exp. Cell Res.* 86:127–35

152. Takagi N, Sasaki M. 1975. Preferential inactivation of the paternally derived X chromosome in the extraembryonic membranes of the mouse. *Nature* 256:640–42

153. Takagi N, Sugawara O, Sasaki M. 1982. Regional and temporal changes in the pattern of X-chromosome replication during the early post-implantation development of the female mouse. *Chromosoma* 85:275–86

154. Tam PP, Williams EA, Tan SS. 1994. Expression of an X-linked HMG-lacZ transgene in mouse embryos: implication of chromosomal imprinting and

lineage-specific X-chromosome activity. *Dev. Genet.* 15:491–503

155. Therman E, Sarto GE, Patau K. 1974. Center for Barr body condensation on the proximal part of the human Xq: a hypothesis. *Chromosoma* 44:361–66

156. Tinker AV, Brown CJ. 1998. Induction of *XIST* expression from the human active X chromosome in mouse/human somatic cell hybrids by DNA demethylation. *Nucleic Acids Res.* 26:2935–40

157. Torchia BS, Call LM, Migeon BR. 1994. DNA replication analysis of FMR1, *XIST*, and factor 8C loci by FISH shows nontranscribed X-linked genes replicate late. *Am. J. Hum. Genet.* 55:96–104

158. van der Vlag J, Otte AP. 1999. Transcriptional repression mediated by the human polycomb-group protein EED involves histone deacetylation. *Nat. Genet.* 23:474–78

159. Wang J, Mager J, Chen Y, Schneider E, Cross JC, et al. 2001. Imprinted X inactivation maintained by a mouse Polycomb group gene. *Nat. Genet.* 28:371–75

160. Warshawsky D, Stavropoulos N, Lee JT. 1999. Further examination of the *Xist* promoter-switch hypothesis in X inactivation: evidence against the existence and function of a P(0) promoter. *Proc. Natl. Acad. Sci. USA* 96:14424–29

161. Webb S, de Vries TJ, Kaufman MH. 1992. The differential staining pattern of the X chromosome in the embryonic and extraembryonic tissues of postimplantation homozygous tetraploid mouse embryos. *Genet. Res.* 59:205–14

162. Willard HF, Salz HK. 1997. Remodelling chromatin with RNA [news; comment]. *Nature* 386:228–29

163. Wolf SF, Jolly DJ, Lunnen KD, Friedmann T, Migeon BR. 1984. Methylation of the hypoxanthine phosphoribosyltransferase locus on the human X chromosome: implications for X-chromosome inactivation. *Proc. Natl. Acad. Sci. USA* 81:2806–10

164. Wutz A, Jaenisch R. 2000. A shift from reversible to irreversible X inactivation is triggered during ES cell differentiation. *Mol. Cell* 5:695–705

165. Wutz A, Rasmussen TP, Jaenisch R. 2002. Chromosomal silencing and localization are mediated by different domains of *Xist* RNA. *Nat. Genet.* 30:167–74

166. Zuccotti M, Boiani M, Ponce R, Guizzardi S, Scandroglio R, et al. 2002. Mouse *Xist* expression begins at zygotic genome activation and is timed by a zygotic clock. *Mol. Reprod. Dev.* 61:14–20

167. Zuccotti M, Monk M. 1995. Methylation of the mouse *Xist* gene in sperm and eggs correlates with imprinted *Xist* expression and paternal X-inactivation. *Nat. Genet.* 9:316–20

Annu. Rev. Genet. 2002. 36:279–303
doi: 10.1146/annurev.genet.36.042602.094806

ORIGINS OF SPONTANEOUS MUTATIONS: Specificity and Directionality of Base-Substitution, Frameshift, and Sequence-Substitution Mutageneses

Hisaji Maki

*Department of Molecular Biology, Graduate School of Biological Sciences,
Nara Institute of Science and Technology, Ikoma, Nara 630-0101, Japan;
e-mail: maki@bs.aist-nara.ac.jp*

Key Words replication fidelity, DNA polymerase, DNA repair, mutator, antimutator, oxygen radicals, repetitive sequence

■ **Abstract** Spontaneous mutations are derived from various sources, including errors made during replication of undamaged template DNA, mutagenic nucleotide substrates, and endogenous DNA lesions. These sources vary in their frequencies and resultant mutations, and are differently affected by the DNA sequence, DNA transactions, and cellular metabolism. Organisms possess a variety of cellular functions to suppress spontaneous mutagenesis, and the specificity and effectiveness of each function strongly affect the pattern of spontaneous mutations. Base substitutions and single-base frameshifts, two major classes of spontaneous mutations, occur non-randomly throughout the genome. Within target DNA sequences there are hotspots for particular types of spontaneous mutations; outside of the hotspots, spontaneous mutations occur more randomly and much less frequently. Hotspot mutations are attributable more to endogenous DNA lesions than to replication errors. Recently, a novel class of mutagenic pathway that depends on short inverted repeats was identified as another important source of hotspot mutagenesis.

CONTENTS

0066-4197/02/1215-0279$14.00

INTRODUCTION

Spontaneous mutations and rearrangements of chromosomes are important cellular processes that lead to alteration of the genome structure and act as engines to drive evolution. Analyses of spontaneous mutations that occur in organisms as various as bacteriophages and human cells have demonstrated that spontaneous mutations do not arise randomly in the genome sequence but rather possess a strong bias in their site distribution and in the types of alteration that they cause within the DNA sequence (19). Each type of mutation shows a pattern of hotspot and cold-spot sites in a given sequence. The site- and type-specificities of spontaneous mutations imply that spontaneous mutagenesis, to a greater or lesser degree, depends on the context of the DNA sequence. Probably, most of the events involved in spontaneous mutagenesis are affected by DNA topology, higher-order structures of DNA, and the extent of DNA transactions, such as transcription, replication, and recombination, on a given DNA sequence. A quantitative theory of directional mutation pressure suggests that the underlying mechanisms leading to mutation bias may vary among bacteria and may operate differently among different parts of the genome in vertebrates (104).

The molecular mechanisms of spontaneous mutagenesis have been elucidated by genetic analyses of mutator and antimutator genes of *Escherichia coli* and by biochemical studies of their gene products (35, 64). It has been also clearly demonstrated that the mechanisms are evolutionarily well conserved among various organisms (21, 70, 83, 84, 93, 102, 111). However, less is known about common and species-specific rules governing the specificity of spontaneous mutations. Although numerous models have been proposed for mutagenic processes that give rise to several types of mutations (24), the factors that determine the specificity and directionality of spontaneous mutations are unknown.

This review summarizes factors determining the specificity of spontaneous mutations, focusing on three major classes of point mutations. The class distribution of spontaneous mutations in several target DNA sequences is also briefly addressed. Finally, possible causes of hotspot mutagenesis are discussed for each type of spontaneous mutation.

PREMUTAGENIC DAMAGE AND PREMUTATIONS INVOLVED IN SPONTANEOUS MUTAGENESIS

As indicated in the mutation theory established from studies of UV- and chemical mutagenesis (24), every mutation is derived from premutagenic damage of DNA, which causes a mutagenic intermediate upon misreplication by the normal replicative apparatus or a special kind of DNA polymerase that participates in translesion DNA synthesis (TLS) (Figure 1). The mutagenic intermediate, termed a premutation in this review, is converted to a mutation at the next round of DNA replication. Premutagenic damage leading to spontaneous mutations is produced by a variety of factors present in normally growing cells. The major causes of premutagenic damage are (*a*) erroneous action of the replicative apparatus during DNA replication with an intact DNA template and the usual dNTPs; (*b*) misinsertion of a mutagenic nucleotide, which has a loose specificity of base-pairing, during DNA replication; and (*c*) chemical reactions by endogenous mutagens, such as active oxygen species, and spontaneous decomposition of DNA bases.

Replication Errors Made During Normal DNA Synthesis

Of the various types of replication errors caused by erroneous action of the replicative apparatus, three types have been well characterized: a single-base mispair leading to a base substitution, a single-base bulge leading to a single-base frameshift, and a multiple-base mismatch leading to a sequence substitution (Figure 2). Despite their low frequencies, these replication errors have been detected in various kinds of in vitro DNA synthesis assays using replicative DNA polymerases (5, 8, 10, 25, 39, 45, 66, 89, 101, 108). Since the replication errors consist only of undamaged nucleotides, these premutagenic damages are referred to as "native replication errors." At the next round of DNA replication, the native replication error would readily be fixed into a mutation. Therefore, the native replication error is a premutation as well as premutagenic damage in itself.

The kinetics of misinsertion that initiates the formation of a mispair has been extensively characterized using the α catalytic subunit alone or the holoenzyme lacking the ε editing subunit of DNA polymerase III of *E. coli* (5, 101). The estimated rate of misinsertion by the polymerase is about 1×10^{-4} and 1×10^{-5}/base/replication for transition and transversion types of terminal mispair,

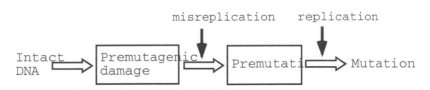

Figure 1   Mutation theory.

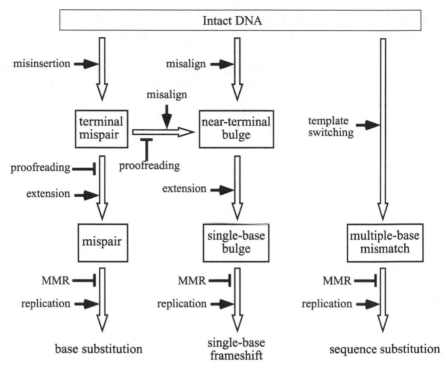

Figure 2   Three mutagenic pathways that begin with native replication errors. MMR; mismatch repair.

respectively. Proper extension of the terminal mispair is an important process for making the mispair effective in mutagenesis, and one that competes with the examination of the newly incorporated nucleotide by the proofreading function (47). Most of the terminal mispairs would not survive during DNA synthesis by the proofreading-proficient replicative DNA polymerase. Thus, the rate of formation of actual mispairs is 50- to 100-fold lower than that for terminal mispairs (5).

Misalignment of the growing chain with the template sequence may occur via simple slippage of the terminus, which promotes the formation of a single-base bulge. Insertion or deletion of a single or multiple unit of dinucleotide repeats may involve the simple slippage event during DNA synthesis, albeit much less frequently than an event that causes a single-base frameshift (46). Simple slippage events may be facilitated by spontaneous breathing of the terminus. Since duplex DNA undergoes such breathing only when the DNA polymerase is dissociated from the primer-template, the breathing-misalignment process seems hardly to occur during the processive DNA synthesis of chromosomes (48). The terminal mispair potentially stimulates misalignment, but most of the misinsertion-misalignment process is strongly suppressed by the proofreading function (2, 5, 81). Another possible source of slippage events is the improper reattachment of the primer terminus back to the template during the course of the proofreading process (25). Evidence for this notion was recently provided from genetic analysis of a novel *pol3* mutant of *Saccharomyces cerevisiae* that shows an antimutator phenotype for a single-base frameshift but not for a base substitution (29). A significant portion of single-base bulges therefore probably results from error in the proofreading process. An extension process for the near-terminal bulge is thought to be important for completing the misalignment (99).

Template switching, another type of erroneous action by the replicative apparatus, potentially generates multiple-base mismatch. This third category of native replication error has been proposed as a mutagenic process leading to sequence substitution, a particular class of spontaneous mutation in which a segment of DNA ranging from 2 to about 20 nucleotides is replaced by a completely different sequence (65, 120) (Figure 3A). The sequence substitution, sometimes

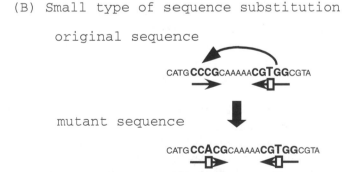

Figure 3 Structural aspects of sequence substitution. Arrows with boxes indicate pseudo-inverted repeats. Open and closed boxes are nonhomologous sequences in the repeats and correspond to replaced and substituted sequences, respectively.

called quasipalindromic-dependent mutation or mistemplated mutation, was first described as a kind of frameshift mutation almost 20 years ago (18). Systematic analyses of many independent sequence substitutions that occur in cells of *E. coli* revealed structural characteristics unique to this class of mutations (120). Based on the definitive requirement of pseudo-inverted repeats for sequence substitution mutagenesis, intra- or interstrand template switching between the repeats was proposed for specific mutagenesis (65, 87, 88, 91, 92, 120) (Figure 4). The involvement of replicative DNA polymerase in sequence substitution mutagenesis both in vivo (65) and in vitro (25) and a significant suppression of mutagenesis by the *mutS*-dependent mismatch repair system (120) support the notion that the multiple-base mismatch is a kind of native replication error. However, the molecular mechanisms of template switching are poorly understood. Note that template switching between pseudo-inverted repeats causes not only sequence substitution but also base substitution and single base frameshift (Figure 3B).

Spontaneous DNA Lesions and Mutagenic Substrate Nucleotides

The nature of native replication errors caused solely by action of the replicative apparatus has been extensively characterized, whereas less is known about spontaneous DNA lesion that potentially induces spontaneous mutations (24). Active oxygen species are produced in aerobically growing cells and attack DNA to produce a wide range of lesions. An estimated 3000–5000 oxidative DNA lesions/cell/generation are produced in cells of *E. coli* under normal aerobic growth conditions (78). Free nucleotides are attacked more efficiently by oxygen radicals than DNA, and in a number of different species, oxidized nucleotides are produced in the cellular nucleotide pool. Among them, 8-oxodGTP (58) and 2-OHdATP (37) possess extraordinarily strong mutagenicity. Considering their high production rate, these mutagenic compounds are potentially the most powerful source of spontaneous mutations. Methylation and hydrolytic decomposition of DNA such as depurination and cytosine deamination cause a variety of endogenous DNA lesions (24). However, the spontaneous frequency of these events is estimated to be much lower than that for the oxidation of DNA. There are only three types of native replication error, essentially no effect on the progression of DNA synthesis, whereas spontaneous DNA lesions contain a wide variety of species that differ substantially in their effects on DNA synthesis. For example, uracil as a consequence of cytosine deamination has no effect on DNA synthesis, whereas abasic DNA damage completely blocks DNA synthesis.

The frequency of mutations produced in in vitro *oriC* plasmid DNA replication, $0.5-0.75 \times 10^{-6}$/bp/replication, may be the best estimate of the net frequency of native replication error made by the replicative apparatus of *E. coli* (25). Based on the rate of native replication error and the size of the genome of *E. coli*, 4 to 6 native replication errors are expected to be produced per chromosome per DNA replication in the organism. Thus, in aerobically growing cells, the extent of

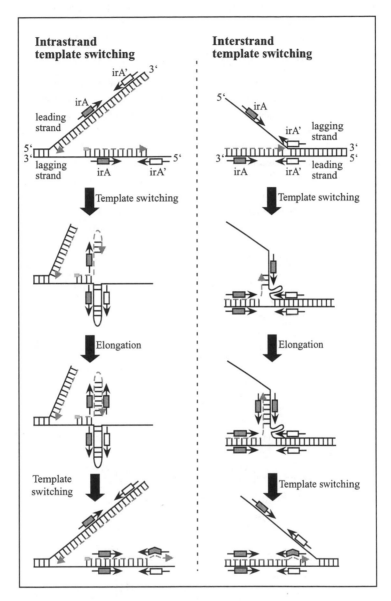

Figure 4   Two models for sequence substitution mutagenesis. Arrows with boxes indicate pseudo-inverted repeats, irA and irA'. Intrastrand template switching occurs when the lagging strand synthesis pauses within the irA', while interstrand template switching occurs when the leading strand synthesis pauses within the irA'. After a short DNA-chain elongation that copies the irA, the growing 3'-OH terminus is positioned back to the authentic template and continues DNA synthesis.

spontaneous DNA lesion seems to be about 1000-fold higher than that of native replication errors.

Replication Errors Induced by Damaged Nucleotides

There are two additional pathways in which replication errors are induced during DNA synthesis by the replicative apparatus (Figure 5). One is the misinsertion of a mutagenic nucleotide. For example, 8-oxodGTP is incorporated opposite either C or A residues of template DNA with almost equal efficiency by DNA polymerase III holoenzyme of *E. coli* (58). In this case, the induced replication error is a mispair containing the damaged nucleotide and would be converted to a premutation at the next round of replication. Fixation of the premutation into a base-substitution mutation requires a further round of DNA replication. The other pathway is the misinsertion of a normal nucleotide opposite a miscoding type of spontaneous DNA lesion. For example, dATP can be misincorporated opposite

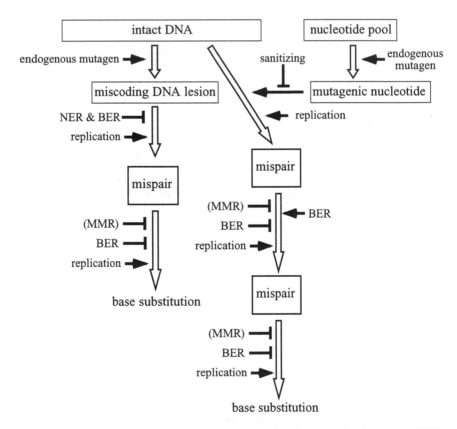

Figure 5   Two mutagenic pathways that begin with induced replication errors. NER, nucleotide excision repair; BER, base excision repair; MMR, mismatch repair.

an 8-oxoG residue in the damaged template DNA (100). The resulting A:8-oxoG mispair is a premutation that leads to a G:C→T:A transversion at the next round of DNA replication. Replication errors occurring in these pathways contain damaged nucleotides and are thus referred to as "induced replication error" in this review. Spontaneous DNA lesions caused by hydrolytic deamination of C or 5-MetC are classified into a special class of miscoding lesions (57). The resulting G:U or G:T mispair invariably induces a G:C → A:T transition mutation at the initial round of DNA replication.

Translesion DNA Synthesis

A significant portion of spontaneous DNA lesions have inhibitory effects on DNA synthesis and generally block the chain-elongation reaction by replicative DNA polymerases. In addition, miscoding-type base-damaged DNA lesions can be processed into replication-blocking type by the action of DNA glycosylases (97). Replication-blocking type DNA lesions are subjected to TLS and produce premutations (1, 6, 23, 27, 31, 56) (Figure 6). Various DNA polymerases participating in TLS have been identified in prokaryotes and eukaryotes (7, 76). *E. coli* possesses three different TLS polymerases: DNA polymerase II, PolII, encoded by *polB*; PolIV (*dinB* product); and PolV (*umuCD* product). Eukaryotic cells also contain multiple species of TLS polymerase. Genetic and biochemical analyses of these TLS polymerases suggest that they differ in lesion specificity and in the accuracy of TLS. TLS polymerases can bypass various DNA lesions containing abasic damage, thymine dimer, and bulky-adduct in multiple ways, depending on the lesion, and produce mispair, single-base bulge, or multiple-base mismatch (3, 11, 26, 32, 33, 38, 41, 50, 51, 59–62, 71, 72, 74, 75, 79, 80, 85, 86, 106, 107, 110, 114, 116). The resulting premutations can be fixed into base substitution, single-base frameshift, and more complex mutations at the next round of DNA replication (Figure 6).

MULTIPLE STEPS OF MUTATION AVOIDANCE MECHANISMS

As predicted from the vast difference between the extent of premutagenic DNA damage and the very low frequency of spontaneous mutation in living cells, cells possess the capacity to suppress the generation of spontaneous mutations with high efficiency. To cope with a wide variety of premutagenic damage and premutations, organisms have evolved many different functions that act to avoid mutation, e.g., to prevent the generation of premutagenic damage, eliminate premutagenic damage, suppress the conversion of premutagenic damage to premutations, and eliminate the premutations before they become fixed as mutations. These mutation-avoidance mechanisms prevent the vast majority of premutagenic damage from actually resulting in mutations. Cells lacking one or more mutation-avoidance functions show a mutator phenotype with an increased frequency of spontaneous mutations.

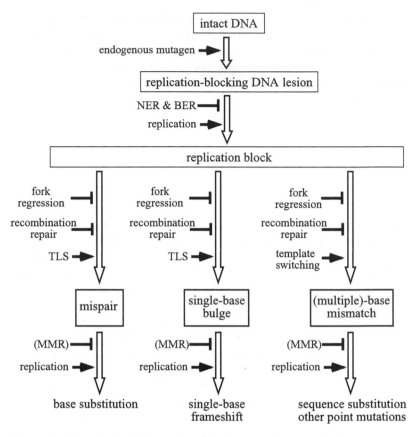

Figure 6   Translesion DNA synthesis leading to various types of spontaneous mutations. NER, nucleotide excision repair; BER, base excision repair; TLS, translesion DNA synthesis; MMR, mismatch repair.

Genetic and biochemical analyses of such mutator cells have helped in identifying the molecular mechanisms of mutation-avoidance functions.

Prevention of Premutagenic Damage

Several cellular functions are involved in preventing premutagenic damage. The base-selection function of replicative DNA polymerase suppresses the generation of terminal single-base mispairs during DNA synthesis with template DNA that either is intact or contains a miscoding type of DNA lesion (47). To some extent, this function also prohibits misinsertion of mutagenic nucleotides, 8-oxodGTP (58) and 2-OHdATP (42). However, the misinsertion of mutagenic nucleotide can be prevented much more efficiently by sanitizing the nucleotide pool in cells (Figure 5). The MutT protein of *E. coli* hydrolyzes 8-oxodGTP to 8-oxodGMP and eliminates the mutagenic nucleotide from the nucleotide pool (58). Similarly, enzymatic detoxification of endogenous mutagens such as active oxygen species is important in preventing the generation of spontaneous lesions on DNA as well as substrate nucleotide (118). Several lines of evidence suggest that a terminal mispair may promote misalignment of the growing chain to generate a single-base bulge (2, 5, 47, 81); the proofreading function of the replicative enzyme that removes the terminal mispair seems to prevent the generation of this kind of slippage error (Figure 2).

Prereplicative and Postreplicative Repair Processes

Four major types of DNA repair mechanisms eliminate premutagenic damage and premutations: the proofreading function built into the replicative apparatus, the mismatch repair system (MMR), the nucleotide excision repair system (NER), and various pathways of base excision repair (BER) (24, 54). Among these, MMR plays a central role in the postreplicative correction that removes premutations before the next round of DNA replication (34, 67, 68). Mutagenesis caused by native replication errors is strongly suppressed by MMR (Figure 2). Mispairs formed during DNA replication with intact template DNA are eliminated very efficiently by the successive actions of proofreading and MMR. Single-base bulges, another type of native replication error, are also corrected very efficiently by MMR. The mismatch repair system is able to eliminate multiple-base mismatches that lead to sequence substitution mutations, but the repair efficiency largely depends on the size of the mismatched segment (120). On the other hand, NER and BER act mostly in the prereplicative elimination of endogenous DNA lesions, and these repair mechanisms cooperate in preventing mutagenesis caused by the miscoding and replication-blocking types of spontaneous DNA lesions (Figures 5, 6).

A key step in mutagenesis caused by replication-blocking DNA lesions is their conversion to premutations by TLS. Other pathways of damage tolerance such as fork regression and recombinational repair compete with TLS for the recovery of stalled replication (14, 49). Since fork regression and recombinational repair are error-free processes, these pathways can suppress the spontaneous mutagenesis that is allowed by TLS. The mechanisms channeling these damage-tolerance pathways, still largely unknown, affect the generation of premutations by TLS. There are several TLS pathways that differ in efficiency and their method of bypassing DNA synthesis on various kinds of DNA lesions. At a given DNA lesion, one TLS pathway may bypass the lesion in a way that leads to a genetic alteration, but "accurate" bypass DNA synthesis may be carried out by another TLS pathway. Therefore, a less error-prone TLS pathway can suppress the conversion of replication-blocking lesions to premutations that may be caused by another more error-prone TLS pathway.

Also important but less well documented is postreplicative correction or elimination of premutations caused by endogenous DNA lesions. The MutY-dependent BER pathway acts in such mutation avoidance, specifically eliminating adenine nucleotide from A:8-oxoG mispair, an induced replication error caused by a template 8-oxoG residue (63). Some kinds of induced replication error caused by miscoding DNA lesions such as O^6-MetG and 8-oxoG are subject to MMR (34, 40, 73). The involvement of MMR in postreplicative suppression of UV-induced mutagenesis (55, 115) suggests that it also plays a role in postreplicative correction of TLS products. G:C → A:T, A:T → G:C, and G:C → T:A mutations that are heavily dependent on aerobic growth conditions were shown in a well-defined *cyc1* reversion mutation assay to be strongly suppressed by MMR in *S. cerevisiae* (20). However, since the damaged nucleotide cannot be eliminated during such a postreplicative correction, the efficiency of the mismatch correction should vary, depending on the nature of the miscoding and replication-blocking DNA lesions.

FACTORS AFFECTING THE SPECIFICITY OF SPONTANEOUS MUTATION

The specificity of spontaneous mutations in a given sequence of chromosomal DNA in normally growing cells is affected by all mutagenic processes that occur within the DNA segment. The frequency and site-distribution of premutagenic damage depend largely on the mechanisms that underlie the generation of each class of premutagenic damage. Biochemical studies of DNA polymerase action indicate that the sequence context of the template as well as the primer DNA could influence the generation of native replication errors (47). In particular, various types of repetitive sequences stimulate misalignment between the primer terminus and the template, which results in the generation of single-base bulges and multiple-base mismatches. On the other hand, the generation of endogenous DNA lesions may be affected by the state of the chromatin structure, the local DNA topology, and the level of transcription within a target DNA sequence, all of which alter the single-strandedness of chromosomal DNA (24). The nonrandom distribution of premutagenic damage in a particular sequence may be an important determinant for the specificity of spontaneous mutations.

However, the pattern of premutagenic damage is unlikely to be well conserved in the final pattern of mutations that arise through each mutagenic pathway. There are at least three reasons for this expectation. First, most premutagenic damage is eliminated by the repair mechanisms specific for each mutagenic pathway, and the efficiency of repair can be influenced by many factors including the sequence context, the state of the chromatin structure, and the level of transcription. Second, only a portion of the endogenous DNA lesions can be converted into premutations by TLS because of a preferential use of other damage-tolerance mechanisms to overcome the blockage of replication (14). The efficiency of conversion is also widely affected by the nature of the DNA lesion and the context of the surrounding sequence. Third, depending on the type of premutagenic DNA damage, the postreplicative correction of the resulting premutations may vary in efficiency.

The pattern of spontaneous mutations is apparently a mixture of the end result of each mutagenic pathway. Therefore, a major determinant in the specificity of spontaneous mutations is the balance of the multiple pathways, beginning with the generation of a specific class of premutagenic DNA damage and ending with the premutations being fixed into mutations. To address the extent to which each mutagenic pathway contributes to the net pattern of spontaneous mutations requires that the characteristics of mutations resulting from each mutagenic pathway be elucidated.

CLASS DISTRIBUTION OF SPONTANEOUS MUTATION

Spectrum analysis of forward mutations that occur in a particular gene has been useful in determining the pattern of spontaneous mutations. To date, a variety of genetic systems have been used. A large number of cell populations grown under controlled conditions are needed in order to determine the very low frequencies of mutations and to collect many independent mutants. Hence, most of the spectra of spontaneous mutations have been obtained using microbial systems, including the *lacI* (95, 96), *rpsL* (65, 120), and *supF* (117) systems of *E. coli* and the *SUP4-o* (43, 90, 119), *URA3* (98, 113), and *CAN1* (73, 109) systems of *S. cerevisiae*.

Caution is advised in comparing the spectrum data from different genetic systems. First, because the mutation assay system is based on the genetic selection of mutants, it is hard to collect all of the sequence alterations that occur in the target gene. Furthermore, the kinds of mutations that can be isolated and the number of sites that can be detected for mutations are largely dependent on the nature of the genetic system. Second, the net frequency of spontaneous mutations is variable and depends on the genetic system used. The rate of mutation per locus differs by up to two orders of magnitude among target sequences, owing partly to differences in the nature of the genetic systems and partly to the nonrandom distribution of spontaneous mutations. The target sequences differ in the number of mutational hotspot sites, and the strength of each hotspot site also varies. These disparities can make it difficult to collect weaker or non-hotspot types of mutations. Thus, relatively rare types of spontaneous mutations can easily be overlooked unless large numbers of independent mutations are analyzed.

Despite the difficulties in addressing the nature of spontaneous mutations, base substitution is probably the most frequent class of spontaneous point mutation and single-base frameshift the second most frequent class of non-hotspot mutation. Spectrum analyses of spontaneous mutations that occur in the protein-coding genes of *E. coli* and *S. cerevisiae* indicate that single-base frameshift occurs, in general, about one to two orders of magnitude less frequently than does base substitution. However, when a target sequence contains a very strong hotspot site for single-base frameshift, the order can be reversed. Another class of spontaneous mutation is sequence substitution. Since most sequence substitutions can be interpreted as composite mutations with one or several base substitutions, insertions, and deletions, they have been classified into "complex mutations" in the literature. Recently, systematic analyses of sequence substitution in wild-type and *dnaE173* mutator strains of *E. coli* established it as a distinct class of spontaneous mutation

(65, 120). However, because of its strong sequence-dependency, the frequency of sequence substitution varies among different target sequences.

BASE SUBSTITUTION

Pattern of Spontaneous Base Substitutions

Large-scale analyses of base-substitution spectra in wild-type cells of *E. coli* have used the *lacI*$^{-d}$ (96) and *rpsL* (120) target sequences, which contain 64 and 93 detectable sites for base substitution, respectively. The base substitutions are non-randomly distributed within each target sequence. Of 293 independent mutations identified within the *lacI*$^{-d}$ sequence on the F plasmid, 63% are located at 19 medium-level hotspot sites. Of 120 base substitutions identified within the *rpsL* sequence on a multicopy plasmid, 63% are located at 9 hotspot sites. Recently, the *rpsL* mutation assay was adapted to analyze mutations in the same target sequence that had become integrated into the chromosome of *E. coli* (K. Yoshiyama, M. Kawano, A. Isogawa & H. Maki, unpublished results). In this case, 70% of 1555 base substitutions examined are confined to only two strong hotspot sites. The remaining 475 mutations are almost evenly distributed at 91 sites within the target sequence. On both the *lacI*$^{-d}$ and *rpsL* target sequences, the rates of the non-hotspot type of base substitutions are calculated to be 10^{-11}–10^{-12}/base/generation, equivalent to the rates of spontaneous base substitutions determined in several reversion mutation assays in *E. coli* (4, 13, 15). The rates of mutations at hotspot sites range widely up to 10^{-8}/base/generation. Spontaneous base substitutions, at least in this organism, likely consist of two distinct types of mutations: hotspot mutation, which is strongly dependent on the context of the DNA sequence, and non-hotspot mutation, which is evenly distributed in the genome sequence at much lower frequency.

Hotspot and non-hotspot mutations may differ in the distribution of the six subtypes of base substitutions classified by mutational directionality. G:C → A:T transitions are three to four times more frequent than A:T → G:C transitions among base substitutions within both the *lacI*$^{-d}$ and *rpsL* sequences on plasmids. This is because both target sequences contain more hotspot sites for G:C → A:T transition than for A:T → G:C transition. In contrast to the plasmid-borne *rpsL*, G:C → T:A and A:T → T:A transversions are the major hotspot-type mutations, occurring at two strong hotspot sites within the chromosomal *rpsL* target sequence. Among 475 non-hotspot types of base substitutions recovered from the chromosomal *rpsL* system, the frequencies of G:C → A:T and A:T → G:C transitions are nearly equal. Furthermore, four subtypes of transversions in non-hotspot-type mutations, representing 211 transversions, occur with virtually identical frequencies. Thus, hotspot-type mutations probably possess a strong bias toward one of the subtypes, depending on the nature of hotspot site and other factors, whereas non-hotspot type mutations are smoothed so as to keep an even frequency for two subtypes of transitions as well as a lower, but constant, frequency for four subtypes of transversions.

A spectrum of 103 base substitutions identified within the *SUP4-o* target sequence in *S. cerevisiae* has been reported (90). Most of the mutations are almost evenly distributed at 48 of 60 detectable sites of base substitution with high mutation rates, about 10^{-8}/base/generation. Since mutation rates determined for the *cyc 1* reversion mutations are 10^{-10}–10^{-11}/base/generation (20), a level similar to the non-hotspot-type base substitutions in *E. coli*, the *SUP4-o* target sequence may contain an unusually high density of hotspot sites or may be in a hypermutable state, for reasons not yet determined. Although it is not possible to depict a reliable site-distribution from relatively small-scale analyses of base substitutions within *URA3* and *CAN1* sequences in *S. cerevisiae*, the mutation rate of base substitutions at a hotspot site in the *URA3* sequence is 2.4×10^{-8}/base/generation and the rate at other sites is possibly much lower (52). This finding may support the notion that spontaneous base substitutions in yeast cells also consist of two distinct types, as observed in cells of *E. coli*.

Differential Contribution of Mutagenic Pathways to Base-Substitution Mutagenesis

There are at least five mutagenic pathways leading to spontaneous base substitutions; two pathways involving native replication errors, base misinsertion, and template switching; two pathways involving the induced replication errors caused by mutagenic nucleotides and miscoding type of endogenous DNA lesions; and a pathway involving TLS caused by replication-blocking–type DNA lesions. The extent to which each pathway has a share in spontaneous base substitutions may vary depending on the target sequences. However, the origins of several hotspot-type base substitutions have been ascertained to be spontaneous DNA lesions. First, deamination of 5-MetC causes G:C → A:T transition at sites containing the methylated cytosine with an extremely high mutation frequency (57). The *dcm*-dependent methylation site is a hotspot site of this mutagenesis in *E. coli* (12), and the high frequency of G:C → A:T mutation at the CpG sites is also due to this specific mutagenesis (17, 44). Second, a few hotspot types of G:C → A:T transitions within the *rpoB* gene of *E. coli* are caused by the *umuCD*-dependent TLS pathway (77), implying that unidentified replication-blocking DNA lesions are responsible for the hotspot mutagenesis. Third, a template-switching mechanism has been suggested to be involved in a strong hotspot type of T:A → A:T transversion within the *thyA* gene of *E. coli* (112). One of the two strong hotspot-type base substitutions within the chromosomal *rpsL* may be caused by the template-switching mechanism (K. Yoshiyama & H. Maki, unpublished results). As in the *thyA* hotspot mutation, the hotspot *rpsL* mutation is resistant to the postreplicative correction by MMR, suggesting that the premutations for these hotspot mutations may contain DNA lesions in the template strand.

By comparing the spectra of base substitutions occurring in wild-type and mutator mutant cells, we may evaluate the contribution of each mutagenic pathway to spontaneous base-substitution mutagenesis. Studies using the *lacI*$^{-d}$ and *rpsL* systems of *E. coli* indicate that the base-misinsertion mutagenesis contributes little to hotspot-type base substitutions. The pattern of base substitutions in MMR-defective cells is substantially different from the spectrum in wild-type cells (94, 96, 120). Most of the hotspot sites found in the MMR-defective spectrum are not hotspots in the wild-type spectrum, indicating a high repair efficiency of MMR. Mutational spectra seen in strains carrying a single or a combination of *mutT*, *mutM*, and *mutY* mutator mutations are not overlapped with the wild-type spectrum, suggesting that 8-oxodGTP and 8-oxoG in DNA do not provide any significant contribution to the final pattern of spontaneous base substitutions (105). A similar conclusion has been reached for the mutagenic pathway of cytosine deamination (22). However, these mutagenic pathways may contribute to some extent to the non-hotspot-type base substitutions.

TLS at endogenous replication-blocking DNA lesions is suggested to play a significant role in the mutagenesis leading to non-hotspot-type base substitutions (4). Cells of *E. coli* that lack a TLS polymerase, PolIV, showed antimutator effects in a reversion assay using five different *trp* alleles. Several different non-hotspot base substitutions can be monitored by the reversion assay. The antimutator effects are strong on A:T → T:A (50-fold reduction) and A:T → C:G (10-fold reduction) transversions, whereas A:T → G:C and G:C → A:T transitions are unaffected in the cells. Therefore, some but not all non-hotspot base substitutions are caused mainly by the TLS pathway. The involvement of oxidative DNA damage in spontaneous TLS mutagenesis has been also suggested by the study. Similar antimutator effects were found in yeast cells that lack a TLS polymerase, Pol ζ (9, 82).

SINGLE-BASE FRAMESHIFT

Because the *lacI*$^{-d}$ system is inadequate for identifying single-base frameshifts, a general view of spontaneous single-base frameshift mutations in *E. coli* has been available only from spectrum analyses using the *rpsL* system (120). The class- and site-distributions of single-base frameshifts are remarkably affected by the directionality of movement of the replication fork relative to the target gene. As indicated in extensive studies using the *lys2ΔBgl* system in *S. cerevisiae* (16, 69), the level of transcription of the target sequence also influences the pattern of single-base frameshift mutations (K. Yoshiyama & H. Maki, unpublished results). These effects mainly alter the frequency of hotspot-type single-base frameshift mutations within the target sequences, suggesting that hotspot and non-hotspot mutations are two distinct types of single-base frameshifts. From the distribution of 42 non-hotspot frameshifts within 375 bp of the *rpsL* target sequence, the following features emerged: −1 and +1 frameshifts occur with almost equal frequencies, 10^{-11}–10^{-12}/bp/replication; about half of the frameshifts are distributed at mononucleotide repeats, and the frequency of frameshift at the repeat depends only slightly on the length of the repeat; one half of the +1 frameshifts are an insertion of the same nucleotide as the 5′- or 3′-neighbor, and the other half are an insertion of a nucleotide unrelated to the neighbor. These characteristics of non-hotspot frameshifts are in good agreement with a dataset from the systematic analysis of single-base frameshifts using the *lys2ΔBgl* system in *S. cerevisiae* (28).

The nature of single-base frameshifts deriving solely from native replication errors has been clearly demonstrated by analyzing *rpsL*⁻ forward mutations induced in a reconstituted system of *oriC* plasmid DNA replication in vitro (25). In contrast to the spontaneous *rpsL*⁻ mutations in vivo, single-base frameshifts occur more frequently than base substitutions in this in vitro system, which is free from mutation avoidance mechanisms other than those of the replicative apparatus. Most frameshifts are distributed at mononucleotide repeats within the target sequence, indicating a very strong dependency on the repetitive sequence. Although the frequencies of both −1 and +1 frameshifts depend on the length of the repeat, −1 frameshifts are predominant at repeats shorter than 5 consecutive residues, whereas +1 frameshifts outnumber −1 frameshift at repeats longer than 5 residues. This pattern of frameshift in vitro corresponds to the single-base frameshifts that occur within the same target sequence in cells of *E. coli* that lack the capacity for MMR. As mentioned above, the spectrum of frameshifts seen in wild-type cells is almost completely unlike that observed in MMR-defective cells. From this difference between the spectra, the repair efficiency of single-base bulges per site is estimated to be 1,000- to 10,000-fold, indicating an extremely high capacity of MMR for postreplicative correction. As a consequence, several mononucleotide repeats longer than 5 nucleotides remain as mild-hotspot sites in the wild-type spectrum.

Although half of the non-hotspot frameshifts are distributed at mononucleotide repeats in the wild-type spectrum, their frequency has a significantly weaker dependence on the length of the repeat. In addition, the occurrence of +1 frameshifts in noniterative sequences is unique to the wild-type spectrum. Therefore, it is conceivable that most spontaneous single-base frameshifts are not derived from single-base bulges caused by the simple slippage of DNA synthesis. As suggested by studies of mutagenesis caused by chemical carcinogens, bulky-adduct DNA lesions can induce frameshifts via TLS (71). Recently, several lines of evidence have suggested the involvement of spontaneous DNA lesions and TLS in spontaneous frameshift mutagenesis in *E. coli* (103) and *S. cerevisiae* (69). However, to what extent the TLS pathway contributes to the net pattern of spontaneous frameshifts is largely unknown.

As observed in the *rpsL* target sequence, long repeats of mononucleotides are hotspot sites for insertion or deletion of the same nucleotide within various target sequences in both prokaryotic and eukaryotic systems. Interestingly, the most frequent hotspot type of frameshift within the *rpsL* sequence is the insertion of a single A residue into a 5′-CC-3′ sequence. The frequency of this frameshift is strongly affected by the directionality of the replication fork movement relative to the sequence and reaches a maximum level of 1.7×10^{-8}/site/replication. This hotspot type of +1 frameshift is caused by a template-switching mechanism that is exclusively dependent on a pair of pseudo-inverted repeats, of which one contains the hotspot site (120) (Figure 3B). However, since the frameshift mutation appears to be resistant to postreplicative correction by MMR, an endogenous DNA lesion seems to be involved in the template switching process and present in the resulting premutation. Within the *rpsL* sequence, more than half of non-hotspot +1 frameshifts, and to a lesser extent −1 frameshifts, were suggested to be caused by the template switching mechanism. Although it is not known whether spontaneous DNA lesions are involved in the non-hotspot-type frameshift mutagenesis, most of the premutations for the non-hotspot frameshifts are probably impervious to MMR.

SEQUENCE SUBSTITUTION

A systematic analysis of the *rpsL* system has revealed common features of sequence substitutions (120). In all of the 64 sequence substitutions examined, a pair of relatively short pseudo-inverted repeats located close to each other with 5–20 intervening nucleotides was converted to a pair of completely inverted repeats (Figure 3A). This characteristic was found in all the sequence substitutions identified within various target sequences in a wide range of species, including T4 bacteriophage (18), bacteria (53, 95), yeast (30, 73, 87, 109), and mammals (17, 36). Within about 400 bp of the *rpsL* target sequence, 17 pseudo-inverted repeats satisfied the common structural requirements for sequence-substitution mutagenesis. However, sequence substitution was confined to 7 of the 17 repeats, and 3 hotspot repeats were found within the target sequence. Therefore some sequence context other than the structure of the pseudo-inverted repeats also appears to be involved in the mutagenesis. The GTG/CAC motif that is always found within or adjacent to the repeat may provide this context (65). Since sequence substitution is strongly dependent on the sequence context and is affected by the directionality of replication fork movements as well as by the level of transcription, the frequencies of sequence substitution found within various target sequences range from less than 1% to about 20% of the total mutations detected in particular target sequences.

As mentioned earlier, the template-switching mechanisms lead not only to sequence substitutions but also to other point mutations, including base substitutions and single-base frameshifts (Figure 3B). However, the occurrence of such point mutations ought to be strongly suppressed by MMR. It has been recently demonstrated that MMR efficiently suppresses sequence substitutions that change in size by one or two nucleotides but not those that change by more than three nucleotides (T. Muroya, T. Ohsu & H. Maki, unpublished results). Thus, in MMR-proficient cells, a multiple-base mismatch containing three or more unpaired nucleotides are only premutations that can survive until the next round of replication.

The findings that strong hotspot-types of base substitution and single-base frameshift within the *rpsL* sequence are resistant to MMR and caused by the template-switching mechanism suggest the possible involvement of spontaneous DNA lesions in hotspot mutagenesis. If so, a portion of spontaneous sequence substitutions may also be induced by such DNA lesions. The presence of hotspot-type sequence substitutions supports this notion. Although many aspects of template-switching mutagenesis remain to be solved, the point mutations caused by the template-switching mechanism clearly play a large role in the spontaneous mutations at hotspot sites.

ACKNOWLEDGMENTS

Work in the author's laboratory is supported by a Grant-in-Aid for Scientific Research on Priority Areas (C) from the Ministry of Education, Culture, Sports, Science and Technology of Japan (12213082).

The *Annual Review of Genetics* is online at http://genet.annualreviews.org

LITERATURE CITED

1. Baynton K, Fuchs RP. 2000. Lesions in DNA: hurdles for polymerases. *Trends Biochem. Sci.* 25:74–79

2. Bebenek K, Roberts JD, Kunkel TA. 1992. The effects of dNTP pool imbalances on frameshift fidelity during DNA replication. *J. Biol. Chem.* 267:3589–96

3. Becherel OJ, Fuchs RP. 2001. Mechanism of DNA polymerase II-mediated frameshift mutagenesis. *Proc. Natl. Acad. Sci. USA* 98:8566–71

4. Bhamre S, Gadea BB, Koyama CA, White SJ, Fowler RG. 2001. An aerobic recA-, umuC-dependent pathway of spontaneous base-pair substitution mutagenesis in *Escherichia coli. Mutat. Res.* 473:229–47

5. Bloom LB, Chen X, Fygenson DK, Turner J, O'Donnell M, Goodman MF. 1997. Fidelity of *Escherichia coli* DNA polymerase III holoenzyme. The effects of beta, gamma complex processivity proteins and epsilon proofreading exonuclease on nucleotide misincorporation efficiencies. *J. Biol. Chem.* 272:27919–30

6. Bridges BA. 1999. DNA repair: Polymerases for passing lesions. *Curr. Biol.* 9:R475–77

7. Burgers PM, Koonin EV, Bruford E, Blanco L, Burtis KC, et al. 2001. Eukaryotic DNA polymerases: proposal for a revised nomenclature. *J. Biol. Chem.* 276:43487–90

8. Carraway M, Rewinski C, Marinus MG. 1990. Mutations produced by DNA polymerase III holoenzyme of *Escherichia coli* after in vitro synthesis in the absence of single-strand binding protein. *Mol. Microbiol.* 4:1645–52

9. Cassier C, Chanet R, Henriques JA, Moustacchi E. 1980. The effects of three PSO genes on induced mutagenesis: a novel class of mutationally defective yeast. *Genetics* 96:841–57

10. Chen X, Zuo S, Kelman Z, O'Donnell M, Hurwitz J, Goodman MF. 2000. Fidelity of eucaryotic DNA polymerase delta holoenzyme from *Schizosaccharomyces pombe. J. Biol. Chem.* 275:17677–82

11. Chiapperino D, Kroth H, Kramarczuk IH, Sayer JM, Masutani C, et al. 2002. Preferential misincorporation of purine nucleotides by human DNA polymerase eta opposite benzo[a]pyrene 7,8-diol 9,10-epoxide deoxyguanosine adducts. *J. Biol. Chem.* 277:11765–71

12. Coulondre C, Miller JH, Farabaugh PJ, Gilbert W. 1978. Molecular basis of base substitution hotspots in *Escherichia coli. Nature* 274:775–80

13. Cox EC, Degnen GE, Scheppe ML. 1972. Mutator gene studies in *Escherichia coli*: the mutS gene. *Genetics* 72:551–67

14. Cox MM. 2001. Recombinational DNA repair of damaged replication forks in

Escherichia coli: questions. *Annu. Rev. Genet.* 35:53–82

15. Cupples CG, Miller JH. 1989. A set of lacZ mutations in *Escherichia coli* that allow rapid detection of each of the six base substitutions. *Proc. Natl. Acad. Sci. USA* 86:5345–49

16. Datta A, Jinks-Robertson S. 1995. Association of increased spontaneous mutation rates with high levels of transcription in yeast. *Science* 268:1616–19

17. de Boer JG, Erfle H, Walsh D, Holcroft J, Provost S, et al. 1998. Spectrum of spontaneous mutations in liver tissue of lacI transgenic mice. *Environ. Mol. Mutagen.* 30:273–86

18. de Boer JG, Ripley LS. 1984. Demonstration of the production of frameshift and base-substitution mutations by quasipalindromic DNA sequences. *Proc. Natl. Acad. Sci. USA* 81:5528–31

19. Drake JW. 1991. Spontaneous mutation. *Annu. Rev. Genet.* 25:125–46

20. Earley MC, Crouse GF. 1998. The role of mismatch repair in the prevention of base pair mutations in *Saccharomyces cerevisiae*. *Proc. Natl. Acad. Sci. USA* 95:15487–91

21. Fishel R, Lescoe MK, Rao MR, Copeland NG, Jenkins NA, et al. 1994. The human mutator gene homolog MSH2 and its association with hereditary nonpolyposis colon cancer. *Cell* 77:167

22. Fix DF, Glickman BW. 1986. Differential enhancement of spontaneous transition mutations in the lacI gene of an Ung- strain of *Escherichia coli*. *Mutat. Res.* 175:41–45

23. Friedberg EC, Fischhaber PL, Kisker C. 2001. Error-prone DNA polymerases: novel structures and the benefits of infidelity. *Cell* 107:9–12

24. Friedberg EC, Walker GC, Siede W. 1995. *DNA Repair and Mutagenesis*. Washington, DC: ASM Press. 698 pp

25. Fujii S, Akiyama M, Aoki K, Sugaya Y, Higuchi K, et al. 1999. DNA replication errors produced by the replicative apparatus of *Escherichia coli*. *J. Mol. Biol.* 289:835–50

26. Gibbs PE, McGregor WG, Maher VM, Nisson P, Lawrence CW. 1998. A human homolog of the *Saccharomyces cerevisiae* REV3 gene, which encodes the catalytic subunit of DNA polymerase zeta. *Proc. Natl. Acad. Sci. USA* 95:6876–80

27. Goodman MF, Tippin B. 2000. Sloppier copier DNA polymerases involved in genome repair. *Curr. Opin. Genet. Dev.* 10:162–8

28. Greene CN, Jinks-Robertson S. 2001. Spontaneous frameshift mutations in *Saccharomyces cerevisiae*: accumulation during DNA replication and removal by proofreading and mismatch repair activities. *Genetics* 159:65–75

29. Hadjimarcou MI, Kokoska RJ, Petes TD, Reha-Krantz LJ. 2001. Identification of a mutant DNA polymerase delta in *Saccharomyces cerevisiae* with an antimutator phenotype for frameshift mutations. *Genetics* 158:177–86

30. Hampsey DM, Koski RA, Sherman F. 1986. Highly mutable sites for ICR-170–induced frameshift mutations are associated with potential DNA hairpin structures: studies with SUP4 and other *Saccharomyces cerevisiae* genes. *Mol. Cell Biol.* 6:4425–32

31. Hanaoka F. 2001. DNA replication. SOS polymerases. *Nature* 409:33–4

32. Haracska L, Prakash S, Prakash L. 2002. Yeast Rev1 protein is a G template specific DNA polymerase. *J. Biol. Chem.* 277:15546–51

33. Haracska L, Unk I, Johnson RE, Johansson E, Burgers PM, et al. 2001. Roles of yeast DNA polymerases delta and zeta and of Rev1 in the bypass of abasic sites. *Genes Dev.* 15:945–54

34. Harfe BD, Jinks-Robertson S. 2000. DNA mismatch repair and genetic instability. *Annu. Rev. Genet.* 34:359–99

35. Horiuchi T, Maki H, Sekiguchi M. 1989. Mutators and fidelity of DNA replication. *Bull. Inst. Pasteur* 87:309–36

36. Ikehata H, Akagi T, Kimura H, Akasaka S, Kato T. 1989. Spectrum of spontaneous mutations in a cDNA of the human hprt gene integrated in chromosomal DNA. *Mol. Gen. Genet.* 219:349–58

37. Inoue M, Kamiya H, Fujikawa K, Ootsuyama Y, Murata-Kamiya N, et al. 1998. Induction of chromosomal gene mutations in *Escherichia coli* by direct incorporation of oxidatively damaged nucleotides. New evaluation method for mutagenesis by damaged DNA precursors in vivo. *J. Biol. Chem.* 273:11069–74

38. Ishikawa T, Uematsu N, Mizukoshi T, Iwai S, Iwasaki H, et al. 2001. Mutagenic and nonmutagenic bypass of DNA lesions by Drosophila DNA polymerases dpoleta and dpoliota. *J. Biol. Chem.* 276:15155–63

39. Izuta S, Roberts JD, Kunkel TA. 1995. Replication error rates for G.dGTP, T.dGTP, and A.dGTP mispairs and evidence for differential proofreading by leading and lagging strand DNA replication complexes in human cells. *J. Biol. Chem.* 270:2595–600

40. Jiricny J. 1998. Replication errors: cha-(lle)nging the genome. *EMBO J.* 17:6427–36

41. Johnson RE, Prakash S, Prakash L. 1999. Efficient bypass of a thymine-thymine dimer by yeast DNA polymerase, Poleta. *Science* 283:1001–4

42. Kamiya H, Maki H, Kasai H. 2000. Two DNA polymerases of *Escherichia coli* display distinct misinsertion specificities for 2-hydroxy-dATP during DNA synthesis. *Biochemistry* 39:9508–13

43. Kang XL, Yadao F, Gietz RD, Kunz BA. 1992. Elimination of the yeast RAD6 ubiquitin conjugase enhances base-pair transitions and G.C–T.A transversions as well as transposition of the Ty element: implications for the control of spontaneous mutation. *Genetics* 130:285–94

44. Kohler SW, Provost GS, Fieck A, Kretz PL, Bullock WO, et al. 1991. Spectra of spontaneous and mutagen-induced mutations in the lacI gene in transgenic mice. *Proc. Natl. Acad. Sci. USA* 88:7958–62

45. Kroutil LC, Frey MW, Kaboord BF, Kunkel TA, Benkovic SJ. 1998. Effect of accessory proteins on T4 DNA polymerase replication fidelity. *J. Mol. Biol.* 278:135–46

46. Kunkel TA. 1993. Nucleotide repeats. Slippery DNA and diseases [news; comment]. *Nature* 365:207–8

47. Kunkel TA, Bebenek K. 2000. DNA replication fidelity. *Annu. Rev. Biochem.* 69:497–529

48. Kunkel TA, Patel SS, Johnson KA. 1994. Error-prone replication of repeated DNA sequences by T7 DNA polymerase in the absence of its processivity subunit. *Proc. Natl. Acad. Sci. USA* 91:6830–34

49. Kunz BA, Straffon AF, Vonarx EJ. 2000. DNA damage-induced mutation: tolerance via translesion synthesis. *Mutat. Res.* 451:169–85

50. Kuraoka I, Robins P, Masutani C, Hanaoka F, Gasparutto D, et al. 2001. Oxygen free radical damage to DNA. Translesion synthesis by human DNA polymerase eta and resistance to exonuclease action at cyclopurine deoxynucleoside residues. *J. Biol. Chem.* 276:49283–88

51. Lawrence CW, Borden A, Woodgate R. 1996. Analysis of the mutagenic properties of the UmuDC, MucAB and RumAB proteins, using a site-specific abasic lesion. *Mol. Gen. Genet.* 251:493–98

52. Lee GS, Savage EA, Ritzel RG, von Borstel RC. 1988. The base-alteration spectrum of spontaneous and ultraviolet radiation-induced forward mutations in the URA3 locus of *Saccharomyces cerevisiae*. *Mol. Gen. Genet.* 214:396–404

53. Levine JG, Schaaper RM, DeMarini DM. 1994. Complex frameshift mutations mediated by plasmid pKM101: mutational mechanisms deduced from 4–aminobiphenyl-induced mutation spectra in Salmonella. *Genetics* 136:731–46

54. Lindahl T, Wood RD. 1999. Quality

control by DNA repair. *Science* 286: 1897–905

55. Liu H, Hewitt SR, Hays JB. 2000. Antagonism of ultraviolet-light mutagenesis by the methyl-directed mismatch-repair system of *Escherichia coli*. *Genetics* 154:503–12

56. Livneh Z. 2001. DNA damage control by novel DNA polymerases: translesion replication and mutagenesis. *J. Biol. Chem.* 276:25639–42

57. Lutsenko E, Bhagwat AS. 1999. Principal causes of hot spots for cytosine to thymine mutations at sites of cytosine methylation in growing cells. A model, its experimental support and implications. *Mutat. Res.* 437:11–20

58. Maki H, Sekiguchi M. 1992. MutT protein specifically hydrolyses a potent mutagenic substrate for DNA synthesis. *Nature* 355:273–75

59. Masutani C, Kusumoto R, Iwai S, Hanaoka F. 2000. Mechanisms of accurate translesion synthesis by human DNA polymerase eta. *EMBO J.* 19:3100–9

60. Masutani C, Kusumoto R, Yamada A, Dohmae N, Yokoi M, et al. 1999. The XPV (xeroderma pigmentosum variant) gene encodes human DNA polymerase eta. *Nature* 399:700–4

61. Matsuda T, Bebenek K, Masutani C, Hanaoka F, Kunkel TA. 2000. Low fidelity DNA synthesis by human DNA polymerase-eta. *Nature* 404:1011–13

62. Matsuda T, Bebenek K, Masutani C, Rogozin IB, Hanaoka F, Kunkel TA. 2001. Error rate and specificity of human and murine DNA polymerase eta. *J. Mol. Biol.* 312:335–46

63. Michaels ML, Miller JH. 1992. The GO system protects organisms from the mutagenic effect of the spontaneous lesion 8-hydroxyguanine (7,8-dihydro-8-oxoguanine). *J. Bacteriol.* 174:6321–25

64. Miller JH. 1996. Spontaneous mutators in bacteria: insights into pathways of mutagenesis and repair. *Annu. Rev. Microbiol.* 50:625–43

65. Mo JY, Maki H, Sekiguchi M. 1991. Mutational specificity of the dnaE173 mutator associated with a defect in the catalytic subunit of DNA polymerase III of *Escherichia coli*. *J. Mol. Biol.* 222:925–36

66. Mo JY, Schaaper RM. 1996. Fidelity and error specificity of the alpha catalytic subunit of *Escherichia coli* DNA polymerase III. *J. Biol. Chem.* 271:18947–53

67. Modrich P. 1991. Mechanisms and biological effects of mismatch repair. *Annu. Rev. Genet.* 25:229–53

68. Modrich P, Lahue R. 1996. Mismatch repair in replication fidelity, genetic recombination, and cancer biology. *Annu. Rev. Biochem.* 65:101–33

69. Morey NJ, Greene CN, Jinks-Robertson S. 2000. Genetic analysis of transcription-associated mutation in *Saccharomyces cerevisiae*. *Genetics* 154:109–20

70. Morrison A, Sugino A. 1994. The 3′–>5′ exonucleases of both DNA polymerases delta and epsilon participate in correcting errors of DNA replication in *Saccharomyces cerevisiae*. *Mol. Gen. Genet.* 242:289–96

71. Napolitano R, Janel-Bintz R, Wagner J, Fuchs RP. 2000. All three SOS-inducible DNA polymerases (Pol II, Pol IV and Pol V) are involved in induced mutagenesis. *EMBO J.* 19:6259–65

72. Nelson JR, Lawrence CW, Hinkle DC. 1996. Thymine-thymine dimer bypass by yeast DNA polymerase zeta. *Science* 272:1646–49

73. Ni TT, Marsischky GT, Kolodner RD. 1999. MSH2 and MSH6 are required for removal of adenine misincorporated opposite 8-oxo-guanine in *S. cerevisiae*. *Mol. Cell* 4:439–44

74. O'Grady PI, Borden A, Vandewiele D, Ozgenc A, Woodgate R, Lawrence CW. 2000. Intrinsic polymerase activities of UmuD′(2)C and MucA′(2)B are responsible for their different mutagenic properties during bypass of a T-T cis-syn cyclobutane dimer. *J. Bacteriol.* 182: 2285–91

75. Ohashi E, Ogi T, Kusumoto R, Iwai S, Masutani C, et al. 2000. Error-prone bypass of certain DNA lesions by the human DNA polymerase kappa. *Genes Dev.* 14:1589–94

76. Ohmori H, Friedberg EC, Fuchs RP, Goodman MF, Hanaoka F, et al. 2001. The Y-family of DNA polymerases. *Mol. Cell* 8:7–8

77. Otterlei M, Kavli B, Standal R, Skjelbred C, Bharati S, Krokan HE. 2000. Repair of chromosomal abasic sites in vivo involves at least three different repair pathways. *EMBO J.* 19:5542–51

78. Park EM, Shigenaga MK, Degan P, Korn TS, Kitzler JW, et al. 1992. Assay of excised oxidative DNA lesions: isolation of 8-oxoguanine and its nucleoside derivatives from biological fluids with a monoclonal antibody column. *Proc. Natl. Acad. Sci. USA* 89:3375–79

79. Pham P, Bertram JG, O'Donnell M, Woodgate R, Goodman MF. 2001. A model for SOS-lesion-targeted mutations in *Escherichia coli. Nature* 409:366–70

80. Pham P, Rangarajan S, Woodgate R, Goodman MF. 2001. Roles of DNA polymerases V and II in SOS-induced error-prone and error-free repair in *Escherichia coli. Proc. Natl. Acad. Sci. USA* 98:8350–54

81. Pham PT, Olson MW, McHenry CS, Schaaper RM. 1999. Mismatch extension by *Escherichia coli* DNA polymerase III holoenzyme. *J. Biol. Chem.* 274:3705–10

82. Quah SK, von Borstel RC, Hastings PJ. 1980. The origin of spontaneous mutation in *Saccharomyces cerevisiae. Genetics* 96:819–39

83. Radicella JP, Dherin C, Desmaze C, Fox MS, Boiteux S. 1997. Cloning and characterization of hOGG1, a human homolog of the OGG1 gene of *Saccharomyces cerevisiae. Proc. Natl. Acad. Sci. USA* 94:8010–15

84. Reenan RA, Kolodner RD. 1992. Isolation and characterization of two *Saccharomyces cerevisiae* genes encoding homologs of the bacterial HexA and MutS mismatch repair proteins. *Genetics* 132:963–73

85. Reuven NB, Arad G, Maor-Shoshani A, Livneh Z. 1999. The mutagenesis protein UmuC is a DNA polymerase activated by UmuD′, RecA, and SSB and is specialized for translesion replication. *J. Biol. Chem.* 274:31763–66

86. Reuven NB, Tomer G, Livneh Z. 1998. The mutagenesis proteins UmuD′ and UmuC prevent lethal frameshifts while increasing base substitution mutations. *Mol. Cell* 2:191–99

87. Ripley LS. 1982. Model for the participation of quasi-palindromic DNA sequences in frameshift mutation. *Proc. Natl. Acad. Sci. USA* 79:4128–32

88. Ripley LS. 1990. Frameshift mutation: determinants of specificity. *Annu. Rev. Genet.* 24:189–213

89. Roberts JD, Nguyen D, Kunkel TA. 1993. Frameshift fidelity during replication of double-stranded DNA in HeLa cell extracts. *Biochemistry* 32:4083–89

90. Roche H, Gietz RD, Kunz BA. 1994. Specificity of the yeast rev3 delta antimutator and REV3 dependency of the mutator resulting from a defect (rad1 delta) in nucleotide excision repair. *Genetics* 137:637–46

91. Rosche WA, Ripley LS, Sinden RR. 1998. Primer-template misalignments during leading strand DNA synthesis account for the most frequent spontaneous mutations in a quasipalindromic region in *Escherichia coli. J. Mol. Biol.* 284:633–46

92. Rosche WA, Trinh TQ, Sinden RR. 1997. Leading strand specific spontaneous mutation corrects a quasipalindrome by an intermolecular strand switch mechanism. *J. Mol. Biol.* 269:176–87

93. Sakumi K, Furuichi M, Tsuzuki T, Kakuma T, Kawabata S, et al. 1993.

Cloning and expression of cDNA for a human enzyme that hydrolyzes 8-oxo-dGTP, a mutagenic substrate for DNA synthesis. *J. Biol. Chem.* 268:23524–30

94. Schaaper RM. 1993. Base selection, proofreading, and mismatch repair during DNA replication in *Escherichia coli*. *J. Biol. Chem.* 268:23762–65

95. Schaaper RM, Danforth BN, Glickman BW. 1986. Mechanisms of spontaneous mutagenesis: an analysis of the spectrum of spontaneous mutation in the *Escherichia coli* lacI gene. *J. Mol. Biol.* 189:273–84

96. Schaaper RM, Dunn RL. 1991. Spontaneous mutation in the *Escherichia coli* lacI gene. *Genetics* 129:317–26

97. Scharer OD, Jiricny J. 2001. Recent progress in the biology, chemistry and structural biology of DNA glycosylases. *BioEssays* 23:270–81

98. Scheller J, Schurer A, Rudolph C, Hettwer S, Kramer W. 2000. MPH1, a yeast gene encoding a DEAH protein, plays a role in protection of the genome from spontaneous and chemically induced damage. *Genetics* 155:1069–81

99. Seki M, Akiyama M, Sugaya Y, Ohtsubo E, Maki H. 1999. Strand asymmetry of +1 frameshift mutagenesis at a homopolymeric run by DNA polymerase III holoenzyme of *Escherichia coli*. *J. Biol. Chem.* 274:33313–19

100. Shibutani S, Takeshita M, Grollman AP. 1991. Insertion of specific bases during DNA synthesis past the oxidation-damaged base 8-oxodG. *Nature* 349:431–34

101. Sloane DL, Goodman MF, Echols H. 1988. The fidelity of base selection by the polymerase subunit of DNA polymerase III holoenzyme. *Nucleic Acids Res.* 16:6465–75

102. Slupska MM, Baikalov C, Luther WM, Chiang JH, Wei YF, Miller JH. 1996. Cloning and sequencing a human homolog (hMYH) of the *Escherichia coli* mutY gene whose function is required for the repair of oxidative DNA damage. *J. Bacteriol.* 178:3885–92

103. Strauss BS, Roberts R, Francis L, Pouryazdanparast P. 2000. Role of the dinB gene product in spontaneous mutation in *Escherichia coli* with an impaired replicative polymerase. *J. Bacteriol.* 182:6742–50

104. Sueoka N. 1988. Directional mutation pressure and neutral molecular evolution. *Proc. Natl. Acad. Sci. USA* 85:2653–57

105. Tajiri T, Maki H, Sekiguchi M. 1995. Functional cooperation of MutT, MutM and MutY proteins in preventing mutations caused by spontaneous oxidation of guanine nucleotide in *Escherichia coli*. *Mutat. Res.* 336:257–67

106. Tang M, Pham P, Shen X, Taylor JS, O'Donnell M, et al. 2000. Roles of *E. coli* DNA polymerases IV and V in lesion-targeted and untargeted SOS mutagenesis. *Nature* 404:1014–18

107. Tang M, Shen X, Frank EG, O'Donnell M, Woodgate R, Goodman MF. 1999. UmuD'(2)C is an error-prone DNA polymerase, *Escherichia coli* pol V. *Proc. Natl. Acad. Sci. USA* 96:8919–24

108. Thomas DC, Roberts JD, Sabatino RD, Myers TW, Tan CK, et al. 1991. Fidelity of mammalian DNA replication and replicative DNA polymerases. *Biochemistry* 30:11751–59

109. Tishkoff DX, Filosi N, Gaida GM, Kolodner RD. 1997. A novel mutation avoidance mechanism dependent on *S. cerevisiae* RAD27 is distinct from DNA mismatch repair [see comments]. *Cell* 88:253–63

110. Tissier A, Frank EG, McDonald JP, Iwai S, Hanaoka F, Woodgate R. 2000. Misinsertion and bypass of thymine-thymine dimers by human DNA polymerase iota. *EMBO J.* 19:5259–66

111. van der Kemp PA, Thomas D, Barbey R, de Oliveira R, Boiteux S. 1996. Cloning and expression in *Escherichia coli* of the OGG1 gene of *Saccharomyces cerevisiae*, which codes for a DNA

glycosylase that excises 7,8-dihydro-8-oxoguanine and 2,6-diamino-4-hydroxy-5-N- methylformamidopyrimidine. *Proc. Natl. Acad. Sci. USA* 93:5197–202

112. Viswanathan M, Lacirignola JJ, Hurley RL, Lovett ST. 2000. A novel mutational hotspot in a natural quasipalindrome in *Escherichia coli. J. Mol. Biol.* 302:553–64

113. von Borstel RC, Ord RW, Stewart SP, Ritzel RG, Lee GS, et al. 1993. The mutator mut7-1 of *Saccharomyces cerevisiae. Mutat. Res.* 289:97–106

114. Wagner J, Gruz P, Kim SR, Yamada M, Matsui K, et al. 1999. The dinB gene encodes a novel *E. coli* DNA polymerase, DNA pol IV, involved in mutagenesis. *Mol. Cell* 4:281–86

115. Wang H, Lawrence CW, Li GM, Hays JB. 1999. Specific binding of human MSH2. MSH6 mismatch-repair protein heterodimers to DNA incorporating thymine- or uracil-containing UV light photoproducts opposite mismatched bases. *J. Biol. Chem.* 274:16894–900

116. Washington MT, Johnson RE, Prakash L, Prakash S. 2002. Human DINB1-encoded DNA polymerase kappa is a promiscuous extender of mispaired primer termini. *Proc. Natl. Acad. Sci. USA* 99:1910–14

117. Watanabe T, van Geldorp G, Najrana T, Yamamura E, Nunoshiba T, Yamamoto K. 2001. Miscoding and misincorporation of 8-oxo-guanine during leading and lagging strand synthesis in *Escherichia coli. Mol. Gen. Genet.* 264:836–41

118. Yamamura E, Nunoshiba T, Kawata M, Yamamoto K. 2000. Characterization of spontaneous mutation in the oxyR strain of *Escherichia coli. Biochem. Biophys. Res. Commun.* 279:427–32

119. Yang Y, Karthikeyan R, Mack SE, Vonarx EJ, Kunz BA. 1999. Analysis of yeast pms1, msh2, and mlh1 mutators points to differences in mismatch correction efficiencies between prokaryotic and eukaryotic cells. *Mol. Gen. Genet.* 261:777–87

120. Yoshiyama K, Higuchi K, Matsumura H, Maki H. 2001. Directionality of DNA replication fork movement strongly affects the generation of spontaneous mutations in *Escherichia coli. J. Mol. Biol.* 307:1195–20

Annu. Rev. Genet. 2002. 36:305–32
doi: 10.1146/annurev.genet.36.052402.152757

GENETICS OF INFLUENZA VIRUSES

David A. Steinhauer
*Department of Microbiology and Immunology, Emory University School
of Medicine, Rollins Research Center, Atlanta, Georgia 30322;
e-mail: steinhauer@microbio.emory.edu*

John J. Skehel
*National Institute for Medical Research, The Ridgeway, Mill Hill, London NW7 1AA
United Kingdom; e-mail: mbrenna@nimr.mrc.ac.uk*

Key Words   mutants, antigenicity, reassortment, recombination, pathogenicity

■ **Abstract**   Influenza A viruses contain genomes composed of eight separate segments of negative-sense RNA. Circulating human strains are notorious for their tendency to accumulate mutations from one year to the next and cause recurrent epidemics. However, the segmented nature of the genome also allows for the exchange of entire genes between different viral strains. The ability to manipulate influenza gene segments in various combinations in the laboratory has contributed to its being one of the best characterized viruses, and studies on influenza have provided key contributions toward the understanding of various aspects of virology in general. However, the genetic plasticity of influenza viruses also has serious potential implications regarding vaccine design, pathogenicity, and the capacity for novel viruses to emerge from natural reservoirs and cause global pandemics.

CONTENTS

0066-4197/02/1215-0305$14.00

INTRODUCTION

Influenza A viruses are lipid-enveloped viruses containing a genome composed of eight strands of negative-sense RNA that encode ten viral proteins. These gene segments are encapsidated in a virally encoded nucleoprotein (NP), and the ribonucleoprotein (RNP) structures are associated with the three subunits of the viral polymerase (PB1, PB2, and PA). Virus particle formation occurs at the surface membrane of infected cells, where budding occurs from regions of the membrane at which the viral glycoproteins, hemagglutinin (HA) and neuraminidase (NA), have accumulated. The viral matrix protein (M1) is the most abundant component of the virion and is thought to play a pivotal role in the process of assembly and budding. Details on the mechanism of assembly are still forthcoming, but it is often assumed that M1 interacts with the RNPs and the cytoplasmic domains of HA, NA, and possibly the third integral membrane protein M2. The two other viral proteins, NS1 and NS2, were initially designated as nonstructural proteins, but there is now evidence for the presence of NS2 in virions.

Depending on the virus strain and passage history, influenza A virions can exhibit a variety of shapes and sizes, ranging from fairly spherical particles of approximately 100 nm in diameter to elongated filamentous forms of the virus (69, 71). Initial human isolates tend to be largely of the filamentous type (26), but upon continuous propagation in the laboratory, viruses with more spherical morphology can be selected (25). Regardless of the virion morphology, the most prominent feature of the virus envelope is the layer of tightly packed HA and NA glycoproteins that project from the viral membrane.

STRUCTURE OF THE INFLUENZA GENOME

Segmented Nature of the Genome and Reassortment

It was not until the 1960s that tissue culture systems for growth and plaque purification of influenza were well established. As a consequence, many of the pioneering studies on influenza were dependent on characteristics such as their ability to grow in allantoic membrane cells of embryonated chicken eggs and their capacity to agglutinate erythrocytes (hemagglutination), which made it possible to approximate virus yields. The early studies of influenza genetics also relied heavily on the availability of genetic markers such as neurovirulence, antigenicity, hemagglutination properties, and virus morphology. These markers were not ideal since the basis of the phenotype was ill defined and often polygenic in nature. Despite this, it was clear that during coinfection with viruses of differing phenotypes, recombination of such markers occurred at frequencies much higher than could be explained by classical microbial genetics. This was first observed by Burnet & Lind in 1949 (19), who reported that antigenicity and neurotropism phenotypes could be segregated following virus infection of mouse brains. The high frequencies of recombination that they noted in this and similar studies led them to speculate that "influenza particles are composite, and on entry into the host cell break down into subunits which replicate independently, giving rise to a pool of virus material from which infectious units can be reconstructed" (20).

In 1951 Henle & Liu (60) showed that UV-inactivated influenza viruses could regain infectivity if cells were infected at a multiplicity greater than one, a phenomenon referred to as "multiplicity reactivation." It was subsequently shown that UV-inactivated virus preparations reactivated by coinfection with infectious viruses can display inheritable characteristics derived from each parent, indicating that the inactivated virus had become infectious by acquiring additional genetic material (5). Single cycle kinetic experiments on multiplicity reactivation led Barry (7) to speculate that the influenza genome might be composed of approximately six independent units as targets of UV inactivation. The numerous observations of high-frequency recombination by influenza viruses compared with other RNA viruses such as poliovirus and Newcastle disease virus, and studies on influenza virus defective interfering (DI) particles that were genetically incomplete, led Hirst to formally propose that the influenza genome was composed of RNA fragments (63). Furthermore, he postulated that for influenza virus there might be "two kinds of recombination," one of which "occurs at high rate and involves the exchange of large pieces." This type of recombination, in which individual gene segments or combinations of segments are exchanged during mixed infections, is referred to as reassortment, and the viruses that result from such genetic exchanges are termed reassortant viruses.

Description of the Gene Segments

With the advent of improved electrophoretic techniques for separating nucleic acids and proteins, it was demonstrated that the influenza A genome consists of eight separate RNA segments (8, 103, 123, 132). In this section we identify the gene segments and give a brief description of the gene products. Details on the RNA structure of the gene segments, the functions of the viral proteins, and an extensive list of relevant references have recently been reviewed (87).

Using the A/Puerto Rico/8/34 virus (PR8) as an example, the influenza A viral gene segments range in size from 890 to 2341 nucleotides and contain from 20 to 45 noncoding nucleotides at the 3' end and 23 to 61 nucleotides at the 5' end, depending on the segment. With the exception of position 4 at the 3' end of viral gene segments, which displays U/C heterogeneity, the 12 nucleotides at the 3' end and the 13 nucleotides at the 5' end are completely conserved for all segments in all strains of influenza A virus. These terminal RNA regions are partially complementary and viral promoter activity has been mapped to these domains, but with the exception of the polyadenylation signal, the functional significance of most of the noncoding sequences beyond the conserved domain remains unresolved. There is common speculation that these sequences are involved in binding of NP and/or polymerase complexes, or that they may contribute to packaging signals.

The three largest gene segments encode the subunits of the viral polymerase, PB2, PB1, and PA, which are so named because of their basic (PB2, PB1) or acidic (PA) properties on isoelectric focusing gels. These are responsible for transcribing messenger RNAs (mRNAs), for synthesizing positive-sense antigenomic template RNAs (cRNAs), and for transcribing the cRNAs into the gene segments (vRNAs) that are incorporated into progeny viruses. Segment 4 encodes the hemagglutinin glycoprotein (HA), which is responsible for binding virus to sialic acid–containing cell-surface receptors and for membrane fusion during virus entry into host cells. It is also the principal target for neutralizing antibodies. The nucleoprotein (NP) is the product of the fifth gene segment. This is the protein that encapsidates cRNAs and vRNAs, which is necessary for them to be recognized as templates for the viral polymerase. Segment 6 encodes the neuraminidase (NA), which cleaves sialic acid from virus and host cell glycoconjugates at the end of the virus life cycle to allow mature virions to be released. Segment 7 generates two gene products, the matrix protein, M1, and the M2 protein. M1 mRNA is a collinear transcript, and its product has a structural role in the virion and it is thought to play a fundamental role in virus assembly. The M2 is a small transmembrane protein derived from spliced mRNA. It has proton channel activity that aids in virus disassembly during the initial stages of infection. The eighth gene segment also encodes two proteins due to alternative splicing. These proteins were originally referred to as NS1 and NS2 because they were thought to be nonstructural, but NS2 has since been shown to be a component of virions. The NS1 has numerous functions. It is a regulator of both mRNA splicing and translation, and it also plays a critical role in the modulation of interferon responses to viral infection. The NS2 functions to mediate the export of newly synthesized RNPs from the nucleus and as such, it is also referred to as the nuclear export protein (NEP).

Replication of the Viral Genome

Transcription of the viral mRNAs and replication of the viral genome both occur in the nucleus of infected cells. Initiation of viral mRNA synthesis is primed by host cell RNA fragments containing an $m^7GpppXm$ cap structure (129, 130). These are generated from cellular pol II mRNA transcripts owing to an endonuclease activity provided by the viral polymerase. The host-derived fragments are between 10 to 13 nucleotides in length and do not hydrogen bond to the vRNAs. Template-directed extension of the primers proceeds from a G residue complementary to the C residue at position 2 of the vRNA 3′ end. Termination of mRNA synthesis occurs due to a polyadenylation (poly-A) signal containing 5 to 7 U residues near the vRNA 5′ end.

The switch from mRNA transcription to replication of antigenomic template RNAs and genomic vRNAs occurs later in infection, as it requires synthesis of viral proteins (6, 58). It is thought that unprimed replication of cRNAs and vRNAs, as well as antitermination at the poly-A signal sequences, are dependent on the presence of newly synthesized soluble NP protein that has been imported back into the nucleus (9, 147). Encapsidated vRNAs are then exported from the nucleus and migrate to the plasma membrane where virion assembly occurs.

Packaging of the Gene Segments During Assembly

In theory, the number of possible reassortant viruses containing eight segments is 256, but in practice this is not observed. Although there are numerous examples in which particular gene segments are segregated randomly as a result of mixed infection, structural or functional constraints imposed on the gene products often result in the co-segregation of two or more segments. A number of genetic studies have indicated that functional cooperativity exists among subsets of viral proteins such as those of the polymerase complex, NP and M, and HA and NA (99). As an example, a balance between HA receptor binding and NA receptor destroying functions is clearly required for optimal replication. Drugs that inhibit neuraminidase function select for mutants not only in the active site of NA where the drug binds, but also in the region of the HA receptor binding site (104). In another example involving a mutant of A/WSN/33 virus containing a deletion in the NA stalk region that does not grow well in eggs, mutants were selected that fell into two categories. Half of these selected for insertions in the NA stalk and the other half involved mutations in HA that decreased sialic acid binding affinity (106). Glycosylation of HA near its receptor binding site and truncation of the NA stalk also combine to influence avian influenza virus replication (4). There are also examples in which the HA and M gene segments require functional compatibility between the HA fusion pH and M2 proton channel properties for efficient replication (50, 51).

The question of how influenza gene segments are incorporated into virions has been the subject of speculation for many years. The two predominant hypotheses involve either random packaging of gene segments or the existence of defined packaging signals for selectively including the different RNPs into virus particles. The true nature of gene segment packaging may lie between the most extreme interpretations of these postulates and involve aspects of each.

If packaging is completely random, the average virion would require incorporation of ten segments or more to ensure that the eight individual segments are present in the appropriate percentage of progeny viruses (37, 86). The pleiomorphic nature of virus morphology indicates that there should not be any strict physical limitations on the number of segments that can be incorporated into virions. Random packaging would likely lead to a high proportion of noninfectious virions being generated, and particle-to-infectivity ratios for influenza viruses have been estimated at around 10:1 (32). However, this is not significantly higher than some estimates for RNA viruses with nonsegmented genomes. Influenza viruses are potentially capable of incorporating more than eight segments. In experiments on reassortant viruses generated by mixed infection of a ts mutant of A/chicken/Rostock/34 and A/chicken/Germany/49, Scholtissek et al. (141) showed that influenza "partial heterozygotes" containing both parental genes for either segment 3 or segment 6 could be maintained during plaque-to-plaque passages. Using reverse genetics, Enami et al. (37) demonstrated that a virus with a ts lesion in the NS1 gene could be complemented with an additional gene segment when grown at nonpermissive temperature. The segment with the NS1 mutation was required for providing NS2 function, whereas the ninth segment was required to express wild-type NS1. There are also numerous studies showing that artificial gene segments encoding reporter genes such as chloramphenicol acetyltransferase (CAT) or green fluorescent protein (GFP) can incorporate into infectious particles as a ninth segment, although the extra segment is usually lost following a few rounds of replication.

Several studies have compared the levels of vRNAs in infected cells with levels in budded virions, but the results of these fail to provide definitive conclusions. Hybridization studies by Smith & Hay (152) showed that in released particles of A/chicken/Rostock/34 virus the eight gene segments existed in equimolar amounts, whereas in infected cells these levels could be variable. However, in a similar study, Enami et al. (35) concluded that in cells infected with A/WSN/34 virus the vRNAs were synthesized coordinately and at equimolar levels. Studies on mutants with reduced levels of vRNA synthesis of individual segments are also inconclusive. A transfectant influenza A virus containing the NA gene flanked by noncoding sequences derived from an influenza B virus NS segment was generated by reverse genetics (110), and this virus showed reduced levels of this particular vRNA both in infected cells and virions. However, in other studies on transfectant viruses with mutations in the NA gene flanking sequences (12), wild-type levels of the segments were detected in virions, despite the observation that the relative amounts of these vRNAs were reduced in infected cells. One caveat to the latter two studies is that potential packaging signals may have been affected by the changes to the gene segment flanking regions.

The arguments for packaging signals include studies on influenza DI particles. These contain deletions in specific gene segments in which coding sequences have been removed and sometimes rearranged, but terminal regions are unaffected. Although they do not encode functional proteins, such segments can nonetheless be incorporated into virus particles. In several (but not all) examples when the truncated segment is incorporated into virions, it specifically interferes with the incorporation of the wild-type segment from which it was derived (2, 33, 112).

Studies on NA deletion mutants selected for growth in the presence of exogenously added neuraminidase, or on cells with depleted cell-surface sialic acid density, suggest that specific packaging signals may exist at terminal sequences of the segment. Liu & Air (93) passaged influenza in medium containing antibodies specific for the viral NA and soluble neuraminidase derived from the bacterium *Micromonosporum viridifaciens* and obtained mutants with large deletions in the gene segment that encodes the viral NA. Although these deletions eliminated the protein coding regions responsible for enzymatic activity, truncated gene segments were maintained upon serial passage. The truncated NA gene segments of mutants selected in this manner contained the noncoding sequences at their 3' and 5' termini (172). In another study, mutant MDCK cell lines were generated by growth in the presence of lectins specific for sialic acid–containing influenza virus receptors. The mutant cell lines that were generated contained significantly reduced levels of cell-surface sialic acid, and virus propagation on such lines was not dependent on NA (72). Similar to the results of Liu & Air (1993), viruses passaged on such mutant MDCK cells gave rise to mutants with truncated NA gene segments containing internal deletions that eliminated NA enzymatic activity, but retained the terminal regions of the segment. In the above examples, the sequences retained at the vRNA 3' end minimally encode the short NA cytoplasmic tail and the hydrophobic segment that serves as both signal sequence and transmembrane anchor domain for the glycoprotein. It is possible that the small protein coding regions are maintained because the truncated proteins serve some function not related to NA activity, such as facilitating virus assembly. Infectious influenza viruses containing deletions of the six-residue cytoplasmic tail have been rescued by reverse genetics, but these replicate with reduced efficiency relative to wild type, and have altered morphology (45, 105). Alternatively, nucleotide sequences at the gene segment termini may contain specific packaging signals that cause the eight individual segments to be selectively incorporated into assembled virions in roughly equimolar quantities. The advent of improved reverse genetics techniques should allow this possibility to be addressed in the near future using reporter genes with alternative end sequences.

MUTANTS

Much of what is known about the functions provided by the individual gene products of influenza results from analyses of mutant viruses. Influenza, like other RNA viruses, displays a high mutation rate due to the error-prone nature of the viral polymerase, so mutant viruses are easy to isolate. Studies have been carried out on naturally occurring variants, spontaneous mutants derived in the laboratory, chemically or physically induced mutants, and mutants selected for a particular phenotype such as host range, drug resistance, or antigenicity. A selection of these, and the functional information obtained from their study, are described below.

Temperature-Sensitive Mutants

In the late 1960s and early 1970s, a large collection of influenza virus temperature-sensitive (ts) mutants, derived from differing strains, was generated in several laboratories worldwide [reviewed in (99)]. Such ts mutants were selected for growth at a permissive temperature (usually between 31°C and 36°C, depending on the study), but were significantly inhibited for replication at a higher nonpermissive temperature (usually between 38°C and 42°C). Simpson & Hirst (149) showed that pair-wise crosses of ts mutants at nonpermissive temperature could give rise to non-ts viruses at frequencies much higher than standard reversions were observed. A series of crosses allowed them to identify five independent complementation groups with their mutants. Similar results with ts mutants were obtained in several other laboratories, and taken cumulatively, it was possible to identify eight separate complementation groups (99). By crossing parental strains with gene segments exhibiting different migration patterns on polyacrylamide gels and using various techniques for selecting reassortants of interest, it was possible to assign each complementation group to an individual gene segment. In some cases the function could be defined based on studies of these ts mutants and in others the results were ambiguous. Here we offer a few examples in which the phenotypes of given ts mutants led to insights on the functions specified by particular gene segments.

For example, studies on segment 1 mutants were useful for identifying the PB2 subunit of the polymerase as the viral component involved in binding to the m^7GpppXm cap structures that are needed to prime viral mRNA synthesis. UV-crosslinking studies of purified polymerase complexes with radiolabeled cap structures showed that the PB2 subunit of wild-type virus could be labeled, but that segment 1 mutants of A/WSN/34 virus (WSN) were inhibited for cap binding function in a temperature-dependent fashion (161). The role for PB2 in cap binding is consistent with results obtained with an A/chicken/Rostock/34 (Rostock virus) mutant that was negative for in vitro transcriptase activity at nonpermissive temperature when rabbit globin RNA was used as primer, but displayed activity when the complementary dinucleotide ApG was used as primer (117). Several other segment 1 mutants also showed defects in RNA polymerase activity in vitro and in infected cells (47, 85, 109, 139, 158).

In vitro transcription assays using a ts mutant of segment 2, which encodes the PB1 protein, suggest that it may be involved in the initiation of transcription (108). This is consistent with transcription assays using radiolabeled nucleoside triphosphates, which showed that the first residue added to the primer bound to the PB1 subunit (68, 161). Assays on infected cells with a WSN PB1 ts mutant suggested that it plays a role in cRNA synthesis as well (85, 122). PB1 contains sequence motifs that are conserved among viral RNA polymerases (131), and mutagenesis studies show that these are required for vRNA synthesis (14), suggesting that PB1 may constitute the catalytic subunit for polymerization. PB1 also contains the nuclease activity required for capped primer formation (89).

Neuraminidase ts mutants provide one of the best examples in which the early studies clearly indicated the protein function. EM studies on WSN virus ts mutants showed that at nonpermissive temperature intact virus particles were produced at the plasma membrane of infected cells, but these aggregated and were inefficiently released (124). When bacterial NA was present during infection, no aggregation occurred and wild-type levels of HA activity was detectable in the infected-cell supernatants. Without functional NA activity to remove sialic acid from the viral and infected-cell surfaces, the interactions between HA receptor binding sites and sialoglycoconjugates prevented virus release.

A study on reassortants of a cold-adapted variant of A/Ann Arbor/6/60 virus used in the development of a live attenuated vaccine (98) provided results consistent with the view that M1 functions in virus assembly (120). At nonpermissive temperature this mutant was capable of synthesizing all viral proteins, but the M1 protein inefficiently associated with the plasma membrane resulting in the reduced production of virus particles. In another study, a ts M1 mutation of A/WSN/33 virus led to nuclear retention of the M1 at nonpermissive temperature (133). This did not affect the capacity for viral RNPs to exit the nucleus, but particle formation was significantly reduced.

Temperature-sensitive mutants of segment 8 have been identified, most of them with changes in the NS1 coding region. These mutants exhibit a variety of effects at nonpermissive temperature. Most display reduced synthesis of the M1, and in some cases the levels of HA, NS proteins, and vRNAs are also reduced [Reference (96) and citations within]. This is consistent with the multifunctional nature of the NS1 protein, which is involved in the regulation of both mRNA splicing and translation (87), as well as being an inhibitor of interferon-mediated antiviral responses (43, 174).

Antibody-Resistant Mutants

Antibodies that neutralize virus infectivity are directed against the surface glycoproteins, in particular the HA. As a consequence, the antigenic properties of the HA and NA change from one year to the next as the viruses evolve to evade existing human immunity. A number of studies have identified and characterized the binding sites for neutralizing antibodies, and much of what is known derives from the analysis of monoclonal antibody-resistant mutants (46). Sequencing studies on mutants selected for growth in the presence of anti-HA neutralizing monoclonal antibodies show that they map to regions on the surface of the globular membrane distal domains of the structure (167). The X-ray crystal structures of the HAs of escape mutants show that only localized changes occur, indicating that the mutated residues reside within the antibody footprints (150). For the HA of the H3 subtype virus A/Aichi/2/68, such studies led to the description of five antigenic regions of the molecule (designated A to E), and the location of the region at which resistant mutants were selected reflected the area of the molecule at which the antibodies were thought to bind based on EM studies of HA-antibody complexes (170). These interpretations have subsequently been confirmed by X-ray crystallography studies on such complexes (16, 40). Sequence changes in antigenic mutants selected by monoclonal antibodies against the H1 subtype have been interpreted similarly (21). The locations of the antigenic regions identified in studies on neutralizing anti-HA monoclonal antibodies are representative of the distribution on the structure of the sites that evolve during antigenic drift. The observation that these sites are proximal to the relatively conserved receptor binding pocket of HA, in conjunction with the structural data on HA-antibody complexes, indicates that interference with virus attachment to host cells may constitute the principal mechanism for antibody-mediated virus neutralization (15).

The selection of escape mutants with monoclonal antibodies directed against the NA also allowed for the identification of the residues where changes were detected (1, 164). As with HA, NA mutations resulting in the loss of reactivity were shown to reside within the antibody footprints (119). The location of such mutations at regions of the NA structure adjacent to the enzyme active site (28) suggests that the antibodies act either by blocking access to the site or by distorting the catalytic domain (164).

Drug-Resistant Mutants

The observation that amantadine (1-aminoadamantane hydrochloride) exhibits anti-influenza activity was first made in the 1960s (30), and it was the first compound made available for clinical use against the virus. Amantadine, and the related compound rimantadine (α-methyl-1-adamantane methylamine hydrochloride), act at the level of virus entry by two separate mechanisms. At concentrations of 100 μM or greater, in vitro, they have broad antiviral activity due to their properties as weak bases, which raises the pH of endosomes to mitigate against the pH-activated HA conformational changes required for membrane fusion (29). Mutants selected under these conditions contain substitutions in HA at various locations in the structure, all of which destabilized the protein as deduced by the elevated pH of fusion each displayed. At concentrations of 5 μM or less these compounds exhibit specific activity against several strains of influenza A viruses, but have no effect on replication of influenza B or influenza C viruses. For many years the mode of action was unclear. In most strains it appeared to be at the stage of virus uncoating early in the replication cycle (18, 75). However, in certain strains of H5 and H7 subtype viruses with HAs that are cleaved intracellularly, an effect in the latter stages of infection was also observed (56). Experiments on reassortant viruses isolated following mixed infection of sensitive and resistant virus strains demonstrated that amantadine was operating on the M gene segment (57, 95). Subsequently, drug-resistant mutants were characterized, and it was found that they contained changes in the transmembrane domain of the M2 protein (59).

Several observations prompted speculation that M2 had a function in modulating pH gradients across membranes. With viruses containing HAs that are cleaved intracellularly, it was shown that the presence of amantadine late in infection caused the HAs to be expressed in the low pH conformation, but that this could be circumvented by the addition of agents that elevate the pH of intracellular compartments (159). It was also found that M2 can form tetramers (66, 160), which can be modeled as a four-helix channel in which residues selected for amantadine resistance oriented toward the interior. Furthermore, an amantadine-resistant Rostock virus mutant was isolated that had no changes in M2, but contained a mutation in the HA that rendered it much more stable to acid pH (154). The concept of M2 as a channel protein was subsequently confirmed by electrophysiological experiments (127), and it was determined that the M2 channel activity is selective for protons (24).

Current models for the mode of action of amantadine reason that, during entry, as endosomes are acidified the M2 allows for the concurrent acidification of the virus interior. This is thought to be required for the dissociation of the viral nucleocapsids from the M1 protein (175), which has been proposed to be important for the subsequent transport of nucleocapsids into the host cell nucleus (18, 101). The function of M2 late in infection is thought to involve the removal of protons from trans-Golgi or post-Golgi vesicles during transport of the viral glycoproteins to the cell surface. This is particularly important for HAs that are cleaved in the Golgi apparatus, as they are more susceptible to acidic conditions that could lead to premature conformational changes.

A number of compounds targeting the NA have been developed with the rationale that if they bind to the active site of NA more tightly than sialic acid, they may have antiviral activity (162). In clinical trials the compounds zanamivir and osteltamivir have proven effective both for prophylactic and therapeutic purposes [reviewed in (53)], although when given at the onset of symptoms they generally reduce the length of illness by only about one day. The drugs appear to act as inhibitors of virus release and dissemination (52) as anticipated based on the knowledge of NA function. Mutants resistant to these drugs are not selected as readily as with amantadine; however, a number of examples have been characterized [reviewed in (104)]. Although mutants with changes in the NA were among those reported, those with changes in the HA were actually more abundant. As expected, the NA mutants contained changes in the active site of the enzyme that involved either residues that play a role in catalysis, or residues involved in structural stability. The HA changes were primarily at residues in, or in close proximity to, the sialic acid–receptor binding pocket, and presumably they alter the HA binding properties. These observations substantiate the concept of interplay between the HA and NA—the loss of NA function can be compensated for by a reduction in binding affinity by the HA. As such, the virus particles can elute from cells despite the lack of NA function caused by the inhibitors.

Receptor-Binding Mutants

Differences in sialic acid binding specificity of influenza virus HAs primarily involve their distinction between sialic acid in $\alpha(2,3)$- and $\alpha(2,6)$-linkages to galactose on carbohydrate side chains. Most HAs of avian and equine influenza viruses preferentially recognize receptors containing the $\alpha(2,3)$ linkage, and those of human viruses the $\alpha(2,6)$-linkage. However, these generalizations should not be cosidered an "all-or-none" phenomenon, as various residues may play a role in specificity and mutants can potentially display intermediate phenotypes. HA mutants of human viruses that recognize $\alpha(2,3)$-linkages can be selected by growth in the presence of glycoproteins rich in $\alpha(2,6)$-linked sialic acids. These mutants defined the location of the receptor binding site on the HA molecule and sequence, and

X-ray crystallographic analyses have indicated the importance of HA1 residue 226 in the site for the differing receptor specificities of viruses from different species (134). The importance for receptor binding generally of conserved residues that form the receptor-binding site has been confirmed by site-specific mutagenesis of expressed HAs and the study of mutants generated by reverse genetics (100).

RNA RECOMBINATION

For influenza viruses there have been relatively few documented cases of true RNA recombination, as opposed to gene segment reassortment. Early sequencing studies on the RNA genomes of influenza revealed examples of inverted sequences within gene segments (38), and that the generation of DI particles resulted from recombination events within or between gene segments (39, 74, 113). In another study, Rohde & Scholtissek (135) observed that following mixed infection of two viral strains, one reassortant contained an NP segment with sequences derived from each parental virus as a result of a recombination event.

There are also examples of viruses using nonhomologous recombination to acquire pathogenic traits relating to HA cleavage, which is required to activate membrane fusion potential (see below). In one example, the nonpathogenic virus A/turkey/Oregon/71 was selected for growth in tissue culture in the absence of a protease to cleave HA. This resulted in the acquisition of a 54-nucleotide insertion derived from 28S ribosomal RNA at the region encoding the HA cleavage site that could be cleaved by intracellular proteases, and the resulting virus was more pathogenic for chickens than the parental virus (81). In a similar study with mutants of A/seal/Massachusetts/1/80 virus, a 20-amino acid insertion of residues derived from the viral NP was found at the cleavage site (121).

Several examples of RNA recombination were also demonstrated in experiments with viruses generated by reverse genetics, including one example of a mosaic gene segment being generated in which sequences derived from the M, PB1, and NA genes of A/PR/8/34 virus were incorporated into a bicistronic gene segment encoding A/WSN/33 NA and a CAT reporter gene (11).

A more recent study illustrated that the frequency at which non-homologous recombination can occur during influenza replication may be higher than once appreciated (106). This study utilized a mutant of WSN virus containing a 24-residue deletion in the stalk region of the viral NA. On MDCK cells this mutant virus grows to titers equivalent to WT, but in embryonated chicken eggs it is severely debilitated. Following continuous passage of the mutant in eggs, 10 independent clones that grew efficiently were isolated. Five of these contained insertions in the NA stalk region, ranging from 10 to 22 amino acids, which were derived from coding sequences for either the PB1, PB2, or the NP proteins. The other five clones displayed reduced HA receptor-binding activity to compensate for the reduction of NA function.

Although examples of true RNA recombination of influenza viruses in nature have been rare, a recent phylogenetic analysis of the HA gene from viruses thought to be responsible for the pandemic of 1918–1919 prompted the investigators to suggest that these viruses may have been HA recombinants (48). They speculated that the HA2 sequences, which compose most of the stalk region of the molecule, may have derived from a human-lineage influenza, whereas the HA1 globular head domain of the molecule, which contains the receptor binding site and antigenic regions, came from swine viruses. However, an alternative interpretation of the data has also been proposed (168).

REVERSE GENETICS

The genetic manipulation of negative strand viruses such as influenza proved to be a challenge, as neither the genomic RNA (vRNA) nor the antigenomic RNA (cRNA) are infectious as naked RNA. For transcription and replication to occur these RNAs must be encapsidated by the NP protein and associated with the proteins of the polymerase complex, PB1, PB2, and PA. In the late 1980s, a major breakthrough for reverse genetics using negative-strand viruses came through studies on influenza virus by Palese and colleagues. They showed that purified NP and polymerase proteins could be used to reconstitute replication-competent in vitro transcribed RNAs (97), and these techniques soon led to the rescue of infectious viruses containing specified mutations (36). The generation of such viruses involved infection of cells by a "helper virus," transfection of the cells with in vitro–generated mutant RNPs, and selection of viruses containing the mutant gene segment away from the pool of viruses with the homologous segment derived from the helper. It subsequently became possible to apply the reverse genetics technology to a broad range of topics including functional studies on several of the viral proteins, examination of the requirements of the noncoding sequences for transcription and replication, the expression of foreign antigens, and the use of manipulated viruses for vaccine studies (44, 114).

The principal limitation on the infection-transfection system was the requirement for a highly efficient selection system. Selection systems based on properties such as host range, antibody reactivity, and drug susceptibility eventually led to the development of rescue systems for six of the eight gene segments, but debilitated viruses could be difficult to rescue, and often the choice of viral strains that were amenable for study was limited. Over several years, many laboratories developed variations of the techniques for virus rescue in attempts to improve efficiency, but none of these eliminated the requirement for helper virus and selection. However, one such alternative developed by Hobom and colleagues ultimately proved to be a significant step forward, as it circumvented the requirement for generating the RNPs in vitro (116). This approach involved the use of cDNAs containing viral gene segments flanked by RNA polymerase I (pol I) promoter and terminator sequences. RNA pol I normally transcribes ribosomal RNAs, which are not capped and contain no signals for 3′ polyadenylation. The pol I promoter and terminator direct defined 5′ transcription start and 3′ stop signals, so it was possible to use these to generate influenza gene segment RNAs with the correct termini. Pleschka et al. (128) then demonstrated the utility of employing constructs such as these for making infectious viruses using the helper virus system.

In 1999, the long-term goal of generating influenza virus entirely from plasmid DNAs was realized (41, 115). In the system developed by Kawaoka and colleagues (115), the eight gene segments of WSN virus were each cloned into plasmids between the human pol I promoter and mouse pol I terminator and these were transfected into human 293T cells along with plasmids encoding the nine influenza structural proteins. Remarkably, within two days in excess of 10^7 plaque-forming units per ml. were recovered from the transfected cell supernatants, and more recently the efficiency has been improved approximately tenfold (114). Using a similar plasmid-only virus rescue system that utilized a pol I promoter to generate $5'$ RNA ends and hepatitis delta virus ribozyme sequences to generate the $3'$ termini of the gene segments, Fodor et al. (41) also demonstrated the rescue of infectious viruses. Subsequently, the pol I system has been modified by Hoffmann et al. (65) to reduce the number of plasmids required to eight. In this system the gene segment RNAs flanked by pol I promoter and terminator sequences are in turn flanked by a CMV promoter (pol II) and polyadenylation signals in the opposite orientation such that viral mRNAs can be transcribed from the same plasmid.

ANTIGENICITY

The genetic properties and ecological diversity of influenza A make it a classic example of a re-emerging virus. Influenza A viruses are renowned for their capacity to cause epidemics on a nearly annual basis due to the continuously evolving nature of their surface glycoproteins, referred to as antigenic drift. At unpredictable intervals viruses with completely different surface antigens are introduced into humans. As large segments of the population have little or no immunity to these strains, they cause global pandemics. This is known as antigenic shift.

Antigenic Shift

Serologically, 15 nonoverlapping subtypes of HA and 9 subtypes of NA have been identified. All of these occur in viruses that circulate in aquatic birds, and these species serve as the natural reservoir for all influenza A viruses (169). Until recently, only H1N1, H2N2, and H3N2 viruses were known to have circulated extensively in humans. In 1918 and 1919, H1N1 viruses were responsible for the pandemic of "Spanish 'flu." This was the worst outbreak of infectious disease in human history, causing an estimated 20 to 40 million deaths. Strains related to the 1918 viruses continued to circulate in humans until 1957, when they were replaced by H2N2 subtype "Asian" strains resulting in the next pandemic. In 1968, another antigenic shift occurred with the emergence of H3N2 subtypes of "Hong Kong 'flu." In this case the HA, but not the NA, was antigenically novel to humans. The fact that a pandemic resulted illustrates the concept that neutralizing antibody responses targeting the HA provide the key component for protection against influenza, although residual immunity to the NA may have reduced the severity of disease resulting from the 1968 viruses.

Descendants of these H3N2 viruses have continued to circulate in humans to this day. However, in 1977 H1N1 subtype viruses that were virtually identical to strains that had previously circulated in 1950 re-emerged in humans (80, 111, 143). The source of the viruses that re-emerged in 1977 is unknown, but it is possible that they had been preserved in a frozen state. Unlike the situation with H2N2 viruses in 1957 and H3N2 viruses in 1968, the 1977 H1N1 strains did not displace the H3N2 viruses. For the past 25 years H3N2 and H1N1 viruses have co-circulated in humans. This has offered the opportunity for these human viruses of different subtypes to reassort with one another (173), and within the past year a number of H1N2 subtype viruses have been isolated from humans (Y. P. Lin & A. Hay, personal communication).

In theory, there are several potential routes by which viruses with novel surface antigens could be introduced into humans. An avian virus could transmit directly to humans, as appears to have been the case with the 1997 H5N1 viruses described below, and with the avian H9N2 subtype viruses that transmitted to humans in 1999 (91). It is also possible that avian-like viruses could infect humans following an intermediate period in an alternative host, as a number of influenza viruses have also been isolated from other mammals. For example, H1N1, H3N2, H9N2, H1N7, and H1N2 subtypes have all been isolated from pigs, and H3N8 and H7N7 subtypes from horses. There have also been sporadic examples of viruses of various subtypes being isolated from seals, whales, and minks (165). The viruses that caused the pandemics of 1957 and 1968 were derived from the reassortment of avian and human strains (77, 144). In 1957, the HA, NA, and PB1 gene segments came from an avian virus and the other segments were obtained from the human H1N1 viruses that were circulating at the time. The H3N2 viruses that emerged in 1968 resulted from the replacement of the HA and PB1 gene segments of circulating human H2N2 viruses with their counterparts from an avian source. The host species in which these reassortment events took place is not known. Avian viruses generally do not replicate well in humans (3), and vice versa (62), so it is possible that reassortment took place in an intermediate host. Pigs have been proposed as a potential "mixing vessel" for reassortment (140), and both avian and human strains can replicate in these hosts (61, 82). One reason for this may be that cells of the pig trachea, where influenza replication occurs, contain the $\alpha(2,3)$-linked sialic acid receptors preferred by avian viruses, as well as those with the $\alpha(2,6)$ linkage favored by human strains (73).

Antigenic Drift

In aquatic birds that provide the natural reservoir for influenza A viruses, phylogenetic analyses show the viruses to be virtually in evolutionary stasis (165). However, when antigenically novel strains are introduced into humans they rapidly and continuously evolve, presumably due to immune selection and various factors that may be involved in adapting to a new host. The glycoproteins in particular exhibit rapid rates of evolution. For the HA the rates of evolution approximate 6.7×10^{-3} substitutions per nucleotide per year and for the NA, 3.2×10^{-3} substitutions per nucleotide per year (151). In the HA most of these occur in the HA1 domain at antigenic sites described above, as a result of immune selection. The genetic drift of the glycoproteins necessitates constant surveillance in order to monitor the antigenicity of circulating strains such that vaccines can be updated when necessary.

If one assumes that humans normally generate polyvalent antibody responses to a given antigen, a question that arises regarding drift concerns the reason that individuals can be re-infected with a given subtype in subsequent years, and how the same subtype can continue to circulate for several decades. One clue, based on the analysis of postinfection human sera, indicates that antibody responses are composed of a limited antibody repertoire (163). As a consequence, only a limited number of changes in the viral glycoproteins may be required to evade the immune response generated by any particular individual to a previously circulating virus.

PATHOGENICITY

For certain avian strains of H5 and H7 subtype viruses the cleavage properties displayed by the HA are the most notable determinant of virus pathogenicity. However, it is also clear that several, if not all eight, gene segments can contribute to the pathogenic phenotype of a virus, depending on considerations such as the genetic background of the virus, the host, and the route of infection.

HA Cleavage as a Major Determinant of Pathogenicity

HA is synthesized as a precursor polypeptide (22) that must be cleaved into the disulfide linked subunits HA1 and HA2 in order to activate membrane fusion function and virus infectivity (84, 88). For the HAs of mammalian and most avian influenza A viruses this cleavage activation step occurs only after the glycoprotein has been transported to the plasma membrane. This restricts the sites of replication in infected animals to organs or tissues where appropriate extracellular proteases are present. In contrast, there are a number of examples of highly pathogenic avian virus strains, in particular among H5 and H7 subtype viruses, which can cause systemic infections in hosts such as chickens and turkeys. The HAs of these viruses are cleaved intracellularly, in the Golgi apparatus, by subtilisin-like proteases such as furin or PC6 (67, 155). These proteases are present in nearly all cell types, and this alleviates one of the restrictions on the potential sites for multicycle replication in the host and facilitates systemic spread of the virus.

In the HA precursor, the cleavage site exists as a loop structure that extends out from the surface of the molecule (22). In nonpathogenic viruses the HA generally contains a single arginine residue at the site of cleavage, which is recognized by extracellular trypsin-like proteases. The HAs of highly pathogenic viruses are often characterized by insertions of additional basic residues at the site of cleavage (83, 153, 166). This serves two functions regarding protease recognition. It extends the site of cleavage further from the surface of the trimer, presumably making it more accessible for activating proteases, and it generates sequence motifs that match or closely resemble the consensus furin recognition sequence R-X-R/K-R. Another factor that can influence cleavage involves the presence or absence of carbohydrate attachment sites adjacent to the cleavage loop. There are examples in which mutations that eliminate such glycosylation sites lead to acquisition of the highly pathogenic phenotype, most likely as the result of an increase in accessibility of the cleavage loop (31, 78, 79).

Of the viruses that have circulated in humans over the past century, there have been no reported examples of strains that contain HA cleavage site features associated with highly pathogenic viruses. However, in 1997 pathogenic avian H5N1 subtype viruses crossed the species barrier and caused a limited outbreak in humans in Hong Kong. Fortunately, these viruses did not develop the capacity for efficient human-to-human transmission, but of the 18 diagnosed cases, 6 were fatal. All of the viruses isolated from H5N1-infected humans were found to contain polybasic sequences at the HA cleavage site that provide furin recognition motifs (10, 27, 148, 156, 157). The viruses isolated were closely related, but in studies using mouse models, nearly all isolates segregated into either high- or low-pathogenicity phenotypes (34, 42, 54, 94, 118). One sequencing study showed that among viral proteins of these isolates, five mutations in four different proteins other than HA correlated with high pathogenicity (76). Two of these were in the PB1 protein, and one each in the PB2, NA, and M1 proteins. An alternative approach using reverse genetics implicated a single mutation in the PB2 protein as the critical determinant of pathogenicity among these isolates in mice, although a notable effect due to heterogeneity of a residue in the HA receptor binding site was also observed (55). Interestingly, the PB2 mutation identified in this study was distinct from the PB2 change observed in the sequencing study. The viruses were not highly pathogenic unless the polybasic sequence at the HA cleavage site was present, regardless of the other mutations, indicating that it is necessary, but not sufficient, for full manifestation of the phenotype.

Multifactorial Nature of Pathogenicity

Examples of studies suggesting that pathogenicity is a polygenic trait date from the 1950s. In experiments that involved mixed infection of the mouse neurovirulent strains WSN or NWS with avirulent strains such as A/Melbourne/35, it was observed that progeny viruses could be generated in which the virulence and antigenic properties were segregated, and that recombinant viruses with varying degrees of virulence could also be obtained (64, 92). In another example, Mayer et al. (102) showed with reciprocal recombinants of NWS virus (H1N1) and the avirulent virus A/Japan/305/57 (H2N2), that both H1N2 and H2N1 viruses were capable of lethally infecting mouse brains, suggesting that neither surface antigen is the exclusive determinant of neurovirulence. Numerous other genetic studies on WSN and NWS strains of influenza have been carried out over the years to indicate that more than one gene influences virulence, and these have been reviewed extensively by Schlesinger et al. (138). However, studies on WSN using reverse genetics technology suggest that a glycosylation site and a C-terminal lysine residue (49, 90) of the viral NA allow it to sequester plasminogen and facilitate HA cleavage (49), and this may have a particular significance for the virulence of this strain. Thus, it seems that when gene segments harbor traits that are known to influence pathogenicity, the genetic background provided by other segments can play a significant role in modulating this phenotype.

Studies on reassortant avian viruses also provided evidence for the polygenic nature of pathogenicity and the complexity of this phenomenon. For example, the viruses A/chicken/Rostock/34 and A/turkey/England/63 both contain polybasic sequences at the HA1-HA2 cleavage site and are highly pathogenic for chickens. However, both nonpathogenic and pathogenic viruses were generated from these strains following co-infection of chick cells, and both HAs were represented among the nonpathogenic reassortants (137). In these viruses, it seemed that the mixing of segments encoding proteins of the polymerase complex was responsible for the loss of pathogenicity. In another study, Rostock virus reassortants were generated containing single gene segment replacements, in which the replaced segments were derived from a variety of different virus strains (142). Among these reassortant viruses, the capacity for modulation of pathogenesis was demonstrated for each of the seven gene segments analyzed.

Many viruses, including influenza, are capable of inducing apoptosis, and this property may contribute to the pathogenicity of these viruses. Different strains of influenza vary in their capacity to induce apoptosis, but detailed mechanisms on how or why this is the case are not yet clear. The viral NA (107, 146) and NS1 (145) proteins have both been implicated in triggering apoptosis, and recently it was reported that the PB1 gene can encode a reasonably well-conserved 87-residue protein from an alternative reading frame, which has the capacity to localize to the mitochondria and induce cell death (23).

INFLUENZA B AND INFLUENZA C VIRUSES

Influenza B and influenza C viruses are orthomyxoviruses that are related to influenza A virus, but intertypic reassortment among the three genera has not been demonstrated. Influenza B virus can cause disease symptoms similar to influenza A, and in some years is probably responsible for more illness than influenza A (169). In general, influenza C virus causes less severe respiratory illness, which rarely progresses to the lower respiratory tract. Although there have been examples of influenza B and C viruses transmitting to other hosts they are essentially human viruses, so unlike influenza A, they do not have a natural reservoir from which they can recruit antigenically novel surface antigens. This may be one factor involved in the different patterns of antigenic drift displayed by influenza B and C viruses by comparison to influenza A. Whereas influenza A viruses generally follow a single evolutionary lineage from one year to the next, antigenically diverse strains of influenza B and C viruses co-circulate (125).

With regard to genome structure, the major difference among influenza A, B, and C viruses pertains to the surface glycoproteins. Influenza A and B viruses have eight gene segments and contain separate attachment and receptor-destroying envelope glycoproteins (HA and NA), whereas influenza C virus has only seven gene segments, with a single glycoprotein, the HEF, providing both of these functions as well as membrane fusion activity. The structures of influenza A HA and influenza C HEF demonstrate that they are largely comparable, but with HEF a domain responsible for the receptor destroying esterase activity is inserted between the stalk region of the protein and the receptor binding domain (136). Other differences among the three genera involve the coding strategies for the gene segment encoding the NA and matrix proteins. Influenza B virus segment six, which encodes the NA, also encodes a protein that may be functionally similar to the influenza A M2 protein. This protein, designated NB, uses an alternative initiation codon to synthesize a 100-residue type III membrane protein that, like M2, is incorporated into virions (13, 17). In addition to the M1 protein, segment seven of influenza B virus also encodes a protein of unknown function, BM2, by utilizing a tandem cistron translational stop-start mechanism (70). Influenza C virus segment six encodes the proteins CM1 and CM2 that may be functionally similar to the influenza A M1 and M2 proteins. CM1 derives from a spliced message (171), whereas CM2 is the proteolytic product of a precursor protein that also produces a rapidly degraded membrane-bound polypeptide of unknown function (126).

Among the influenza A, B, and C viruses, proteins that appear to have similar functions have evolved strategically different mechanisms for their expression during virus replication. From an evolutionary standpoint, it will be interesting to analyze and compare the structural and functional properties of the proteins encoded by these viruses, to attempt to relate these to the various ecological niches that these viruses inhabit, and to understand how and why these viruses can establish themselves in a particular host while possibly maintaining the genetic means to transmit to another.

The *Annual Review of Genetics* is online at http://genet.annualreviews.org

LITERATURE CITED

1. Air GM, Els MC, Brown LE, Laver WG, Webster RG. 1985. Location of antigenic sites on the three-dimensional structure of the influenza N2 virus neuraminidase. *Virology* 145:237–48
2. Akkina RK, Chambers TM, Nayak DP. 1984. Expression of defective-interfering influenza virus-specific transcripts and polypeptides in infected cells. *J. Virol.* 51: 395–403
3. Baere AS, Webster RG. 1991. Replication of avian influenza viruses in humans. *Arch. Virol.* 119:37–42
4. Baigent SJ, McCauley JW. 2001. Glycosylation of haemagglutinin and stalk-length of neuraminidase combine to regulate

the growth of avian influenza viruses in tissue culture. *Virus Res.* 79:177–85

5. Baron S, Jensen KE. 1955. Evidence for genetic interaction between non-infectious and infectious influenza A viruses. *J. Exp. Med.* 102:677–97

6. Barrett T, Wolstenholme AJ, Mahy BW. 1979. Transcription and replication of influenza virus RNA. *Virology* 98:211–25

7. Barry RD. 1961. The multiplication of influenza virus. II. Multiplicity reactivation of ultraviolet irradiated virus. *Virology* 14:398–405

8. Bean WJ Jr, Simpson RW. 1976. Transcriptase activity and genome composition of defective influenza virus. *J. Virol.* 18:365–69

9. Beaton AR, Krug RM. 1986. Transcription antitermination during influenza viral template RNA synthesis requires the nucleocapsid protein and the absence of a 5′ capped end. *Proc. Natl. Acad. Sci. USA* 83:6282–86

10. Bender C, Hall H, Huang J, Klimov A, Cox N, et al. 1999. Characterization of the surface proteins of influenza A (H5N1) viruses isolated from humans in 1997–1998. *Virology* 254:115–23

11. Bergmann M, Garcia-Sastre A, Palese P. 1992. Transfection-mediated recombination of influenza A virus. *J. Virol.* 66: 7576–80

12. Bergmann M, Muster T. 1995. The relative amount of an influenza A virus segment present in the viral particle is not affected by a reduction in replication of that segment. *J. Gen. Virol.* 76:3211–15

13. Betakova T, Nermut MV, Hay AJ. 1996. The NB protein is an integral component of the membrane of influenza B virus. *J. Gen. Virol.* 77:2689–94

14. Biswas SK, Nayak DP. 1994. Mutational analysis of the conserved motifs of influenza A virus polymerase basic protein 1. *J. Virol.* 68:1819–26

15. Bizebard T, Barbey-Martin C, Fleury D, Gigant B, Barrere B, et al. 2001. Structural studies on viral escape from antibody neutralization. *Curr. Top. Microbiol. Immunol.* 260:55–64

16. Bizebard T, Gigant B, Rigolet P, Rasmussen B, Diat O, et al. 1995. Structure of influenza virus haemagglutinin complexed with a neutralizing antibody. *Nature* 376:92–94

17. Brassard DL, Leser GP, Lamb RA. 1996. Influenza B virus NB glycoprotein is a component of the virion. *Virology* 220: 350–60

18. Bukrinskaya AG, Vorkunova NK, Kornilayeva GV, Narmanbetova RA, Vorkunova GK. 1982. Influenza virus uncoating in infected cells and effect of rimantadine. *J. Gen. Virol.* 60:49–59

19. Burnet FM, Lind PE. 1949. Recombination of characters between two influenza virus strains. *Aust. J. Sci.* 12:109–10

20. Burnet FM, Lind PE. 1951. A genetic approach to variation in influenza viruses. 3. Recombination of characters in influenza virus strains used in mixed infections. *J. Gen. Microbiol.* 5:59–66

21. Caton AJ, Brownlee GG, Yewdell JW, Gerhard W. 1982. The antigenic structure of the influenza virus A/PR/8/34 hemagglutinin (H1 subtype). *Cell* 31:417–27

22. Chen J, Lee KH, Steinhauer DA, Stevens DJ, Skehel JJ, Wiley DC. 1998. Structure of the hemagglutinin precursor cleavage site, a determinant of influenza pathogenicity and the origin of the labile conformation. *Cell* 95:409–17

23. Chen W, Calvo PA, Malide D, Gibbs J, Schubert U, et al. 2001. A novel influenza A virus mitochondrial protein that induces cell death. *Nat. Med.* 7:1306–12

24. Chizhmakov IV, Geraghty FM, Ogden DC, Hayhurst A, Antoniou M, Hay AJ. 1996. Selective proton permeability and pH regulation of the influenza virus M2 channel expressed in mouse erythroleukaemia cells. *J. Physiol.* 494:329–36

25. Choppin PW, Murphy JS, Tamm I. 1960. Studies of two kinds of virus particles which comprise influenza A2 strains. III.

Morphological characteristics: independence of morphological and functional traits. *J. Exp. Med.* 112:945–52

26. Chu CM, Dawson IM, Elford WJ. 1949. Filamentous forms associated with newly isolated influenza virus. *Lancet* i:602–3

27. Claas EC, Osterhaus AD, van Beek R, De Jong JC, Rimmelzwaan GF, et al. 1998. Human influenza A H5N1 virus related to a highly pathogenic avian influenza virus. *Lancet* 351:472–77

28. Colman PM, Varghese JN, Laver WG. 1983. Structure of the catalytic and antigenic sites in influenza virus neuraminidase. *Nature* 303:41–44

29. Daniels RS, Downie JC, Hay AJ, Knossow M, Skehel JJ, et al. 1985. Fusion mutants of the influenza virus hemagglutinin glycoprotein. *Cell* 40:431–39

30. Davies WL, Grunert RR, Haff RF, McGahen JW, Neumayer EM, et al. 1964. Antiviral activity of 1-adamantanamine (amantadine). *Science* 144:862–63

31. Deshpande KL, Fried VA, Ando M, Webster RG. 1987. Glycosylation affects cleavage of an H5N2 influenza virus hemagglutinin and regulates virulence. *Proc. Natl. Acad. Sci. USA* 84:36–40

32. Donald HB, Issacs A. 1954. Counts of influenza virus particles. *J. Gen. Microbiol.* 10:457–64

33. Duhaut SD, McCauley JW. 1996. Defective RNAs inhibit the assembly of influenza virus genome segments in a segment-specific manner. *Virology* 216:326–37

34. Dybing JK, Schultz-Cherry S, Swayne DE, Suarez DL, Perdue ML. 2000. Distinct pathogenesis of Hong Kong-origin H5N1 viruses in mice compared to that of other highly pathogenic H5 avian influenza viruses. *J. Virol.* 74:1443–50

35. Enami M, Fukuda R, Ishihama A. 1985. Transcription and replication of eight RNA segments of influenza virus. *Virology* 142:68–77

36. Enami M, Luytjes W, Krystal M, Palese P. 1990. Introduction of site-specific mutations into the genome of influenza virus. *Proc. Natl. Acad. Sci. USA* 87:3802–5

37. Enami M, Sharma G, Benham C, Palese P. 1991. An influenza virus containing nine different RNA segments. *Virology* 185:291–98

38. Fields S, Winter G. 1981. Nucleotide-sequence heterogeneity and sequence rearrangements in influenza virus cDNA. *Gene* 15:207–14

39. Fields S, Winter G. 1982. Nucleotide sequences of influenza virus segments 1 and 3 reveal mosaic structure of a small viral RNA segment. *Cell* 28:303–13

40. Fleury D, Barrere B, Bizebard T, Daniels RS, Skehel JJ, Knossow M. 1999. A complex of influenza hemagglutinin with a neutralizing antibody that binds outside the virus receptor binding site. *Nat. Struct. Biol.* 6:530–34

41. Fodor E, Devenish L, Engelhardt OG, Palese P, Brownlee GG, Garcia-Sastre A. 1999. Rescue of influenza A virus from recombinant DNA. *J. Virol.* 73:9679–82

42. Gao P, Watanabe S, Ito T, Goto H, Wells K, et al. 1999. Biological heterogeneity, including systemic replication in mice, of H5N1 influenza A virus isolates from humans in Hong Kong. *J. Virol.* 73:3184–89

43. Garcia-Sastre A. 2001. Inhibition of interferon-mediated antiviral responses by influenza A viruses and other negative-strand RNA viruses. *Virology* 279:375–84

44. Garcia-Sastre A, Palese P. 1993. Genetic manipulation of negative-strand RNA virus genomes. *Annu. Rev. Microbiol.* 47:765–90

45. Garcia-Sastre A, Palese P. 1995. The cytoplasmic tail of the neuraminidase protein of influenza A virus does not play an important role in packaging this protein into viral envelopes. *Virus Res.* 37:37–47

46. Gerhard W. 2001. The role of the antibody response in influenza virus infection. *Curr. Top. Microbiol. Immunol.* 260:171–90

47. Ghendon YZ, Markushin SG, Klimov AI, Hay AJ. 1982. Studies of fowl plague virus temperature-sensitive mutants with

defects in transcription. *J. Gen. Virol.* 63: 103–11

48. Gibbs MJ, Armstrong JS, Gibbs AJ. 2001. Recombination in the hemagglutinin gene of the 1918 "Spanish flu." *Science* 293:1842–45

49. Goto H, Kawaoka Y. 1998. A novel mechanism for the acquisition of virulence by a human influenza A virus. *Proc. Natl. Acad. Sci. USA* 95:10224–28

50. Grambas S, Bennett MS, Hay AJ. 1992. Influence of amantadine resistance mutations on the pH regulatory function of the M2 protein of influenza A viruses. *Virology* 191:541–49

51. Grambas S, Hay AJ. 1992. Maturation of influenza A virus hemagglutinin-estimates of the pH encountered during transport and its regulation by the M2 protein. *Virology* 190:11–18

52. Gubareva LV, Bethell R, Hart GJ, Murti KG, Penn CR, Webster RG. 1996. Characterization of mutants of influenza A virus selected with the neuraminidase inhibitor 4-guanidino-Neu5Ac2en. *J. Virol.* 70:1818–27

53. Gubareva LV, Kaiser L, Hayden FG. 2000. Influenza virus neuraminidase inhibitors. *Lancet* 355:827–35

54. Gubareva LV, McCullers JA, Bethell RC, Webster RG. 1998. Characterization of influenza A/HongKong/156/97 (H5N1) virus in a mouse model and protective effect of zanamivir on H5N1 infection in mice. *J. Infect. Dis.* 178:1592–96

55. Hatta M, Gao P, Halfmann P, Kawaoka Y. 2001. Molecular basis for high virulence of Hong Kong H5N1 influenza A viruses. *Science* 293:1840–42

56. Hay A, Zambon MC, Wolstenholme AJ, Skehel JJ, Smith MH. 1986. Molecular basis of resistance of influenza A viruses to amantadine. *J. Antimicrobiol. Chemother.* 18(Suppl. B):19–29

57. Hay AJ, Kennedy NC, Skehel JJ, Appleyard G. 1979. The matrix protein gene determines amantadine-sensitivity of influenza viruses. *J. Gen. Virol.* 42:189–91

58. Hay AJ, Skehel JJ, McCauley J. 1982. Characterization of influenza virus RNA complete transcripts. *Virology* 116:517–22

59. Hay AJ, Wolstenholme AJ, Skehel JJ, Smith MH. 1985. The molecular basis of the specific anti-influenza action of amantadine. *EMBO J.* 4:3021–24

60. Henle W, Liu OC. 1951. Studies on host-virus interactions in the chick embryo influenza virus system. *J. Exp. Med.* 94: 305–22

61. Hinshaw VS, Webster RG, Easterday BC, Bean WJ. 1981. Replication of avian influenza A viruses in mammals. *Infect. Immunol.* 34:354–61

62. Hinshaw VS, Webster RG, Naeve CW, Murphy BR. 1983. Altered tissue tropism of human-avian reassortant influenza viruses. *Virology* 128:260–63

63. Hirst GK. 1962. Genetic recombination with Newcastle disease virus, poliovirus, and influenza. *Cold Spring Harbor Symp. Quant. Biol.* 27:303–8

64. Hirst GK, Gotlieb T. 1955. The experimental production of combination forms of virus. V. Alterations in the virulence of neurotropic influenza virus as a result of mixed infection. *Virology* 1:221–35

65. Hoffmann E, Neumann G, Kawaoka Y, Hobom G, Webster RG. 2000. A DNA transfection system for generation of influenza A virus from eight plasmids. *Proc. Natl. Acad. Sci. USA* 97:6108–13

66. Holsinger LJ, Lamb RA. 1991. Influenza virus M2 integral membrane protein is a homotetramer stabilized by formation of disulfide bonds. *Virology* 183:32–43

67. Horimoto T, Nakayama K, Smeekens SP, Kawaoka Y. 1994. Proprotein-processing endoproteases PC6 and furin both activate hemagglutinin of virulent avian influenza viruses. *J. Virol.* 68:6074–78

68. Horisberger MA. 1982. Identification of a catalytic activity of the large basic P polypeptide of influenza virus. *Virology* 120:279–86

69. Horne RW, Waterson AP, Wildy P, Farnham AP. 1960. The structure and composition of myxoviruses. I. Electron microscope studies of the structure of myxovirus particles by negative staining techniques. *Virology* 11:79–98

70. Horvath CM, Williams MA, Lamb RA. 1990. Eukaryotic coupled translation of tandem cistrons: identification of the influenza B virus BM2 polypeptide. *EMBO J.* 9:2639–47

71. Hoyle L, Horne RW, Waterson AP. 1961. The structure and composition of the myxoviruses II. Components released from the influenza virus particle by ether. *Virology* 13:448–59

72. Hughes MT, McGregor M, Suzuki T, Suzuki Y, Kawaoka Y. 2001. Adaptation of influenza A viruses to cells expressing low levels of sialic acid leads to loss of neuraminidase activity. *J. Virol.* 75:3766–70

73. Ito T, Couceiro JN, Kelm S, Baum LG, Krauss S, et al. 1998. Molecular basis for the generation in pigs of influenza A viruses with pandemic potential. *J. Virol.* 72:7367–73

74. Jennings PA, Finch JT, Winter G, Robertson JS. 1983. Does the higher order structure of the influenza virus ribonucleoprotein guide sequence rearrangements in influenza viral RNA? *Cell* 34:619–27

75. Kato N, Eggers HJ. 1969. Inhibition of uncoating of fowl plague virus by 1-adamantanamine hydrochloride. *Virology* 37:632–41

76. Katz JM, Lu X, Tumpey TM, Smith CB, Shaw MW, Subbarao K. 2000. Molecular correlates of influenza A H5N1 virus pathogenesis in mice. *J. Virol.* 74:10807–10

77. Kawaoka Y, Krauss S, Webster RG. 1989. Avian-to-human transmission of of the PB1 gene of influenza A virus in the 1957 and 1968 pandemics. *J. Virol.* 63:4603–8

78. Kawaoka Y, Naeve CW, Webster RG. 1984. Is virulence of H5N2 influenza viruses in chickens associated with loss of carbohydrate from the hemagglutinin? *Virology* 139:303–16

79. Kawaoka Y, Webster RG. 1989. Interplay between carbohydrate in the stalk and the length of the connecting peptide determines the cleavability of influenza virus hemagglutinin. *J. Virol.* 63:3296–300

80. Kendal AP NG, Skehel JJ, Dowdle WR. 1978. Antigenic similarity of influenza A (H1N1) viruses from epidemics in 1977–1978 to "Scandinavian" strains isolated in epidemics of 1950–1951. *Virology* 89:632–36

81. Khatchikian D, Orlich M, Rott R. 1989. Increased viral pathogenicity after insertion of a 28S ribosomal RNA sequence into the haemagglutinin gene of an influenza virus. *Nature* 340:156–57

82. Kida HIT, Yasuda J, Shimizu Y, Itakura C, Shortridge KF, et al. 1994. Potential for transmission of avian influenza viruses to pigs. *J. Gen. Virol.* 75:2183–88

83. Klenk HD, Garten W. 1994. Host cell proteases controlling virus pathogenicity. *Trends Microbiol.* 2:39–43

84. Klenk HD, Rott R, Orlich M, Blodorn J. 1975. Activation of influenza A viruses by trypsin treatment. *Virology* 68:426–39

84a. Knipe DM, Howley PM, eds. 2001. *Fields Virology.* Philadelphia: Lippincott, Williams & Wilkins

85. Krug RM, Ueda M, Palese P. 1975. Temperature-sensitive mutants of influenza WSN virus defective in virus-specific RNA synthesis. *J. Virol.* 16:790–96

86. Lamb RA, Choppin PW. 1983. The gene structure and replication of influenza virus. *Annu. Rev. Biochem.* 52:467–506

87. Lamb RA, Krug RM. 2001. Orthomyxoviridae: the viruses and their replication. See Ref. 84a, pp. 1487–531

88. Lazarowitz SG, Choppin PW. 1975. Enhancement of the infectivity of influenza A and B viruses by proteolytic cleavage of the hemagglutinin polypeptide. *Virology* 68:440–54

89. Li ML, Rao P, Krug RM. 2001. The

active sites of the influenza cap-dependent endonuclease are on different polymerase subunits. *EMBO J.* 20:2078–86

90. Li S, Schulman J, Itamura S, Palese P. 1993. Glycosylation of neuraminidase determines the neurovirulence of influenza A/WSN/33 virus. *J. Virol.* 67:6667–73

91. Lin YP, Shaw M, Gregory V, Cameron K, Lim W, et al. 2000. Avian-to-human transmission of H9N2 subtype influenza A viruses: relationship between H9N2 and H5N1 human isolates. *Proc. Natl. Acad. Sci. USA* 97:9654–58

92. Lind PE, Burnet FM. 1957. Recombination between virulent and non-virulent strains of influenza virus. 2. The behavior of virulence markers on recombination. *Aust. J. Exp. Biol. Med. Sci.* 35:67–78

93. Liu C, Air GM. 1993. Selection and characterization of a neuraminidase-minus mutant of influenza virus and its rescue by cloned neuraminidase genes. *Virology* 194:403–7

94. Lu X, Tumpey TM, Morken T, Zaki SR, Cox NJ, Katz JM. 1999. A mouse model for the evaluation of pathogenesis and immunity to influenza A (H5N1) viruses isolated from humans. *J. Virol.* 73:5903–11

95. Lubeck MD, Schulman JL, Palese P. 1978. Susceptibility of influenza A viruses to amantadine is influenced by the gene coding for M protein. *J. Virol.* 28:710–16

96. Ludwig S, Vogel U, Scholtissek C. 1995. Amino acid replacements leading to temperature-sensitive defects of the NS1 protein of influenza A virus. *Arch. Virol.* 140:945–50

97. Luytjes W, Krystal M, Enami M, Pavin JD, Palese P. 1989. Amplification, expression, and packaging of foreign gene by influenza virus. *Cell* 59:1107–13

98. Maassab HF, Bryant ML. 1999. The development of live attenuated cold-adapted influenza virus vaccine for humans. *Rev. Med. Virol.* 9:237–44

99. Mahy BWJ. 1983. Mutants of influenza virus. In *Genetics of Influenza Viruses*, ed.

DW Kingsbury, P Palese, pp. 192–254. Berlin: Springer-Verlag

100. Martin J, Wharton SA, Lin YP, Takemoto DK, Skehel JJ, et al. 1998. Studies of the binding properties of influenza hemagglutinin receptor-site mutants. *Virology* 241:101–11

101. Martin K, Helenius A. 1991. Nuclear transport of influenza virus ribonucleoproteins: the viral matrix protein (M1) promotes export and inhibits import. *Cell* 67:117–30

102. Mayer V, Schulman JL, Kilbourne ED. 1973. Nonlinkage of neurovirulence exclusively to viral hemagglutinin or neuraminidase in genetic recombinants of A/NWS (H0N1) influenza virus. *J. Virol.* 11:272–78

103. McGeoch D, Fellner P, Newton C. 1976. The influenza virus genome consists of eight distinct RNA species. *Proc. Natl. Acad. Sci. USA* 73:3045–49

104. McKimm-Breschkin JL. 2000. Resistance of influenza viruses to neuraminidase inhibitors—a review. *Antiviral Res.* 47:1–17

105. Mitnaul LJ, Castrucci MR, Murti KG, Kawaoka Y. 1996. The cytoplasmic tail of influenza A virus neuraminidase (NA) affects NA incorporation into virions, virion morphology, and virulence in mice but is not essential for virus replication. *J. Virol.* 70:873–79

106. Mitnaul LJ, Matrosovich MN, Castrucci MR, Tuzikov AB, Bovin NV, et al. 2000. Balanced hemagglutinin and neuraminidase activities are critical for efficient replication of influenza A virus. *J. Virol.* 74:6015–20

107. Morris SJ, Price GE, Barnett JM, Hiscox SA, Smith H, Sweet C. 1999. Role of neuraminidase in influenza virus-induced apoptosis. *J. Gen. Virol.* 80:137–46

108. Mowshowitz SL. 1978. P1 is required for initiation of cRNA synthesis in WSN influenza virus. *Virology* 91:493–95

109. Mukaigawa J, Hatada E, Fukuda R, Shimizu K. 1991. Involvement of the

influenza A virus PB2 protein in the regulation of viral gene expression. *J. Gen. Virol.* 72:2661–70

110. Muster T, Subbarao EK, Enami M, Murphy BR, Palese P. 1991. An influenza A virus containing influenza B virus 5' and 3' noncoding regions on the neuraminidase gene is attenuated in mice. *Proc. Natl. Acad. Sci. USA* 88:5177–81

111. Nakajima K, Desselberger U, Palese P. 1978. Recent human influenza A (H1N1) viruses are closely related genetically to strains isolated in 1950. *Nature* 274:334–39

112. Nakajima K, Ueda M, Sugiura A. 1979. Origin of small RNA in von Magnus particles of influenza virus. *J. Virol.* 29:1142–48

113. Nayak DP, Chambers TM, Akkina RK. 1985. Defective-interfering (DI) RNAs of influenza viruses: origin, structure, expression, and interference. *Curr. Top. Microbiol. Immunol.* 114:103–51

114. Neumann G, Kawaoka Y. 2001. Reverse genetics of influenza virus. *Virology* 287:243–50

115. Neumann G, Watanabe T, Ito H, Watanabe S, Goto H, et al. 1999. Generation of influenza A viruses entirely from cloned cDNAs. *Proc. Natl. Acad. Sci. USA* 96:9345–50

116. Neumann G, Zobel A, Hobom G. 1994. RNA polymerase I-mediated expression of influenza viral RNA molecules. *Virology* 202:477–79

117. Nichol ST, Penn CR, Mahy BW. 1981. Evidence for the involvement of influenza A (fowl plague Rostock) virus protein P2 in ApG and mRNA primed in vitro RNA synthesis. *J. Gen. Virol.* 57:407–13

118. Nishimura H, Itamura S, Iwasaki T, Kurata T, Tashiro M. 2000. Characterization of human influenza A (H5N1) virus infection in mice: neuro-, pneumo- and adipotropic infection. *J. Gen. Virol.* 81:2503–10

119. Nuss JM, Whitaker PB, Air GM. 1993. Identification of critical contact residues in the NC41 epitope of a subtype N9 influenza virus neuraminidase. *Proteins* 15:121–32

120. Odagiri T, Tanaka T, Tobita K. 1987. Temperature-sensitive defect of influenza A/Ann Arbor/6/60 cold-adapted variant leads to a blockage of matrix polypeptide incorporation into the plasma membrane of the infected cells. *Virus Res.* 7:203–18

121. Orlich M, Gottwald H, Rott R. 1994. Non-homologous recombination between hemagglutinin gene and the nucleoprotein gene of an influenza virus. *Virology* 204:462–65

122. Palese P, Ritchey MB, Schulman JL. 1977. P1 and P3 proteins of influenza virus are required for complementary RNA synthesis. *J. Virol.* 21:1187–95

123. Palese P, Schulman JL. 1976. Differences in RNA patterns of influenza A viruses. *J. Virol.* 17:876–84

124. Palese P, Tobita K, Ueda M, Compans RW. 1974. Characterization of temperature sensitive influenza virus mutants defective in neuraminidase. *Virology* 61:397–410

125. Palese P, Young JF. 1982. Variation of influenza A, B, and C viruses. *Science* 215:1468–74

126. Pekosz A, Lamb RA. 2000. Identification of a membrane targeting and degradation signal in the p42 protein of influenza C virus. *J. Virol.* 74:10480–88

127. Pinto LH, Holsinger LJ, Lamb RA. 1992. Influenza virus M2 protein has ion channel activity. *Cell* 69:517–28

128. Pleschka S, Jaskunas R, Engelhardt OG, Zurcher T, Palese P, Garcia-Sastre A. 1996. A plasmid-based reverse genetics system for influenza A virus. *J. Virol.* 70:4188–92

129. Plotch SJ, Bouloy M, Krug RM. 1979. Transfer of 5'-terminal cap of globin mRNA to influenza viral complementary RNA during transcription in vitro. *Proc. Natl. Acad. Sci. USA* 76:1618–22

130. Plotch SJ, Bouloy M, Ulmanen I, Krug RM. 1981. A unique cap(m7GpppXm)-dependent influenza virion endonuclease

cleaves capped RNAs to generate the primers that initiate viral RNA transcription. *Cell* 23:847–58

131. Poch O, Blumberg BM, Bougueleret L, Tordo N. 1990. Sequence comparison of five polymerases (L proteins) of unsegmented negative-strand RNA viruses: theoretical assignment of functional domains. *J. Gen. Virol.* 71:1153–62

132. Pons MW. 1976. A re-examination of influenza single-and double-stranded RNAs by gel electrophoresis. *Virology* 69:789–92

133. Rey O, Nayak DP. 1992. Nuclear retention of M1 protein in a temperature-sensitive mutant of influenza (A/WSN/33) virus does not affect nuclear export of viral ribonucleoproteins. *J. Virol.* 66:5815–24

134. Rogers GN, Paulson JC, Daniels RS, Skehel JJ, Wilson IA, Wiley DC. 1983. Single amino acid substitutions in influenza haemagglutinin change receptor binding specificity. *Nature* 304:76–78

135. Rohde W, Scholtissek C. 1980. On the origin of the gene coding for an influenza A virus nucleocapsid protein. *Arch. Virol.* 64:213–23

136. Rosenthal PB, Zhang X, Formanowski F, Fitz W, Wong CH, et al. 1998. Structure of the haemagglutinin-esterase-fusion glycoprotein of influenza C virus. *Nature* 396:92–96

137. Rott R, Orlich M, Scholtissek C. 1979. Correlation of pathogenicity and gene constellation of influenza A viruses. III. Non-pathogenic recombinants derived from highly pathogenic parent strains. *J. Gen. Virol.* 44:471–77

138. Schlesinger RW, Husak PJ, Bradshaw GL, Panayotov PP. 1998. Mechanisms involved in natural and experimental neuropathology of influenza viruses: evidence and speculation. *Adv. Virus Res.* 50:289–379

139. Scholtissek C, Bowles AL. 1975. Isolation and characterization of temperature-sensitive mutants of fowl plague virus. *Virology* 67:576–87

140. Scholtissek C, Burger H, Kistner O, Shortridge KF. 1985. The nucleoprotein as a possible major factor in determining host specificity of influenza H3N2 viruses. *Virology* 147:287–94

141. Scholtissek C, Rohde W, Harms E, Rott R, Orlich M, Boschek CB. 1978. A possible partial heterozygote of an influenza A virus. *Virology* 89:506–16

142. Scholtissek C, Rott R, Orlich M, Harms E, Rohde W. 1977. Correlation of pathogenicity and gene constellation of an influenza A virus (fowl plague). I. Exchange of a single gene. *Virology* 81:74–80

143. Scholtissek C, von Hoyningen V, Rott R. 1978. Genetic relatedness between the new 1977 epidemic strains (H1N1) of influenza and human influenza strains isolated between 1947 and 1957 (H1N1). *Virology* 89:613–17

144. Scholtissek C, von Hoyningen V, Rott R. 1978. On the origin of the human influenza virus subtypes H2N2 and H3N2. *Virology* 87:13–20

145. Schultz-Cherry S, Dybdahl-Sissoko N, Neumann G, Kawaoka Y, Hinshaw VS. 2001. Influenza virus ns1 protein induces apoptosis in cultured cells. *J. Virol.* 75:7875–81

146. Schultz-Cherry S, Hinshaw VS. 1996. Influenza virus neuraminidase activates latent transforming growth factor beta. *J. Virol.* 70:8624–29

147. Shapiro GI, Krug RM. 1988. Influenza virus RNA replication in vitro: synthesis of viral template RNAs and virion RNAs in the absence of an added primer. *J. Virol.* 62:2285–90

148. Shortridge KF, Zhou NN, Guan Y, Gao P, Ito T, et al. 1998. Characterization of avian H5N1 influenza viruses from poultry in Hong Kong. *Virology* 252:331–42

149. Simpson RW, Hirst GK. 1968. Temperature-sensitive mutants of influenza A virus: isolation of mutants and preliminary observations on genetic recombination and complementation. *Virology* 35:41–49

150. Skehel JJ, Wiley DC. 2000. Receptor

binding and membrane fusion in virus entry: the influenza hemagglutinin. *Annu. Rev. Biochem.* 69:531–69

151. Smith FL, Palese P. 1989. Variation in influenza virus genes: epidemiology, pathogenic, and evolutionary consequenses. In *The Influenza Viruses*, ed. RM Krug, pp. 319–59. New York: Plenum

152. Smith GL, Hay AJ. 1982. Replication of the influenza virus genome. *Virology* 118: 96–108

153. Steinhauer DA. 1999. Role of hemagglutinin cleavage for the pathogenicity of influenza virus. *Virology* 258:1–20

154. Steinhauer DA, Wharton SA, Skehel JJ, Wiley DC, Hay AJ. 1991. Amantadine selection of a mutant influenza virus containing an acid-stable hemagglutinin glycoprotein: evidence for virus-specific regulation of the pH of glycoprotein transport vesicles. *Proc. Natl. Acad. Sci. USA* 88:11525–29

155. Stieneke-Grober A, Vey M, Angliker H, Shaw E, Thomas G, et al. 1992. Influenza virus hemagglutinin with multibasic cleavage site is activated by furin, a subtilisin-like endoprotease. *EMBO J.* 11: 2407–14

156. Suarez DL, Perdue ML, Cox N, Rowe T, Bender C, et al. 1998. Comparisons of highly virulent H5N1 influenza A viruses isolated from humans and chickens from Hong Kong. *J. Virol.* 72:6678–88

157. Subbarao K, Klimov A, Katz J, Regnery H, Lim W, et al. 1998. Characterization of an avian influenza A (H5N1) virus isolated from a child with a fatal respiratory illness. *Science* 279:393–96

158. Sugiura A, Ueda M, Tobita K, Enomoto C. 1975. Further isolation and characterization of temperature-sensitive mutants of influenza virus. *Virology* 65:363–73

159. Sugrue RJ, Bahadur G, Zambon MC, Hall-Smith M, Douglas AR, Hay AJ. 1990. Specific structural alteration of the influenza haemagglutinin by amantadine. *EMBO J.* 9:3469–76

160. Sugrue RJ, Hay AJ. 1991. Structural char-

acteristics of the M2 protein of influenza A viruses: evidence that it forms a tetrameric channel. *Virology* 180:617–24

161. Ulmanen I, Broni BA, Krug RM. 1981. Role of two of the influenza virus core P proteins in recognizing cap 1 structures (m7GpppNm) on RNAs and in initiating viral RNA transcription. *Proc. Natl. Acad. Sci. USA* 78:7355–59

162. von Itzstein M, Wu WY, Kok GB, Pegg MS, Dyason JC, et al. 1993. Rational design of potent sialidase-based inhibitors of influenza virus replication. *Nature* 363:418–23

163. Wang ML, Skehel JJ, Wiley DC. 1986. Comparative analyses of the specificities of anti-influenza hemagglutinin antibodies in human sera. *J. Virol.* 57:124–28

164. Webster RG, Air GM, Metzger DW, Colman PM, Varghese JN, et al. 1987. Antigenic structure and variation in an influenza virus N9 neuraminidase. *J. Virol.* 61:2910–16

165. Webster RG, Bean WJ, Gorman OT, Chambers TM, Kawaoka Y. 1992. Evolution and ecology of influenza A viruses. *Microbiol. Rev.* 56:152–79

166. Webster RG, Rott R. 1987. Influenza virus A pathogenicity: the pivotal role of hemagglutinin. *Cell* 50:665–66

167. Wiley DC, Skehel JJ. 1987. The structure and function of the hemagglutinin membrane glycoprotein of influenza virus. *Annu. Rev. Biochem.* 56:365–94

168. Worobey M, Rambaut A, Pybus OG, Robertson DL. 2002. Questioning the evidence for genetic recombination in the 1918 "Spanish flu" virus. *Science* 296:211

169. Wright PF, Webster RG. 2001. Orthomyxoviruses. See Ref. 84a, pp. 1533–79

170. Wrigley NG, Brown EB, Daniels RS, Douglas AR, Skehel JJ, Wiley DC. 1983. Electron microscopy of influenza haemagglutinin-monoclonal antibody complexes. *Virology* 131:308–14

171. Yamashita M, Krystal M, Palese P. 1988. Evidence that the matrix protein of

influenza C virus is coded for by a spliced mRNA. *J. Virol.* 62:3348–55

172. Yang P, Bansal A, Liu C, Air GM. 1997. Hemagglutinin specificity and neuraminidase coding capacity of neuraminidase-deficient influenza viruses. *Virology* 229:155–65

173. Young JF, Palese P. 1979. Evolution of human influenza A viruses in nature: recombination contributes to genetic variation of H1N1 strains. *Proc. Natl. Acad. Sci. USA* 76:6547–51

174. Yuan W, Krug RM. 2001. Influenza B virus NS1 protein inhibits conjugation of the interferon (IFN)-induced ubiquitin-like ISG15 protein. *EMBO J.* 20:362–71

175. Zhirnov OP. 1990. Solubilization of matrix protein M1/M from virions occurs at different pH for orthomyxo- and paramyxoviruses. *Virology* 176:274–79

Annu. Rev. Genet. 2002. 36:333–60
doi: 10.1146/annurev.genet.36.043002.091635

ALLOSTERIC CASCADE OF SPLICEOSOME ACTIVATION

David A. Brow

*Department of Biomolecular Chemistry, University of Wisconsin-Madison,
Madison, Wisconsin 53706-1532; e-mail: dabrow@facstaff.wisc.edu*

Key Words   splicing, snRNP, RNA, intron, pre-mRNA

■ **Abstract**   Introns are removed from precursor messenger RNAs in the cell nucleus by a large ribonucleoprotein complex called the spliceosome. The spliceosome contains five subcomplexes called snRNPs, each with one RNA and several protein components. Interactions of the snRNPs with each other and the intron are highly dynamic, changing in an ordered progression throughout the splicing process. This allosteric cascade of interactions is programmed into the RNA and protein components of the spliceosome, and is driven by a family of DExD/H-box RNA-dependent ATPases. The dependence of cascade progression on multiple intron-recognition events likely serves to enforce the accuracy of splicing. Here, the progression of the allosteric cascade from the first recognition event to the first catalytic step of splicing is reviewed.

CONTENTS

0066-4197/02/1215-0333$14.00

INTRODUCTION

Eukaryotic protein-coding genes are riddled with sequences that are faithfully transcribed by RNA polymerase II, but then are spliced out of the primary transcript and so do not appear in the mature messenger RNA (mRNA). These intervening sequences, or introns, can comprise up to 99% of the primary transcript (pre-mRNA). An extreme example, the dystrophin gene, encodes a >2 million nucleotide pre-mRNA, which is processed into a 14,000 nucleotide mRNA by the removal of 78 introns (118). Since each of the 79 exons (retained sequences) is translated, each splice junction must be correct or else the protein sequence will be altered, which in the case of this mRNA could result in muscular dystrophy.

Pre-mRNA splicing entails the identification of the precise junctions between an intron and its flanking exons, namely the 5′ and 3′ splice sites, and removal of the intervening sequence via a two-step chemical reaction (Figure 1). This process is catalyzed by the spliceosome, a multi-megaDalton ribonucleoprotein complex

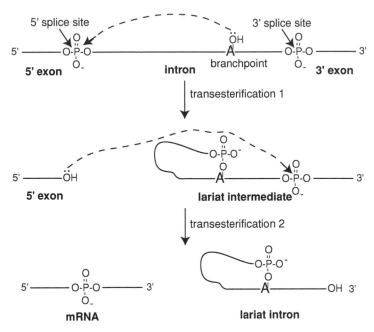

Figure 1 Schematic of the chemical reactions catalyzed by the spliceosome. At top is diagrammed a pre-mRNA with a single intron. In the first transesterification, the 2′ oxygen of a particular adenosine (A) residue of the intron forms a bond with the phosphorus atom that links the 5′ exon and the intron, displacing the bond to the 5′ exon and forming a "lariat" intron/3′ exon intermediate. In the second transesterification, the oxygen atom of the nascent 3′ hydroxyl group of the 5′ exon forms a bond with the phosphorus atom that links the intron and 3′ exon, displacing the bond to the lariat intron and joining the 5′ and 3′ exons to form the mRNA.

assembled from five subcomplexes called snRNPs (18, 40, 51). This review focuses primarily on the spliceosome from the budding yeast *Saccharomyces cerevisiae*, on which extensive genetic studies have been done, but relevant findings from the biochemical analysis of human spliceosomes is also discussed. All evidence suggests that the two human spliceosomes, both the major (U2 RNA-dependent) and minor (U1 2RNA-dependent or "ATAC") forms, are highly similar to the *S. cerevisiae* spliceosome in fundamental aspects of structure and function (20). Nevertheless, there are important differences in the details of the splicing mechanism in humans and *S. cerevisiae*, in part due to the much greater size and abundance of introns in humans. In this regard, characterization of the splicing machinery from the fission yeast *Schizosaccharomyces pombe* will also be of significant interest, since at least some aspects of fission yeast splicing is more human-like than budding yeast splicing (62; see below). Genetic analysis of the fission yeast spliceosome will be facilitated by the recently completed genome sequence of *S. pombe* (162). Here I use the term "yeast" to refer to *S. cerevisiae* and "fission yeast" to refer to *S. pombe*.

OVERVIEW OF THE SPLICEOSOME

Years of effort by many investigators have recently yielded the essentially complete enumeration of spliceosome components (20, 71, 113, 144). Each of the five spliceosomal snRNPs comprises one small nuclear RNA and several proteins (Table 1). The spliceosomal RNAs are called U1, U2, U4, U5, and U6 (U3 is

TABLE 1   Protein components of the yeast spliceosomal snRNPs

U1[a]	U2	U4	U5	U6	U4/U6	U4/U5/U6
Sm B-G	Sm B-G	Sm B-G	Sm B-G	Lsm 2-8	Prp6	Prp38
70K/Snp1[b]	A'/Lea1	Prp3	Dib1/Snu16	Prp24	Prp31	Snu23
A/Mud1	B''/Yib9/Msl1	Prp4	Prp8			Snu 66
C/Yhc1	Cus1	Snu13	Prp28			Spp381
Luc7	Cus2		Prp44/Brr2			(Prp18)[c]
Nam8/Mud15	Hsh49		Snu40			
Prp39	Hsh155		Snu114			
Prp40	Prp9					
Prp42/Mud16	Prp11					
Snu56/Mud10	Prp21					
Snu71	Rse1					
	Snu17					

[a]Each column represents the snRNP containing the snRNA(s) listed at the top. Proteins found only in multi-snRNP complexes are listed in the two right-hand columns. (See text for penta-snRNP-specific proteins.)

[b]Protein synonyms are separated by a slash.

[c]Prp18 dissociates from the tri-snRNP on glycerol gradients (47).

involved in ribosomal RNA processing). U1–U5 spliceosomal RNAs each bind seven common "Sm" proteins, designated B, D1, D2, D3, E, F, and G. U6 RNA binds a homologous group of seven "Lsm" (like Sm) proteins, designated 2–8. In addition, each RNA binds a group of snRNP-specific proteins. In yeast, many of these proteins are named "Prp" proteins, to indicate their function in Pre-mRNA processing. In addition to the snRNP proteins listed in Table 1, there is a group of splicing factors associated specifically with a "penta-snRNP" complex isolated under low salt conditions (144). These include the eight proteins of the Prp19 complex (27), Prp19, Ntc20, Ntc25/Snt309, Ntc30/Isy1, Ntc31/Sfy2, Ntc77/Syf3/Clf1, Ntc85/Cef1, and Ntc90/Syf1, as well as several other proteins: Cwc2, Ecm2/Slt11, Prp45, Prp46/Cwc1, and Sad1.

A number of non-snRNP proteins are also required for splicing, most notably a family of DExD/H-box RNA helicases (146) (Table 2). (Note that two members of this family, Prp28 and Prp44/Brr2, are U5 snRNP proteins.) Other non-snRNP splicing factors include the two pre-mRNA binding proteins Bbp1/Msl5/Ysf1 and Mud2, the two cap-binding complex subunits Cbc1/Sto1 and Cbc2/Mud13, and five other proteins: Aar2, Exo84, Prp17/Slu4, Slu7, and Spp2. Thus, more than 75 protein splicing factors exist in yeast. Many of these proteins are discussed in more detail below; information on and references for all can be obtained from the Saccharomyces Genome Database: http://genome-www.stanford.edu/Saccharomyces/. Of course, knowing the components of the spliceosome doesn't tell us how it works. The knowledge does, however, provide an experimental basis for studying spliceosome function and some clues on where to focus one's efforts. In particular, the requirement for eight different DExD/H-box proteins in a single process underscores the importance of RNA structural transitions in pre-mRNA splicing (5, 94, 107, 142).

TABLE 2   Spliceosomal DExD/H-box proteins

DExD/H-box protein	Human homolog	Stage in splicing cycle	Likely substrate	Cs allele
Sub2	UAP56	Pre-spliceosome	Intron b.p./BBP/Mud2	*sub2-5*
Prp5	KIAA0801[a]	Pre-spliceosome	U2 snRNP	?
Prp28	U5-100 kD	Early step I activation	U1/5'ss/U1-C	*prp28-1*
Prp44/Brr2	U5-200 kD	Early step I activation	U4/U6	*brr2-1*
Prp2	KIAA0577[b]	Late step I activation	?	?
Prp16	hPrp16	Step II activation	U2/U6	*prp16-302*
Prp22	Hrh1	mRNA release	?	*prp22-2*
Prp43	mDEAH9	Intron release	?	?

[a]Charles Query, personal communication.

[b]Assignment of human homolog currently based only on sequence similarity.

Spliceosome Assembly: The Holospliceosome Hypothesis

Incubation of synthetic pre-mRNA with yeast or human cell extract in the presence of ATP results in the production of splicing intermediates and products (55, 70, 86, 111). When splicing reactions containing radioactively labeled pre-mRNA are subjected to nondenaturing gel electrophoresis, at least four distinct complexes can be resolved (29, 68, 117, 129, 137), here called the commitment complex, pre-spliceosome, complete spliceosome, and active spliceosome. These complexes differ in their snRNP composition and order of appearance. All contain pre-mRNA, but only active spliceosomes contain products of the first and second catalytic steps of splicing. Based on these and other findings, a sequential model of spliceosome assembly was developed (Figure 2*a*). In this model, U1 snRNP binding to the 5′ splice site initiates the commitment of an intron to splicing. U2 RNA binds subsequently to the branchpoint, followed by docking of a preassembled U4/U5/U6 tri-snRNP with the pre-spliceosome. Significantly, U1 and U4 RNAs are seen to exit the complete spliceosome before it becomes catalytically active, which requires the disruption of base-pairs formed earlier between U1 RNA and the 5′ splice site, and between U4 and U6 RNAs. Unwinding of these helices is thought to be required for subsequent rearrangements that lead to formation of the active site. The Prp19 protein complex joins the spliceosome at about the time the U1 and U4 snRNPs leave.

However, not all data agree with the tri-snRNP addition model described above. In the absence of pre-mRNA a portion of the U2 snRNP present in cell extracts is found to be associated with the U4/U6/U5 tri-snRNP (47, 69, 121). Psoralen crosslinking of RNAs in HeLa cell nuclear extract suggests that the majority of U4/U6 RNA complex is also complexed with U2 RNA (156), consistent with evidence that pairing of the 5′ end of U2 RNA with the 3′ end of U6 RNA (53) is required for stable pairing of human U4 and U6 RNAs (19). Furthermore, U2 snRNP was found to be associated with E complex (the mammalian commitment

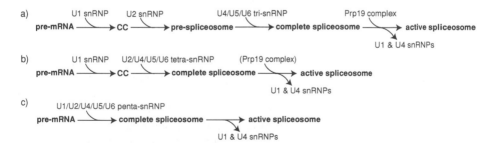

Figure 2   Three different models for the spliceosome assembly pathway. (*a*) The tri-snRNP model. (*b*) The tetra-snRNP model. (*c*) The penta-snRNP or holospliceosome model. CC, commitment complex. Note that the Prp19 protein complex is a component of the penta-snRNP, but not the tri-snRNP. It is not known if the Prp19 complex is in the tetra-snRNP.

complex) isolated under gentle conditions (33). Very recently, a U1/U2/U4/U5/U6 penta-snRNP complex that also contains the Prp19 complex was isolated from yeast cell extract at low salt concentration (144). These results suggest the existence of a "holospliceosome" complex that binds as a unit to an intron (Figure 2c). In the holospliceosome model, the sequential complexes resolved on native gels reflect the ordered modulation of specific contacts between different snRNPs, and between snRNPs and the pre-mRNA, rather than the stepwise assembly of the spliceosome on the pre-mRNA. Because the U1 snRNP is the most loosely associated component of the penta-snRNP (144), an alternative version of the model suggests that a U2/U4/U5/U6 tetra-snRNP binds to the commitment complex (Figure 2b). Regardless, the native gel data clearly indicate that the interactions between spliceosomal components change in an ordered fashion throughout the splicing cycle.

Programmed Conformational Switches: The Allosteric Cascade Hypothesis

The splicing reaction itself is essentially isoenergetic: Two transesterifications result in no net change in the number of phosphoester bonds (Figure 1). However, the lariat intron is subsequently debranched by enzymatic hydrolysis of the $2'-5'$ phosphoester bond (64), and then the intron is presumably degraded to mononucleotides by nucleases, greatly increasing the entropy and decreasing the enthalpy of the system. Thus, splicing is expected to be strongly exothermic. Nevertheless, at several points in the splicing cycle ATP must be hydrolyzed, and at each of these points a DExD/H-box RNA-dependent ATPase is required.

One can view spliceosome interaction with the pre-mRNA as an allosteric cascade in which early recognition steps (e.g., U1 snRNP binding to the 5' splice site) induce conformational changes that are required for subsequent recognition steps, which in turn are required for catalytic activation. In this manner, the endpoint of a catalytically active spliceosome will not be reached unless all of the landmarks of an authentic intron are present and properly positioned. The DExD/H box proteins presumably provide the energy to fuel the allosteric switches. When combined with the holospliceosome hypothesis, the allosteric cascade model makes it unnecessary to divide the early splicing pathway into "assembly" and "activation" stages, as is typically done. Instead, one can imagine a fully assembled spliceosome encountering an intron and undergoing an ordered series of interactions and allosteric conformational changes that eventually result in the catalysis of the first transesterification reaction. This review describes what is known about the spliceosomal allosteric cascade up to the point of branchpoint attack on the 5' splice site. It does not cover subsequent steps in the allosteric cascade that are necessary for the second transesterification (149) and spliceosome disassembly (6, 97).

Because of this allosteric cascade, the spliceosome is a highly dynamic complex and therefore difficult to study. For biochemical analyses it is desirable to trap and accumulate spliceosomes at a particular point in the splicing cycle. A variety of means have been developed to arrest splicing in vitro and in vivo, including factor depletion or inactivation, cofactor (magnesium, ATP) depletion, and intron mutation. Ideally, one would like a reversible treatment, which allows one to prove that the arrested complex is an active intermediate and not a dead-end product. A very useful biochemical and genetic tool in this regard is worth mentioning at the outset: conditional mutations in RNA that result in a cold-sensitive (cs) growth defect. As detailed below, cs phenotypes of RNA mutations appear to be due to the inability of secondary structures stabilized (directly or indirectly) by the mutation to melt at an adequate rate at the restrictive temperature (32, 39, 81, 139, 143, 174). Not only do cs mutations provide a means to isolate a relatively homogeneous population of reversibly arrested "frozen" spliceosomes in vitro (74, 143), isolation of suppressors or enhancers of cs spliceosomal RNA mutations in vivo provides an efficient means for identifying factors that influence specific conformational switches (74, 75, 139, 153, 159). Indeed, mutations in the spliceosomal DExD/H box proteins often confer a cs phenotype (108) (Table 2), presumably due to defects in unwinding RNA helices.

THE ALLOSTERIC CASCADE DRIVING SPLICEOSOME ACTIVATION

An extreme form of the allosteric cascade model is that each spliceosomal interaction (except the very first) is strictly dependent upon a previous interaction, i.e., all interactions are strictly ordered. This is certainly not true. For example, early interactions at the branchpoint/3' splice site can occur independently of early interactions at the 5' splice site (101). Some spliceosomal interactions exhibit cooperativity with no evidence of allostery. Only a highly detailed understanding of the dynamics of spliceosome structure will allow the full extent of allosteric regulation of splicing to be determined. Nevertheless, several examples of allostery are well established, particularly among the spliceosomal RNAs, and more are likely to emerge. Below, the known interactions in the splicing pathway are described, and their interdependence is considered.

Recognition of the 5' Splice Site by the U1 snRNP

Because the 5' splice site is the first portion of an intron to emerge from RNA polymerase II during synthesis of a pre-mRNA, recognition of the 5' splice site is likely to precede other snRNP-pre-mRNA interactions on a given intron. An important component of this recognition event is binding of the U1 snRNP in the ATP-independent process of commitment complex formation (127, 130). In part, U1 snRNP binding is mediated by base-pairing of the 5' end of U1 RNA with the conserved sequences immediately downstream of the 5' splice site (Figure 3a). Pairing typically involves the first six intron nucleotides (127), which are highly conserved in yeast but less so in humans (20, 88, 141). However, removal of the sequences in U1 RNA that are complementary to the 5' end of the intron does not abolish binding of the U1 snRNP to the 5' splice site (34). Furthermore, in vitro

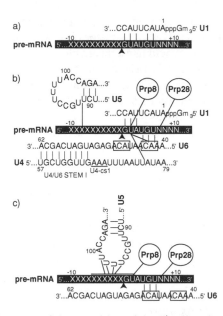

Figure 3   Model for sequential recognition of the 5′ splice site. The pre-mRNA is in inverse font. Ten nucleotides upstream and downstream of the 5′ splice site (*arrowhead*) are shown. X, nonconserved exon nucleotides. N, nonconserved intron nucleotides. (*a*) In the commitment complex the 5′ end of U1 RNA base-pairs with conserved residues at the 5′ end of the intron. "Gm₃" indicates the 2,2,7-trimethylguanosine cap found at the 5′ end of U1-U5 RNAs. See text for a description of the protein crosslinks that are observed in the commitment complex. (*b*) In the complete spliceosome, U1 RNA/5′ splice site pairing is at least partially disrupted and U6 RNA base-pairs to the 5′ splice site via the most 5′ of two ACA trinucleotides (*boxed*). Long lines indicate UV light-induced crosslinks that form between spliceosome components and pre-mRNA substituted with 4-thioU at positions -8 and $+2$. Additional U6/pre-mRNA crosslinks are described in the text. The underlined nucleotides in U4 are changed to 5′UUG3′ by the U4-cs1 mutation. (*c*) In the active spliceosome, U4 has unwound from U6, and U6/5′ splice site pairing has shifted to the downstream ACA trinucleotide. The apical U residues of U5 loop I crosslink to pre-mRNA residues -6 to $+2$.

selection for efficient 5′ splice site sequences in the absence of the complementary sequences in U1 RNA yielded a nucleotide frequency distribution very similar to the consensus 5′ splice site sequence (89), indicating that something other than the 5′ end of U1 RNA recognizes this sequence.

Several protein-RNA interactions stabilize U1 snRNP binding. Seven U1 snRNP proteins can be UV crosslinked to the 5′ splice site region in commitment complexes: U1-70K/Snp1, U1-C/Yhc1, Nam8/Mud15, Snu56/Mud10, and the Sm proteins B, D1 and D3 (119, 176). Although U1-70K has not been directly implicated in commitment complex assembly, the remaining six proteins have. Depletion of

U1-C inhibits commitment complex formation without altering U1 snRNP levels (145). Nam8/Mud15 is required for efficient splicing of nonconsensus 5′ splice sites, and for consensus 5′ splice sites in the presence of the U1-4U mutation, which weakens pairing of U1 RNA with the 5′ splice site, or in the absence of a cap on the 5′ exon (48, 119). Snu56/Mud10 binds to immunopurified cap-binding complex (CBC; see below), and so may link the U1 snRNP to the pre-mRNA 5′ cap (38). Sm proteins B, D1 and D3 have extended C-terminal tails, deletion of which diminishes commitment complex stability (175).

The heterodimeric nuclear yeast cap binding complex, comprised of Cbc1/Sto1 and Cbc2/Mud13, also appears to participate in commitment complex formation, since Cbc1/Sto1 crosslinks to the 5′ exon in the commitment complex (176) and depletion of Cbc2/Mud13 inhibits commitment complex formation in vitro (31, 80). Furthermore, deletion of the *CBC1* and *CBC2* genes is lethal in combination with mutations in the genes encoding several U1 snRNP proteins, including U1-A/Mud1, Nam8/Mud15, Snu56/Mud10, Snu71, Luc7, and SmD3 (38). Indeed, Luc7 was discovered in this synthetic lethal screen, and was subsequently shown to be a U1 snRNP protein required for commitment complex assembly in vitro (37). These results are consistent with the finding that temperature sensitive mutations in the gene for the yeast capping enzyme, *CEG1*, confer a pre-mRNA splicing defect (41, 136). Thus, the U1 snRNP is anchored to the 5′ splice site by multiple interactions, the individual strengths of which may vary depending on distance of the intron from the 5′ cap and the nature of the sequences encompassing the 5′ splice site.

Recognition of the 5′ Splice Site by the U4/U5/U6 tri-snRNP

Native gel analysis provides no evidence for interaction of the tri-snRNP with the 5′ splice site prior to binding of the U2 snRNP to the branchpoint. However, RNA crosslinking studies reveal an early interaction of U5 RNA with the 5′ splice site region while the U1 snRNP is bound there (105, 164). Recently, Maroney et al. (96) identified a U2 snRNP-independent interaction of the human and nematode U5 snRNP with the 5′ splice site region, which was detected by UV crosslinking of the U5 snRNP protein Prp8 to the GU dinucleotide at the 5′ end of the intron (124). Unlike commitment complex formation, U5 snRNP binding requires ATP as well as the U5 RNA loop 1 sequences (Figure 3*b*), the U6 RNA 5′ splice site pairing region, and intact U4 RNA, indicating that it is actually the U4/U5/U6 tri-snRNP, rather than free U5 snRNP, that is binding. Tri-snRNP binding probably accounts for the previously observed early interaction between the residues U91 and U92 of the yeast U5 RNA loop 1 and the 5′ exon 8 nuclotides upstream of the 5′ splice site (105) (Figure 3*b*).

Interactions between U1 snRNP proteins and tri-snRNP proteins that may stabilize tri-snRNP binding at the 5′ splice site have been identified by two-hybrid assays. In particular, U1-70K/Snp1 interacts with the U5 snRNP proteins Prp44/Brr2 and Prp8 (10, 43, 152) and with Exo84, a recently identified splicing factor (10).

Exo84, in turn, interacts with Prp8 (73), which also interacts with U1 snRNP proteins Prp39 and Prp40 (2, 152). Indeed, a U1/U5 bi-snRNP has been identified in yeast cell extract (104, 129), although this complex lacks many of the U5 snRNP proteins and may be an intermediate in U5 snRNP biogenesis (46). Consistent with early U1/U5 snRNP interaction at the 5′ splice site, a psoralen crosslink between human U1 and U5 RNAs is observed at very early times in a splicing reaction (9). While the U1 snRNP alone protects about 14 nucleotides of the intron and 5 nucleotides of 5′ exon from micrococcal nuclease digestion, addition of the tri-snRNP extends the protection 5′ to include another 10 nucleotides of 5′ exon (96). It is thus likely that tri-snRNP binding and U1 snRNP binding are cooperative prior to unwinding of the U4/U6 base-pairing.

Interestingly, when U1 RNA/5′ splice site complementarity exceeds the usual 6–9 base-pairs, stable U4/U5/U6 tri-snRNP interaction with the pre-mRNA is inhibited (89, 143). This inhibition results in a cold-sensitive splicing defect that is enhanced (i.e., exacerbated) by a mutation in the helicase domain of the U5 snRNP-associated DExD/H-box protein Prp28, *prp28-1*, that alone also confers cold-sensitivity (143). Conversely, the U1-4U mutation in U1 RNA that decreases complementarity with the consensus 5′ splice site suppresses the cold-sensitivity of *prp28-1*. Indeed, the U1-4U mutation eliminates the requirement for the normally essential *PRP28* gene, as does a mutation in the gene for the U1 snRNP protein U1-C (28). The same mutation in U1-C protein suppresses the splicing defect caused by increasing the length of the U1 RNA/5′ splice site helix. These results indicate that Prp28 destabilizes the U1 RNA/5′ splice site helix, while U1-C stabilizes it. The former conclusion is further supported by crosslinking of human Prp28 to the 5′ splice site at the time of tri-snRNP addition (58), and the latter conclusion is consistent with earlier biochemical studies on human U1-C (54) as well as strong crosslinking of yeast U1-C to position +6 of the rp51A intron (176). Thus, the ATP-dependence of tri-snRNP binding to the 5′ splice site region may reflect the requirement for at least partial displacement of U1 RNA from the 5′ splice site by Prp28. In support of the tetra-snRNP or holospliceosome addition models, disruption of U1/5′ splice site pairing by the U1-4U mutation allows weak ATP-independent formation of pre-spliceosomes (82), suggesting that the observed ATP-dependence of stable U2 snRNP binding to the branchpoint (see below) is at least partly due to the ATP-dependence of tri-snRNP binding to the 5′ splice site.

Further evidence for coupling of U1 snRNP and U4/U5/U6 tri-snRNP recognition of the 5′ splice site comes from the U4-cs1 mutation (81). This triple substitution extends base-pairing between U4 and U6 RNAs (Figure 3b), masking the region in U6 that ultimately pairs with the 5′ splice site, and conferring a cold-sensitive block to splicing prior to catalysis. U4-cs1 spliceosomes arrested at 16°C have high levels of U1 snRNP, which is rapidly released, along with U4 snRNP, upon shift to 30°C (74). Furthermore, the U4-cs1 growth defect is enhanced (exacerbated) by the *prp28-1* mutation (75). Both results suggest coupling between U1 RNA displacement from, and U6 pairing with, the 5′ splice site. In addition, mutations that increase the pairing potential between U6 RNA and the 5′ splice site suppress the splicing defect caused by U1 RNA/5′ splice site hyperstabilization (143).

The location in U6 RNA of a subset of *trans*-acting suppressors of U4-cs1 (81) suggests that U6 RNA may first recognize the 5′ splice site with sequences upstream of the ACAGA-box element that pairs with the 5′ splice site in the active spliceosome (60, 79) (Figure 3c). This upstream pairing is supported by two UV-crosslinking studies (59, 135), the more recent of which also found evidence of U4 RNA contact with the 5′ splice site in a branchpoint-independent, ATP-dependent complex (Figure 3b). A third crosslinking study is consistent with either register of U6/5′ splice site pairing (63). The early contacts between the U4/U6 snRNP and the 5′ splice site may be the reason that the tri-snRNP appears to be more stably bound to U4-cs1-arrested spliceosomes than to U1/5′ splice site hyperstabilized spliceosomes, when each is subjected to native gel electrophoresis (74, 143). There is some evidence of cooperativity of U2 snRNP and tri-snRNP binding even at this early stage, since U1 snRNP but not tri-snRNP association with pre-mRNA is observed on native gels in the absence of U2 RNA (129). Unwinding of the U4/U6 RNA pairing and rearrangement of contacts at the 5′ splice site to those shown in Figure 3c apparently requires U2 snRNP bound to the branchpoint.

Recognition of the Branchpoint/3′ Splice Site by BBP/SF1 and Mud2/U2AF

Seconds to minutes after a 5′ splice site emerges from a transcribing RNA polymerase, the branchpoint and 3′ splice site appear. In human pre-mRNAs, which tend to have multiple, long introns and short exons, branchpoint/3′ splice site recognition is often coupled to U1 snRNP binding to the next downstream 5′ splice site. This mechanism is known as exon definition (14, 123) and involves accessory splicing factors called SR proteins (23, 52). Neither SR proteins nor exon definition are known to exist in yeast, so this mechanism is not discussed here. Even fission yeast, which has SR protein homologs (62), appears to use intron definition rather than exon definition (125), perhaps due to the short average length of its introns [81 nucleotides (162)]. Discussion below focuses on direct recognition of the branchpoint/3′ splice site region and cross-intron bridging of complexes bound at the ends of a single intron.

Like the 5′ splice site, the sequence encompassing the branchpoint adenylate residue is highly conserved in yeast, but less so in humans (20, 88, 141). The consensus branchpoint sequence is UACUA<u>A</u>C, where the underlined "A" is the branch nucleotide (Figure 4). In yeast, this sequence is often preceded by two U residues and followed by an A residue (88, 141). The 3′ splice site consensus is much more limited, consisting of YAG/ in both yeast and mammals, where Y is either pyrimidine base and the slash indicates the 3′ splice site. The spacing between the branch nucleotide and 3′ splice site is quite variable, ranging in yeast from 10

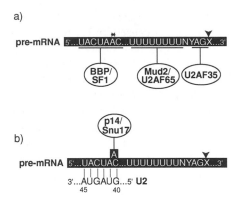

Figure 4 Model for sequential recognition of the branchpoint/3′ splice site region. Shown are the sequences immediately upstream of the 3′ splice site (*arrowhead*). Y, pyrimidine base. The asterisk marks the branchpoint adenosine residue. (*A*) Protein factors that recognize the conserved sequences in the commitment complex are indicated with labeled ovals. (*B*) In the pre-spliceosome and complete spliceosome, U2 RNA pairs with sequences encompassing the branchpoint, and p14 contacts the branchpoint adenosine. See text for additional interactions.

to 155 nucleotides, with a mean of 39 nucleotides (141). In addition, many yeast introns have a pyrimidine-rich (predominantly uracil) tract of 8–12 nucleotides that ends one nucleotide upstream of the 3′ splice site. The pyrimidine-rich tract is more ubiquitous and generally longer in human introns, perhaps to compensate for the weak branchpoint consensus (see below).

The branchpoint sequence is first recognized by a protein called BBP (or Bbp1) in yeast and SF1 in humans (7, 16). The KH domain of BBP/SF1 specifically binds the branch residue and flanking nucleotides (17, 87, 115, 122). Yeast BBP was discovered in a synthetic lethal screen with a deletion of *MUD2* (2). Mud2 and the other *MUD* gene products (Table 1) were identified in a synthetic lethal screen with a partial deletion allele of *SNR19*, the U1 RNA gene (83). Mud2 is the yeast homolog of the large subunit of U2AF, a heterodimeric human splicing factor that binds to the polypyrimidine tract and 3′ splice site (173). The small subunit of U2AF (U2AF35) directly contacts the AG dinucleotide at the 3′ splice site (100, 163, 178), while the large subunit (U2AF65) binds to the polypyrimidine tract (Figure 4*a*) (140). Budding yeast has no apparent U2AF35 homolog and does not require the 3′ splice site for the first catalytic step of splicing (134), whereas fission yeast does have a U2AF35 homolog (160) and does require the 3′ splice site for the first catalytic step (126).

The synthetic lethality of mutations in BBP in the absence of Mud2 is easily understood in light of the cooperative binding displayed by their mammalian homologs SF1 and U2AF. Either protein alone binds a branchpoint/polypyrimidine tract RNA with an apparent K_d of 1–10 μM, while presence of the partner protein increases the apparent affinity 5- to 20-fold (15). Binding cooperativity of a similar magnitude was inferred from an in vivo reporter system (115). Cooperative binding is mediated by protein-protein interactions between BBP/SF1 and Mud2/U2AF65 (2, 15, 122).

Cross-intron bridging of complexes at the 5′ splice site and branchpoint may first occur in the commitment complex. Two forms of the commitment complex, named CC1 and CC2, are resolved on native gels (138). While both complexes contain the U1 snRNP and are disrupted by mutations in the 5′ splice site, only CC2 is disrupted by mutations in the branchpoint sequence (112, 138). This fact and the slower mobility of CC2 on native gels suggested that CC2 contains additional factors that bind the branchpoint sequence. Indeed, both BBP and Mud2 are components of CC2, and SF1 and U2AF are present in the homologous mammalian E complex (1, 2, 133). Full stability of the commitment complex requires the conserved sequences at both the branchpoint and 5′ splice site (78, 130), suggesting that the U1 snRNP may bind cooperatively with BBP + Mud2 to the intron. Bridging of the complexes assembled at the 5′ splice site and branchpoint in CC2 may be facilitated by protein-protein interactions between BBP and the U1 snRNP protein Prp40 (2, 61). In addition, the Prp19 complex protein Ntc77/Syf3/Clf1 interacts with both Prp40 and Mud2, and appears to stabilize the tri-snRNP in the spliceosome (30). Finally, interaction of Mud2 with the cap-binding complex may also contribute to cross-intron bridging (38).

Recognition of the Branchpoint by the U2 snRNP

After it is bound by BBP, the branchpoint is recognized by base-pairing with U2 RNA. Although recent evidence suggests that the U2 snRNP may be a component of the ATP-independent mammalian commitment (E) complex (33), stable association of the U2 snRNP with the branchpoint requires ATP. There are two DExD/H-box proteins that are required at this point in the allosteric cascade: Sub2/UAP56 and Prp5. The human protein UAP56 was discovered by virtue of its interaction with U2AF65, and is required for in vitro splicing and association of U2 snRNP with pre-mRNA (36). The yeast homolog, Sub2, was identified as a high-copy suppressor of cold-sensitive growth caused either by the snRNP biogenesis mutant *brr1-1* (67), or by deletion of the *NAM8* gene in the presence of HA-tagged Prp40 (84). Sub2 is 63% identical to UAP56, and expression of UAP56 in yeast complements the lethal deletion of *SUB2* (177). Similar to the genetic interaction between U1-C and Prp28, deletion of the nonessential *MUD2* gene suppresses the lethality of deletion of the *SUB2* gene (67). This result suggests that the essential function of Sub2 is the ATP-dependent removal of Mud2 and/or BBP from the branchpoint to allow binding of the U2 snRNP. However, depletion or inactivation of Sub2 in yeast cell extract leads variously to decreased formation of CC2 (177), pre-spliceosomes (84), or complete spliceosomes (67). Thus, Sub2 may have multiple functions early in the allosteric cascade.

To bind productively to the branchpoint region, the U2 RNA must be in the correct conformation and be associated with the proper proteins (Table 1). Biochemical studies identified human SF3a (3 subunits) and SF3b (5 subunits) as U2 snRNP-associated splicing factors essential for pre-spliceosome formation (71, 72, 161). SF3 is also present in yeast: SF3a is composed of Prp9, Prp11, and Prp21 (71, 131), while SF3b comprises Cus1, Hsh49, Hsh155, Rse1, and Snu17 (25, 45, 57, 155). In addition, the U2A′/Lea1 and U2B″/Yib9 proteins are also important for pre-spliceosome formation in vitro (24).

The importance of U2 RNA structure is illustrated by the fact that destabilizing mutations in stem IIa of yeast U2 can induce formation of a competing stem that inhibits U2 snRNP binding to the pre-mRNA, and results in cold-sensitive growth (174). A selection for suppressors of one such mutation, U2-G53A, yielded the gene for the SF3b subunit Cus1 (114, 159), as well as a novel U2 snRNP protein, Cus2 (169). Cus2 is nonessential, has RNA-binding activity, and appears to facilitate a conformational change in U2 RNA required for pre-spliceosome assembly (116). The DExD/H-box protein Prp5, acting in conjunction with SF3a and SF3b, seems to participate in a subsequent conformational change, which increases the accessibility of U2 RNA's branchpoint-binding sequences (109, 158). Strikingly, in the absence of Cus2, pre-spliceosome assembly in vitro is ATP-independent (116). This result and genetic interactions between *CUS2* and *PRP5* suggest that Prp5 displaces Cus2 from U2 RNA and/or disrupts a Cus2-stabilized U2 RNA structure (116). However, unlike the case of Mud2 and Sub2, deletion of *CUS2* does not render Prp5 dispensable, so Prp5 likely has a Cus2-independent function as well.

The U2 snRNP is anchored to the branchpoint both by base-pairing between U2 RNA and the intron (Figure 4*b*) (94), and by binding of SF3a and SF3b to flanking sequences primarily upstream of the branchpoint (49, 50). In addition, one human SF3b subunit, p14, contacts the branchpoint directly (161). Snu17 is the likely yeast homolog of p14 (45), although Snu17 has not yet been shown to bind the branchpoint. Binding of human U2 snRNP to a short RNA containing just the branchpoint sequence and a polypyrimidine tract is ATP-independent, but addition of 18 nucleotides 5′ of the branchpoint sequence renders U2 snRNP binding ATP-dependent, suggesting that the accommodation of upstream sequences by SF3 requires a conformational change (106). Although the ATP-dependence of U2 snRNP binding appears to be sequence nonspecific with respect to the upstream RNA, in a related study purine-rich sequences upstream of the branchpoint were found to alter the conformation of the U2 snRNP and its interaction with U6 RNA (8). Yeast U2 snRNP binding to a 17-nucleotide 2′ O-methyl branchpoint oligonucleotide was analyzed in a similar fashion (3). In this case, with only 4 nucleotides upstream of the branchpoint and no polypyrimidine tract, binding was stimulated by ATP in a manner dependent on the presence of functional Prp5. Indeed, Prp5 was found to be associated with the U2 snRNP in the presence or absence of branchpoint oligonucleotide (3). These results support the hypothesis that Prp5 alters U2 snRNP structure to promote pairing with the branchpoint sequence.

Unwinding of U4/U6 and the Switch of U6 for U1 at the 5′ Splice Site

Up to this point in the splicing cycle, "communication" between the 5′ splice site and branchpoint/3′ splice site appears not to be essential; interactions at the two ends of the intron are able to form independently. Progression of the allosteric cascade beyond this point, however, seems to require cross-intron interactions. In the classical model of the splicing cycle, this point is when spliceosome assembly is considered to be complete and catalytic activation of the spliceosome begins. In the holospliceosome model, however, this is simply a stage in the ongoing allosteric cascade during which all five snRNPs are tightly bound together. The next major event is the complete displacement of U1 RNA from the 5′ splice site, establishment of a new pairing of U6 RNA with the 5′ splice site, and unwinding of the extensive base-pairing between U4 and U6 RNAs (5, 94, 107, 142) (see Figure 3c). At about the same time, the interaction of U5 loop 1 with the 5′ exon is also altered (4). These conformational changes appear to require the branchpoint sequence, presumably because the "activated" U2 snRNP has a role in the process.

Displacement of U1 and U4 RNA, and likely the other conformational changes that occur at this point, are inhibited at low temperature by the U4-cs1 mutation (Figure 3b) (74). U4-cs1 cold-arrested spliceosomes can be isolated by affinity chromatography and chased by raising the temperature in the presence of ATP, whereupon U1 and U4 snRNPs rapidly dissociate (74). The DExD/H-box protein Prp44, also called Brr2, Rss1, Slt22, and Snu246 (77, 85, 108, 168), has been implicated in unwinding the U4/U6 helices (65, 76, 121) and likely confers the ATP requirement. A cold-sensitive mutation (brr2-1) in the helicase domain of Prp44 that inhibits U4/U6 unwinding in vitro (121) is synthetic lethal with U4-cs1 (73), consistent with having defects in the same process. However, U4-cs1 is also synthetic lethal with conditional mutations in Prp28 and in the U6 snRNP protein Prp24 (73, 75). Thus, U4-cs1 may affect several steps in splicing, or Prp28 and Prp24 may directly or indirectly influence U4/U6 unwinding. Previously identified genetic interactions of Prp24 with mutations in U4 and U6 RNAs favor the latter hypothesis (139, 153), but an in vitro study does not support an important function of Prp24 in the spliceosome (120).

Mutations in at least five distinct regions of the 270-kDa U5 snRNP protein Prp8 suppress the cold-sensitive phenotype of U4-cs1 (73). One such mutation in the N-terminal region also suppresses prp28-1, while a cluster of mutations near the center of Prp8 suppress the brr2-1 mutation (75). These interactions suggest that Prp8 may regulate the activity of both DExD/H-box proteins. Certain mutations in the U4 snRNP protein Prp4 block the allosteric cascade at a very similar point as U4-cs1, just prior to U4/U6 unwinding (11). Prp4 interacts genetically with the U4 snRNP protein Prp3 and with Prp24 (92, 131), and interacts physically with Prp3 (12). In addition, the small, acidic tri-snRNP proteins Prp38 and Spp381 appear to have an important function in U4/U6 unwinding (91, 165).

A remarkable convergence of mapping studies on three disease genes has helped to draw together these complex genetic interactions. Three genes responsible for autosomal dominant retinitis pigmentosa were found to encode the human homologs of Prp3, Prp8, and Prp31 (26, 98, 154). Prp31 is a U4/U6 snRNP protein implicated in stabilizing the tri-snRNP and its association with the pre-spliceosome (95, 157). A temperature-sensitive mutation in yeast *PRP31* is synthetic lethal with U4-cs1 at 30°C (75). Thus, all three proteins are directly or indirectly implicated in U4/U6 unwinding, along with Prp4, Prp24, Prp28, and Prp44. Why a defect in this process should result in the progressive degeneration of the retina is a mystery. However, it is notable that the phenotypic effects of certain mutations in general yeast splicing factors have in at least two cases been traced to the altered splicing of a single pre-mRNA (21, 61).

Pairing of U2 and U6 RNAs to form the Catalytic Core

The intermediate structure(s) of U6 RNA in the transition from the U4/U6 pairing in the complete spliceosome to the U2/U6 pairing in the active spliceosome (Figure 5) are currently unclear. Genetic evidence suggests that U6 RNA may transiently exist as a free RNA, forming a 3′ intramolecular stem-loop (ISL) that is thought to persist in the U2/U6 pairing, and a potential pseudoknot structure called the telestem (153). On the other hand, UV-crosslinking studies on the human U12-dependent spliceosome suggest that U2/U6 helix I forms even before the U4/U6 pairing is completely unwound and the U6 ISL has formed (42). A screen for mutations synthetically lethal with U2 RNA substitutions that disrupt an intramolecular stem as well as U2/U6 helix II yielded alleles of *PRP44* and *PRP8* (168), suggesting that U4/U6 RNA unwinding may be coupled with U2/U6 helix II formation. Furthermore, a novel splicing factor identified in the screen, Slt11, binds in vitro to a U4/U6 RNA mimic and promotes annealing of an RNA oligonucleotide to form a helix analogous to U2/U6 helix II (167). However, other mutations obtained in this screen were in genes for factors involved in the second catalytic step of splicing, including Prp17 and Slu7 (166), so the involvement of Prp44, Prp8, and Slt11 could be due to a function subsequent to U4/U6 unwinding.

Prp19 is an essential splicing factor that associates with the spliceosome around the time U4 is unwound from U6 RNA, but prior to Prp2 function (147). Prp19 is complexed with at least seven other "Ntc" (Nineteen complex) proteins (27). Three of these proteins, Ntc90/Syf1, Ntc31/Syf2, and Ntc77/Clf1/Syf3, were previously identified in a synthetic lethal screen using a deletion of the *PRP17* gene (13). Strikingly, there is significant overlap between genes identified in this (*syf*) synthetic lethal screen and the U2 RNA synthetic lethal screen (*slt*) described above, including *PRP8*, *SLT11*, and the U2 RNA gene itself. Very recently, Prp19 and the Ntc proteins have been shown to be components of a much larger complex that contains Prp17, Slt11, and components of the U2, U5, and U6 snRNPs (110). The fission yeast homolog of this complex contains the Cdc5 protein, which is homologous to Ntc85/Cef1, and mutations in at least some of the components of the fission and budding yeast complexes result in cell cycle defects (110, 132). The

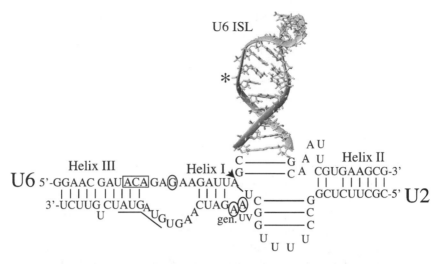

Figure 5 Model of the U2/U6 RNA base-pairing interaction in the active spliceosome. Shown are residues 33–95 of human U6 RNA and 3–46 of human U2 RNA. The lowest free-energy solution structure of the yeast U6 ISL replaces the equivalent nucleotides in human U6 RNA, which are expected to adopt a similar structure (S. Butcher, personal communication). The asterisk marks the phosphate that binds a magnesium ion essential for the first transesterification. U2/U6 helices I, II, and III are indicated, and are not drawn to scale with the U6 ISL. Alternative pairings are possible at the junction of helices I and II with the U6 and U2 intramolecular stems. The boxed residues in U6 RNA base-pair with the 5′ splice site (Figure 3c). The underlined residues in U2 RNA base-pair with the branchpoint (Figure 4b). The circled G residue in U6 RNA forms a tertiary interaction with the circled A residues in U2 RNA, as inferred from genetic interaction in yeast (gen.) or UV crosslinking of the human RNAs in vitro (UV). The arrowhead indicates the site of phosphotriester formation with the branchpoint adenosine in the in vitro ribozyme reaction. See text for discussion and references.

Cdc5/Cef1 complex could be a spliceosome that has completed splicing, given that it corresponds closely to the penta-snRNP complex lacking the U1 and U4 snRNPs (144). Conceivably, this post-splicing complex participates in a cell cycle checkpoint [but see (21)].

The only DExD/H box protein known to act subsequent to U4/U6 unwinding but prior to the catalysis of the first transesterification is Prp2 (29, 170). Spliceosomes assembled in Prp2-inactivated extract can be isolated free of U1 and U4 RNA and then chased through the first transesterification by the addition of purified Prp2 and a partially purified, heat-stable, snRNP-free fraction called "HP," demonstrating that U1 and U4 RNAs are not required for catalysis (66, 170, 171). Dominant-negative mutant forms of Prp2 remain associated with arrested spliceosomes and can be UV crosslinked to the pre-mRNA, suggesting that the pre-mRNA may be the target of Prp2 (35, 148). However, since both U2 and U6 RNAs are based-paired to the pre-mRNA at this point, they may also be the target of Prp2. The essential splicing factor Spp2, identified as a high-copy suppressor of *prp2* mutants, functions at the same time as Prp2 and interacts with Prp2 in a two-hybrid assay (128). Both proteins appear to leave the spliceosome after ATP hydrolysis by Prp2. There are no known subsequent rearrangements prior to the attack of the branchpoint adenosine on the 5' splice site.

Catalysis of the First Transesterification

Because some group II eubacterial/organellar introns self-splice in the absence of proteins using a two-step transesterifcation reaction identical to that of the spliceosome (102), it is suspected that the spliceosome is fundamentally a ribozyme. Further support for this hypothesis comes from the common metal ion dependence of both the first and second transesterification reactions in group II and spliceosomal introns (44). The most direct evidence that the spliceosome is a ribozyme was recently obtained by Valadkhan & Manley (151), who observed weak catalytic activity analogous to the first transesterification in a trimolecular complex consisting of fragments of U6, U2, and branchpoint RNAs. The fragments of U2 and U6 RNAs that were used are capable of all known intermolecular base-pairing interactions between these RNAs (helices I, II, and III), and the U6 fragment includes the ISL (Figure 5). The branchpoint adenosine of the substrate oligonucleotide apparently forms an unusual 2',3',5'-phosphotriester linkage between the A and G residues of the invariant AGCA sequence at the 5' end of the U6 ISL (Figure 5). Interestingly, mutations flanking the AGCA sequence partially suppress the in vivo splicing inhibition caused by intron branchpoint mutations, which was interpreted to suggest that this region of U6 RNA binds to the catalytic core of the spliceosome (99). The maximal extent of the in vitro reaction is only about 0.15% of input branchpoint oligonucleotide in 24 hours at saturating U2/U6 concentrations (151). Nevertheless, this is the first splicing-related ribozyme activity observed with wild-type U2 and U6 RNA sequences, and it provides hope that a more authentic and efficient RNA-catalyzed splicing reaction may be achieved.

UV crosslinking of the catalytically active human U2/U6 RNA complex confirms the existence of a tertiary contact inferred from genetic evidence to form in vivo (93, 150) (Figure 5). Otherwise, no structural analysis of the complex has been reported. However, the solution structure of the yeast U6 ISL has recently been solved (56) (Figure 5). Consistent with the ribozyme model for U6 RNA function, NMR analysis of the yeast U6 ISL indicates that the U80 Sp phosphate oxygen that appears to bind a magnesium ion essential for the first transesterification in the context of the spliceosome (172) also exhibits stereospecific metal binding in the isolated U6 ISL (56). Furthermore, an adjacent A-C base-pair is protonated at physiological pH, and the pKa of this base-pair becomes more acidic upon metal binding. Protonated bases are present near the active sites of several ribozymes (22), and the HDV ribozyme in particular exhibits an acidic pKa shift upon metal binding (90, 103). Thus, the recent biochemical and biophysical analyses of U2 and U6 RNAs strongly support the hypothesis that the spliceosome is a ribozyme.

CONCLUSIONS

We are at a very exciting point in the ongoing investigation of the spliceosome. The molecular parts list for this ribonucleoprotein machine is substantially complete, although there are sure to be a few surprises left. We have some idea of how the parts fit together on a gross scale, but very little is available in the way of atomic resolution structures. Sequence motifs provide clues to the functions of some spliceosomal proteins, e.g., the DExD/H box family members, while the molecular mechanisms of others, such as Prp8, are as yet a complete mystery. The reward for the hard work of divining the molecular details of the splicing pathway will be an understanding of how RNA and protein can collaborate to accomplish a highly exacting biochemical task. The strategies used by the spliceosome are expected to provide an interesting and informative contrast to those of its fellow ribonucleoprotein machine, the ribosome.

Genetic approaches have been very important in identifying the spliceosomal components and building a basic map of interactions. The emphasis is now shifting to the atomic scale, for example, by saturation mutagenesis of specific domains of splicing factors and isolation of allele-specific suppressors. Information from these studies will be invaluable for developing detailed models of in vivo spliceosome function, which can then be tested by in vitro biochemical techniques. Comparative analysis of yeast and human spliceosomes, and those of other organisms, will continue to be useful for distinguishing aspects of the splicing mechanism that are hard-wired into the machinery from those that are more adaptable to the specific requirements of a species.

The way that we think about the spliceosome has been heavily influenced by the biochemical techniques with which it was first discovered, most notably native gel electrophoresis. New approaches are now challenging the classical model of spliceosome assembly, and the concept of a pre-assembled holospliceosome must be given serious consideration. The relationship between the penta-snRNP and the Cdc5/Cef1 complex is an intriguing mystery and underscores how little we know about recycling of the spliceosome after catalysis. Finally, the unexpected discovery that dominant mutations in genes encoding three different essential spliceosomal proteins can result in a relatively common form of human blindness serves as a reminder of the complex biological implications of the splicing process.

ACKNOWLEDGMENTS

I thank Andreas Kuhn, Eric Steinmetz, Sharon Kwan, and Christine Guthrie for critical reading of the manuscript, and Sam Butcher for help with Figure 5. I apologize to colleagues whose work was not cited due to space constraints. Work on pre-mRNA splicing in my lab is supported by NIH grant GM54018.

The *Annual Review of Genetics* is online at http://genet.annualreviews.org

LITERATURE CITED

1. Abovich N, Liao XC, Rosbash M. 1994. The yeast MUD2 protein: an interaction with PRP11 defines a bridge between commitment complexes and U2 snRNP addition. *Genes Dev.* 8:843–54

2. Abovich N, Rosbash M. 1997. Cross-intron bridging interactions in the yeast commitment complex are conserved in mammals. *Cell* 89:403–12

3. Abu Dayyeh BK, Quan TK, Castro M, Ruby SW. 2002. Probing interactions between the U2 snRNP and the DEAD-box protein Prp5. *J. Biol. Chem.* 277:20221–33

4. Alvi RK, Lund M, O'Keefe RT. 2001. ATP-dependent interaction of yeast U5 snRNA loop 1 with the 5′ splice site. *RNA* 7:1013–23

5. Ares M Jr, Weiser B. 1995. Rearrangement of snRNA structure during assembly and function of the spliceosome. *Prog. Nucleic Acids Res. Mol. Biol.* 50:131–59

6. Arenas JE, Abelson JN. 1997. Prp43: An RNA helicase-like factor involved in spliceosome disassembly. *Proc. Natl. Acad. Sci. USA* 94:11798–802

7. Arning S, Grüter P, Bilbe G, Krämer A. 1996. Mammalian splicing factor SF1 is encoded by variant cDNAs and binds to RNA. *RNA* 2:794–810

8. Ast G, Pavelitz T, Weiner AM. 2001. Sequences upstream of the branch site are required to form helix II between U2 and U6 snRNA in a *trans*-splicing reaction. *Nucleic Acids Res.* 29:1741–49

9. Ast G, Weiner AM. 1997. A novel U1/U5 interaction indicates proximity between U1 and U5 snRNAs during an early step of mRNA splicing. *RNA* 3:371–81

10. Awashti S, Palmer R, Castro M, Mobarak CD, Ruby SW. 2001. New roles for the Snp1 and Exo84 proteins in yeast pre-mRNA splicing. *J. Biol. Chem.* 276:31004–15

11. Ayadi L, Miller M, Banroques J. 1997. Mutations within the yeast U4/U6 snRNP protein Prp4 affect a late stage of spliceosome assembly. *RNA* 3:197–209

12. Ayadi L, Callebaut I, Saguez C, Villa T, Mornon J-P, Banroques J. 1998. Functional and structural characterization of the Prp3 binding domain of the yeast Prp4 splicing factor. *J. Mol. Biol.* 284:673–87

13. Ben-Yehuda S, Dix I, Russell CS, McGarvey M, Beggs JD, Kupiec M. 2000. Genetic and physical interaction between factors involved in both cell cycle progression and pre-mRNA splicing in *Saccharomyces cerevisiae*. *Genetics* 156:1503–17

14. Berget SM. 1995. Exon recognition in vertebrate splicing. *J. Biol. Chem.* 270:2411–14

15. Berglund JA, Abovich N, Rosbash M. 1998. A cooperative interaction between U2AF65 and mBBP/SF1 facilitates branchpoint region recognition. *Genes Dev.* 12:858–67

16. Berglund JA, Chua K, Abovich N, Reed R, Rosbash M. 1997. The splicing factor BBP interacts specifically with the pre-mRNA branchpoint sequence UACUAAC. *Cell* 89:781–87

17. Berglund JA, Fleming ML, Rosbash M. 1998. The KH domain of the branchpoint sequence binding protein determines specificity for the pre-mRNA branchpoint sequence. *RNA* 4:998–1006

18. Brody E, Abelson J. 1985. The "spliceosome": yeast pre-messenger RNA associates with a 40S complex in a splicing-dependent reaction. *Science* 228:963–67

19. Brow DA, Vidaver RM. 1995. An element in human U6 RNA destabilizes the U4/U6 spliceosomal RNA complex. *RNA* 1:122–31

20. Burge CB, Tuschl TA, Sharp PA. 1999. Splicing of precursors to mRNAs by

the spliceosomes. In *RNA World*, ed. RF Gesteland, TR Cech, JF Atkins, pp. 525–60. Cold Spring Harbor, NY: Cold Spring Harbor Lab. Press

21. Burns CG, Ohi R, Mehta S, O'Toole ET, Winey M, et al. 2002. Removal of a single alpha-tubulin gene intron suppresses cell cycle arrest phenotypes of splicing factor mutations in *Saccharomyces cerevisiae. Mol. Cell. Biol.* 22:801–15

22. Butcher SE. 2001. Structure and function of the small ribozymes. *Curr. Opin. Struct. Biol.* 11:315–20

23. Cáceres JF, Kornblihtt AR. 2002. Alternative splicing: multiple control mechanisms and involvement in human disease. *Trends Genet.* 18:186–93

24. Caspary F, Séraphin B. 1998. The yeast U2A′/U2B″ complex is required for pre-spliceosome formation. *EMBO J.* 17.6348–58

25. Caspary F, Shevchenko A, Wilm M, Séraphin B. 1999. Partial purification of the yeast U2 snRNP reveals a novel yeast pre-mRNA splicing factor required for pre-spliceosome formation. *EMBO J.* 18:3463–74

26. Chakarova CF, Hims MM, Bolz H, Abu-Safieh L, Patel RJ, et al. 2002. Mutations in HPRP3, a third member of pre-mRNA splicing factor genes, implicated in autosomal dominant retinitis pigmentosa. *Hum. Mol. Genet.* 11:87–92

27. Chen C-H, Yu W-C, Tsao TY, Wang L-Y, Chen H-R, et al. 2002. Functional and physical interactions between components of the Prp19p-associated complex. *Nucleic Acids Res.* 30:1029–37

28. Chen JY, Stands L, Staley JP, Jackups RR, Latus LJ, Chang T-H. 2001. Specific alterations of U1-C protein or U1 small nuclear RNA can eliminate the requirement of Prp28p, an essential DEAD box splicing factor. *Mol. Cell* 7:227–32

29. Cheng S-C, Abelson J. 1987. Spliceosome assembly in yeast. *Genes Dev.* 1:1014–27

30. Chung S, McLean MR, Rymond BC.

1999. Yeast ortholog of the *Drosophila* crooked neck protein promotes spliceosome assembly through stable U4/U6.U5 snRNP addition. *RNA* 5:1042–54

31. Colot HV, Stutz F, Rosbash M. 1996. The yeast splicing factor Mud13p is a commitment complex component and corresponds to CBP20, the small subunit of the nuclear cap-binding complex. *Genes Dev.* 10:1699–708

32. Dammel C, Noller H. 1993. A cold-sensitive mutation in 16S rRNA provides evidence for helical switching in ribosome assembly. *Genes Dev.* 7:660–70

33. Das R, Zhou Z, Reed R. 2000. Functional association of U2 snRNP with the ATP-independent spliceosomal complex E. *Mol. Cell* 5:779–87

34. Du H, Rosbash M. 2001. Yeast U1 snRNP–pre-mRNA complex formation without U1 snRNA–pre-mRNA base pairing. *RNA* 7:133–42

35. Edwalds-Gilbert G, Kim D-H, Kim S-H, Tseng Y-H, Yu Y, Lin R-J. 2000. Dominant negative mutants of the yeast splicing factor Prp2 map to a putative cleft region in the helicase domain of DExD/H-box proteins. *RNA* 6:1106–19

36. Fleckner J, Zhang M, Valcárcel J, Green MR. 1997. U2AF65 recruits a novel human DEAD box protein required for the U2 snRNP-branchpoint interaction. *Genes Dev.* 11:1864–72

37. Fortes P, Bilbao-Cortés D, Fornerod M, Rigaut G, Raymond W, et al. 1999. Luc7p, a novel yeast U1 snRNP protein with a role in 5′ splice site recognition. *Genes Dev.* 13:2425–38

38. Fortes P, Kufel J, Fornerod M, Polycarpou-Schwarz M, Lafontaine D, et al. 1999. Genetic and physical interactions involving the yeast nuclear cap-binding complex. *Mol. Cell. Biol.* 19:6543–53

39. Fortner DM, Troy RG, Brow DA. 1994. A stem-loop in U6 RNA defines a conformational switch required for pre-messenger RNA splicing. *Genes Dev.* 8:221–33

40. Frendewey D, Keller W. 1985. Stepwise

assembly of a pre-mRNA splicing complex requires U-snRNPs and specific intron sequences. *Cell* 42:355–67

41. Fresco LD, Buratowski S. 1996. Conditional mutants of the yeast mRNA capping enzyme show that the cap enhances, but is not required for, mRNA splicing. *RNA* 2:584–96

42. Frilander MJ, Steitz JA. 2001. Dynamic exchanges of RNA interactions leading to catalytic core formation in the U12-dependent spliceosome. *Mol. Cell* 7:217–26

43. Fromont-Racine M, Rain J-C, Legrain P. 1997. Toward a functional analysis of the yeast genome through exhaustive two-hybrid screens. *Nat. Genet.* 16:277–82

44. Gordon PM, Sontheimer EJ, Piccirilli JA. 2000. Metal ion catalysis during the exon-ligation step of nuclear pre-mRNA splicing: extending the parallels between the spliceosome and group II introns. *RNA* 6:199–205

45. Gottschalk A, Bartels C, Neubauer G, Lührmann R, Fabrizio P. 2001. A novel yeast U2 snRNP protein, Snu17, is required for the first catalytic step of splicing and for progression of spliceosome assembly. *Mol. Cell. Biol.* 21:3037–46

46. Gottschalk A, Kastner B, Lührmann R, Fabrizio P. 2001. The yeast U5 snRNP coisolated with the U1 snRNP has an unexpected protein composition and includes the splicing factor Aar2p. *RNA* 7: 1554–66

47. Gottschalk A, Neubauer G, Banroques J, Mann M, Lührmann R, Fabrizio P. 1999. Identification by mass spectrometry and functional analysis of novel proteins of the yeast [U4/U6 · U5] tri-snRNP. *EMBO J.* 18:4535–48

48. Gottschalk A, Tang J, Puig O, Salgado J, Neubauer G, et al. 1998. A comprehensive biochemical and genetic analysis of the yeast U1 snRNP reveals five novel proteins. *RNA* 4:374–93

49. Gozani O, Feld R, Reed R. 1996. Evidence that sequence-independent binding of highly conserved U2 snRNP proteins upstream of the branch site is required for assembly of spliceosomal complex A. *Genes Dev.* 10:233–43

50. Gozani O, Potashkin J, Reed R. 1998. A potential role for U2AF-SAP 155 interactions in recruiting U2 snRNP to the branch site. *Mol. Cell. Biol.* 18:4752–60

51. Grabowski PJ, Seiler SR, Sharp PA. 1985. A multicomponent complex is involved in the splicing of messenger RNA precursors. *Cell* 42:345–53

52. Gravely BR. 2000. Sorting out the complexity of SR protein function. *RNA* 6:1197–211

53. Hausner T-P, Giglio LM, Weiner AM. 1990. Evidence for base-pairing between mammalian U2 and U6 small nuclear ribonucleoprotein particles. *Genes Dev.* 4:2146–56

54. Heinrichs V, Bach M, Winkelmann G, Lührmann R. 1990. U1-specific protein C needed for efficient complex formation of U1 snRNP with a 5′ splice site. *Science* 247:69–72

55. Hernandez N, Keller W. 1983. Splicing of in vitro synthesized messenger RNA precursors in HeLa cell extracts. *Cell* 35:89–99

56. Huppler A, Nikstad LJ, Allmann AM, Brow DA, Butcher SE. 2002. Metal binding and base ionization in the U6 RNA intramolecular stem-loop structure. *Nat. Struct. Biol.* 9:431–35

57. Igel H, Wells S, Perriman R, Ares M Jr. 1998. Conservation of structure and subunit interactions in yeast homologs of splicing factor 3b (SF3b) subunits. *RNA* 4:1–10

58. Ismaïli N, Sha M, Gustafson EH, Konarska MM. 2001. The 100-kDa U5 snRNP protein (hPrp28p) contacts the 5′ splice site through its ATPase site. *RNA* 7:182–93

59. Johnson TL, Abelson J. 2001. Characterization of U4 and U6 interactions with the 5′ splice site using a *S. cerevisiae* in vitro *trans*-splicing system. *Genes Dev.* 15:1957–70

60. Kandels-Lewis S, Seraphin B. 1993. Involvement of U6 snRNA in 5′ splice site selection, *Science* 262:2035–39

61. Kao H-Y, Siliciano PG. 1996. Identification of Prp40, a novel essential yeast splicing factor associated with the U1 small nuclear ribonucleoprotein particle. *Mol. Cell. Biol.* 16:960–67

62. Käufer NF, Potashkin J. 2000. Analysis of the splicing machinery in fission yeast: a comparison with budding yeast and mammals. *Nucleic Acids Res.* 28:3003–10

63. Kim CH, Abelson J. 1996. Site-specific crosslinks of yeast U6 snRNA to the pre-mRNA near the 5′ splice site. *RNA* 2:995–1010

64. Kim JW, Kim HC, Kim GM, Yang JM, Boeke JD, Nam K. 2000. Human RNA lariat debranching enzyme cDNA complements the phenotypes of *Saccharomyces cerevisiae* dbr1 and *Schizosaccharomyces pombe* dbr1 mutants. *Nucleic Acids Res.* 28:3666–73

65. Kim D-H, Rossi JJ. 1999. The first ATPase domain of the yeast 246-kDa protein is required for in vivo unwinding of the U4/U6 duplex. *RNA* 5:959–71

66. Kim S-H, Lin R-J. 1996. Spliceosome activation by PRP2 ATPase prior to the first transesterification reaction of pre-mRNA splicing. *Mol. Cell. Biol.* 16:6810–19

67. Kistler AL, Guthrie C. 2001. Deletion of *MUD2*, the yeast homolog of U2AF65, can bypass the requirement for Sub2, an essential spliceosomal ATPase. *Genes Dev.* 15:42–49

68. Konarska MM, Sharp PA. 1986. Electrophoretic separation of complexes involved in the splicing of precursors to mRNA. *Cell* 46:845–55

69. Konarska MM, Sharp PA. 1988. Association of U2, U4, U5, and U6 small nuclear ribonucleoproteins in a spliceosome-type complex in absence of precursor mRNA. *Proc. Natl. Acad. Sci. USA* 85:5459–62

70. Krainer ΛR, Maniatis T, Ruskin B, Green MR. 1984. Normal and mutant human beta-globin pre-mRNAs are faithfully and efficiently spliced in vitro. *Cell* 36:993–1005

71. Krämer A. 1996. The structure and function of proteins involved in mammalian pre-mRNA splicing. *Annu. Rev. Biochem.* 65:367–409

72. Krämer A, Grüter P, Gröning K, Kastner B. 1999. Combined biochemical and electron microscopic analyses reveal the architecture of the mammalian U2 snRNP. *J. Cell Biol.* 145:1355–68

73. Kuhn AN, Brow DA. 2000. Suppressors of a cold-sensitive mutation in yeast U4 RNA define five domains in the splicing factor Prp8 that influence spliceosome activation. *Genetics* 155:1667–82

74. Kuhn AN, Li Z, Brow DA. 1999. Splicing factor Prp8 governs U4/U6 unwinding during activation of the spliceosome. *Mol. Cell* 3:65–75

75. Kuhn AN, Reichl EM, Brow DA. 2002. Distinct domains of splicing factor Prp8 mediate different aspects of spliceosome activation. *Proc. Natl. Acad. Sci. USA* 99:9145–49

76. Laggerbauer B, Achsel T, Lührmann R. 1998. The human U5-200kD DEXH-box protein unwinds U4/U6 RNA duplices in vitro. *Proc. Natl. Acad. Sci. USA* 95:4188–92

77. Lauber J, Fabrizio P, Teigelkamp S, Lane WS, Hartmann E, Lührmann R. 1996. The HeLa 200 kDa U5 snRNP-specific protein and its homologue in *Saccharomyces cerevisiae* are members of the DEXH-box protein family of putative RNA helicases. *EMBO J.* 15:4001–15

78. Legrain P, Séraphin B, Rosbash M. 1988. Early commitment of yeast pre-mRNA to the spliceosome pathway. *Mol. Cell. Biol.* 8:3755–60

79. Lesser CF, Guthrie C. 1993. Mutations in U6 snRNA that alter splice site specificity: implications for the active site. *Science* 262:1982–88

80. Lewis JD, Görlich D, Mattaj IW. 1996. A yeast cap binding protein complex

(yCBC) acts at an early step in pre-mRNA splicing. *Nucleic Acids Res.* 24:3332–6

81. Li Z, Brow DA. 1996. A spontaneous duplication in U6 spliceosomal RNA uncouples the early and late functions of the ACAGA element in vivo. *RNA* 2:879–94

82. Liao XC, Colot HV, Wang Y, Rosbash M. 1992. Requirements for U2 snRNP addition to yeast pre-mRNA. *Nucleic Acids Res.* 20:4237–45

83. Liao XC, Tang J, Rosbash M. 1993. An enhancer screen identifies a gene that encodes the yeast U1A snRNP protein: implications for snRNP protein function in pre-mRNA splicing. *Genes Dev.* 7:419–28

84. Libri D, Graziani N, Saguez C, Boulay J. 2001. Multiple roles for the yeast SUB2/yUAP56 gene in splicing. *Genes Dev.* 15:36–41

85. Lin J, Rossi JJ. 1996. Identification and characterization of yeast mutants that overcome an experimentally introduced block to splicing at the 3' splice site. *RNA* 2:835–48

86. Lin RJ, Newman AJ, Cheng S-C, Abelson J. 1985. Yeast mRNA splicing in vitro. *J. Biol. Chem.* 260:14780–92

87. Liu Z, Luyten I, Bottomley MJ, Messias AC, Houngninou-Molango S, et al. 2001. Structural basis for recognition of the intron branch site RNA by Splicing Factor 1. *Science* 294:1098–102

88. Lopez PJ, Séraphin B. 1999. Genomic-scale quantitative analysis of yeast pre-mRNA splicing: implications for splice-site recognition. *RNA* 5:1135–37

89. Lund M, Kjems J. 2002. Defining a 5' splice site by functional selection in the presence and absence of U1 snRNA 5' end. *RNA* 8:166–79

90. Lupták A, Ferré-D'Amaré AR, Zhou K, Zilm KW, Doudna JA. 2001. Direct pKa measurement of the active-site cytosine in a genomic hepatitis delta virus ribozyme. *J. Am. Chem. Soc.* 123:8447–52

91. Lybarger S, Beickman K, Brown V, Dembla-Rajpal N, Morey K, et al. 1999. Elevated levels of a U4/U6.U5 snRNP-associated protein, Spp381p, rescue a mutant defective in spliceosome maturation. *Mol. Cell. Biol.* 19:577–84

92. Maddock JR, Weidenhammer EM, Adams CC, Lunz RL, Woolford JL Jr. 1994. Extragenic suppressors of *Saccharomyces cerevisiae prp4* mutations identify a negative regulator of *PRP* genes. *Genetics* 136:833–47

93. Madhani HD, Guthrie C. 1994. Randomization-selection analysis of snRNAs in vivo: evidence for a tertiary interaction in the spliceosome. *Genes Dev.* 8:1071–86

94. Madhani HD, Guthrie C. 1994. Dynamic RNA-RNA interactions in the spliceosome. *Annu. Rev. Genet.* 28:1–26

95. Makarova OV, Makarov EM, Liu S, Vornlocher H-P, Lührmann R. 2002. Protein 61K, encoded by a gene (*PRPF31*) linked to autosomal dominant retinitis pigmentosa, is required for U4/U6.U5 tri-snRNP formation and pre-mRNA splicing. *EMBO J.* 21:1148–57

96. Maroney PA, Romfo CM, Nilsen TW. 2000. Functional recognition of the 5' splice site by U4/U6.U5 tri-snRNP defines a novel ATP-dependent step in early splicesome assembly. *Mol. Cell* 6:317–28

97. Martin A, Schneider S, Schwer B. 2002. Prp43 is an essential RNA-dependent ATPase required for release of lariat-intron from the spliceosome. *J. Biol. Chem.* 277:17743–50

98. McKie AB, McHale JC, Keen TJ, Tarttelin EE, Goliath R, et al. 2001. Mutations in the pre-mRNA splicing factor gene PRPC8 in autosomal dominant retinitis pigmentosa (RP13). *Hum. Mol. Genet.* 10:1555–62

99. McPheeters DS. 1996. Interactions of the yeast U6 RNA with the pre-mRNA branch site. *RNA* 2:1110–23

100. Merendino L, Guth S, Bilbao D, Martinez C, Valcárcel J. 1999. Inhibition of *msl-2* splicing by Sex-lethal reveals

interaction between U2AF35 and the 3' splice site AG. *Nature* 402:838–41

101. Michaud S, Reed R. 1993. A functional association between the 5' and 3' splice sites is established in the earliest prespliceosome complex (E) in mammals. *Genes Dev.* 7:1008–20

102. Michel F, Ferat J-L. 1995. Structure and activities of group II introns. *Annu. Rev. Biochem.* 64:435–61

103. Nakano S, Chadalavada DM, Bevilacqua PC. 2000. General acid-base catalysis in the mechanism of a hepatitis delta virus ribozyme. *Science* 287:1493–97

104. Neubauer G, Gottschalk A, Fabrizio P, Séraphin B, Lührmann R, Mann M. 1997. Identification of the proteins of the yeast U1 small nuclear ribonucleoprotein complex by mass spectrometry. *Proc. Natl. Acad. Sci. USA* 94:385–90

105. Newman AJ, Teigelkamp S, Beggs JD. 1995. snRNA interactions at the 5' and 3' splice sites monitored by photoactivated crosslinking in yeast splicesomes. *RNA* 1:968–80

106. Newnham CM, Query CC. 2001. The ATP requirement for U2 snRNP addition is linked to the pre-mRNA region 5' to the branch site. *RNA* 7:1298–309

107. Nilsen TW. 1998. RNA-RNA interactions in nuclear pre-mRNA splicing. In *RNA Structure and Function*, ed. R. Simons, M. Grunberg-Manago, pp. 279–307. Cold Spring Harbor, NY: Cold Spring Harbor Lab. Press

108. Noble SM, Guthrie C. 1996. Identification of novel genes required for yeast pre-mRNA splicing by means of cold-sensitive mutations. *Genetics* 143:67–80

109. O'Day CL, Dalbadie-McFarland G, Abelson J. 1996. The *Saccharomyces cerevisiae* Prp5 protein has RNA-dependent ATPase activity with specificity for U2 small nuclear RNA. *J. Biol. Chem.* 271:33261–67

110. Ohi MD, Link AJ, Ren L, Jennings JL, McDonald WH, Gould K. 2002. Proteomics analysis reveals stable multi-protein complexes in both fission and budding yeasts containing Myb-related Cdc5p/Cef1p, novel pre-mRNA splicing factors, and snRNAs. *Mol. Cell. Biol.* 22:2011–24

111. Padgett RA, Hardy SF, Sharp PA. 1983. Splicing of adenovirus RNA in a cell-free transcription system. *Proc. Natl. Acad. Sci. USA* 80:5230–34

112. Pascolo E, Séraphin B. 1997. The branch-point residue is recognized during commitment complex formation before being bulged out of the U2 snRNA–pre-mRNA duplex. *Mol. Cell. Biol.* 17:3469–76

113. Patterson B, Guthrie C. 1988. Spliceosomal snRNAs. *Annu. Rev. Genet.* 22:387–419

114. Pauling MH, McPheeters DS, Ares M Jr. 2000. Functional Cus1p is found with Hsh155p in a multiprotein splicing factor associated with U2 snRNA. *Mol. Cell. Biol.* 20:2176–85

115. Peled-Zehavi H, Berglund A, Rosbash M, Frankel AD. 2001. Recognition of RNA branch point sequences by the KH domain of splicing factor 1 (mammalian branch point binding protein) in a splicing factor complex. *Mol. Cell. Biol.* 21:5232–41

116. Perriman R, Ares M Jr. 2000. ATP can be dispensable for prespliceosome formation in yeast. *Genes Dev.* 14:97–107

117. Pikielny CW, Rymond BC, Rosbash M. 1986. Electrophoresis of ribonucleoproteins reveals an ordered assembly of yeast splicing complexes. *Nature* 324:341–45

118. Pozzoli U, Sironi M, Cagliani R, Comi GP, Bardoni A, Bresolin N. 2002. Comparative analysis of the human Dystrophin and Utrophin gene structures. *Genetics* 160:793–98

119. Puig O, Gottschalk A, Fabrizio P, Séraphin B. 1999. Interaction of the U1 snRNP with nonconserved intronic sequences affects 5' splice site selection. *Genes Dev.* 13:569–80

120. Raghunathan PL, Guthrie C. 1998. A spliceosomal recycling factor that reanneals U4 and U6 small nuclear

ribonucleoprotein particles. *Science* 279: 857–60

121. Raghunathan PL, Guthrie C. 1998. RNA unwinding in U4/U6 snRNPs requires ATP hydrolysis and the DEIH-box splicing factor Brr2. *Curr. Biol.* 8:847–55

122. Rain J-C, Rafi Z, Rhani Z, Legrain P, Krämer A. 1998. Conservation of functional domains involved in RNA binding and protein-protein interactions in human and *Saccharomyces cerevisiae* pre-mRNA splicing factor SF1. *RNA* 4: 551–65

123. Reed R. 1996. Initial splice-site recognition and pairing during pre-mRNA splicing. *Curr. Opin. Genet. Dev.* 6:215–20

124. Reyes JL, Kois P, Konforti B, Konarska MM. 1996. The canonical GU dinucleotide at the 5′ splice site is recognized by p220 of the U5 snRNP within the spliceosome. *RNA* 2:213–25

125. Romfo CM, Alvarez CJ, van Heeckeren WJ, Webb CJ, Wise JA. 2000. Evidence for splice site pairing via intron definition in *Schizosaccharomyces pombe*. *Mol. Cell. Biol.* 20:7955–70

126. Romfo CM, Wise JA. 1997. Both the polypyrimidine tract and the 3′ splice site function prior to the first step of splicing in fission yeast. *Nucleic Acids Res.* 25:4658–65

127. Rosbash M, Séraphin B. 1991. Who's on first? The U1snRNP-5′ splice site interaction and splicing. *Trends Biochem. Sci.* 16:187–90

128. Roy J, Kim K, Maddock JR, Anthony JG, Woolford JL Jr. 1995. The final stages of spliceosome maturation require Spp2p that can interact with the DEAH box protein Prp2p and promote step 1 of splicing. *RNA* 1:375–90

129. Ruby SW. 1997. Dynamics of the U1 small nuclear ribonucleoprotein during yeast spliceosome assembly. *J. Biol. Chem.* 272:17333–41

130. Ruby SW, Abelson J. 1988. An early hierarchic role of U1 small nuclear ribonu-cleoprotein in spliceosome assembly. *Science* 242:1028–35

131. Ruby SW, Chang T-H, Abelson J. 1993. Four yeast spliceosomal proteins (PRP5, PRP9, PRP11 and PRP21) interact to promote U2 snRNP binding to pre-mRNA. *Genes Dev.* 7:1909–25

132. Russell CS, Ben-Yehuda S, Dix I, Kupiec M, Beggs JD. 2000. Functional analyses of interacting factors involved in both pre-mRNA splicing and cell cycle progression in *Saccharomyces cerevisiae*. *RNA* 6:1565–72

133. Rutz B, Séraphin B. 1999. Transient interaction of BBP/ScSF1 and Mud2 with the splicing machinery affects the kinetics of spliceosome assembly. *RNA* 5:819–31

134. Rymond BC, Rosbash M. 1985. Cleavage of 5′ splice site and lariat formation are independent of 3′ splice site in yeast mRNA splicing. *Nature* 317:735–37

135. Sawa H, Abelson J. 1992. Evidence for a base-pairing interaction between U6 small nuclear RNA and 5′ splice site during the splicing reaction in yeast. *Proc. Natl. Acad. Sci. USA* 89:11269–73

136. Schwer B, Shuman S. 1996. Conditional inactivation of mRNA capping enzyme affects yeast pre-mRNA splicing in vivo. *RNA* 2:574–83

137. Séraphin B, Rosbash M. 1989. Identification of functional U1 snRNA-pre-mRNA complexes committed to spliceosome assembly and splicing. *Cell* 59:349–58

138. Séraphin B, Rosbash M. 1991. The yeast branchpoint sequence is not required for the formation of a stable U1 snRNA-pre-mRNA complex and is recognized in the absence of U2 RNA. *EMBO J.* 10:1209–16

139. Shannon KW, Guthrie C. 1991. Suppressors of a U4 snRNA mutation define a novel U6 snRNP protein with RNA-binding motifs. *Genes Dev.* 5:773–85

140. Singh R, Banerjee H, Green MR. 2000. Differential recognition of the poly-pyrimidine-tract by the general splicing

factor U2AF65 and the splicing repressor sex-lethal. *RNA* 6:901–11

141. Spignola M, Grate L, Haussler D, Ares M. 1999. Genome-wide bioinformatic and molecular analysis of introns in *Saccharomyces cerevisiae*. *RNA* 5:221–34

142. Staley JP, Guthrie C. 1998. Mechanical devices of the spliceosome: motors, clocks, springs, and things. *Cell* 92:315–26

143. Staley JP, Guthrie C. 1999. An RNA switch at the 5′ splice site requires ATP and the DEAD box protein Prp28p. *Mol. Cell* 3:55–64

144. Stevens SW, Ryan DE, Ge HY, Moore RE, Young MK, Lee TD, Abelson JA. 2002. Composition and functional characterization of the yeast spliceosomal penta-snRNP. *Mol. Cell* 9:31–44

145. Tang J, Abovich N, Fleming ML, Séraphin B, Rosbash M. 1997. Identification and characterization of a yeast homolog of U1 snRNP-specific protein C. *EMBO J.* 16:4082–91

146. Tanner NK, Linder P. 2001. DExD/H box RNA helicases: from generic motors to specific dissociation functions. *Mol. Cell* 8:251–62

147. Tarn W-Y, Lee K-R, Cheng S-C. 1993. Yeast precursor mRNA processing protein PRP19 associates with the spliceosome concomitant with or just after dissociation of U4 small nuclear RNA. *Proc. Natl. Acad. Sci. USA* 90:10821–25

148. Teigelkamp S, McGarvey M, Plumpton M, Beggs JD. 1994. The splicing factor PRP2, a putative RNA helicase, interacts directly with pre-mRNA. *EMBO J.* 13:888–97

149. Umen JG, Guthrie C. 1995. The second catalytic step of splicing. *RNA* 1:869–85

150. Valadkhan S, Manley JL. 2000. A tertiary interaction detected in a human U2/U6 snRNA complex assembled in vitro resembles a genetically proven interaction in yeast. *RNA* 6:206–19

151. Valadkhan S, Manley JL. 2001. Splicing-related catalysis by protein-free snRNAs. *Nature* 413:701–7

152. van Nues RW, Beggs JD. 2001. Functional contacts with a range of splicing proteins suggest a central role for Brr2p in the dynamic control of the order of events in spliceosomes of *Saccharomyces cerevisiae*. *Genetics* 157:1451–67

153. Vidaver RM, Fortner DM, Loos-Austin LS, Brow DA. 1999. Multiple functions of *Saccharomyces cerevisiae* splicing protein Prp24 in U6 RNA structural rearrangements. *Genetics* 153:1205–18

154. Vithana EN, Abu-Safieh L, Allen MJ, Carey A, Papaioannou M, et al. 2001. A human homolog of yeast pre-mRNA splicing gene, PRP31, underlies autosomal dominant retinitis pigmentosa on chromosome 19q13.4 (RP11). *Mol. Cell* 8:375–81

155. Wang C, Chua K, Seghezzi W, Lees E, Gozani O, Reed R. 1998. Phosphorylation of spliceosomal protein SAP 155 coupled with splicing catalysis. *Genes Dev.* 12:1409–14

156. Wassarman DA, Steitz JA. 1992. Interactions of small nuclear RNA's with precursor messenger RNA during in vitro splicing. *Science* 257:1918–25

157. Weidenhammer EM, Ruiz-Noriega M, Woolford JL Jr. 1997. Prp31p promotes the association of the U4/U6 · U5 tri-snRNP with pre-spliceosomes to form spliceosomes in *Saccharomyces cerevisae*. *Mol. Cell. Biol.* 17: 3580–88

158. Weist DK, O'Day CL, Abelson J. 1996. *In vitro* studies of the Prp9 · Prp11 · Prp21 complex indicate a pathway for U2 small nuclear ribonucleoprotein activation. *J. Biol. Chem.* 271:33268–76

159. Wells SE, Neville M, Haynes M, Wang J, Igel H, Ares M Jr. 1996. *CUS1*, a suppressor of cold-sensitive U2 snRNA mutations, is a novel yeast splicing factor homologous to human SAP 145. *Genes Dev.* 10:220–32

160. Wentz-Hunter K, Potashkin J. 1996. The small subunit of the splicing factor U2AF

is conserved in fission yeast. *Nucleic Acids Res.* 24:1849–54

161. Will CL, Schneider C, MacMillan AM, Katopodis NF, Neubauer G, et al. 2001. A novel U2 and U11/U12 snRNP protein that associates with the pre-mRNA branch site. *EMBO J.* 20:4536–46

162. Wood V, Gwilliam R, Rajandream M-A, Lyne M, Stewart A, et al. 2002. The genome sequence of *Schizosaccharomyces pombe. Nature* 415:871–80

163. Wu S, Romfo CM, Nilsen TW, Green MR. 1999. Functional recognition of the 3′ splice site AG by the splicing factor U2AF35. *Nature* 402:832–35

164. Wyatt JR, Sontheimer EJ, Steitz JA. 1992. Site-specific cross-linking of mammalian U5 snRNP to the 5′ splice site before the first step of pre-mRNA splicing. *Genes Dev.* 6:2542–53

165. Xie J, Beickman K, Otte E, Rymond BC. 1998. Progression through the spliceosome cycle requires Prp38p function for U4/U6 snRNA dissociation. *EMBO J.* 17:2938–46

166. Xu D, Field DJ, Tang S-J, Moris A, Bobechko BO, Friesen JD. 1998. Synthetic lethality of yeast *slt* mutations with U2 small nuclear RNA mutations suggest functional interactions between U2 and U5 snRNPs that are important for both steps of pre-mRNA splicing. *Mol. Cell. Biol.* 18:2055–66

167. Xu D, Friesen JD. 2001. Splicing factor Slt11p and its involvement in formation of U2/U6 helix II in activation of the yeast spliceosome. *Mol. Cell. Biol.* 21:1011–23

168. Xu D, Nouraini S, Field D, Tang S-J, Friesen JD. 1996. An RNA-dependent ATPase associated with U2/U6 snRNAs in pre-mRNA splicing. *Nature* 381:709–13

169. Yan D, Perriman R, Igel H, Howe KJ, Neville M, Ares M Jr. 1998. Cus2, a yeast homolog of human Tat-SF1, rescues function of misfolded U2 through an unusual RNA recognition motif. *Mol. Cell. Biol.* 18:5000–9

170. Yean S-L, Lin R-J. 1991. U4 small nuclear RNA dissociates from a yeast spliceosome and does not participate in the subsequent splicing reaction. *Mol. Cell. Biol.* 11:5571–77

171. Yean S-L, Lin R-J. 1996. Analysis of small nuclear RNAs in a precatalytic spliceosome. *Gene Express.* 5:301–13

172. Yean S-L, Wuenschell G, Termini J, Lin R-J. 2000. Metal-ion coordination by U6 small nuclear RNA contributes to catalysis in the spliceosome. *Nature* 408:881–84

173. Zamore PD, Green MR. 1989. Identification, purification, and biochemical characterization of U2 small nuclear ribonucleoprotein auxiliary factor. *Proc. Natl. Acad. Sci. USA* 86:9243–47

174. Zavanelli MI, Britton JS, Igel AH, Ares M Jr. 1994. Mutations in an essential U2 small nuclear RNA structure cause cold-sensitive U2 small nuclear ribonucleoprotein function by favoring competing alternative U2 RNA structures. *Mol. Cell. Biol.* 14:1689–97

175. Zhang D, Abovich N, Rosbash M. 2001. A biochemical function for the Sm complex. *Mol. Cell* 7:319–29

176. Zhang D, Rosbash M. 1999. Identification of eight proteins that cross-link to pre-mRNA in the yeast commitment complex. *Genes Dev.* 13:581–92

177. Zhang M, Green MR. 2001. Identification and characterization of yUAP/Sub2p, a yeast homolog of the essential human pre-mRNA splicing factor hUAP56. *Genes Dev.* 15:30–35

178. Zorio DAR, Blumenthal T. 1999. Both subunits of U2AF recognize the 3′ splice site in *Caenorhabditis elegans. Nature* 402:835–38

Annu. Rev. Genet. 2002. 36:361–88
doi: 10.1146/annurev.genet.36.061102.093104

Genetic Engineering Using Homologous Recombination[1]

Donald L. Court, James A. Sawitzke, and Lynn C. Thomason

Gene Regulation and Chromosome Biology Laboratory, Center for Cancer Research, National Cancer Institute at Frederick, Frederick, Maryland 21702; e-mail: court@ncifcrf.gov, sawitzke@ncifcrf.gov, lthomason@ncifcrf.gov

Key Words DNA replication forks, strand annealing, in vivo cloning, oligo recombination, recombineering

■ **Abstract** In the past few years, in vivo technologies have emerged that, due to their efficiency and simplicity, may one day replace standard genetic engineering techniques. Constructs can be made on plasmids or directly on the *Escherichia coli* chromosome from PCR products or synthetic oligonucleotides by homologous recombination. This is possible because bacteriophage-encoded recombination functions efficiently recombine sequences with homologies as short as 35 to 50 base pairs. This technology, termed recombineering, is providing new ways to modify genes and segments of the chromosome. This review describes not only recombineering and its applications, but also summarizes homologous recombination in *E. coli* and early uses of homologous recombination to modify the bacterial chromosome. Finally, based on the premise that phage-mediated recombination functions act at replication forks, specific molecular models are proposed.

CONTENTS

INTRODUCTION

Recombinant DNA constructs made in vivo via homologous recombination have been a fundamental analytical tool used by bacterial geneticists. Homologous recombination is the process of exchanging DNA between two molecules through regions of identical sequence. In this way, it ensures precise exchange and joining of two DNA molecules with the limits of the exchange events defined by the homologies between molecules. Homologous DNA recombination systems are extremely useful for moving mutations into and out of the bacterial chromosome. However, these manipulations require extensive in vitro engineering of plasmids or phages in the initial stages of the protocol. Thus, the ability to create specific changes on the chromosome of *E. coli* has always been time-consuming and in certain instances very difficult. We touch on homologous recombination in *E. coli*, and ways in which it has been used to modify the bacterial chromosome. In addition, this review describes phage-encoded homologous recombination systems including very recent technological advances that eliminate the need for restriction enzymes and DNA ligase for modifying or subcloning DNA, thereby eliminating many of the time-consuming in vitro steps of genetic engineering. The phage systems described here share properties with the yeast double-strand break repair system (107, 143) that is able to generate recombinants between linear duplex DNA and the yeast chromosome through very short (<50-bp) regions of DNA identity (6). Because of this ability, yeast researchers have had an advantage that until recently was not shared by the *E. coli* geneticist. Genetic engineering with phage-encoded recombination functions that utilize short homologies has been named "recombineering," a convenient term to describe homology-dependent, recombination-mediated, genetic engineering (29, 38).

The discovery of restriction enzymes more than 30 years ago and their use with DNA ligase to cleave and join novel combinations of DNA molecules in vitro revolutionized molecular biology and led to the advent of genetic engineering. Since then, many advances in genetic engineering have occurred that have allowed the technology to keep up with the rapid expansion of the field of molecular biology. We are now in a new era of biology. Genomic sequencing has provided the complete genome information for many bacterial and eukaryotic organisms. At the same time, *E. coli* vectors have been developed that accommodate clones containing 100s of kb of foreign DNA, thus enabling full-length eukaryotic genes to be isolated and studied with their regulatory regions. Bacterial artificial chromosomes, BACs, or P1 artificial chromosomes, PACs, are single-copy cloning vectors derived from F or P1 plasmids, respectively (58, 128, 141). They contain the F or P1 replication and partitioning systems that ensure low copy number and faithful segregation. The complete genome of herpes virus has been cloned in one BAC (11), and complete genomic libraries from many eukaryotic organisms are represented in BAC vectors (128). Making these large clones in *E. coli* is only the beginning of the manipulations and functional analyses that are being attempted with eukaryotic genes. The full gamut of genetic tests that were once reserved for *E. coli* and yeast studies are now commonplace in the mammalian model, mouse. Mouse genomic clones modified in *E. coli* can be reintroduced into the mouse genome as random transgenic events or more specifically by homologous recombination as replacements of the native segments. Subtle modifications, like point mutations, and more complex changes, like insertions, often need to be made to the BAC clones before reintroduction into their original chromosomal location. It is at this point that classical genetic engineering has become the rate-limiting step in the functional analyses of many of these large clones. The problems encountered in trying to engineer nearly megabase-size clones are the same as those faced by microbial geneticists in modifying the 4.6-megabase bacterial chromosome.

Precision in generating recombinant DNA molecules by standard genetic engineering techniques with restriction enzymes and DNA ligase is lost when working with large DNA molecules. Most cloning techniques depend upon unique restriction sites. With large DNA molecules, even rare cutters, such as the *Not*I enzyme, have many sites of action. Additionally, large DNA can be difficult to work with in vitro because it is prone to breakage. Homologous recombination in vivo is a more versatile and precise way to engineer large DNA molecules (6, 98, 155). Phage-mediated homologous recombination systems, such as the bacteriophage λ Red system, have recently been exploited for these purposes (78, 94, 99, 156, 159). These systems have critical differences from the standard *E. coli* RecA-dependent recombination pathways. The phage systems have a unique advantage in that they can catalyze efficient recombination with very short regions of sequence homology (<50-bp). Importantly, they function even in the absence of RecA, a protein essential for *E. coli* homologous recombination. This is an advantage since RecA action can lead to unwanted recombination and rearrangement of large genomic clones on BACs (29, 98, 128).

In contrast to classical genetic engineering techniques, recombineering does not require construction of plasmid or phage DNA intermediates containing the appropriately pre-engineered homology segments. All that is required in vitro is the synthesis of standard oligonucleotides (oligos) that provide the homology. These oligos can be used directly for recombineering or for construction of PCR products that are used for recombineering. For effective gene replacements, the PCR products are generated with ~50-bp ends that are homologous to sequence targets in the genome.

This review describes the functions of the λ Red recombination system, with emphasis on their use in homologous recombination and in recombineering. Technological advances of recombineering are but one aspect of the review; the recombination of linear DNA by the λ Red system has provided new insights into the mechanisms of homologous recombination. Based on the results of recombineering studies, molecular models describing the λ Red-mediated recombination process are proposed.

RECOMBINATION FUNCTIONS IN *E. COLI*

Homologous recombination has been studied extensively in *E. coli*. Although investigated originally because of its usefulness in genetics, a primary role of homologous recombination in the cell is almost certainly the repair of DNA damage (14). A major portion of this repair occurs at the replication fork itself, and homologous recombination is now known to be a major factor in re-establishing a stalled or disrupted replication fork (32, 57, 73). Skalka first invoked the interplay of replication and recombination processes using phage λ as a model (134, 135), but it is only recently that a direct involvement of recombination functions in the formation and/or activation of a replication fork have been demonstrated (4, 66, 74, 123).

In *E. coli*, most recombination depends on RecA. RecA protein binds to single-strand segments of DNA, forming DNA-protein filaments that have the ability to search other DNA molecules for sequence homology. Once homology is found, the RecA filament pairs and can exchange strands with the homologous segment (Figure 1) (74, 120). The RecBCD and RecF recombination systems operate in *E. coli* and both require RecA function (25, 56). The RecBCD enzyme initiates recombination at double-strand DNA (dsDNA) ends by generating 3' single-strand DNA (ssDNA) overhangs (69, 74, 100). The enzyme also aids RecA in binding to this ssDNA, allowing RecA to promote strand exchange and subsequent

Figure 1   RecA-mediated single-strand invasion and Beta-mediated single-strand annealing.

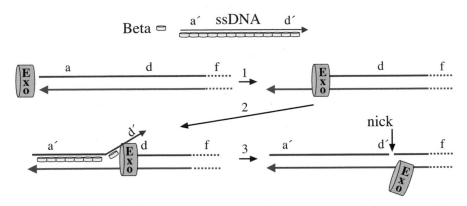

Figure 3 Single-strand annealing and assimilation by λ Exo and Beta. Exo is shown loading at the end of a DNA molecule. Arrows labeled 1, 2, 3 represent the progression of events. A ssDNA is shown bound with Beta protein before annealing and as it anneals to a ssDNA overhang. Exo falls off as annealing is completed to generate a nick in the DNA.

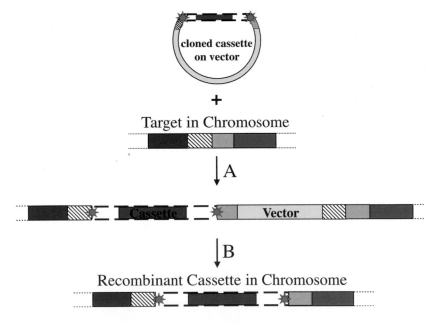

Figure 4 Homologous recombination-mediated insertion and excision of vector DNA for allelic replacement. A DNA cassette (red) is cloned into a vector (yellow) between two DNA segments that have homologies to target sites in the bacterial chromosome. Stars represent restriction sites. The homologies are represented by striped and gray boxes on the vector and chromosome. Insertion (*A*) occurs by RecA-mediated recombination between the striped homology segment on the chromosome and vector. Excision (*B*) of the vector leaves behind the cassette as the product of a second recombination between the gray segments. The vector described here could be either a phage λ or a plasmid that is conditionally defective for replication.

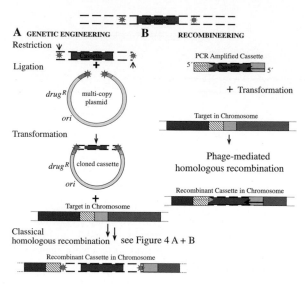

Figure 5 A comparison of steps required to create a recombinant DNA in the chromosome using standard genetic engineering (*A*) or recombineering (*B*). The cassette is indicated in its native site at the top flanked by restriction sites (stars). Figure 4 describes other details of the figure. The horizontal arrows in part *B* indicate the primers used to PCR amplify the cassette with flanking homologies.

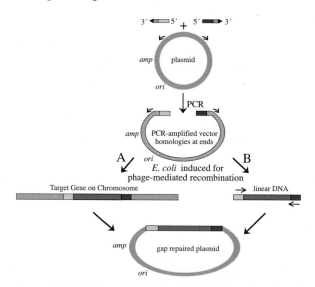

Figure 6 Cloning DNA by gap repair. Two procedures are described to repair a gapped plasmid using phage recombination functions. The colored segments with 5′ and 3′ ends represent 70-nt primers with 5′ homology (yellow or red) segments and 3′ ends (arrows) to amplify the plasmid vector to a linear form. (*A*) represents events used to retrieve a target gene from the bacterial chromosome by gap repair following electroporation of the linear vector. (*B*) represents cloning by co-electroporation of a linear vector with a linear PCR-amplified DNA having the same flanking homologies.

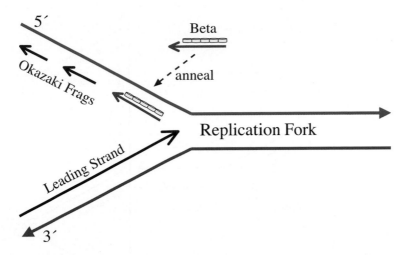

Figure 7 Annealing of the ssDNA oligo to the replication fork by Beta. A ssDNA oligo is shown bound by Beta. Beta anneals the ssDNA to the lagging strand gap at the replication fork.

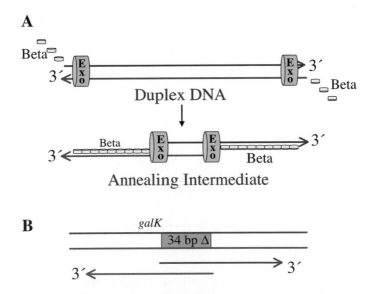

Figure 8 The dsDNA intermediate with 3′ single-strand overhangs. (*A*) λ Exo and Beta generate an annealing intermediate from linear dsDNA. The 3′ ends are indicated. (*B*) Recombination between a synthetic annealing intermediate and the bacterial chromosome at the *galK* gene. The *galK* gene contains a 34-bp deletion as indicated. In the annealing intermediate, the 3′ overhangs are 36 nt long and the duplex segment is 34 bp in length.

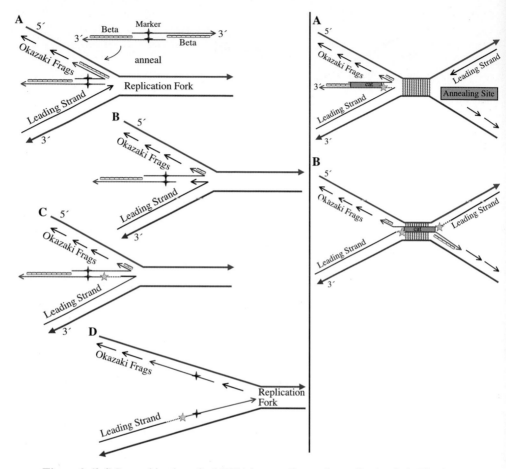

Figure 9 (*left*) Recombination of a dsDNA intermediate at the replication fork. The dsDNA intermediate described in Figure 8*A* is shown with 3′ overhangs. A genetic marker is indicated by blue stars. (*A*) The annealing of one 3′ overhang with the leading strand gap. (*B*) Arrest and backtracking of the replication fork stimulated by Beta-mediated annealing of the ssDNA end with the leading strand. (*C*) PolA-catalzyed polymerization at the 3′ end of the leading strand (blue dashes) and joining of that strand to the intermediate by DNA ligase (yellow star). (*D*) The re-established replication fork after branch migration to the right.

Figure 10 (*right*) Recombination of a dsDNA containing large internal nonhomologies. In this model, the *cat* cassette represents the nonhomologous marker and is shown in place of the blue marker described in Figure 9. The nonhomology prevents branch migration to the right and the fork becomes moribund. (*A*) The second *E. coli* replication fork arriving at the position of the stalled fork. Vertical lines indicate base pairing between the unreplicated parental strands. The annealing site in the gap of the lagging strand at the second replication fork is complementary to the free 3′ end of the dsDNA intermediate. (*B*) The structure after annealing of the free intermediate strand and joining of the leading strand to the *cat* cassette. Note that the original parental segment shown with base pairing (vertical lines) has not been replicated, and the *cat* substitution has been folded over during annealing and ligation, leaving cross-over junctions at each end that must still be resolved.

recombination (3). The RecF pathway also initiates recombination at dsDNA ends, but at a much reduced (100-fold) frequency; this represents the residual recombination activity in a *recBC* mutant (39). Perhaps the most important function of the RecF pathway is to repair defective replication forks (32, 74). Although the RecF pathway requires a broader set of proteins to carry out the recombination reactions than the RecBCD pathway, similar enzymatic functions are needed. For example, RecQ and RecJ process dsDNA to generate 3′ ssDNA overhangs for RecA binding (30). The RecO, RecR, and RecF proteins enhance binding of RecA protein to these single-strand substrates in the presence of single-strand DNA-binding protein, Ssb (147). Once RecA is bound to ssDNA, homology search and strand exchange generate recombination intermediates that may require DNA synthesis to fill any gaps. Recombination intermediates (Holliday structures) in both pathways can be resolved by the process of branch migration catalyzed by RuvAB or RecG, and endonucleolytic cleavage catalyzed by RuvC (127). DNA ligase seals any nicks remaining after resolution.

Additionally, phage-mediated recombination systems can be provided in *E. coli* by the Rac prophage functions, RecE and RecT, or the λ phage Red functions, Exo and Beta. Under some conditions, the phage systems use RecA function (139, 140); however, the phage systems can also generate recombinants in the complete absence of RecA (10, 129).

Strand Invasion Versus Single-Strand Annealing

A primary difference between RecA-dependent recombination and the RecA-independent phage-mediated recombination is the way in which homologous pairing and strand exchange occurs. The RecA-independent Red-mediated recombination is comparatively simple. As defined both in vivo and in vitro, the Red functions Exo and Beta generate recombinants by a process called single-strand annealing (Figure 1) (17, 55, 139, 140). If two homologous DNAs each receive a double-strand break at different points, Exo can degrade the 5′ ended strands exposing 3′ overhangs that Beta binds. Once the complementary sequence is exposed, Beta can anneal the two strands to generate recombinants. In comparison, RecA-mediated exchange can also progress by single-strand annealing, but often occurs through a strand invasion mechanism. RecA bound to the 3′ end of ssDNA can find homologous unbroken DNA and invade, generating the recombination intermediate shown in Figure 1 (74). As described above, additional functions can complete the recombination. Thus, the mechanics of strand invasion promoted by RecA and strand annealing promoted by Red are two well-defined alternatives to generate recombination products. A point relative to the discussion of recombineering is that the Red system cannot generate recombinants between a linear DNA duplex and a nonreplicating DNA circle in the absence of RecA (112, 140). However, the Red system can generate recombinants between linear DNA and the bacterial chromosome in the absence of RecA (97, 156), presumably because the circular chromosome is replicating.

THE BACTERIOPHAGE λ RED FUNCTIONS: EXO, BETA, AND GAM

Mutation of *recA* (24) eliminates recombination mediated by the two bacterial systems (22, 56); however, recombination between phage λ genomes is nearly as efficient in *recA* mutants as in wild-type *E. coli* (10) because λ encodes its own recombination functions. The *exo* (*redα*) and *bet* (*redβ*) genes of phage λ are defined as mutations that eliminate λ homologous recombination in a *recA* mutant strain (36, 41, 130). Some of the *bet* mutations not only cause a Beta defect, but also are defective for λ exonuclease (Exo) activity (129, 157). Polarity is not the only cause of this double defect; in some cases a defective interaction between mutant Beta protein and Exo is thought to reduce the exonuclease activity (129). The notion that the two proteins interact is further supported by the copurification of Beta and Exo in a complex (118).

A third gene, *gam*, provides the full recombination potential to λ. Genetic studies first indicated that RecBCD was a target for Gam (161), and that Gam binds to the RecB subunit (87, 122). Although the RecBCD enzyme is present at only 10 molecules per cell (145), it aggressively destroys most cellular linear dsDNA. By binding, Gam protein inhibits this potent nuclease (63, 93). Other genetic studies indicated that Gam inhibited a second function of *E. coli*, the SbcCD endonuclease (20, 44, 70). Sequence comparison studies show that SbcC and RecB are derived from a common ancestral protein (102), thus a common binding site may remain. In vitro studies with purified SbcCD demonstrate an endonuclease activity targeted to DNA palindromes, which is accompanied by a processive dsDNA-dependent 3′ to 5′ exonuclease activity (27, 28). In vivo, SbcCD repairs double-strand breaks on the bacterial chromosome through recombination with a sister chromosome (33). Thus, Gam inhibits two nucleases, RecBCD and SbcCD, both involved in double-strand break-dependent recombination.

The *exo*, *bet*, and *gam* genes are located next to each other in the p_L operon of phage λ (Figure 2) and are expressed early following infection by the phage or after induction of the prophage. In the prophage, the p_L promoter is directly controlled by the CI repressor, and even following removal of the repressor the expression of the *exo*, *bet*, and *gam* genes is initially prevented by transcription termination. Ultimately, λ N function modifies RNA polymerase to prevent

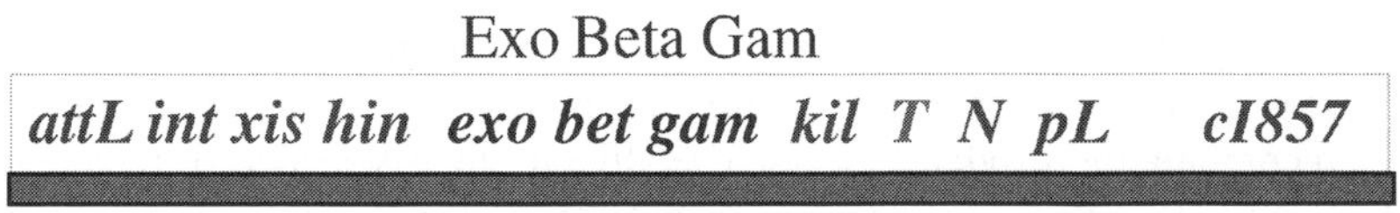

Exo Beta Gam

| *attL int xis hin* | *exo bet gam* | *kil* | *T* | *N* | *pL* | *cI857* |

Figure 2   The p_L operon of prophage λ. The p_L promoter transcribes the genes to its left in the map. *T* represents a transcription terminator. The recombination functions Exo, Beta, and Gam are shown above their genes. Other genes are described elsewhere (31).

transcription termination and allow expression of the recombination functions (31), thereby coordinately activating all the genes in the p_L operon.

A cryptic λ-like prophage, called Rac, is found in the genome of some strains of *E. coli*. This prophage contains genes, *recE* and *recT*, encoding homologous recombination functions that are analogous to *exo* and *bet* (48, 60, 61). Mutants in Rac called *sbcA* (underline{s}uppressor of *recBC*) have been selected in which expression of these genes has been activated resulting in increased recombination in the absence of RecBCD (5). Our discussion focuses primarily on the λ Red gene functions, but they are very similar in activity and function to RecE and RecT. In fact, RecE and RecT can substitute for Exo and Beta on λ (47).

λ Exo: A 5′ to 3′ dsDNA–Dependent Exonuclease

λ Exo has a subunit molecular weight of 24 kDa and degrades linear dsDNA in a 5′ to 3′ direction processively at a rate of 1000 bases per second in vitro (16, 80, 88). Single nucleotides are removed processively leaving long, 3′ ssDNA overhangs (Figure 3, see color insert) that can reach almost half the length of the original duplex DNA (16, 55, 80). Exo requires a dsDNA end to begin digestion and remains bound to the dsDNA as it degrades one strand; it does not initiate at nicks or gaps in the DNA (17, 18). The active form of the protein is a trimer that has a central hole (68, 150). While the entrance to the hole accommodates a dsDNA, the exit diameter is the size of ssDNA. Thus, Exo binds a dsDNA end, slides down the 3′-ended strand, and cleaves mononucleotides from the 5′ strand, leaving behind the intact 3′ overhang (68). Exo also has a much weaker 5′ exonuclease activity on short ssDNA oligos (138).

A novel activity of Exo is displayed during strand assimilation (Figure 3, see color insert). In situations where Exo is degrading a dsDNA from its 5′ end, a complementary single-strand DNA may be annealed to the 3′ single-strand overhang being generated by degradation. Exo degradation of the 5′ strand is prevented as soon as the strand annealing is complete, leaving a nick at that point (18). A nick may also be generated through the combined strand assimilation actions of Exo and Beta as shown in Figure 3 (79).

λ Beta: A ssDNA-Binding Protein that Anneals Complementary DNA Strands

The λ Beta protein has a subunit molecular weight of 25.8 kDa. When isolated from the cell, Beta copurifies with λ Exo and two other proteins until the final phosphocellulose step. Even after this step, most of the Beta protein remains associated in a high-molecular-weight complex with the two other proteins, the S1 ribosomal protein and the NusA transcription elongation factor (91, 151). It is not known whether this interaction occurs in vivo or whether NusA and/or S1 affect recombination.

The Beta protein binds stably to ssDNA (118) greater than 35 nucleotides in length (101), and protects the DNA from single-strand nuclease attack (62, 92).

Beta promotes pairing or annealing between complementary ssDNAs, (Figure 3) (62, 65, 92). Although Beta does not bind directly to dsDNA (62, 92), after it anneals complementary ssDNAs, it remains tightly bound to the annealed duplex (62, 79, 92). This annealed dsDNA-Beta complex is resistant to DNase I and is much more stable than the ssDNA-Beta complex (79). Thus, Beta is bound to duplex DNA but only as the nascent product of strand annealing.

Beta spontaneously assembles into a ring structure in solution and when bound to ssDNA, as it anneals strands it forms a filament on dsDNA (109). Similar ring and filament structures have been found for several other proteins that bind and anneal ssDNA to a complementary sequence, including the yeast RAD52 protein (109). Beta enters the ssDNA at the 3' end and loads in a polarized manner, binding from 3' to 5' (62, 79). Since Exo and Beta are thought to form a complex in vivo (18, 129), it is reasonable that they act cooperatively such that Exo degrades the 5' ends of duplex DNA revealing 3' ssDNA to which Beta can bind. For RecE and RecT, such a combined activity has been demonstrated in vitro (49).

In addition to promoting strand annealing, Beta can also promote a limited type of strand exchange. It can displace a strand of a DNA duplex but must have an adjacent single-stranded gap for initiating the annealing reaction; Beta can then promote displacement using only the energy gained by the adjacent annealing reaction. Beta cannot directly invade duplex DNA with a homologous ssDNA as RecA does (79). RecT has been shown to carry out a similar single-strand displacement of duplex DNA adjacent to gaps (49). RecT-mediated strand invasion into supercoiled circular DNA has been demonstrated in vitro under conditions of low salt and in the absence of divalent cations (103). It remains an open question as to whether strand invasion occurs in vivo during Beta- and/or RecT-dependent recombination events in the absence of RecA. The phage systems have been proposed to cause strand invasion in the absence of RecA (71, 83, 133, 144), However, Kuzminov has explained the same events by single-strand annealing reactions (74).

λ Gam: A Modifier of RecBCD and SbcCD

The λ Gam protein binds stochiometrically to the RecBCD enzyme forming the RecBCD-Gam complex (63, 93). In this complex, several activities, including the nuclease activities, of RecBCD are inhibited (63, 93, 148). A more recently defined activity of RecBCD has not been tested for Gam inhibition, the loading of RecA protein onto ssDNA (2, 3, 21).

If Gam completely inactivates RecBCD, bacterial strains expressing Gam should have the same phenotypes as a RecBCD null mutant. Indeed, both cell types exhibit similar phenotypes; they allow DNA concatemers to form during λ replication (42, 114), generate concatemer–like multimers during ColE1 plasmid replication (26, 40, 131, 132), are UV sensitive, and show reduced cell viability in culture (42, 93).

One major difference between the two cell types is that the strain expressing Gam maintains a high level of recombination activity, arguing against the idea that

Gam completely inactivates RecBCD. In the presence of Gam, recombination is retained for both Hfr crosses and P1 transductions (51, 93, 116) and is actually stimulated for the repair of double-strand breaks caused by X rays (146). Moreover, recombination in phage λ crosses in which Gam (but not Exo or Beta) is present is also efficient (129). These different types of recombination still require RecBCD function since *recBC* or *recD* mutants are defective for recombination or repair in the presence of Gam (93, 116, 146, 148). Thus, Gam does not inhibit a RecBCD-dependent recombination activity despite greatly altering the activity of the RecBCD enzyme. Gam does, however, create new requirements for this altered RecBCD recombination. Several studies show that Gam-inhibited RecBCD recombination and repair requires RecA, RecJ, RecQ, and RecN of the RecF pathway but not RecF itself (82, 93, 108, 146). These proteins likely provide the RecBCD enzymatic activities inhibited by Gam. RecF, a function involved in loading RecA onto ssDNA (147), is not required, indicating that RecBCD-Gam may retain this activity (108). Thus, the RecBCD-Gam complex retains some RecBCD function but requires components of the RecF pathway. It remains a possibility that the SbcCD recombination function (33) retains a residual activity when complexed with Gam and that the two complexes, RecBCD-Gam and SbcCD-Gam, both contribute to recombination.

GENE MODIFICATION AND REVERSE GENETICS IN *E. COLI*

The plasmid and phage vectors, so well studied in *E. coli*, provide great advantages for cloning, amplification, and manipulation of foreign DNA. These methods depend upon classical genetic manipulations as well as standard recombinant DNA techniques such as cutting DNA with restriction endonucleases, purifying DNA fragments, making novel DNA joints with DNA ligase, and transforming the clones into competent cells. Recombineering can replace many of these tedious manipulations.

Classical Genetics and Early Genetic Engineering Methods

In *E. coli*, the earliest systems for manipulating genes and mutant alleles involved moving them from one strain to another by conjugational mating of an integrated F (Hfr crosses) or by F' elements (F plasmids that carry specific segments of bacterial DNA). Transduction by phages that carry bacterial DNA was also used to transfer markers from one strain to another.

As cloning of genes came into vogue, bacteriophage λ vectors were used to clone nearly every gene of *E. coli* (67). These vectors were deleted for λ recombination functions. Genetic modifications of the cloned genes could be engineered on λ and then exchanged from λ to the chromosome (52). Allelic exchange of this type was usually accomplished by forcing the λ to integrate and later excise using homology shared between the cloned gene and the chromosome (Figure 4, see color insert).

These two events, integration and excision, are dependent upon RecA (46, 84), and occur at frequencies approaching 1 in 1000 cells.

Similar RecA-dependent homologous recombination methods have been used to exchange alleles between plasmid clones and the bacterial chromosome (119). Generally, the system is set up so that the plasmid is conditionally defective for replication and can only be maintained if it is integrated via homologous recombination between its cloned gene and the bacterial chromosome (50, 105, 136, 155). These integrants are normally selected via a drug resistance conferred by the plasmid. These plasmid and phage systems are ideal for transferring specific modifications made in vitro back onto the chromosome (Figure 4, see color insert). However, they rely on cloning the homologous regions onto the vectors and creating the desired changes in vitro, which may entail additional cloning and testing stages (Figure 5A, see color insert). Another limitation with these integration and resolution systems is that the integrated form may cause polarity on downstream genes within an operon. Thus, integrations cannot easily be made and tested where essential genes are involved.

Transformation and Recombination with Linear DNA

Homologous recombination can be used to recombine a linear dsDNA fragment into the genome. This type of allelic replacement was first commonly utilized in *Saccharomyces cerevisiae*. In yeast, because double-strand break repair recombination is very proficient and is stimulated by linear DNA ends (143), recombinants are readily created between homologous segments of the chromosome and transformed DNA. The homology segments at the ends of the linear DNA can be as short as 50-bp. Such linear DNAs flanked by short homologies can be generated directly by PCR using primers carrying the 50-bp homologies (6, 75). This recombination technology is extremely useful because it allows direct in vivo engineering of the chromosomes and plasmids in yeast without the time-consuming and cumbersome efforts required in generating clones and modifications of clones in vitro using restriction enzymes and DNA ligase. Figure 5B illustrates the steps involved in constructing chromosomal allelic replacements using these short homologies on linear DNA. Yet, there are serious limitations in using yeast for more general recombinant DNA procedures. Yeast does not have vectors for shuttling engineered DNA to organisms other than *E. coli*, and it is difficult to produce and isolate sufficient levels of plasmid from yeast cells. Finally, the recombination activities in yeast are intrinsic and are not controllable for genetic engineering. For these reasons, a regulated system in *E. coli* that mimics the powerful genetic techniques provided by yeast homologous recombination would be an invaluable tool. Early attempts to develop such bacterial or phage systems, described below, were limited in their usefulness because of their requirement for long homologies and their poor efficiencies. More recently, a phage-encoded system that allows direct modification of the bacterial chromosome as well as *E. coli* vectors including BACs that can contain large DNA inserts (~300 kb) has been developed

(29, 98, 111). This latest technology, called recombineering, has advantages over all previous systems.

Bacterial-Encoded Systems for Linear DNA Recombination

In wild-type *E. coli*, unlike yeast, linear dsDNA is rapidly degraded by nucleases. *E. coli* mutants that are defective for the main nuclease, *recB* or *recC*, do not degrade linear DNA as rapidly (8, 9), are defective for recombination, and grow very poorly, producing up to 80% nonviable cells (15). Suppressor mutations, *sbcB sbcC*, restore recombination activity and viability to these *recBC* mutant strains (5, 72, 81). The *sbcB* mutation is a special allele of the *xonA* gene, which affects the 3′ to 5′ exonuclease activity of Exo I. As mentioned above, *sbcC* encodes part of another RecBCD-like nuclease. Linear DNA is stable in these suppressor strains and can undergo homologous recombination with the chromosome using the RecF pathway. This recombination is completely dependent upon RecA (23). In *recBC sbcB sbcC* mutant cells, recombination requires very long regions (~1000 bp) of flanking homology that must be engineered by classical cloning techniques and isolated as linear DNA by restriction digestion. Despite the long homologies, the frequency of recombinants is low and requires microgram amounts of transformed DNA (59, 86).

The RecBCD functions can also be used for linear DNA recombination under specific conditions. Mutations in the *recD* gene inactivate the RecBCD nuclease but not recombination activities and thus, linear DNA is preserved and can be recombined with the chromosome (121). Dabert & Smith (34) used another approach to recombine linear DNA in wild-type *E. coli* containing the RecBCD nuclease. Special sites were engineered in the linear DNA causing RecBCD that entered these DNAs to lose its destructive nuclease activity and to become recombinogenic. In fact, electroporation itself has been suggested to reduce DNA degradation by RecBCD nuclease and allow recombination with linear dsDNA (37). Thus, several different strategies have been utilized to allow linear DNA recombination in *E.coli*; all require RecA and either the RecF or RecBCD recombination pathway functions. Unfortunately, all are very inefficient, with only a few recombinants found per transformation. In addition, thousands of base pairs of homology and high DNA concentrations are required to generate these rare recombinants.

Phage-Encoded Systems for Linear DNA Recombination

Another class of *recBC* suppressor mutations, *sbcA*, that expresses recombination functions RecE and RecT (48) from the Rac prophage, also generates rare recombinants between linear DNA and the host chromosome or plasmids (152). A similar system, Red, is encoded by phage λ. Murphy (94) developed the λ system, which enhanced the efficiency of linear DNA recombination at least 50-fold compared with previous systems. Moreover, this λ recombination is functional in most strains, not just *recBC* or *recD* mutants. In this Red system, the λ *exo*, *bet*, and *gam* genes are under *lac* promoter control on a multicopy plasmid. Because Gam function inhibits

RecBCD nuclease, and because Exo and Beta provide recombination activity, linear DNA is not degraded but recombines with the circular bacterial chromosome. In the wild-type strain, a similar plasmid that expressed only Exo and Beta (no Gam) did not generate recombinants under the same conditions, demonstrating again the inhibitory effect of RecBCD nuclease. Murphy (94) also demonstrated that the phage P22 recombination functions could promote linear recombination but at a reduced efficiency relative to the λ functions.

When the λ *gam* gene is expressed, ColE1 plasmids fail to replicate as circles and generate linear multimers (26, 114, 132). These multimers can inhibit recombination between linear transformed DNA and the chromosome, perhaps because they compete for the Exo and Beta products (94). The plasmid concatemers may also be toxic to cells containing them (42). For these reasons, Murphy (94) replaced the chromosomal *recBCD* operon with the λ *exo* and *bet* genes under *lac* promoter control. Conveniently, in this *recBCD* replacement, Gam was not required to inactivate the RecBCD nuclease. Despite being expressed from a single copy in the chromosome, Exo and Beta functions increased linear recombination several-fold over the plasmid-induced level. The Poteete and Murphy laboratories (94, 95, 113, 115) have shown that linear recombination in these *recBCD::P_{lac} bet exo* cells is completely dependent upon the λ Exo and Beta functions. Recombination levels are also dependent upon RecA as they are reduced nearly 100-fold in *recA* mutants. In addition, several recombination genes, *recQ, recO, recR, recF*, and *ruvC*, were required, but *recJ* and *recG* were not. Thus, recombination generated in this *recBCD* deletion mutant background uses a combination of the Red and RecF pathway functions and is largely dependent upon RecA. In these studies, long homologies (>1000 bp) between the linear DNA and the chromosome were used. Recombination was enhanced by increasing the time of Exo and Beta induction (94) and by increasing the linear DNA concentration (95). At very high concentrations of linear DNA (~30 micrograms per electroporation), recombinants were found in 1% of the cells surviving electroporation.

Genetic Engineering with Short DNA Homologies: Recombineering

Homologous recombination studies and in vivo genetic engineering were taken to yet another level by Francis Stewart and colleagues (159). They found that short DNA homologies, 42 to 50 bp in length, generated recombinants that depended upon expression of the RecE and RecT functions induced in the *recBC sbcA* mutant strain described above. Although this RecET-mediated recombination is not very efficient, the crucial advantage is that it could use short homologies (159). This technology is extremely useful for genetic engineering of BAC or PAC clones and the chromosome of *E. coli* (99, 159). The homology required could be incorporated in the primers used to PCR amplify the drug cassette, thereby eliminating multiple steps: namely, the need for in vitro construction of plasmids containing the long (>1000 bp) flanking homologies (Figure 5, see color insert).

The Stewart laboratory (159) created a portable system by cloning *recE, recT*, and the λ *gam* genes under control of three separate promoters in a ColE1-type plasmid. Although transferable, this plasmid-based system is poorly maintained and somewhat toxic due to *gam* expression. Of the three genes, only the *recE* gene was placed under a regulated promoter, the arabinose *pBAD* promoter; *gam* and *recT* were placed under constitutive promoters. Expression of Gam inhibited RecBCD nuclease, allowing linear DNA to survive and the short homologies at the DNA ends to be used as substrates for the recombination functions of RecET. The plasmid construct generated ~threefold more recombinants than the chromosomal *recBC sbcA* mutant (159). Thus, a flexible system that could generate linear DNA recombinants in most strains using short homologies was created. Another very similar plasmid construct was created that substituted the λ *exo* and *bet* genes for *recE* and *recT*, respectively. This Red system worked as well if not better than the RecET system (99). These same plasmid sets have also been designed without *gam* and have been used by the Stewart laboratory (97) in *recBC* mutant strains, while other plasmid replicons derived to express the *gam*, *bet*, and *exo* genes include pSC101 (35) and pR6K116 containing the R6K gamma origin of replication (160).

A defective λ prophage-based system has also been developed to express *gam*, *bet*, and *exo* genes in their natural context. Here the genes are carried as a single copy on the bacterial genome and expressed from the powerful λ p_L promoter. Expression is tightly regulated by the temperature-sensitive λ CI857 repressor (Figure 2); at 32°C the repressor blocks the p_L promoter. Inactivating the repressor by a temperature shift to 42°C turns on the promoter, allowing coordinated expression of *gam*, *bet*, and *exo* genes (156).

In plasmid and prophage recombineering systems that use short homologies, linear DNA recombination requires both *exo* (*recE*) and *bet* (*recT*) expression; *gam* is also required in *recBCD+* cells (97, 156). Muyrers et al. (97) demonstrated that RecE works only with RecT and that Exo works only with Beta. Mutation of *recA* only reduced this recombination a few fold (97, 156). This differs from the Murphy and Poteete laboratories' Red-mediated system that uses long homologous ends. In that system, the absence of RecA causes up to a 100-fold reduction in recombination (94, 95, 113, 115). This discrepancy is not understood (111); however, homology length does not appear to be the cause (97, 156).

The plasmid- and prophage-based recombineering systems each have their advantages and disadvantages. Generally, the plasmid system is more mobile and easily transferred among *E. coli* strains and even to other bacterial species (117, 149). Recombination functions in the prophage system are tightly controlled and coordinately expressed. Three problems are avoided by being able to coordinately induce the prophage recombination functions for a brief (<15 min) time: (*a*) Leaky expression of recombination functions leads to unwanted recombination products. This is more of a problem with BAC (or PAC) clones carrying genomic DNA from eukaryotes that possess many long repetitive sequences. (*b*) Constitutive expression of Gam function inactivates RecBCD leading to plasmid instability (26, 40,

93) and cell toxicity (42, 126). (*c*) Altered expression ratios of RecE and RecT functions affect recombination efficiency (97).

Coordinate expression of the Exo and Beta functions from their natural context in the prophage generates especially high recombination levels, which can be a great advantage. In fact, Lee et al. (78) were able to identify recombinants by colony hybridization without selection. In their experiment, a 24-bp sequence encoding the FLAG TAG epitope had been targeted by flanking homology to a gene in a BAC clone in a *recA* mutant background. Seven colonies in 4200 examined from the electroporation hybridized to the 24-bp probe, and all seven were found to be correct by sequence analysis.

APPLICATIONS OF RECOMBINEERING WITH LINEAR dsDNA

Targeting the Chromosome

Antibiotic cassettes with appropriate flanking homology can be conveniently used to target specific genes or regions of the chromosome for replacement. Such replacements are directly selected as drug resistant. The junctions between the homology arms and the start of the resistance cassette define the ends of the deleted region (see Figure 5*B*, see color insert). This technology has been used to insert cassettes between two adjacent base pairs without deleting any bases (38, 159) and to replace as much as 70-kb with the cassette (54).

In addition to the selectable marker, other DNA sites or coding sequences can be incorporated on the same fragments as the resistance cassette and recombined jointly with it. Examples include *lacZ* fusions, GFP fusions, and His-tags. The limiting condition in these cases is usually the size of the DNA elements to be amplified and the fidelity of the PCR amplification of those elements. Targets for site-specific recombinases, such as *loxP* or *frt* sites, can be added on the flanks of the drug cassette, allowing subsequent removal of the cassette by activating expression of site-specific recombinases Cre or Flp that act on *loxP* and *frt* sites, respectively (35, 99, 159). Counter-selectable genes such as *sacB* can also be recombined along with the selectable drug marker. The SacB function, once established in a cell, converts sucrose to a toxic form and kills *E. coli* (43). A second round of linear recombination can be used to delete the drug cassette and *sacB* by plating the recombination mixture on agar containing sucrose to select for these recombinants (78, 95, 96). This counter-selection technique can generate a perfect deletion of the drug cassette and *sacB*, whereas the *loxP* or *frt* recombination always leaves behind a scar of the remaining *loxP* or *frt* site. The counter-selection technique can also be used to insert other nonselectable markers such as fusions or tags or even point mutations at the original site of recombination. Thus, a targeted region can be replaced by *cat-sacB* in the first step; then DNAs containing various mutations can be introduced in the second step, allowing selection of numerous site-specific mutations in a gene of interest.

Technically, the positive selection with a drug marker is generally free of background resistant colonies, and all resistant colonies are likely to be correct. This is not the case during counter-selection against *sacB* as spontaneous sucrose-resistant mutations, inactivated for *sacB*, occur at a frequency of about 1 in 10,000 cells. Therefore, the desired recombinants must be detected from among this spontaneous background of sucrose-resistant colonies. Screening among sucrose-resistant colonies for those that have also lost the drug cassette detects the desired recombinants. Hence, when recombination is high, fewer sucrose-resistant colonies need to be screened.

Targeting Plasmids

Just as linear DNA can be targeted to the chromosome, it can also be targeted to plasmids either already resident in the cell, or co-electroporated with linear DNA (156, 159). Both of these procedures require that the plasmid DNA be retransformed to generate pure clones of recombinant plasmids. Therefore, a selectable marker on the incoming linear fragment is useful. Co-electroporation is preferable for ColE1-type plasmids because an already established plasmid generates multimers when Gam is expressed (26). These multimers contain both recombinant and parental segments that must be separated in vitro (156).

In vivo DNA Retrieval by Gap-Repair

Recombineering can be used to subclone DNA directly into a linear plasmid vector backbone without restriction enzymes or DNA ligase. Nearly any region from the bacterial chromosome, a plasmid, or a BAC clone can be retrieved into an appropriate vector. The precision of the technology allows fusions to be made between the retrieved gene and regulatory elements like promoters and translational signals on the vector.

Figure 6A (see color insert) describes the generation of a linear vector with ends that contain short homology segments to a target in the cell. Minimally, the vector needs a selectable drug marker and an active origin of replication. Gap-repair of the plasmid by recombination with the target circularizes the vector, allowing selection for the drug marker. Two plasmids, p15A and ColE1, with different replication origins, have worked well in gap-repair cloning (78, 160). A third plasmid, pSC101, did not generate gap-repaired recombinants in similar crosses (E. Lee & D. Court, unpublished data). Possibly the type of replicon involved determines whether linear DNA recombination can be completed. Linear DNA recombination and replication may be closely coupled events (discussed below).

The use of ColE1 plasmids presents another potential problem. High-copy, pUC-type, pBluescript vectors were used to subclone fragments up to ~25 kb, but larger fragments were much more difficult to clone. Yet, with a lower-copy pBR322 vector, fragments as large as 80-kb could be subcloned in one step (78). The high-copy vector with large inserts overtaxes the capacity of the cell for DNA synthesis.

Unwanted drug-resistant colonies can also be generated without retrieving the target in the cell. This occurs by directly joining ends via repeats longer than 5-bp, which are located at or near the end. Removal of these short repeats reduces this "end-joining" reaction dramatically (160). This type of end-joining reaction occurs rarely in normal *E. coli* strains and appears to be elevated to a higher level by the presence of the RecET and Red functions.

Earlier reports had implicated RecET function in elevating RecA-independent intrachromosomal deletions between short (~7-bp) repeats (64). The frequency of deletion formation is reduced by overexpression of the 3′ to 5′ exonuclease, Exo I, encoded by *xonA* (153). Likewise, a *xonA* null mutation caused an increased deletion frequency in normal cells without RecET (1). This suggests that RecE (or λ Exo) generates excess 3′ ends, favoring end joining of linear vectors and intrachromosomal deletions, whereas Exo I removes 3′ ends reducing end joining and intrachromosomal deletion formation. End joining may depend on annealing of 3′ single-strand overhangs just as recombination mediated by RecET and Red functions does. SbcCD function in concert with single-strand annealing has also been implicated in the generation of intrachromosomal deletions through short repeats (13).

In vivo Cloning by Gap-Repair

A technology called in vivo cloning (see Figure 6*B*, see color insert) has been used in *recBC sbcBC* mutants of *E. coli* since 1993 (12, 106) and is based on single-strand annealing. The efficiency of the procedure is enhanced dramatically by the Red and RecET systems (97). Again, a linear plasmid vector is used, but in this instance it is designed to retrieve DNA from a co-electroporated DNA fragment whose ends are homologous to the vector ends. Exo (RecE) degrades the 5′-ended strands of each co-electroporated linear DNA. Beta (RecT) binds to the ssDNA ends and anneals them to their complementary strands, where they are covalently joined by DNA ligase generating the drug-resistant plasmid clone.

A modification of the in vivo cloning protocol may allow direct cloning of a fragment from a complete genomic DNA mixture from almost any organism. This type of technology was developed in yeast where it works with homologies as short as 60-bp (77, 104). A similar technology is being developed in *E. coli* (160). Purified genomic DNA from any source is fragmented and co-electroporated with a linear vector. Homologies on the ends of the vector are used to find and retrieve the target DNA from the complex mixture in the cells. Each cell receives a portion of the genomic DNA fragments, and the complexity of the genome determines the efficiency of the retrieved target (77, 160).

RECOMBINEERING WITH ssDNA

In yeast, transformation with ssDNA has been used for recombination to generate mutations (89, 90, 154). Recent studies demonstrate that in *E. coli*, ssDNA is also recombinogenic when using λ Red (38, 142). The efficiency of recombination with ssDNA is dramatically higher than that obtained with dsDNA. Whereas the function(s) responsible for ssDNA recombination in yeast is not known, in *E. coli* the only λ function required is Beta. Exo and Gam, required for dsDNA recombination, are not required for ssDNA. There is also no dependence on RecA for the ssDNA recombination tested (38).

The ssDNA can be supplied as synthetic oligos to make a single base change. An oligo length of 30-nt generates recombinants but a large increase is obtained with 40-nt long oligos (38). The dramatic increase in recombination efficiency between 30- and 40-nt is similar to the increase observed using flanking homologies in dsDNA (156). This critical length dependence correlates well with a length dependence of 36-nt for tight binding by Beta protein in vitro to ssDNA (62, 101). In this regard, a ssDNA oligo of 70-nt increased recombination fivefold over that for 40-, 50-, or 60-nt long oligos (38). The 70mer may provide two tight binding domains for Beta protein. Longer ssDNA oligos have also been used; a 164-nt oligo was used to introduce a 24-nt FLAG TAG sequence into a precise position in the Brca2 mouse gene on a BAC clone (142). However, a much longer ssDNA (>1000 nt) generated from denaturation of a PCR product with 50 nt of homology on the ends was extremely inefficient for recombination (D. Yu & D. Court, unpublished data). This may be explained by the 3' to 5' polarity of Beta binding; any nuclease that attacks the 5' end, like RecJ, could eliminate homology from that end before Beta could bind and protect it (62, 79).

In *E. coli*, ssDNA oligos have been used to create point mutations, to repair mutations, to create deletions, and to create small insertions on the chromosome and in BAC clones (38, 142). Two *galK* defects, a point mutation and a 3.3-kb insertion at the same site as the point mutation, are corrected to *galK*$^{+}$ at the same frequency by an identical 70-nt oligo. Thus, a single base substitution and a 3.3-kb deletion were generated with equal efficiency. Other 70-nt ssDNAs have been used to cure five different *Tn10* elements from the chromosome with recombination efficiencies that reached 6% of cells surviving electroporation (38).

Recombineering with ssDNA is more efficient than with dsDNA and is the method of choice to create point mutations and other changes in a single step. Recombinants from ssDNA recombineering are so frequent that they can be screened directly from among total cells in an electroporation. In fact, a special PCR amplification screen can be used to detect the single base change of a recombinant (19, 142). In these cases, point mutations are created at frequencies approaching 1% of electroporated cells, even in the absence of RecA activity (38, 142). At these high recombination efficiencies, colony hybridization has also been used to screen for unselected recombinants, which have insertions or deletions of multiple bases (N. Costantino, L. Thomason & D. Court, unpublished data).

Recombineering with ssDNA may work in many bacterial species and even in eukaryotes under appropriate conditions, since Beta is the only λ function required and the proposed mechanism (see below) seems likely to be universal. Because this system is so simple and requires only Beta, it seemed possible that the Beta protein could be co-electroporated with ssDNA to generate recombinants not only in *E. coli*, but other organisms as well. Protein-nucleic acid co-electroporation has worked to generate in vivo products with other DNA-binding proteins (45, 76). However, initial attempts to generate recombinants by co-electroporating Beta protein bound to ssDNA into *E. coli* cells have failed (H. Ellis, K. Murphy & D. Court, unpublished data).

Chromosome Recombineering with ssDNA Oligos: A Model

For any particular allele, recombinants can be generated with oligos corresponding to either of the two complementary DNA strands that include that allele. Invariably, one strand is found to recombine more frequently than the other. For markers located in different segments of the *E. coli* chromosome, the more efficient strand correlates with the direction of replication through the region being tested. In the seven cases examined, the more efficient strand corresponded to the lagging strand of DNA replication (38). This strand bias has been explained by a model in which the ssDNA oligo is annealed by Beta to the gaps present in the lagging strand as the replication fork (Figure 7, see color insert) passes through the targeted region (29, 74).

The ssDNA gaps at the replication fork are coated by Ssb protein (85, 125). Therefore, Beta must either anneal the strands with Ssb still bound, or Ssb must be displaced. An Ssb displacement activity has been postulated for proteins similar to Beta (7). Since both complementary strands generate recombinants (38), ssDNA gaps may also occur in leading strand synthesis, during transcription processes, by supercoiling, and at DNA repair events. Beta may anneal the ssDNA oligos at any of these places.

RECOMBINEERING WITH dsDNA: AN INTERMEDIATE

When dsDNA is used for linear recombination, *exo*, *bet*, and *gam* gene products are required for efficient recombination (156). These results and the known biochemical properties of the Exo and Beta proteins suggested the DNA structure shown in Figure 8A (see color insert). In this model, a linear dsDNA with flanking 3' ssDNA overhangs is generated as an annealing intermediate. Such a structure made in vitro and electroporated into a cell should be recombinogenic, and recombination should not require Exo.

Muyrers et al. (97) constructed substrates with variable length 3' overhangs flanking a drug-resistance marker. Regardless of the length of 3' overhangs, recombinants were fully dependent on Exo in addition to Beta. Even with Exo and Beta present, however, recombination efficiency was low in these experiments, perhaps because of how the substrate was constructed or because Gam was not present (97). Yu et al. (D. Yu, J. Sawitzke, H. Ellis & D. Court, manuscript in preparation) have performed similar experiments with differing results. In this case, the 3' overhangs were constructed by co-electroporating two 70-nt oligos. The oligos were such that 34-nt at the 5' ends were complementary and contained the DNA missing within a 34-bp *galK* deletion in the recipient strain (Figure 8*B*, see color insert). The 36-nt at each 3′ end remained unpaired and were homologous to DNA flanking the 34-bp deletion. As predicted by the model, this substrate recombined and, more importantly, required only Beta. The presence or absence of Exo had no effect. This recombination was also found to be *recA* independent (D. Yu, J. Sawitzke, H. Ellis & D. Court, manuscript in preparation). Because ssDNA ends are involved, this intermediate, like ssDNA, may initiate recombination at a DNA replication fork.

Chromosome Recombineering with Linear dsDNA: A Model

A model is described in Figure 9 (see color insert) to account for linear dsDNA recombination using λ Red. As this recombination can be RecA independent (97, 156), we propose it occurs at the DNA replication fork. It has been suggested that recombination in the absence of RecA is catalyzed by Beta-dependent annealing (139, 140) of complementary 3′ ssDNA. In this way, the dsDNA intermediate (Figure 8*A*, see color insert) anneals at the fork and triggers a block to fork progression (32). As drawn, this structure looks like a precursor to a four-way Holliday junction (Figure 9*A*, see color insert). Stalled DNA replication forks are known to backtrack (32). During backtracking (Figure 9*B*), the leading strand is transferred and annealed to the 3′ ssDNA of the dsDNA intermediate to create a chicken foot-like structure (53, 110). DNA polymerase I (PolA) can initiate repair synthesis on the backtracked leading strand (Figure 9*C*), using the dsDNA intermediate as a template; DNA ligase can then covalently join the leading strand to the intermediate. The replication fork is reestablished by branch migration as shown in Figure 9*D* such that each strand of the original linear dsDNA intermediate forms the new strand at the regenerated replication fork.

Blocking Branch Migration by a Large Nonhomologous Region: A Modified Model

RecA-independent recombination can occur between the chromosome and linear DNA containing a nonhomologous cassette flanked by homologies to the target (156). We propose that with these substrates the 3′ single-strand end of the linear duplex also anneals at the replication fork and generates the same intermediates as described in Figure 9*C*. However, branch migration, like that in Figure 9*D*, is blocked by the large nonhomology of the *cat* cassette (Figure 10*A*, see color insert) stalling the replication fork and preventing completion of the recombination event. However, both processes can be rescued by the second *E. coli* replication fork that traverses the chromosome in the opposite direction (Figure 10*A*). The second fork must pass through the *E. coli* terminus and proceed to the stalled fork. It is not clear how long this would take or how efficient it is. Preliminary studies indicate that other λ functions in the p_L operon prevent new rounds of DNA replication from initiating (K. Surgueev, D. Court & S. Austin, submitted). Once at the stalled fork, the second fork can provide an annealing site that is complementary to the 3' single-strand at the other side of the drug cassette. These two single-strand regions can be annealed by Beta, and as described in Figure 9C, PolA catalyzes replication, and DNA ligase can then repair the leading strand gap joining it to the *cat* cassette (Figure 10B). The daughter chromosomes remain linked at the *cat* cassette and resolution of the junctions will presumably require RuvC (127) or perhaps a topoisomerase (158). The end result is that two chromosomes are generated, a parental and recombinant with the cassette replacement.

RECOMBINEERING: TAPPING ITS POTENTIAL

Currently, the phage-encoded requirements for recombineering are well defined and the process is very efficient even in *recA* mutant cells; however, other host requirements have not been determined. Our models that describe Red-mediated recombination at the replication fork suggest candidate functions that act during DNA replication. Replication functions associated with λ Red-mediated processes are DNA polymerase I and DNA ligase. Phage λ growth in *lig* or *polA* mutants is defective in the absence of *exo*, *bet*, or *gam* (134, 137), indicating the interactions of these functions.

Recombineering is so efficient that it can provide the substrates for in vivo biochemistry. For example, insertion of modified nucleotides (biotinylated, fluorescent, phosphothiolated, etc.) directly into the chromosome can be used to study the processes of DNA replication, recombination, and chromosome segregation. One potential use of such modified nucleotides is to identify the site of Red-mediated recombination in the cell and the structure of the DNA at that site. For example, a biotin tag could be used to isolate a fragment of the chromosome containing recombination intermediates to help determine their structure.

Although recombineering in *E. coli* has been studied for only four years, it is being used by large numbers of researchers in both prokaryotic and eukaryotic applications. Using oligo mutagenesis, any base on the chromosome can be changed in one step without selection. With appropriate controls, it should be possible to determine which changes affect viability of the cell. Because the recombineering technology is well defined, it may be possible to adapt it to other organisms (117, 124). In organisms where Red is not active, analogs that can carry out the same functions will undoubtedly be found in the viruses of these organisms, be they prokaryotic or eukaryotic.

ACKNOWLEDGMENTS

We thank Nina Costantino, David Friedman, and Carolyn McGill for critical reading and discussion of the manuscript. We thank Marg Mills for preparing the manuscript and editorial comments. We also thank Daiguan Yu, Hillary Ellis, and E-Chiang Lee for stimulating discussions on these subjects.

The *Annual Review of Genetics* is online at http://genet.annualreviews.org

LITERATURE CITED

1. Allgood ND, Silhavy TJ. 1991. *Escherichia coli xonA (sbcB)* mutants enhance illegitimate recombination. *Genetics* 127: 671–80

2. Amundsen SK, Taylor AF, Smith GR. 2000. The RecD subunit of the *Escherichia coli* RecBCD enzyme inhibits RecA loading, homologous recombination, and DNA repair. *Proc. Natl. Acad. Sci. USA* 97:7399–404

3. Anderson DG, Kowalczykowski SC. 1997. The translocating RecBCD enzyme stimulates recombination by directing RecA protein onto ssDNA in a chi-regulated manner. *Cell* 90:77–86

4. Asai T, Bates DB, Kogoma T. 1994. DNA replication triggered by double-stranded breaks in *E. coli*: dependence on homologous recombination functions. *Cell* 78:1051–61

5. Barbour SD, Nagaishi H, Templin A, Clark AJ. 1970. Biochemical and genetic studies of recombination proficiency in *Escherichia coli*. II. Rec⁺ revertants caused by indirect suppression of rec⁻ mutations. *Proc. Natl. Acad. Sci. USA* 67: 128–35

6. Baudin A, Ozier-Kalogeropoulos O, Denouel A, Lacroute F, Cullin C. 1993. A simple and efficient method for direct gene deletion in *Saccharomyces cerevisiae*. *Nucleic Acids Res.* 21:3329–30

7. Beernink HT, Morrical SW. 1999. RMPs: recombination/replication mediator proteins. *Trends Biochem. Sci.* 24:385–89

8. Benzinger R, Enquist LW, Skalka A. 1975. Transfection of *Escherichia coli* spheroplasts. V. Activity of RecBC nuclease in rec⁺ and rec⁻ spheroplasts measured with different forms of bacteriophage DNA. *J. Virol.* 15:861–71

9. Brcic-Kostic K, Stojiljkovic I, Salaj-Smic E, Trgovcevic Z. 1989. The *recB* gene product is essential for exonuclease V-dependent DNA degradation *in vivo*. *Mutat. Res.* 227:247–50

10. Brooks K, Clark AJ. 1967. Behavior of λ bacteriophage in a recombination-deficient strain of *Escherichia coli*. *J. Virol.* 1:283–93

11. Brune W, Messerle M, Koszinowski UH. 2000. Forward with BACs: new tools for herpesvirus genomics. *Trends Genet.* 16:254–59

12. Bubeck P, Winkler M, Bautsch W. 1993. Rapid cloning by homologous recombination in vivo. *Nucleic Acids Res.* 21:3601–2

13. Bzymek M, Lovett ST. 2001. Evidence for two mechanisms of palindrome-stimulated deletion in *Escherichia coli*: single-strand annealing and replication slipped mispairing. *Genetics* 158:527–40

14. Campbell A. 1984. Types of recombination: common problems and common strategies. *Cold Spring Harb. Symp. Quant. Biol.* 49:839–44

15. Capaldo FN, Barbour SD. 1975. The role of the *rec* genes in the viability of *Escherichia coli* K12. *Basic Life Sci.* 5A:405–18

16. Carter DM, Radding CM. 1971. The role of exonuclease and β protein of phage λ in genetic recombination. II. Substrate specificity and the mode of action of lambda exonuclease. *J. Biol. Chem.* 246:2502–12

17. Cassuto E, Lash T, Sriprakash KS, Radding CM. 1971. Role of exonuclease and β protein of phage λ in genetic recombination. V. Recombination of λ DNA in vitro. *Proc. Natl. Acad. Sci. USA* 68:1639–43

18. Cassuto E, Radding CM. 1971. Mechanism for the action of λ exonuclease in genetic recombination. *Nat. New Biol.* 229:13–16

19. Cha RS, Zarbl H, Keohavong P, Thilly WG. 1992. Mismatch amplification mutation assay (MAMA): application to the

c-H-ras gene. *PCR Methods Appl.* 2:14–20

20. Chalker AF, Leach DR, Lloyd RG. 1988. *Escherichia coli sbcC* mutants permit stable propagation of DNA replicons containing a long palindrome. *Gene* 71:201–5

21. Churchill JJ, Anderson DG, Kowalczykowski SC. 1999. The RecBC enzyme loads RecA protein onto ssDNA asymmetrically and independently of chi, resulting in constitutive recombination activation. *Genes Dev.* 13:901–11

22. Clark AJ. 1973. Recombination deficient mutants of *E. coli* and other bacteria. *Annu. Rev. Genet.* 7:67–86

23. Clark AJ. 1991. *rec* genes and homologous recombination proteins in *Escherichia coli.* *Biochimie* 73:523–32

24. Clark AJ, Margulies AD. 1965. Isolation and characterization of recombination-deficient mutants of *Escerichia coli* K12. *Proc. Natl. Acad. Sci. USA* 53:451–59

25. Clark AJ, Sandler SJ. 1994. Homologous genetic recombination: the pieces begin to fall into place. *Crit. Rev. Microbiol.* 20:125–42

26. Cohen A, Clark AJ. 1986. Synthesis of linear plasmid multimers in *Escherichia coli* K-12. *J. Bacteriol.* 167:327–35

27. Connelly JC, de Leau ES, Leach DR. 1999. DNA cleavage and degradation by the SbcCD protein complex from *Escherichia coli.* *Nucleic Acids Res.* 27:1039–46

28. Connelly JC, Kirkham LA, Leach DR. 1998. The SbcCD nuclease of *Escherichia coli* is a structural maintenance of chromosomes (SMC) family protein that cleaves hairpin DNA. *Proc. Natl. Acad. Sci. USA* 95:7969–74

29. Copeland NG, Jenkins NA, Court DL. 2001. Recombineering: a powerful new tool for mouse functional genomics. *Nat. Rev. Genet.* 2:769–79

30. Courcelle J, Hanawalt PC. 1999. RecQ and RecJ process blocked replication forks prior to the resumption of replication

in UV-irradiated *Escherichia coli. Mol. Gen. Genet.* 262:543–51

31. Court D, Oppenheim AB. 1983. Phage lambda's accessory genes. See Ref. 51a, pp. 251–77

32. Cox MM. 2001. Recombinational DNA repair of damaged replication forks in *Escherichia coli*: questions. *Annu. Rev. Genet.* 35:53–82

33. Cromie GA, Millar CB, Schmidt KH, Leach DR. 2000. Palindromes as substrates for multiple pathways of recombination in *Escherichia coli. Genetics* 154:513–22

34. Dabert P, Smith GR. 1997. Gene replacement with linear DNA fragments in wild-type *Escherichia coli*: enhancement by Chi sites. *Genetics* 145:877–89

35. Datsenko KA, Wanner BL. 2000. One-step inactivation of chromosomal genes in *Escherichia coli* K-12 using PCR products. *Proc. Natl. Acad. Sci. USA* 97:6640–45

36. Echols H, Gingery R. 1968. Mutants of bacteriophage λ defective in vegetative genetic recombination. *J. Mol. Biol.* 34:239–49

37. El Karoui M, Amundsen SK, Dabert P, Gruss A. 1999. Gene replacement with linear DNA in electroporated wild-type *Escherichia coli. Nucleic Acids Res.* 27:1296–99

38. Ellis HM, Yu D, DiTizio T, Court DL. 2001. High efficiency mutagenesis, repair, and engineering of chromosomal DNA using single-stranded oligonucleotides. *Proc. Natl. Acad. Sci. USA* 98:6742–46

39. Emmerson PT. 1968. Recombination deficient mutants of *Escherichia coli* K12 that map between *thyA* and *argA. Genetics* 60:19–30

40. Feiss M, Siegele DA, Rudolph CF, Frackman S. 1982. Cosmid DNA packaging in vivo. *Gene* 17:123–30

41. Franklin NC. 1967. Extraordinary recombinational events in *Escherichia coli.* Their independence of the *rec$^+$* function. *Genetics* 55:699–707

42. Friedman SA, Hays JB. 1986. Selective inhibition of *Escherichia coli recBC* activities by plasmid-encoded GamS function of phage λ. *Gene* 43:255–63

43. Gay P, Le Coq D, Steinmetz M, Ferrari E, Hoch JA. 1983. Cloning structural gene *sacB*, which codes for exoenzyme levansucrase of *Bacillus subtilis*: expression of the gene in *Escherichia coli*. *J. Bacteriol.* 153:1424–31

44. Gibson FP, Leach DRF, Lloyd RG. 1992. Identification of *sbcD* mutations as cosuppressors of *recBC* that allow propagation of DNA palindromes in *Escherichia coli* K-12. *J. Bacteriol.* 174:1222–28

45. Goryshin IY, Jendrisak J, Hoffman LM, Meis R, Reznikoff WS. 2000. Insertional transposon mutagenesis by electroporation of released *Tn5* transposition complexes. *Nat. Biotechnol.* 18:97–100

46. Gottesman ME, Yarmolinsky MB. 1968. Integration-negative mutants of bacteriophage λ. *J. Mol. Biol.* 31:487–505

47. Gottesman MM, Gottesman ME, Gottesman S, Gellert M. 1974. Characterization of bacteriophage λ reverse as an *Escherichia coli* phage carrying a unique set of host-derived recombination functions. *J. Mol. Biol.* 88:471–87

48. Hall SD, Kane MF, Kolodner RD. 1993. Identification and characterization of the *Escherichia coli* RecT protein, a protein encoded by the *recE* region that promotes renaturation of homologous single-stranded DNA. *J. Bacteriol.* 175:277–87

49. Hall SD, Kolodner RD. 1994. Homologous pairing and strand exchange promoted by the *Escherichia coli* RecT protein. *Proc. Natl. Acad. Sci. USA* 91:3205–9

50. Hamilton CM, Aldea M, Washburn BK, Babitzke P, Kushner SR. 1989. New method for generating deletions and gene replacements in *Escherichia coli*. *J. Bacteriol.* 171:4617–22

51. Hays JB, Smith TA, Friedman SA, Lee E, Coffman GL. 1984. RecF and RecBC function during recombination of nonreplicating, UV-irradiated phage λ DNA and during other recombination processes. *Cold Spring Harb. Symp. Quant. Biol.* 49:475–83

51a. Hendrix RW, Roberts JW, Stahl FW, Weisberg RA, eds. 1983. *Lambda II*. Cold Spring Harbor, NY: Cold Spring Harbor Lab.

52. Henry MF, Cronan JE Jr. 1991. Direct and general selection for lysogens of *Escherichia coli* by phage λ recombinant clones. *J. Bacteriol.* 173:3724–31

53. Higgins NP, Kato K, Strauss B. 1976. A model for replication repair in mammalian cells. *J. Mol. Biol.* 101:417–25

54. Hill F, Benes V, Thomasova D, Stewart AF, Kafatos FC, Ansorge W. 2000. BAC trimming: minimizing clone overlaps. *Genomics* 64:111–13

55. Hill SA, Stahl MM, Stahl FW. 1997. Single-strand DNA intermediates in phage λ's Red recombination pathway. *Proc. Natl. Acad. Sci. USA* 94:2951–56

56. Horii Z, Clark AJ. 1973. Genetic analysis of the RecF pathway to genetic recombination in *Escherichia coli* K12: isolation and characterization of mutants. *J. Mol. Biol.* 80:327–44

57. Horiuchi T, Fujimura Y. 1995. Recombinational rescue of the stalled DNA replication fork: a model based on analysis of an *Escherichia coli* strain with a chromosome region difficult to replicate. *J. Bacteriol.* 177:783–91

58. Ioannou PA, Amemiya CT, Garnes J, Kroisel PM, Shizuya H, et al. 1994. A new bacteriophage P1-derived vector for the propagation of large human DNA fragments. *Nat. Genet.* 6:84–89

59. Jasin M, Schimmel P. 1984. Deletion of an essential gene in *Escherichia coli* by site-specific recombination with linear DNA fragments. *J. Bacteriol.* 159:783–86

60. Joseph JW, Kolodner R. 1983. Exonuclease VIII of *Escherichia coli*. I. Purification and physical properties. *J. Biol. Chem.* 258:10411–17

61. Joseph JW, Kolodner R. 1983. Exonuclease VIII of *Escherichia coli*. II. Mechanism of action. *J. Biol. Chem.* 258:10418–24

62. Karakousis G, Ye N, Li Z, Chiu SK, Reddy G, Radding CM. 1998. The β protein of phage λ binds preferentially to an intermediate in DNA renaturation. *J. Mol. Biol.* 276:721–31

63. Karu AE, Sakaki Y, Echols H, Linn S. 1975. The gamma protein specified by bacteriophage λ. Structure and inhibitory activity for the RecBC enzyme of *Escherichia coli*. *J. Biol. Chem.* 250:7377–87

64. Keim P, Lark KG. 1990. The RecE recombination pathway mediates recombination between partially homologous DNA sequences: structural analysis of recombination products. *J. Struct. Biol.* 104:97–106

65. Kmiec E, Holloman WK. 1981. β protein of bacteriophage λ promotes renaturation of DNA. *J. Biol. Chem.* 256:12636–39

66. Kogoma T. 1997. Stable DNA replication: interplay between DNA replication, homologous recombination, and transcription. *Microbiol. Mol. Biol. Rev.* 61:212–38

67. Kohara Y, Akiyama K, Isono K. 1987. The physical map of the whole *E. coli* chromosome: application of a new strategy for rapid analysis and sorting of a large genomic library. *Cell* 50:495–508

68. Kovall R, Matthews BW. 1997. Toroidal structure of lambda-exonuclease. *Science* 277:1824–27

69. Kowalczykowski SC. 2000. Initiation of genetic recombination and recombination-dependent replication. *Trends Biochem. Sci.* 25:156–65

70. Kulkarni SK, Stahl FW. 1989. Interaction between the *sbcC* gene of *Escherichia coli* and the *gam* gene of phage λ. *Genetics* 123:249–53

71. Kusano K, Sunohara Y, Takahashi N, Yoshikura H, Kobayashi I. 1994. DNA double-strand break repair: genetic determinants of flanking crossing-over. *Proc. Natl. Acad. Sci. USA* 91:1173–77

72. Kushner SR, Nagaishi H, Clark AJ. 1972. Indirect suppression of *recB* and *recC* mutations by exonuclease I deficiency. *Proc. Natl. Acad. Sci. USA* 69:1366–70

73. Kuzminov A. 1995. Collapse and repair of replication forks in *Escherichia coli*. *Mol. Microbiol.* 16:373–84

74. Kuzminov A. 1999. Recombinational repair of DNA damage in *Escherichia coli* and bacteriophage λ. *Microbiol. Mol. Biol. Rev.* 63:751–813

75. Lafontaine D, Tollervey D. 1996. One-step PCR mediated strategy for the construction of conditionally expressed and epitope tagged yeast proteins. *Nucleic Acids Res.* 24:3469–71

76. Lamberg A, Nieminen S, Qiao M, Savilahti H. 2002. Efficient insertion mutagenesis strategy for bacterial genomes involving electroporation of *in vitro*-assembled DNA transposition complexes of bacteriophage mu. *Appl. Environ. Microbiol.* 68:705–12

77. Larionov V. 1999. Direct isolation of specific chromosomal regions and entire genes by TAR cloning. *Genet. Eng.* 21:37–55

78. Lee EC, Yu D, Martinez de Velasco J, Tessarollo L, Swing DA, et al. 2001. A highly efficient *Escherichia coli*–based chromosome engineering system adapted for recombinogenic targeting and subcloning of BAC DNA. *Genomics* 73:56–65

79. Li Z, Karakousis G, Chiu SK, Reddy G, Radding CM. 1998. The β protein of phage λ promotes strand exchange. *J. Mol. Biol.* 276:733–44

80. Little JW. 1967. An exonuclease induced by bacteriophage λ. II. Nature of the enzymatic reaction. *J. Biol. Chem.* 242:679–86

81. Lloyd RG, Buckman C. 1985. Identification and genetic analysis of *sbcC* mutations in commonly used *recBC sbcB* strains of *Escherichia coli* K-12. *J. Bacteriol.* 164:836–44

82. Lovett ST, Kolodner RD. 1989. Identification and purification of a single-stranded-DNA-specific exonuclease encoded by the

recJ gene of *Escherichia coli. Proc. Natl. Acad. Sci. USA* 86:2627–31

83. Luisi-DeLuca C, Kolodner RD. 1992. Effect of terminal non-homology on intramolecular recombination of linear plasmid substrates in *Escherichia coli. J. Mol. Biol.* 227:72–80

84. Manly KF, Signer ER, Radding CM. 1969. Nonessential functions of bacteriophage λ. *Virology* 37:177–88

85. Marians KJ. 1996. Replication fork propagation. In *Escherichia coli and Salmonella*, ed. FC Neidhardt, R Curtiss III, JL Ingraham, ECC Lin, KB Low, et al., pp. 749–63. Washington, DC: ASM Press

86. Marinus MG, Carraway M, Frey AZ, Brown L, Arraj JA. 1983. Insertion mutations in the *dam* gene of *Escherichia coli* K-12. *Mol. Gen. Genet.* 192:288–89

87. Marsic N, Roje S, Stojiljkovic I, Salaj-Smic E, Trgovcevic Z. 1993. In vivo studies on the interaction of RecBCD enzyme and λ Gam protein. *J. Bacteriol.* 175:4738–43

88. Matsuura S, Komatsu J, Hirano K, Yasuda H, Takashima K, et al. 2001. Real-time observation of a single DNA digestion by λ exonuclease under a fluorescence microscope field. *Nucleic Acids Res.* 29:E79

89. Moerschell RP, Das G, Sherman F. 1991. Transformation of yeast directly with synthetic oligonucleotides. *Methods Enzymol.* 194:362–69

90. Moerschell RP, Tsunasawa S, Sherman F. 1988. Transformation of yeast with synthetic oligonucleotides. *Proc. Natl. Acad. Sci. USA* 85:524–28

91. Muniyappa K, Mythili E. 1993. Phage λβ protein, a component of general recombination, is associated with host ribosomal S1 protein. *Biochem. Mol. Biol. Int.* 31:1–11

92. Muniyappa K, Radding CM. 1986. The homologous recombination system of phage λ. Pairing activities of β protein. *J. Biol. Chem.* 261:7472–78

93. Murphy KC. 1991. λ Gam protein inhibits the helicase and chi-stimulated re-combination activities of *Escherichia coli* RecBCD enzyme. *J. Bacteriol.* 173:5808–21

94. Murphy KC. 1998. Use of bacteriophage λ recombination functions to promote gene replacement in *Escherichia coli. J. Bacteriol.* 180:2063–71

95. Murphy KC, Campellone KG, Poteete AR. 2000. PCR-mediated gene replacement in *Escherichia coli. Gene* 246:321–30

96. Muyrers JP, Zhang Y, Benes V, Testa G, Ansorge W, Stewart AF. 2000. Point mutation of bacterial artificial chromosomes by ET recombination. *EMBO Rep.* 1:239–43

97. Muyrers JP, Zhang Y, Buchholz F, Stewart AF. 2000. RecE/RecT and Redα/Redβ initiate double-stranded break repair by specifically interacting with their respective partners. *Genes Dev.* 14:1971–82

98. Muyrers JP, Zhang Y, Stewart AF. 2001. Techniques: recombinogenic engineering—new options for cloning and manipulating DNA. *Trends Biochem. Sci.* 26:325–31

99. Muyrers JP, Zhang Y, Testa G, Stewart AF. 1999. Rapid modification of bacterial artificial chromosomes by ET-recombination. *Nucleic Acids Res.* 27:1555–57

100. Myers RS, Stahl FW. 1994. Chi and the RecBC D enzyme of *Escherichia coli. Annu. Rev. Genet.* 28:49–70

101. Mythili E, Kumar KA, Muniyappa K. 1996. Characterization of the DNA-binding domain of β protein, a component of phage λ Red-pathway, by UV catalyzed cross-linking. *Gene* 182:81–87

102. Naom IS, Morton SJ, Leach DR, Lloyd RG. 1989. Molecular organization of *sbcC*, a gene that affects genetic recombination and the viability of DNA palindromes in *Escherichia coli* K-12. *Nucleic Acids Res.* 17:8033–45

103. Noirot P, Kolodner RD. 1998. DNA strand invasion promoted by *Escherichia coli* RecT protein. *J. Biol. Chem.* 273:12274–80

104. Noskov VN, Koriabine M, Solomon G, Randolph M, Barrett JC, et al. 2001. Defining the minimal length of sequence homology required for selective gene isolation by TAR cloning. *Nucleic Acids Res.* 29:E32

105. O'Connor M, Peifer M, Bender W. 1989. Construction of large DNA segments in *Escherichia coli. Science* 244:1307–12

106. Oliner JD, Kinzler KW, Vogelstein B. 1993. In vivo cloning of PCR products in *E. coli. Nucleic Acids Res.* 21:5192–97

107. Orr-Weaver TL, Szostak JW, Rothstein RJ. 1981. Yeast transformation: a model system for the study of recombination. *Proc. Natl. Acad. Sci. USA* 78:6354–58

108. Paskvan I, Salaj-Smic E, Ivancic-Bace I, Zahradka K, Trgovcevic Z, Brcic-Kostic K. 2001. The genetic dependence of RecBCD-Gam mediated double strand end repair in *Escherichia coli. FEMS Microbiol. Lett.* 205:299–303

109. Passy SI, Yu X, Li Z, Radding CM, Egelman EH. 1999. Rings and filaments of β protein from bacteriophage λ suggest a superfamily of recombination proteins. *Proc. Natl. Acad. Sci. USA* 96:4279–84

110. Postow L, Crisona NJ, Peter BJ, Hardy CD, Cozzarelli NR. 2001. Topological challenges to DNA replication: conformations at the fork. *Proc. Natl. Acad. Sci. USA* 98:8219–26

111. Poteete AR. 2001. What makes the bacteriophage λ Red system useful for genetic engineering: molecular mechanism and biological function. *FEMS Microbiol. Lett.* 201:9–14

112. Poteete AR, Fenton AC. 1993. Efficient double-strand break-stimulated recombination promoted by the general recombination systems of phages λ and P22. *Genetics* 134:1013–21

113. Poteete AR, Fenton AC. 2000. Genetic requirements of phage λ Red-mediated gene replacement in *Escherichia coli* K-12. *J. Bacteriol.* 182:2336–40

114. Poteete AR, Fenton AC, Murphy KC. 1988. Modulation of *Escherichia coli* RecBCD activity by the bacteriophage λ Gam and P22 Abc functions. *J. Bacteriol.* 170:2012–21

115. Poteete AR, Fenton AC, Murphy KC. 1999. Roles of RuvC and RecG in phage λ Red-mediated recombination. *J. Bacteriol.* 181:5402–8

116. Poteete AR, Volkert MR. 1988. Activation of RecF-dependent recombination in *Escherichia coli* by bacteriophage λ and P22-encoded functions. *J. Bacteriol.* 170:4379–81

117. Price-Carter M, Tingey J, Bobik TA, Roth JR. 2001. The alternative electron acceptor tetrathionate supports B12-dependent anaerobic growth of *Salmonella enterica* serovar typhimurium on ethanolamine or 1,2-propanediol. *J. Bacteriol.* 183:2463–75

118. Radding CM, Rosenzweig J, Richards F, Cassuto E. 1971. Separation and characterization of exonuclease, β protein, and a complex of both. *J. Biol. Chem.* 246:2510–12

119. Raibaud O, Mock M, Schwartz M. 1984. A technique for integrating any DNA fragment into the chromosome of *Escherichia coli. Gene* 29:231–41

120. Roca AI, Cox MM. 1997. RecA protein: structure, function, and role in recombinational DNA repair. *Prog. Nucleic Acid Res. Mol. Biol.* 56:129–223

121. Russell CB, Thaler DS, Dahlquist FW. 1989. Chromosomal transformation of *Escherichia coli recD* strains with linearized plasmids. *J. Bacteriol.* 171:2609–13

122. Salaj-Smic E, Dermic D, Brcic-Kostic K, Cajo GC, Trgovcevic E. 2000. In vivo studies of the *Escherichia coli* RecB polypeptide lacking its nuclease center. *Res. Microbiol.* 151:769–76

123. Sandler SJ, Marians KJ. 2000. Role of PriA in replication fork reactivation in *Escherichia coli. J. Bacteriol.* 182:9–13

124. Scheirer KE, Higgins NP. 2001. Transcription induces a supercoil domain barrier in bacteriophage Mu. *Biochimie* 83:155–59

125. Sekimizu K, Bramhill D, Kornberg A. 1988. Sequential early stages in the in vitro initiation of replication at the origin of the *Escherichia coli* chromosome. *J. Biol. Chem.* 263:7124–30

126. Sergueev K, Yu D, Austin S, Court D. 2001. Cell toxicity caused by products of the p_L operon of bacteriophage lambda. *Gene* 272:227–35

127. Sharples GJ, Ingleston SM, Lloyd RG. 1999. Holliday junction processing in bacteria: insights from the evolutionary conservation of RuvABC, RecG, and RusA. *J. Bacteriol.* 181:5543–50

128. Shizuya H, Birren B, Kim UJ, Mancino V, Slepak T, et al. 1992. Cloning and stable maintenance of 300-kilobase-pair fragments of human DNA in *Escherichia coli* using an F-factor-based vector. *Proc. Natl. Acad. Sci. USA* 89:8794–97

129. Shulman MJ, Hallick LM, Echols H, Signer ER. 1970. Properties of recombination-deficient mutants of bacteriophage λ. *J. Mol. Biol.* 52:501–20

130. Signer ER, Weil J. 1968. Recombination in bacteriophage λ. I. Mutants deficient in general recombination. *J. Mol. Biol.* 34:261–71

131. Silberstein Z, Cohen A. 1987. Synthesis of linear multimers of OriC and pBR322 derivatives in *Escherichia coli* K-12: role of recombination and replication functions. *J. Bacteriol.* 169:3131–37

132. Silberstein Z, Maor S, Berger I, Cohen A. 1990. λ Red-mediated synthesis of plasmid linear multimers in *Escherichia coli* K12. *Mol. Gen. Genet.* 223:496–507

133. Silberstein Z, Tzfati Y, Cohen A. 1995. Primary products of break-induced recombination by *Escherichia coli* RecE pathway. *J. Bacteriol.* 177:1692–98

134. Skalka A. 1974. A replicator's view of recombination (and repair). In *Mechanisms in Recombination*, ed. RF Grell, pp. 421–32. New York: Plenum

135. Skalka AM. 1977. DNA replication—bacteriophage λ. *Curr. Top. Microbiol. Immunol.* 78:201–37

136. Slater S, Maurer R. 1993. Simple phage-mid-based system for generating allele replacements in *Escherichia coli*. *J. Bacteriol.* 175:4260–62

137. Smith GR. 1983. General recombination. See Ref. 51a, pp. 175–209

138. Sriprakash KS, Lundh N, Huh M-O, Radding CM. 1975. The specificity of λ exonuclease. Interactions with single-stranded DNA. *J. Biol. Chem.* 250:5438–45

139. Stahl FW. 1998. Recombination in phage λ: one geneticist's historical perspective. *Gene* 223:95–102

140. Stahl MM, Thomason L, Poteete AR, Tarkowski T, Kuzminov A, Stahl FW. 1997. Annealing vs. invasion in phage λ recombination. *Genetics* 147:961–77

141. Sternberg NL. 1992. Cloning high molecular weight DNA fragments by the bacteriophage P1 system. *Trends Genet.* 8:11–16

142. Swaminathan S, Ellis HM, Waters LS, Yu D, Lee E-C, et al. 2001. Rapid engineering of bacterial artificial chromosomes using oligonucleotides. *Genesis* 29:14–21

143. Szostak JW, Orr WT, Rothstein RJ, Stahl FW. 1983. The double-strand-break repair model for recombination. *Cell* 33:25–35

144. Takahashi NK, Kusano K, Yokochi T, Kitamura Y, Yoshikura H, Kobayashi I. 1993. Genetic analysis of double-strand break repair in *Escherichia coli*. *J. Bacteriol.* 175:5176–85

145. Taylor A, Smith GR. 1980. Unwinding and rewinding of DNA by the RecBC enzyme. *Cell* 22:447–57

146. Trogovcevic Z, Rupp WD. 1975. Lambda bacteriophage gene products and X-ray sensitivity of *Escherichia coli*: comparison of *red*-dependent and *gam*-dependent radioresistance. *J. Bacteriol.* 123:212–21

147. Umezu K, Chi N-W, Kolodner RD. 1993. Biochemical interaction of the *Escherichia coli* RecF, RecO, and RecR proteins with RecA protein and single-stranded DNA binding protein. *Proc. Natl. Acad. Sci. USA* 90:3875–79

148. Unger RC, Clark AJ. 1972. Interaction of the recombination pathways of bacteriophage λ and its host *Escherichia coli* K12: effects on exonuclease V activity. *J. Mol. Biol.* 70:539–48

149. Uzzau S, Figueroa-Bossi N, Rubino S, Bossi L. 2001. Epitope tagging of chromosomal genes in Salmonella. *Proc. Natl. Acad. Sci. USA* 98:15264–69

150. van Oostrum J, White JL, Burnett RM. 1985. Isolation and crystallization of λ exonuclease. *Arch. Biochem. Biophys.* 243:332–37

151. Venkatesh TV, Radding CM. 1993. Ribosomal protein S1 and NusA protein complexed to recombination protein β of phage λ. *J. Bacteriol.* 175:1844–46

152. Winans SC, Elledge SJ, Krueger JH, Walker GC. 1985. Site-directed insertion and deletion mutagenesis with cloned fragments in *Escherichia coli*. *J. Bacteriol.* 161:1219–21

153. Yamaguchi H, Hanada K, Asami Y, Kato JI, Ikeda H. 2000. Control of genetic stability in *Escherichia coli*: the SbcB 3′-5′ exonuclease suppresses illegitimate recombination promoted by the RecE 5′-3′ exonuclease. *Genes Cells* 5:101–9

154. Yamamoto T, Moerschell RP, Wakem LP, Ferguson D, Sherman F. 1992. Parameters affecting the frequencies of transformation and co-transformation with synthetic oligonucleotides in yeast. *Yeast* 8:935–48

155. Yang XW, Model P, Heintz N. 1997. Homologous recombination based modification in *Escherichia coli* and germline transmission in transgenic mice of a bacterial artificial chromosome. *Nat. Biotechnol.* 15:859–65

156. Yu D, Ellis HM, Lee EC, Jenkins NA, Copeland NG, Court DL. 2000. An efficient recombination system for chromosome engineering in *Escherichia coli*. *Proc. Natl. Acad. Sci. USA* 97:5978–83

157. Zagursky RJ, Hays JB. 1983. Expression of the phage λ recombination genes *exo* and *bet* under *lac*PO control on a multicopy plasmid. *Gene* 23:277–92

158. Zechiedrich EL, Cozzarelli NR. 1995. Roles of topoisomerase IV and DNA gyrase in DNA unlinking during replication in *Escherichia coli*. *Genes Dev.* 9:2859–69

159. Zhang Y, Buchholz F, Muyrers JP, Stewart AF. 1998. A new logic for DNA engineering using recombination in *Escherichia coli*. *Nat. Genet.* 20:123–28

160. Zhang Y, Muyrers JP, Testa G, Stewart AF. 2000. DNA cloning by homologous recombination in *Escherichia coli*. *Nat. Biotechnol.* 18:1314–17

161. Zissler J, Signer E, Schaefer F. 1971. The role of recombination in growth of bacteriophage λ. I. The *gamma* gene. In *The Bacteriophage Lambda*, ed. AD Hershey, pp. 455–68. Cold Spring Harbor, NY: Cold Spring Harbor Lab.

Annu. Rev. Genet. 2002. 36:389–410
doi: 10.1146/annurev.genet.36.040202.092802

CHROMOSOME REARRANGEMENTS AND TRANSPOSABLE ELEMENTS

Wolf-Ekkehard Lönnig and Heinz Saedler

Max-Planck-Institut für Züchtungsforschung, Carl-von-Linné-Weg 10, D-50829 Köln, Germany; e-mail: loennig@mpiz-koeln.mpg.de

Key Words   Barbara McClintock, chromosome rearrangements, synteny, living fossils, stasis

■ **Abstract**   There has been limited corroboration to date for McClintock's vision of gene regulation by transposable elements (TEs), although her proposition on the origin of species by TE-induced complex chromosome reorganizations in combination with gene mutations, i.e., the involvement of both factors in relatively sudden formations of species in many plant and animal genera, has been more promising. Moreover, resolution is in sight for several seemingly contradictory phenomena such as the endless reshuffling of chromosome structures and gene sequences versus synteny and the constancy of living fossils (or stasis in general). Recent wide-ranging investigations have confirmed and enlarged the number of earlier cases of TE target site selection (hot spots for TE integration), implying preestablished rather than accidental chromosome rearrangements for nonhomologous recombination of host DNA. The possibility of a partly predetermined generation of biodiversity and new species is discussed. The views of several leading transposon experts on the rather abrupt origin of new species have not been synthesized into the macroevolutionary theory of the punctuated equilibrium school of paleontology inferred from thoroughly consistent features of the fossil record.

CONTENTS

INTRODUCTION

The phenomenon of variegation among living organisms, namely, the mosaic appearance of some individuals, has long been a source of fascination. As early as 1588, Jacob Theodor von Bergzabern provided a lively description of the multicolored and variegated kernels of maize from the New World (101), and in 1623, Basilius Besler published an impressive figure of flower variation in *Mirabilis* in his *Hortus Eystettensis* (18). However, the concept of transposon activities for living organisms, first described by Barbara McClintock in 1944 and in the years following, was that of Ac/Ds (activator/dissociation) involved in the breakage-fusion-bridge cycle (23, 24, 62, 91). Her depiction of a phenomenon directly involved in gross chromosome rearrangements mediated by transposable elements (TEs) marked the onset of transposon research. However, since this general definition of chromosomal rearrangements as "rearrangements of the linear sequence of chromosomes including transposition, duplication, deletion, inversion or translocation of nucleic acid segments" (57), extensive molecular research has shown that large chromosome rearrangements are only the tip of the iceberg.

Throughout the existence of most plant and animal species, both gross and small quantitative chromosome reshufflings have often added up to enormous chromosomal changes—even only considering the overall mass of DNA—so that, for example, about 70% of the chromosomes of McClintock's model plant *Zea mays* is reported to consist of DNA-sequences due to TE-activities alone. The large majority of transposons of *Z. mays*, in addition to the examples described below, consists of class I element families, i.e., the retrotransposons. Similar conclusions have been reached for most other organisms studied to date, from *Arabidopsis thaliana* (14%) to *Vicia faba* (some 90%), *Drosophila melanogaster* (15%) to humans (45%).

As models of transposition and mechanisms of gross and small chromosome rearrangements mediated by TEs have already been extensively reviewed (3, 4, 11, 17, 27, 57, 67–70, 100, 101, 112, 127–129, 149), we refer to them only insofar as they relate to the problem of "what, if anything, the masses of TE-caused chromosome rearrangements biologically are good for in so many plant and animal species." Answering this question remains central to TE research; several hypothetical answers, often contradictory, have been advanced over the past 50 years (3, 4, 69, 84). We also focus on an interdisciplinary approach between TE-genetics (TE-induced gross and small genome reorganizations in time) and some particularly intriguing aspects of the fossil record (the abrupt appearance and stasis of most new life forms). We start the discussion with some of the main hypotheses advanced by the discoverer of TE-caused chromosome rearrangements, Barbara McClintock.

SOME NEW EVIDENCE FOR MCCLINTOCK'S VIEW ON THE FUNCTIONS OF TE-MEDIATED CHROMOSOME REARRANGEMENTS

Ontogeny

McClintock's hypotheses regarding a system of functions for TEs, both ontogenetically and evolutionarily, are still eminently thought-provoking and problematic (24, 69, 90, 91, 99, 101). She was struck by the gross chromosome rearrangements that are induced especially in connection with the "breakage-fusion-bridge-cycle," as well as a wide array of small to large effects of the *Ac/Ds*, *En/Spm* and *Dt* systems on gene expression, and the effects of transposons on the *R* and *B* loci and the stress activations of TEs, which were discovered later. Let us first briefly focus on her hypotheses on TEs and ontogeny. Arguing repeatedly that "the real point is control" for transposons, "she believed that precise transpositions of controlling elements successively inactivated suites of genes, thus executing the developmental program by which an embryo becomes a plant" (23, see also 24, 91, 124).

New evidence that may corroborate McClintock's views on the importance of TEs in ontogeny comes from two discoveries, albeit not in plants: TEs are involved in the unusual organization of telomeres in *Drosophila melanogaster* (43, 107) and in the parasitic protozoan *Giardia lamblia* (1, 108). In these organisms, TEs seem to contribute to chromosome stability by expanding the buffer between the ends of the chromosomes and the genes. Moreover, *Giardia* chromosomes display a developmentally impressive variation in length in reaction to environmental stress (TEs are also involved); for instance, chromosome 1 can vary between 1.1 and 1.9 Mb, i.e., an overall increase of about 73% (1.1 = 100%) or decrease to 58%, respectively (1, 108). Although chromosome stability is essential for development, it is still not clear whether these TEs have any function in normal development other than that of contributing to stability. Nevertheless, as the first results of regular TE functions in development, these discoveries may be viewed as a first step in the direction of the functional relevance that McClintock envisioned originally.

However, if we have learned anything over the past two decades about animal and plant development, it is that there are as yet no clearcut results to support the proposition that any of the major ontogenetic steps are caused and governed by any known transposable elements, i.e., by means of transposon movements accompanied by small TE-mediated chromosome rearrangements that successively inactivate or activate different suites of genes. To the contrary, this task has been largely assigned to gene classes such as the homeobox, MADS-box, and other regulatory gene families. Current data indicate that, during development, these classes of "controlling elements" do not normally move from one place to another within a given set of chromosomes.

Moreover, the telomeric functions referred to above are probably only substitutions of functional sequences that were originally equally useful but now differ from those found in the telomeres of most other known species (1). If, on the other hand, a selective advantage is assumed, why did TEs not long ago overtake all of the telomeric sequences in the majority of life forms on Earth?

Due to insertional preferences to subtelomeric and telomeric regions, TEs have been detected in many species, although usually only as a minority of the overall sequences [reviewed in (1)]. It is assumed that they cannot do much damage in these regions and take on the buffer function referred to above. It is not known whether TEs are able to extend the buffer function during ontogeny, perhaps adapting chromosomal functions to stress in the species where they have been found.

However, active TEs can indeed modulate and modify plant and animal development, as demonstrated in many examples from the work of McClintock and more recent investigators. Yet, the ontogenetic modifications observed and identified to date are not the causes of normal ontogeny; rather they appear to constitute a set of factors that occur more or less concomitantly (and often even disruptive) and are tolerated during development as long as they do not become too heavy a burden for the affected organisms [for details, see (69, 84)]. Nevertheless, the outcome of TE activities may produce beautiful results, like the flower color variation shown in Figure 1 (see color insert), taken from Basilius Besler's *Herbarium*, published in 1613 (18). This is probably a depiction of some effects of TEs on normal pigment formation, whereby gene functions of the anthocyanine pathway are disrupted and released within certain developmental boundaries. Normal ontogenetic regulation and expression of the anthocyanine pathway, however, does not depend on the activities of transposable elements (ongoing rounds of excisions and insertions).

Origin of Species

In McClintock's view, phenotypic changes and the origin of new species by TE-induced macromutations, i.e., by abrupt and more or less simultanenous sequence changes in chromosomal organization and in a greater number of genes were the key to the problem of the origin of life in all its forms. Moreover, "late in life, she synthesized her life's work into a vision of the genome as a sensitive organ of the cell, capable of rearranging itself in response to environmental cues" (23), with TEs again playing the decisive role. New evidence appears to be somewhat stronger in support of this function generating variation for morphological species formation than for McClintock's views on TEs and ontogeny.

Among McClintock's favorite examples was the muntjac deer, one of the prime examples in this context. The diploid chromosome number of the Chinese *Muntiacus reevesi* is 46, but that of the closely related Indian dwarf *Muntiacus muntjak* is 6 in the female and 7 in the male deer, which is "the lowest chromosome number in any vertebrate" (64). As McClintock commented, "Observations of the chromosomes in the hybrid between these two species strongly supports chromosome fusion as the mechanism of origin of the reduced number and huge size of the Indian muntjac chromosomes" (90). However, the muntjac case may even be more significant and revealing than McClintock envisioned some 20 years ago. Recent investigations suggest that the Indian muntjac karyotype may be derived "directly from 2n = 70 original karyotype rather than from an intermediate 2n = 46 Chinese muntjac-like karyotype" and that the 5 "old" as well as the several recently discovered muntjac chromosome races and species likely obtained their different chromosome numbers by parallel chromosomal fusions rather than by a simple line of descent (143, see also 73, 150).

M. muntjak displays a 20% difference in DNA content due to loss of middle repetitive sequences in comparison with *M. reevesi* (61). TEs are putatively ideal candidates to be involved in both the inferred chromosome rearrangements as well as in such gross differences in DNA content. Whether these DNA changes and the many other chromosome rearrangements that have been found are really causally related to TE activities in species formation in the genus *Muntiacus* (or in *Sorex, Mus, Rattus*, among others, and in plants; see below) has not been determined (69).

Some examples where TEs play a role in chromosome rearrangements in natural populations have been identified in *Drosophila*. In Hawaiian natural populations of *D. melanogaster*, for example, the transposable element *hobo* is associated with both cytological breakpoints of three endemic inversions, and a fourth such inversion has a solitary *hobo* insert at one breakpoint, all in clear contrast to cosmopolitan inversions in the same chromosome (86).

Another report has shown the involvement of TEs in chromosome rearrangements of the *virilis* group of *Drosophila*, in which species differ by multiple chromosome fusions and inversions, among other features (42). Although the causal relationship between these rearrangements and species formation has not been elucidated, the authors found that the *Penelope* and *Ulysses* transposons are "nonrandomly distributed in 12 strains of *D. virilis*" and that the insertion sites display a statistically relevant linkage with the breakpoints of several inversions detected in different species of the *virilis* aggregate (over half of 12 hybrid-dysgenesis-induced inversions, 2 translocations, and 2 deletions and others, totaling 16 rearrangements, have been reported as showing breakpoints in association with *Penelope* and *Ulysses* that are identical to those found in natural populations and species). An extrapolation to additional species of the group appears reasonable.

The most important problems still to be resolved on this topic have to do with causal relationships between chromosome rearrangements and species formations in the wild, selection, and the possibilities and limits of extrapolating from existing results.

1. What is the ratio between "normal" and TE-induced chromosome reshuffling in species in the wild?

2. To what extent is selection involved in the birth of chromosome races or are the majority of these phenomena simply the result of genetic drift (further points below)?

3. What are the exact possibilities and limits of the origin of species by TE-mediated gross and small chromosome rearrangements?

Before extending this topic at the end of our review, we first take a closer look at some seemingly contradictory phenomena whose unraveling appears to point to certain constraints of transposon activities in the development and the origin of species.

"DYNAMIC CHROMOSOMES," SYNTENY, AND LIVING FOSSILS

Synteny

Molecular analyses have disclosed extensive syntenic regions not only in closely related species (often already considered to be millions of years apart from each other), but also in practically all the distinct genera within different plant and animal families thus far investigated, and even between higher systematic categories, to a certain extent. Hence, some strongly functional constraints have been postulated for chromosome rearrangements (75–79, 80, 109, 114, 115). Despite exceptions at the microlinearity level (often also TE caused) (10), the general phenomenon of synteny is so well established that reliable predictions can be made even for species and genera of a plant or animal family that have not yet been analyzed (34, 39, 47, 117). The chromosomes of man, cattle, and muntjac, even with the many chromosome reshufflings mentioned above, still display extensive homoeologies (46, 151). However, in plants as distinct as rice (monocot) and *Arabidopsis* (dicot), so far synteny appears to be hardly detectable (34, 35, 88, 117).

The significance of selection in the regular processes of chromosome rearrangements, with the "tendency for similar structural changes to establish themselves in one member of the karyotype after another" (146) in wild-type species, was the subject of lively debate in the 1970s and 1980s [reviewed in (80)]. Selection-directed (145, 146) versus autonomous chromosome rearrangements being selectively neutral (75–80, 109, 114, 115)—or perhaps even slightly disadvantageous—were the two most diametrically opposing views at that time, and the argument has still not been resolved. Despite some attractive examples that might be interpreted as being due to selection, no clearcut selective values for the chromosome rearrangements observed in the wild have ever been obtained, to the authors' knowledge. Since even such established textbook examples for natural selection as that of the peppered moth (26, 58a, 82) and the neck of the giraffe (53) have fallen into disrepute recently or at least have proved to be much more intricate than once was believed, Lima-de-Faria's words on chromosome reshuffling are particularly pertinent in the current debate (76): "What I am trying to convey is that due to the absence of knowledge of molecular mechanisms, selection has been employed like a kind of general remedy by the biologist. Every time a phenomenon appeared in biology, and one obviously ignored its mechanism, selection was invoked as an explanation and the matter was settled."

Whatever the preponderant causes prove to be, the commonality of chromosome rearrangements as well as synteny are equally true for both the animal and plant kingdoms. In general, the plant kingdom even appears to surpass the animal kingdom in the frequencies and latitudes of chromosome rearrangements, especially in combination with different forms of polyploidy.

In the plant kingdom, chromosome rearrangements that often accompany chromosome and genome duplications are so common that fixed chromosome numbers and strict gene sequences alone have long been considered insufficient to identify and define a plant species (50). Literally thousands of species and genera display extensive gross chromosome rearrangements (in part, probably TE-induced), and widely different chromosome numbers have been detected all within a species and/or genus (for examples displaying both phenomena, see Table 1). Because synteny is equally well established among plant species and genera as well as some

TABLE 1   Selected examples of different chromosome numbers often accompanied by chromosome rearrangements within plant species and genera[a]

Family	Species	Chromosome numbers *within* the species
Nymphaeaccac	*Nymphaea alba*	2n = 48, 64, 84, 105, 112
Ranunculaceae	*Caltha palustris*	2n = 16, 32, 56, 64, 80
	Anemone nemorosa	2n = 16, 24, 30, 45
Moraceae	*Morus nigra*	2n = 26, 28, 30–308
Caryophyllaceae	*Arenaria multicaulis*	2n = 80, 120, 160, 200
	Arenaria biflora	2n = 20, 22
	Stellaria media	2n = 40, 42, 44
	Cerastium pumilum	2n = 72, 90, 94, 96,100
	Cerastium cerastoides	2n = 36, 38, 40
Polygonaceae	*Polygonum bistorta*	2n = 24, 44, 46, 48, 120
	Polygonum viviparum	2n = 88, 100, 110, 132
Salicaceae	*Salix helvetica*	2n = 36, 38, 39
Brassicaceae	*Cardamine pratensis*	2n = 16, 28, 30, 40, 48
	Cardamine palustris	2n = 56–96
	Cardaminopsis arenosa	2n = 16, 20, 24, 32, 39, 40
	Erophila verna	2n = 14–64
	Diplotaxis muralis	2n = 20, 22, 42, 44
Resedaceae	*Reseda luteola*	2n = 24, 26, 28
Primulaceae	*Primula auricula*	2n = 62, 63, 64, 66
	Primula integrifolia	2n = 62, 66, 68, 70
	Lysimachia nummularia	2n = 32, 36, 43, 45
Crassulaceae	*Sedum montanum*	2n = 34–136
	Sedum acre	2n = 16, 24, 48, 60, 80
Saxifragaceae	*Saxifraga bulbifera*	2n = 26, 28
Rosaceae	*Fragaria moschata*	2n = 28, 35, 42, 56

[a]See References (20, 48, 72).

higher systematic categories, both these phenomena, albeit seemingly contradictory, are almost all-pervading for most life forms on Earth.

To put this problem in the context of this nearly paradigmatic table: It will not be easy to establish selective advantages for all the chromosome rearrangements seen within the many plant species. How then is synteny to be explained in the face of the apparently endless chromosome reshufflings found in the history of nearly all forms of life on Earth?

According to Lima de Faria, there are a range of essentially nonfunctional and thus "forbidden" chromosome rearrangements (disruptive and destructive translocations). Thus the chromosome mutations are potentially channeled in certain directions, and the conservative reshufflings [reciprocal symmetrical, centric fusion (Robertsonian), and dissociation] are found most frequently in natural populations (76, 80, 114, 115).

As mentioned above, extensive new investigations have corroborated the fact that there are hot spots of TE integrations (1–4, 11, 19, 22, 36, 42, 74, 86, 106, 112, 125, 149), so that TE-mediated gross chromosome rearrangements participate in forming a predestined path of possible breakage and fusions, although the extent is still to be determined. Combining the chromosome field theory with TE-predestined nonhomologous or illegitimate recombination of host DNA segments could provide another explanation for synteny in the face of the seemingly endless chromosome reshufflings that are widely found in plants and animals alike (for further discussion of site-specific transposon integration, see below).

Extending the discussion of phenomena that at first sight appear to be mutually exclusive, yet on closer inspection display some promise of a solution, let us turn to the topic of TE-mediated chromosome rearrangements and the numerous occurrences of living fossils, or the phenomenon of paleontological stasis in general.

TE-Mediated Chromosome Rearrangements and Living Fossils

Longstanding differences between molecular biology and paleontology as to the phylogenetic relationships and times of origin of many different plant and animal groups (14) underlie the reluctance among molecular biologists for interdisciplinary discussion. However, avoidance does not solve the problem. We should differentiate between rather tentative hypotheses and clearcut results that have been independently corroborated and consolidated by diverse research groups over the years—in the following case even over some 200 years.

The paleontological point in focus here is the abrupt appearance and stasis (morphological constancy) for the overwhelming majority of life forms in Earth's history (13, 31, 40, 41, 51, 55, 56, 65, 66, 80, 81, 85, 89, 97, 111, 113, 116, 132, 133, 139). Originally proposed by Cuvier in the 1790s (31, 113, 131) and subsequently reworked and reelaborated (40, 41, 55, 56, 96, 116, 133), the phenomenon led in the 1970s to the theory of punctuated equilibrium and in the 1990s, to its consolidation

Figure 1 *Mirabilis jalapa* L. published by Basilius Besler in 1613 in his *Hortus Eystettensis*. In plants, transposable elements (TEs) reveal their presence by generating variegation patterns as in the candidate depicted here. In these cases, each patch of variegation means that a TE has left the locus so that the wild-type function of a respective gene involved in flower color synthesis is restored. The earlier the event, the more widespread the resulting cell line and the larger the restored color dot or sector.

as a widely acknowledged paleontological view of life in the western world (40, 55, 56). This view, which is based on some 200 million catalogued fossils in museums worldwide but is not consistent with the strict neo-Darwinian theory of essentially continuous evolution, seems to be in agreement, at least phenotypically, with the McClintock hypothesis of saltational generation of new species.

Hence, the following discussion on TEs, "dynamic chromosomes," and living fossils [i.e., any life forms existing essentially unchanged over geologic times up to the present] with a brief consideration of stasis in general, [for further discussion, see (80)] consists of two phenomena, stasis and genetic flux. Each is equally well established in its respective discipline, yet is seemingly irreconcilably at odds with the other: How is it possible that the genomes have always been in a permanent state of flux possibly by TE-mediated and other small- to gross-chromosome rearrangements (11, 12, 44, 45, 58, 59, 60, 91, 95, 99, 101, 123, 126–130, 137) or other "normal" mutations as well as molecular drive (38), yet produce life forms that are so regularly morphologically and anatomically constant for enormous periods of geological time?

This conflict between an all-pervasive genetic flux and overwhelming evidence of morphological constancy and stasis in the fossil record has led to questioning of the quality of the fossil evidence. However, the richness and often extraordinary quality of the fossil record (15, 37, 63, 80, 116, 139), e.g., the rich fossil amber floras and faunas of the world (66, 111, 144) show all relevant morphological and anatomical features of the entrapped organisms in microscopic detail, or the silica-embedded flora of the Rhynie cherts to the fossilized fauna of Grube Messel, near Darmstadt, Germany, leave no reasonable doubt in the minds of most qualified observers as to the existence of stasis. This conclusion has found increasing acceptance even among staunch neo-Darwinians: "Stasis is a real perspective and a fascinating one" (89).

The following points may help solve this dilemma.

1. TE-induced and other gross chromosome rearrangements can lead to postzygotic isolating mechanisms that result in almost total cross-fertilization barriers between different lines of the same species in experimental organisms in a relatively short time period, as, for instance, in *Pisum sativum* (71). So-called twin species, i.e., "species" with no corresponding morphologic differences to distinguish one systematic species from another (80), have regularly been generated experimentally. This lack of phenotypic differences prevents twin species from being discriminated in the fossil record.

2. Due to TE-induced changes in DNA mass (see Introduction), the amounts of DNA in the haploid genomes of closely related species that are hardly distinguishable from each other morphologically can differ enormously (the C-value paradox). Species of the genus *Vicia*, for example, vary between 1.8 and 13.3 pg (95, 103, 104) of DNA per haploid genome. Even within the same nonpolyploid plant species, the C-value can vary significantly, even though some original examples for this phenomenon were attributable to technical

problems (7, 8). For such "selfish" TE-proliferations (3, 4, 7–9, 59, 69, 87, 98) in combination with chromosome reshuffling, both gross chromosome rearrangements and strong differences in haploid DNA mass (pg) need not be correlated to morphological disparities. The same conclusion likely holds for many reshufflings in microcolinearity. Assuming that the more or less erratic molecular clocks (80, 81) will run on and that molecular drive will continue, larger changes can also be expected to accumulate on the level of the functional nucleotide sequences. For a living fossil or some of the many fossil genera that existed essentially unchanged anatomically throughout millions of years, from their abrupt appearance in the paleobiological record until their usually equally abrupt extinctions, this could mean enormous changes on the levels of chromosome rearrangements, gene sequences, and DNA mass despite eons of morphological stasis. In short, this could perhaps be referred to as the constancy of the vessel (morphology), versus changes in contents (genetics). Of the many possibly neutral sequence rearrangements accumulating over geologic time, none would be detectable by an anatomical comparison between a living organism and its morphologically corresponding "ancestor" species in the fossil record.

3. However, McClintock and other TE geneticists discussed the phenomenon of TE activation in the wake of stress, all the way from radiation-caused induction in the vicinity of nuclear testing in the 1950s to recent tissue culture experiments (69, 90, 91, 101, 110). These observations led to the hypothesis that TE activities can remain dormant over eons of time, usually in a methylated state of inactivity, and be "awakened" in new stressful environmental situations to produce the genetic flexibility necessary for further adaptations (90, 91, 131). This hypothesis brings us back to some of the basic ideas on periodic induction of mutations formulated by Hugo de Vries in the early twentieth century (24, 140, 141). Relatively short intervals of genetic instabilities in restricted areas of the distribution of a species generating new forms alternate with longer periods of low or no changes in the constitution of the species. If true, to what magnitude could the rise of such new species be predetermined by constraints of TE activities and the genetic potential of their host?

HOT SPOTS OF TE INSERTION AND THE PREDETERMINED ORIGIN OF SPECIES

Sequence-specific target site selection can be due to the sequence specificity of the integrase, which is an intrinsic feature of the endonuclease cutting the acceptor locus for class I elements such as LINEs and SINEs, both of which are non-LTR retrotransposons. Further elements prefer certain types of genes: *Ty1*, an LTR retrotransposon, preferentially integrates near tRNA or other genes transcribed by RNA pol III. Most likely, this is caused by the interactions necessary to transcribe these genes. Sequence preference can also be observed for *Ty1* that displays nonrandom integration in N A/T A/T A/T N sites. Insertion site preference has also been reported for the *P* elements in *Drosophila* (class II, i.e., DNA transposon). A 14-bp palindrome containing the sequence of the 8-bp target duplication can be identified for this element (74). Furthermore, the state of the chromosomal DNA can apparently limit the access for a transposon to a locus. The maize *Ac* transposon, for example, preferentially transposes into transcribed regions, perhaps because of loose packaging (strongly acetylated and less tightly bound histones) of the chromosomal DNA at these positions (3, 4).

New results on transposon target site selection, i.e., hot spots for TE integration (1, 2, 11, 19, 22, 36, 42, 74, 86, 106, 112, 125, 149), combined with the theory of chromosome field (see above), suggest that there is an appropriate degree of channeling of the future course of chromosome diversification, corresponding to something like a constraint and predestination concerning biodiversity. Hence, a largely TE- (and otherwise) preestablished chromosome rearrangement potential might eventually replace the idea of an infinite array of purely random shufflings of chromosomal nucleotide sequences following TE host nonhomologous recombinations. Moreover, retrotransposons inserting in members of the same class and the same or even different family (19) could increase the possibilty of illegitimate recombination of host DNA sequences due to longer stretches of TE sequence similarities in different parts of the genome. However, the exact extent of the constraints and the ratio of random to nonrandom chromosome shuffling have not yet been determined.

The synthesis of a partially predetermined generation of biodiversity seems to be reinforced by some observations on the level of small nucleic acid segment rearrangements, i.e., on the gene level. Our current understanding is that multiple, independently arisen TE footprints (123, 124) result in differently altered host sequences that do not usually restore the wild types, in agreement with most observations. A mechanism for TE-generated footprints has been suggested by Nevers & Saedler (100). Indeed, Scott et al., in their extensive investigations of transposon footprints in one locus (125), found that over 90% of >800 analyzed *Ds* excision products of maize *Waxy*-alleles showed mutant sequences that did not restore the wild type. The sequence deviations proved, however, to be "surprisingly nonrandom." Depending on the allele and the insertion site, between 37% and 88% carry a predominant footprint, and even the less prevalent footprints are often similar to the prevalent ones. Moreover, in 1% to 6% of the excisions, no footprint at all was formed. This implies that only the rest appears to consist of random sequences.

Since the vast majority of TE insertions into gene sequences can be classified as loss-of-function mutations, the first step of such TE activities constitutes a partially predestined method for essentially one aim: to switch off the function of a specific gene more or less permanently. The second step, excision with footprint formation in Class II elements, leads either to reversion to approximately wild-type gene functions or to a certain amount of foreordained variation of promoter and/or gene functions. Both phenomena could be significant in the context of morphological species formation, especially regressive evolution including resistances and the origin of ecotypes (see below).

As discussed elsewhere (69, 80, 84), the mechanisms in TE gene inactivation could be relevant to the origin of cultivated plants and domestic animals, as well as to regressive evolution in general. Indeed, a few cases have already been definitely established in plants (69). Moreover, ectopic gene expression and thus the generation of dominant traits have been reported in one case of domestication: *Zea mays* (142, 147), although it is still unclear whether TEs were involved. Nonetheless, genetic redundancy in combination with nonrandom gene inactivation and release mechanisms could also be pertinent for explosive species radiations such as those found in the cichlids in East African lakes. Here, in time lapses of merely a few hundred to several thousand years, there occurred a huge amount of regular morphological and ethological convergences in four independent radiations (93). Certain gene sets of the redundant potential may be switched off and others switched on. It is conceivable that there are still other mechanisms such as gene methylation (28), differential splicing (118), alternative promoters (83), or accelerated formation of alleles of regulatory and other genes, perhaps even caused by TEs (102, 148). As in regular chromosome rearrangements in the wild (see above), selection versus autonomous species formation and combinations of both have been widely postulated in the cichlid example, termed "one of the most spectacular examples of convergence in all evolutionary biology" (94). On one decisive point, however, there is widespread agreement: The time spans for conventional point mutations in combination with gene duplications are too short to give rise to novel genes and gene reaction chains to explain the phenomena (93). An inclusive genetic potential appears to be a more realistic approach to the problem. But this could open up another question, the origin of the genetic redundancy itself (83).

If the phenomenon of regressive evolution is applied to the structures and functions of TEs themselves, it is plausible that mechanisms for stress activation, target site selection, as well as gene inactivation and release originally were more effective in several transposon lines and organisms. Conversely, target site selection has also been found for heterochromatic and other regions with low or no gene content (1, 11, 103, 147), and this kind of selectivity may be part of the survival strategy of selfish TEs. They can proliferate there with minimal harm to their hosts and thus to themselves.

A phenomenon that may also be relevant in the context of possible transposon functions is that of the flax genotrophs (29, 30, 105, 119). The process of environmentally induced changes in the flax genomes "does not appear to be the generation of random variation" (30). Cullis et al. postulate that the heritable changes in flax are due to specific rearrangements at distinct positions of the genome. Low-copy-number, middle-repetitive, and highly repetitive sequences have been shown to be involved in the polymorphisms detected, and sequence alterations of specific subsets of 5SrDNA have been identified (119). TEs may also be involved in these rearrangements. If so, this evidence could be added validation for McClintock's view on the functions of TE-mediated chromosome rearrangements in ontogeny. In this case, transposon-mediated nonrandom genomic reshuffling would have some direct effects on gene regulation in ontogeny and, due to its heritability, on biodiversity also.

However, a discriminating, in-depth discussion and final evaluation of these topics call for much more evidence. Future investigations will require precise determination of (*a*) the overall ratio of nonrandom TE-induced sequence changes of relatively small nucleotide segments to the rest, which can only be obtained by concentrating on statistically relevant numbers of independent TE visits and their footprints in individual loci; (*b*) the causes and percentages of TE target site selections in relation to accidental insertions, especially for larger chromosome rearrangements; (*c*) the amount of genetic redundancy and its potential for generating morphological and other differences; and (*d*) the magnitude, if any, of TE involvement in the heritable genomic changes of the flax genotrophs mentioned above.

At present, the limited evidence for constraints caused by chromosome rearrangements consisting of TE target site selection (largely predetermining gross chromosome rearrangements), TE-induced nonrandom sequence deviations on the gene level, and the theory of the chromosome field are in agreement with the suggestion of a nonselection-driven, autonomous origin of a part of biodiversity and chromosomal species (29, 49, 75–79, 80–84, 90, 91, 93, 97, 109, 114–116, 119, 121, 122, 135, 140, 141).

TE-MEDIATED CHROMOSOME REARRANGEMENTS AND THE POSSIBILITY OF ABRUPT BOOSTING OF BIODIVERSITY AND SPECIES FORMATION

The following section focuses on possible TE-mediated chromosome rearrangements and the likelihood of an abrupt appearance of biodiversity and new life forms.

In agreement with McClintock, Syvanen (137) remarked, "I believe that transposons have the potential to induce highly complex changes in a single event." Shapiro (128) agreed that "there must exist mechanisms for large-scale, rapid reorganisations of diverse sequence elements into new configurations" for the integrated mosaic genome to make evolutionary sense [for further discussion, see (69)]. As noted above, this view seems to accord, at least phenotypically, with the findings of paleontology over the past 200 years. The question then centers on the extent to which TE-mediated chromosomal rearrangments can really explain the abrupt appearance not only of species but also of higher systematic categories in the face of the typical stasis of new life forms that so consistently characterizes the fossil record, i.e., the theory of punctuated equilibrium. Particularly apposite here are the following words of "so tough and influential a man" (52) as the German paleontologist Otto H. Schindewolf (116):

"According to Darwin's theory, evolution takes place exclusively by way of slow, continuous formation and modification of species: the progressive addition of ever newer differences at the species level results in increasing divergence and leads to the formation of genera, families, and higher taxonomic and phylogenetic units. Our experience, gained from the observation of fossil material, directly contradicts this interpretation. We found that the organizing structure of a family or an order did not arise as the result of continuous modification in a long chain of species, but rather by means of *a sudden, discontinuous direct refashioning of the type complex from family to family, from order to order, from class to class*. The characters that account for the distinctions among species are completely different from those that distinguish one type from another." (Italics by Schindewolf).

The existence in many animal groups such as foraminifers, brachiopods, corals, and others of an overwhelmingly rich fossil material consisting of literally millions (and generally in micropaleontology, billions) of individual exemplars led many paleontologists to emphasize that the objection most often raised against this view, i.e., the imperfection of the fossil record, is no longer valid [(15, 16, 37, 63, 116, 133, 139); see also discussion in (80, 85)]. Although Schindewolf's causal explanations of evolution from inner compulsion are not accepted by most contemporary paleontologists, there is general agreement on his factual representation of the punctuated mode of appearance of new organisms in the fossil record. Given that Schindewolf's description is phenotypically essentially correct, albeit with some exceptions (15, 56, 133), to what extent could TE-mediated small and gross chromosome rearrangements cope with the problem raised by the origin of families, orders, and classes?

In regard to the question of species formation in muntjac deer and cyclops, McClintock (90) suggested, "it is difficult to resist concluding that some specific "genomic shock" was responsible for origins of new species" in these genera. Despite the reservations and open questions noted above, her hypothesis is intrinsically attractive and a promising possibilty that warrants further investigation. However, if all the proof that is still lacking to substantiate her view on the origin of species were available, would that also give us the mode of origin of the higher systematic categories and types of life referred to by Schindewolf? To be more specific: If so, to what extent can any of the TE-incited rearrangements contribute to the origin of novel genes and new gene reaction chains as well as the genesis of irreducibly complex structures? All three of these may be especially relevant for the origin of higher systematic categories (3–5, 33, 69, 80–85, 121, 122, 130).

We could apply these same questions to a well-known problem of the fossil record: To what extent could small and gross chromosome rearrangements mediated by TEs and other factors solve the Cambrian explosion (6, 13, 14, 25, 51, 80, 92, 120, 132, 133)? (*a*) Even if one does not accept the assumption of the rather abrupt appearance of approximately 100 new animal Baupläne (51, 92), suggesting instead that lower systematic categories make their debut there, at least in part; (*b*) assuming further that some limited links to the Ediacara Fauna are probably established (25); and (*c*) accepting the hypothetical relevance of some fossilized Precambrian embryos (21, 54, 120, 152) as a tiny step in the origin of at least some

450 unexpected animal families, including many novel phyla, Darwin's 1859 verdict (32) still seems to have some force for the uncommitted observer: "The case at present must remain inexplicable; and may be truly urged as a valid argument against the views here entertained."

If we reject Darwin's idea of a strictly gradual evolution of most life forms, to what extent could TEs contribute to a resolution of such problems as the Cambrian explosion and the origin of higher systematic categories in general?

An exact evaluation of the possibilities and limits of TE-mediated contributions to the origin of species and especially higher systematic categories is only in its early stages. Extensive research on these and related questions is still needed.

Still pertinent to the state of experimentally induced species formation at the beginning of the twenty-first century is the view of McClintock's personal friend and intellectual soulmate, Richard Goldschmidt (49), on the abrupt origin of species by sudden genome reorganizations: "Needless to say, I did not succeed in producing a higher category in a single step; but it must be kept in mind that neither have the Neo-Darwinians ever built up as much as the semblance of a new species by recombination of micromutations. In such well-studied organisms as *Drosophila*, in which numerous visible and, incidentally, small invisible mutations have been recombined, never has even the first step in the direction of a new species been accomplished, not to mention higher categories."

CONCLUSIONS AND PERSPECTIVE

The Darwinian centennial in 1959 (138) provided the occasion for many voices to declare that all the basic problems of the origin of species and higher systematic categories had fully been solved by the modern synthesis. The speakers and authors of the centennial were, of course, aware of some outstanding problems, but it would be only a matter of time and effort before these were also resolved within the frame of the synthetic theory. In any case, there was nothing of substance to be really worried about. This was in stark contrast to the many opposing views advanced in 1909 at the 50th anniversary celebration of the publication of Darwin's *Origin of Species*. TEs had no part in these discussions.

Extrapolating from the wide range of current opinions (3–5, 14, 29, 30, 33, 38, 40, 55, 56, 64, 69, 71, 75–85, 89–93, 101, 109, 121, 122, 126–130, 132–138), we might safely predict that if similar meetings are held in 2009, the climate of opinion will be much closer to that of the 1909 semicentennial than to that of the 1959 centennial, and transposable elements will have played a special part in discussions on the origin of systematic species, from small to large chromosome rearrangements in muntjac deer to many different plant chromosome lines.

However, in the face of the numerous scientific problems still unresolved in the context of the origin of species and higher systematic categories, we would probably be well advised to continue to welcome the plethora of different and diverging ideas and hypotheses on the origin of life in all its forms as well as to remain open-minded on real results of investigations, wherever they may lead.

The *Annual Review of Genetics* is online at http://genet.annualreviews.org

LITERATURE CITED

1. Arkhipova IR, Morrison HG. 2001. Three retrotransposon families in the genome of *Gardia lamblia*: two telomeric, one dead. *Proc. Natl. Acad. Sci. USA* 98:14497–502
2. Beal EL, Rio DC. 1997. *Drosophila* P-element transposase is a novel site-specific endonuclease. *Genes Dev.* 11: 2137–51
3. Becker H-A, Lönnig W-E. 2002. Transposons, eukaryotic. In *Encyclopedia of Life Sciences*, 18:529–39. London: Nature Publ. Group/MacMillan
4. Becker H-A, Saedler H, Lönnig W-E. 2002. Transposable elements in plants. *Encyclopedia of Genetics*, ed. S Brenner, JH Miller, 4:2020–33. San Diego: Academic
5. Behe M. 1996. *Darwin's Black Box*. New York: Free Press. 307 pp.
6. Bengtson S. 1991. Oddballs from the Cambrian start to get even. *Nature* 351: 184–85
7. Bennett MD, Leitch IJ. 1995. Nuclear DNA amounts in angiosperms. *Ann. Bot.* 76:113–76
8. Bennett MD, Leitch IJ. 1997. Nuclear DNA amounts in angiosperms—583 new estimates. *Ann. Bot.* 80:169–96
9. Bennetzen JL. 1998. The structure and evolution of angiosperm nuclear genomes. *Curr. Opin. Plant. Biol.* 1:103–8
10. Bennetzen JL. 2000. Comparative sequence analysis of plant nuclear genomes: microcolinearity and its many exceptions. *Plant Cell* 12:1021–29
11. Bennetzen JL. 2000. Transposable element contributions to plant gene and genome evolution. *Plant Mol. Biol.* 42: 251–69
12. Bennetzen JL, Kellogg EA. 1997. Do plants have a one-way ticket to genome obesity? *Plant Cell* 9:1509–14
13. Benton MJ. 1993. *The Fossil Record 2*. London: Chapman & Hall. 835 pp.

14. Benton MJ. 2001. Finding the tree of life: matching phylogenetic trees to the fossil record through the 20th century. *Proc. R. Soc. London Ser. B* 268:2123–30
15. Benton MJ, Pearson PN. 2001. Speciation in the fossil record. *Trends Ecol. Evol.* 16:405–11
16. Benton MJ, Wills MA, Hitchin R. 2000. Quality of the fossil record through time *Nature* 403:534–37
17. Berg DE, Howe MM. 1989. *Mobile DNA*. Washington, DC: Am. Soc. Microbiol. 972 pp.
18. Besler B. 1613 (1997). *Der Garten von Eichstätt (Hortus Eystettensis). Das grosse Herbarium des Basilius Besler von 1613. Mit einem Vorwort von D Vogellehner und botanische Erläuterungen von GG Aymonin*. München: Schirmer/Mosel Verlag. 367 pp.
19. Cantrell MA, Filanoski BJ, Ingermann AR, Olsson K, DiLuglio N, et al. 2001. An ancient retrovirus-like element contains hot spots for SINE insertion. *Genetics* 158:769–77
20. Cave MS, ed. 1958–1964. *Plant Chromosome Numbers*. Chapel Hill: Univ. N.C. Press. 700 pp.
21. Chen J-Y, Oliveri P, Li C-W, Zhou G-Q, Gao F, et al. 2000. Special feature: Precambrian animal diversity: putative phosphatized embryos from the Doushantuo Formation of China. *Proc. Natl. Acad. Sci.USA* 97:4457–62
22. Colot V, Haedens V, Rossignol JL. 1998. Extensive, nonrandom diversity of excision footprints generated by *Ds*-like transposon *Ascot-1* suggests new parallels with V(D)J recombination. *Mol. Cell. Biol.* 18:4337–46
23. Comfort NC. 1999. "The real point is control": the reception of Barbara McClintock's controlling elements. *J. Hist. Biol.* 32:133–62

24. Comfort NC. 2001. *The Tangled Field: Barbara McClintock's Search for the Patterns of Genetic Control.* Cambridge, MA: Harvard Univ. Press. 337 pp.

25. Conway Morris S. 1998. *The Crucible of Creation. The Burgess Shale and the Rise of Animals.* Oxford: Oxford Univ. Press. 242 pp.

26. Coyne JA. 1998. Not black and white. Review of *Melanism*, MEN Majerus (1998). *Nature* 396:35–36

27. Craig NL, Craigie R, Gellert M, Lambowitz AM, ed. 2002. *Mobile DNA II.* Washington, DC: Am. Soc. Microbiol. Press. 1232 pp.

28. Cubas P, Vincent C, Coen E. 1999. An epigenetic mutation responsible for natural variation in floral symmetry. *Nature* 401:157–61

29. Cullis CA. 1999. Environmental stress—a generator of adaptive variation? In *Plant Adaptations to Stress Environments*, ed. HR Lerner, pp. 149–60. New York: Marcel Dekker

30. Cullis CA, Song Y, Swami S. 1999. RAPD polymorphisms in flax genotrophs. *Plant Mol. Biol.* 41:795–800

31. Cuvier G. 1830. Discours sur les révolutions de la surface du globe, et sur les changements qu'elles sont produits dans le règne animal. Sixième èdition francaise. Paris: Heideloff. 400 pp. (See Refs. 113, 131)

32. Darwin C. 1859 (1967 reprint of 1872 ed.). *On the Origin of Species by Means of Natural Selection, or the Preservation of Favoured Races in the Struggle for Life.* London: John Murray. 488 pp.

33. Dembski WA. 2002. *No Free Lunch: Why Specified Complexity Cannot Be Purchased without Intelligence.* Lanham, MD: Rowman & Littlefield. 404 pp.

34. Devos KM, Beales J, Nagamura Y, Sasaki T. 1999. *Arabidopsis*-rice: Will colinearity allow gene prediction across the eudicot-monocot divide? *Genome Res.* 9:825–29

35. Devos KM, Gale MD. 2000. Genome relationships: the grass model in current research. *Plant Cell* 12:637–46

36. Dietrich CR, Cui F, Packila ML, Li J, Ashlock DA, et al. 2002. Maize *Mu* transposons are targeted to the 5' untranslated region of the gl8 gene and sequences flanking *Mu* target-site duplications exhibit nonrandom nucleotide composition throughout the genome. *Genetics* 160: 697–716

37. Donovan SK, Paul CRC, ed. 1998. *The Adequacy of the Fossil Record.* Chichester: Wiley. 312 pp.

38. Dover G. 2000. *Dear Mr Darwin. Letters on the Evolution of Life and Human Nature.* Berkeley: Univ. Calif. Press. 268 pp.

39. Dubcovsky J, Ramakrishna W, SanMiguel P, Busso CS, Yan L, et al. 2001. Comparative sequence analysis of colinear barley and rice bacterial artificial chromosomes. *Plant Physiol.* 125:1342–53

40. Eldredge N. 1999. *The Pattern of Evolution.* New York: Freeman. 219 pp.

41. Eldredge N, Stanley SM, eds. 1984. *Living Fossils.* New York: Springer Verlag. 291 pp.

42. Evgen'ev MB, Zelentsova H, Poluectova H, Lyozin GT, Veleikodvorskaja V, et al. 2000. Mobile elements and chromosomal evolution in the virilis group of Drosophila. *Proc. Natl. Acad. Sci. USA* 97:11337–42

43. Fanti L, Pimpinelli S. 1999. The peculiar organization of telomeres in *Drosophila melanogaster. Gene Ther. Mol. Biol.* 4:1–10

44. Fedoroff N, Botstein D, eds. 1992. *The Dynamic Genome.* Plainview, NY: Cold Spring Harbor Lab. Press. 422 pp.

45. Flavell RB, Bennett MD, Smith JB, Smith DB. 1974. Genome size and the proportion of repeated nucleotide sequence DNA in plants. *Biochem. Genet.* 12:257–69

46. Fronicke L, Chowdhary BP, Scherthan H. 1997. Segmental homology among cattle (*Bos taurus*), Indian muntjac (*Muntiacus muntjak vaginalis*) and Chinese muntjac

(*M. reevesi*) karyotypes. *Cytogenet. Cell Genet.* 77:223–27

47. Gale MD, Devos KM. 1998. Plant comparative genetics after 10 years. *Science* 282:656–59

48. Goldblatt P, Johnson DE, eds. 1994. *Index to Plant Chromosome Numbers,* Vol. 51. St. Louis: Mo. Bot. Gard. 267 pp.

49. Goldschmidt R. 1948. Ecotype, ecospecies, and macroevolution. In *Richard Goldschmidt—Controversial Geneticist and Creative Biologist,* ed. LK Piternick, pp. 140–53. Basel: Birkhäuser Verlag. 153 pp.

50. Gottschalk W. 1976. *Die Bedeutung der Polyploidie für die Evolution der Pflanzen.* Stuttgart: Gustav Fischer Verlag. 501 pp.

51. Gould SJ. 1989. *Wonderful Life. The Burgess Shale and the Nature of History.* New York: Norton. 347 pp.

52. Gould SJ. 1993. Foreword to Schindewolf OH, *Basic Questions in Paleontology,* pp. IX–XIV. Chicago: Univ. Chicago Press. 467 pp.

53. Gould SJ. 1996. The tallest tale. Is the textbook version of the giraffe evolution a bit of a stretch? *Nat. Hist.* 5:18–23, 54–57

54. Gould SJ. 2000. *The Lying Stones of Marrakesh.* New York: Three Rivers Press. 371 pp.

55. Gould SJ. 2002. *The Structure of Evolutionary Theory.* Cambridge, MA: Belknap/Harvard Univ. Press. 1464 pp.

56. Gould SJ, Eldredge N. 1993. Punctuated equilibrium comes of age. *Nature* 366:223–27

57. Gray YHM. 2000. It takes two transposons to tango. *Trends Genet.* 16:461–68

58. Hohn B, Dennis ES, eds. 1985. *Genetic Flux in Plants.* Wien: Springer-Verlag. 253 pp.

58a. Hooper J. 2002. *Of Moths and Men: Intrigue, Tragedy & the Peppered Moth.* London: Fourth Estate. 377 pp.

59. Hurst GDD, Werren JH. 2001. The role of selfish genetic elements in eucaryotic evolution. *Nat. Rev. Genet.* 2:597–606

60. Ingham LD, Hanna WW, Baier JW, Hannah LC. 1993. Origin of the main class of repetitive DNA within selected Pennisetum species. *Mol. Gen. Genet.* 238:350–56

61. Johnston FP, Church RB, Lin CC. 1982. Chromosome rearrangement between the Indian muntjac and Chinese muntjac as accompanied by a deletion of middle repetitive DNA. *Can. J. Biochem.* 60:497–506

62. Keller EF. 1993. *A Feeling for the Organism: The Life and Work of Barbara McClintock.* New York: Freeman. 235 pp. 2nd ed.

63. Kerr RA. 1991. Old bones aren't so bad after all. *Science* 252:32–33

64. King M. 1993. *Species Evolution.* Cambridge: Cambridge Univ. Press. 336 pp.

65. Kleesattel W. 2001. *Die Welt der Lebenden Fossilien.* Darmstadt: Wiss. Buchges. 192 pp.

66. Krumbiegel G, Krumbiegel B. 1994. *Bernstein—Fossile Harze aus aller Welt.* Weinstadt: Goldschneck-Verlag. 110 pp.

67. Kumar A, Bennetzen JL. 1999. Plant retrotransposons. *Annu. Rev. Genet.* 33:479–532

68. Kunze R. 1996. The *Activator (Ac)* element of *Zea mays* L. In *Transposable Elements,* ed. H Saedler, A Gierl, pp. 161–94. Heidelberg: Springer-Verlag

69. Kunze R, Saedler H, Lönnig W-E. 1997. Plant transposable elements. *Adv. Bot. Res.* 27:331–470

70. Kunze R, Weil CF. 2002. The *hAT* and CACTA superfamilies of plant transposons. In *Mobile DNA II,* ed. NL Craig, R Craigie, M Gellert, AM Lambowitz, pp. 565–610. Washington, DC: Am. Soc. Microbiol. Press

71. Lamprecht H. 1974. *Monographie der Gattung Pisum.* Graz: Steiermärk. Landesdruck. 655 pp.

72. Lauber K, Wagner G. 2001. *Flora Helvetica.* Bern, Switz: Verlag Paul Haupt. 1615 pp. 3rd. ed., rev.

73. Li YC, Lee C, Sanoudou D, Hsu TH,

Li SY, Lin CC. 2000. Interstitial colocalization of two cervid satellite DNAs involved in the genesis of the Indian muntjac karyotype. *Chromosome Res.* 8:363–73

74. Liao GC, Rehm EJ, Rubin GM. 2000. Insertion site preference of the P transposable element in *Drosophila melanogaster. Proc. Natl. Acad. Sci. USA* 97:3347–51

75. Lima-de-Faria A. 1980. Classification of genes, rearrangements and chromosomes according to the chromosome field (study of over 700 species, from algae to humans) *Hereditas* 93:1–46

76. Lima-de-Faria A. 1986. *Molecular Evolution and Organization of the Chromosome.* Amsterdam: Elsevier. 1186 pp.

77. Lima-de-Faria A. 1988. *Evolution without Selection. Form and Function by Autoevolution.* Amsterdam: Elsevier. 372 pp.

78. Lima-de-Faria A. 1999. The chromosome field theory confirmed by DNA and hybridization. *Riv. Biol.* 92:513–15

79. Lima-de-Faria A, Arnason U, Widegren B, Isaksson M, Essen-Moller J, Jaworska H. 1986. DNA cloning and hybridization in deer species supporting the chromosome field theory. *Biosystems* 19:185–212

80. Lönnig W-E. 1993. *Artbegriff, Evolution und Schöpfung.* Köln: Naturwiss. Verlag. 622 pp. Internet ed. 2002

81. Lönnig W-E. 1999. *Johann Gregor Mendel: Why His Discoveries Were Ignored for 35 (72) Years.* Köln: Naturwiss. Verlag and http://www.mpiz.mpg.de/~loennig/mendel/mendel.htm (In German with English summary and note on Mendel's integrity). 70 pp.

82. Lönnig W-E. 2001. Natural selection. In *The Corsini Encyclopedia of Psychology and Behavioral Sciences,* ed. WE Craighead, CB Nemeroff, 3:1008–16. New York: Wiley. 3rd ed.

83. Lönnig W-E. 2002. *Das Gesetz der rekurrenten Variation.* Köln: Naturwiss. Verlag. 48 pp.

84. Lönnig W-E, Saedler H. 1997. Plant trans-

posons: contributors to evolution? *Gene* 205:245–53

85. Lönnig W-E, Wittlich K. 2002. *Kann der Neodarwinismus durch biologische Tatsachen widerlegt werden?* Köln: Naturwiss. Verlag. 42 pp. 3rd ed.

86. Lyttle TW, Haymer DS. 1992. The role of the transposable element hobo in the origin of endemic inversions in wild populations of *Drosophila melanogaster. Genetica* 86:113–26

87. Maside X, Bartolomé C, Assimacopoulos S, Charlesworth B. 2001. Rates of movement and distribution of transposable elements in *Drosophila melanogaster*: in situ hybridization vs Southern blotting data. *Genet. Res. Camb.* 78:121–36

88. Mayer K, Murphy G, Tarchini R, Wambutt R, Volckaert G, et al. 2001. Conservation of microstructure between a sequenced region of the genome of rice and multiple segments of the genome of *Arabidopsis thaliana. Genome Res.* 11:1167–74

89. Maynard Smith J. 1988. Punctuation in perspective. *Nature* 332:311–12

90. McClintock B. 1983. *The Significance of Responses of the Genome to Challenge.* Nobel lecture, 8 Dec. Cold Spring Harbor, NY: Cold Spring Harbor Lab. See also *Science* 226:792–801

91. McClintock B. 1987. The discovery and characterization of transposable elements. In *Genes, Cells and Organisms. Great Books in Experimental Biology,* ed. JA Moore. New York: Garland

92. McMenamin MAS, McMenamin DLS. 1990. *The Emergence of Animals. The Cambrian Breakthrough.* New York: Columbia Univ. Press. 217 pp.

93. Menting G. 2001. Explosive Artbildung bei ostafrikanischen Buntbarschen. *Naturwiss. Rundsch.* 54:401–10

94. Meyer A. 2001. Evolutionary celebrities. Review of *The Cichlid Fishes: Nature's Grand Experiment in Evolution,* GW Barlow. *Nature* 410:17–18

95. Murray MG, Peters DL, Thompson WF. 1981. Ancient repeated sequences in the

pea and mung bean genomes and implications for genome evolution. *J. Mol. Evol.* 17:31–42

96. Nelson G. 1994. Older than that. *Nature* 367:108

97. Nilsson H. 1953. *Synthetische Artbildung.* Lund: Gleerups. 1300 pp.

98. Neumann P, Nouzová M, Macas J. 2001. Molecular and cytogenetic analysis of repetitive DNA in pea (*Pisum sativum* L.). *Genome* 44:716–28

99. Nevers P, Saedler H. 1977. Transposable genetic elements as agents of gene instability and chromosomal rearrangements. *Nature* 268:109–15

100. Nevers P, Saedler H. 1985. Transposition in plants: a molecular model. *EMBO J.* 4:585–90

101. Nevers P, Shepherd NS, Saedler H. 1986. Plant transposable elements. *Adv. Bot. Res.* 12:103–203

102. Nordborg M, Walbot V. 1995. Estimating allelic diversity generated by excision of different transposon types. *Theor. Appl. Genet.* 90:771–75

103. Nouzová M, Kubaláková M, Zelová MD, Koblítzková A, Neumann P, et al. 1999. Cloning and characterization of new repetitive sequences in field bean (*Vicia faba* L.). *Ann. Bot.* 83:535–41

104. Nouzová M, Neumann P, Navrátilová A, Galbraith DW, Macas J. 2000. Microarray-based survey of repetitive genomic sequences in *Vicia* spp. *Plant Mol. Biol.* 45:229–44

105. Oh T, Gorman M, Cullis CA. 2000. RFLP and RAPD mapping in flax (*Linum usitatissimum*). *Theor. Appl. Genet.* 101: 590–93

106. Ono T, Kondoh Y, Kagiyama N, Sonta S, Yoshida MC. 2001. Genomic organization and chromosomal distribution of rat ID elements. *Genes Genet. Syst.* 76:213–20

107. Pardue M-L, DeBaryshe PG. 1999. Drosophila telomeres: two transposable elements with important roles in chromosomes. *Genetica* 107:189–96

108. Pardue M-L, DeBaryshe PG, Lowenhaupt K. 2001. Another protozoan contributes to understanding telomeres and transposable elements. *Proc. Natl. Acad. Sci. USA* 98:14195–97

109. Peters GB. 1982. The recurrence of chromosome fusion in inter-population hybrids of the grasshopper *Atractomorpha similis. Chromosoma* 85:323–47

110. Peterson P. 1953. A mutable pale green locus in maize. *Genetics* 45:115–33

111. Poinar G Jr, Poinar R. 1999. *The Amber Forest. A Reconstruction of a Vanished World.* Princeton, NJ: Princeton Univ. Press. 239 pp.

112. Ros F, Kunze R. 2001. Regulation of *Activator/Dissociation* transposition by replication and DNA methylation. *Genetics* 157:1723–33

113. Rudwick MJS. 1997. *Georges Cuvier, Fossil Bones, and Geological Catastrophes.* Chicago: Univ. Chicago Press. 301 pp.

114. Scherthan H, Arnason U, Lima-de-Faria A. 1987. The chromosome field theory tested in muntjac species by DNA cloning and hybridization. *Hereditas* 107:175–84

115. Scherthan H, Arnason U, Lima-de-Faria A. 1990. Localization of cloned, repetitive DNA sequences in deer species and its implications for maintenance of gene territory. *Hereditas* 112:13–20

116. Schindewolf OH. 1993. *Basic Questions in Paleontology.* Chicago: Univ. Chicago Press. 467 pp.

117. Schmidt R. 2000. Synteny: recent advances and future prospects. *Curr. Opin. Plant Biol.* 3:97–102

118. Schmucker D, Clemens JC, Shu H, Worby CA, Xiao J, et al. 2000. Drosophila Dscam is an axon guidance receptor exhibiting extraordinary molecular diversity. *Cell* 101:671–84

119. Schneeberger RG, Cullis CA. 1991. Specific DNA alterations associated with the environmental induction of heritable changes in flax. *Genetics* 128:619–30

120. Schopf JW. 1999. *Cradle of Life: The*

Discovery of Earth's Earliest Fossils. Princeton, NJ: Princeton Univ. Press. 367 pp.

121. Schwabe C. 2001. *The Genomic Potential Hypothesis: A Chemist's View of the Origins, Evolution and Unfolding of Life.* Georgetown, TX: Landes Biosci. 114 pp.

122. Schwabe C, Büllesbach EE. 1998. *Relaxin and the Fine Structure of Proteins.* Berlin: Springer. 200 pp.

123. Schwarz-Sommer Z, Gierl A, Cuypers H, Peterson PA, Saedler H. 1985. Plant transposable elements generate the DNA sequence diversity needed in evolution. *EMBO J.* 4:591–97

124. Schwarz-Sommer Z, Saedler H. 1987. Can plant transposable elments generate novel regulatory units? *Mol. Gen. Genet.* 209:207–9

125. Scott L, LaFoe D, Weil CF. 1996. Adjacent sequences influence DNA repair accompanying transposon excision in maize. *Genetics* 142:237–46

126. Shapiro JA. 1991. Genomes as smart systems. *Genetica* 84:3–4

127. Shapiro JA. 1993. Natural genetic engineering in evolution. In *Transposable Elements and Evolution*, ed. JF McDonald, pp. 325–37. Dordrecht: Kluwer

128. Shapiro JA. 1995. The discovery and significance of mobile genetic elements. In *Mobile Genetic Elements*, ed. DJ Sherratt, pp. 1–13. Oxford: IRL Press/Oxford Univ. Press

129. Shapiro JA. 1997. Genome organization, natural genetic engineering and adaptive mutation. *Trends Genet.* 13:98–104

130. Shapiro JA. 2000. Transposable elements as the key to a 21st century view of evolution. *Genetica* 107:171–79

131. Smith JC. 1993. *Georges Cuvier. An Annotated Bibliography of His Published Works.* Washington, DC: Smithonian Inst. Press. 251 pp.

132. Stanley SM. 1981. *The New Evolutionary Timetable: Fossils, Genes, and the Origin of Species.* New York: Basic. 222 pp.

133. Stanley SM. 1998. *Macroevolution. Pattern and Process.* Baltimore: John Hopkins Univ. Press. 332 pp.

134. Starlinger P. 1993. What do we still need to know about transposable elements Ac? *Gene* 135:251–55

135. Steenis CGGJ van. 1981. *Rheophytes of the World. An Account of the Flood-Resistant Flowering Plants and Ferns and the Theory of Autonomous Evolution.* Alphen aan den Rijn: Sijthoff & Noordhoff. 407 pp.

136. Sterelny K. 2001. *Dawkins vs. Gould. Survival of the Fittest.* Duxford, Camb.: Icon Books. 156 pp.

137. Syvanen M. 1984. The evolutionary implications of mobile genetic elements. *Annu. Rev. Genet.* 18:271–93

138. Tax S. 1960. *Evolution after Darwin.* Chicago: Univ. Chicago Press. Vol. 1, 629 pp.; Vol. 2, 473 pp.; Vol. 3, 310 pp.

139. Valentine JW. 1989. How good was the fossil record? Clues from the California Pleistocene. *Paleobiology* 15:83–94

140. Vries H de. 1901/1903. *Die Mutationstheorie (Versuche und Beobachtungen über die Entstehung von Arten im Pflanzenreich).* Leipzig: Verlag von Veit. Vol. 1, 648 pp., Vol. 2, 752 pp.

141. Vries H de. 1906. *Arten und Varietäten und ihre Entstehung durch Mutation.* Berlin: Gebrüder Bornträger. 365 pp.

142. Wang RL, Stec A, Hey J, Lukens L, Doebley J. 1999. The limits of selection during maize domestication. *Nature* 398:236–39

143. Wang W, Lan H. 2000. Rapid and parallel chromosomal number reductions in muntjac deer inferred from mitochondrial DNA phylogeny. *Mol. Biol. Evol.* 17:1326–33

144. Weitschat W, Wichard W. 1998. *Atlas der Pflanzen und Tiere im Baltischen Bernstein.* München: Verlag Dr. Friedrich Pfeil. 256 pp.

145. White MJD. 1973. *Animal Cytology and Evolution.* Cambridge: Cambridge Univ. Press. 3rd. ed.

146. White MJD. 1975. Chromosomal repatterning—regularities and restrictions. *Genetics* 79:63–72

147. White S, Doebley J. 1998. Of genes and genomes and the origin of maize. *Trends Genet.* 14:327–32

148. Wisman E, Hartmann U, Sagasser M, Baumann E, Palme K, et al. 1998. *Proc. Natl. Acad. Sci. USA* 95:12432–37

149. Xiao Y-L, Li X, Peterson T. 2000. A c insertion site affects the frequency of transposon-induced homologous recombination at the maize *p1* locus. *Genetics* 156:2007–17

150. Yang F, Obrien PCM, Wienberg J, Neitzel H, Lin CC, Fergusonsmith MA. 1997. Chromosomal evolution of the Chinese muntjac (*Muntiacus reevesi*). *Chromosoma* 106:37–43

151. Yang FT, Muller S, Just R, Ferguson-smith MA, Wienberg J. 1997. Comparative chromosome painting in mammals—human and the Indian muntjac (*Muntiacus muntjak vaginalis*). *Genomics* 39:396–401

152. Zhang Y, Yuan X, Yin L, Li C, Chen J, Hua T. 1998. Interpreting Late Precambrian microfossils. *Science* 282:1783A–83

Annu. Rev. Genet. 2002. 36:411–53
doi: 10.1146/annurev.genet.36.061802.101708

GENETICS OF SENSORY MECHANOTRANSDUCTION

Glen G. Ernstrom and Martin Chalfie
*Department of Biological Sciences, 1012 Fairchild Center, Columbia University,
1212 Amsterdam Avenue, New York, NY 10027; e-mail: ge29@columbia.edu,
mc21@columbia.edu*

Key Words mechanosensation, touch, balance, hearing, proprioception

■ **Abstract** The molecular mechanisms for the transduction of light and chemical signals in animals are fairly well understood. In contrast, the processes by which the senses of touch, balance, hearing, and proprioception are transduced are still largely unknown. Biochemical approaches to identify transduction components are difficult to use with mechanosensory systems, but genetic approaches are proving more successful. Genetic research in several organisms has demonstrated the importance of cytoskeletal, extracellular, and membrane components for sensory mechanotransduction. In particular, researchers have identified channel proteins in the DEG/ENaC and TRP families that are necessary for signaling in a variety of mechanosensory cells. Proof that these proteins are components of the transduction channel, however, is incomplete.

CONTENTS

INTRODUCTION

Sensory mechanotransduction is the transformation of mechanical force into electrical signals by specialized cells that communicate with the rest of the nervous system. Depending on the level of complexity of the mechanosensory cell—the mechanoreceptor—animals can respond to a touch or the high frequency deflections caused by sound. In addition, mechanoreceptors sense mechanical forces generated by the animal's own movement and activity.

Although anatomists and physiologists have described a variety of mechanoreceptors, little is known about the molecules utilized by any of these cells to transduce mechanical signals. Indeed, the holy grail of research in this field is the identification of the molecule or molecules that underlie mechanosensory transduction. Electrophysiological studies of vestibular hair cells in vertebrates and, more recently, fruit fly mechanosensing bristles demonstrate that mechanotransduction can be so rapid that the use of second messengers is unlikely (35, 36, 200). These observations, especially those on the vertebrate hair cells, have greatly influenced thinking in the field and led to models in which mechanosensory transduction is mediated by physical manipulation of a sensory channel or channel complex (135).

The identification of components needed to sense mechanical signals using biochemical or molecular biological approaches is complicated by the lack of suitable sources and of suitable assays. Mechanotransduction channels are found in dispersed or difficult-to-access tissues and are typically low in number. The human retina contains approximately 10^9 rods, each with 10^7 molecules of rhodopsin (163). In contrast, each of the several thousand hair cells in the vertebrate vestibular and auditory systems has approximately 100 mechanotransduction channels (37, 43, 93, 97). Each of the 14–16 mechanoreceptors in the slit-sense organ, a stretch-sensor at leg joints, of the tropical wandering spider has about 250 transduction channels (87). In addition, mechanosensory cells in skin, while concentrated in some regions of the body, are often associated with other types of mechanoreceptors and support cells. Enrichment for a particular transduction complex would be challenging.

Unlike receptors or channels that are stimulated by light or chemicals, mechanosensory channels must be gated by stimuli that are not readily administered in reconstitution experiments. Because of this difficulty, functional screening of mRNA populations, which identified, for example, the glutamate and capsaicin receptors (19, 88), is not feasible. A further complication is the uncertainty of whether a mechanotransduction channel would function normally outside of its normal cellular context. Many models of mechanotransduction channels posit a receptor complex containing a channel closely associated with other cellular components that enable it to respond to physical manipulation (36, 79, 150, 200). Channels in the absence of other components might remain in sustained opened or closed states, so that they are no longer mechanically sensitive.

A more fruitful approach has been the identification of components genetically using phenotypes associated with defects in mechanosensory systems. In general behaviors associated with specific mechanosensing cells have been characterized; loss of these behaviors has been used to identify mutants. This approach can potentially identify genes needed for mechanosensory transduction, but it will also produce genes required for the development of the mechanosensing cells or for the function (and development) of more downstream cells in their neural circuits. Once distinguished, however, the mechanosensory genes can be used for further study to examine their role in mechanosensation and to identify related genes in other organisms.

In this review, we describe how genetics has contributed to our current molecular understanding of sensory mechanotransduction in animals. We have concentrated on studies investigating the function of mechanosensing cells mediating touch, hearing, balance, and proprioception (position sense), although mechanotransduction, in its broadest sense, can include the processes by which cellular stress and strain regulate the growth and development of nonsensory cells. The latter topic has been reviewed elsewhere (3). We describe research using animals in which genetic screens have uncovered mutants defective in mechanosensation: nematodes, flies, zebrafish, and mice. Human genetics has also contributed to our knowledge of mechanosensation, but is not reviewed here because the genetics of hearing loss in humans was reviewed last year in this series (149; see also www.uia.ac.be/dnalab/hhh/) and because known genes underlying human peripheral sensory neuropathies do not appear to contribute directly to transduction. Mutation in peripheral myelin protein 22 gene (PMP22) [e.g. (160, 194)] and the myelin protein P_0 (MPZ) gene [e.g. (85, 86, 114)], for instance, lead to gradual demyelination of peripheral nerves. Cutaneous sensory loss is only one of several symptoms that are associated with these gene lesions. Mutation of the gene that encodes Connexin 32 is linked to another peripheral neuropathy, the X-linked Charcot-Marie-Tooth syndrome. (12). Hereditary sensory neuropathy type I is linked to a mutation in a ubiquitously expressed enzyme, serine palmitoyltransferase (42).

The Variety of Mechanoreceptor Cells

Animals usually have several types of morphologically distinct cells that sense mechanical signals, suggesting that the particular morphology is important for their function as mechanoreceptors. Figures 1 to 4 diagram the types of cells that are discussed in this review. Although we discuss the function of each cell type below, we comment here on several general features they possess. First, many of the cells have distinctive cytoskeletal structures. These structures include the large diameter microtubules of the *Caenorhabditis elegans* receptors for gentle touch, the ciliated structures of mechanoreceptors in *C. elegans* and Drosophila, and the actin-based filament bundles of vertebrate hair cells. Second, many of the cells have distinctive extracellular features that are associated with the cells. These features include the prominent extracellular material associated with the *C. elegans* receptors for gentle touch, structures that appear to cap the ciliated endings of Drosophila Type

I mechanosensory cells, and the tip-links that appear to join the hairs of the vertebrate hair cells to each other. Because these features appear to be localized to the supposed sites of transduction, they support a model of sensory transduction in which the transduction channel is regulated by extracellular and intracellular components. Direct demonstration of these associations, however, has not been made. Third, many of the cells have specialized support cells, such as the socket and sheath cells of many ciliated mechanosensory organs in invertebrates and the capsules encasing particular mammalian cutaneous receptors like the Pacinian corpuscles and Meissner's corpuscles. In some cases, these support cells produce a specialized extracellular environment in which the sensing cells act.

Not all mechanosensing cells have ultrastructurally distinct features. Many cells innervating the skin are said to have free-nerve endings, i.e., they are not associated with obvious support cells. Some cells, like the Rohon-Beard cells in *Xenopus* larvae, have termini with varicosities and swollen endings (162); others, such as some cells innervating mammalian skin, having endings with no apparent specializations (178). Such processes may still be associated with extracellular matrix; the processes of the multiple dendritic neurons in Drosophila, for example, appear to be attached to the epidermis (see below).

Defining a Sensory Mechanotransduction Channel

An important goal of most of the genetic studies described here is the discovery of the transduction mechanism or mechanisms underlying the sensing of mechanical stimuli. To be the basis of mechanosensory signaling, a molecule or molecules should satisfy several criteria. For simplicity we assume that the mechanosensory molecule is itself a channel. This assumption, which has not been proven, is accepted by many in the field because of the rapid response time of vertebrate hair cells (35) and Drosophila mechanosensitive bristles (200). First, the channel should be localized to the correct cells and to the correct position within the cells. Second, the channel must be necessary for the electrical response of the sensory cell to the mechanical stimulus and not for the subsequent activity of the cell. Third, mechanical forces must gate the channel expressed in a heterologous cell or lipid bilayer. Fourth, the gating should recapitulate the mechanically gated current observed in its native environment. The last two criteria are the most difficult to demonstrate, and have not been established for any of the candidate mechanosensory proteins identified in animals.

The mechanosensitive channel MscL in *Escherichia coli*, however, has most fully met these criteria [reviewed in (136)]. The MscL channel, which acts as a safety valve to release ions and macromolecules in hypo-osmotic environments, was the first molecularly identified mechanosensitive channel. This channel is expressed in the periplasmic membrane, is necessary for the large mechanically gated conductance characterized in spheroblasts, and can recapitulate the mechanically gated activity when reconstituted in lipid bilayers. Although tension in the lipid bilayer gates MscL, the prevailing view is that the cytoskeletal and/or extracellular proteins directly tether and gate sensory mechanotransduction channels in invertebrates and vertebrates (36, 79, 150, 200). Thus, mechanical force could be transmitted to the channel either through the lipid bilayer or through extracellular and cytoplasmic attachments [see (81) for a recent discussion]. The first mechanism is much easier to demonstrate by controlled suction through a patch-clamp pipette. However, some voltage-gated channels can be activated in this manner. Morris and colleagues, for example, found that the voltage-gated Shaker K^+ channel—a channel well known to shape the time course of action potentials—can be activated by pipette suction (78). Thus, channel complexes may be susceptible to activation by mechanical perturbations of the bilayer but this response may not be physiologically relevant.

SENSORY MECHANOTRANSDUCTION IN *C. ELEGANS*

Despite having a nervous system with only 302 neurons, the *Caenorhabditis elegans* hermaphrodite is sensitive to a variety of mechanical stimuli and has several different types of mechanosensory receptor cells. Different sets of cells respond to gentle touch along the body (22) [or to tapping of the plate on which the animals are grown (158)], to harsh body touch (21, 205), to touch at the nose (105), to touch at the side of the head (83), and to the texture of the substrate on which the animals move (171). In addition, *C. elegans* may also detect its own movement using stretch receptors in the undifferentiated process endings of its ventral cord motor neurons (188). Such self-generated mechanical stimuli may be important for coordinating locomotion.

Cell ablation studies have demonstrated the involvement of particular sensory neurons in sensing mechanical stimuli (22, 26, 83, 105, 171, 205, 211). In each of the mechanosensory behaviors studied, genes have been identified that are expressed in the essential mechanoreceptor(s) and can be mutated to cause a mechanosensory deficit. In addition to the identification of different specialized mechanoreceptors, genetic studies have identified different genes that are needed for these mechanosensory behaviors.

Gentle Body Touch

When the worm is stroked across its head with a fine hair, the animal moves backward; when it is stroked across its tail, it moves forward. This response, which can be elicited by 10 μN of force (I. Chin, M.B. Goodman & M. Chalfie, unpublished data), requires the function of the six mechanoreceptors for gentle touch. These cells send processes along the body just below the cuticle and synapse onto command interneurons that regulate locomotion (22, 23). Three cells (ALML, ALMR, and AVM) sense touch in the anterior half of the animals; two other cells (PLML and PLMR) sense posterior touch. Touch perceived through these cells has several other effects, including the regulation of egg-laying, pharyngeal pumping, and defecation rhythm (23, 121, 190). Several of these behaviors are mediated through other synapses made by these sensory cells (23). A sixth cell with similar morphology and gene expression (PVM) does not appear to be needed for touch-mediated movement (22), but is thought to be involved in other touch-mediated circuits (23).

The processes of all six touch receptors have two characteristic features that distinguish them from other cells and that may be important for mechanosensation (Figure 1, see color insert). One feature is a bundle of large-diameter microtubules that packs the process (24). These microtubules, which contain 15 protofilaments (25), are found only in these cells. The second feature is a prominent extracellular matrix that is particularly extensive between the touch receptor neurons and the overlaying cuticle (23). The extracellular matrix is not uniformly distributed along the processes, but periodically swells. This swelling is most easily seen in the processes of the AVM and PVM cells.

MUTANTS INSENSITIVE TO GENTLE TOUCH   The first screen for touch-insensitive animals, searching for animals with normal movement but that failed to move when stroked with an eyebrow hair, was done in *C. elegans* and yielded genes that affected the development and function of the six receptor neurons for gentle touch (22, 184). Subsequent mutageneses have identified over 450 independent mutant lines, defining 18 genes, that are insensitive to gentle touch (21, 79). Most of these genes are named *mec* for *mec*hanosensory abnormal. Because multiple alleles have been identified for all but one gene, *mec-17*, which has a weak touch-insensitivity phenotype, these screens are probably saturated for genes whose loss of function is restricted to touch insensitivity.

Several of the genes are represented by many (>50) alleles. These alleles include deletions, loss-of-functions alleles, hypomorphic missense mutations (including temperature-sensitive alleles), and unusual, neomorphic alleles. This collection of alleles has been useful in several ways. For example, although animals sense touch at hatching, temperature-sensitive alleles have been used to demonstrate that many of the function *mec* genes must be expressed throughout larval development for the adults to be touch sensitive (21). This result suggests that the genes are needed for the function and not the development of the cells. The temperature-sensitive alleles have also been used to identify dominant enhancing mutations to demonstrate gene interactions among the *mec* genes (52, 79). Rare mutations in the *mec-4* gene produced an unusual phenotype; the touch receptor neurons vacuolate and lyse (21, 22), suggesting an ionic imbalance in the cells. This phenotype provided not only a model for neurodegenerative disease, but also the first hint that the MEC-4 protein might be a channel protein (48).

Several genes are required for the normal development of these touch receptors. For example, the *mec-3* gene, encoding a LIM-homeodomain transcription factor that is expressed after the receptor neurons arise, is needed for their differentiation (204, 205). *mec-3* animals generate cells that would normally differentiate into touch cells, but the cells fail to make the large-diameter microtubules or extracellular matrix or to express several other *mec* genes. Twelve other genes (*mec-1, 2,*

4–7, 9, 10, 12, 14, 15, and *18*) mutate to touch-insensitivity, yet result in differentiated touch receptor neurons. These genes affect the function of these cells and are good candidates for genes needed for mechanosensory transduction. Eleven of the 12 *mec* function genes (all but *mec-15*) have been cloned and their expression patterns have been determined. Ten of the 11 genes are expressed in the touch receptor neurons (*mec-5*, encoding a collagen, is made by the surrounding hypodermis). Two genes (*mec-4* and *mec-18*) are found exclusively in these cells, and one (*mec-6*) is more generally expressed. The remaining genes are found in one or a few additional cell types. Except for *mec-6*, which is also needed for coordinated movement (see below), the loss-of-function phenotype for these genes is touch insensitivity.

A MODEL FOR MECHANOSENSATION   The sequences of the function genes and gene interactions deduced from studies of suppressing and enhancing mutations have suggested a model for mechanosensation in which a mechanosensory complex is formed by a channel, associated extracellular and cytoskeletal proteins, and modulating proteins (52, 79, 95). In this model the relative movement of its attachments to the extracellular matrix and the microtubule cytoskeleton gates the channel. Support for several aspects of the model have come from recent biochemical and electrophysiological studies.

The channel complex   At the core of the putative mechanotransduction complex are the pore-forming channel subunits encoded by the *mec-4* and *mec-10* genes. Both genes encode similar but nonredundant proteins, called degenerins. The degenerins are members of the DEG/ENaC superfamily of amiloride-sensitive, sodium channel subunits (Figure 5) (48, 74, 95, 117). Consistent with their role as channel subunits, dominant gain-of-function mutations cause cell swelling and lysis (21, 22, 95), probably due to an abnormally large increase in ion flux. Channel activity can be detected in Xenopus oocytes expressing MEC-4 or both MEC-4 and MEC-10 with the gain-of-function mutation, but not in their wild-type form (74). Although the mutant proteins form channels in frog oocytes, they have not been shown to respond to mechanical stimuli. For example, superfusion of hypoosmotic solution does not increase channel activity.

Several studies of mutated and chimeric proteins have provided insights into how the structure of DEG/ENaC channels relates to their function [reviewed in (4)]. The degenerins and other DEG/ENaC proteins have two transmembrane domains, a large extracellular loop, and small, cytoplasmic termini. The second transmembrane domain and the adjacent amino acids preceding it are the most conserved portion of these channels. These residues are critical for sodium ion selectivity, conductance, and for the micromolar affinity of the pore-blocker amiloride. Additional conserved regions in all DEG/ENaC proteins include cytoplasmic residues immediately preceding the first transmembrane domain and two extracellular cysteine-rich domains. The *C. elegans* degenerins have an additional cysteine-rich domain and a 22-amino-acid regulatory sequence in the extracellular loop (71).

Extensive mutational analysis of *mec-4* has suggested that all of these domains are important for degenerin function (90, 91). Gain-of-function mutations, which cause neurodegeneration, occur at an alanine located near the external entry of the pore of both *mec-4* and *mec-10* (48, 95). Mutations within the conserved 22-amino-acid sequence also give this gain-of-function phenotype (71). These results suggest

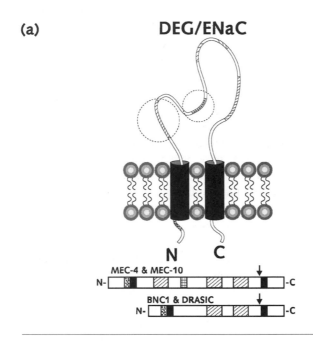

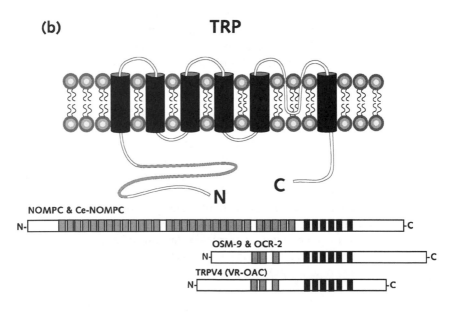

that these regions are important for controlling ion flux. These results have been supported by expression studies in heterologous cells of MEC-4 and MEC-10 (74) and of another *C. elegans* degenerin, UNC-105 (69). Several otherwise recessive mutations in *mec-4* and one in *mec-10* act as dominant negative suppressors of the gain-of-function mutations (90, 95). This suppression suggests that each gene probably contributes more than one subunit to the degenerin channel (90).

Two other proteins, the products of the touch function genes *mec-2* and *mec-6*, appear to form a channel complex with MEC-4 and MEC-10 (74) (D. Chelur, S. Zhang & M. Chalfie, unpublished). Heterologously expressed MEC-2 and MEC-6 increase the MEC-4/MEC-10 currents 30-fold and 45-fold, respectively, without increasing the amount of degenerin protein at the oocyte surface. These effects are synergistic; coexpression of all four proteins results in 200-fold higher current compared to those produced by MEC-4 and MEC-10. MEC-2 and MEC-6 also change the ion selectivity of the channel. The mechanism whereby these modulatory components enhance channel activity is not known, but may be elucidated by single channel studies. In addition, these proteins coimmunoprecipitate the degenerins expressed in heterologous cells and colocalize with MEC-4 in the touch receptors. The localization and/or stability of MEC-4 in these cells also require wild-type MEC-6 function.

MEC-2 is a membrane protein with *C. elegans*-specific termini and a central 280-amino-acid region that is 64% identical with stomatin, a putative regulator of ion flux in humans (96, 118). Both proteins are monotopic proteins with a membrane-associated domain that does not appear to span, but remains on the cytosolic side of the lipid bilayer (96, 167). Human stomatin is found associated with lipid rafts (168), suggesting that MEC-2 may be similarly localized. The additional N and C termini MEC-2 are needed for the MEC-2 enhancement of MEC-4 activity in Xenopus oocytes, suggesting that these termini may be important in regulating MEC-4 gating (74).

mec-6 mutations suppress the degenerations or dominant phenotypes caused by gain-of-function mutations in *mec-4*, *mec-10*, and other *C. elegans* degenerins (26, 95, 188; D. Chelur & M. Chalfie, unpublished data). This result initially suggested that MEC-6 might be another degenerin subunit [making the touch receptor channel similar to the vertebrate epithelial sodium channel, which has three similar

---

Figure 5 Membrane topology and domain structure of ion channel subunits implicated in mechanotransduction. (*a*) DEG/ENaC subunits have two transmembrane domains (*black*), a large extracellular loop, and intracellular termini. Dotted regions are not found in all members of the superfamily. MEC-4 and MEC-10 are representative of a branch of the superfamily that has three cysteine-rich domains (*diagonal stripes*) and a conserved extracellular regulatory domain (*checkered*). BNC1 and DRASIC have two cysteine-rich domains. A conserved N-terminal region is dotted and the position of the degeneration-inducing residue is indicated by an arrow. (*b*) TRP ion channel subunits have six membrane-spanning domains (*black*), a pore-forming P-loop between the fifth and sixth segments, and a variable number of ankyrin repeats (*gray*).

subunits (18)]. *mec-6*, however, encodes a single transmembrane domain protein that is almost entirely extracellular and that is similar to vertebrate paraoxonases (D. Chelur & M. Chalfie, unpublished data). The paraoxonase proteins receive their name because the first of three homologous human proteins to be discovered (PON1) degrades paraoxon, a metabolite of the insecticide parathion and potent inhibitor of acetylcholinesterase (64). PON1 and the rabbit homolog of PON3 are also found in serum associated with the high-density lipoprotein (HDL) particles (47). In vitro, HDL-associated paraoxonase protects low-density lipoprotein (LDL) particles from oxidation (131, 132, 174). This function in vivo may prevent artherosclerosis. PON2 also has antioxidant properties, but it is ubiquitously expressed and is not associated with HDL particles (143). The physiological role of the non-HDL-associated paraoxoanase protein is not known, although the results with the *mec-6* protein suggest that it may affect channel function.

Is the degenerin channel complex a mechanosensory transduction complex? The evidence is incomplete. Although the components are needed for touch sensitivity and are expressed along the length of the sensory process [the animals are sensitive to touch along the entire process (22)], we do not know whether these channels underlie a mechanosensory current in these cells or whether the heterologously expressed channels are mechanosensitive. If forces acting on the large extracellular domains of MEC-4 and MEC-10 activate the channels, methods are required to manipulate these domains while allowing for simultaneous voltage-clamp recording. Antibody attachment of magnetic beads [as in magnetic twisting cytometry (201)] or laser-tweezers might be adapted for this purpose.

Other channels may affect touch sensitivity in these receptor neurons. A gain-of-function mutation of *egl-2*, which encodes an ether-a-go-go-like K^+ channel, causes anterior touch insensitivity, presumably by preventing depolarization (207). The role of this protein in touch sensitivity is unclear, although it may affect the responsiveness of the cells to mechanical stimuli.

Extracellular matrix   Three touch function genes, *mec-1*, *mec-5*, and *mec-9*, encode components of the extracellular matrix. *mec-5* encodes a collagen that is produced by the surrounding hypodermis (52). MEC-5 is unique; it only shares similarity with other collagens in the Gly-X-Y repeat domain. The protein appears to be specifically used by the touch receptors for gentle touch, since no other mutant phenotypes are produced by null alleles.

Two other genes, *mec-1* and *mec-9*, are expressed in the touch receptor neurons and encode proteins that contain multiple epidermal growth factor (EGF) domains and multiple Kunitz protease inhibitor domains (52; L. Emtage, G. Gu & M. Chalfie, unpublished data). Both domains are probably protein-binding domains; the Kunitz domains are unlikely to inhibit proteases, since they lack the appropriate P1 residue needed for this activity. Both MEC-1 and MEC-9 appear to be secreted from the touch receptor cells. Loss of *mec-9* function does not result in visible alterations in the extracellular matrix. In contrast, loss of *mec-1* results in the loss of most, if not all, of the extracellular matrix and a concomitant loss of attachment of the processes near the body surface (22). This attachment, which requires at least some of the Kunitz domains of MEC-1, is not essential for touch sensitivity. Animals with nonsense mutations that result in products with most, but not all, of the Kunitz domains do not have the attachment defect, but are touch insensitive (L. Emtage, G. Gu & M. Chalfie, unpublished data). In contrast, *him-4* mutations disrupt the attachment of the touch receptor neurons to the body wall (causing them to appear as they do in *mec-1* animals), but do not result in touch insensitivity (196). *him-4* encodes a muscle-derived protein, hemicentrin, that is located in the extracellular matrix of the touch receptor neurons.

Two lines of evidence suggest that the putative mechanosensory channel interacts with components within the extracellular matrix. First, normally recessive mutations in several *mec* genes, including *mec-2*, *mec-4*, *mec-6*, *mec-7*, *mec-9*, *mec-10*, and *mec-12*, are dominant enhancers of temperature-sensitive *mec-5* mutations (52, 79). Since this enhancement is not allele-specific, these results suggest that *mec-5* can be made rate limiting for touch sensitivity, but do not determine whether any of the proteins interact. In this context, however, we note that mutation in another collagen gene, *let-2*, genetically suppresses the hypercontraction caused by the semidominant mutation in another *C. elegans* DEG/ENaC channel subunit expressed in muscle, UNC-105 (69, 126). Second, although MEC-5::GFP is distributed continuously along the touch receptor processes in wild-type animals, it is distributed in a punctate pattern that often corresponds with that of MEC-2 (as a marker of the channel complex) in *him-4* mutants (L. Emtage & M. Chalfie, unpublished data).

Cytoskeleton   The most prominent and unique features of the receptors for gentle touch are the 15-protofilament microtubules that fill the cell processes (24, 25). Reconstruction of the processes from serial electron micrographs reveals that the microtubules increase in number and length throughout most of larval growth, are relatively short compared to the length of the process (about $1/20^{th}$ of the length), and are apparently crossbridged to each other. The proximal ends of the microtubules can be anywhere within the microtubule bundle, but the distal ends are always on the outside of the bundle, i.e., in a position where they could contact the plasma membrane. In addition, these ends are frequently associated with electron-diffuse material that appears to contact the membrane. The *mec-7* and *mec-12* genes encode β and α tubulins, respectively, that are required for the formation of the 15-protofilament microtubules and for touch sensitivity (63, 170). A special function in mechanosensation is suggested not only by the unique occurrence of these microtubules in these cells, but also by the finding that *mec-7* and *mec-12* null mutations result in cells that contain smaller diameter microtubules that permit the apparently normal outgrowth of the cells

(22). The role played by the microtubules in these cells, however, is unclear. The location of the distal ends near the plasma membrane suggests a direct association with the channel complex. In partial support of this role, microtubules are needed for the distribution of *mec-2* and *mec-18* in the touch cell processes (96; G. Gu & M. Chalfie, unpublished data). Alternatively, the microtubules may play a more passive structural role and are not directly associated with the channel complexes. The punctate localization of MEC-4 and the other components of the channel complex correspond to the periodicity in the extracellular matrix, not the distribution of the distal ends of the microtubules. Electron microscopic localization of the MEC proteins should clarify the role of the microtubules in sensory transduction.

Other factors   Two other genes, *mec-14* and *mec-18*, encode proteins with probable enzymatic activities that may alter the function of the channel. *mec-14* encodes a protein in the aldo-keto reductase superfamily (N. Hom, S. Gangadharan, Y. Tu, L. Chen & M. Chalfie, unpublished data). Mutations of *mec-14*, which is exclusively expressed in the touch cells (G. Caldwell & M. Chalfie, unpublished data), partially suppress the cell death caused by a gain-of-function mutation of *mec-10* (96). This suppression suggests that wild-type MEC-14 activates the degenerin channels. That MEC-14 could be a channel regulator is weakly supported by the fact that β-subunits of Shaker potassium channels are also members of the same protein superfamily (28, 137).

Unlike mutations in *mec-6* and *mec-4*, which abolish the degenerations produced by a gain-of-function mutation in *mec-10*, or mutations in *mec-2*, *mec-14*, *mec-12*, and *mec-5*, which reduced the number of deaths, mutations in *mec-18* increase the number. This enhancement suggests that *mec-18* may negatively regulate the degenerin channel in the touch receptor neurons (95). *mec-18* encodes a protein with similarity over its entire length with firefly luciferase and the plant protein 4-coumarate coA ligase (G. Gu & M. Chalfie, unpublished data). Both of these latter proteins form activated carboxyl groups while converting ATP to AMP and PP_i. The target or targets of MEC-18 activity are not known.

Mutations in the *egl-3* gene also abolish sensitivity to gentle touch (106). *egl-3* encodes a putative polyprotein convertase that is expressed in the touch receptor cells and downstream interneurons as well as several other cells. Because expression of the wild-type gene in the interneurons does not rescue the body touch insensitivity, Kass et al. (106) suggested that it must act in the receptors for gentle touch. Its role in these cells is unclear.

Harsh Body Touch

When the touch receptors for gentle touch are inactivated by mutation or ablation, the animals can still respond to the stronger stimulus of a prod with a platinum wire. This harsher touch at the tail requires the PVC interneurons, since the degeneration of these cells in *deg-1* mutants (26) makes the animals incapable of responding to the prod. The role of these interneurons in this response is unclear. They may sense the stronger stimulus or may be required for the circuit of a sensory cell that does sense this stimulus [some additional putative sensory cells synapse onto the PVC cells (209)]. If the PVC cells have a sensory function, the *deg-1* gene, which encodes a degenerin like MEC-4 and MEC-10, may be involved. However, since loss of *deg-1* activity does not give a mutant phenotype (26), the gene may be redundant or not important for PVC function.

An additional response to harsh touch, typically a backward movement, occurs when animals lacking the receptors for gentle touch are prodded midway along their lengths. This response depends on the two PVD neurons but not the touch cells needed for gentle touch (205) (Figure 1, see color insert). These cells arise in the second of the four larval stages and extend along the midline of the body and a shorter process that synapses onto command interneurons that control locomotion (209).

Although the mechanosensory function of the PVD cells was identified first by the response to harsh touch, the cells may also modify the function of other touch receptors. Specifically, the worm's response to a plate-tapping stimulus depends on both the receptors for gentle touch and the PVD cells (211). Wild-type animals respond to tapping on the Petri plate on which they grow by moving backward. If animals lack the anterior receptors for gentle touch (the ALM cells), the frequency and magnitude of the reversals are reduced. If the posterior receptor cells (the PLM cells) are killed, then the animals tend to move forward. If the PVD cells are ablated, otherwise intact animals reverse less often and the responses of both ALM and PLM ablated animals are lessened. These results suggest that the PVD cells may modulate the sensitivity of ventral cord interneurons to input from the receptors for gentle touch (211).

PVD cells express MEC-3, the transcription factor that specifies the fate of the receptors for gentle touch, and *mec-3* mutants do not respond to harsh touch (205). Although most proteins needed for gentle touch are not expressed in PVD, MEC-10 is (95). An intriguing possibility is that a different degenerin complex detects higher threshold mechanical stimuli in the PVD cells. *mec-10* mutants are not, however, insensitive to harsh touch. Possibly other degenerin genes are expressed in these cells.

Nose Touch

Even without the receptors for gentle touch, *C. elegans* backs away from objects it bumps with its head (39, 105). This nose touch response requires the function of several ciliated sensory neurons in the head (105). Laser ablation of the ASH (Figure 1, see color insert) neurons reduces the ability of the animals to respond by 45%, and killing the FLP neurons reduces it by 30%. The remainder of the response is eliminated when the OLQ neurons and the ALM neurons are also killed (killing these cells on their own has no effect on the nose touch response) (49, 105).

The two ASH neurons are needed not only for mechanosensation, but also for osmosensation and chemosensation (9, 82, 105). These different responses can be separated behaviorally and genetically. Habituating animals to nose touch leaves the osmotic and chemical avoidance responses intact, and habituating the animals to a noxious chemical does not affect nose touch (82). In addition, mutation of the novel protein gene *osm-10* eliminates the response to high osmolarity, but not the other behaviors (82). Mutations in the *glr-1* gene abolish the nose touch response, but not the responses of ASH to the other stimuli (83, 134). Somewhat surprisingly, *glr-1* encodes an AMPA-type glutamate receptor that is not expressed in the ASH cells, but in interneurons onto which they synapse. Mutation of *lin-10*, which encodes a PDZ protein needed for the localization of GLR-1, affects ASH nose touch, but not its other activities (164). These results suggest that different senses use different signaling systems in this cell.

Several other genes can be mutated to give animals with nose touch defects, but all of these also affect the other responses mediated by the ASH cells. These genes include *eat-4* (11), which encodes a vesicular glutamate transporter (10, 121, 186), and *odr-3*, which encodes a Gα protein (161).

In contrast to their effect on gentle touch, mutations of *egl-3* suppress the nose touch insensitivity of *glr-1* and *lin-10* mutants, but not *eat-4* mutants (106). Although *egl-3* is expressed in the ASH cells, this suppression requires *egl-3* expression in the interneurons. The putative polyprotein convertase encoded by *egl-3* could be needed for the production of a neuropeptide that regulates synaptic function. Kass et al. (106) speculate that this function may be needed for habituation.

Mutation of *osm-9* abolishes nose touch as well as other sensory reception in the ASH cells (32). OSM-9, a TRP (transient receptor potential)–like channel subunit, is expressed in ASH as well as several other ciliated cells in the head and tail. TRP ion channels have six membrane-spanning domains and intracellular termini that are expressed in a variety of excitable and nonexcitable cells (Figure 5) (30). The N terminus often has a variable number of ankyrin repeats that probably act as protein-binding domains. The possible role of a TRP channel in the mechanosensory function of these cells is intriguing given the finding of a TRP channel in Drosophila that may also transduce mechanical signals (see below). Recently, a second TRP-like subunit, the product of the *ocr-2* gene, is expressed in the ASH cells and is needed for all of its sensory responses (193). Moreover, both OSM-9 and OCR-2 are needed for their appropriate localization at the ciliated endings. Electrical recording from the ASH cells should determine whether these various channel proteins are needed for mechanosensory currents.

Two other channel proteins, the degenerin protein products of the *deg-1* and *unc-8* genes, are expressed in the ASH cells (80, 188). The role of these genes in the ASH cells is unclear. Although gain-of-function mutation of *deg-1* causes the death of the ASH cells, loss-of-function alleles of deg-*1* and *unc-8* do not yield a detectable mutant phenotype (82).

Texture

An additional set of ciliated neurons that have termini adjacent to the cuticle (148, 208) also appear to respond to mechanical stimuli. Laser ablation of these cells [the four CEP neurons (Figure 1, see color insert), the two deirid (ADE) neurons, and the two postdeirid (PDE) neurons] abolishes the normal slowing that is seen when animals enter a bacterial lawn on an agar medium (171). Since wild type, but not animals with these cells ablated, also slow their movement when they encounter appropriately sized Sephadex beads, the slowing behavior is thought to depend on mechanical rather than chemical properties of the bacteria. These cells contain the neurotransmitter dopamine (184), and mutations that interfere with the production and function of dopamine mimic the effects of these ablations (171). The CEP and ADE cells express the closest nematode homolog of the NOMPC protein, a candidate mechanotransduction channel in Drosophila [(200), see below]. The expression in the PDE cells was not reported.

The function of these cells appears to be modulated in two ways. First, the rate at which the animal slows depends on whether they have been well-fed or starved (171). This behavioral modulation, mediated by serotonin, provides an interesting system to study the role of neuromodulation on mechanosensation. Second, the receptors for gentle touch chemically synapse onto these cells (23). Since a gentle touch stimulus can prevent animals from stopping when they encounter bacteria, these synapses may serve to down regulate the dopaminergic cells.

Proprioception

The sinusoidal motion of *C. elegans* requires several sets of motor neurons that are distributed along the ventral cord motor neurons (208). The cells in each set have processes that overlap and are often joined by gap junctions. Chemical synapses are located near the cell bodies leaving more distal portions bare. This morphology led R. Russell (personal communication) to suggest that the bare processes of the cells served as stretch sensors that helped coordinate locomotion. As such, these cells could be acting as dual function sensory and motor neurons [a precedent for dual function neurons exists in *C. elegans* where the IL1 sensory cells are also motor neurons (203)]. Loss of the *unc-8* gene, which encodes a degenerin channel subunit related to MEC-4 and MEC-10, causes a mild uncoordinated phenotype (188). Like the receptors for gentle touch, the motor neurons coexpress a second degenerin, DEL-1 (188); a stomatin-like protein, UNC-1 (157); and MEC-6 (D. Chelur & M. Chalfie, unpublished data). Moreover, *unc-1* and *unc-8* interact genetically to modulate sensitivity to volatile anesthetics (157), and *mec-6* mutation causes a slight locomotor defect (188) and greatly suppresses the uncoordination caused by a gain-of-function mutation in *unc-8* (175). These results suggest that a mechanosensory complex similar to that in the receptors for gentle touch functions in motor neurons.

DROSOPHILA SENSORY MECHANOTRANSDUCTION

The mechanosensory cells of Drosophila are classified into two groups depending on whether they have one (Type I) or more than one (Type II) sensory dendrite [reviewed in (103)] (Figure 2, see color insert). The Type I neurons have ciliated sensory endings and are associated with support cells that form small mechanosensory sensilla. Some sensilla (the external sensory organs) have external bristles or cuticular domes (the campaniform sensilla); others simply attach to apparently unspecialized cuticle through support cells (the chordotonal organs). A chordotonal organ is composed of a sensory unit called a scolopidium: a ciliated dendrite inserted into a scolopale support cell (see Figure 1, see color insert). Chordotonal organs in larvae skin have individual scolopidia, whereas the adult Johnston's organ, a structure between the second and third segments of the antenna, has hundreds of scolopidia (57). The Type II or multiple dendritic cells innervate the epidermis. It is likely that there are several classes of cells within the dendritically aborizing multidendritic neurons of Drosophila as four distinct morphological subclasses have been described (77). The ultrastructure of the dendritic endings of the aborizing multidendritic cells has not been examined, but the branched dendrites appear to be attached to the epidermis because they resist being pulled away from it (W.D. Tracey, personal communication). Consistent with their being attached, the endings move with the overlying epidermis in living larvae (66). More direct evidence that Type II neurons are mechanoreceptors comes from electrophysiological studies that show some of the homologous multidendritic neurons in the *Manduca* moth respond to mechanical stimulation (76).

Type I Touch Sensitivity

Removal of most external sensory sensilla by mutation of the *achaete-scute* genes produces larval touch insensitivity (the failure of larvae to move away from the touch of a hair) and adult uncoordination and lethality (108). These latter phenotypes suggest that external sensory mechanosensation is essential in the adult. Kernan et al. (108) used these phenotypes to identify candidate mechanosensory mutants. A screen for touch-insensitive larvae identified X-linked mutations in two previously identified genes, *uncoordinated* (*unc*) and *uncoordinated-like* (*uncl*), and in four new mapped genes, *touch insensitive larva A–D* (*tilA–tilD*). Additional mutations, some resulting in changes in bristle morphology, were also identified but not mapped. A separate screen for uncoordinated adults (with early onset lethality) identified ten mutations in seven complementation groups on the second chromosome. Subsequent mutageneses have identified other mechanosensation genes on the second chromosome (M. Kernan, D.M. Cowan & R. Walker, personal communication); these additional genes are listed in FlyBase (www.flybase.org) (33).

Many of the mutations giving this uncoordinated phenotype probably produce defects in the mechanosensory sensilla and not in downstream cells. For the *unc* gene, the focus of action in the bristles has been demonstrated in mosaic animals (108). In decapitated mosaic animals, stimulation of wild-type bristles on the dorsal abdomen elicited movement of the third leg, whereas stimulation of *unc* mutant bristles in nearby mosaic patches did not.

In other mutant strains, electrophysiological recording has indicated defects in the sensory sensilla. By placing an electrode on the cut end of a bristle, Kernan et al. (108) measured the resting transepithelial potential and the voltage that resulted when the bristle was displaced (called the mechanoreceptor potential or MRP). The bristles from *unc* and *uncl* adults, as well as those from *no mechanoreceptor receptor potential A–D* (*nompA–nompD*) mutants (these genes are on the second chromosome), had normal transepithelial potentials, but had greatly reduced or absent MRPs, suggesting that these genes are good candidates for encoding transduction components. The single *nompD* mutation eliminated the MRP except in bristles of the dorsal humerus. This result may indicate that these bristles require different gene products or that the mutation is hypomorphic. Mutation of one other second chromosome gene, *reduced mechanoreceptor potentials* (*remp*, now called *rempA*) had reduced transepithelial potentials and MRPs whereas the second chromosome genes, *uncoordinated with mechanoreceptor potentials A* and *B* (*umpA* and *umpB*) had normal mechanoreceptor potentials but were still uncoordinated. The presence of normal potentials generated in the sensilla in *umpA* and *umpB* mutants suggests these genes act in downstream cells that coordinate locomotion. The *tilB* mutant appears to affect one class of Type I mechanoreceptors because external sensory bristle mechanoreceptor potentials are normal, but the chordotonal organs do not function normally (see below).

Genes that are needed for the mechanosensory potential are important candidates for encoding components needed for mechanosensory transduction. The *unc*, *nompA*, *nompB*, and *nompC* genes have been characterized molecularly and appear to fulfill this promise. Two genes, *unc* and *nompB*, are needed for the proper production of the ciliary axoneme. *unc* encodes a novel coiled-coil protein that is associated with the basal body of the ciliary axoneme of sensory neurons in Type I sensilla and is needed for the correct production of these axonemes (and those of sperm) (M. Kernan, personal communication). *nompB* is the Drosophila homolog of the mouse Tg737/orpk, nematode *osm-5*, and Chlamydomonas IFT88 genes. NOMPB is a tetratricopeptide repeat protein that is part of a conserved intraflagellar transport complex required to construct and maintain cilia and flagella. *nompB* mutants have truncated sensory cilia, but unlike *unc* mutants have normal, functional sperm (Y.G. Han & M. Kernan, personal communication).

nompA encodes an extracellular protein that has a zona pellucida domain and several plasminogen N-terminal domains, suggesting that it is an extracellular matrix protein (29). The protein is expressed in the support cells of chordotonal and extracellular sensilla, but not in cells surrounding the multidendritic mechanosensory cells. In the Type I sensilla the ciliated nerve endings insert into dendritic caps. NOMPA is located in these caps in wild-type animals, but the caps are disrupted or fail to attach to the dendritic cilia in *nompA* mutants. Since many scolopidia have caps in *nompA* mutants, the protein could tether the mechanosensory channel to the cap and, thus, connect the dendrite to the cap, tether another component (or components) of the dendrite to the cap, or maintain the integrity of the cap. Whatever the mechanism, NOMPA is another example of an extracellular protein needed for normal mechanoreceptor function.

Perhaps most important is the *nompC* gene, which encodes a putative channel protein of the TRP family with a large number of ankyrin repeats (Figure 5) (200). When recordings are made from the cut end of a wild-type bristle in the voltage-clamp mode, displacement of the bristle produces a rapidly increasing current that soon decays to a plateau current that lasts for the duration of the displacement. In putative null *nompC* mutants, the peak current is absent, but the plateau remains. A hypomorphic mutation, *nompC*[4], causes the peak to adapt faster than in wild type.

The expression pattern for this gene has not been pinpointed to individual sensory cells, but in situ hybridization reveals that the gene is expressed in the bristle sensory organs and other mechanosensory organs. A similar gene, *Ce-nompC*, is expressed in the CEP and ADE sensory neurons of *C. elegans* (200). These cells have ciliated endings embedded in the cuticle (148) and appear to respond to the texture of the bacteria on which the nematodes are grown (171). Interestingly, the *Ce-nompC::gfp* construct, which contains the first four ankyrin repeats, was localized to the sensory ending of the CEP neurons. These results suggest that the ankyrin repeats are important for the localization, and thus function, of the protein. Confirmation of this localization of a full-length construct would be desirable. The *gfp* construct was also expressed in two interneurons, the DVA and DVC cells. This nonsensory expression could indicate a different role for the gene in these cells or could reflect an artifact because of the relatively large 5' sequence included in the construct. Localization using an antibody specific to the *C. elegans* protein should resolve this issue. Alternatively, these cells might have a previously unrecognized mechanosensory function. Wicks & Rankin (211) found that ablation of the DVA cells reduced the sensitivity of the plate-tapping response; perhaps DVA also transduces mechanical signals.

Is NOMPC a mechanotransduction channel in Drosophila? The evidence is tantalizing but incomplete. The localization of the protein in Drosophila to the mechanosensory organs is supportive, but not conclusive. Localization of the protein to the sensory neurons and to the ciliated endings in Drosophila remains to be established. The loss of the peak in the mechanoreceptor current and the concomitant loss of mechanosensory activity certainly indicate that the current provided by this channel is important for the function of the mechanoreceptors. We do not know, however, if this current (assuming that the channels are expressed in the sensory cells) is the sole transduction current. As discussed by Walker et al. (200), the residual current may be an independent mechanotransduction current. Alternatively, the residual current may be a transduction current that precedes the NOMPC-dependent current. In wild-type cells (200) the latency is very rapid (200 μs). If the initial transduction current depends on NOMPC, the latency of the

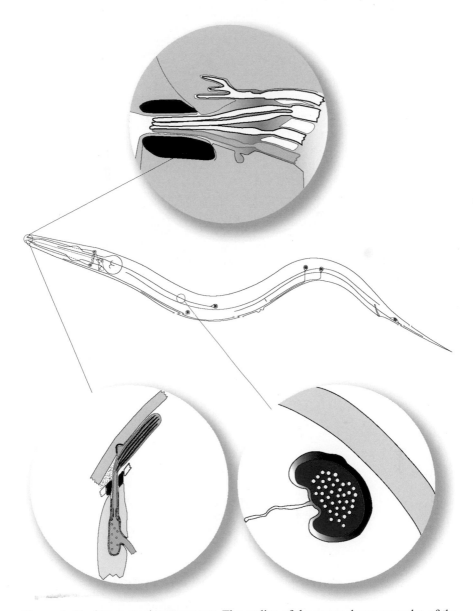

Figure 1 *C. elegans* mechanoreceptors. The outline of the worm shows examples of the ASH cells (green), which respond to nose touch, the receptors for gentle touch (red), the CEP cells (blue), which respond to texture, and the PVD cells (purple), which respond to harsh body touch. Enlargements show (*clockwise from the top*): the ciliated ending of an ASH cell in the amphid at the nose of the animal, a cross-section of an ALM receptor for gentle touch showing its prominent bundle of microtubules and extracellular matrix (dark gray), and the ciliated ending of a CEP neuron and its association with the cuticle (gray). Pictures have been drawn with reference to (22, 148).

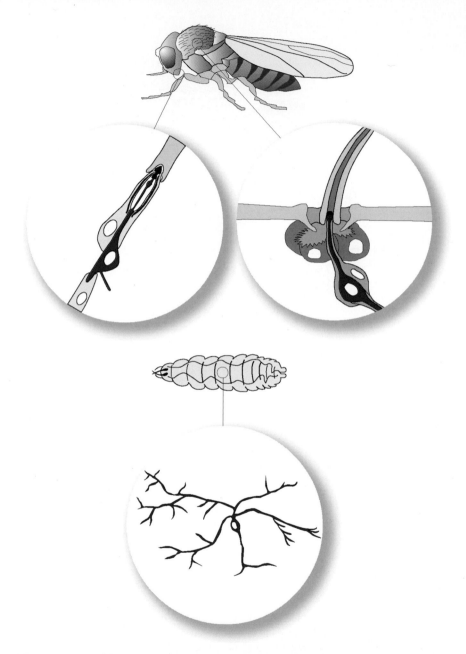

Figure 2 Drosophila mechanoreceptors. In the adult (*top*) Type I sensory organs include the chordotonal organ (*left*) and the external bristle. The sensory neuron is red; support cells are light gray. A Type II multidendritic cell (red) is diagramed from a larva (*bottom*). Pictures have been drawn adapted from (29, 103).

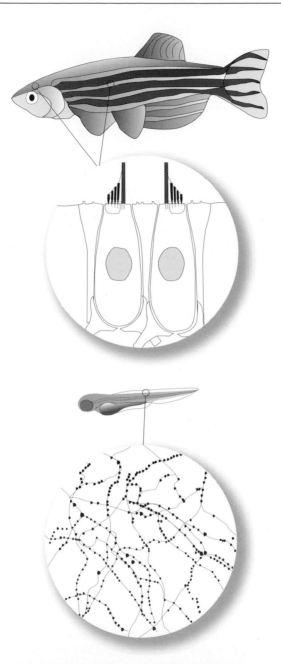

Figure 3 Fish mechanoreceptors. Hair cells in the adult (*top*) with the stereocilia indicated in red and the tall kinocilia in blue. Sensory processes from Rohon-Beard cell in the skin of larvae (*bottom*) are depicted in the lower enlargement. Pictures have been drawn with reference to Flock (60a) for hair cell and Svoboda et al. (Svoboda KR, Linares AE, Ribera AB. 2001. Activity Regulates programmed cell death of zebrafish Rohon-Beard neurons. *Development*. 128:3511–20) for Rohon-Beard cells.

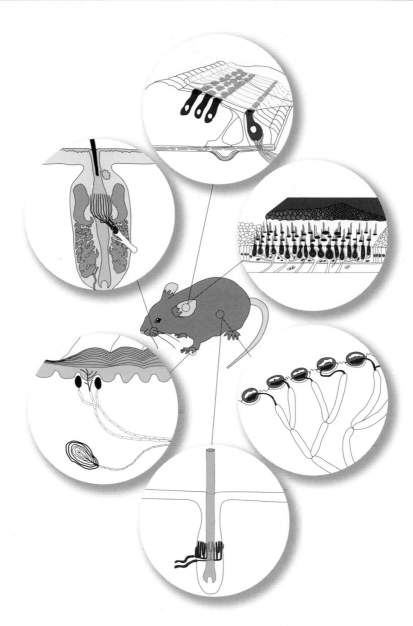

Figure 4 Mammalian mechanoreceptors. Enlargements (*clockwise from top*) show the inner (red) and outer (blue) hair cells of the organ of Corti; the hair cells (red) and the otolithic membrane (blue) from the sacculus; the Merkel cells (blue) and Merkel disks (red) in touch domes from hairy skin; the palisade endings (red) around a body hair follicle; Meissner-like corpuscles (red), Pacinian-like corpuscle (black), and intraepidermal ending (blue) in glabrous skin; and a lanceolatae ending (red) innervating a whisker. Pictures have been drawn with reference to (96a) for the organ of Corti, (123a) for the sacculus; (101) for Merkel cells, (72) for Palisade endings, (62) for the whisker mechanoreceptor, and (72a) for glabrous skin receptors..

residual current in the *nompC* mutants may be longer. This measurement has not been reported.

Johnston's Organ

Loss of chordotonal organs (including those of the Johnston's organs) by mutations in the *atonal* bHLH transcription factor gene results in animals that do not respond to sound. Using this phenotype, Eberl et al. (56) obtained 15 mutant lines in which the adults were insensitive to sound. Two of the mutations caused generalized neuronal degeneration. Of the remaining mutations, only one failed to produce sound-evoked potentials (57). This latter gene was named *beethoven*. The sound-evoked potential was also absent (*unc, uncl, nompA, B, D, F*, and *rempA*) or almost always absent (*nompE, I*, and *J*) in several of the mutants affected for bristle mechanosensation, demonstrating that many genes needed for the function of external sensory organs are also required for chordotonal mechanosensory organs (57). A strong *tilB* mutation also eliminated the sound-evoked potential even though it did not eliminate the MRP in the bristles. Ultrastructural analysis suggests that both *beethoven* and *tilB* might affect axonemal structure in the chordotonal sensory neurons.

Surprisingly, mutation of *nompC* results in a relatively weak defect in the chordotonal sensilla needed to detect sound. Although the strong *nompC* mutations eliminate the MRP (108), it only reduced the sound-evoked potential from the Johnston's organ by about 60% (57). These results suggest that other channel proteins contribute to the electrical response in Johnston's organ.

Type II Touch Sensitivity

nompC and the other cloned *nomp* genes are not expressed in the multidendritic cells, suggesting that different proteins may be needed for mechanosensation by Type II cells. Two groups have identified a DEG/ENaC-encoding gene [*pickpocket* (*ppk*) (1); *drosophila multidendritic neuron Na channel* (*dmdNaC1*) (41)] that appears to be exclusively expressed in some of the multidendritic cells. Using an antibody against PPK, Adams et al. (1) localized the protein to the cell bodies and dendritic arbors of the multidendritic cells. Neither group demonstrated electrical activity of this protein in frog oocytes, even with the equivalent of the *mec-4* degeneration-causing mutation. Adams et al., however, showed that the protein could coimmunoprecipitate with and interfere with the current produced by an activated mammalian DEG/ENaC protein, BNC1 (1). Although this protein was hypothesized to be needed for the mechanosensory response, *ppk* mutants have no detectable loss in mechanosensation (W. Johnson, personal communication). Some of the multidendritic neurons may function as nociceptors as other genes affecting the function of the multidendritic cells have been identified in screens for larvae that fail to respond to noxious heat and mechanical stimuli (W.D. Tracey & S. Benzer, personal communication). The analysis of these mutations should provide insights into the sensation of nociceptive stimuli.

ZEBRAFISH BALANCE AND TOUCH

Two types of mechanoreceptors in fish, hair cells and Rohon-Beard cells, have been affected by mutations (Figure 3, see color insert). The detection of sound and balance rely on specialized epithelial cells, the hair cells in the inner ears of vertebrates. In fish, additional hair cells project directly into the external environment as part of the lateral line organ. These cells sense waterborne mechanical stimuli originating near the animal (152). In contrast, the touch sensing cells in fish are neuronal cells with endings embedded in the skin. In larval fish and amphibians, the Rohon-Beard cells detect gentle, mechanical stimuli applied to the trunk (31, 159, 162). The Rohon-Beard cells eventually degenerate and are replaced by dorsal root ganglion sensory cells in the adult.

The structure and function of hair cells have been studied in a wide variety of vertebrates. Although hair cells display some species-specific differences, they share a similar mechanoelectrical transduction mechanism (54, 60). Transduction in hair cells occurs in the apical, actin filament–containing extensions of the main cell body—the stereocilia (61, 173, 191). In addition to attachments near their base, each stereocilium is tethered to an adjacent taller stereocilium by a thin extracellular filament, the tip-link (150). The kinocilium, a true cilium, is typically found toward the tall end of the hair cell bundle. The kinocilium is often attached to an overlying membrane, and thus mechanically couples the stereociliary bundle to motion in the surrounding medium. Transduction molecules probably do not act in kinocilia, since the generation of normal mechanotransduction potentials does not require kinocilia displacement in vitro (99). Displacement of the hair cell bundle toward the kinocilium leads to an inward flow of a current through a nonselective cation channel. This current in vivo is carried primarily by potassium and perhaps in part by calcium (34, 98, 129). The mechanoreceptor current in vestibular hair cells is extraordinarily fast ($\sim$10 μs at 37°C) and sensitive (mammalian hair cells can respond to $\sim$1 nm displacements) (35, 98, 166). In addition, these cells respond in a direction-dependent manner (98). Moving the bundle toward the tallest stereocilium in the staircase array of stereocilia enhances the transduction current, whereas moving it toward the shortest stereocilium decreases the current. According to the gating-spring model, changing tension in an elastic element—perhaps the tip-link—gates tethered transduction channels (36, 151). Forward movement stretches the tip-links and therefore pulls the transduction gate open, whereas backward movement slackens the tip-links, allowing the channel gate to close. Although the manner of tip-link function in this model has been challenged (139), the tip-links are essential for hair cell mechanotransduction (6).

Adaptation, the reduced response of sensory cells to sustained stimulation, occurs in hair cells when the stereociliary bundle is displaced for prolonged periods of time [reviewed in (54)]. One way hair cells adapt to a continuous displacement is a reduction in transduction channel activity (55). Both rapid and slow adaptation of mechanotransduction has been observed. Rapid adaptation (within a few milliseconds) is a process whereby calcium entering the transduction channel acts on or near the channel to inhibit its activity (37, 93). Slow adaptation (occurring in tens of milliseconds) depends on myosin-actin interactions (5, 89, 92). This adaptation is explained by myosin-mediated adjustment of the tension of the gating spring linked to the transduction channel: During a prolonged stimulus, the tension-generating myosin motor slips along the actin core, dragging the tip-link attachment and the associated channel toward the base of the bundle. This movement reduces the tension and resets the responsiveness of the channel to displacement.

The larval Rohon-Beard cells, as described in Xenopus (162), have unmyelinated nerve endings with varicosities that are embedded in the skin. These relatively unbranched neurites, from cell bodies close to but not part of the developing spinal cord, innervate the skin of the larvae. The varicose endings, embedded in invaginated pits in the superficial layers of skin, are often found with many mitochondria, tubules, and vesicles (162). Gentle stroking of the skin with a hair triggers spikes in these otherwise quiescent cells (31). The spiking activity rapidly adapts during prolonged skin displacements, suggesting that these cells sense transient mechanical stimuli and are insensitive to pressure.

Circler Mutants

Two phenotypes have been used to identify mutant larvae with potential mechanosensory defects in the zebrafish, *Danio rerio* (75). One phenotype is a movement abnormality in which the fish swim in circles (the circler phenotype). Since upright swimming requires an intact vestibular system in the inner ear, a circular swimming pattern could indicate abnormalities in the hair cells of this sensory organ. The other phenotype is an insensitivity to the touch of a probe to the tail in animals that have otherwise normal movement. Because they move normally, these mutants are presumably defective in the Rohon-Beard cells of the embryos, the dorsal root ganglion neurons of the larvae, or interneurons that relay their signals.

The circler mutants have normal touch sensitivity, and the tactile mutants swim in an upright position and do not circle. That these other responses are intact suggests a defect at or near the sensory ending, not in more centrally located neurons. Furthermore, although the screens are not completely saturated, these results suggest that different sets of genes are needed for these different mechanosensory responses.

Seventeen mutant lines were initially recovered that had the circler phenotype without gross defects in the inner ear (75). [Mutations in at least 20 additional complementation groups produce defects in the development of the inner ear; mutation of some of these genes causes circling (210).] Two mutations were alleles of the *nevermind* gene; both resulted in abnormal retinotectal projections. One mutation caused subtle developmental defects in the epithelial columns, which join to create the semicircular canals (T. Nicolson, personal communication). The remaining 14 mutations were subsequently analyzed and were shown to constitute 8 complementation groups (144). Mutations in these genes do not produce gross defects in the inner ear or lateral line or in neuronal projections. Additional mutageneses have enlarged this mutant collection, resulting in the identification of at least 14 new circler genes (T. Nicolson, personal communication). Of the 22 genes, 10 have multiple alleles and in some cases, several genes have 6 or more alleles.

Many of the circler mutants have defects in the function of the hair cells of the inner ear and/or the lateral line. Analysis of the circler mutants has suggested further subdivision of the genes, particularly on the basis of whether animals respond normally to an acoustic stimulus, hair cell morphology is normal, and lateral line hair cells produce extracellular voltage changes upon mechanical stimulation (microphonic potentials). Zebrafish larvae carrying mutations in two genes, *sputnik* and *mariner*, have splayed hair cell bundles, do not respond to acoustic stimuli, and have no microphonic potentials. These mutants point to the importance of hair cell bundle integrity for their function. The *mariner* gene encodes Myosin VIIA (58). Mutation of orthologous genes in mouse (see below) and human [Usher syndrome type 1B (206)] similarly affect hair cell bundle morphology and result in vestibular and hearing defects. The *sputnik* gene encodes Cadherin 23, an extracellular adhesion protein that appears to be needed for the proper formation of the tip-links, which are abnormal in the mutants (T. Nicolson, personal communication). As indicated below, mutant *cadherin 23* is also the basis of deafness in mouse (*waltzer*) and humans [Usher syndrome type 1D (16)].

Mutations in two other genes—*mercury* and *orbiter*—also fail to produce microphonic potentials. Since these latter mutants have morphologically normal hair cell bundles, these genes are good candidates for components of the mechanosensing apparatus. These genes have not been identified molecularly. The two mutant alleles of a third gene, *gemini*, which encodes an L-type calcium channel protein, result in a reduction of the microphonic potential (144; T. Nicolson, personal communication). The styryl dye FM1-43 is thought to enter mouse hair cells through the transduction channel (65). Since lateral line hair cells in *gemini* mutants display normal dye uptake (172), this calcium channel is probably not directly involved in transduction. Similarly, the products of *astronaut* and *cosmonaut*, genes represented by single mutant alleles that result in circlers with normal microphonic potentials, are not required for sensory transduction. The importance of an additional circler gene, *skylab*, is difficult to assess, because the two alleles (one recessive, the other semidominant) result in the degeneration of the hair cells (144; T. Nicolson, personal communication). Semidominant and recessive gain-of-function mutations in the *C. elegans* degenerin gene *deg-1* also cause neurodegeneration, whereas loss-of-function mutations do not produce any detectable phenotype (26, 71). If the *skylab* gene acts similarly, it may be a candidate for a component of the mechanosensory apparatus.

Touch Mutants

Eight mutations in six genes produce the touch-insensitive phenotype (75). Mutants defective in three of the touch-sensitivity genes, *alligator*, *macho*, and *steifftier* have been examined more extensively by Ribera & Nüsslein-Volhard (159). In wild-type animals larvae become responsive to touch at 27 h after fertilization. A couple of hours before this time, the Rohon-Beard cells display action potentials with small overshoots. Larger action potentials are seen with the advent of touch sensitivity. At this time, the cells show an increase in the voltage-dependent potassium current (I_{Kv}) and the voltage-activated sodium current (I_{Na}). In *macho* animals, the large overshoot action potentials do not appear and the increase in I_{Na} is not seen. The change in I_{Kv}, however, is seen, and the resting potential is normal. The overshoot of the action potentials is reduced but not absent in *alligator* and *steifftier* mutants. In contrast, *touch-down* mutants, which show a delayed onset of touch sensitivity and have melanophore defects (107), have normal action potentials. Thus, the mutants examined so far appear to be needed for the development of the cells or for the electrical properties of the cells at a step following mechanosensory transduction. Mutations in the *crocodile* and *schlaffi* genes cause late-onset touch insensitivity, and have little (*crocodile*) or no (*schlaffi*) effect on Rohon-Beard membrane properties. These results may indicate that these genes are more important for the function of the dorsal root ganglion neurons than the Rohon-Beard cells.

MOUSE HEARING, BALANCE, AND TOUCH

Mammals have a wide variety of mechanosensory cells that transduce external or internal mechanical stimuli (Figure 4, see color insert). Mechanoreceptors that transduce external signals include touch receptors in the skin and hair cells of the inner ear. Internal mechanoreceptors sense blood vessel and internal organ stretch while mechanoreceptors innervating muscle tendon organs and joints provide the position sense, or proprioception, of the limbs. Here we review recent advances made with transmission and molecular genetic approaches toward our understanding of the composition and function of inner ear mechanoreceptors and cutaneous mechanoreceptors.

Mechanosensory Genes of the Inner Ear

Different sets of hair cells in the inner ear transduce mechanical stimuli originating from sound, linear head acceleration, or angular head acceleration (97). Sound is sensed by hair cells in the organ of Corti in the snail-shaped cochlea of the inner ear of mammals. The organ of Corti contains inner hair cells, outer hair cells, the overlying tectorial membrane, and the underlying basilar membrane. Based on their predominant sensory afferent innervation, inner hair cells appear to be the primary sensory cells (179). The electromotile outer hair cells (169), which are sparsely innervated by afferents and highly innervated by efferent neurons, appear to regulate inner hair cell sensitivity (40). The basilar membrane vibrates in response to sound-generated pressure waves in the fluid-filled compartments of the cochlea, and the hair cell bundles riding on top of this membrane approach the tectorial membrane as the basilar membrane rises. The shearing forces that result displace the bundle, leading to the generation of mechanoreceptor current in the hair cells.

The utricule, saccule, and semicircular canals are separate compartments of the inner ear that sense horizontal, vertical, and angular acceleration of the head, respectively. The kinocilia of the hair cell bundles are physically attached to an overlying gelatinous membrane (35). As the head moves, extracellular fluid deflects these membranes, in turn displacing the kinocilia and associated stereociliary bundle; bundle displacement produces an inward mechanosensory current. The gelatinous otolithic membrane on the top of the hair cells of the utricule and saccule contain otoconia, particles comprised of calcium carbonate crystals and protein (187, 202), which add to the mechanical load of the hair cells. In contrast, the thick, gelatinous membrane of the hair cells of the semicircular canals, which is known as the cupula, lacks otoconia.

An ever-increasing number of genes required for hearing and balance are being identified. These mouse genes, which can affect the development and function of hair cells and other components of the vestibular and auditory systems, are listed at www.ihr.mrc.ac.uk/hereditary/MutantsTable.shtml and www.jax.org/research/hhim/documents/models.html. Steel & Kros have recently compared genes needed for mouse and human hearing (181). In this review, we focus mainly on those genes needed for the function, not development, of the hair cells.

Mutations resulting in deafness and/or imbalance have arisen spontaneously or from systematic mutageneses (94, 145). In addition, some genes have been targeted for knockout because mutation of similar human genes causes deafness [e.g. (59, 138)]. Deaf mice lack a startle response (the folding back of ear pinna or jumping) to sound, and mice with balance defects constantly circle, waddle, and/or toss their heads.

NON-HAIR CELL MUTATIONS   Mutations affecting cells other than the hair cells are important because they enable researchers to gain insights into the function of difficult to access components of the inner ear. Some mutations affect the organization of the regulatory membranes; others affect the constitution of the K^+-rich extracellular fluid (the endolymph) that bathes the hair cells.

Mutation of the *otogelin* gene (mutated in twister) causes both hearing and vestibular defects and disrupts the organization of the tectorial and otolithic membranes and the cupula (176, 177). The head tilt, tilted, and tilted-head mutants, which are defective in three different genes, have vestibular defects and lack otoconia (13, 124, 147). Targeted deletion of α-tectorin, a component of the tectorial membrane, causes a malformed membrane that results in inner hair cells tuned to particular frequencies but that lack the active feedback to enable them to respond to lower intensity sounds (122).

Potassium secretion is needed to form the endolymph. When endolymph is not made, the Reissner's membrane (the membrane that partially bounds the hair cell compartment) collapses onto the organ of Corti. Loss of isk (a.k.a. minK), an auxiliary channel subunit of the KvLQT1 K^+ channel, decreases potassium secretion and collapses the Reissner's membrane (195). Mutation of the genes for the KvLQT1 (KCNQ1) channel and the Na-K-Cl co-transporter (encoded by *Slc12a2*, one of the genes mutated in shaker-with-syndactylism animals) produces a similar phenotype (46, 120). Disruption of *Kcc4*, which encodes a K-Cl cotransporter, results in deaf mice whose outer hair cells degenerate (15). Because the Reissner's membrane does not collapse in *Kcc4* mutants, this transporter is not essential for endolymph production. The localization of the Kcc4 protein to basolateral support cells suggests that it clears K^+ from the basolateral side of the outer hair cells.

HAIR CELL MUTATIONS

Extracellular matrix   Mutations disrupting the function of the hair cells affect extracellular, cytoskeletal, and membrane proteins. Mutations in two genes for extracellular cell adhesion molecules cause deafness and circling. The hair bundles in these mutants become disorganized and gradually degenerate. One gene, *Cdh23*, encoding Cadherin-23, is mutated in waltzer mice (44). These mutant mice show no electrical activity in the downstream auditory brainstem neurons prior to hair cell degeneration, suggesting that they are functionally compromised before hair cell structure is disrupted. Mutations in this gene lead to hair cell dysfunction in zebrafish and humans as well. The second gene, *Pcdh15*, encoding Protocadherin-15, is mutated in Ames waltzer mice (2). An attractive hypothesis for the function of these proteins is that they are components of the tip-links that join the hairs of the hair cells, and that the splaying of the hair cell bundles occurs because of the loss of these connections (2, 44). Direct proof for this hypothesis, however, is lacking. Antibody localization of these proteins to the tip-links remains to be done.

Cytoskeleton   The stereocilia are filled with actin filaments cross-linked, in part, with the protein fimbrin (192). As a consequence of actin cross-linking, the stereocilia are rigid structures that pivot at their base (38). Positional cloning of the gene mutated in jerker mice, which are defective in hearing and balance, identified an actin-bundling protein, epsin (217). Unlike other actin-bundling proteins, epsin lacks calcium-dependent inhibition, a possibly important property since calcium flows through the transduction channel (128, 217).

The actin bundles also participate in adaptation, in which movement of the transduction channel through connections to the actin filaments is thought to alter the tension of the gating spring [reviewed in (54)]. The role of one myosin, Myosin 1c, in this process has been tested in an elegant genetic experiment: Holt et al. (89) mutated the mouse gene for Myosin 1c (changing the tyrosine at position 61 to a glycine) so that it could uniquely bind and be inhibited by N^6 (2-methyl butyl) ADP. Expression of the mutant gene in utricular hair cells of transgenic mice had a dominant negative phenotype, but only in the presence of the modified ADP. Adaptation, but not transduction, was blocked in the presence of the analogue; neither was blocked in its absence.

Myosin 1c is not the only myosin needed for hair cell function. Mutation of three unconventional myosin genes [*Myosin 7A* (mutant in shaker-1 mice), *Myosin 15* (mutant in shaker-2 mice), and *Myosin 6* (mutant in Snell's waltzer mice)] disrupt the structural integrity, and therefore the function, of the hair cells (7, 73, 156). shaker-2 mutants have unusually small stereocilia; thus, this myosin may function in the maturation of the stereocilia (156). Because unconventional myosins may bind to molecules other than actin, the function of these proteins is unclear. In two-hybrid screens, the tail of the human Myosin 7A interacted with a novel cell adhesion protein, vezatin. Vezatin may provide support at the base of the stereocilia by interacting with one of the cadherins (116). Myosin 7A may contribute to adaptation. Although greater displacement is needed to open the transduction channels in shaker-1 mutants, the channels adapt more quickly and more completely than in wild-type animals. Unlike Myosin 1c, which is restricted to each end of the tip-links (67), Myosin 7A is expressed throughout the stereocilia (84). Myosin 7A, by maintaining membrane attachments to the actin core, may slow the movement of these tethered components during adaptation (112).

Membrane proteins   Sensory transduction and signaling in the hair cells use calcium. Calcium enters the transduction channels (43, 128), but because stereocilia lack internal compartments to sequester calcium, it must be extruded through the large number of plasma membrane calcium ATPases (PMCAs) that produce a substantial outward calcium flux near the tips of the stereocilia (215). The production of a local rise in external calcium concentration may feed back on the calcium permeable transduction channel to regulate its sensitivity. At the basolateral membrane, calcium regulates synaptic vesicle release and the calcium-dependent potassium channels needed for the tuning of auditory hair cells. The knockout of the gene for the plasma membrane calcium ATPase type 2 (PMCA2) (111) or its mutation in deafwaddler mice (182) results in auditory and vestibular defects. PMCA protein, identified by an antibody recognizing all four PMCA isotypes, is greatly reduced in deafwaddler mice in both apical and basal portions of auditory and vestibular hair cells (182). The localization of splice variant products of PMCA2 has not been reported in mice. However, one splice variant, PMCA2a, is restricted to the apical portions of the hair cells in the bullfrog sacculus (53). Thus, PMCA pumps may play different roles in signal transduction and processing depending on where they are expressed in the hair cells. Electrophysiological analysis of mutants that target individual PMCA genes may help to identify how different forms contribute to hair cell function.

Another protein, a small, single-pass membrane protein encoded by the *Transmembrane inner ear (Tmie)* gene, is mutant in spinner mice (140). The recessive spinner mutations lead to vestibular and deafness defects and to the eventual degeneration of hair cells. Prior to degeneration the kinocilia of some inner hair cells are abnormally retained whereas some stereocilia of outer hair cells are shortened or missing. Thus, *Tmie* appears important for the maturation of hair bundles. *Tmie* transcripts have been localized to the inner ear, but the cellular site of expression is not known.

At the moment, mouse mutations have not identified the hair cell transduction channel. The recent identification of the *Transmembrane cochlear protein 1* (*Tmc1*) gene, first identified as the basis of a dominantly inherited deafness in humans (115), however, provides an intriguing candidate. The *Tmc1* gene encodes a previously uncharacterized membrane protein with from 6 to 11 predicted membrane-spanning regions. In mice the *Tmc1* gene is mutated in the deafness and Beethoven mutants (115, 197). The recessive deafness mutation affects cochlear hair cell function but the hair bundles develop normally (14). Beethoven, a semidominant deafness mutation, causes late-onset degeneration primarily of the inner hair cells of the organ of Corti; the mutants do not have a vestibular phenotype (197). The *Tmc1* message is expressed in inner and outer hair cells of the inner ear (197). A clearer understanding of this protein will come with more detailed protein localization, in situ physiological studies, and heterologous studies.

Other genes   Several other genes affect the development and function of the hair cells; many, but not all, of these mutations are associated with late-onset degeneration of the hair cells (180). Some genes encode transcription factors that regulate late aspects of hair cell development, e.g., the POU homeobox gene *Pou4f3* (*Brn-3C*) (212, 213) and the thyroid hormone receptor β receptor gene *Thrb* (165). Other genes encode cytoplasmic proteins that are important for hair cell function, e.g., the AP-3 adaptor protein known to be involved in endocytosis (mutated in mocha mice) and the endopeptidase-like product of the *Pex* gene that is mutated in Gyro mice and is needed for normal renal phosphate metabolism (104, 119, 130, 183, 185). Although these genes are important for hair cell function, they are unlikely to contribute directly to transduction.

Mechanosensory Genes of the Skin

Neurons innervating the skin have cell bodies in the dorsal root ganglia. Different cutaneous sensory endings, characterized by their morphology (100), are responsible for transducing mechanical, thermal, and painful stimuli. Most of the ultrastructural studies have characterized skin mechanosensory terminals in mammals other than mice, but similar endings have been seen in mice in the light microscope. Such mechanosensory endings include palisade hair follicle endings, neural Merkel disks and the closely associated nonneural Merkel cells, Meissner-like corpuscles, Pacinian-like corpuscles, and branching intraepidermal endings. Hair follicles are associated with nerve endings (palisade endings) that ascend the follicle, while whiskers have thick nerve endings (lanceolate endings) associated with the finger-like projections of Schwann cells. Hair follicle movement can stimulate these nerve endings (17). Neural Merkel disks and the closely associated nonneural Merkel cells are also found near hair follicles, but do not respond to hair follicle movement directly. These complexes respond in hairy and nonhairy (glabrous) skin to local skin indentation (101). In other regions of hairy skin, Merkel cell/neurite complexes are concentrated in touch-domes, or small areas of raised skin (101). Merkel cells virtually envelop the nerve terminal, and they contain putative transmitter vesicles. Researchers disagree on whether the Merkel cells directly transduce mechanical stimuli, modulate sensitivity, or serve a developmental, rather than functional role [see e.g., (45, 102, 109, 146, 189)]. Meissner-like corpuscles are sensory endings in glabrous skin. The otherwise myelinated fiber innervating the corpuscle is ensheathed at its unmyelinated ending by collagenous lamellae (20). As has been demonstrated for the similar Pacinian corpuscle (127), the encapsulating structure of Meissner-like endings probably functions to allow the unmyelinated nerve ending to preferentially detect vibration. Other myelinated, mechanically sensitive endings in glabrous skin include intraepidermal nerve endings that tend to respond only to high threshold stimuli and are not associated with elaborate auxiliary structures.

Sensory nerve fibers leading away from these terminals have been characterized electrophysiologically by how fast they conduct action potentials and how quickly they adapt. In the mouse, fast conducting fibers, which are large-diameter myelinated fibers, can be either slowly adapting (SA) or rapidly adapting (RA) (110). RA fibers signal at the onset and offset of a skin displacement and are therefore associated with transmitting vibratory stimuli, or transient motion of hair follicles. SA fibers signal throughout skin displacement and preferentially encode texture and pressure. In mice, more moderately conducting (D-hair) fibers are very sensitive to small forces and are associated with stimulation of the down hair (110). Other moderately conducting (AM) fibers are much less sensitive and require large forces to evoke a response (110). Both of these moderately conducting fibers are myelinated but have smaller diameters than RA and SA fibers. The slowest conducting fibers are thin, unmyelinated fibers that primarily conduct signals produced by painful and/or thermal stimuli.

Most studies associating nerve terminals with specific nerve fibers use cats and primates, not mice. With this caveat in mind, SA fibers end with Merkel disks (101), RA fibers end with palisade endings (141) and Meissner-like corpuscles (142), and high-threshold AM fibers are likely to have the sprouting, intraepidermal nerve endings (113). The endings associated with D-hair fibers are uncertain.

As with other sensory mechanotransduction systems, the molecular identity of the transducer proteins in cutaneous touch receptors remains uncertain. Although no large-scale genetic screens for touch-insensitive mice have been reported, transgenic mice have been used to test whether DEG/ENaC channels contribute to touch sensitivity. Two genes that encode DEG/ENaC proteins—BNC1 (also called BNaC1, ASIC2, and MDEG1) and DRASIC (also called ASIC3)—have been of particular interest because they are expressed in many dorsal root ganglion neurons in an overlapping pattern (68, 155, 198, 199).

Both wild-type proteins produce tiny or no amiloride-sensitive current in Xenopus oocytes at physiological pH (155), but are activated by acid. BNC1 is maximally activated at a pH near 4.0 (27). DRASIC shows a biphasic response to acid; lowering the pH to 6.5 produces a rapid transient current, whereas lowering the pH to 4.0 produces an additional longer-lasting current (198). DRASIC and BNC1a (one of two isoforms) form a heteromeric channel with different kinetic properties

(8). DRASIC and BNC1b (the other isoform) form a channel with altered ion selectivity that differs from the channels produced when each protein is expressed alone (125).

Both BNC1 and DRASIC are expressed in many, if not all, of the mechanosensory endings in skin. García-Añoveros et al. (72) found that BNC1a (BNaC1α) is transported to most, if not all, cutaneous mechanoreceptor endings in glabrous and hairy skin (72). BNC1a was not detected in the smaller diameter nociceptive fibers. Using in situ hybridization, Price et al. (153) found that mRNA for BNC1b was expressed at higher levels and in most of the large-diameter (mechanosensory) fibers from the dorsal root ganglia. They also found protein in the nerve endings that surround hair follicles using an isoform-nonspecific antibody. DRASIC is also localized to most, if not all, mechanoreceptor nerve terminals (although localization in Merkel cells and Merkel disks could not be distinguished) (154).

Although BNC1 and DRASIC are expressed in many types of dorsal root sensory neurons, gene disruption selectively alters, but does not abolish, touch sensitivity in a few types of mechanoreceptors. RA and SA fiber touch sensitivity, assessed by analyzing the action potential frequency as a function of skin displacement in in vitro skin-nerve preparations (110), is reduced in BNC1 knockouts, whereas D-hair and AM fibers are unaffected (153). The response of cultured neurons to current injection is normal, suggesting that transduction, and not membrane or conduction properties, is affected (153). DRASIC knockouts show an increase in RA fiber sensitivity and a decrease in SA and AM fiber sensitivity (154). The different cellular phenotypes in the two knockout strains may reflect different channel properties that result when only one of the two proteins is present. Consistent with the presence of a heteromeric channel in mechanosensory neurons is the alteration of proton-gated current kinetics in cultured, large diameter dorsal root ganglion neurons from DRASIC knockout mice (214). Double mutants, therefore, may produce stronger phenotypes.

Expression of *C. elegans* MEC-2 greatly enhances MEC-4/MEC-10 channel activity in frog oocytes (74). A much smaller, but significant, enhancement is produced when the stomatin-like central portion of MEC-2 or human stomatin is used. Since stomatin is expressed in mammalian dorsal root sensory cells and the lanceolate endings innervating whiskers (62, 133), it, or another homologue, may be important for mechanosensation. A stomatin knockout mouse has been produced (218), but its mechanosensory properties have not been reported. The examination of this mouse and of mice that are also mutated in the BNC1 and DRASIC genes may help determine whether these proteins are required for mechanosensory function.

SUMMARY AND FUTURE DIRECTIONS

Despite the considerable work to date, the genetic analysis of mechanosensory receptors is a very young field. Genetic screens are incomplete, and the molecular identity of most of the genes has not been determined. Nonetheless, the analysis of mechanosensory defective mutants has identified a variety of proteins needed for the development and function of several different types of mechanoreceptors.

Although incomplete, these genetic studies suggest two broad conclusions. First, sensory mechanotransduction appears to be mediated in several different ways. In worms, flies, and zebrafish, different sets of genes appear to be needed for the function of different mechanosensory receptors. Mutations in members of a given set of genes affect one mechanosensory modality but not others. Since the nature of the mechanosensing complex is not known for any cell, we are still left with the possibility that a more generally expressed component that is modified by special factors underlies mechanosensory transduction. Second, the molecular identities of the genes investigated so far support the importance of membrane components (especially channel proteins), the cytoskeleton, extracellular matrix, and extracellular milieu in the function of many mechanosensory receptors. These studies support a model in which a mechanosensory complex involving a channel coupled to extracellular and intracellular proteins is at the heart of mechanosensory transduction. In many cases, however, the role of these components is not clear. One goal for the future is to establish how many of these components are directly involved in sensory mechanotransduction.

Perhaps the most important result from genetic studies in several animals has been the identification of several candidates for components of the mechanosensory channels. The two best-studied candidates are members of the DEG/ENaC family and of the TRP family (Figure 5). The evidence that either is the transduction channel is incomplete. One of the greatest challenges facing the field is the establishment of proof that any or all of these candidates are needed for mechanosensory transduction.

DEG/ENaC proteins have been implicated in mechanosensation in worms and mammals, but their roles are not fully understood. The *C. elegans* DEG/ENaC proteins MEC-4 and MEC-10, for example, form amiloride-sensitive channels that are expressed in mechanosensory cells and are needed for mechanosensitivity. Neither the requirement for these channels for the in vivo mechanosensory current, nor the mechanical gating of heterologously expressed channels has been shown. Another *C. elegans* degenerin, UNC-8, is thought to be involved in proprioception, but its properties have been even less well studied. The mammalian BNC1 and DRASIC proteins are expressed in cutaneous mechanoreceptors, form channels, and are needed for the normal touch responsiveness in skin. Touch sensitivity is not abolished in BNC1 and DRASIC mutants, however, and these proteins are not known to be mechanosensitive. Moreover, mechanosensory currents or potentials have not been measured in the mechanosensory endings of wild-type and mutant mice. Finally, in addition to their expression in kidney, lung, and intestinal epithelia, the ENaC channel subunits are expressed in several sites of mechanosensation: rat paw (50), trigeminal sensory neurons (62), and baroreceptors (51). The function of the ENaC subunits in these cells, however, is not known.

Several TRP-like channel proteins have been implicated in mechanosensation. One, the Drosophila NOMP-C protein, is required for a large component of the mechanosensory current in the mechanosensory bristles. Although this protein is needed for most of the mechanosensory current, it has not been shown to be gated mechanically. The *C. elegans* homolog of *nompC* is expressed in mechanoreceptors that detect texture, but a requirement for this gene in these cells has not been demonstrated. Two *C. elegans* TRP-like genes, *osm-9* and *ocr-2*, are needed for detection of nose touch and other sensory functions of the ASH neurons. Their channel properties have not been investigated. Finally, the *Trpv4* gene encodes a mammalian TRP-like protein, TRPV4 (previously named VR-OAC), that is expressed in hair cells and is stimulated by hypoosmotic swelling when expressed in heterologous cells (123); the contribution of VR-OAC to hair cell function awaits analysis of the knockout mice.

Future experiments will determine whether any other members of these gene families are also involved in mechanosensory transduction. In addition, other unrelated channel proteins may have roles in the response of sensory cells to mechanical stimuli. One intriguing candidate is the product of the *Tmc1* gene, which is needed for hearing in mice and humans. The analysis of the properties of this protein, and of the cells that express it should indicate whether it might be a mechanosensory transducer. The analysis of *Tmc1* homologues, which are found in *C. elegans*, will also help elucidate the function of this protein.

Another goal for the future is to establish whether these channel proteins as well as associated proteins directly transduce mechanical signals or provide permissive conditions for mechanotransduction. The parallel approaches of patch-clamp recording in mechanoreceptors of genetic mutants and reconstitution experiments in heterologous cells are needed to establish the roles of these proteins.

Although the genetic analysis of mechanosensory processes is incomplete, the near future should bring the discovery and molecular characterization of many more genes needed for the mechanical senses. The discovery of additional players cannot be accomplished using classic genetics alone; classic genetic screens for mechanosensory defective mutants are useful for obtaining candidate genes, but they have inherent limitations. A major problem is the inability to easily identify redundant genes, pleiotropic genes, genes that subtly modify mechanosensory behaviors, and genes that down-regulate mechanosensory activity (i.e., genes whose loss produces animals that are supersensitive to mechanical stimuli). Other methods, such as yeast two-hybrid screens, coimmunoprecipitation and mass spectroscopy analysis of protein complexes, and DNA microarray experiments, will also add greatly to our understanding of mechanosensory receptors. A recent experiment may serve as an example of what may be expected in the near future: Microarray experiments identified 71 candidate genes that required *mec-3* for expression in the *C. elegans* receptors for gentle touch (216). These genes included most of the known *mec* genes (the epidermally expressed *mec-5* gene and the ubiquitously expressed *mec-6* gene were not included) and many more genes that had not been identified by the genetic screens that appeared to be at or near saturation. Presumably, many more genes required in mechanosensing cells will be discovered by similar methods; but as this example demonstrates, they may not identify all the genes needed for mechanosensation. In addition, these new methods, for example the combination of DNA microarrays and RNA interference, will allow the study of the molecular basis of mechanosensation in organisms in which classical genetic studies have not been done. The Rohon-Beard cells of Xenopus, for example, could be studied by these methods.

A combination of biochemical, molecular, electrophysiological, and genetic approaches will be needed to understand the molecular basis of mechanosensory transduction. As has already happened with the study of the DEG/ENaC genes, TRP genes, and stomatin genes, discoveries made with one organism will influence work with others. The molecular genetic study of mechanosensation in multiple organisms will also continue to provide insights and tools, both experimental and theoretical, needed to unravel the puzzle of metazoan mechanosensation.

ACKNOWLEDGMENTS

We thank our colleagues for helpful discussions and sharing unpublished data. We especially thank Bob O'Hagan, Teresa Nicolson, Maurice Kernan, Dan Tracey, and Peter Gillespie for comments on the manuscript and Carolyn Crowley for preparation of the figures. Research from our lab is supported by NIH grant GM30997 to M.C.

The *Annual Review of Genetics* is online at http://genet.annualreviews.org

LITERATURE CITED

1. Adams CM, Anderson MG, Motto DG, Price MP, Johnson WA, Welsh MJ. 1998. Ripped pocket and pickpocket, novel Drosophila DEG/ENaC subunits expressed in early development and in mechanosensory neurons. *J. Cell Biol.* 140:143–52

2. Alagramam KN, Murcia CL, Kwon HY, Pawlowski KS, Wright CG, Woychik RP. 2001. The mouse Ames waltzer hearing-loss mutant is caused by mutation of *Pcdh15*, a novel protocadherin gene. *Nat. Genet.* 27:99–102

3. Alenghat FJ, Ingber DE. 2002. Mechanotransduction: all signals point to cytoskeleton, matrix, and integrins. *Sci. Signal Transduct. Knowl. Environ.* 2002: PE6

4. Alvarez de la Rosa D, Canessa CM, Fyfe GK, Zhang P. 2000. Structure and regulation of amiloride-sensitive sodium channels. *Annu. Rev. Physiol.* 62:573–94

5. Assad JA, Corey DP. 1992. An active motor model for adaptation by vertebrate hair cells. *J. Neurosci.* 12:3291–309

6. Assad JA, Shepherd GM, Corey DP. 1991. Tip-link integrity and mechanical transduction in vertebrate hair cells. *Neuron* 7:985–94

7. Avraham KB, Hasson T, Steel KP, Kingsley DM, Russell LB, et al. 1995. The mouse *Snell's waltzer* deafness gene encodes an unconventional myosin required for structural integrity of inner ear hair cells. *Nat. Genet.* 11:369–75

8. Babinski K, Catarsi S, Biagini G, Séguéla P. 2000. Mammalian ASIC2a and ASIC3 subunits co-assemble into heteromeric proton-gated channels

sensitive to Gd3$^+$. *J. Biol. Chem.* 275: 28519–25

9. Bargmann CI, Thomas JH, Horvitz HR. 1990. Chemosensory cell function in the behavior and development of *Caenorhabditis elegans*. *Cold Spring Harbor Symp. Quant. Biol.* 55:529–38

10. Bellocchio EE, Reimer RJ, Fremeau RT Jr, Edwards RH. 2000. Uptake of glutamate into synaptic vesicles by an inorganic phosphate transporter. *Science* 289:957–60

11. Berger AJ, Hart AC, Kaplan JM. 1998. Gα_s-induced neurodegeneration in *Caenorhabditis elegans*. *J. Neurosci.* 18:2871–80

12. Bergoffen J, Scherer SS, Wang S, Scott MO, Bone LJ, et al. 1993. Connexin mutations in X-linked Charcot-Marie-Tooth disease. *Science* 262:2039–42

13. Bergstrom RA, You Y, Erway LC, Lyon MF, Schimenti JC. 1998. Deletion mapping of the head tilt (*het*) gene in mice: a vestibular mutation causing specific absence of otoliths. *Genetics* 150:815–22

14. Bock GR, Steel KP. 1983. Inner ear pathology in the deafness mutant mouse. *Acta Otolaryngol.* 96:39–47

15. Boettger T, Hübner CA, Maier H, Rust MB, Beck FX, Jentsch TJ. 2002. Deafness and renal tubular acidosis in mice lacking the K-Cl co-transporter Kcc4. *Nature* 416:874–8

16. Bolz H, von Brederlow B, Ramírez A, Bryda EC, Kutsche K, et al. 2001. Mutation of *CDH23*, encoding a new member of the cadherin gene family, causes Usher syndrome type 1D. *Nat. Genet.* 27:108–12

17. Brown AG, Iggo A. 1967. A quantitative study of cutaneous receptors and afferent fibers in the cat and rabbit. *J. Physiol.* 193:707–33

18. Canessa CM, Schild L, Buell G, Thorens B, Gautschi I, et al. 1994. Amiloride-sensitive epithelial Na$^+$ channel is made of three homologous subunits. *Nature* 367:463–67

19. Caterina MJ, Schumacher MA, Tominaga M, Rosen TA, Levine JD, Julius D. 1997. The capsaicin receptor: a heat-activated ion channel in the pain pathway. *Nature* 389:816–24

20. Cauna N, Ross LL. 1960. The fine structure of Meissner's touch corpuscles of human fingers. *J. Biophys. Biochem. Cytol.* 8:472–82

21. Chalfie M, Au M. 1989. Genetic control of differentiation of the *Caenorhabditis elegans* touch receptor neurons. *Science* 243:1027–33

22. Chalfie M, Sulston J. 1981. Developmental genetics of the mechanosensory neurons of *Caenorhabditis elegans*. *Dev. Biol.* 82:358–70

23. Chalfie M, Sulston JE, White JG, Southgate E, Thomson JN, Brenner S. 1985. The neural circuit for touch sensitivity in *Caenorhabditis elegans*. *J. Neurosci.* 5:956–64

24. Chalfie M, Thomson JN. 1979. Organization of neuronal microtubules in the nematode *Caenorhabditis elegans*. *J. Cell Biol.* 82:278–89

25. Chalfie M, Thomson JN. 1982. Structural and functional diversity in the neuronal microtubules of *Caenorhabditis elegans*. *J. Cell Biol.* 93:15–23

26. Chalfie M, Wolinsky E. 1990. The identification and suppression of inherited neurodegeneration in *Caenorhabditis elegans*. *Nature* 345:410–16

27. Champigny G, Voilley N, Waldmann R, Lazdunski M. 1998. Mutations causing neurodegeneration in *Caenorhabditis elegans* drastically alter the pH sensitivity and inactivation of the mammalian H$^+$-gated Na$^+$ channel MDEG1. *J. Biol. Chem.* 273:15418–22

28. Chouinard SW, Wilson GF, Schlimgen AK, Ganetzky B. 1995. A potassium channel β subunit related to the aldo-keto reductase superfamily is encoded by the *Drosophila* hyperkinetic locus. *Proc. Natl. Acad. Sci. USA* 92:6763–67

29. Chung YD, Zhu J, Han Y, Kernan MJ.

2001. *nompA* encodes a PNS-specific, ZP domain protein required to connect mechanosensory dendrites to sensory structures. *Neuron* 29:415–28

30. Clapham DE, Runnels LW, Strubing C. 2001. The TRP ion channel family. *Nat. Rev. Neurosci.* 2:387–96

31. Clarke JD, Hayes BP, Hunt SP, Roberts A. 1984. Sensory physiology, anatomy and immunohistochemistry of Rohon-Beard neurones in embryos of *Xenopus laevis*. *J. Physiol.* 348:511–25

32. Colbert HA, Smith TL, Bargmann CI. 1997. OSM-9, a novel protein with structural similarity to channels, is required for olfaction, mechanosensation, and olfactory adaptation in *Caenorhabditis elegans*. *J. Neurosci.* 17:8259–69

33. Consortium TF. 2002. The FlyBase database of the *Drosophila* genome projects and community literature. *Nucleic Acids Res.* 30:106–8

34. Corey DP, Hudspeth AJ. 1979. Ionic basis of the receptor potential in a vertebrate hair cell. *Nature* 281:675–77

35. Corey DP, Hudspeth AJ. 1979. Response latency of vertebrate hair cells. *Biophys. J.* 26:499–506

36. Corey DP, Hudspeth AJ. 1983. Kinetics of the receptor current in bullfrog saccular hair cells. *J. Neurosci.* 3:962–76

37. Crawford AC, Evans MG, Fettiplace R. 1991. The actions of calcium on the mechano-electrical transducer current of turtle hair cells. *J. Physiol.* 434:369–98

38. Crawford AC, Fettiplace R. 1985. The mechanical properties of ciliary bundles of turtle cochlear hair cells. *J. Physiol.* 364:359–79

39. Croll N. 1976. When *C. elegans* (Nematoda: Rhabditidae) bumps into a bead. *Can. J. Zool.* 54:566–70

40. Dallos P. 1992. The active cochlea. *J. Neurosci.* 12:4575–85

41. Darboux I, Lingueglia E, Pauron D, Barbry P, Lazdunski M. 1998. A new member of the amiloride-sensitive sodium channel family in *Drosophila melanogaster* peripheral nervous system. *Biochem. Biophys. Res. Commun.* 246:210–16

42. Dawkins JL, Hulme DJ, Brahmbhatt SB, Auer-Grumbach M, Nicholson GA. 2001. Mutations in *SPTLC1*, encoding serine palmitoyltransferase, long chain base subunit-1, cause hereditary sensory neuropathy type I. *Nat. Genet.* 27:309–12

43. Denk W, Holt JR, Shepherd GM, Corey DP. 1995. Calcium imaging of single stereocilia in hair cells: localization of transduction channels at both ends of tip links. *Neuron* 15:1311–21

44. Di Palma F, Holme RH, Bryda EC, Belyantseva IA, Pellegrino R, et al. 2001. Mutations in *Cdh23*, encoding a new type of cadherin, cause stereocilia disorganization in waltzer, the mouse model for Usher syndrome type 1D. *Nat. Genet.* 27:103–7

45. Diamond J, Mills LR, Mearow KM. 1988. Evidence that the Merkel cell is not the transducer in the mechanosensory Merkel cell-neurite complex. *Prog. Brain Res.* 74:51–56

46. Dixon MJ, Gazzard J, Chaudhry SS, Sampson N, Schulte BA, Steel KP. 1999. Mutation of the Na-K-Cl co-transporter gene *Slc12a2* results in deafness in mice. *Hum. Mol. Genet.* 8:1579–84

47. Draganov DI, Stetson PL, Watson CE, Billecke SS, La Du BN. 2000. Rabbit serum paraoxonase 3 (PON3) is a high density lipoprotein-associated lactonase and protects low density lipoprotein against oxidation. *J. Biol. Chem.* 275: 33435–42

48. Driscoll M, Chalfie M. 1991. The *mec-4* gene is a member of a family of *Caenorhabditis elegans* genes that can mutate to induce neuronal degeneration. *Nature* 349:588–93

49. Driscoll M, Kaplan J. 1997. Mechanotransduction. In *C.elegans II*, ed. DL Riddle, T Blumenthal, BJ Meyer, JR

Preiss, pp. 645–77. Cold Spring Harbor, NY: Cold Spring Harbor Lab. Press

50. Drummond HA, Abboud FM, Welsh MJ. 2000. Localization of β and γ subunits of ENaC in sensory nerve endings in the rat foot pad. *Brain Res.* 884:1–12

51. Drummond HA, Price MP, Welsh MJ, Abboud FM. 1998. A molecular component of the arterial baroreceptor mechanotransducer. *Neuron* 21:1435–41

52. Du H, Gu G, William CM, Chalfie M. 1996. Extracellular proteins needed for C. elegans mechanosensation. *Neuron* 16:183–94

53. Dumont RA, Lins U, Filoteo AG, Penniston JT, Kachar B, Gillespie PG. 2001. Plasma membrane Ca^{2+}-ATPase isoform 2a is the PMCA of hair bundles. *J. Neurosci.* 21:5066–78

54. Eatock RA. 2000. Adaptation in hair cells. *Annu. Rev. Neurosci.* 23:285–314

55. Eatock RA, Corey DP, Hudspeth AJ. 1987. Adaptation of mechanoelectrical transduction in hair cells of the bullfrog's sacculus. *J. Neurosci.* 7:2821–36

56. Eberl DF, Duyk GM, Perrimon N. 1997. A genetic screen for mutations that disrupt an auditory response in *Drosophila melanogaster. Proc. Natl. Acad. Sci. USA* 94:14837–42

57. Eberl DF, Hardy RW, Kernan MJ. 2000. Genetically similar transduction mechanisms for touch and hearing in *Drosophila. J. Neurosci.* 20:5981–88

58. Ernest S, Rauch GJ, Haffter P, Geisler R, Petit C, Nicolson T. 2000. *Mariner* is defective in *myosin VIIA*: a zebrafish model for human hereditary deafness. *Hum. Mol. Genet.* 9:2189–96

59. Everett LA, Belyantseva IA, Noben-Trauth K, Cantos R, Chen A, et al. 2001. Targeted disruption of mouse *Pds* provides insight about the inner-ear defects encountered in Pendred syndrome. *Hum. Mol. Genet.* 10:153–61

60. Fettiplace R, Fuchs PA. 1999. Mechanisms of hair cell tuning. *Annu. Rev. Physiol.* 61:809–34

60a. Flock A. 1965. Transducing mechanisms in lateral line canal organ receptors. *Cold Spring Harbor Symp.* 30:133

61. Flock A, Cheung HC, Flock B, Utter G. 1981. Three sets of actin filaments in sensory cells of the inner ear. Identification and functional orientation determined by gel electrophoresis, immunofluorescence and electron microscopy. *J. Neurocytol.* 10:133–47

62. Fricke B, Lints R, Stewart G, Drummond H, Dodt G, et al. 2000. Epithelial Na^+ channels and stomatin are expressed in rat trigeminal mechanosensory neurons. *Cell Tissue Res.* 299:327–34

63. Fukushige T, Siddiqui ZK, Chou M, Culotti JG, Gogonea CB, et al. 1999. MEC-12, an α-tubulin required for touch sensitivity in *C. elegans. J. Cell Sci.* 112:395–403

64. Furlong CE, Li WF, Brophy VH, Jarvik GP, Richter RJ, et al. 2000. The PON1 gene and detoxication. *Neurotoxicology* 21:581–87

65. Gale JE, Marcotti W, Kennedy HJ, Kros CJ, Richardson GP. 2001. FM1–43 dye behaves as a permanent blocker of the hair-cell mechanotransducer channel. *J. Neurosci.* 21:7013–25

66. Gao FB, Kohwi M, Brenman JE, Jan LY, Jan YN. 2000. Control of dendritic field formation in Drosophila: the roles of flamingo and competition between homologous neurons. *Neuron* 28:91–101

67. García JA, Yee AG, Gillespie PG, Corey DP. 1998. Localization of myosin-Iβ near both ends of tip links in frog saccular hair cells. *J. Neurosci.* 18:8637–47

68. García-Añoveros J, Derfler B, Neville-Golden J, Hyman BT, Corey DP. 1997. BNaC1 and BNaC2 constitute a new family of human neuronal sodium channels related to degenerins and epithelial sodium channels. *Proc. Natl. Acad. Sci. USA* 94:1459–64

69. García-Añoveros J, García JA, Liu JD, Corey DP. 1998. The nematode degenerin UNC-105 forms ion channels

that are activated by degeneration-
or hypercontraction-causing mutations.
Neuron 20:1231–41

70. Deleted in proof

71. García-Añoveros J, Ma C, Chalfie M.
1995. Regulation of *Caenorhabditis elegans* degenerin proteins by a putative
extracellular domain. *Curr. Biol.* 5:441–
48

72. García-Añoveros J, Samad TA, Zuvela-
Jelaska L, Woolf CJ, Corey DP. 2001.
Transport and localization of the DEG/
ENaC ion channel BNaC1α to peripheral
mechanosensory terminals of dorsal root
ganglia neurons. *J. Neurosci.* 21:2678–
86

72a. Gardner EP, Martin JH, Jessell TM. The
bodily senses. In *Principles of Neuroscience*, ed. ER Kandel, JH Schwartz,
TM Jessell, 22:430–50. New York:
McGraw-Hill. 4th ed.

73. Gibson F, Walsh J, Mburu P, Varela A,
Brown KA, et al. 1995. A type VII
myosin encoded by the mouse deafness
gene *shaker-1*. *Nature* 374:62–64

74. Goodman MB, Ernstrom GG, Chelur
DS, O'Hagan R, Yao CA, Chalfie M.
2002. MEC-2 regulates *C. elegans* DEG/
ENaC channels needed for mechanosensation. *Nature* 415:1039–42

75. Granato M, van Eeden FJ, Schach U,
Trowe T, Brand M, et al. 1996. Genes
controlling and mediating locomotion
behavior of the zebrafish embryo and
larva. *Development* 123:399–413

76. Grueber WB, Graubard K, Truman JW.
2001. Tiling of the body wall by multidendritic sensory neurons in Manduca sexta. *J. Comp. Neurol.* 440:271–
83

77. Grueber WB, Jan LY, Jan YN. 2002.
Tiling of the Drosophila epidermis by
multidendritic sensory neurons. *Development* 129:2867–78

78. Gu CX, Juranka PF, Morris CE. 2001.
Stretch-activation and stretch-inacti-
vation of Shaker-IR, a voltage-gated K+
channel. *Biophys. J.* 80:2678–93

79. Gu G, Caldwell GA, Chalfie M. 1996.
Genetic interactions affecting touch sen-
sitivity in *Caenorhabditis elegans*. *Proc.
Natl. Acad. Sci. USA* 93:6577–82

80. Hall DH, Gu G, García-Añoveros J,
Gong L, Chalfie M, Driscoll M. 1997.
Neuropathology of degenerative cell
death in *Caenorhabditis elegans*. *J. Neurosci.* 17:1033–45

81. Hamill OP, Martinac B. 2001. Molecular
basis of mechanotransduction in living
cells. *Physiol Rev.* 81:685–740

82. Hart AC, Kass J, Shapiro JE, Kaplan JM.
1999. Distinct signaling pathways mediate touch and osmosensory responses in
a polymodal sensory neuron. *J. Neurosci.*
19:1952–58

83. Hart AC, Sims S, Kaplan JM. 1995.
Synaptic code for sensory modalities revealed by *C. elegans* GLR-1 glutamate
receptor. *Nature* 378:82–85

84. Hasson T, Gillespie PG, Garcia JA, Mac-
Donald RB, Zhao Y, et al. 1997. Unconventional myosins in inner-ear sensory
epithelia. *J. Cell Biol.* 137:1287–307

85. Hayasaka K, Himoro M, Sato W, Takada
G, Uyemura K, et al. 1993. Charcot-
Marie-Tooth neuropathy type 1B is associated with mutations of the myelin P_0
gene. *Nat. Genet.* 5:31–34

86. Hayasaka K, Himoro M, Sawaishi Y,
Nanao K, Takahashi T, et al. 1993. *De
novo* mutation of the myelin P_0 gene in
Dejerine-Sottas disease (hereditary mo-
tor and sensory neuropathy type III). *Nat.
Genet.* 5:266–68

87. Hoger U, French AS. 1999. Estimated
single-channel conductance of mechani-
cally-activated channels in a spider
mechanoreceptor. *Brain Res.* 826:230–
35

88. Hollmann M, O'Shea-Greenfield A, Rogers SW, Heinemann S. 1989. Cloning
by functional expression of a member
of the glutamate receptor family. *Nature*
342:643–48

89. Holt JR, Gillespie SK, Provance DW,
Shah K, Shokat KM, et al. 2002. A

chemical-genetic strategy implicates myosin-1c in adaptation by hair cells. *Cell* 108:371–81

90. Hong K, Driscoll M. 1994. A transmembrane domain of the putative channel subunit MEC-4 influences mechanotransduction and neurodegeneration in *C. elegans. Nature* 367:470–73

91. Hong K, Mano I, Driscoll M. 2000. *In vivo* structure-function analyses of *Caenorhabditis elegans* MEC-4, a candidate mechanosensory ion channel subunit. *J. Neurosci.* 20:2575–88

92. Howard J, Hudspeth AJ. 1987. Mechanical relaxation of the hair bundle mediates adaptation in mechanoelectrical transduction by the bullfrog's saccular hair cell. *Proc. Natl. Acad. Sci. USA* 84: 3064–68

93. Howard J, Hudspeth AJ. 1988. Compliance of the hair bundle associated with gating of mechanoelectrical transduction channels in the bullfrog's saccular hair cell. *Neuron* 1:189–99

94. Hrabe de Angelis MH, Flaswinkel H, Fuchs H, Rathkolb B, Soewarto D, et al. 2000. Genome-wide, large-scale production of mutant mice by ENU mutagenesis. *Nat. Genet.* 25:444–47

95. Huang M, Chalfie M. 1994. Gene interactions affecting mechanosensory transduction in *Caenorhabditis elegans. Nature* 367:467–70

96. Huang M, Gu G, Ferguson EL, Chalfie M. 1995. A stomatin-like protein necessary for mechanosensation in *C. elegans. Nature* 378:292–95

96a. Hudspeth AJ. 2000. Hearing. See Ref. 72a, 30:590–613

97. Hudspeth AJ. 1989. How the ear's works work. *Nature* 341:397–404

98. Hudspeth AJ, Corey DP. 1977. Sensitivity, polarity, and conductance change in the response of vertebrate hair cells to controlled mechanical stimuli. *Proc. Natl. Acad. Sci. USA* 74:2407–11

99. Hudspeth AJ, Jacobs R. 1979. Stereocilia mediate transduction in vertebrate hair

cells. *Proc. Natl. Acad. Sci. USA* 76: 1506–9

100. Iggo A, Andres KH. 1982. Morphology of cutaneous receptors. *Annu. Rev. Neurosci.* 5:1–31

101. Iggo A, Muir AR. 1969. The structure and function of a slowly adapting touch corpuscle in hairy skin. *J. Physiol.* 200:763–96

102. Ikeda I, Yamashita Y, Ono T, Ogawa H. 1994. Selective phototoxic destruction of rat Merkel cells abolishes responses of slowly adapting type I mechanoreceptor units. *J. Physiol.* 479(Pt 2):247–56

103. Jan YN, Jan LY. 1993. The peripheral nervous system. In *The Development of Drosophila melanogaster*, ed. M Bate, AM Arias, pp. 1207–44. Plainview, NY: Cold Spring Harbor Lab. Press

104. Kantheti P, Qiao X, Diaz ME, Peden AA, Meyer GE, et al. 1998. Mutation in AP-3δ in the *mocha* mouse links endosomal transport to storage deficiency in platelets, melanosomes, and synaptic vesicles. *Neuron* 21:111–22

105. Kaplan JM, Horvitz HR. 1993. A dual mechanosensory and chemosensory neuron in *Caenorhabditis elegans. Proc. Natl. Acad. Sci. USA* 90:2227–31

106. Kass J, Jacob TC, Kim P, Kaplan JM. 2001. The EGL-3 proprotein convertase regulates mechanosensory responses of *Caenorhabditis elegans. J. Neurosci.* 21: 9265–72

107. Kelsh RN, Brand M, Jiang YJ, Heisenberg CP, Lin S, et al. 1996. Zebrafish pigmentation mutations and the processes of neural crest development. *Development* 123:369–89

108. Kernan M, Cowan D, Zuker C. 1994. Genetic dissection of mechanosensory transduction: mechanoreception-defective mutations of Drosophila. *Neuron* 12: 1195–206

109. Kinkelin I, Stucky CL, Koltzenburg M. 1999. Postnatal loss of Merkel cells, but not of slowly adapting mechanoreceptors

in mice lacking the neurotrophin receptor p75. *Eur. J. Neurosci.* 11:3963–69

110. Koltzenburg M, Stucky CL, Lewin GR. 1997. Receptive properties of mouse sensory neurons innervating hairy skin. *J. Neurophysiol.* 78:1841–50

111. Kozel PJ, Friedman RA, Erway LC, Yamoah EN, Liu LH, et al. 1998. Balance and hearing deficits in mice with a null mutation in the gene encoding plasma membrane Ca^{2+}-ATPase isoform 2. *J. Biol. Chem.* 273:18693–96

112. Kros CJ, Marcotti W, van Netten SM, Self TJ, Libby RT, et al. 2002. Reduced climbing and increased slipping adaptation in cochlear hair cells of mice with *Myo7a* mutations. *Nat. Neurosci.* 5:41–47

113. Kruger L, Perl ER, Sedivec MJ. 1981. Fine structure of myelinated mechanical nociceptor endings in cat hairy skin. *J. Comp. Neurol.* 198:137–54

114. Kulkens T, Bolhuis PA, Wolterman RA, Kemp S, te Nijenhuis S, et al. 1993. Deletion of the serine 34 codon from the major peripheral myelin protein P0 gene in Charcot-Marie-Tooth disease type 1B. *Nat. Genet.* 5:35–39

115. Kurima K, Peters LM, Yang Y, Riazuddin S, Ahmed ZM, et al. 2002. Dominant and recessive deafness caused by mutations of a novel gene, *TMC1*, required for cochlear hair-cell function. *Nat. Genet.* 30:277–84

116. Küssel-Andermann P, El-Amraoui A, Safieddine S, Nouaille S, Perfettini I, et al. 2000. Vezatin, a novel transmembrane protein, bridges myosin VIIA to the cadherin-catenins complex. *EMBO J.* 19:6020–29

117. Lai CC, Hong K, Kinnell M, Chalfie M, Driscoll M. 1996. Sequence and transmembrane topology of MEC-4, an ion channel subunit required for mechanotransduction in *Caenorhabditis elegans. J. Cell Biol.* 133:1071–81

118. Lande WM, Thiemann PV, Mentzer WC Jr. 1982. Missing band 7 membrane protein in two patients with high Na, low K erythrocytes. *J. Clin. Invest.* 70:1273–80

119. Lane PW, Deol MS. 1974. Mocha, a new coat color and behavior mutation on chromosome 10 of the mouse. *J. Hered.* 65:362–64

120. Lee MP, Ravenel JD, Hu RJ, Lustig LR, Tomaselli G, et al. 2000. Targeted disruption of the *Kvlqt1* gene causes deafness and gastric hyperplasia in mice. *J. Clin. Invest.* 106:1447–55

121. Lee RY, Sawin ER, Chalfie M, Horvitz HR, Avery L. 1999. EAT-4, a homolog of a mammalian sodium-dependent inorganic phosphate cotransporter, is necessary for glutamatergic neurotransmission in *Caenorhabditis elegans. J. Neurosci.* 19:159–67

122. Legan PK, Lukashkina VA, Goodyear RJ, Kössi M, Russell IJ, Richardson GP. 2000. A targeted deletion in α-tectorin reveals that the tectorial membrane is required for the gain and timing of cochlear feedback. *Neuron* 28:273–85

123. Liedtke W, Choe Y, Marti-Renom MA, Bell AM, Denis CS, et al. 2000. Vanilloid receptor-related osmotically activated channel (VR-OAC), a candidate vertebrate osmoreceptor. *Cell* 103:525–35

123a. Lindeman HH. 1973. Anatomy of the otolith organs. *Adv. Oto-Rhino-Laryng.* 20:405–33

124. Lim DJ, Erway LC, Clark DL. 1978. Tilted-head mice with genetic otoconial anomaly. Behavioral and morphological correlates. In *Vestibular Mechanisms in Health and Disease*, ed. JD Hood, pp. 195–206. London: Academic

125. Lingueglia E, de Weille JR, Bassilana F, Heurteaux C, Sakai H, et al. 1997. A modulatory subunit of acid sensing ion channels in brain and dorsal root ganglion cells. *J. Biol. Chem.* 272:29778–83

126. Liu J, Schrank B, Waterston RH. 1996. Interaction between a putative mechanosensory membrane channel and a collagen. *Science* 273:361–64

127. Loewenstein WR, Mendelson M. 1965. Components of receptor adaptation in a Pacinian corpuscle. *J. Physiol.* 177:377–97

128. Lumpkin EA, Hudspeth AJ. 1995. Detection of Ca^{2+} entry through mechanosensitive channels localizes the site of mechanoelectrical transduction in hair cells. *Proc. Natl. Acad. Sci. USA* 92: 10297–301

129. Lumpkin EA, Marquis RE, Hudspeth AJ. 1997. The selectivity of the hair cell's mechanoelectrical-transduction channel promotes Ca^{2+} flux at low Ca^{2+} concentrations. *Proc. Natl. Acad. Sci. USA* 94:10997–1002

130. Lyon MF, Scriver CR, Baker LR, Tenenhouse HS, Kronick J, Mandla S. 1986. The *Gy* mutation: another cause of X-linked hypophosphatemia in mouse. *Proc. Natl. Acad. Sci. USA* 83:4899–903

131. Mackness MI, Arrol S, Abbott C, Durrington PN. 1993. Protection of low-density lipoprotein against oxidative modification by high-density lipoprotein associated paraoxonase. *Atherosclerosis* 104:129–35

132. Mackness MI, Arrol S, Durrington PN. 1991. Paraoxonase prevents accumulation of lipoperoxides in low-density lipoprotein. *FEBS Lett.* 286:152–54

133. Mannsfeldt AG, Carroll P, Stucky CL, Lewin GR. 1999. Stomatin, a MEC-2 like protein, is expressed by mammalian sensory neurons. *Mol. Cell Neurosci.* 13: 391–404

134. Maricq AV, Peckol E, Driscoll M, Bargmann CI. 1995. Mechanosensory signalling in *C. elegans* mediated by the GLR-1 glutamate receptor. *Nature* 378: 78–81

135. Markin VS, Hudspeth AJ. 1995. Modeling the active process of the cochlea: phase relations, amplification, and spontaneous oscillation. *Biophys. J.* 69:138–47

136. Martinac B. 2001. Mechanosensitive channels in prokaryotes. *Cell Physiol. Biochem.* 11:61–76

137. McCormack T, McCormack K. 1994. Shaker K$^+$ channel β subunits belong to an NAD(P)H-dependent oxidoreductase superfamily. *Cell* 79:1133–35

138. McGuirt WT, Prasad SD, Griffith AJ, Kunst HP, Green GE, et al. 1999. Mutations in *COL11A2* cause non-syndromic hearing loss (DFNA13). *Nat. Genet.* 23: 413–19

139. Meyer J, Furness DN, Zenner HP, Hackney CM, Gummer AW. 1998. Evidence for opening of hair-cell transducer channels after tip-link loss. *J. Neurosci.* 18: 6748–56

140. Mitchem KL, Hibbard E, Beyer LA, Bosom K, Dootz GA, et al. 2002. Mutation of the novel gene *Tmie* results in sensory cell defects in the inner ear of spinner, a mouse model of human hearing loss DFNB6. *Hum. Mol. Genet.* 11:1887–98

141. Munger BL, Ide C. 1988. The structure and function of cutaneous sensory receptors. *Arch. Histol. Cytol.* 51:1–34

142. Munger BL, Page RB Jr, Pubols BH. 1979. Identification of specific mechanosensory receptors in glabrous skin of dorsal root ganglioectomized primates. *Anat. Rec.* 193:630–31

143. Ng CJ, Wadleigh DJ, Gangopadhyay A, Hama S, Grijalva VR, et al. 2001. Paraoxonase-2 is a ubiquitously expressed protein with antioxidant properties and is capable of preventing cell-mediated oxidative modification of low density lipoprotein. *J. Biol. Chem.* 276: 44444–49

144. Nicolson T, Rusch A, Friedrich RW, Granato M, Ruppersberg JP, Nüsslein-Volhard C. 1998. Genetic analysis of vertebrate sensory hair cell mechanosensation: the zebrafish circler mutants. *Neuron* 20:271–83

145. Nolan PM, Peters J, Strivens M, Rogers D, Hagan J, et al. 2000. A systematic, genome-wide, phenotype-driven mutagenesis programme for gene function

studies in the mouse. *Nat. Genet.* 25: 440–43

146. Ogawa H. 1996. The Merkel cell as a possible mechanoreceptor cell. *Prog. Neurobiol.* 49:317–34

147. Ornitz DM, Bohne BA, Thalmann I, Harding GW, Thalmann R. 1998. Otoconial agenesis in *tilted* mutant mice. *Hear Res.* 122:60–70

148. Perkins LA, Hedgecock EM, Thomson JN, Culotti JG. 1986. Mutant sensory cilia in the nematode *Caenorhabditis elegans*. *Dev. Biol.* 117:456–87

149. Petit C, Levilliers J, Hardelin JP. 2001. Molecular genetics of hearing loss. *Annu. Rev. Genet.* 35:589–646

150. Pickles JO, Comis SD, Osborne MP. 1984. Cross-links between stereocilia in the guinea pig organ of Corti, and their possible relation to sensory transduction. *Hear Res.* 15:103–12

151. Pickles JO, Corey DP. 1992. Mechanoelectrical transduction by hair cells. *Trends Neurosci.* 15:254–59

152. Popper AN, Platt C. 1993. Inner ear and lateral line. In *The Physiology of Fishes*, ed. DH Evan, pp. 99–136. Ann Arbor: CRC Press

153. Price MP, Lewin GR, McIlwrath SL, Cheng C, Xie J, et al. 2000. The mammalian sodium channel BNC1 is required for normal touch sensation. *Nature* 407:1007–11

154. Price MP, McIlwrath SL, Xie J, Cheng C, Qiao J, et al. 2001. The DRASIC cation channel contributes to the detection of cutaneous touch and acid stimuli in mice. *Neuron* 32:1071–83

155. Price MP, Snyder PM, Welsh MJ. 1996. Cloning and expression of a novel human brain Na$^+$ channel. *J. Biol. Chem.* 271:7879–82

156. Probst FJ, Fridell RA, Raphael Y, Saunders TL, Wang A, et al. 1998. Correction of deafness in *shaker-2* mice by an unconventional myosin in a BAC transgene. *Science* 280:1444–47

157. Rajaram S, Spangler TL, Sedensky MM, Morgan PG. 1999. A stomatin and a degenerin interact to control anesthetic sensitivity in *Caenorhabditis elegans*. *Genetics* 153:1673–82

158. Rankin CH, Beck CD, Chiba CM. 1990. *Caenorhabditis elegans*: a new model system for the study of learning and memory. *Behav. Brain Sci.* 37:89–92

159. Ribera AB, Nüsslein-Volhard C. 1998. Zebrafish touch-insensitive mutants reveal an essential role for the developmental regulation of sodium current. *J. Neurosci.* 18:9181–91

160. Roa BB, Dyck PJ, Marks HG, Chance PF, Lupski JR. 1993. Dejerine-Sottas syndrome associated with point mutation in the peripheral myelin protein 22 (PMP22) gene. *Nat. Genet.* 5:269–73

161. Roayaie K, Crump JG, Sagasti A, Bargmann CI. 1998. The Gα protein ODR-3 mediates olfactory and nociceptive function and controls cilium morphogenesis in *C. elegans* olfactory neurons. *Neuron* 20:55–67

162. Roberts A, Hayes BP. 1977. The anatomy and function of 'free' nerve endings in an amphibian skin sensory system. *Proc. R. Soc. London Ser. B Biol. Sci.* 196:415–29

163. Rodieck R. 1998. *The First Steps in Seeing.* Sunderland, MA: Sinauer

164. Rongo C, Whitfield CW, Rodal A, Kim SK, Kaplan JM. 1998. LIN-10 is a shared component of the polarized protein localization pathways in neurons and epithelia. *Cell* 94:751–59

165. Rüsch A, Erway LC, Oliver D, Vennström B, Forrest D. 1998. Thyroid hormone receptor β-dependent expression of a potassium conductance in inner hair cells at the onset of hearing. *Proc. Natl. Acad. Sci. USA* 95:15758–62

166. Russell IJ, Richardson GP, Cody AR. 1986. Mechanosensitivity of mammalian auditory hair cells in vitro. *Nature* 321:517–19

167. Salzer U, Ahorn H, Prohaska R. 1993. Identification of the phosphorylation site on human erythrocyte band 7 integral

membrane protein: implications for a monotopic protein structure. *Biochim. Biophys. Acta* 1151:149–52

168. Salzer U, Prohaska R. 2001. Stomatin, flotillin-1, and flotillin-2 are major integral proteins of erythrocyte lipid rafts. *Blood* 97:1141–43

169. Santos-Sacchi J, Dilger JP. 1988. Whole cell currents and mechanical responses of isolated outer hair cells. *Hear Res.* 35:143–50

170. Savage C, Hamelin M, Culotti JG, Coulson A, Albertson DG, Chalfie M. 1989. *mec-7* is a β-tubulin gene required for the production of 15-protofilament microtubules in *Caenorhabditis elegans*. *Genes Dev.* 3:870–81

171. Sawin ER, Ranganathan R, Horvitz HR. 2000. *C. elegans* locomotory rate is modulated by the environment through a dopaminergic pathway and by experience through a serotonergic pathway. *Neuron* 26:619–31

172. Seiler C, Nicolson T. 1999. Defective calmodulin-dependent rapid apical endocytosis in zebrafish sensory hair cell mutants. *J. Neurobiol.* 41:424–34

173. Shepherd GM, Barres BA, Corey DP. 1989. "Bundle blot" purification and initial protein characterization of hair cell stereocilia. *Proc. Natl. Acad. Sci. USA* 86:4973–77

174. Shih DM, Gu L, Xia YR, Navab M, Li WF, et al. 1998. Mice lacking serum paraoxonase are susceptible to organophosphate toxicity and atherosclerosis. *Nature* 394:284–87

175. Shreffler W, Magardino T, Shekdar K, Wolinsky E. 1995. The *unc-8* and *sup-40* genes regulate ion channel function in *Caenorhabditis elegans* motorneurons. *Genetics* 139:1261–72

176. Simmler MC, Cohen-Salmon M, El-Amraoui A, Guillaud L, Benichou JC, et al. 2000. Targeted disruption of *Otog* results in deafness and severe imbalance. *Nat. Genet.* 24:139–43

177. Simmler MC, Zwaenepoel I, Verpy E,

Guillaud L, Elbaz C, et al. 2000. Twister mutant mice are defective for otogelin, a component specific to inner ear acellular membranes. *Mamm. Genome* 11:960–66

178. Sinclair D. 1981. *Mechanisms of Cutaneous Sensation*. Oxford, England: Oxford Univ. Press. 363 pp.

179. Spoendlin H. 1979. Innervation of the cochlear receptor. In *Basic Mechanisms in Hearing*, ed. A Moller, pp. 185–230. New York: Academic

180. Steel KP. 1995. Inherited hearing defects in mice. *Annu. Rev. Genet.* 29:675–701

181. Steel KP, Kros CJ. 2001. A genetic approach to understanding auditory function. *Nat. Genet.* 27:143–49

182. Street VA, McKee-Johnson JW, Fonseca RC, Tempel BL, Noben-Trauth K. 1998. Mutations in a plasma membrane Ca^{2+}-ATPase gene cause deafness in deafwaddler mice. *Nat. Genet.* 19:390–94

183. Strom TM, Francis F, Lorenz B, Böddrich A, Econs MJ, et al. 1997. *Pex* gene deletions in *Gy* and *Hyp* mice provide mouse models for X-linked hypophosphatemia. *Hum. Mol. Genet.* 6:165–71

184. Sulston J, Dew M, Brenner S. 1975. Dopaminergic neurons in the nematode *Caenorhabditis elegans*. *J. Comp. Neurol.* 163:215–26

185. Swank RT, Reddington M, Howlett O, Novak EK. 1991. Platelet storage pool deficiency associated with inherited abnormalities of the inner ear in the mouse pigment mutants muted and mocha. *Blood* 78:2036–44

186. Takamori S, Rhee JS, Rosenmund C, Jahn R. 2000. Identification of a vesicular glutamate transporter that defines a glutamatergic phenotype in neurons. *Nature* 407:189–94

187. Takemura T, Sakagami M, Nakase T, Kubo T, Kitamura Y, Nomura S. 1994. Localization of osteopontin in the otoconial organs of adult rats. *Hear Res.* 79:99–104

188. Tavernarakis N, Shreffler W, Wang S,

Driscoll M. 1997. *unc-8*, a DEG/ENaC family member, encodes a subunit of a candidate mechanically gated channel that modulates *C. elegans* locomotion. *Neuron* 18:107–19

189. Tazaki M, Suzuki T. 1998. Calcium inflow of hamster Merkel cells in response to hyposmotic stimulation indicate a stretch activated ion channel. *Neurosci. Lett.* 243:69–72

190. Thomas JH. 1990. Genetic analysis of defecation in *Caenorhabditis elegans*. *Genetics* 124:855–72

191. Tilney LG, Derosier DJ, Mulroy MJ. 1980. The organization of actin filaments in the stereocilia of cochlear hair cells. *J. Cell Biol.* 86:244–59

192. Tilney MS, Tilney LG, Stephens RE, Merte C, Drenckhahn D, et al. 1989. Preliminary biochemical characterization of the stereocilia and cuticular plate of hair cells of the chick cochlea. *J. Cell Biol.* 109:1711–23

193. Tobin DM, Madsen DM, Kahn-Kirby A, Peckol EL, Moudler G, et al. 2002. Combinatorial expression of TRPV channel proteins defines their sensory functions and subcellular localization in *C. elegans* neurons. *Neuron.* 35:307–18

194. Valentijn LJ, Baas F, Wolterman RA, Hoogendijk JE, van den Bosch NH, et al. 1992. Identical point mutations of PMP-22 in Trembler-J mouse and Charcot-Marie-Tooth disease type 1A. *Nat. Genet.* 2:288–91

195. Vetter DE, Mann JR, Wangemann P, Liu J, McLaughlin KJ, et al. 1996. Inner ear defects induced by null mutation of the isk gene. *Neuron* 17:1251–64

196. Vogel BE, Hedgecock EM. 2001. Hemicentin, a conserved extracellular member of the immunoglobulin superfamily, organizes epithelial and other cell attachments into oriented line-shaped junctions. *Development* 128:883–94

197. Vreugde S, Erven A, Kros CJ, Marcotti W, Fuchs H, et al. 2002. Beethoven, a mouse model for dominant, progressive hearing loss DFNA36. *Nat. Genet.* 30:257–58

198. Waldmann R, Bassilana F, de Weille J, Champigny G, Heurteaux C, Lazdunski M. 1997. Molecular cloning of a non-inactivating proton-gated Na⁺ channel specific for sensory neurons. *J. Biol. Chem.* 272:20975–78

199. Waldmann R, Champigny G, Voilley N, Lauritzen I, Lazdunski M. 1996. The mammalian degenerin MDEG, an amiloride-sensitive cation channel activated by mutations causing neurodegeneration in *Caenorhabditis elegans*. *J. Biol. Chem.* 271:10433–36

200. Walker RG, Willingham AT, Zuker CS. 2000. A *Drosophila* mechanosensory transduction channel. *Science* 287:2229–34

201. Wang N, Ingber DE. 1995. Probing transmembrane mechanical coupling and cytomechanics using magnetic twisting cytometry. *Biochem. Cell Biol.* 73:327–35

202. Wang Y, Kowalski PE, Thalmann I, Ornitz DM, Mager DL, Thalmann R. 1998. Otoconin-90, the mammalian otoconial matrix protein, contains two domains of homology to secretory phospholipase A2. *Proc. Natl. Acad. Sci. USA* 95:15345–50

203. Ward S, Thomson N, White JG, Brenner S. 1975. Electron microscopical reconstruction of the anterior sensory anatomy of the nematode *Caenorhabditis elegans*. *J. Comp. Neurol.* 160:313–37

204. Way JC, Chalfie M. 1988. *mec-3*, a homeobox-containing gene that specifies differentiation of the touch receptor neurons in *C. elegans*. *Cell* 54:5–16

205. Way JC, Chalfie M. 1989. The *mec-3* gene of *Caenorhabditis elegans* requires its own product for maintained expression and is expressed in three neuronal cell types. *Genes Dev.* 3:1823–33

206. Weil D, Blanchard S, Kaplan J, Guilford P, Gibson F, et al. 1995. Defective myosin VIIA gene responsible for Usher syndrome type 1B. *Nature* 374:60–61

207. Weinshenker D, Wei A, Salkoff L, Thomas JH. 1999. Block of an ether-a-go-go-like K$^+$ channel by imipramine rescues *egl-2* excitation defects in *Caenorhabditis elegans*. *J. Neurosci.* 19: 9831–40

208. White JG, Southgate E, Thomson JN, Brenner S. 1976. The structure of the ventral nerve cord of *Caenorhabditis elegans*. *Philos. Trans. R. Soc. London Ser. B. Biol. Sci.* 275:327–48

209. White JG, Southgate E, Thompson JN, Brenner S. 1986. The structure of the nervous system of the nematode *Caenorhabditis elegans*. *Philos. Trans. R. Soc. London Ser. B. Biol. Sci.* 314:1–340

210. Whitfield TT, Granato M, van Eeden FJ, Schach U, Brand M, et al. 1996. Mutations affecting development of the zebrafish inner ear and lateral line. *Development* 123:241–54

211. Wicks SR, Rankin CH. 1995. Integration of mechanosensory stimuli in *Caenorhabditis elegans*. *J. Neurosci.* 15:2434–44

212. Xiang M, Gan L, Li D, Chen ZY, Zhou L, et al. 1997. Essential role of POU-domain factor Brn-3c in auditory and vestibular hair cell development. *Proc. Natl. Acad. Sci. USA* 94:9445–50

213. Xiang M, Gao WQ, Hasson T, Shin JJ. 1998. Requirement for Brn-3c in maturation and survival, but not in fate determination of inner ear hair cells. *Development* 125:3935–46

214. Xie J, Price MP, Berger AL, Welsh MJ. 2002. DRASIC contributes to pH-gated currents in large dorsal root ganglion sensory neurons by forming heteromultimeric channels. *J. Neurophsyiol.* 87:2835–43

215. Yamoah EN, Lumpkin EA, Dumont RA, Smith PJ, Hudspeth AJ, Gillespie PG. 1998. Plasma membrane Ca^{2+}-ATPase extrudes Ca^{2+} from hair cell stereocilia. *J. Neurosci.* 18:610–24

216. Zhang Y, Ma C, Delohery T, Nasipak B, Foat BC, et al. 2002. Identification of genes expressed in *C. elegans* touch receptor neurons. *Nature*. 418:331–35

217. Zheng L, Sekerkova G, Vranich K, Tilney LG, Mugnaini E, Bartles JR. 2000. The deaf jerker mouse has a mutation in the gene encoding the espin actin-bundling proteins of hair cell stereocilia and lacks espins. *Cell* 102:377–85

218. Zhu Y, Paszty C, Turetsky T, Tsai S, Kuypers FA, et al. 1999. Stomatocytosis is absent in "stomatin"-deficient murine red blood cells. *Blood* 93:2404–10

NOTE ADDED IN PROOF

The research in our laboratory on the *mec-6* gene cited as unpublished (pp. 419, 420, 425 in text) is now in press: Chelur DS, Ernstrom GG, Goodman MB, Yao CA, Chen L, O'Hagan R, Chalfie M. 2002. The mechanosensory protein MEC-6 is a subunit of the *C. elegans* touch-cell degerin channel. *Nature*. In press.

Annu. Rev. Genet. 2002. 36:455–88
doi: 10.1146/annurev.genet.36.052802.114101

UNDERSTANDING THE FUNCTION OF ACTIN-BINDING PROTEINS THROUGH GENETIC ANALYSIS OF *DROSOPHILA* OOGENESIS

Andrew M. Hudson[1] and Lynn Cooley[1, 2]

*Departments of Genetics[1] and Cell Biology[2], Yale University School of Medicine,
P.O. Box 208005, New Haven, Connecticut 06520-8005; e-mail: lynn.cooley@yale.edu,
andrew.hudson@yale.edu*

Key Words ring canal, actin regulation, actin crosslinker, follicle cell, migration

■ **Abstract** Much of our knowledge of the actin cytoskeleton has been derived from biochemical and cell biological approaches, through which actin-binding proteins have been identified and their in vitro interactions with actin have been characterized. The study of actin-binding proteins (ABPs) in genetic model systems has become increasingly important for validating and extending our understanding of how these proteins function. New ABPs have been identified through genetic screens, and genetic results have informed the interpretation of in vitro experiments. In this review, we describe the molecular and ultrastructural characteristics of the actin cytoskeleton in the *Drosophila* ovary, and discuss recent genetic analyses of actin-binding proteins that are required for oogenesis.

CONTENTS

0066-4197/02/1215-0455$14.00

INTRODUCTION

The production of a *Drosophila* egg is accomplished through the close association between the developing oocyte and 15 interconnected nurse cells. Transcriptional products from the oocyte nucleus contribute little to the RNA and protein that will sustain early embryonic development. Instead, the nurse cells, and to a lesser extent the follicle cells, synthesize and transport the cytoplasmic components that will support the development of the oocyte and later, the embryo (84).

The mechanism by which cytoplasm moves from the nurse cells to the oocyte has been probed using both cell biological [e.g., (13, 51)] and genetic [e.g., (24, 153)] approaches. Cell biological studies revealed an elaborate actin cytoskeleton in developing egg chambers including extended F-actin bundles and F-actin–rich intercellular bridges [e.g., (109, 145)]. Genetic analysis underscored the functional relevance of the actin cytoskeleton in this process: The majority of mutations characterized at the molecular level that affect nurse cell cytoplasm transport alter ABPs.

Somatic follicle cells surround the germline cysts in each egg chamber, and the follicle cells contribute key signals to determine oocyte and embryo polarity (31). In addition, polarized follicle cells secrete the proteins making up the oocyte vitelline membrane and eggshell late in oogenesis. To carry out these tasks, follicle cells must form and maintain a polarized epithelium, undergo changes in cell shape, and carry out several cell migrations. All of these processes are associated with dynamic changes in the follicle cell actin cytoskeleton.

Thus, the *Drosophila* ovary provides a system in which the functions of actin-binding proteins (ABPs) can be dissected in a metazoan context. In this review, we focus on recent work on the genetic analysis of actin-binding protein function during *Drosophila* oogenesis. [For a review of earlier work, see (115).]

Overview of Oogenesis

As in other species, the development of an egg in *Drosophila* involves complex interactions between germ cells and somatic cells (90, 107). Drosophila ovaries are organized into approximately 15 ovarioles, which are tubular structures that contain progressively maturing egg chambers. Egg chambers initiate development in the anterior of the ovariole, in a region termed the germarium (Figure 1, see color insert). Here, each germline stem cell divides asymmetrically, generating a daughter cell (called the cystoblast) and another stem cell. A cystoblast undergoes four rounds of cell division without completing cytokinesis, generating a cyst of 16 interconnected germ cells. The arrested cleavage furrows in this cluster are subsequently transformed into ring canals, the stable intercellular bridges that facilitate cytoplasm transport throughout oogenesis (112). Interactions between the germ cells and follicle cells in the germarium are complex and not fully understood, but follicle cells are required to maintain germline stem cell identity (151), and also regulate germline stem cell proliferation (152). As the 16-cell cyst moves posteriorly in the germarium, interactions with somatic follicle cells continue. Two follicle cell stem cells located at the periphery of the germarium provide a source of follicle cells that encapsulate the germ cells (88). Encapsulation involves complex migratory movements by the follicle cells in the germarium, as the follicle cells send long extensions between adjacent germ cell clusters, resulting in the complete envelopment of the germ cell cluster by a layer of follicle cells (125, 135) (Figure 1*B–E*). Once this process is complete, egg chambers bud from the germarium and continue to move posteriorly through the ovariole as oogenesis proceeds.

From the cluster of 16 germ cells, one is chosen to become the oocyte. The pattern of synchronized germ cell division in the germarium results in a stereotypical pattern of interconnections, such that only two germ cells have four ring canals, and the oocyte is always selected from one of these two [discussed in (20)]. The oocyte is then localized to the posterior of the egg chamber in a cell-sorting process involving differential cadherin expression in the nurse cells, the oocyte, and the follicle cells (46).

As oogenesis proceeds, the nurse cell nuclei become highly polyploid, whereas the oocyte nucleus remains arrested in meiotic prophase and is transcriptionally quiescent. The increase in nurse cell ploidy facilitates high levels of biosynthetic activity in the nurse cells, and the cytoplasmic products of this activity begin to move into the growing oocyte. Initial transport is slow and selective from stage 1, the point where the egg chamber buds off from the germarium, until stage 10, where the oocyte has grown to a size equal to that of the nurse cell cluster. During stage 10, the egg chambers undergo preparation to rapidly transfer the remaining cytoplasmic contents; this rapid transport phase is initiated by a cue to enter a modified apoptotic program [reviewed in (17)].

The apoptotic cue leads to a number of striking downstream events (17). First, a network of actin bundles forms in the nurse cells, creating a halo of F-actin around each of the large polyploid nurse cell nuclei. This is followed by the disassembly of the nurse cell nuclear envelopes, which allows the nurse cell nuclear contents to mix with cytoplasm. Finally, the nurse cells actively contract, expelling their remaining cytoplasmic contents into the oocyte. The cytoplasmic bundles are required to restrain the nurse cell nuclei during rapid transport, as mutations that compromise bundle formation still contract, but fail to transfer cytoplasm due to the movement of the large, polyploid nuclei into the ring canals, blocking further transport [e.g., (24)]. Based on several observations, the contractile force appears to be generated by cytoplasmic myosin II. Germline clones of *sqh*, the gene encoding the myosin II regulatory light chain, fail to transport due to a failure in contraction, and Myosin II is localized to the nurse cell cortices (34, 149).

Follicle cells undergo a precisely regulated series of migrations during oogenesis [reviewed in (31)]. Beginning at stage 9, most follicle cells migrate as an epithelial sheet posteriorly toward the oocyte. About 50 follicle cells remain covering the nurse cell cluster and become stretched into a squamous epithelium. During the posterior migration of the follicular epithelium, a special population of follicle cells, the border cells, delaminate from the anterior of the egg chamber and move through the nurse cell cluster to the anterior margin of the oocyte. These cells are required to form a functional micropyle, a structure that facilitates sperm entry [reviewed in (126)].

The Actin Cytoskeleton in the Germline

The organization of the actin cytoskeleton in the *Drosophila* ovary has been extensively characterized in terms of ultrastructural organization and molecular composition. We briefly review what is known about the ovarian actin cytoskeleton, and compare the actin-rich structures in the ovary with common organizational features of the mammalian actin cytoskeleton. In ovarian germ cells, two striking populations of actin filaments exist: ring canal actin, and stage-specific cytoplasmic actin bundles.

The cytoplasmic actin bundles that form just prior to the rapid phase of cytoplasm transport bear a number of striking similarities to parallel actin bundles in other systems, including brush border microvilli and hair cell stereocilia [discussed in (5, 29)]. The cytoplasmic bundles are composed of unipolar, hexagonally packed actin filaments with their fast-growing barbed ends facing the membrane (Figure 2A and C, F and G, see color insert) (49). At least two filament crosslinking proteins are required to organize these bundles. Quail, a relative of brush border villin (85, 91), appears to loosely bundle filaments, while fascin, the product of the *singed* gene (19), organizes these filaments into maximally crosslinked, hexagonally packed filament bundles (19). Actin bundles exhibit a striated appearance when visualized with fluorescent phalloidin (49, 109) (see Figure 2C). The striated appearance likely reflects their organization, as the bundles, which extend up to 40 μm in length, are composed of a series of overlapping 2–3 μm modules. These modules appear to be attached to each other through lateral associations, and the gaps in fluorescence likely reflect regions of little overlap of adjacent bundles (49) (Figure 2G).

Fluorescent-phalloidin staining of ring canals reveals a tight band of F-actin (Figure 2B). Images of ring canals in cross section obtained by thin section EM show an electron-dense outer rim adjacent to the ring canal membrane and an electron opaque inner rim that corresponds to the F-actin visible by phalloidin staining [e.g., (140)] (Figure 2D, E). By EM, the F-actin at the ring canal appears to be organized into loosely packed parallel bundles throughout most of oogenesis

(140) (Figure 2E). A number of proteins have been found by immunofluorescence to localize to ring canals, and a distinction can be made between ring canal proteins that colocalize with F-actin in the inner rim and proteins that are more closely associated with the membrane in the ring canal outer rim.

It is difficult to make a perfect analogy between the ring canal and other F-actin structures. However, an intriguing comparison can also be made between ring canals and the leading edge in a migrating cell. Perhaps the most telling similarity is the high rate of actin filament turnover in both ring canals (69) and leading edge lamellipodia [e.g., (136)]. This high rate of F-actin turnover is in marked contrast to the F-actin associated with stress fibers [(158) and references therein]. In addition, ring canals share several molecular components with lamellipodia, including filamin and the Arp2/3 complex (discussed below).

The Actin Cytoskeleton in Follicle Cells

Ovarian somatic follicle cells adopt markedly different morphologies and functions during oogenesis, providing an excellent model for a number of actin-dependent processes. The follicle cells that surround the young egg chambers and cover the oocyte in later egg chambers provide a beautiful example of a polarized epithelium (Figure 3A, see color insert). This includes what appears to be an actin-rich adhesion belt near the apical surface and clear apico-basal polarity [reviewed in (100)].

The follicular epithelial cells also have an array of actin bundles associated with their basal plasma membranes, which is in contact with an overlying basement membrane (52, 53) (Figure 3D, E). These bundles may be analogous to stress fibers, and they are required for egg chamber shape. Late in oogenesis, egg chambers change in shape from round to oblong. This shape change is preceded by a reorganization of the basal actin bundles: Basal bundles in younger egg chambers appear randomly oriented from cell to cell, but by stage seven become aligned such that bundles in all cells lie perpendicular to the anterior-posterior axis of the egg chamber (40). In mutants where the bundles are disorganized, such as *kugelei*, the resulting egg chambers fail to elongate properly and the resulting eggs are round and sterile (53). Given this phenotype, it seems plausible that the bundles are contractile, and that they are thus bona fide stress fibers. However, it has not been determined whether myosin is a component of these bundles, nor has the polarity of the filaments in the bundles been determined.

Near the end of oogenesis, the border cells begin their migration as a group of 8 to 10 follicle cells at the anterior of the egg chamber that delaminate from the follicular epithelium (Figure 3G). In the course of their migration, the border cells exhibit all of the stereotypical features associated with migrating cells: lamella and lamellipodium extension, filopodial extension (see Figure 3G), and cell translocation. Thus, border cell migration presents the opportunity to study the actin cytoskeleton in cells migrating in situ in a genetically tractable system.

PROTEINS AFFECTING THE POLYMERIZATION STATE OF ACTIN

Understanding how cells rapidly reorganize their actin cytoskeleton requires the identification and characterization of proteins that influence actin polymerization dynamics. In vitro, the initiation of actin polymerization (nucleation) is an inherently slow process; however, once polymerization is initiated, it proceeds rapidly until a steady state between G-actin and F-actin pools is achieved. Proteins that initiate, terminate, or otherwise affect the rate of actin polymerization are likely to be important regulators of actin polymerization in vivo. In recent years, the study of additional proteins involved in actin polymerization dynamics in *Drosophila* has added to our understanding of the actin cytoskeleton.

The Arp2/3 Complex and its Regulators, Scar and Wasp

The Arp2/3 complex is a complex of seven proteins first identified as a profilin binding complex in *Acanthamoeba* (78). It consists of the actin-related proteins Arp2 and Arp3, a WD-repeat protein Arpc1, and four other proteins, Arpc2–5 (79). The Arp2/3 complex is the only actin nucleation factor thus far characterized that creates new actin filaments that elongate from their barbed ends (101). As most actin polymerization within a cell involves growth at the faster-growing barbed end, the Arp2/3 complex is positioned to be an important regulator of actin polymerization. The Arp2/3 complex requires two cofactors for efficient nucleation: an activating protein of the Wasp/Scar family and preexisting actin filaments. Current data indicate that the Arp2/3 complex binds to an activating protein and to the side of an actin filament, and this activated complex mimics the structure of a barbed end of an actin filament (117, 144). The binding to the side of a filament is thought to be obligatory for Arp2/3 nucleation (3), and this leads to the polymerization of new filaments at a characteristic 70° angle from the mother filament (101). In agreement with this, the Arp2/3 complex has been localized to 70° filament branches in lamellipodia of cultured cells (37, 132).

Scar and Wasp activation are downstream of the Rho-family of small GTPases. In the case of Scar, there is evidence that the Rho family GTPase Rac causes changes in the actin cytoskeleton through Scar (80, 96), and that Scar in turn activates the Arp2/3 complex via an intermediary protein, IRSp53 (97). However, the best-documented signaling relationship involves actin polymerization downstream of the small GTPase Cdc42. Cdc42 binds and activates Wasp, which in turn activates the Arp2/3 complex, leading to new actin polymerization [for review, see (59)]. Scar and Wasp are homologous in their C-terminal regions that mediate interaction with the Arp2/3 complex, but have divergent N-terminal domains that interact with upstream signaling molecules (Figure 4, see color insert).

Recent isolation of mutations in the *Drosophila* homologs of Scar, Wasp, and Arp2/3 complex components has offered insights into how these proteins function to shape the actin cytoskeleton. Mutations in genes encoding the two Arp2/3 complex subunits, Arp3 and Arpc1 (the WD-repeat protein), have been recently characterized (62). All alleles of these mutations are lethal, but germline clones reveal defects in cytoplasm transport due to defects in actin accumulation and organization in the ring canal.

The ring canal defects associated with Arp2/3 mutants offer insights into both ring canal assembly and Arp2/3 complex function. While at late stages, the defects are quite severe and transport through the rings is presumably compromised, at earlier stages, ring canals appear less affected (62). This may reflect a mechanistic difference in the way F-actin is recruited to the ring canal at different stages. During mid-oogenesis, there is a large increase in the amount of actin that is recruited to the ring canal to support the expansion of the ring canal diameter (140), and this coincides with the onset of the Arp2/3 ring canal defect. Thus, the Arp2/3 complex may be required for this late growth phase. Interestingly, the *Drosophila* genome encodes two Arpc3 subunits, and one of these, designated Arpc3B, appears to be specific to the female germline and highly enriched at ring canals (62). It is possible that Arpc3B forms part of a specialized ring canal Arp2/3 complex at the ring canal, perhaps with a modified nucleation activity suited to the requirements of the ring canal actin cytoskeleton.

Mutations in *Drosophila* homologs of the Arp2/3 complex regulators Scar and Wasp have also been reported (9, 155). Like the Arp2/3 complex mutations, germline clones of complete loss-of-function *SCAR* alleles exhibit severe defects in the ring canal actin cytoskeleton (155). The similarity between the phenotypes associated with Arp2/3 complex mutations and *SCAR* mutations suggests that *SCAR* is the primary regulator of the Arp2/3 complex during oogenesis. In contrast, germline clones of strong mutations in the sole *Drosophila* Wasp homolog exhibit no apparent defects in ring canals (155), and indeed, these germline clones produce viable, fertile eggs. Thus it appears that SCAR and the Arp2/3 complex cooperate to create a pool of F-actin to support ring canal growth, whereas the relationship between the Arp2/3 complex and Wasp during oogenesis is not important. In flies, the co-requirements for Wasp and the Arp2/3 complex appear to be restricted to certain types of cell fate decisions in neurogenesis (134).

In contrast to ring canals, the parallel cytoplasmic actin bundles form in egg chambers mutant for the Arp2/3 complex, *SCAR*, and *Wasp* (62; A.M. Hudson & L. Cooley, unpublished observations). The simplest interpretation of this result is that the formation of at least some types of parallel actin bundles does not depend on the Arp2/3 complex and its regulators. These results are consistent with the finding that N-WASP–deficient cells form filopodia (77). In addition, examination of Arp2/3 phenotypes in other tissues and other organisms supports an Arp2/3-independent mechanism for parallel F-actin structures; Arp2/3 mutants cells form extensive parallel actin bundles in developing bristles (62), and actin cables in yeast do not require the Arp2/3 complex (35, 121, 150).

spire

spire (*spir*) was isolated as a maternal effect gene in *Drosophila* that affects axial patterning in the early embryo (87). In *spir* mutants, the localization of molecules essential for anterior-posterior and dorsal-ventral patterning is lost. This phenotype is associated with a phenomenon called premature microtubule streaming, where a normal microtubule-driven swirling of the oocyte contents occurs prematurely, resulting in the delocalization of embryonic determinants (137). Despite the apparent dysfunction of the microtubule cytoskeleton, a number of observations suggest that the primary defect is a disruption of the actin cytoskeleton. Mutations in other genes encoding actin-binding proteins, including *chickadee* (profilin) and *cappuccino* (a formin homology protein), result in a similar premature microtubule phenotype (86, 137). In addition, treatment of cultured egg chambers with cytochalasin D also results in premature microtubule streaming. However, the defect in the actin cytoskeleton that leads to premature microtubule streaming is unknown.

spire was recently cloned, and the encoded protein SPIR possesses several domains that suggest a function in regulating the actin cytoskeleton (147) (Figure 4A). SPIR contains two tandem Wasp-homology 2 (WH2) domains; WH2 domains bind G-actin in several other proteins, and SPIR protein fragments containing the WH2 domains also bind to G-actin. In addition, two-hybrid interaction tests indicated that the N terminus of SPIR binds to wild-type and dominant-negative Rho1, Rac1, and Cdc42, suggesting that SPIR may link signaling from Rho-family GTPases to the actin cytoskeleton.

SPIR was also recently identified in the yeast two-hybrid system as a protein interacting with DJNK, the *Drosophila* homolog of Jun N-terminal kinase (105). In this study, SPIR induced actin polymerization upon transfection into mammalian cells, suggesting that SPIR proteins can promote actin polymerization in vivo. In addition, SPIR was phosphorylated by DJNK in vitro, suggesting that multiple signaling inputs may impinge on SPIR.

SPIR and Wasp/SCAR proteins share several features: actin monomer-binding WH2 domains, the ability to induce actin polymerization in cultured cells, and a Rho-family GTPase interaction domain. However, SPIR does not contain the characteristic sequences that mediate interaction between Wasp/SCAR proteins and the Arp2/3 complex (147). Still, the similarities between SPIR and Wasp/SCAR proteins raise the possibility that they play similar roles in regulating the actin cytoskeleton. Perhaps SPIR relays signals from Rho-family GTPases to an as-yet-unidentified cytoskeletal effector molecule. Alternatively, SPIR may itself be able to effect rearrangements of the actin cytoskeleton in response to upstream signals.

CAP

CAP, or cyclase-associated protein, was first identified based on its association with adenylyl cyclase, and was implicated in organizing the actin cytoskeleton through genetic interactions with profilin in yeast [reviewed in (61)]. CAP has an N-terminal domain that interacts with adenylate cyclase, a linker region containing a polyproline region, and a C-terminal domain that binds G-actin and mediates its cytoskeletal function (Figure 4A). Subsequent studies revealed CAP to be an actin monomer-binding protein and, like other monomer proteins, CAP proteins were found to inhibit actin polymerization due to actin monomer sequestration (38, 42, 47). However, recent mammalian cell culture studies using antibody inhibition suggest a requirement for CAP in the formation of stress fibers, and a GFP-CAP fusion protein localizes to stress fibers; moreover, injection of purified CAP protein stimulated F-actin formation (39). Thus, the normal cellular role of CAP is unclear.

Several labs have recently reported the isolation of mutations in the *Drosophila* CAP homolog, referred to as *capulet* (*cap*) or *act up* (*acu*) (7, 10), and the results have offered substantial insights into CAP function. Clones of cells mutant for *cap/acu* exhibit excess actin polymerization in the ovarian germline (7), the columnar ovarian follicle cells (8), and in epithelial cells in the developing eye (10). This phenotype is consistent with CAP normally functioning to inhibit actin polymerization through actin monomer sequestration. However, additional data suggest the relationship between CAP and actin polymerization may be more complex. First, when the *cap/acu* mutant phenotype was compared to the phenotype caused by the mutation in *tsr*, which encodes the *Drosophila* cofilin homolog, the *cap/acu* phenotype was distinct in that ectopic actin only accumulated in discrete locations, whereas in *tsr* mutants ectopic actin was seen throughout the mutant cells (7, 8). This suggests that *cap/acu* may be required to specifically regulate polarized F-actin. Second, *chickadee* mutations were genetically epistatic to *cap/acu* mutations, indicating that the ectopic actin polymerization seen in CAP-deficient cells is dependent on profilin function (8, 10). Further, Baum & Perrimon (8) showed that mutations in *enabled* (*ena*), a *Drosophila* VASP relative, are also epistatic to *cap/acu* mutations. Ena/VASP proteins are modular proteins that can link cellular signaling components to the actin cytoskeleton by their ability to interact with cellular signaling components and their ability to bind F-actin. Taken together, these results suggest that loss of CAP function may result in an increase in F-actin by perturbing the regulation of polarized F-actin assembly.

Cofilin/ADF

Observations of actin filament turnover in cells indicate that F-actin depolymerizes at much higher rates in vivo than experiments with purified F-actin would predict [reviewed in (158)]. Proteins of the cofilin/ADF (Actin-depolymerizing factor) family are important for accelerating the disassembly of F-actin in vivo [reviewed in (4)], for example, in motile cells [reviewed in (108)]. In vitro, cofilin/ADF proteins bind to both F-actin and G-actin, and under many conditions dramatically accelerate actin filament depolymerization. ADF/cofilin has been reported to possess two activities that could account for its depolymerizing activity: filament severing and the ability to increase the actin monomer off-rate [reviewed in (82)]. Which one of these activities is more relevant to cofilin's depolymerization activity is somewhat controversial, but it is possible that both activities contribute to cofilin-accelerated depolymerization [discussed in (82)]. Although most data support a primary role for cofilin in depolymerizing F-actin, recent data from cultured cell experiments suggest that cofilin severing activity is required to generate free barbed ends to generate new actin polymerization sites, possibly in cooperation with the Arp2/3 complex (63, 157).

Mutations in cofilin/ADF were isolated in *Drosophila* as mutants defective in cell division. Among the defects observed was a failure of meiotic spindle asters to migrate, leaving the two asters adjacent to one another, which inspired the name *twinstar* (*tsr*) (50). In addition to defects in aster migration, *tsr* mutants were defective in cytokinesis, and cells accumulated F-actin aggregates. In studies of *cap/acu* function in the ovary, Baum and colleagues also examined both germline (7) and follicle cell (8) clones of *tsr*. In both instances, *tsr* mutant cells accumulated F-actin aggregates. It was not clear, however, whether specific F-actin structures such as ring canals or nurse cell cytoplasmic bundles were affected.

A genetic analysis of tissue morphogenesis in ovarian development identified a novel *tsr* mutation, designated *tsr^{ntf}* (ntf = no terminal filament), that specifically affects an ovarian structure called the terminal filament (22). The terminal filament consists of a stack of follicle cells located at the anterior end of the germarium (45), and these cells appear to be important in regulating germline proliferation (66). The terminal filament forms during ovarian development by a complex cell sorting and intercalation process (45). In *tsr^{ntf}* mutants, terminal filament cells are present, but fail to undergo the rearrangements and movements required to form the terminal filament structure. In addition, when *tsr^{ntf}* was combined with other *tsr* alleles, defects in border cell migration resulted. When F-actin distribution was examined in developing *tsr^{ntf}* ovaries, extremely high levels of F-actin were evident in the terminal filament cells, consistent with previous results. The tissue specificity of the *tsr^{ntf}* mutation is likely due to the presence of a Gypsy transposable element in the *tsr* gene; Gypsy elements frequently block interactions between tissue-specific enhancer elements and their promoters (25).

Taken together, results from the genetic analysis of *tsr* provide evidence that cofilin/ADF proteins are required for F-actin depolymerizing in vivo; this finding is consistent with genetic analysis of cofilin in *Saccharomyces cerevisiae*. In addition, these results provide the first genetic evidence that cofilin/ADF activity is required for cell migratory movements. These results are fully consistent with a requirement for cofilin/ADF in depolymerizing F-actin during cell migration. However, cofilin/ADF-mediated F-actin severing may also be required to support new F-actin assembly during migration, as suggested by work on cultured cells [(157); discussed in (23)].

ACTIN CROSSLINKING PROTEINS

F-Actin Bundling Proteins that Promote Parallel Bundle Formation

Null mutations in the genes *singed* and *quail* result in female sterility due to a specific defect in the formation of the parallel cytoplasmic actin bundles that form in stage 10 nurse cells. The molecular characterization of *singed* showed that it encodes the *Drosophila* homolog of the actin-bundling protein fascin (15, 19, 106), while *quail* encodes a protein in the villin family of actin-binding proteins (85).

The actin-binding properties of fascin and villin have been extensively characterized. Fascin is a ~55-kDa globular, monomeric actin-bundling protein [recently reviewed in (73)]. In vitro, fascin bundles actin filaments into highly ordered arrays, with all filaments in the bundle exhibiting the same polarity. The presence of one F-actin binding site has been mapped to a C-terminal proteolytic fragment of the fascin (104), and a second actin-binding site may reside in a highly conserved N-terminal region that is similar to the F-actin binding site in the MARCKS protein (104). Interestingly, the MARCKS homology sequence contains a consensus PKC phosphorylation site that can be phosphorylated in vivo and is phosphorylated by PKC in vitro (104). Phosphorylation of the serine at this site inhibits F-actin binding, suggesting that the bundling activity of fascin may be regulated. However, mutation of the homologous serine in *Drosophila* fascin does not appear to interfere with fascin function in the ovary (K. Ayers & L. Cooley, unpublished observations).

Villin was isolated as an actin-binding protein constituent of the intestinal brush border microvilli [reviewed in (98)]. Structurally, villin shares extensive homology to the actin capping/severing protein gelsolin, as the N-terminal portions of each protein consist of six homologous repeats. However, villin differs from gelsolin in organization by having a C-terminal ~8-kDa extension termed the headpiece; the headpiece contains a second F-actin binding site that is required for F-actin crosslinking (Figure 4A). In vitro, villin exhibits a diverse range of actin-modulating activities that are regulated by levels of free Ca^{++}. At submicromolar Ca^{++}, villin bundles F-actin and has no other effect on actin dynamics. At micromolar Ca^{++} concentrations and above, villin can sever preexisting actin filaments, cap the fast-growing barbed ends of actin filaments, and also nucleate actin filament assembly. In contrast to filaments nucleated by the Arp2/3 complex, filaments nucleated by villin elongate from their slow-growing (pointed) ends (44).

In many systems, F-actin bundles in vivo are associated with more than one actin-bundling protein (5, 29). Whether multiple actin-bundling proteins provide essential functions or are redundant can be tested genetically in *Drosophila*. As mentioned above, mutations in either *singed* or *quail* fail to form cytoplasmic actin bundles, indicating that these proteins are not functionally redundant. The relative requirements of *quail* and *singed* have been examined further through a series of experiments altering the expression levels of these two genes (18). First, a partial loss-of-function mutation in *quail* was found to dominantly enhance a weak *singed* allele with reduced fertility, again demonstrating a requirement for both proteins. It was found that a weak *singed* mutation could be completely rescued by doubling the copy number of wild-type *quail*, and these nurse cells produced robust actin bundles. However, doubling the dosage of *quail* in a *singed* null background could only partially rescue the *singed* fertility defect. In these partially rescued nurse cells, cytoplasmic actin bundles appeared to initiate assembly based on immunofluorescence, but subtle ultrastructural defects were apparent. The short bundle modules possibly did not adhere to one another as they do in wild type to form long actin bundles. The reverse experiment has also been done. Doubling the copy number of wild-type *singed* genes in a *quail* mutant background does not rescue the *quail* phenotype (K. Ayers & L. Cooley, unpublished observations). These results suggest that Quail is perhaps the more important protein for initiating the assembly of the actin bundles, and that fascin provides a distinct function in maintaining their integrity, and perhaps also participates in linking bundle modules together to form long bundles.

The effect of free Ca^{++} on *Drosophila* Quail has been carefully explored (91). In contrast to villin, Quail bound and bundled F-actin independent of Ca^{++} concentration, producing tight bundles of actin filaments highly similar to villin-bundled F-actin. A possible reason for this functional divergence between Quail and villin was revealed when the distribution of free Ca^{++} in live, late-stage egg chambers was examined (91). Free Ca^{++} levels in the nurse cell cytoplasm rise dramatically at stage 11 as the nurse cells permeabilize their nuclear membranes during apoptosis. Thus, the presence of a Ca^{++}-dependent filament severing activity could have disastrous consequences for the integrity of the nurse cell cytoplasmic actin bundles. Indeed, when human villin was expressed in mutant nurse cells with reduced Quail activity, human villin appeared to have a dominant-negative effect consistent with a Ca^{++}-activated severing activity. While actin bundles began to form at stage 10, and human villin initially localized to these bundles, bundles were absent at late stages when free cytoplasmic Ca^{++} increased, suggesting that filaments bundled by human villin were severed as free Ca^{++} levels became elevated.

The detailed analysis of genetic phenotypes and biochemical behavior of both Quail and villin has been instructive. While sequence similarity of Quail to villin initially suggested mechanisms by which Quail could participate in bundle formation, in vitro experiments and in vivo observations were essential to developing an accurate understanding of Quail function. Thus, as new proteins are identified based on sequence similarity to villin [e.g., (70)], ascribing in vivo function will often require further investigation. Villin can be viewed as a gelsolin-like core to which an F-actin binding headpiece has been appended, creating an F-actin-bundling protein. The particular features of gelsolin [e.g., severing, capping, Ca^{++}-sensitivity] that are preserved or discarded through evolution may then determine the in vivo function of the protein. Indeed, careful sequence analysis of Quail revealed that at least one Ca^{++} binding site is poorly conserved in Quail, consistent with its Ca^{++} insensitivity (91).

F-Actin Crosslinking Proteins Associated with Ring Canal Actin

Screens for mutations that affect nurse cell cytoplasm transport led to the identification of genes encoding F-actin crosslinking proteins. These include *cheerio* (*cher*), which encodes a *Drosophila* filamin/ABP-280 molecule, and *kelch*, which encodes a protein with homology to the F-actin crosslinking protein scruin. In addition, HtsRC, a novel 60-kDa protein product of the *hu-li-tai shao* (*hts*) gene is required for F-actin accumulation at ring canals (111, 154). Of these proteins, the functions of filamin and kelch are the best characterized.

KELCH Kelch (the German word for goblet; *kelch* mutant females lay small, cup-like eggs) was the founding member of a family of proteins containing a series of ~50 amino acid repeats, designated kelch repeats. The crystal structure of one kelch-repeat protein, galactose oxidase (64), revealed that the kelch repeats form a β-barrel structure, similar to WD-repeat proteins. Based on this structure and conserved sequence motifs, other kelch-repeat proteins have been predicted to adopt a similar β-barrel structure (14). Like WD-repeat proteins, kelch-repeat proteins participate in a diverse array of cellular processes; thus, the mere presence of kelch repeats does not suggest function [for detailed review of Kelch proteins, see (1)]. The domain structure of *Drosophila* Kelch also includes a BTB dimerization domain near the N terminus of the protein, and this domain is preceded by a short, glutamine rich N-terminal sequence (Figure 4A) (113, 153).

Several kelch-repeat proteins interact directly with actin. These include α-scruin from *Limulus*, which crosslinks actin filaments in the sperm acrosomal process, IPP, and Mayven from human. The phenotype of *kelch* mutant egg chambers suggests that Kelch could also be an actin-binding protein. In *kelch* mutant ring canals F-actin is abundant but highly disorganized, and the ring canal lumens become partially occluded with the disorganized F-actin (111, 140). Structure-function studies on *Drosophila* Kelch led to the model that Kelch is recruited to ring canals where it functions as a dimer to form low-affinity F-actin crosslinks that both organize F-actin and allow actin filament sliding to accommodate ring canal expansion (113).

In a recent study specific elements of this model were tested (69), and additional significant insights were provided into the mechanism of ring canal growth. The model proposed by Robinson & Cooley (113) assumed that the Kelch-repeat domain contains an F-actin binding site, but this had not been demonstrated. Using purified recombinant protein, Kelch was shown to bind to F-actin, and through purification of individual kelch repeats, the actin-binding domain was mapped to kelch repeat 5. This repeat corresponds to the region in α-scruin that binds to F-actin (131). In addition, purified full-length Kelch was shown to bundle actin filaments, suggesting that the general model for actin crosslinking by Kelch was correct.

Further experiments revealed that the affinity of Kelch for F-actin is regulated by phosphorylation. In vivo, Kelch was shown to exist in both tyrosine-phosphorylated and non-phosphorylated forms, and the phosphorylation site was also mapped to repeat 5. Both endogenous phosphorylated Kelch, as well as recombinant Kelch containing a phosphomimetic mutation, displayed a markedly reduced affinity for F-actin in vitro. When a non-phosphorylatable Kelch (Kel[Y627A]) was expressed in the ovary in place of wild-type Kelch, several defects in ring canal growth and organization were observed. First, electron microscopy of Kel[Y627A]-expressing egg chambers revealed that ring canal F-actin was unable to reorganize into a number of subbundles, as occurs late in wild-type oogenesis. This defect was accompanied by an increase in the number of filaments observed in cross-section, as well as a decrease in ring canal diameter relative to wild type. Further, an analysis of actin turnover at ring canals using FRAP revealed that GFP-actin recovery rates were significantly lower in egg chambers expressing Kel^{Y627A} than in wild type. These results suggest that Kelch phosphorylation is required for Kelch to properly organize ring canal F-actin, and underscores the importance of Kelch in accommodating the orderly addition of F-actin to the ring canal. The phenotypes associated with expression of Kel^{Y627A} are remarkably similar to *Src64*, a mutant for which the only described phenotype is a defect in ring canal growth (32, 69); this observation is discussed further below.

FILAMIN   Filamin/ABP-280 (hereafter referred to as FLN) was initially isolated as an actin-binding protein from chick [the history, structure, and function of FLNs have been recently reviewed in (129)]. FLNs are extended proteins containing an N-terminal actin-binding domain followed by a series of 20 to 24 repeats, each of which folds into a compact structure termed a beta-sandwich (Figure 4A). The repeats are interrupted by two linker sequences, termed hinges, which confer flexibility to the molecule. FLNs form dimers through a domain located in their C-terminal repeat, and extended, V-shaped FLN dimers can be observed crosslinking actin filaments by electron microscopy. FLNs are extremely potent crosslinkers of F-actin, and the immunolocalization of FLN to orthogonal F-actin arrays in vivo underscores the relevance of this function. At high FLN:actin ratios, FLNs have also been reported to organize F-actin into parallel bundles in vitro (55), although the physiological significance of this observation is not clear. A large number of molecules have been identified as FLN binding partners, many binding to the C-terminal repeats. Many of these proteins are transmembrane proteins and signal transduction molecules, and it has been hypothesized that FLNs link the F-actin cytoskeleton to the plasma membrane, and also that FLNs may serve as scaffolds for signal transduction components.

The *cheerio* mutation was isolated as a female-sterile mutant that exhibited defective ring canals. In *cheerio* mutant egg chambers, F-actin does not accumulate at ring canals, although other ring canal markers, such as phosphotyrosine, are still present (75, 116). *cheerio* was found to encode a *Drosophila* FLN protein, and immunolocalization of FLN revealed that it was highly enriched at ring canals (75, 124). At least two protein products are produced from the *cheerio* locus. FLN1-20 is similar in organization to vertebrate FLNs, although it contains four fewer repeats. FLN12-20 is a 90-kDa protein consisting of only the C-terminal nine repeats, but does not contain the N-terminal actin-binding domain (Figure 4A). FLN12-20 expression in the ovary is restricted to the somatic muscle sheath that surrounds each ovariole. However, truncation mutations that eliminate FLN12-20 and the C-terminal domain of FLN1-20 are viable and do not display defects other than ring canal defects associated with loss of FLN1-20.

The analysis of *cheerio* in the ovarian germline provides several general insights into FLN function. First, FLN has been best characterized as an F-actin crosslinking protein that produces orthogonal networks of filaments. Although it is difficult to know precisely how ring canal actin filaments are organized, they have been described as loosely packed, circumferentially oriented, parallel filaments based on transmission electron microscopy (140). The requirement for FLN for ring canal F-actin organization suggests that the ability of FLN to form parallel F-actin bundles under some conditions in vitro (55) may have physiological relevance. Second, the strict requirement for full-length FLN1-20 in the germline indicates that the C-terminal repeats are essential for ring canals, even as they appear otherwise dispensable in the fly (discussed below). As the binding sites for many transmembrane proteins are found in the C-terminal domain of FLN, it is tempting to speculate that at the ring canal, FLN may function in part to link the ring canal inner rim F-actin to the membrane-associated outer rim. Consistent with this idea, FLN was shown to be a component of both the actin-rich inner rim and the membrane-associated outer rim of ring canals (124). In addition, FLN localization to ring canals follows the localization of D-mucin, a transmembrane glycoprotein that is the earliest known marker of ring canals.

F-Actin-Binding Proteins Involved in Follicle Cell Morphogenesis

A number of recent studies have investigated the role of actin crosslinking proteins in the shaping the somatic follicle cells. The follicular epithelium has been used as one model to examine the requirements for spectrins in tissue morphogenesis, the establishment and maintenance of apical-basal polarity, and cell migration. In addition, filamin mutations that specifically affect follicle cell morphogenesis have been characterized, providing an important genetic model for understanding how filamin proteins participate in cell movements.

SPECTRIN   Spectrins are actin crosslinking proteins that were first characterized as components of the erythrocyte cytoskeleton (11). Two basic types of spectrin exist, α-spectrin and β-spectrin; each consists of a tandem array of 106 amino acid spectrin repeats, but they exhibit differences at their termini. β-spectrin possesses an actin-binding domain at its N terminus, whereas the α-spectrin C-terminus contains an EF-hand Ca^{++} binding domain (Figure 4A). These molecules form antiparallel heterodimers that associate in a head-to-head fashion to form elongated tetramers. The β-spectrin actin-binding domains are located at either end of the tetramer, creating an extended F-actin crosslinking molecule (Figure 4B). β-spectrin molecules also possess a C-terminal pleckstrin homology domain, as well as an internal binding site for the membrane protein ankyrin, both of which may mediate the association of spectrin with membranes. Spectrins are thus thought to participate in the organization and formation of the membrane cytoskeleton.

Studies in *Drosophila* led to the identification of a novel β-spectrin isoform termed β_{Heavy}-spectrin (β_H-spectrin) (33). The β_H-spectrin isoform is larger than conventional β-spectrin (430 kDa vs. 230 kDa) and forms an elongated tetramer with α-spectrin to form a longer (240 nm) tetrameric F-actin crosslinker. Immunolocalization studies revealed that in the polarized epithelium of the embryo, β-spectrin and β_H-spectrin localize to distinct membrane domains: β-spectrin is found on lateral membranes, and β_H-spectrin is restricted to apical membranes (138). Each is presumably associated with α-spectrin, which localizes to both domains (Figure 4B). Recently, a mammalian ortholog of β_H-spectrin was identified, though it does not appear to be differentially localized in most polarized epithelia (128).

In egg chambers, the three spectrin isoforms exhibit a similar distribution in the follicular epithelium: α-spectrin is localized to both lateral and apical cortices, β-spectrin is restricted to lateral membranes, and β_H-spectrin is localized to the apical domain (74, 156). Analysis of follicle cells mutant for either α-spectrin (α-spec) (74) or karst (kar), the gene encoding β_H-spectrin (156), indicates that these two proteins are dependent on one another for their localization to the apical domain. In α-spec mutants, β-spectrin persists on lateral membranes, albeit at reduced levels (74). In the germline, α-spectrin is required for regulating germline cell division in the germarium, and both α-spectrin and β-spectrin localize to the fusome, a vesicular organelle involved in regulating germline proliferation [(27); reviewed in (115, 127)].

The partitioning of β-spectrin and β_H-spectrin into distinct cellular domains raises the question of whether the spectrins participate in the establishment or maintenance of epithelial polarity. Phenotypic analyses of mutations in α-spec (74) and kar (156) indicate that spectrins are not essential to establish apical-basal polarity. In kar mutant females a number of markers for apical-basal polarity were examined, and all remained properly localized throughout oogenesis (156). In α-spec mutant clones that covered the entire follicular epithelium, markers for apical-basal polarity were appropriately localized in follicle cells in younger egg chambers, indicating that α-spectrin is not required to establish apico-basal polarity. However, in elongating stage 9 egg chambers, α-spec mutant follicle cells exhibited phenotypes associated with a loss of polarity. In these egg chambers, follicle cells at the posterior and occasionally the anterior poles lost their monolayer character and formed a disorganized multilayered epithelium. Markers for both the apical and lateral membrane domains became uniformly distributed in cells exhibiting this phenotype. These results suggest that the spectrin cytoskeleton is required to maintain apical polarity in cells such as the posterior follicle cells, which are likely to undergo mechanical stress to accommodate the growing oocyte. The observation that kar mutant cells can maintain apico-basal polarity, even as all detectable spectrin is absent from the apical domains, suggests that the lateral spectrin cytoskeleton may be more important in maintaining polarity in these cells. Alternatively, spectrin function at either the apical or lateral domain could be sufficient to maintain apical-basal polarity. Since β-spectrin is specifically localized to the lateral membranes, analysis of mutants in β-spectrin could resolve this issue, as only lateral spectrin should be affected.

Although the apical-basal polarity of follicle cells lacking β_H-spectrin was apparently unaffected, *kar* mutant follicle cells exhibit other phenotypes that indicate the structural integrity of the apical domain was compromised (156). In *kar* mutants, the stage 9 migration of main body follicle cells to the posterior is significantly delayed relative to germline development and border cell migration. The migration of the main body follicle cells to the posterior of the egg chamber is accompanied by a change in their shape from cuboidal to columnar, and the apical surfaces of these cells constrict as the cells concentrate at the posterior. In *kar* mutants, follicle cells remained somewhat cuboidal, and the apical regions did not constrict as they do in wild type. Indeed, after the migration was completed, the apical regions of *kar* mutant follicle cells actually expanded, whereas wild-type follicle cell apices retain their shape. Further evidence of compromised apical integrity was apparent from the disrupted adherens junction in these cells. How apical spectrin contributes to apical constriction is not clear, but the disruption of the adherens junction in these cells may affect the contractile actin adhesion belt that is associated with the adherens junction.

FILAMIN   In addition to *cheerio* mutations that affect ring canals, a second class of *cher* mutation that specifically disrupts follicle cell morphogenesis has been identified (125). In this class, exemplified by the *cher^{EP}* mutation, egg chamber formation in the germarium is severely affected at the stage when follicle cells surround germline clusters to form a follicular epithelium. In wild type, follicle cells that surround the germarium extend processes between adjacent germline cysts (see Figure 1*C*). This is followed by migration of the follicle cells between cysts, leading to the encapsulation of each cyst and the formation of an interfollicular stalk that separates adjacent cysts. In *cher^{EP}* mutants, the follicle cells in the germarium are unable to complete this process, resulting in the fusion of adjacent germline cysts. In addition to encapsulation defects, these *cher^{EP}* mutations exhibit a delay in border cell migration, further implicating FLN in somatic cell movements.

An analysis of the FLN proteins expressed in *cher* mutants, as well as their tissue distribution, led to surprising conclusions about the in vivo requirements for FLN in *Drosophila*. The *cher^{EP}* allele lacked FLN expression in the somatic follicle cells, whereas germline FLN expression was readily detectable. In contrast, *cher* mutations that specifically disrupted ring canals were often chemically induced mutations that produced truncated FLN proteins. Strikingly, FLN proteins consisting of only the N-terminal actin-binding domain and 5 filamin repeats appeared sufficient to carry out FLN function in follicle cells. To confirm that the N-terminal half of FLN was sufficient for follicle cell morphogenesis, *cher^{EP}* mutations were rescued by a FLN transgene expressing a truncated FLN in the somatic cells.

Together, these mutations have been useful in dissecting the tissue-specific requirements for FLN function in the ovary. Full-length FLN1-20 is expressed in the follicle cells and the nurse cells, and mutations that produce truncated FLN1-20 molecules, or have reduced levels of germline expression, exhibit defective ring canals. In contrast, truncated FLN molecules that presumably lack the FLN dimerization domain can fully support follicle cell function (125). Since the amino-half resembles *Dictyostelium* ABP 120, it is possible that the function of FLN in cell migration is the more ancient role of FLN. Consistent with this, expression of a C-terminally truncated FLN can also rescue the morphological phenotype associated with loss of FLN in human cells (95). Whether monomeric FLN molecules are sufficient for migration in follicle cells, or whether the truncated FLN proteins are still able to dimerize and thus crosslink F-actin, is unclear. In flies, full-length FLN that includes the C-terminal dimerization domain is required only for ring canals. Determining cell types that require full-length FLN in vertebrates will be of great interest.

MYOSIN MOTORS

Myosins are molecular motors that use the energy from ATP hydrolysis to move along actin filaments. Since the discovery of classical muscle myosin (myosin II), a diverse collection of related motors have been discovered, collectively referred to as unconventional myosins [reviewed in (94, 122)]. The in vivo functions of many of these myosins are only beginning to be understood. Two classes of myosin have been examined in *Drosophila* oogenesis: cytoplasmic myosin II and myosin VI.

Myosin II

Myosin II, or conventional myosin, is a filament-forming myosin that mediates contractile sliding of actin filaments. Myosin II is termed a two-headed myosin, as it consists of a dimer of two myosin heavy chain motor domains. Each heavy chain is associated with two light chains, an essential light chain (EMLC) and a regulatory light chain (RMLC). Myosin II activity is regulated primarily through phosphorylation of the RMLC by a Ca^{++}/calmodulin-activated myosin light chain kinase (MLCK). Phosphorylation of specific N-terminal serine and threonine residues in the RMLC stimulates the motor activity of myosin II, leading to contraction (68). Cytoplasmic myosin II has been implicated in a diverse array of actin-based processes, including cytokinesis, cell motility, and cell spreading [reviewed in (81)].

Myosin II function in *Drosophila* oogenesis has been examined through the genetic analysis of the RMLC, encoded by the *spaghetti squash* (*sqh*) gene. Previously, it was shown that reducing, but not eliminating, *sqh* activity resulted in a failure to complete cytoplasm transport to the oocyte due to a specific defect in contraction; actin bundles and ring canals appeared unaffected (34, 149). The nature of myosin II regulation in the germline has been examined further, through the analysis of the N-terminal phosphorylation sites of the RMLC (67). Although the importance of RMLC phosphorylation for the activation of myosin II has been well established for smooth muscle (141), the role of phosphorylation in non-muscle cells is less clear. The two phosphorylated residues in vertebrate RMLCs are conserved in the *Drosophila* RMLC, and these were mutated either singly or in combination. Changing both phosphorylation sites (Thr20, Ser21) resulted in a lethal phenotype indistinguishable from a *sqh* null mutation, indicating that the phosphorylation sites are essential.

In flies expressing only the mutant RMLC proteins in the ovarian germline, defects in cytokinesis and germline contraction were apparent (67), as had been previously observed in loss-of-function *sqh* germline clones (34, 149). However, in contrast to the earlier studies of *sqh* partial loss-of-function mutations, moderately penetrant defects in ring canal morphology were observed in egg chambers expressing only non-phosphorylatable or phosphomimetic RMLC proteins (67). In addition, similar defects were observed when a null allele of *sqh* was examined. However, these defects were not severe enough to compromise cytoplasm transport. While these data suggest that myosin II may function in maintaining ring canal integrity, the authors note that the defects could also be an indirect consequence of assembling ring canals at the site of a defective contractile ring.

Myosin VI

Myosin VI is distinct among all known myosins in that is moves along actin filaments toward the pointed, or slow-growing end of filaments (148). As all other myosins examined move toward the barbed end, the implications of a pointed end–directed myosin are intriguing [discussed in (26)]. First, only cargo carried by a myosin VI motor will be able to track along unipolar actin filaments in a direction opposite that of other cargo carrying myosins. Second, since myosin VI is predicted to dimerize and form a two-headed myosin (57), myosin VI dimers could interact with bipolar actin filaments to generate an expansive, pushing force instead of the contractile force associated with conventional myosin II.

Mutations in myosin VI genes are associated with deafness in mice (2) and humans (92), and these mutations lead to a degeneration of the hair cell stereocilia, the specialized microvilli that mediate auditory mechanotransduction [reviewed in (43)]. Immunolocalization of myosin VI in hair cells revealed it to be localized to the cuticular plate of inner ear hair cells. The cuticular plate is an actin-rich structure linking the stereocilia of the hair bundle to the hair cell body, suggesting that myosin VI functions to anchor the actin bundles of the stereocilia at the cuticular plate (56). In addition, myosin VI has been reported to be recruited to lamellipodia, membrane ruffles, and Golgi complex in cultured cells stimulated with growth factor, suggesting a role in dynamic actin rearrangements (16). In *Drosophila*, myosin VI is encoded by *jaguar* (21, 58). Existing *jaguar* mutants are male-sterile, and myosin VI expression is specifically affected in spermatogenesis, where myosin VI is required for spermatid individualization (58). Antibody inhibition studies in *Drosophila* embryos have implicated myosin VI in vesicle movement (93).

The function of myosin VI in oogenesis has been addressed in two recent studies. By injecting fluorescently labeled myosin VI monoclonal antibodies into nurse cells, numerous labeled particles were observed moving vectorially toward the oocyte (12). When an inhibitory myosin VI polyclonal antibody was injected, vectorial particle movement ceased. These results suggest that myosin VI mediates vectorial particle movement in *Drosophila* nurse cells, as has been shown in embryos (93). In addition, these observations suggest that a polarized population of F-actin exists in nurse cells to accommodate this transport; however, such filaments have not been characterized.

In an additional observation made during this study, myosin VI monoclonal antibody labeled ring canals in live egg chambers in a manner that suggested localization to the actin-rich inner rim (12). In considering the possible functions of a pointed end–directed motor, Cramer (26) proposed that myosin VI might slide filaments apart during ring canal expansion, and the localization of myosin VI to ring canals is consistent with this idea. However, ring canal localization was not reported when the same monoclonal antibody was used to examine the distribution of myosin VI in fixed egg chambers (28). The use of other antibodies and fixation techniques could resolve the issue of myosin VI localization at ring canals. Determining whether myosin VI has a function in ring canal morphogenesis may require the identification of stronger alleles of the myosin VI gene, *jaguar*, so that ovarian phenotypes can be examined.

In a second study, an enhancer-trap transposon insertion was isolated that expressed high levels of a reporter gene in migrating cells during oogenesis (28). Subsequent analysis revealed that the transposon was inserted into the myosin VI gene. To examine the functional requirements of myosin VI in oogenesis, an antisense expression system was established that resulted in a reduction of myosin VI protein levels. In oogenesis, the antisense RNA was highly expressed in the main body follicle cells that migrate posteriorly at stage 8/9. In these egg chambers, the follicle cells adopted aberrant morphologies, sometimes lost monolayer character, and often failed to migrate to the posterior, implicating myosin VI in migration and maintaining cell morphology. Antisense myosin VI RNA was not present at high levels in either the border cells or the germline cells, and therefore examination of phenotypes in these tissues was not possible.

SIGNAL TRANSDUCTION PROTEINS AND THE ACTIN CYTOSKELETON

Rho Family GTPases

The Rho family of small GTPases, consisting of Rho, Rac, and Cdc42, are important regulators of the actin cytoskeleton. Numerous experiments implicate these proteins as molecular switches to orchestrate rearrangements of the actin cytoskeleton, and cell culture studies indicate that Rho induces stress fiber formation, Rac promotes lamellipodia and membrane ruffling, and Cdc42 initiates filopodia and microspikes [reviewed in (159)]. In addition, the Rho family GTPases have also been implicated in signal transduction events leading to changes in gene transcription, principally through the Jun N-terminal Kinase and Map Kinase signaling pathways [for review, see (133)].

With the completion of the *Drosophila* genome sequence, it appears that there are seven Rho family members in flies. Notable among them are three genes closely related to *Rac* (*Rac1*, *Rac2*, and *Mtl*), as well as three genes highly homologous to *Rho* (*Rho1*, *RhoL*, and *RhoBTB*). All but *RhoBTB* have been previously characterized. *RhoBTB* represents the *Drosophila* member of a recently described conserved class of *Rho*-related genes (110). The *RhoBTB* genes are predicted to encode proteins with an N terminus highly homologous to *Rho*, with an extended C terminus that contains a proline-rich domain and two BTB domains arranged in tandem, followed by a conserved C-terminal domain. Although these genes are conserved among humans, *Dictyostelium*, and *Drosophila*, nothing is known about their function.

Previously, the functions of the Rho family GTPases during oogenesis were examined using transgenic expression of dominant-negative and constitutively active Rho family GTPases (102). In mid-oogenesis, alterations in *Cdc42* or *RhoL* activity resulted in destabilization of cortical membranes, and in the case of *RhoL*, defects in adhesion between germ cells and follicle cells. Expression of dominant-negative *Cdc42*, *Rac1*, or *RhoL* prior to stage 10 disrupted cytoplasmic bundle formation, suggesting a requirement for each of these genes in cytoplasmic actin bundle assembly. In follicle cells, expression of dominant-negative *Rac1*, but not *Cdc42* or *RhoL*, strongly inhibited border cell migration.

Recently, loss-of-function mutations in *Cdc42* (36), *Rho1* (83, 130), and the three *Rac* genes (54, 103) have been reported. These mutations have been useful in more carefully dissecting the in vivo functions of these genes.

The genetics of the *Drosophila Rac* genes *Rac1*, *Rac2*, and *Mtl* are complex (54, 103). Null mutants for any one of these genes are viable and fertile, as are double mutants for *Rac2* and *Mtl*; however, all other double mutant combinations are inviable. Analysis of various combinations of Rac mutations revealed differential requirements for these three genes in specific tissues and processes (54, 103). In an effort to remove the maternal contribution of the three *Rac* genes from embryos, Hakeda-Suzuki et al. (54) engineered an elegant strategy to produce *Rac1*, *Rac2*, and *Mtl* triple-mutant germline clones. Embryos were not recovered, however, indicating that Rac function is required in the female germline for completion of oogenesis. Given the ring canal phenotype of *SCAR* and Arp2/3 complex mutations, it will be interesting to determine whether Rac function is also required for ring canal actin accumulation, or whether Rac is required for cytoplasmic actin bundle formation, as suggested by expression of dominant-negative *Rac1*.

A study of the developmental requirements of *Rho1* revealed possible functions for *Rho1* during oogenesis (83). Germline clones of homozygous *Rho1^{null}* cells were not recovered, indicating that *Rho1* is necessary for cell viability or proliferation (83, 130). To circumvent this problem, *Rho1^{null}* was made doubly heterozygous with $RpII140^{wimp}$, an allele of an RNA polymerase subunit (83). The $RpII140^{wimp}$ background, which presumably reduces global transcription levels, provided a sensitized background in which $Rho1^{null}/+$ could be examined for effects. While $Rho1^{null}/+$; $RpII140^{wimp}/+$ females were fertile, the ovaries exhibited subtle defects in the germline actin cytoskeleton. Ring canals, while apparently stable and of normal size, appeared to have a peripheral ring of actin bundles emanating from the surrounding nurse cell membranes, giving the ring canals a fuzzy appearance at low magnification. In addition, the cortical actin in the oocytes appeared to be more loosely organized than in wild type. These results suggest that $Rho1$ may function to organize the ovarian F-actin cytoskeleton, but it is not clear how. Other methods that reduce $Rho1$ function in the ovary, such as clonal analysis using weaker alleles of $Rho1$ (130), may provide more revealing phenotypes.

Loss-of-function mutations in the *Drosophila* homolog of CDC42 have recently been characterized (36, 41). In the germline, loss of $Cdc42$ appears to have relatively little effect early in oogenesis, and the cortical membrane instability associated with dominant-negative Cdc42 expression was not evident (41). Consistent with what was observed with dominant-negative Cdc42 expression, a reduction in cytoplasmic actin bundle assembly was reported in late-stage germline clones. In addition, a myc-tagged Cdc42 protein expressed in the ovary appeared to be associated with the cytoplasmic bundles near the cortical membranes where the bundles initiate, suggesting a function for Cdc42 in cytoplasmic bundle assembly. This result is intriguing, as *Drosophila Wasp* and the Arp2/3 complex, the best-characterized downstream cytoskeletal effectors of Cdc42, are not required for actin bundle assembly, as discussed above. These results are consistent with recent work suggesting the existence of cytoskeletal effectors other than the Arp2/3 complex being downstream of Cdc42 (72).

In somatic follicle cells, $Cdc42$ mutant clones exhibit a more dramatic phenotype (41). Follicle cells that normally adopt a highly polarized, columnar cell shape late in oogenesis failed to do so in mutant clones, and instead often appeared irregular in shape. In addition, $Cdc42$ mutant follicle cells were no longer maintained as a monolayer. By morphology alone, $Cdc42$ mutant follicle cells appeared to have lost polarity, consistent with a well-established role for CDC42 in cell polarity in other systems (65). However, it was reported that several molecular markers for apico-basal polarity, including the adherens junction protein armadillo/β-catenin, were properly localized in $Cdc42$ mutant cells.

The Src and Tec Protein Tyrosine Kinases

Src family protein tyrosine kinases were first identified as transforming proteins in avian tumor viruses. Intensive study over the past 25 years has implicated Src kinases in diverse array of cellular functions, including regulation of proliferation, differentiation, and F-actin organization [reviewed in (139)]. Ample evidence implicates Src family kinases in regulating actin cytoskeletal dynamics. Src kinases localize to many sites where F-actin and actin-binding proteins are localized, and

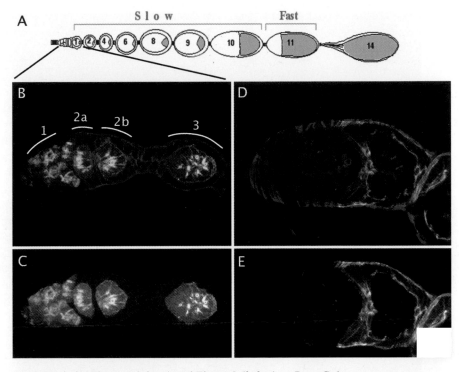

See legends for Figure 1 (*above*) and Figure 3 (*below*) on Page C-4

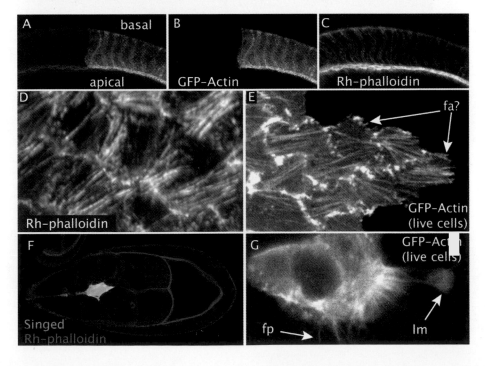

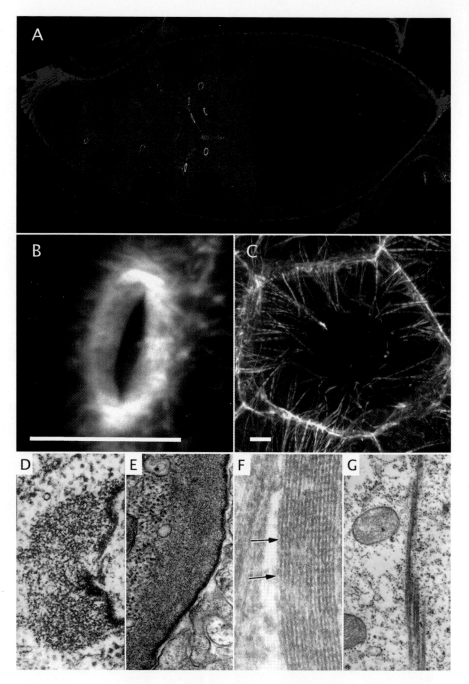

See legends for Figure 2 (*above*) and Figure 4 (*opposite*) on Page C-4

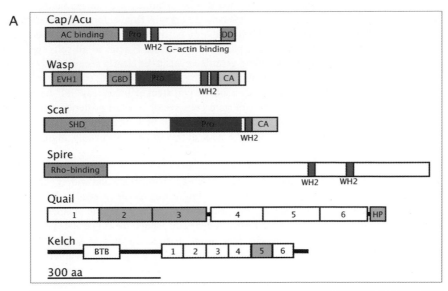

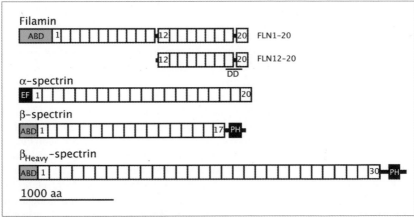

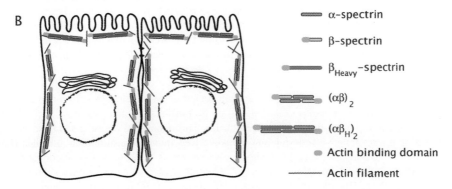

Figure 1 Overview of oogenesis. (*A*) Schematic of an ovariole, with the germarium at the anterior on the left. Progressively maturing egg chambers are arrayed from anterior to posterior; gray cell in each egg chamber is the oocyte. (*B, C*) Close-up of a germarium expressing GFP-Actin in the germ cells (green) and stained for rhodamine-phalloidin to reveal F-actin in all cells (red). GFP-Actin expression allows the shape of developing germline cysts to be clearly seen in *C*. (*D, E*) Germarium stained for rhodamine phalloidin (red) and containing a clone of follicle cells expressing GFP-Actin (green). The follicle cell clone is probably a follicle cell stem cell clone, as GFP-Actin expression is observed beginning where the follicle cell stem cells reside, and all posterior cells are expressing GFP-Actin. The GFP-Actin highlights the follicle cells as they migrate between germ cell clusters to encapsulate a cyst; note the high level of F-actin at tips of the centripetally migrating follicle cells (*E*).

Figure 2 Germline actin structures. (*A*) Stage 10B egg chamber stained with rhodamine phalloidin to reveal F-actin (red), and HtsRC antibody to highlight ring canals (green; HtsRC colocalizes with F-actin, producing yellow ring canals). Both ring canals and cytoplasmic bundles are apparent. (*B*) Close-up view of a ring canal; F-actin forms a tight band around the ring canal, as revealed by phalloidin staining. (*C*) Phalloidin-stained cytoplasmic F-actin bundles at stage 10B; note striated appearance. (*D, E*) Ring canals viewed in cross section (*D*) and grazing section (*E*) by thin section electron microscopy [adapted from (140)]. Note the electron-dense outer rim on the right in *D* and *E*, and the loosely bundled actin filaments that line the outer rim. Actin filaments in the ring canal do not appear to be highly ordered. (*F, G*) Thin section electron micrographs of cytoplasmic actin bundles [adapted from (49)]. Note the transverse banding of filaments in *F*; this is indicative of maximally crosslinked, hexagonally packed actin filaments. In *G*, three F-actin bundle modules can be seen overlapping; the regions of overlap are likely to create the brightly fluorescent regions observed in the striated bundles shown in *C*. Scale bar in B and C: 10 μm.

Figure 3 F-actin in follicle cells. (*A–C*) F-actin (red in *A* and white in *C*) is polarized in the columnar follicular epithelial cells that overlie the oocyte late in oogenesis. A clone of follicle cells expressing GFP-Actin is shown in green in *A* and white in *B*, and allows the apical F-actin in the follicle cells to be distinguished from the cortical oocyte actin. (*D, E*) Basal actin bundles in columnar follicle cells visualized with rhodamine-phalloidin in *D* and as a clone of GFP-Actin in live cells in *E*. The clone of cells expressing GFP-Actin in *E* allows the shape of the GFP-Actin expressing cells to be clearly seen, and the basal F-actin bundles appear to terminate in structures that interdigitate into neighboring cells. These terminal structures may be similar to focal adhesions (fa?), but are also reminiscent of the "adhesion zippers" observed in Reference (143). (*F*) Migrating border cells are marked by their strong expression of singed, the *Drosophila* fascin homolog. The border cells are a cluster of 8–10 cells that migrate from the anterior of the egg chamber through the nurse cells to the oocyte. (*G*) A clone of live border cells expressing GFP-Actin as they migrate. Finer F-actin structures such as filopodia (fp) and lamellipodia (lm) are more easily visualized in live cells than in fixed tissue.

Figure 4 (*A*) Schematic depicting the domain organization of selected actin-binding proteins discussed in this review. Lengths of schematics are proportional to molecular mass; note that proteins in each boxed region are drawn to different scales. Throughout the figure, green boxes indicate predicted or demonstrated F-actin-binding domains. Red indicates proline-rich stretches (Pro), blue indicates WH2 domains (wasp-homology 2, also known as verprolin homology domains), and yellow indicates Arp2/3 complex-binding CA (central–acidic) domains. Other abbreviations: AC, adenylyl cyclase; DD, dimerization domain; HP, headpiece; BTB, BTB-type dimerization domain; EF, EF-hand domain; PH, pleckstrin homology domain. (*B*) Cartoon of epithelial cells depicting subcellular localization of different *Drosophila* spectrin heterotetramers in a polarized epithelium.

a number of Src substrates and binding proteins are known actin-binding proteins. In addition, osteoclasts derived from $src^{-/-}$ mice exhibit defects in their F-actin cytoskeleton, and exhibit reduced motility.

Mutations in a *Drosophila* Src homolog, *Src64*, have been identified (32). All mutants thus far identified are fully viable, and the only detectable phenotype is female sterility. The *Src64* locus produces several transcripts, and the lesions associated with female sterility affect a distal promoter, raising the possibility that the *Src64* mutations isolated thus far may affect a germline-specific transcript in a manner similar to female-sterile profilin mutations.

The female sterility of *Src64* mutant flies is due to a specific defect in ring canal growth. Ring canals from *Src64* mutant egg chambers were on average 40% smaller than wild-type ring canals (32). In addition, the intense phosphotyrosine labeling observed in wild-type ring canals was eliminated in *Src64* mutants. All other aspects of ring canal morphology appeared to be normal in *Src64* mutants. The ring canals appeared to be normally shaped, and the cytoskeletal components Kelch, HtsRC (32), filamin (124), and the Arp2/3 complex component Arpc3B (A.M. Hudson & L. Cooley, unpublished observations) all localized to ring canals, as in wild type. Antibodies raised against the Src64 protein label cortical F-actin in the germline and appear somewhat enriched at ring canals (32).

The involvement of *Src64* in ring canal growth presented the opportunity to examine the links between Src activity and cytoskeletal reorganization. Two groups recently reported on the interaction between *Src64* and the *Drosophila* Tec family protein tyrosine kinase *Tec29* [for review of Tec family kinases, see (123)]. In one approach, a weakly fertile *Src64* mutant was used as a sensitized genetic background to screen for mutations that either dominantly enhanced or suppressed the *Src64* fertility defect. One mutation that enhanced the *Src64* phenotype was revealed to be a lethal mutation in the *Tec29* gene (48). In addition to enhancing *Src64*, *Tec29* germline clones produced a small ring canal phenotype similar to *Src64*. A second group simultaneously isolated lethal *Tec29* mutations through a reverse genetic approach and described similar clonal phenotypes and *Src64* interactions (120). Antibodies raised against Tec29 protein specifically localize to ring canal. Moreover, Tec29 ring canal localization was dependent on Src64 function: in *Src64* mutants, Tec29 fails to localize, whereas the localization of Src64 in *Tec29* germline clones was unaffected (48, 120). These results indicate that Tec29 is downstream of Src64 in a signaling pathway required for ring canal growth.

A link between the Src/Tec signaling and a cytoskeletal effector molecule was uncovered through analysis of the Kelch protein (69). As discussed earlier, the actin-binding affinity of Kelch is regulated by phosphorylation, and the expression of a non-phosphorylatable Kelch (Kel^{Y627A}) in place of wild-type Kelch caused a phenotype identical to *Src64*. Furthermore, when the phosphorylation state of Kelch in *Src64* mutant ovaries was examined, only the non-phosphorylated form was detected (69). Two conclusions from these results can be drawn. First, Kelch phosphorylation is downstream of Src64 activity, either directly or indirectly.

Second, Kelch appears to be the only significant cytoskeletal effector of Src64 at ring canals, as expression of Kel^{Y627A} reproduces all of the phenotypes associated with loss of *Src64* function. The similar phenotype of *Tec29* germline clones suggests that it could function as a signaling intermediate between Src64 and Kelch; however, this possibility requires further investigation. An additional important question for future research is the identification of the phosphatase(s) responsible for balancing the phosphorylation state of these proteins.

The Lar Receptor Protein Tyrosine Phosphatase

Receptor tyrosine phosphatases (RPTPs) have been well characterized as important molecules in axon guidance and neural development [reviewed in (142)]. The receptor protein tyrosine phosphatase Lar was originally identified as a Leukocyte antigen-related protein. It contains an extracellular domain similar to cell adhesion molecules, containing both immunoglobulin and fibronectin type-II repeats, and an intracellular domain consisting of two tandem phosphatase domains. Many RPTPS are highly enriched in neuronal tissue (142), and homologs of Lar and several other RPTPs are required for axon guidance in *Drosophila* (30, 71); however, less is known about the role of RPTPs in other tissues.

Two groups recently identified a function for *Drosophila Lar* in organizing F-actin in the follicle cells (6, 40). In both cases, loss of *Lar* function in the follicle cells resulted in a round egg phenotype. Further, the round egg phenotype was associated with a loss of polarity in the basal actin bundles in the follicle cells. In wild type, these bundles become uniformly oriented perpendicular to the A/P axis of the egg chamber (discussed above; see Figure 2*C*,*D*). However, in *Lar* mutants, the basal actin bundles appeared to be randomly oriented from cell to cell. These results suggest that LAR is involved in a signaling pathway required to polarize the actin cytoskeleton in the follicular epithelium.

Both groups found that *Lar* was required in the somatic follicle cells (and not the germline) to polarize the basal bundles; however, the two groups arrived at differing conclusions about how *Lar* participates in this process. In one study, it was proposed that *Lar* is required for the basal bundle polarization based on several observations (6). First, LAR protein was enriched at the basal surfaces of follicle cells where the basal actin bundles terminate during stages 7 to 9, the time that polarization occurs. In addition, follicle cell clones mutant for a null allele of *Lar* had a loss-of-bundle polarity, although the polarization defect was not cell autonomous. Furthermore, other proteins known to interact with LAR in vertebrates, including α- and β-integrin subunits, exhibited a similar subcellular distribution during stages 7–9, and follicle cell clones of these mutants also resulted in a round egg phenotype. Finally, they found that expression of a *Lar* cDNA construct in late stage follicle cells was sufficient to rescue the *Lar* round egg phenotype.

In the second study, it was suggested that the bundle polarity phenotype was a secondary consequence of an earlier defect in cell fate specification (40). This hypothesis was proposed after a more extensive clonal analysis using a null *Lar* allele revealed no particular correlation between the location of *Lar* clones and the presence of actin bundle polarization defects (40). Strikingly, large *Lar* mutant clones were observed in which the basal actin bundles were appropriately polarized. Additional observations on how the basal actin bundles become polarized led to the proposal that *Lar* functions early in oogenesis, perhaps in the germarium, to specify cell fate. First, in describing how the basal actin bundles become polarized during mid-oogenesis, it was noted that polarization first becomes apparent near the anterior and posterior poles of an egg chamber. Second, *Lar* mutant egg chambers often had defects affecting the polar cells, two pairs of specialized cells located at the anterior and posterior poles. In *Lar* mutants, extra polar cells were often specified, and ectopic polar cell clusters were sometimes observed. Moreover, the ectopic polar cell clusters sometimes induced basal actin cell polarization in neighboring follicle cells. Together, these data suggested that the polar cells produce an instructive cue that polarizes the basal actin bundles in the follicular epithelium. To explain the *Lar* phenotype, it was proposed that *Lar* might be required for the proper specification of the polar cells, which takes place in the germarium (88). This possibility could not be tested directly, as markers for polar cells at this early stage were not available. However, in germaria from *Lar* mutant females, defects in follicle cell interactions and morphogenesis were observed, consistent with a possible function in mediating follicle cell fate decisions (40).

Together, these reports provide evidence for a role of a receptor tyrosine phosphatase in organizing the actin cytoskeleton in a nonneuronal tissue. Whether *Lar* functions directly in follicle cells undergoing bundle polarization or indirectly through the specification of polar cells is not completely clear. However, the finding that follicle cell clones mutant for a null *Lar* allele can have properly polarized basal actin bundles strongly suggests that Lar is not required in follicle cells at the time when the bundles become polarized. In addition, the Gal4 expression system used to rescue the *Lar* phenotype in late-stage *Lar* egg chambers (6) has also been reported to drive expression in follicle cells in the germarium (76), consistent with a requirement for *Lar* to specify polar cell fate early in oogenesis. Regardless of the mechanism, these findings demonstrate a requirement for RPTP function in a nonneuronal context, and provide an additional system in which the link between RPTP signaling and cytoskeletal organization can be further investigated.

FUTURE DIRECTIONS

The study of the actin cytoskeleton through genetic analysis of the *Drosophila* ovary should benefit from a number of recent technological advances. New methods of screening for genes will certainly identify novel actin-binding proteins as well as proteins that regulate their function. One screening technique involves overexpression screens, where large numbers of genes can be surveyed for gain-of-function phenotypes associated with overexpression (89, 118). A second screening technique relies on GFP-based protein trapping, where proteins associated with the actin cytoskeleton can be identified based on a direct examination of their subcellular localization (99); this method has already been used to identify novel genes whose products colocalize with F-actin in the ovary (M. Buszczak, R. Kelso, A. Quiñones, L. Cooley, unpublished observations). Both of these screening techniques will identify genes with redundant functions that are otherwise refractory to loss-of-function genetic screens. In addition, the recent development of a transgenic system capable of driving expression in the ovarian germline (119) will greatly facilitate the in vivo analysis of actin-binding proteins. Finally, live in situ imaging of GFP-tagged proteins, whether engineered or generated through protein trapping, is providing new insights into how actin-binding proteins influence dynamic actin processes (69), and will continue to do so.

ACKNOWLEDGMENTS

We thank the members of the Cooley lab for many discussions about the results and ideas presented in this review, as well as for specific comments on this manuscript. Research in the Cooley laboratory is supported by grants from the National Institutes of Health.

The *Annual Review of Genetics* is online at http://genet.annualreviews.org

LITERATURE CITED

1. Adams J, Kelso R, Cooley L. 2000. The kelch repeat superfamily of proteins: propellers of cell function. *Trends Cell Biol.* 10:17–24

2. Avraham KB, Hasson T, Steel KP, Kingsley DM, Russell LB, et al. 1995. The mouse *Snell's waltzer* deafness gene encodes an unconventional myosin required for structural integrity of inner ear hair cells. *Nat. Genet.* 11:369–75

3. Bailly M, Ichetovkin I, Grant W, Zebda N, Machesky LM, et al. 2001. The F-actin side binding activity of the Arp2/3 complex is essential for actin nucleation and lamellipod extension. *Curr. Biol.* 11:620–25

4. Bamburg JR. 1999. Proteins of the ADF/cofilin family: essential regulators of actin dynamics. *Annu. Rev. Cell Dev. Biol.* 15:185–230

5. Bartles JR. 2000. Parallel actin bundles and their multiple actin-bundling proteins. *Curr. Opin. Cell Biol.* 12:72–78

6. Bateman J, Reddy RS, Saito H, Van Vactor D. 2001. The receptor tyrosine phosphatase Dlar and integrins organize actin filaments in the *Drosophila* follicular epithelium. *Curr. Biol.* 11:1317–27

7. Baum B, Li W, Perrimon N. 2000. A cyclase-associated protein regulates actin and cell polarity during *Drosophila* oogenesis and in yeast. *Curr. Biol.* 10:964–73

8. Baum B, Perrimon N. 2001. Spatial control of the actin cytoskeleton in *Drosophila* epithelial cells. *Nat. Cell Biol.* 3:883–90

9. Ben-Yaacov S, Le Borgne R, Abramson I, Schweisguth F, Schejter ED. 2001. *Wasp*, the *Drosophila* Wiskott-Aldrich syndrome gene homologue, is required for cell fate decisions mediated by *Notch* signaling. *J. Cell Biol.* 152:1–13

10. Benlali A, Draskovic I, Hazelett DJ,

Treisman JE. 2000. *act up* controls actin polymerization to alter cell shape and restrict Hedgehog signaling in the *Drosophila* eye disc. *Cell* 101:271–81

11. Bennett V. 1990. Spectrin: a structural mediator between diverse plasma membrane proteins and the cytoplasm. *Curr. Opin. Cell Biol.* 2:51–56

12. Bohrmann J. 1997. *Drosophila* unconventional myosin VI is involved in intra- and intercellular transport during oogenesis. *Cell Mol. Life Sci.* 53:652–62

13. Bohrmann J, Biber K. 1994. Cytoskeleton-dependent transport of cytoplasmic particles in previtellogenic to mid-vitellogenic ovarian follicles of *Drosophila*: time-lapse analysis using video-enhanced contrast microscopy. *J. Cell Sci.* 107:849–58

14. Bork P, Doolittle RF. 1994. *Drosophila* kelch motif is derived from a common enzyme fold. *J. Mol. Biol.* 236:1277–82

15. Bryan J, Edwards R, Matsudaira P, Otto J, Wulfkuhle J. 1993. Fascin, an echinoid actin-bundling protein, is a homolog of the *Drosophila singed* gene product. *Proc. Natl. Acad. Sci. USA* 90:9115–19

16. Buss F, Kendrick-Jones J, Lionne C, Knight AE, Cote GP, Paul Luzio J. 1998. The localization of myosin VI at the golgi complex and leading edge of fibroblasts and its phosphorylation and recruitment into membrane ruffles of A431 cells after growth factor stimulation. *J. Cell Biol.* 143:1535–45

17. Buszczak M, Cooley L. 2000. Eggs to die for: cell death during *Drosophila* oogenesis. *Cell Death Differ.* 7:1071–74

18. Cant K, Knowles BA, Mahajan-Miklos S, Heintzelman M, Cooley L. 1998. *Drosophila* fascin mutants are rescued by overexpression of the villin-like protein, quail. *J. Cell Sci.* 111:213–21

19. Cant K, Knowles BA, Mooseker MS, Cooley L. 1994. *Drosophila singed*, a fascin homolog, is required for actin bundle formation during oogenesis and bristle extension. *J. Cell Biol.* 125:369–80

20. Carpenter AT. 1994. Egalitarian and the choice of cell fates in *Drosophila melanogaster* oogenesis. *Ciba Found. Symp.* 182:223–46

21. Castrillon DH, Gonczy P, Alexander S, Rawson R, Eberhart CG, et al. 1993. Toward a molecular genetic analysis of spermatogenesis in *Drosophila melanogaster*: characterization of male-sterile mutants generated by single P element mutagenesis. *Genetics* 135:489–505

22. Chen J, Godt D, Gunsalus K, Kiss I, Goldberg M, Laski FA. 2001. Cofilin/ADF is required for cell motility during *Drosophila* ovary development and oogenesis. *Nat. Cell Biol.* 3:204–9

23. Condeelis J. 2001. How is actin polymerization nucleated in vivo? *Trends Cell Biol.* 11:288–93

24. Cooley L, Verheyen E, Ayers K. 1992. *chickadee* encodes a profilin required for intercellular cytoplasm transport during *Drosophila* oogenesis. *Cell* 69:173–84

25. Corces VG, Geyer PK. 1991. Interactions of retrotransposons with the host genome: the case of the gypsy element of *Drosophila*. *Trends Genet.* 7:86–90

26. Cramer LP. 2000. Myosin VI. Roles for a minus end-directed actin motor in cells. *J. Cell Biol.* 150:121–26

27. de Cuevas M, Lee JK, Spradling AC. 1996. alpha-Spectrin is required for germline cell division and differentiation in the *Drosophila* ovary. *Development* 122:3959–68

28. Deng W, Leaper K, Bownes M. 1999. A targeted gene silencing technique shows that *Drosophila* myosin VI is required for egg chamber and imaginal disc morphogenesis. *J. Cell Sci.* 112:3677–90

29. DeRosier DJ, Tilney LG. 2000. F-actin bundles are derivatives of microvilli: What does this tell us about how bundles might form? *J. Cell Biol.* 148:1–6

30. Desai CJ, Gindhart JG Jr, Goldstein LS, Zinn K. 1996. Receptor tyrosine phosphatases are required for motor axon

guidance in the *Drosophila* embryo. *Cell* 84:599–609

31. Dobens LL, Raftery LA. 2000. Integration of epithelial patterning and morphogenesis in *Drosophila* ovarian follicle cells. *Dev. Dyn.* 218:80–93

32. Dodson GS, Guarnieri DJ, Simon MA. 1998. *Src64* is required for ovarian ring canal morphogenesis during *Drosophila* oogenesis. *Development* 125:2883–92

33. Dubreuil RR, Byers TJ, Stewart CT, Kiehart DP. 1990. A beta-spectrin isoform from *Drosophila* (beta H) is similar in size to vertebrate dystrophin. *J. Cell Biol.* 111:1849–58

34. Edwards KA, Kiehart DP. 1996. *Drosophila* nonmuscle myosin II has multiple essential roles in imaginal disc and egg chamber morphogenesis. *Development* 122:1499–511

35. Evangelista M, Pruyne D, Amberg DC, Boone C, Bretscher A. 2002. Formins direct Arp2/3-independent actin filament assembly to polarize cell growth in yeast. *Nat. Cell Biol.* 4:32–41

36. Fehon RG, Oren T, LaJeunesse DR, Melby TE, McCartney BM. 1997. Isolation of mutations in the *Drosophila* homologues of the human neurofibromatosis 2 and yeast *CDC42* genes using a simple and efficient reverse-genetic method. *Genetics* 146:245–52

37. Flanagan LA, Chou J, Falet H, Neujahr R, Hartwig JH, Stossel TP. 2001. Filamin A, the Arp2/3 complex, and the morphology and function of cortical actin filaments in human melanoma cells. *J. Cell Biol.* 155:511–17

38. Freeman NL, Chen Z, Horenstein J, Weber A, Field J. 1995. An actin monomer binding activity localizes to the carboxyl-terminal half of the *Saccharomyces cerevisiae* cyclase-associated protein. *J. Biol. Chem.* 270:5680–85

39. Freeman NL, Field J. 2000. Mammalian homolog of the yeast cyclase associated protein, CAP/Srv2p, regulates actin fil-

ament assembly. *Cell Motil. Cytoskelet.* 45:106–20

40. Frydman HM, Spradling AC. 2001. The receptor-like tyrosine phosphatase Lar is required for epithelial planar polarity and for axis determination within *Drosophila* ovarian follicles. *Development* 128:3209–20

41. Genova JL, Jong S, Camp JT, Fehon RG. 2000. Functional analysis of *Cdc42* in actin filament assembly, epithelial morphogenesis, and cell signaling during *Drosophila* development. *Dev. Biol.* 221:181–94

42. Gieselmann R, Mann K. 1992. ASP-56, a new actin sequestering protein from pig platelets with homology to CAP, an adenylate cyclase-associated protein from yeast. *FEBS Lett.* 298:149–53

43. Gillespie PG, Walker RG. 2001. Molecular basis of mechanosensory transduction. *Nature* 413:194–202

44. Glenney JR Jr, Kaulfus P, Weber K. 1981. F actin assembly modulated by villin: Ca^{++}-dependent nucleation and capping of the barbed end. *Cell* 24:471–80

45. Godt D, Laski FA. 1995. Mechanisms of cell rearrangement and cell recruitment in *Drosophila* ovary morphogenesis and the requirement of *bric a brac*. *Development* 121:173–87

46. Godt D, Tepass U. 1998. *Drosophila* oocyte localization is mediated by differential cadherin-based adhesion. *Nature* 395:387–91

47. Gottwald U, Brokamp R, Karakesisoglou I, Schleicher M, Noegel AA. 1996. Identification of a cyclase-associated protein (CAP) homologue in *Dictyostelium discoideum* and characterization of its interaction with actin. *Mol. Biol. Cell* 7:261–72

48. Guarnieri DJ, Dodson GS, Simon MA. 1998. SRC64 regulates the localization of a Tec-family kinase required for *Drosophila* ring canal growth. *Mol. Cell* 1:831–40

49. Guild GM, Connelly PS, Shaw MK,

Tilney LG. 1997. Actin filament cables in *Drosophila* nurse cells are composed of modules that slide passively past one another during dumping. *J. Cell Biol.* 138:783–97

50. Gunsalus KC, Bonaccorsi S, Williams E, Verni F, Gatti M, Goldberg ML. 1995. Mutations in *twinstar*, a *Drosophila* gene encoding a cofilin/ADF homologue, result in defects in centrosome migration and cytokinesis. *J. Cell Biol.* 131:1243–59

51. Gutzeit HO. 1986. The role of microfilaments in cytoplasmic streaming in *Drosophila* follicles. *J. Cell Sci.* 80:159–69

52. Gutzeit HO. 1990. The microfilament pattern in the somatic follicle cells of mid-vitellogenic ovarian follicles of *Drosophila*. *Eur. J. Cell Biol.* 53:349–56

53. Gutzeit HO, Eberhardt W, Gratwohl E. 1991. Laminin and basement membrane-associated microfilaments in wild-type and mutant *Drosophila* ovarian follicles. *J. Cell Sci.* 100:781–88

54. Hakeda-Suzuki S, Ng J, Tzu J, Dietzl G, Sun Y, et al. 2002. Rac function and regulation during *Drosophila* development. *Nature* 416:438–42

55. Hartwig JH, Stossel TP. 1981. Structure of macrophage actin-binding protein molecules in solution and interacting with actin filaments. *J. Mol. Biol.* 145:563–81

56. Hasson T, Gillespie PG, Garcia JA, MacDonald RB, Zhao Y, et al. 1997. Unconventional myosins in inner-ear sensory epithelia. *J. Cell Biol.* 137:1287–307

57. Hasson T, Mooseker MS. 1994. Porcine myosin-VI: characterization of a new mammalian unconventional myosin. *J. Cell Biol.* 127:425–40

58. Hicks JL, Deng WM, Rogat AD, Miller KG, Bownes M. 1999. Class VI unconventional myosin is required for spermatogenesis in *Drosophila*. *Mol. Biol. Cell* 10:4341–53

59. Higgs HN, Pollard TD. 2001. Regulation of actin filament network formation through the Arp2/3 complex: activation

by a diverse array of proteins. *Annu. Rev. Biochem.* 70:649–76

60. Huang C, Ni Y, Wang T, Gao Y, Haudenschild CC, Zhan X. 1997. Down-regulation of the filamentous actin cross-linking activity of cortactin by Src-mediated tyrosine phosphorylation. *J. Biol. Chem.* 272:13911–15

61. Hubberstey AV, Mottillo EP. 2002. Cyclase-associated proteins: CAPacity for linking signal transduction and actin polymerization. *FASEB J.* 16:487–99

62. Hudson AM, Cooley L. 2002. A subset of dynamic actin rearrangements in *Drosophila* requires the Arp2/3 complex. *J. Cell Biol.* 156:677–87

63. Ichetovkin I, Grant W, Condeelis J. 2002. Cofilin produces newly polymerized actin filaments that are preferred for dendritic nucleation by the Arp2/3 complex. *Curr. Biol.* 12:79–84

64. Ito N, Phillips SE, Stevens C, Ogel ZB, McPherson MJ, et al. 1991. Novel thioether bond revealed by a 1.7 Å crystal structure of galactose oxidase. *Nature* 350:87–90

65. Johnson DI. 1999. Cdc42: an essential Rho-type GTPase controlling eukaryotic cell polarity. *Microbiol. Mol. Biol. Rev.* 63:54–105

66. Jones L. 2001. Stem cells: So what's in a niche? *Curr. Biol.* 11:R484–86

67. Jordan P, Karess R. 1997. Myosin light chain-activating phosphorylation sites are required for oogenesis in *Drosophila*. *J. Cell Biol.* 139:1805–19

68. Kamm KE, Stull JT. 2001. Dedicated myosin light chain kinases with diverse cellular functions. *J. Biol. Chem.* 276:4527–30

69. Kelso RJ, Hudson AM, Cooley L. 2002. *Drosophila* Kelch regulates actin organization via Src64-dependent tyrosine phosphorylation. *J. Cell Biol.* 156:703–13

70. Klahre U, Friederich E, Kost B, Louvard D, Chua NH. 2000. Villin-like actin-binding proteins are expressed

ubiquitously in Arabidopsis. *Plant Physiol.* 122:35–48

71. Krueger NX, Van Vactor D, Wan HI, Gelbart WM, Goodman CS, Saito H. 1996. The transmembrane tyrosine phosphatase DLAR controls motor axon guidance in *Drosophila. Cell* 84:611–22

72. Krugmann S, Jordens I, Gevaert K, Driessens M, Vandekerckhove J, Hall A. 2001. Cdc42 induces filopodia by promoting the formation of an IRSp53:Mena complex. *Curr. Biol.* 11:1645–55

73. Kureishy N, Sapountzi V, Prag S, Anilkumar N, Adams JC. 2002. Fascins and their roles in cell structure and function. *BioEssays* 24:350–61

74. Lee JK, Brandin E, Branton D, Goldstein LS. 1997. alpha-Spectrin is required for ovarian follicle monolayer integrity in *Drosophila melanogaster. Development* 124:353–62

75. Li MG, Serr M, Edwards K, Ludmann S, Yamamoto D, et al. 1999. Filamin is required for ring canal assembly and actin organization during *Drosophila* oogenesis. *J. Cell Biol.* 146:1061–74

76. Liu Y, Montell DJ. 1999. Identification of mutations that cause cell migration defects in mosaic clones. *Development* 126:1869–78

77. Lommel S, Benesch S, Rottner K, Franz T, Wehland J, Kuhn R. 2001. Actin pedestal formation by enteropathogenic *Escherichia coli* and intracellular motility of *Shigella flexneri* are abolished in N-WASP-defective cells. *EMBO Rep.* 2:850–57

78. Machesky LM, Atkinson SJ, Ampe C, Vandekerckhove J, Pollard TD. 1994. Purification of a cortical complex containing two unconventional actins from *Acanthamoeba* by affinity chromatography on profilin-agarose. *J. Cell Biol.* 127:107–15

79. Machesky LM, Gould KL. 1999. The Arp2/3 complex: a multifunctional actin organizer. *Curr. Opin. Cell. Biol.* 11:117–21

80. Machesky LM, Insall RH. 1998. Scar1

and the related Wiskott-Aldrich syndrome protein, WASP, regulate the actin cytoskeleton through the Arp2/3 complex. *Curr. Biol.* 8:1347–56

81. Maciver SK. 1996. Myosin II function in non-muscle cells. *BioEssays* 18:179–82

82. Maciver SK. 1998. How ADF/cofilin depolymerizes actin filaments. *Curr. Opin. Cell Biol.* 10:140–44

83. Magie CR, Meyer MR, Gorsuch MS, Parkhurst SM. 1999. Mutations in the Rho1 small GTPase disrupt morphogenesis and segmentation during early *Drosophila* development. *Development* 126:5353–64

84. Mahajan-Miklos S, Cooley L. 1994. Intercellular cytoplasm transport during *Drosophila* oogenesis. *Dev. Biol.* 165:336–51

85. Mahajan-Miklos S, Cooley L. 1994. The villin-like protein encoded by the *Drosophila quail* gene is required for actin bundle assembly during oogenesis. *Cell* 78:291–301

86. Manseau L, Calley J, Phan H. 1996. Profilin is required for posterior patterning of the *Drosophila* oocyte. *Development* 122:2109–16

87. Manseau LJ, Schupbach T. 1989. *cappuccino* and *spire*: two unique maternal-effect loci required for both the antero-posterior and dorsoventral patterns of the *Drosophila* embryo. *Genes Dev.* 3:1437–52

88. Margolis J, Spradling A. 1995. Identification and behavior of epithelial stem cells in the *Drosophila* ovary. *Development* 121:3797–807

89. Mata J, Curado S, Ephrussi A, Rorth P. 2000. Tribbles coordinates mitosis and morphogenesis in *Drosophila* by regulating string/CDC25 proteolysis. *Cell* 101:511–22

90. Matova N, Cooley L. 2001. Comparative aspects of animal oogenesis. *Dev. Biol.* 231:291–320

91. Matova N, Mahajan-Miklos S, Mooseker MS, Cooley L. 1999. *Drosophila* quail, a

villin-related protein, bundles actin filaments in apoptotic nurse cells. *Development* 126:5645–57

92. Melchionda S, Ahituv N, Bisceglia L, Sobe T, Glaser F, et al. 2001. MYO6, the human homologue of the gene responsible for deafness in *Snell's waltzer* mice, is mutated in autosomal dominant nonsyndromic hearing loss. *Am. J. Hum. Genet.* 69:635–40

93. Mermall V, McNally JG, Miller KG. 1994. Transport of cytoplasmic particles catalysed by an unconventional myosin in living *Drosophila* embryos. *Nature* 369: 560–62

94. Mermall V, Post PL, Mooseker MS. 1998. Unconventional myosins in cell movement, membrane traffic, and signal transduction. *Science* 279:527–33

95. Meyer SC, Zuerbig S, Cunningham CC, Hartwig JH, Bissell T, et al. 1997. Identification of the region in actin-binding protein that binds to the cytoplasmic domain of glycoprotein IB alpha. *J. Biol. Chem.* 272:2914–19

96. Miki H, Suetsugu S, Takenawa T. 1998. WAVE, a novel WASP-family protein involved in actin reorganization induced by Rac. *EMBO J.* 17:6932–41

97. Miki H, Yamaguchi H, Suetsugu S, Takenawa T. 2000. IRSp53 is an essential intermediate between Rac and WAVE in the regulation of membrane ruffling. *Nature* 408:732–35

98. Mooseker MS. 1985. Organization, chemistry, and assembly of the cytoskeletal apparatus of the intestinal brush border. *Annu. Rev. Cell Biol.* 1:209–41

99. Morin X, Daneman R, Zavortink M, Chia W. 2001. A protein trap strategy to detect GFP-tagged proteins expressed from their endogenous loci in *Drosophila*. *Proc. Natl. Acad. Sci. USA* 98:15050–55

100. Muller HA. 2000. Genetic control of epithelial cell polarity: lessons from *Drosophila*. *Dev. Dyn.* 218:52–67

101. Mullins RD, Heuser JA, Pollard TD. 1998. The interaction of Arp2/3 complex with

actin: nucleation, high affinity pointed end capping, and formation of branching networks of filaments. *Proc. Natl. Acad. Sci. USA* 95:6181–86

102. Murphy AM, Montell DJ. 1996. Cell type-specific roles for Cdc42, Rac, and RhoL in *Drosophila* oogenesis. *J. Cell Biol.* 133: 617–30

103. Ng J, Nardine T, Harms M, Tzu J, Goldstein A, et al. 2002. Rac GTPases control axon growth, guidance and branching. *Nature* 416:442–47

104. Ono S, Yamakita Y, Yamashiro S, Matsudaira PT, Gnarra JR, et al. 1997. Identification of an actin binding region and a protein kinase C phosphorylation site on human fascin. *J. Biol. Chem.* 272:2527–33

105. Otto IM, Raabe T, Rennefahrt UE, Bork P, Rapp UR, Kerkhoff E. 2000. The p150-Spir protein provides a link between c-Jun N-terminal kinase function and actin reorganization. *Curr. Biol.* 10:345–48

106. Paterson J, O'Hare K. 1991. Structure and transcription of the *singed* locus of *Drosophila melanogaster*. *Genetics* 129: 1073–84

107. Pepling ME, de Cuevas M, Spradling AC. 1999. Germline cysts: a conserved phase of germ cell development? *Trends Cell Biol.* 9:257–62

108. Pollard TD, Blanchoin L, Mullins RD. 2000. Molecular mechanisms controlling actin filament dynamics in nonmuscle cells. *Annu. Rev. Biophys. Biomol. Struct.* 29:545–76

109. Riparbelli MG, Callaini G. 1995. Cytoskeleton of the *Drosophila* egg chamber: new observations on microfilament distribution during oocyte growth. *Cell Motil. Cytoskelet.* 31:298–306

110. Rivero F, Dislich H, Glockner G, Noegel AA. 2001. The *Dictyostelium discoideum* family of Rho-related proteins. *Nucleic Acids Res.* 29:1068–79

111. Robinson DN, Cant K, Cooley L. 1994. Morphogenesis of *Drosophila* ovarian ring canals. *Development* 120:2015–25

112. Robinson DN, Cooley L. 1996. Stable intercellular bridges in development: the cytoskeleton lining the tunnel. *Trends Cell Biol.* 6:474–79

113. Robinson DN, Cooley L. 1997. *Drosophila* kelch is an oligomeric ring canal actin organizer. *J. Cell Biol.* 138:799–810

114. Robinson DN, Cooley L. 1997. Examination of the function of two kelch proteins generated by stop codon suppression. *Development* 124:1405–17

115. Robinson DN, Cooley L. 1997. Genetic analysis of the actin cytoskeleton in the *Drosophila* ovary. *Annu. Rev. Cell Dev. Biol.* 13:147–70

116. Robinson DN, Smith-Leiker TA, Sokol NS, Hudson AM, Cooley L. 1997. Formation of the *Drosophila* ovarian ring canal inner rim depends on *cheerio*. *Genetics* 145:1063–72

117. Robinson RC, Turbedsky K, Kaiser DA, Marchand JB, Higgs HN, et al. 2001. Crystal structure of Arp2/3 complex. *Science* 294:1679–84

118. Rorth P. 1996. A modular misexpression screen in *Drosophila* detecting tissue-specific phenotypes. *Proc. Natl. Acad. Sci. USA* 93:12418–22

119. Rorth P. 1998. Gal4 in the *Drosophila* female germline. *Mech. Dev.* 78:113–18

120. Roulier EM, Panzer S, Beckendorf SK. 1998. The Tec29 tyrosine kinase is required during *Drosophila* embryogenesis and interacts with Src64 in ring canal development. *Mol. Cell* 1:819–29

121. Sagot I, Klee SK, Pellman D. 2002. Yeast formins regulate cell polarity by controlling the assembly of actin cables. *Nat. Cell Biol.* 4:42–50

122. Sellers JR. 2000. Myosins: a diverse superfamily. *Biochim. Biophys. Acta* 1496:3–22

123. Smith CI, Islam TC, Mattsson PT, Mohamed AJ, Nore BF, Vihinen M. 2001. The Tec family of cytoplasmic tyrosine kinases: mammalian Btk, Bmx, Itk, Tec, Txk and homologs in other species. *BioEssays* 23:436–46

124. Sokol NS, Cooley L. 1999. *Drosophila* filamin encoded by the *cheerio* locus is a component of ovarian ring canals. *Curr. Biol.* 9:1221–30

125. Sokol NS, Cooley L. 2002. *Drosophila* Filamin is required for follicle cell motility during oogenesis. Submitted

126. Spradling AC. 1993. Developmental genetics of oogenesis. In *The Development of Drosophila melanogaster*, ed. M Bate, A. Martinez-Arias, pp. 1–70. Plainview, NY: Cold Spring Harbor Press

127. Spradling AC, de Cuevas M, Drummond-Barbosa D, Keyes L, Lilly M, et al. 1997. The *Drosophila* germarium: stem cells, germ line cysts, and oocytes. *Cold Spring Harb. Symp. Quant. Biol.* 62:25–34

128. Stabach PR, Morrow JS. 2000. Identification and characterization of beta V spectrin, a mammalian ortholog of *Drosophila* beta H spectrin. *J. Biol. Chem.* 275:21385–95

129. Stossel TP, Condeelis J, Cooley L, Hartwig JH, Noegel A, et al. 2001. Filamins as integrators of cell mechanics and signalling. *Nat. Rev. Mol. Cell Biol.* 2:138–45

130. Strutt DI, Weber U, Mlodzik M. 1997. The role of *RhoA* in tissue polarity and Frizzled signalling. *Nature* 387:292–95

131. Sun S, Footer M, Matsudaira P. 1997. Modification of Cys-837 identifies an actin-binding site in the beta-propeller protein scruin. *Mol. Biol. Cell.* 8:421–30

132. Svitkina TM, Borisy GG. 1999. Arp2/3 complex and actin depolymerizing factor/cofilin in dendritic organization and treadmilling of actin filament array in lamellipodia. *J. Cell Biol.* 145:1009–26

133. Symons M. 1996. Rho family GTPases: the cytoskeleton and beyond. *Trends Biochem. Sci.* 21:178–81

134. Tal T, Vaizel-Ohayon D, Schejter ED. 2002. Conserved interactions with cytoskeletal but not signaling elements are

an essential aspect of *Drosophila* Wasp function. *Dev. Biol.* 243:260–71

135. Tanentzapf G, Smith C, McGlade J, Tepass U. 2000. Apical, lateral, and basal polarization cues contribute to the development of the follicular epithelium during *Drosophila* oogenesis. *J. Cell Biol.* 151: 891–904

136. Theriot JA, Mitchison TJ. 1991. Actin microfilament dynamics in locomoting cells. *Nature* 352:126–31

137. Theurkauf WE. 1994. Premature microtubule-dependent cytoplasmic streaming in *cappuccino* and *spire* mutant oocytes. *Science* 265:2093–96

138. Thomas GH, Kiehart DP. 1994. Beta heavy-spectrin has a restricted tissue and subcellular distribution during *Drosophila* embryogenesis. *Development* 120: 2039–50

139. Thomas SM, Brugge JS. 1997. Cellular functions regulated by Src family kinases. *Annu. Rev. Cell Dev. Biol.* 13:513–609

140. Tilney LG, Tilney MS, Guild GM. 1996. Formation of actin filament bundles in the ring canals of developing *Drosophila* follicles. *J. Cell Biol.* 133:61–74

141. Tohtong R, Yamashita H, Graham M, Haeberle J, Simcox A, Maughan D. 1995. Impairment of muscle function caused by mutations of phosphorylation sites in myosin regulatory light chain. *Nature* 374:650–53

142. Van Vactor D. 1998. Protein tyrosine phosphatases in the developing nervous system. *Curr. Opin. Cell Biol.* 10:174–81

143. Vasioukhin V, Bauer C, Yin M, Fuchs E. 2000. Directed actin polymerization is the driving force for epithelial cell-cell adhesion. *Cell* 100:209–19

144. Volkmann N, Amann KJ, Stoilova-McPhie S, Egile C, Winter DC, et al. 2001. Structure of Arp2/3 complex in its activated state and in actin filament branch junctions. *Science* 293:2456–59

145. Warn RM, Gutzeit HO, Smith L, Warn A. 1985. F-actin rings are associated with the ring canals of the *Drosophila* egg chamber. *Exp. Cell Res.* 157:355–63

146. Weed SA, Parsons JT. 2001. Cortactin: coupling membrane dynamics to cortical actin assembly. *Oncogene* 20:6418–34

147. Wellington A, Emmons S, James B, Calley J, Grover M, et al. 1999. Spire contains actin binding domains and is related to *ascidian* posterior end mark-5. *Development* 126:5267–74

148. Wells AL, Lin AW, Chen LQ, Safer D, Cain SM, et al. 1999. Myosin VI is an actin-based motor that moves backwards. *Nature* 401:505–8

149. Wheatley S, Kulkarni S, Karess R. 1995. *Drosophila* nonmuscle myosin II is required for rapid cytoplasmic transport during oogenesis and for axial nuclear migration in early embryos. *Development* 121:1937–46

150. Winter DC, Choe EY, Li R. 1999. Genetic dissection of the budding yeast Arp2/3 complex: a comparison of the in vivo and structural roles of individual subunits. *Proc. Natl. Acad. Sci. USA* 96:7288–93

151. Xie T, Spradling AC. 1998. *decapentaplegic* is essential for the maintenance and division of germline stem cells in the *Drosophila* ovary. *Cell* 94:251–60

152. Xie T, Spradling AC. 2000. A niche maintaining germ line stem cells in the *Drosophila* ovary. *Science* 290:328–30

153. Xue F, Cooley L. 1993. *kelch* encodes a component of intercellular bridges in *Drosophila* egg chambers. *Cell* 72:681–93

154. Yue L, Spradling AC. 1992. *hu-li tai shao*, a gene required for ring canal formation during *Drosophila* oogenesis, encodes a homolog of adducin. *Genes Dev.* 6:2443–54

155. Zallen JA, Cohen Y, Hudson AM, Cooley L, Wieschaus E, Schejter ED. 2002. SCAR is a primary regulator of Arp2/3-dependent morphological events in Drosophila. *J. Cell Biol.* 156:689–701

156. Zarnescu DC, Thomas GH. 1999. Apical spectrin is essential for epithelial morphogenesis but not apicobasal polarity in *Drosophila. J. Cell Biol.* 146:1075–86

157. Zebda N, Bernard O, Bailly M, Welti S, Lawrence DS, Condeelis JS. 2000. Phosphorylation of ADF/cofilin abolishes EGF-induced actin nucleation at the leading edge and subsequent lamellipod extension. *J. Cell Biol.* 151:1119–28

158. Zigmond SH. 1993. Recent quantitative studies of actin filament turnover during cell locomotion. *Cell Motil. Cytoskelet.* 25:309–16

159. Zigmond SH. 1996. Signal transduction and actin filament organization. *Curr. Opin. Cell Biol.* 8:66–73

Annu. Rev. Genet. 2002. 36:489–519
doi: 10.1146/annurev.genet.36.043002.091619

THE GENETICS OF RNA SILENCING

Marcel Tijsterman, René F. Ketting, and Ronald H. A. Plasterk

Hubrecht Laboratory, Center for Biomedical Genetics, Uppsalalaan 8, 3584 CT, Utrecht, The Netherlands; e-mail: tijsterman@niob.knaw.nl; ketting@niob.knaw.nl; plasterk@niob.knaw.nl.

Key Words RNA silencing, RNAi, Post-Transcriptional Gene Silencing (PTGS), cosuppression, siRNAs

■ **Abstract** Although initially recognized as a handy tool to reduce gene expression, RNA silencing, triggered by double-stranded RNA molecules, is now recognized as a mechanism for cellular protection and cleansing: It defends the genome against molecular parasites such as viruses and transposons, while removing abundant but aberrant nonfunctional messenger RNAs. The underlying mechanisms in distinct gene silencing phenomena in different genetic systems, such as cosuppression in plants and RNAi in animals, are very similar. There are common RNA intermediates, and similar genes are required in RNA silencing pathways in protozoa, plants, fungi, and animals, thus indicating an ancient pathway. This chapter gives an overview of both biochemical and genetic approaches leading to the current understanding of the molecular mechanism of RNA silencing and its probable biological function.

CONTENTS

INTRODUCTION

Posttranscriptional Gene Silencing in Plants

In an attempt to deepen the purple color of petunia plants, the groups of Jorgensen and Mol introduced extra copies of the pigment-producing genes: dihydroflavonol-4-reductase or chalcone synthase. Rather than increasing flower pigmentation as expected, the opposite result was observed in many plants: variegated or even completely white flowers (76, 112). These observations [and (68, 101)], the first manifestation of RNA silencing, were originally termed cosuppression because of the apparent communication between unlinked but homologous loci: RNAs derived from both the transgenes and the homologous endogenous genes were degraded, resulting in the loss of pigmentation phenotype. Cosuppression has subsequently been found to occur in many species of plants, fungi, and animals (see below). The silencing does not depend on the presence of endogenously encoded homologous loci; silencing can occur between two related transgenes, and it is not only triggered by transgenes but can also be initiated by viruses (8), known as VIGS, for virus induced gene silencing. Cosuppression can operate at the transcriptional level (TGS for transcriptional gene silencing), in which it appears to involve alterations at the DNA level, e.g., DNA methylation (114), and at the posttranscriptional level: Nuclear run-on experiments demonstrate that the homologous transcripts are produced, but are rapidly degraded in the cytoplasm (28, 111). This process does not affect transcription of the endogenous locus, hence the name posttranscriptional gene silencing (PTGS). In this chapter we focus on the posttranscriptional part of RNA silencing; for studies on TGS, the reader is directed to recent reviews (69, 116) and references therein.

Over the past decade much research has been done to unravel the PTGS phenomenon, and at least two observations suggest that RNA molecules constitute the sequence-specific trigger of PTGS: (*a*) Transcriptionally active genes are better inducers of PTGS than transcriptionally inactive genes (88, 117), and (*b*) RNA viruses can induce silencing of homologous genes encoded by the host plant (8, 56, 93). Numerous models have been proposed, often invoking a causal presence of single-stranded RNAs that result from aberrant processing of the transgenes, hence the name aberrant RNAs (10). Such RNAs, but with antisense polarity, were hypothesized to account for the sequence-specificity and posttranscriptional nature of PTGS by the ability to pair with the target mRNA, leading to its subsequent destruction. Such antisense species, however, were never detected by conventional RNA analyses. In perhaps one of the biggest breakthroughs in PTGS (and RNA silencing in general), Hamilton & Baulcombe discovered an unexpectedly short RNA species of around 25 nt, corresponding to both the sense and antisense sequence of the targeted mRNAs, that had a strict correlation with the occurrence of PTGS (44). These species, later designated siRNAs, were found in a broad spectrum of plant systems undergoing PTGS, including transgene and virus-induced silencing, but were not detected in the absence of PTGS. The accumulation of both sense and the antisense siRNAs suggests that double-stranded RNA (dsRNA) is produced prior to formation of such species. The importance of dsRNA in triggering PTGS was also proposed by Waterhouse et al. who demonstrated that constructs producing RNAs capable of duplex formation, either via hairpin/panhandle constructs or by crossing sense- and antisense-producing clones, are more potent inducers of gene silencing than constructs that produce RNAs of only sense or antisense polarity (123). Invoking a dsRNA species in PTGS (see Figure 1) can now explain why silencing is observed more frequently where the introduced transgenic copies are arranged as inverted repeats, compared to transgenes in a direct repeat orientation [(e.g., 88, 102, 103)].

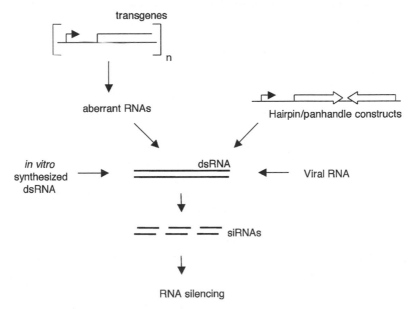

Figure 1   RNA silencing directed by foreign dsRNA. In this model, all triggers that induce RNA silencing operate through a dsRNA intermediate giving rise to the formation of siRNAs. While inversely orientated transgenes, in vitro prepared dsRNA, and viruses are direct sources of dsRNA, an additional step is needed to explain the production of dsRNA from multicopy transgenic arrays or highly transcribed single-copy transgenes. This might occur via read-through transcription, either from an endogenous promoter at the site of integration or as a result of head-to-head organization of transcription units in the multicopy array, giving rise to dsRNA directly or to antisense RNA that can pair with the sense to form dsRNA. Alternatively, sense (and possibly antisense) aberrant RNAs are converted into dsRNA via de novo RNA synthesis, reminiscent of virus replication.

RNAi in *Caenorhabditis elegans*

The first direct evidence that dsRNA could lead to RNA silencing came from work in the nematode *C. elegans*. While using an antisense approach to inhibit gene function in *C. elegans*, Guo & Kemphues surprisingly found that molecules with sense polarity were as effective in knocking down target gene expression as molecules with antisense polarity (43). This paradox was resolved by the Fire and Mello labs; they demonstrated that dsRNA is actually the potent inducer of RNA interference (RNAi) (38) and that the outcome of RNAi experiments with single-stranded (antisense) RNAs can largely be explained by modest amounts of dsRNA that contaminate in vitro prepared single-stranded RNA populations. By contrast, pure single-stranded RNA of sense polarity was unable to trigger gene silencing.

Subsequent experiments suggested that RNAi targets a posttranscriptional event in gene expression (38, 71). By increasing the rate of mRNA turnover in a sequence-specific manner, RNAi leads to phenotypes that are either identical to genetic null mutations, or resemble allelic series of mutants. The notion that only a few molecules of dsRNA per cell are sufficient to completely interfere with gene expression suggests a catalytic or amplification component in RNAi (38, 70). dsRNA targets mRNA both in the cytoplasm and in the nucleus of exposed cells, but acts post splicing: Individual mRNAs that are polycistronically transcribed can be targeted separately, and dsRNA directed against promoter sequences and introns are not effective at inducing RNAi. The sequence of the dsRNA is not necessarily identical to the target: dsRNAs that were 88% identical to the target mRNA triggered RNAi, but these triggers still contained significant stretches (of up to 41 nt) that perfectly matched the target mRNA (84). Although dsRNA is required for triggering RNAi, the chemical composition of the antisense strand is more important than the sense strand. Numerous chemical modifications on the dsRNA trigger, either directly on the bases or on the RNA backbone, preferentially block RNAi when carried out on the antisense strand, suggesting that the two strands have different functions in RNAi (84).

Analogous to PTGS in plants, RNAi in *C. elegans* is associated with the formation of small RNAs of 20–25 nt (siRNAs). The formation of siRNAs from dsRNA precursors was demonstrated by injecting radio-labeled dsRNA into the syncytial gonad of the nematode, thus using the reproductive organ as an in vivo test tube (83, 84). Furthermore, RNase protection assays on RNA isolated from animals fed on bacteria that generate dsRNA homologous to a specific *C. elegans* gene also revealed small RNA molecules of 20–25 nt in length (98).

RNA SILENCING EFFECTOR MOLECULES: siRNAs

Important insight into the generation of siRNAs and their function in RNA silencing came from the biochemical dissection of RNAi when in vitro systems that recapitulate the features of RNAi observed in vivo were developed. In a cell-free system derived from *Drosophila* embryo, dsRNA targets a corresponding mRNA for degradation during which time RNA species of 21-23 nt are formed that are processed from the input dsRNA (109, 128). Independently, it was found that *Drosophila* S2 cells supported sequence-specific mRNA degradation upon transfection of homologous dsRNA (45). A sequence-specific nuclease activity was partially purified from these cells, which also contained RNA fragments of approximately 25 nt in length, corresponding to the input dsRNA. This led to the hypothesis that such species, by providing sequence-specificity, guide the nuclease to the substrate mRNA (45). This proposed function was further supported by the observation that in the cell-free system, substrate mRNAs are cleaved in regular intervals (of 21–23 nt) but only in the region that is covered by the dsRNA (128). Taken together, these data led to a simple two-step model to explain RNAi: (*a*) long dsRNA molecules are diced into 21-23nt siRNAs by a dsRNA-specific nuclease; (*b*) these siRNAs guide a nuclease-containing protein complex, designated RISC (for RNAi-induced silencing complex), to the substrate through conventional base-pairing interactions of the antisense strand of the siRNA to the mRNA, triggering its subsequent destruction (Figure 2). Although it was noted early on that the RNA component of RISC was essential for its activity (45), formal proof that siRNAs are the intermediates in the RNAi reaction and mediate sequence-specific mRNA degradation was only recently obtained by showing that chemically synthesized RNA duplexes with sizes similar to siRNAs can guide target cleavage in the *Drosophila* in vitro system described above (34, 35).

The discrete size of siRNAs hinted at the involvement of an RNAse III or related enzyme because these cleave dsRNA substrates (like hairpin structures) into products with a defined length [(e.g., (1)]. In a directed approach, Hannon and coworkers tested *Drosophila* RNAse III family members for their ability to generate siRNAs from input dsRNA and found that dsRNA dicing activity could indeed be immunoprecipitated with antibodies raised against one particular member, subsequently designated Dicer (14). To address the involvement of Dicer in RNAi in living cells, these authors performed RNAi on a gene whose function is required for the interference. Transfection of *Drosophila* S2 cells with dsRNA homologous to Dicer significantly reduced the ability of these cells to silence a GFP reporter via RNAi.

Dicer is evolutionarily well conserved with homologs present in fungi (*Neurospora crassa*, *Saccharomyces pombe*), plants (*Arabidopsis*), and animals, including *C. elegans*, *Drosophila*, and mammals. Functional conservation of this family of proteins and their requirement in RNAi came from the demonstration that human Dicer was also found to dice dsRNA into siRNA (14), and that *C. elegans* mutants with a genetic defect in the Dicer ortholog (DCR-1) were resistant to RNAi induced by dsRNA (41, 52, 55). It is thus reasonable to assume that all systems that support dsRNA-induced gene silencing depend on a member of this protein family to convert dsRNA into siRNA intermediates that target the mRNA for destruction.

With the rapid progress of biochemical and genetic approaches, a minimal set of proteins that are sufficient to carry out the RNAi reaction in the test tube should soon be identified (see also below). Some of these proteins are or will

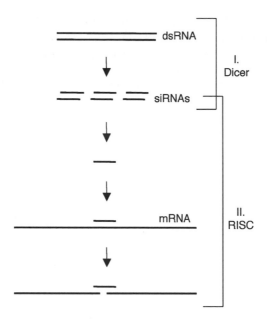

Figure 2   Two-step model to explain RNAi. In the first step, dsRNA is diced by an ATP-dependent ribonuclease (later found to be Dicer) into short interfering RNAs (siRNAs): duplexes of 21–23 nucleotides bearing two-nucleotide 3′ overhanging ends. These siRNAs are subsequently transferred to a second enzyme complex, designated RISC for RNAi-induced silencing complex, which contains an endoribonuclease that is distinct from Dicer. The siRNA guides the endonuclease—the antisense strand of the siRNA is perfectly complementary to the target mRNA—to the target mRNA leading to its destruction. The position of the cleavage site in the target is within the sequences covered by the siRNA (near the center) and is determined by the 5′ end of the guiding molecules (34, 35).

be identified using genetic screens aimed at isolating mutants that are defective in RNA silencing, followed by positional cloning of the mutated loci. Here, we discuss genetic approaches used successfully in several systems (*C. elegans*, *Arabidopsis*, *N. crassa*) that exhibit different manifestations of RNA silencing (RNAi, PTGS, transgene silencing).

GENETIC ANALYSIS OF RNA SILENCING

Quelling in *Neurospora crassa*

RNA silencing, manifested as transgene-induced gene silencing, has also been described in the filamentous fungus *Neurospora crassa*, and has been termed quelling (21, 92). Tandemly inserted transgenic copies containing part of the albino-1 (*al-1*) gene, essential for the biosynthesis of carotenoids, can invariably silence the endogenous *al-1* locus, leading to an albino phenotype instead of the orange color seen when the *al-1* gene is expressed at wild-type levels. This simple visual reporter system was used to identify components required for silencing, by isolating quelling-deficient (qde) mutants followed by the identification of the responsible gene defects (22). All recessive qde mutants that were isolated fall into three complementation groups, qde-1, qde-2, and qde-3 (loss of transgenic copies also leads to the mutant phenotype). In addition to the release of *al-1* silencing in these mutants, the transgene-encoded selectable marker, which is required for construction purposes, was expressed at elevated levels, suggesting that this transgene was partially silenced in the starting Neurospora strain (22).

As in RNAi in animals and PTGS in plants, quelling is associated with the formation of siRNAs derived from the transgenic copies of the silenced locus (19). [Note that the existence of a dominant diffusible factor in quelling was proposed early on, based on the observation that heterokaryons containing a quelled and nonquelled nucleus exhibit a quelled phenotype (21)]. qde-1 and qde-3 mutants fail to produce siRNAs, suggesting that these gene products are involved either in the generation of these silencing intermediates or in steps upstream in the silencing (19). qde-3 encodes a DNA helicase (24) belonging to the RecQ family of DNA helicases that generally function in DNA repair and recombination; this family includes the human genes involved in Bloom's and Werner's syndromes. It has therefore been proposed that QDE-3 operates at the DNA level, perhaps by remodeling the chromatin status of the tandemly organized transgenic copies to allow the production of the silencing RNA signal. RNase protection assays, however, indicate that sense RNA is still abundantly produced from the transgenic copies in a qde-3 genetic background (22).

qde-1 strains also fail to produce siRNAs (19) as a result of mutations in a gene with sequence homology to a putative RNA-dependent RNA polymerase isolated from tomato (23, 95, 96). This discovery strengthened the idea that RNA silencing induced by transgenes involves the generation of dsRNA using overexpressed or improperly expressed mRNAs (also termed aberrant RNAs) as a template. The notion that qde-1 encodes a protein that could, in principle, fulfill this function supports such models, and may explain why siRNAs are not detected: The precursor long dsRNAs are not generated. Thus far, long dsRNA molecules have not been detected in Neurospora, and biochemical proof is still lacking that the putative RdRPs, required for RNA silencing, catalyze de novo RNA synthesis on an RNA template.

qde-2 encodes a PAZ/PIWI domain containing protein, and belongs to the Argonaute family of proteins (18). Although the function of these evolutionarily well-conserved protein motifs is not known, their key importance in RNA silencing is illustrated by the notion that several members of this family are implicated in RNA silencing in distinct biological systems (see Table 1). Transgene-dependent siRNA species are still produced in qde-2 mutants, indicating that QDE-2 acts in a downstream step of RNA silencing perhaps in sequence-specific mRNA

TABLE 1 Proteins implicated in RNA silencing

Domain structure	Protein	Organism	Silencing	Mutant phenotype	Putative function
PAZ- and C-terminal PIWI domain	RDE-1	*Caenorhabditis elegans*	RNAi	RNAi resistant; not required for cosuppression	Initiation of RNAi, downstream of siRNA production;
	QDE-2	*Neurospora crassa*	Quelling	Quelling defective	Initiation of silencing
	AGO1	*Arabidopsis thaliana*	PTGS	PTGS deficient, developmental defects	Initiation of silencing
	Ago1	*Drosophila melanogaster*	RNAi	RNAi deficient	RNA silencing downstream of siRNA production
	Ago2	*Drosophila melanogaster*	RNAi	—	Component of RISC
	Aubergine	*Drosophila melanogaster*	Stellate silencing	Failure to silence Stellate locus; developmental defects	Translational repressor
RNA-dependent RNA polymerase	QDE-1	*Neurospora crassa*	Quelling	Quelling defective	Generation of dsRNA from aberrant RNAs
	EGO-1	*Caenorhabditis elegans*	Germline RNAi	RNAi defective for germline genes; germline-development defects	Generation of dsRNA (germline specific)
	RRF-1	*Caenorhabditis elegans*	Somatic RNAi	RNAi defective for somatic genes	Generation of dsRNA; secondary siRNA production
	RRF-3	*Caenorhabditis elegans*	RNAi	Increased sensitivity to RNAi	Dominantly interfering with EGO-1/RRF-1
	SGS2/SDE1	*Arabidopsis thaliana*	PTGS	PTGS deficient; VIGS proficient; abnormal leaf development	Generation of dsRNA;
	RrpA	*Dictyostelium discitium*	RNAi	RNAi defective	Generation of dsRNA

Domain	Gene	Organism	Process	Phenotype	Function
RNA helicase-, PAZ-, RNAse III- and dsRNA-binding-domains	Dicer	Drosophila melanogaster	RNAi	—	Dicing long dsRNA into siRNAs; miRNAs production
		Homo sapiens	RNAi	—	Generation of siRNAs and miRNAs
	DCR-1	Caenorhabditis elegans	RNAi	RNAi defective; developmental timing defects; sterile	Dicing long dsRNA into siRNAs; stRNAs and miRNAs production
dsRNA-binding	RDE-4	Caenorhabditis elegans	RNAi	RNAi defective	Initiation of RNAi; generation of siRNAs
Putative RNA-helicase domains (various types)	Mut6	Chlamydomonas reinhardtii		Deficient in transgene silencing; transposon activation; failure to degrade aberrant RNAs	RNA-unwinding
	SDE3	Arabidopsis thaliana		PTGS deficient; VIGS proficient	RNA-unwinding
	SMG-2	Caenorhabditis elegans	RNAi	Failure to sustain RNAi after initiation	RNA-unwinding
	MUT-14	Caenorhabditis elegans	RNAi	RNAi deficient for germline-expressed genes	RNA-unwinding
	Spinde-E	Drosophila melanogaster	Stellate silencing	Failure to silence Stellate locus; developmental defects; activation of retrotransposons	RNA-unwinding
Rnase D domain	MUT-7	Caenorhabditis elegans	RNAi	RNAi deficient for germline-expressed genes; cosuppression defective	RNA-degradation
RecQ DNA helicase	QDE-3	Neurospora crassa	Quelling	Quelling defective	Generation of aberrant RNAs

degradation (19). In support, QDE-2 was found to copurify with siRNAs. This situation resembles RNAi in *Drosophila*, in which the QDE-2 homolog AGO-2 is part of an siRNA containing RNA-directed ribonuclease complex, termed RISC, that targets mRNA destruction (45, 46). RISC in *Drosophila* is a multiprotein complex, including an AGO-2 unrelated nuclease activity. Such a degree of similarity between different systems suggests that QDE-2 is unlikely to be the only protein required for siRNA-induced mRNA degradation in *Neurospora*. Biochemical approaches are needed to reveal their existence, as the genetic screens suggest that only null alleles of qde-1, qde-2, and qde-3 are compatible with life.

Genetic Dissection of PTGS in Plants

Several groups performed genetic screens in Arabidopsis to identify components of the PTGS mechanism. Vaucheret and colleagues used a 35S-*GUS* (encoding β-glucuronidase) transgenic system to isolate mutants in at least three genetic loci, *sgs1*, *sgs2*, and *sgs3* (*sgs*: suppressor of gene silencing) (36). Nuclear run-on experiments verified that the defect is in PTGS, not in TGS: Transcriptional activity of the transgene was not affected. Importantly, whereas the defect in PTGS was not limited to the exogenous 35S-*GUS* transgene (transgene-induced silencing of endogenous *Nia* genes was also affected), PTGS in these mutants could still be triggered by transgenes designed to produce dsRNA (13, 75). It has therefore been proposed that the role of these proteins in PTGS is to turn RNA triggers, possibly aberrant RNAs that are produced from certain PTGS-inducing transgenes, into dsRNA. This hypothesis is fueled by the notion that the *SGS2* gene encodes a protein that is homologous to the RNA-dependent RNA polymerases (75) previously found to be mutated in *Neurospora* PTGS and *C. elegans* RNAi (see below and Table 1).

The *SGS3*-encoded protein has no significant similarity to other known proteins in plants or other kingdoms (75), nor does the amino acid sequence provide any insight into its possible function because of a lack of known protein motifs other than a coiled-coil domain, which could point to protein-protein interactions. Both *sgs2* and *sgs3* mutant plants show enhanced susceptibility to cucumovirus (CMV), providing proof for the hypothesis that PTGS reflects an antiviral defense mechanism (75, 113, 118). The observation that a plant-encoded RdRP is able to counteract virus infection may indicate that PTGS targets the virus when it has a single-stranded conformation or that ssRNAs produced from the virus induce PTGS.

Another gene required for PTGS was isolated in a similar but more controlled screen in order to isolate mutants that in addition to this defect might also be compromised in certain stages of development (37). This gene encodes AGO1, which was previously identified to control development—mutations in *AGO1* pleiotropically affect general plant architecture (15, 72)—and is homologous to QDE-2 and *RDE-1*, which are required for quelling in *Neurospora* and RNAi in *C. elegans*, respectively.

The putative RdRP-encoding gene *SGS2*, described above, was independently found by Baulcombe and colleagues (26), who used the following experimental approach to screen for mutants in Arabidopsis PTGS. Two different GFP-encoding transgenes were used: (*a*) an initiator of PTGS, a potato virus X:GFP vector-encoding transgene (35S-*PVX:GFP*) that is able to mediate weak PTGS, and (*b*) a recipient of PTGS: a 35S-*GFP*-encoding transgene that, when present alone, is not silenced. When combined in one plant, strong PTGS is observed (25, 26). Isolated mutant alleles (*sde* for silencing defective) were categorized into four complementation groups: *sde1* and *sde2*, which result in the complete loss of PTGS, giving rise to a full green phenotype; *sde3*, which shows a delayed loss of PTGS; and *sde4*, (only one allele in a collection of 64 mutants), which shows transient loss of PTGS only at newly emerging leaves. *SDE1* was cloned and found to encode the identical protein as SGS2, described above. The analysis of PTGS-associated siRNAs in these mutant and wild-type transgenic plants revealed some striking clues (26). First, siRNAs detected in the double transgenic wild-type plants could be fully attributed to the target (35S-*GFP*) and not the initiator of PTGS (35S-*PVX:GFP* or replicating *PVX:GFP* RNA), arguing that siRNAs of the targeted transcript are also generated as a consequence of PTGS and not just as a source to trigger PTGS. These data also suggest that siRNAs, or a major fraction in this system, accumulate in a target-dependent fashion. Second, siRNAs were less abundant in *sde1* plants, but in this genetic background, siRNAs were derived from the replicating initiator *PVX:GFP* RNA that was present at elevated levels in *sde1* plants. This suggested that SDE1 is not required for PTGS induced by a virus. Indeed, when tested directly, GFP expression in *sde1* plants, although resistant to PTGS mediated by a transgene, is still sensitive to silencing triggered by a *Tobacco rattle virus* vector construct (27). This sensitivity was also found for plants defective for *SDE3*, which encodes a putative RNA helicase with sequence similarity to RNA helicase-like proteins conserved in all kingdoms (27).

These observations led to a model in which PTGS is triggered through dsRNA intermediates that are either directly supplied (e.g., by virus infection) or made from a transgene in a process that requires at least SDE1/SGS2, AGO1, SGS3, and SDE3, thus placing the action of these proteins upstream of siRNA production (115). This hypothesis was strengthened by the recent finding that *sgs2, sgs3*, and *ago1* mutants (isolated because of their resistance to PTGS induced by highly transcribed sense transgenes) are proficient in RNA silencing when triggered by transgene loci producing dsRNA (13, 75).

Transgene and Transposon Silencing in *Chlamydomonas reinhardtii*

In the unicellular green alga *Chlamydomonas reinhardtii*, loss of a DEAH-Box RNA helicase, *Mut6*, not only results in the loss of PTGS (of a transgene) but also leads to enhanced transposition of both the retroelement TOC1 and the DNA-mediated transposon Gulliver (126). The levels of RNAs corresponding to the transgene and the retroelement TOC1 are markedly increased in the Mut-6 genetic background (in a wild-type genetic background, transgenically derived RNAs are not observed). By inhibiting de novo RNA synthesis, while endogenous mature mRNAs are degraded normally in Mut-6, TOC1 transcripts become stable, suggesting that the degradation of these RNAs depends on the *Mut6* protein. Interestingly, the plasmid insertion into the *Mut-6* locus, causing the defect, results in improperly processed *Mut-6* transcripts with defects in splicing and polyadenylation. These aberrant RNAs disappeared when the Mut-6 defect was complemented by the introduction of a single copy of the wild-type *Mut-6* allele. This could mean that specifically aberrant RNAs are degraded in a *Mut6*-dependent process that can also silence transgenes, whereas correctly processed mRNAs are left untouched.

Genetic Screens to Identify Components in *C. elegans* RNAi

To identify factors involved in *C. elegans* RNAi, genetic screens were performed that made use of the systemic nature of RNAi, i.e., RNAi can also be triggered by soaking animals in a solution containing dsRNA (104) or by feeding them bacteria that are genetically modified to produce dsRNA homologous to a worm gene (107, 108). Thus, mutagenized animal populations were fed bacteria that produce dsRNA homologous to an essential worm gene, and mutants were selected on their ability to grow on such media. A subset of mutants that were also resistant to injection of dsRNA directly into the animal was further analyzed to determine the molecular nature of the RNAi defect (105). Several complementation groups have been identified this way: *rde-1*, *rde-3*, and *rde-4* are completely RNAi defective, and *rde-2* has a defect in RNAi that appears to be specific for RNAi directed to germline-expressed (or maternally provided) mRNA. Somatically expressed target genes are fully sensitive to dsRNA triggers.

Another genetic screen that yielded RNAi-defective mutants was initially aimed at identifying components of the mechanism that silences DNA transposition in the germline of *C. elegans* (53). Although transposition can readily be detected in somatic tissues of the nematode, the germline is protected from these mutagenic events; worms can be cultured for many years without a single case of transposition. A large fraction of these mutator (*mut*) mutants—loss of silencing leads to a mutator phenotype as a result of frequent gene interruptions by transposon insertions—also have a defect in RNAi (53). Many of these are either allelic to or have similar RNAi characteristics as *rde-2* or *rde-3*, as was also suggested by the reciprocal observation that *rde-2* and *rde-3* lost their defense against DNA transposition. The notion that transposon silencing depends on factors that are also required for dsRNA-induced gene silencing led to the speculation that these DNA elements are tamed via a dsRNA intermediate. This hypothesis, however, is difficult to reconcile with the observation that completely RNAi-deficient *rde-1* and *rde-4* mutant animals (also when the dsRNA is expressed cell autonomously) do not display the mutator phenotype (105).

On the other hand, it is difficult to place the action of the RDE-1 and RDE-4 proteins as central to RNA silencing: Although essential for RNAi, these gene products are not required for transgene-induced cosuppression (29, 54) or RNA silencing via direct administration of short antisense RNAs (106), while for both manifestations of RNA silencing, functionality of the mutator/RNAi type of genes, e.g., *mut-7*, is crucial. Further evidence that the RNAi-specific genes *rde-1* and *rde-4* can be mechanistically separated from the mutator/RNAi genes came from elegantly designed experiments that addressed the inheritance of RNAi (42). It has long been recognized that for a number of target genes, the RNAi pheno-type was not restricted to the exposed animals but could also be transmitted to the next generation (F1); in most cases, the F2 generation reverted to a wild-type phenotype. Mello and coworkers investigated this inheritability of RNAi and found clues for the existence of a dominant extragenic agent: For some sensitive targets—genes expressed in the maternal germline—the interference is sustained for more generations and can be passed on to subsequent generations even in the absence of the endogenous gene that is the target of silencing. The authors uncoupled the initiation of this inferred inherited species (in the mother exposed to dsRNA triggers) from the execution step in RNAi (interference of the target mRNA in the F2 offspring), and subsequently analyzed the genetic requirements of both events. They found that the formation of the inherited agent depends on the RNAi genes *rde-1* and *rde-4* but not on the RNAi/mutator genes *rde-2* and *mut-7*. In contrast, execution of silencing in the descendants of the exposed an-imals required RDE-2 and MUT-7 but did not depend on the wild-type activity of *rde-1* and *rde-4* within these animals. Taken together, these data suggest an order in the mechanism of RNAi: (*a*) the initiation of RNAi by the generation of a heritable extragenic sequence-specific intermediate; requirement for RDE-1 and RDE-4, followed by (*b*) the execution of interference downstream of initiation, which depends, at least, on the *rde-2* and *mut-7* genes (42).

$$\text{dsRNA} \xrightarrow[\substack{\text{(exposed animal)}}]{\text{RDE-1, RDE-4}} \substack{\text{Inherited}\\\text{extragenic}\\\text{species}} \xrightarrow[\substack{\text{(progeny of}\\\text{exposed animal)}}]{\text{RDE-2, MUT-7}} \substack{\text{mRNA}\\\text{destruction}}$$

$$\text{dsRNA} \xrightarrow[\text{Dicer reaction}]{} \text{siRNAs} \xrightarrow[\text{RISC}]{} \substack{\text{mRNA}\\\text{destruction}}$$

One obvious class of candidate molecules to constitute the extragenic inherited agents are siRNA species (see the parallels between the *C. elegans* genetic data and the model to explain RNAi in vitro in the scheme depicted above). Fire and coworkers demonstrated that siRNAs are produced upon injection of dsRNA into the gonadal syncytium, and although wild-type animals were able to dice the dsRNA into molecules of sizes reminiscent of siRNAs, this activity was almost completely abolished in *rde-4* mutant animals, which suggests that RDE-4 is di-rectly involved in this processing step (83). In contrast, phenotypically similar *rde-1* mutant animals were fully proficient in dicing dsRNA, suggesting that this pro-tein acts downstream of siRNA production. In agreement with this interpretation, direct injection of short dsRNA molecules into the animal could bypass the requirement for RDE-4, but not RDE-1, in triggering gene silencing (83).

The involvement of these genes in the Dicer reaction was also analyzed in vitro by assaying *C. elegans* extracts prepared from wild-type and various mutant animals for their ability to produce siRNAs from long double-stranded molecules, an activity attributed to *C. elegans* DCR-1, the ortholog of the *Drosophila* RNAse III enzyme Dicer; siRNA-producing activity can be immunoprecipitated with antisera raised against DCR-1 (52). Although extracts from the RNAi mutants *rde-1*, *mut-7*, and *mut-14* were fully proficient (106), extract prepared from *rde-4* animals failed to generate siRNAs (unpublished observations), supporting the proposed role for RDE-4 in the first step of RNAi: dicing dsRNA into siRNAs. Using RNase protection assays, this study also analyzed the production of siRNAs in vivo in wild-type and mutants animals that were exposed to dsRNA by feeding on bacteria that generate dsRNA homologous to a specific *C. elegans* gene. Whereas in wild-type animals, RNA molecules of 20–25 nt could easily be detected, most mutants, including those that are proficient in siRNA production in vitro, failed to produce detectable levels of small RNA species (106). The observation that only RNA species of antisense polarity, not of sense polarity, are detected in wild-type animals suggests that siRNAs that are directly diced of the input dsRNA were not visualized; if so, one would expect species of both sense and antisense polarity, as was found in the *C. elegans* in vitro assay. The species that are detected in a wild-type genetic background more likely result either from stabilization (presumably on the target mRNA, which would explain why only antisense siRNAs are seen) and/or amplification. These more downstream steps in the RNAi reaction probably require the action of several genes that have been isolated in mutant screens.

mRNA DESTRUCTION

As mentioned above, biochemical experiments in *Drosophila* and mammalian cells demonstrated that siRNAs are the effector molecules in RNAi (33, 34). However, most of the genes isolated in the genetic screens described above can be placed either upstream of the formation of dsRNA or in the process of dicing the dsRNA into siRNAs; relatively little is known about the protein components involved in the execution step of RNAi: destruction of the mRNA. The genetic data from *C. elegans* suggest that the mutator/RNAi genes, e.g., *mut-7* and *rde-2*, might be involved, but biochemical support is lacking. In the short term, further identification of proteins that copurify with RISC activity should help to elucidate this issue; identification of additional RISC components is under way in several laboratories. The only RISC factor thus far published is *Drosophila* AGO2 (46), a homolog of RDE-1, QDE-2, and Arabidopsis Ago1, which are required for RNA silencing in genetic systems (20). Another *Drosophila* member of this family, AGO1, was recently found also to be required for RNAi in vivo (124). Remarkably, RDE-1 and Arabidopsis Ago1 are thought to operate in initiating stages of RNA silencing, and not in the more downstream steps. In these systems, other family members, not yet identified genetically, [*C. elegans* encodes at least 24 members of this family (41, 97)] may be required, perhaps because these are essential for viability. Support for a scenario in which downstream components in RNA silencing, possibly of RISC, might not be isolated in genetic screens (as a result of an associated essential function) comes from the notion that most, if not all, PTGS mutants isolated in plant systems are still proficient when RNA silencing is triggered directly by dsRNA. This suggests that only upstream components can be found.

Alternatively, the RNA silencing pathways might be more diverged between species than initially anticipated: In *Drosophila* and also in mammalian cells, RISC activity in RNA silencing has been well established, including the ability to bypass the upstream part of RNAi by direct administration of siRNAs in vivo and in vitro. However, in *Neurospora*, plants, and *C. elegans*, evidence for RISC activity has not been found to date.

Secondary siRNAs Produced by Transitive RNAi

Some light on possible differences in executing mRNA degradation was shed recently when it was found that in addition to siRNAs that could be attributed to the input dsRNA, siRNAs were also detected from sequences outside of the targeted area (98). This could mean that siRNAs are also produced from the targeted mRNA transcript (and not just from the input dsRNA), as secondary products of mRNA degradation. These secondary siRNAs must be produced by de novo RNA synthesis because the 20–25-nt molecules of antisense polarity that comprise one strand of the double-stranded siRNAs are not present in unexposed animals nor in the trigger dsRNA molecules. Subsequent analysis indicates that secondary siRNAs are functional. An endogenous gene could be targeted with noncomplementary dsRNA as long as the animals also express an engineered transcript that has sequences of the target mRNA fused to sequences that are homologous to the input dsRNA (see also Figure 3). In agreement with the polarity seen in the production of secondary siRNAs, in these experiments RNAi transits (hence called transitive RNAi) in one direction. Interference of the endogenous locus was observed only when the homologous sequence was positioned 5′ of the sequences corresponding to the dsRNA input. These data suggest that mRNAs are not only targets but also amplifiers of the original signal, which may explain earlier observations that only traces of dsRNA can wipe out an excess of mRNAs, and the fact that RNAi can persist for more than one generation.

In plants such a unidirectional generation of secondary siRNAs is not seen: Secondary siRNAs are also detected for sequences downstream of the input trigger and their generation is dependent on a putative RdRP protein. Possible explanations include (*a*) the presence of undetectable levels of full-length antisense transcripts: The sense strand of the siRNA can then prime RNA synthesis identical to its antisense counterpart on the sense mRNA, thus giving rise to siRNAs that can be assigned to both upstream and downstream sequences. (*b*) The sense transcript

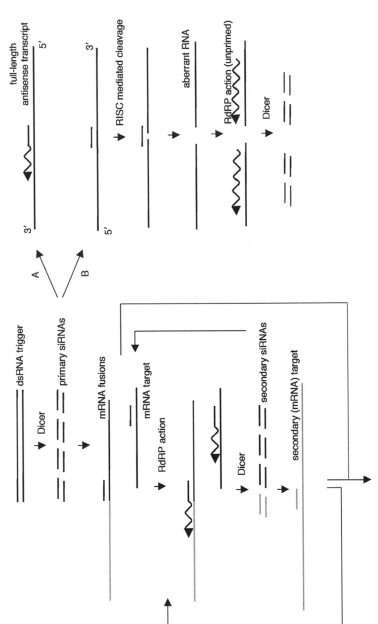

is cleaved by the primary siRNAs, but the mRNA degradation products are seen as aberrant RNAs that are turned into dsRNA in a process that likely involves primer-independent RdRP action. At present, biochemical data exist only from a member of this family isolated from Tomato: This protein can produce RNA products by extending primers on RNA templates but is also capable of initiating RNA synthesis in the absence of a primer.

Generation of secondary siRNAs requires the action of a protein capable of de novo RNA synthesis using the mRNA as a template, which would result in dsRNA that can be a substrate for Dicer-mediated cleavage. Likely candidates for this function are RNA-dependent RNA polymerases (RdRPs), especially since in *C. elegans* putative RNA-dependent RNA polymerases have also been implicated in gene silencing phenomena. Based on the observation that gene silencing in *Neurospora* requires a putative RdRP (Qde-1), Maine and coworkers investigated RNAi in animals mutated in *ego-1*, one of the four homologous genes encoded by the *C. elegans* genome, and found that this protein is involved in RNAi, but the requirement appears specific for germline tissue (100). Recently, Sijen et al. demonstrated that loss of *rrf-1*, a second member of this family of four (and in the genome positioned immediately downstream of *ego-1*), also results in an RNAi-defective phenotype, but in this case, only somatic cells are resistant to RNAi (98). This could indicate that both proteins fulfill the same function but in different tissues.

Additional support for the involvement of RdRP action in *C. elegans* RNAi comes from the observation that short RNA of antisense polarity can also trigger RNA silencing when administered in close proximity to the target mRNA (106). These antisense species have a less strict size requirement than double-stranded

---

←

Figure 3   RNAi amplification, transitive RNAi, and the generation of secondary siRNAs. The left part of the figure depicts a model to explain how transitive RNAi can lead to amplification of RNAi in *C. elegans*: Primary siRNAs diced from the input dsRNA anneal with RNA molecules that contain complementary sequences. This primer-template substrate can then be used to support de novo RNA synthesis by an RNA-dependent RNA polymerase resulting in dsRNA 5′ to the sequences that are within the input dsRNA. The detection of siRNAs corresponding to such upstream regions suggests that Dicer cuts the dsRNA leading to "secondary" siRNAs that, in a second round, can anneal to a novel target that does not share any sequence homology with the initial input dsRNA, but shares homology with the intermediate bridging RNA molecules. Another round can now, in principle, be initiated, degrading the mRNA in the process, thus completing RNAi. Note that this mechanism does not require, nor exclude, the action of a RISC activity: cutting the mRNA within the siRNA-covered region. The notion that RNAi is not initiated at all in RdRP mutants suggests that either mRNA is degraded via this mechanism or that an amplification of the original signal is required to completely wipe out mRNA.

siRNAs—siRNAs only work when they are of 21–23 nt in length (35, 78)—suggesting that these molecules do not enter RISC-mediated mRNA degradation directly. Instead, as RNAs might directly act as RNA primers allowing extension of the 3′ end by an RdRP, thereby creating dsRNA that can subsequently be a substrate for Dicer action (106). Together, these observations led to the following model (Figure 3) (2, 86, 127). Although the model includes the involvement of a RISC complex in degradation of the mRNA, the presumed action of an RdRP on the mRNA target resulting in dicer substrates will automatically lead to its destruction thus creating the amplification products. In such a scenario, a separate parallel RISC activity is superfluous.

Mutations in a third member of this RdRP gene family, *rrf-3*, make *C. elegans* supersensitive to RNAi (98), a phenotype that might be explained if the RFF-3 protein dominantly interferes with the action of EGO-1 and RRF-1.

Although the term transitive RNAi was first coined in *C. elegans*, the phenomenon was initially observed for PTGS in plants. In *N. benthamiana* plants that express full-length GFP (as a presumed bridging molecule), GFP DNA fragments containing only part of the GFP sequence can "transitively" silence the expression of nonoverlapping GFP sequences that are independently expressed from a virus-based vector (122). A putative RdRP (SGS2/SDE1) is also required, although in contrast to *C. elegans*, no directionality is observed: siRNAs are produced both 5′ and 3′ of the original trigger, indicating bidirectional spreading from the initiator region into adjacent regions of the target gene (Figure 3) (110).

Independently, Nellen and coworkers, in *Dictyostelium discoideum*, showed that hairpin constructs can silence expression of homologous transgenes or endogenous genes. Here too, silencing required the action of an enzyme with presumed RdRP activity (67). Of the three RdRPs encoded by the *Dictyostelium* genome, RrpA, RrpB, and RrpC, only RrpA was required to induce RNA silencing; however, RrpA was also essential for the detection of siRNAs, despite the fact that extracts of rrpA in vitro were proficient in dicing dsRNA into siRNAs. This could indicate that secondary siRNAs rather than primary siRNAs (coming from the initiating dsRNA hairpin) are visualized in the in vivo assay. This hypothesis is further supported by the observation that siRNAs were not detected in the absence of a target locus, suggesting that the target gene is involved in siRNA generation and amplification. This situation resembles observations made in *C. elegans* in which most siRNAs that are detected in vivo are thought to be secondary (as discussed above).

Does this mean that in these species RNA silencing operates via an RdRP dependent mechanism, whereas *Drosophila* and mammals execute RNAi via RISC? Although homologs of the RdRP genes described above have not been found in BLAST searches in the *Drosophila* and human genome, in vitro experiments using *Drosophila* cell extracts suggest that siRNAs might prime RdRP action on complementary RNA targets (63). Perhaps the most convincing finding against a role for obligatory polymerase action on the siRNA is that in human cells, siRNAs that cannot act as primers because their 3′ terminus is blocked by a FITC moiety are still potent inducers of RNA silencing (47).

SYSTEMIC NATURE OF RNA SILENCING

Systemic Propagation of PTGS

An amplification step has also been implicated in the systemic spread of RNA silencing throughout exposed organisms. RNA silencing is not restricted to individual cells but can spread from the site of infection to more distant tissues. The systemic nature of PTGS in plants has been termed SAS, for systemic acquired silencing, and was most convincingly demonstrated in grafting experiments (81). Sequence-specific silencing was transmitted unidirectionally from silenced stocks (lower tissues and root system) to target scions (vegetative upper tissues), suggesting the presence of a diffusible silencing agent. Noteworthy in these experiments is that the expression level of the target gene needed to be elevated compared to wild-type plants to observe systemic PTGS (82). Non-silenced scions that either contain extra transgene-encoded copies or are overexpressing the endogenous target gene as a result of metabolic derepression, are sensitive to PTGS upon grafting, although systemic PTGS was not observed when wild-type scions were grafted onto silenced stocks. This could mean that only a limited number of triggers reach the distant tissues and some sort of target-dependent amplification is required to completely degrade all mRNAs. The notion that transmission of PTGS also occurred between a silenced stock and an engrafted scion through a stem of a non-transgenic wild-type plant suggests that such amplification is not required during long-distance transport of the proposed, but as yet unidentified, RNAi signaling molecule.

The systemic propagation of PTGS was also demonstrated by Voinnet et al. (119, 122). These authors show that in non-silenced GFP transgenic tobacco plants, silencing can be initiated by localized introduction of an additional GFP transgene, either via *Agrobacterium* infiltration or biolistic transformation. Although GFP silencing is initially localized to the exposed tissues, after some time silencing starts to spread—long-distance movement of the signal occurs through the phloem, and cell-to-cell movement is through plasmodesmata—eventually leading to PTGS even in the upper leaves of the plant. A system that amplifies the initiating silencing event has been proposed in this case also.

Systemic RNAi in *C. elegans*

In *C. elegans*, silencing is also not restricted to those cells that are exposed to the original dsRNA trigger. GFP expression can be silenced in the germline and in almost all somatic cells by feeding animals on dsRNA directed against GFP (50, 107, 108). Apparently, the dsRNA is taken up from the environment and the signal spreads throughout the organism to exert its effect in differentiated tissues. Only neuronal cells appear to be less sensitive to incoming triggers. RNAi works when the dsRNA is expressed within these cells, but silencing is not complete when the dsRNA is administered via food. A possible explanation is that neurons are less proficient in the uptake of such nucleic acids from their surroundings.

Because genetic screens to isolate mutants in the mechanism of RNAi made use of this systemic nature of RNAi, mutants with defects in the uptake of dsRNA or transport of RNAi signals have also been found. Mutagenized animals were selected on bacteria producing dsRNA homologous to genes essential for embryonic development (thus resulting in dead embryos), which are expressed in the maternal germline. Thus, animals with a defect in the uptake of dsRNA or in the transport of trigger molecules to the target tissue would survive this selection. To obtain true RNAi-deficient (*rde*) mutants, a second filter was applied: Only those mutants that were also RNAi defective when the dsRNA was directly injected into the target tissue (thus bypassing the routing of RNAi signals) were further analyzed.

Genetic screens have recently been performed directly aimed at identifying components required for systemic RNAi (125). Mutants were isolated that lost dsRNA-induced silencing of a GFP marker gene but only when the dsRNA triggers come from outside the silenced cells: to control for an intact RNAi machinery, GFP was also silenced by transgene-induced expression of GFP dsRNA in a limited number of cells. Mutants that fail to silence GFP triggered by environmentally produced dsRNA, but that had a wild-type response to dsRNA expressed cell autonomously, were further analyzed and three complementation groups were identified: *sid-1*, *sid-2*, and *sid-3* (for systemic RNAi deficient). *sid-1* has been cloned and encodes a transmembrane domain-containing protein that is required cell autonomously, perhaps for the uptake of dsRNA or RNAi intermediates.

Notably, in this experimental system, some systemic spread is seen from cells that express GFP dsRNA to cells that only have the recipient GFP mRNA target, but this spread was very limited. One explanation might lie in a limited intercellular transport between the tissues that were analyzed (muscle cells). However, expression of hairpin constructs within intestinal cells (which take up food from the lumen of the gut and are thus the likely entry point of environmentally provided dsRNA) silences GFP expression within these cells but does not trigger silencing throughout the organism (our unpublished observations). Perhaps environmentally provided dsRNA (i.e., by feeding or soaking) is immediately endocytosed into specialized cellular compartments for further transport. It is unclear whether there are distinct routes by which RNAi signaling or effecter molecules travel through the organism or whether these molecules are hitchhiking along existing pathways to take up nutrients. Many mutants that show some sort of defect in transitive RNA are viable and do not have a significant growth delay, which a priori would not be expected if the defect compromises the intake of food.

BIOLOGICAL FUNCTIONS OF RNA SILENCING

Interplay Between PTGS and RNA Viruses

The discovery that mutant plants defective in PTGS are also hypersensitive to the infection of certain viruses strengthened the hypothesis that PTGS participates in a defense mechanism to protect the plant (and other species) against infectious agents. Several observations had pointed to a complex interplay between PTGS and virus infection. For example, plants that show transgene-induced PTGS of a virus sequence are immune to the cognate virus [e.g., (66)]. Also in some cases that involve wild-type plants, recovery from virus infection is associated with specific degradation of virus RNA (3, 90); these wild-type plants have now become immune to that virus and to viruses that have been genetically modified to contain part of the virus used for the initial infection (89, 90).

Not surprisingly from an evolutionary point of view, in response to this defense strategy, plant viruses produce proteins that counteract PTGS. This idea followed from the observed synergistic effects in disease symptoms when plants were infected with two viruses. Because the helper component proteinase (Hc-Pro) from potyvirus suppressed the host defense mechanism, the pathogenicity and accumulation of two other viruses, *Cucumber mosaic virus* and *Tobacco mosaic virus*, were enhanced (87). Subsequently, it was shown that HC-Pro and also the 2b protein of the CMV virus function as effective suppressors of PTGS (7, 16, 51), further strengthening the proposed role of PTGS as a natural mechanism against viruses. Several virally encoded suppressors of PTGS have subsequently been identified that target PTGS at different stages: initiation, systemic spread, or maintenance (12, 121). For example, HC-Pro inhibits PTGS in all symptomatic tissues, and this suppression appears to act at a step preceding accumulation of the PTGS-associated siRNAs: No siRNAs were observed in plants in which PTGS has been suppressed by HC-Pro (64). Remarkably, overexpression of a tobacco gene, *rgs-CaM*, which was identified because the encoding protein interacts with Hc-Pro, also inhibits PTGS (6). Possibly, this type of virus-induced suppression of PTGS is established through activation of the plant-encoded *rgs-CaM* gene: *rgs-CaM* mRNA levels are increased by virus infection or in plants expressing HC-Pro. The biological function of this calmodulin-related protein in wild-type plants is not yet known.

Other types of virally encoded suppressors, including the CMV 2b and TBSV 19K proteins, have a more restricted inhibitory effect: Initiation of PTGS was inhibited only at the growing points of the plants, but no effect was observed in tissues where PTGS has already been established (16). In yet another interesting case, the 25-kDa viral movement protein (p25) of *Potato virus X*, although originally thought not to interfere with PTGS (16), was shown to target the spread of silencing. Grafting experiments showed that systemic PTGS was absent in the presence of functional p25 (120).

Virally encoded suppressors of RNA silencing are not restricted to plants. In animal cells, RNA silencing can also act as an antiviral defense: *Flock house virus* (FHV), an RNA virus that can infect both vertebrate and invertebrate hosts and is both an initiator and target of RNA silencing in *Drosophila* host cells, encodes a protein, B2, that also operates as an RNA-silencing suppressor (61). Interestingly, this protein was investigated not because of sequence similarity to known suppressors, but because of its similar location in the viral genome to the CMV RNA silencing-suppressor 2b (62).

RNAi and Repression of DNA Transposition

In addition to protecting organisms against viruses, RNA silencing can also protect the organisms from DNA transposition, as observed for *C. elegans* and *Chlamydomonas reinhardtii* in which mutations in RNA silencing can also result in increased rates of transposition (53, 126). In another example of coordinate loss of RNA silencing and loss of transposition repression, Aravin et al. observed that mutations in the RNA helicase spindle-E not only relieved silencing of the Stellate genes—Stellate expression is sensitive to dsRNA-mediated silencing—but also resulted in derepression of retrotransposition in the *Drosophila* germline (9). A further indication that the RNA silencing machinery might directly target retrotransposon transcripts comes from the detection of retrotransposon-derived siRNAs in *Trypanosoma brucei*, where dsRNA-induced RNA silencing has been shown (77), indicating that cognate dsRNA intermediates are generated in this parasite that can be processed through Dicer action (30).

Cleaning Up Aberrant RNAs

Yet another function of the RNA silencing mechanism, or at least of some of its components, might be to remove aberrant nonfunctional RNAs from the nuclear or cellular pool of RNAs to fulfill a role in mRNA surveillance. If aberrant RNAs are turned into dsRNA to trigger RNA silencing, they are destroyed in the process. This would impose a serious problem to the cell if a single aberrant molecule is sufficient to trigger the reaction, sweeping the cell clean of any homologous mRNA. One would assume that only if these aberrant RNAs are produced above a certain threshold level, RNA silencing is triggered. Strong evidence for a threshold level in RNA silencing is seen in *Drosophila* undergoing PTGS (80). Although the strongest support comes from the work in *C. reinhardtii* (11, 126), a possible role for RNA silencing components in mRNA surveillance is also fueled by the discovery that mutations in certain *smg* genes, which are required for mRNA surveillance in *C. elegans*, show a defect in the attenuation of RNAi (31). Mutant animals recover much faster from dsRNA treatment than the wild-type controls [see also (65) for a comprehensive review on the RNAi/mRNA surveillance connection]. In addition, the Arabidopsis gene SDE3, which when mutated renders plants resistant to PTGS, is homologous to the *C. elegans smg-2* gene (27). Such implications could undermine the interpretation of mutants defective in the RNAi response: Any mutation resulting in the abundant production of aberrant RNAs might saturate the RNA silencing machinery and falsely be considered as defective in the RNAi pathway. Goldstein and coworkers (32) raised this possibility as a possible explanation for why mutations in chromatin condensation factors also resulted in a diminished capacity to respond to dsRNA.

RNA Silencing and Development: siRNAs, stRNAs, miRNAs

Some components of the silencing machinery are interconnected with gene regulatory mechanisms that control developmental programs. Following up on in vitro data in *Drosophila* demonstrating that the RNAse III-type nuclease DICER is responsible for processing dsRNA into siRNA species, several research groups independently reported the involvement of the orthologous gene, *dcr-1*, in *C. elegans* RNAi (41, 52, 55). In addition to an RNAi-defective phenotype, i.e., proof for a role of dicer in RNAi in vivo, DCR-1-defective animals are sterile; their oocytes are abnormal and no fertilized eggs were detected. They also display several developmental abnormalities during larval growth, including a protruding vulva and a tendency to burst at the vulval region before reaching the adult stage. In addition, seam cells fail to fuse during the L4-to-adult transition. In some cases, additional rounds of cell division are observed that result in animals without adult-specific alae (41, 52). These developmental defects are reminiscent of phenotypes resulting from mutations in the heterochronic genes *lin-4* and *let-7* (60, 91). These genes produce noncoding RNAs that regulate the timing of developmental events (hence named small temporal RNAs or stRNAs) via regulation of translation of downstream target genes. This overlap in phenotypes together with the notion that stRNAs, which are of similar length as siRNAs, can be produced by processing of larger precursors predicted to form a dsRNA hairpin structure, triggered investigation of the role of *dcr-1* in the generation of such stRNAs. Loss of DCR-1 led to accumulation of the *let-7* and *lin-4* "unprocessed" precursor molecules and reduced levels of mature stRNAs (41, 52). Dicer fulfills this function in other animal systems, where the *let-7* stRNA is conserved across much of animal phylogeny with a 100% sequence conservation between *C. elegans*, *Drosophila*, and humans (85): Immunopurified *Drosophila* DICER as well as *Drosophila* embryo extracts could convert the *let-7* precursor RNA into 21-nt-sized stRNA (48, 52), and downregulation of human Dicer in Hela cells results in the accumulation of *let-7* RNA precursor (48).

The *let-7* and *lin-4* stRNAs are just the tip of the iceberg. Several laboratories have recently cloned many additional stRNA-like RNAs from *C. elegans*, *Drosophila*, and human cells, subsequently termed micro-RNAs or miRNAs (57–59, 74). Like the stRNAs, all of these miRNAs are predicted to be processed from duplexed precursor molecules and are therefore likely processed by Dicer, which was indeed shown for two of them (59). The eminent role of the dicer enzyme in catalyzing the maturation of such a plethora of miRNAs might explain why development in *C. elegans dcr-1* animals is completely compromised. The fact that first-generation *dcr-1* homozygotes are viable (but subsequently fail to produce progeny) is likely the result of maternal contribution of DCR-1. Also, dramatic developmental defects are observed in Arabidopsis that are mutated in the Dicer homolog CARPEL FACTORY (49).

None of the miRNAs is perfectly complementary to any known protein-encoding transcripts in the genome of the cognate species, which suggests that miRNAs do not negatively regulate expression via the RNAi pathway, i.e., destruction of the homologous transcript. Also, vice versa, animals defective for the RNAi protein RDE-4, which is thought to cooperate with Dicer in siRNA production, do not display developmental phenotypes. Instead, miRNAs may interfere with the translation machine, as for *lin-4* and *let-7* stRNAs, by binding the 3'UTR of the downstream target genes (4, 99). The regulatory potential of miRNAs is further illustrated by the observation that many miRNAs, including those found in *Drosophila* and mammalian cells, have a very strict time- or tissue-specific expression pattern. No function has yet been ascribed to any of the novel miRNAs. A big challenge for today's molecular biologist is to define the potential role of miRNAs (in gene regulation) and identify their target genes.

FUTURE DIRECTIONS

The tremendous pace in which our understanding of RNA silencing progressed over the past four years, despite many unanswered questions, makes it likely that most protein factors required for RNA silencing will be identified through combined biochemical and genetic approaches, and that their specific function in the silencing mechanism will be addressed. A stronger link is needed between the biochemical and genetic data, e.g., by setting up in vitro systems for the genetic systems. Although the outline of the RNA silencing pathways is coming into focus, much interest will drift away from the core silencing mechanism into related areas of research. As an example, the recent identification of a mechanistic link between RNA silencing and gene regulation via stRNAs/miRNAs, together with the notion that such miRNAs are abundantly present throughout the animal kingdom, opened up a completely new area of research into how and to what extent these small noncoding RNA molecules regulate cellular or developmental processes (5, 73, 94).

One of the most interesting aspects of understanding how RNA silencing works is its practical use: the ability to reduce gene expression in a manner that is highly sequence specific as well as technologically facile and cheap. In the *C. elegans* community, RNAi is now routinely used in reverse genetic approaches studying particular gene functions. Moreover, large-scale RNAi screens have been performed in *C. elegans* using RNAi as a forward mutagenesis tool (39, 40). Currently, several laboratories are scanning the complete genome for specific phenotypes using libraries of bacteria that produce dsRNA directed against each ORF encoded by the *C. elegans* genome. Similar tools are about to be applied to higher eukaryotes, including human cells. Upon the demonstration that siRNAs can be used to reduce gene expression in mammalian cells (33), bypassing the PKR response that is triggered when larger dsRNA molecules enter the cells, many laboratories designed vector-based DNA systems that express siRNAs or siRNA-precursor hairpin cell-autonomously, thus providing a stable source of RNA silencing molecules [e.g. (17, 79)]. This tool can be easily scaled to set up a library to systematically target all ORFs encoded by the human genome. However, caution is needed here. Although thus far siRNAs have been shown to be potent, not all siRNAs are equally efficient. In practice, this could mean that several siRNAs must be tested empirically to find one that reduces mRNA levels sufficiently to draw definite conclusions in analysis of gene function. For *C. elegans*, not all genes are targeted efficiently; what constitutes the parameters for success is not yet known.

The *Annual Review of Genetics* is online at http://genet.annualreviews.org

LITERATURE CITED

1. Abou Elela S, Ares M Jr. 1998. Depletion of yeast RNase III blocks correct U2 3′ end formation and results in polyadenylated but functional U2 snRNA. *EMBO J.* 17:3738–46

2. Ahlquist P. 2002. RNA-dependent RNA polymerases, viruses, and RNA silencing. *Science* 296:1270–73

3. Al-Kaff NS, Covey SN, Kreike MM, Page AM, Pinder R, Dale PJ. 1998. Transcriptional and posttranscriptional plant gene silencing in response to a pathogen. *Science* 279:2113–15

4. Ambros V. 2000. Control of developmental timing in *Caenorhabditis elegans*. *Curr. Opin. Genet. Dev.* 10:428–33

5. Ambros V. 2001. MicroRNAs: tiny regulators with great potential. *Cell* 107:823–26

6. Anandalakshmi R, Marathe R, Ge X, Herr JM Jr, Mau C, et al. 2000. A calmodulin-related protein that suppresses posttranscriptional gene silencing in plants. *Science* 290:142–44

7. Anandalakshmi R, Pruss GJ, Ge X, Marathe R, Mallory AC, et al. 1998. A viral suppressor of gene silencing in plants. *Proc. Natl. Acad. Sci. USA* 95:13079–84

8. Angell SM, Baulcombe DC. 1997. Consistent gene silencing in transgenic plants expressing a replicating potato virus X RNA. *EMBO J.* 16:3675–84

9. Aravin AA, Naumova NM, Tulin AV, Vagin VV, Rozovsky YM, Gvozdev VA. 2001. Double-stranded RNA-mediated silencing of genomic tandem repeats and transposable elements in the *D. melanogaster* germline. *Curr. Biol.* 11:1017–27

10. Baulcombe DC. 1996. RNA as a target and an initiator of post-transcriptional gene silencing in transgenic plants. *Plant Mol. Biol.* 32:79–88

11. Baulcombe DC. 2001. Unwinding RNA silencing. *Science* 290:1108

12. Beclin C, Berthome R, Palauqui JC, Tepfer M, Vaucheret H. 1998. Infection of tobacco or Arabidopsis plants by CMV counteracts systemic post-transcriptional silencing of nonviral (trans)genes. *Virology* 252:313–17

13. Beclin C, Boutet S, Waterhouse P, Vaucheret H. 2002. A branched pathway for transgene-induced RNA silencing in plants. *Curr. Biol.* 12:684–88

14. Bernstein E, Caudy AA, Hammond SM, Hannon GJ. 2001. Role for a bidentate ribonuclease in the initiation step of RNA interference. *Nature* 409:363–66

15. Bohmert K, Camus I, Bellini C, Bouchez D, Caboche M, Benning C. 1998. AGO1 defines a novel locus of Arabidopsis controlling leaf development. *EMBO J.* 17: 170–80

16. Brigneti G, Voinnet O, Li WX, Ji LH, Ding SW, Baulcombe DC. 1998. Viral pathogenicity determinants are suppressors of transgene silencing in *Nicotiana benthamiana*. *EMBO J.* 17:6739–46

17. Brummelkamp TR, Bernards R, Agami R. 2002. A system for stable expression of short interfering RNAs in mammalian cells. *Science* 296:550–53

18. Catalanotto C, Azzalin G, Macino G, Cogoni C. 2000. Gene silencing in worms and fungi. *Nature* 404:245

19. Catalanotto C, Azzalin G, Macino G, Cogoni C. 2002. Involvement of small RNAs and role of the qde genes in the gene silencing pathway in Neurospora. *Genes Dev.* 16:790–95

20. Cerutti L, Mian N, Bateman A. 2000. Domains in gene silencing and cell differentiation proteins: the novel PAZ domain and redefinition of the Piwi domain. *Trends Biochem. Sci.* 25:481–82

21. Cogoni C, Irelan JT, Schumacher M, Schmidhauser TJ, Selker EU, Macino G. 1996. Transgene silencing of the *al-1*

gene in vegetative cells of Neurospora is mediated by a cytoplasmic effector and does not depend on DNA-DNA interactions or DNA methylation. *EMBO J.* 15: 3153–63

22. Cogoni C, Macino G. 1997. Isolation of quelling-defective (*qde*) mutants impaired in posttranscriptional transgene-induced gene silencing in *Neurospora crassa. Proc. Natl. Acad. Sci. USA* 94: 10233–38

23. Cogoni C, Macino G. 1999. Gene silencing in *Neurospora crassa* requires a protein homologous to RNA-dependent RNA polymerase. *Nature* 399:166–69

24. Cogoni C, Macino G. 1999. Posttranscriptional gene silencing in Neurospora by a RecQ DNA helicase. *Science* 286:2342–44

25. Dalmay T, Hamilton A, Mueller E, Baulcombe DC. 2000. Potato virus X amplicons in Arabidopsis mediate genetic and epigenetic gene silencing. *Plant Cell* 12:369–79

26. Dalmay T, Hamilton A, Rudd S, Angell S, Baulcombe DC. 2000. An RNA-dependent RNA polymerase gene in *Arabidopsis* is required for posttranscriptional gene silencing mediated by a transgene but not by a virus. *Cell* 101:543–53

27. Dalmay T, Horsefield R, Braunstein TH, Baulcombe DC. 2001. SDE3 encodes an RNA helicase required for post-transcriptional gene silencing in Arabidopsis. *EMBO J.* 20:2069–78

28. de Carvalho F, Gheysen G, Kushnir S, Van Montagu M, Inzé D, Castresana C. 1992. Suppression of beta-1,3-glucanase transgene expression in homozygous plants. *EMBO J.* 11:2595–602

29. Dernburg AF, Zalevsky J, Colaiacovo MP, Villeneuve AM. 2000. Transgene-mediated cosuppression in the *C. elegans* germline. *Genes Dev.* 14:1578–83

30. Djikeng A, Shi H, Tschudi C, Ullu E. 2001. RNA interference in *Trypanosoma brucei*: cloning of small interfering RNAs provides evidence for retroposon-derived 24–26-nucleotide RNAs. *RNA* 7:1522–30

31. Domeier ME, Morse DP, Knight SW, Portereiko M, Bass BL, Mango SE. 2000. A link between RNA interference and nonsense-mediated decay in *Caenorhabditis elegans. Science* 289:1928–31

32. Dudley NR, Labbe JC, Goldstein B. 2002. Using RNA interference to identify genes required for RNA interference. *Proc. Natl. Acad. Sci. USA* 99:4191–96

33. Elbashir SM, Harborth J, Lendeckel W, Yalcin A, Weber K, Tuschl T. 2001. Duplexes of 21-nucleotide RNAs mediate RNA interference in cultured mammalian cells. *Nature* 411:494–98

34. Elbashir SM, Lendeckel W, Tuschl T. 2001. RNA interference is mediated by 21- and 22-nucleotide RNAs. *Genes Dev.* 15:188–200

35. Elbashir SM, Martinez J, Patkaniowska A, Lendeckel W, Tuschl T. 2001. Functional anatomy of siRNAs for mediating efficient RNAi in *Drosophila melanogaster* embryo lysate. *EMBO J.* 20: 6877–88

36. Elmayan T, Balzergue S, Beon F, Bourdon V, Daubremet J, et al. 1998. Arabidopsis mutants impaired in cosuppression. *Plant Cell* 10:1747–58

37. Fagard M, Boutet S, Morel JB, Bellini C, Vaucheret H. 2000. AGO1, QDE-2, and RDE-1 are related proteins required for post-transcriptional gene silencing in plants, quelling in fungi, and RNA interference in animals. *Proc. Natl. Acad. Sci. USA* 97:11650–54

38. Fire A, Xu S, Montgomery MK, Kostas SA, Driver SE, Mello CC. 1998. Potent and specific genetic interference by double-stranded RNA in *Caenorhabditis elegans. Nature* 391:806–11

39. Fraser AG, Kamath RS, Zipperlen P, Martinez-Campos M, Sohrmann M, Ahringer J. 2000. Functional genomic analysis of *C. elegans* chromosome I by systematic RNA interference. *Nature* 408:325–30

40. Gonczy P, Echeverri G, Oegema K, Coulson A, Jones SJ, et al. 2000. Functional genomic analysis of cell division in *C. elegans* using RNAi of genes on chromosome III. *Nature* 408:331–36

41. Grishok A, Pasquinelli AE, Conte D, Li N, Parrish S, et al. 2001. Genes and mechanisms related to RNA interference regulate expression of the small temporal RNAs that control *C. elegans* developmental timing. *Cell* 106:23–34

42. Grishok A, Tabara H, Mello CC. 2000. Genetic requirements for inheritance of RNAi in *C. elegans*. *Science* 287:2494–97

43. Guo S, Kemphues KJ. 1995. *par-1*, a gene required for establishing polarity in *C. elegans* embryos, encodes a putative Ser/Thr kinase that is asymmetrically distributed. *Cell* 81:611–20

44. Hamilton AJ, Baulcombe DC. 1999. A species of small antisense RNA in posttranscriptional gene silencing in plants. *Science* 286:950–52

45. Hammond SM, Bernstein E, Beach D, Hannon GJ. 2000. An RNA-directed nuclease mediates post-transcriptional gene silencing in Drosophila cells. *Nature* 404:293–96

46. Hammond SM, Boettcher S, Caudy AA, Kobayashi R, Hannon GJ. 2001. Argonaute2, a link between genetic and biochemical analyses of RNAi. *Science* 293:1146–50

47. Holen T, Amarzguioui M, Wiiger MT, Babaie E, Prydz H. 2002. Positional effects of short interfering RNAs targeting the human coagulation trigger Tissue Factor. *Nucleic Acids Res.* 30:1757–66

48. Hutvagner G, McLachlan J, Pasquinelli AE, Balint E, Tuschl T, Zamore PD. 2001. A cellular function for the RNA-interference enzyme Dicer in the maturation of the let-7 small temporal RNA. *Science* 293:834–38

49. Jacobsen SE, Running MP, Meyerowitz EM. 1999. Disruption of an RNA helicase/RNAse III gene in Arabidopsis causes unregulated cell division in floral meristems. *Development* 126:5231–43

50. Kamath RS, Martinez-Campos M, Zipperlen P, Fraser AG, Ahringer J. 2000. Effectiveness of specific RNA-mediated interference through ingested double-stranded RNA in *Caenorhabditis elegans*. *Genome Biol.* 2:1–10

51. Kasschau KD, Carrington JC. 1998. A counterdefensive strategy of plant viruses: suppression of posttranscriptional gene silencing. *Cell* 95:461–70

52. Ketting RF, Fischer SE, Bernstein E, Sijen T, Hannon GJ, Plasterk RH. 2001. Dicer functions in RNA interference and in synthesis of small RNA involved in developmental timing in *C. elegans*. *Genes Dev.* 15:2654–59

53. Ketting RF, Haverkamp TH, van Luenen HG, Plasterk RH. 1999. Mut-7 of *C. elegans*, required for transposon silencing and RNA interference, is a homolog of Werner syndrome helicase and RNaseD. *Cell* 99:133–41

54. Ketting RF, Plasterk RH. 2000. A genetic link between co-suppression and RNA interference in *C. elegans*. *Nature* 404:296–98

55. Knight SW, Bass BL. 2001. A role for the RNase III enzyme DCR-1 in RNA interference and germ line development in *Caenorhabditis elegans*. *Science* 293:2269–71

56. Kumagai MH, Donson J, della-Cioppa G, Harvey D, Hanley K, Grill LK. 1995. Cytoplasmic inhibition of carotenoid biosynthesis with virus-derived RNA. *Proc. Natl. Acad. Sci. USA* 92:1679–83

57. Lagos-Quintana M, Rauhut R, Lendeckel W, Tuschl T. 2001. Identification of novel genes coding for small expressed RNAs. *Science* 294:853–58

58. Lau NC, Lim LP, Weinstein EG, Bartel DP. 2001. An abundant class of tiny RNAs with probable regulatory roles in *Caenorhabditis elegans*. *Science* 294:858–62

59. Lee RC, Ambros V. 2001. An extensive

class of small RNAs in *Caenorhabditis elegans*. *Science* 294:862–64

60. Lee RC, Feinbaum RL, Ambros V. 1993. The *C. elegans* heterochronic gene *lin-4* encodes small RNAs with antisense complementarity to *lin-14*. *Cell* 75:843–54

61. Li H, Li WX, Ding SW. 2002. Induction and suppression of RNA silencing by an animal virus. *Science* 296:1319–21

62. Li HW, Lucy AP, Guo HS, Li WX, Ji LH, et al. 1999. Strong host resistance targeted against a viral suppressor of the plant gene silencing defence mechanism. *EMBO J.* 18:2683–91

63. Lipardi C, Wei Q, Paterson BM. 2001. RNAi as random degradative PCR: siRNA primers convert mRNA into dsRNAs that are degraded to generate new siRNAs. *Cell* 107:297–307

64. Mallory AC, Ely L, Smith TH, Marathe R, Anandalakshmi R, et al. 2001. HC-Pro suppression of transgene silencing eliminates the small RNAs but not transgene methylation or the mobile signal. *Plant Cell* 13:571–83

65. Mango SE. 2001. Stop making nonSense: the *C. elegans smg* genes. *Trends Genet.* 17:646–53

66. Marathe R, Anandalakshmi R, Smith TH, Pruss GJ, Vance VB. 2000. RNA viruses as inducers, suppressors and targets of post-transcriptional gene silencing. *Plant Mol. Biol.* 43:295–306

67. Martens H, Novotny J, Oberstrass J, Steck TL, Postlethwait P, Nellen W. 2002. RNAi in *Dictyostelium*: the role of RNA-directed RNA polymerases and double-stranded RNase. *Mol. Biol. Cell* 13:445–53

68. Matzke MA, Primig M, Trnovsky J, Matzke AJM. 1989. Reversible methylation and inactivation of marker genes in sequentially transformed tobacco plants. *EMBO J.* 8:643–49

69. Mittelsten Scheid O, Paszkowski J. 2000. Transcriptional gene silencing mutants. *Plant Mol. Biol.* 43:235–41

70. Montgomery MK, Fire A. 1998. Double-stranded RNA as a mediator in sequence-specific genetic silencing and co-suppression. *Trends Genet.* 14:255–58

71. Montgomery MK, Xu S, Fire A. 1998. RNA as a target of double-stranded RNA-mediated genetic interference in *Caenorhabditis elegans*. *Proc. Natl. Acad. Sci. USA* 95:15502–7

72. Morel JB, Godon C, Mourrain P, Beclin C, Boutet S, et al. 2002. Fertile hypomorphic ARGONAUTE (*ago1*) mutants impaired in post-transcriptional gene silencing and virus resistance. *Plant Cell* 14:629–39

73. Moss EG. 2000. Non-coding RNA's: lightning strikes twice. *Curr. Biol.* 10: R436–39

74. Mourelatos Z, Dostie J, Paushkin S, Sharma A, Charroux B, et al. 2002. miRNPs: a novel class of ribonucleoproteins containing numerous microRNAs. *Genes Dev.* 16:720–28

75. Mourrain P, Beclin C, Elmayan T, Feuerbach F, Godon C, et al. 2000. Arabidopsis SGS2 and SGS3 genes are required for posttranscriptional gene silencing and natural virus resistance. *Cell* 101:533–42

76. Napoli C, Lemieux C, Jorgensen RA. 1990. Introduction of a chimeric chalcone synthase gene into *Petunia* results in reversible co-suppression of homologous genes in *trans*. *Plant Cell* 2:279–89

77. Ngo H, Tschudi C, Gull K, Ullu E. 1998. Double-stranded RNA induces mRNA degradation in *Trypanosoma brucei*. *Proc. Natl. Acad. Sci. USA* 95:14687–92

78. Nykanen A, Haley B, Zamore PD. 2001. ATP requirements and small interfering RNA structure in the RNA interference pathway. *Cell* 107:309–21

79. Paddison PJ, Caudy AA, Hannon GJ. 2002. Stable suppression of gene expression by RNAi in mammalian cells. *Proc. Natl. Acad. Sci. USA* 99:1443–48

80. Pal-Bhadra M, Bhadra U, Birchler JA. 2002. RNAi related mechanisms affect both transcriptional and posttranscriptional transgene silencing in Drosophila. *Mol. Cell* 9:315–27

81. Palauqui JC, Elmayan T, Pollien JM, Vaucheret H. 1997. Systemic acquired silencing: transgene-specific post-transcriptional silencing is transmitted by grafting from silenced stocks to non-silenced scions. *EMBO J.* 16:4738–45

82. Palauqui JC, Vaucheret H. 1998. Transgenes are dispensable for the RNA degradation step of cosuppression. *Proc. Natl. Acad. Sci. USA* 95:9675–80

83. Parrish S, Fire A. 2001. Distinct roles for RDE-1 and RDE-4 during RNA interference in *Caenorhabditis elegans. RNA* 7:1397–402

84. Parrish S, Fleenor J, Xu S, Mello C, Fire A. 2000. Functional anatomy of a dsRNA trigger: differential requirement for the two trigger strands in RNA interference. *Mol. Cell* 6:1077–87

85. Pasquinelli AE, Reinhart BJ, Slack F, Martindale MQ, Kuroda MI, et al. 2000. Conservation of the sequence and temporal expression of *let-7* heterochronic regulatory RNA. *Nature* 408:86–89

86. Plasterk RH. 2002. RNA silencing: the genome's immune system. *Science* 296: 1263–65

87. Pruss G, Ge X, Shi XM, Carrington JC, Vance VB. 1997. Plant viral synergism: the potyviral genome encodes a broad-range pathogenicity enhancer that transactivates replication of heterologous viruses. *Plant Cell* 9:859–68

88. Que Q, Wang H-Y, English JJ, Jorgensen RA. 1997. The frequency and degree of co-suppression by sense chalcone synthase transgenes are dependent on transgene promoter strength and are reduced by premature nonsense codons in the transgene coding sequence. *Plant Cell* 9:1357–68

89. Ratcliff F, Harrison BD, Baulcombe DC. 1997. A similarity between viral defense and gene silencing in plants. *Science* 276:1558–60

90. Ratcliff FG, MacFarlane SA, Baulcombe DC. 1999. Gene silencing without DNA.

RNA-mediated cross-protection between viruses. *Plant Cell* 11:1207–16

91. Reinhart BJ, Slack FJ, Basson M, Pasquinelli AE, Bettinger JC, et al. 2000. The 21-nucleotide *let-7* RNA regulates developmental timing in *Caenorhabditis elegans. Nature* 403:901–6

92. Romano N, Macino G. 1992. Quelling: transient inactivation of gene expression in *Neurospora crassa* by transformation with homologous sequences. *Mol. Microbiol.* 6:3343–53

93. Ruiz MT, Voinnet O, Baulcombe DC. 1998. Initiation and maintenance of virus-induced gene silencing. *Plant Cell* 10:937–46

94. Ruvkun G. 2001. Molecular biology. Glimpses of a tiny RNA world. *Science* 294:797–99

95. Schiebel W, Haas B, Marinkovic S, Klanner A, Sanger HL. 1993. RNA-directed RNA polymerase from tomato leaves. I. Purification and physical properties. *J. Biol. Chem.* 268:11851–57

96. Schiebel W, Haas B, Marinkovic S, Klanner A, Sanger HL. 1993. RNA-directed RNA polymerase from tomato leaves. II. Catalytic in vitro properties. *J. Biol. Chem.* 268:11858–67

97. Schwarz DS, Zamore PD. 2002. Why do miRNAs live in the miRNP? *Genes Dev.* 16:1025–31

98. Sijen T, Fleenor J, Simmer F, Thijssen KL, Parrish S, et al. 2001. On the role of RNA amplification in dsRNA-triggered gene silencing. *Cell* 107:465–76

99. Slack F, Ruvkun G. 1997. Temporal pattern formation by heterochronic genes. *Annu. Rev. Genet.* 31:611–34

100. Smardon A, Spoerke JM, Stacey SC, Klein ME, Mackin N, Maine EM. 2000. EGO-1 is related to RNA-directed RNA polymerase and functions in germline development and RNA interference in *C. elegans. Curr. Biol.* 10:169–78

101. Smith CJ, Watson CF, Bird CR, Ray J, Schuch W, Grierson D. 1990. Expression of a truncated tomato polygalacturonase

gene inhibits expression of the endogenous gene in transgenic plants. *Mol. Gen. Genet.* 224:477–81

102. Stam M, de Bruin R, Kenter S, van der Hoorn RA, van Blokland R, et al. 1997. Post-transcriptional silencing of chalcone synthase in *Petunia* by inverted transgene repeats. *Plant J.* 12:63–82

103. Stam M, de Bruin R, van Blokland R, van der Hoorn RA, Mol JN, Kooter JM. 2000. Distinct features of post-transcriptional gene silencing by antisense transgenes in single copy and inverted T-DNA repeat loci. *Plant J.* 21:27–42

104. Tabara H, Grishok A, Mello CC. 1998. RNAi in *C. elegans*: soaking in the genome sequence. *Science* 282:430–31

105. Tabara H, Sarkissian M, Kelly WG, Fleenor J, Grishok A, et al. 1999. The *rde-1* gene, RNA interference, and transposon silencing in *C. elegans*. *Cell* 99:123–32

106. Tijsterman M, Ketting RF, Okihara KL, Sijen T, Plasterk RH. 2002. RNA helicase MUT-14-dependent gene silencing triggered in *C. elegans* by short antisense RNAs. *Science* 295:694–97

107. Timmons L, Court DL, Fire A. 2001. Ingestion of bacterially expressed dsRNAs can produce specific and potent genetic interference in *Caenorhabditis elegans*. *Gene* 263:103–12

108. Timmons L, Fire A. 1998. Specific interference by ingested dsRNA. *Nature* 395:854

109. Tuschl T, Zamore PD, Lehmann R, Bartel DP, Sharp PA. 1999. Targeted mRNA degradation by double-stranded RNA in vitro. *Genes Dev.* 13:3191–97

110. Vaistij FE, Jones L, Baulcombe DC. 2002. Spreading of RNA targeting and DNA methylation in RNA silencing requires. Transcription of the target gene and a putative RNA-dependent RNA polymerase. *Plant Cell* 14:857–67

111. van Blokland R, van der Geest N, Mol JN, Kooter JM. 1994. Transgene-mediated suppression of chalcone synthase expression in *Petunia hybrida* results from an increase in RNA turnover. *Plant J.* 6:861–77

112. van der Krol AR, Mur LA, Beld M, Mol JN, Stuitje AR. 1990. Flavonoid genes in *Petunia*: addition of a limited number of gene copies may lead to a suppression of gene expression. *Plant Cell* 2:291–99

113. Vance V, Vaucheret H. 2001. RNA silencing in plants—defense and counter-defense. *Science* 292:2277–80

114. Vaucheret H, Beclin C, Elmayan T, Feuerbach F, Godon C, Morel JB, et al. 1998. Transgene-induced gene silencing in plants. *Plant J.* 16:651–59

115. Vaucheret H, Beclin C, Fagard M. 2001. Post-transcriptional gene silencing in plants. *J. Cell Sci.* 114:3083–91

116. Vaucheret H, Fagard M. 2001. Transcriptional gene silencing in plants: targets, inducers and regulators. *Trends Genet.* 17:29–35

117. Vaucheret H, Nussaume L, Palauqui J-C, Quillere I, Elmayan T. 1997. A transcriptionally active state is required for post-transcriptional silencing (co-suppression) of nitrate reductase host genes and transgenes. *Plant Cell* 9:1495–504

118. Voinnet O. 2001. RNA silencing as a plant immune system against viruses. *Trends Genet.* 17:449–59

119. Voinnet O, Baulcombe DC. 1997. Systemic signalling in gene silencing. *Nature* 389:553

120. Voinnet O, Lederer C, Baulcombe DC. 2000. A viral movement protein prevents spread of the gene silencing signal in *Nicotiana benthamiana*. *Cell* 103:157–67

121. Voinnet O, Pinto YM, Baulcombe DC. 1999. Suppression of gene silencing: a general strategy used by diverse DNA and RNA viruses of plants. *Proc. Natl. Acad. Sci. USA* 96:14147–52

122. Voinnet O, Vain P, Angell S, Baulcombe DC. 1998. Systemic spread of sequence-specific transgene RNA degradation in

plants is initiated by localized introduction of ectopic promoterless DNA. *Cell* 95:177–87

123. Waterhouse PM, Graham MW, Wang MB. 1998. Virus resistance and gene silencing in plants can be induced by simultaneous expression of sense and antisense RNA. *Proc. Natl. Acad. Sci. USA* 95:13959–64

124. Williams RW, Rubin GM. 2002. ARGONAUTE1 is required for efficient RNA interference in Drosophila embryos. *Proc. Natl. Acad. Sci. USA* 99:6889–94

125. Winston WM, Molodowitch C, Hunter CP. 2002. Systemic RNAi in *C. elegans*

requires the putative transmembrane protein SID-1. *Science* 295:2456–59

126. Wu-Scharf D, Jeong B, Zhang C, Cerutti H. 2000. Transgene and transposon silencing in *Chlamydomonas reinhardtii* by a DEAH-box RNA helicase. *Science* 290:1159–62

127. Zamore PD. 2002. Ancient pathways programmed by small RNAs. *Science* 296:1265–69

128. Zamore PD, Tuschl T, Sharp PA, Bartel DP. 2000. RNAi: double-stranded RNA directs the ATP-dependent cleavage of mRNA at 21 to 23 nucleotide intervals. *Cell* 101:25–33

Annu. Rev. Genet. 2002. 36:521–56
doi: 10.1146/annurev.genet.36.060402.100441

TRANSVECTION EFFECTS IN DROSOPHILA

Ian W. Duncan

*Department of Biology, Washington University, Campus Box 1229, St. Louis,
Missouri 63130; e-mail: duncan@biology.wustl.edu*

Key Words transvection, zeste, bithorax complex, somatic pairing,
enhancer-promoter interactions

■ **Abstract**   An unusual feature of the Diptera is that homologous chromosomes are
intimately synapsed in somatic cells. At a number of loci in Drosophila, this pairing
can significantly influence gene expression. Such influences were first detected within
the bithorax complex (BX-C) by E.B. Lewis, who coined the term transvection to describe them. Most cases of transvection involve the action of enhancers in *trans*. At
several loci deletion of the promoter greatly increases this action in *trans*, suggesting
that enhancers are normally tethered in *cis* by the promoter region. Transvection can
also occur by the action of silencers in *trans* or by the spreading of position effect
variegation from rearrangements having heterochromatic breakpoints to paired unrearranged chromosomes. Although not demonstrated, other cases of transvection may
involve the production of joint RNAs by *trans*-splicing. Several cases of transvection
require Zeste, a DNA-binding protein that is thought to facilitate homolog interactions
by self-aggregation. Genes showing transvection can differ greatly in their response to
pairing disruption. In several cases, transvection appears to require intimate synapsis
of homologs. However, in at least one case (transvection of the *iab-5,6,7* region of the
BX-C), transvection is independent of synapsis within and surrounding the interacting gene. The latter example suggests that transvection could well occur in organisms
that lack somatic pairing. In support of this, transvection-like phenomena have been
described in a number of different organisms, including plants, fungi, and mammals.

CONTENTS

0066-4197/02/1215-0521$14.00

INTRODUCTION

The term transvection was introduced by E.B. Lewis in 1954 to describe cases in which gene activity is influenced by homologous pairing (86). Although treated initially as a curiosity relevant to only a few genes, in recent years transvection has been found at a rapidly growing number of loci. Most of these cases are in Drosophila, where homologs are intimately synapsed in somatic cells, and involve the action of enhancers in *trans*. However, several transvection-like phenomena have been described in other organisms as well, suggesting that homolog interactions are of general importance.

The first part of this review covers examples of transvection in which enhancers act positively in *trans*. The focus is on the bithorax complex (BX-C), which is the most thoroughly studied locus with respect to transvection. In the second part, a heterogeneous group of cases that involve repressive interactions in *trans* is described. At the end, potential examples of transvection in other organisms are touched upon. For a different perspective, the reader is referred to several other reviews on transvection (63, 64, 71, 77, 118, 146, 149, 150, 152).

POSITIVE ACTION OF ENHANCERS IN *TRANS*

Transvection within the Bithorax Complex (BX-C): The *Ultrabithorax* Gene

The BX-C contains three homeobox genes, *Ultrabithorax* (*Ubx*), *abdominal-A* (*abd-A*), and *Abdominal-B* (*Abd-B*), that control the identities of segments in the abdomen and posterior thorax [see (39) for review]. These genes are activated within parasegments, anatomical units consisting of the posterior compartment of one segment and the anterior compartment of the next most posterior segment. Together, the three BX-C genes control the identities of parasegments 5 through 14. Most of the BX-C DNA consists of a series of regulatory domains that serve to activate transcription of one of the protein-coding genes within specific parasegments. Remarkably, these regulatory domains are in the same order as the parasegments each controls. The best studied of the BX-C genes with respect to transvection is *Ubx*. *Ubx* is the left-most of the BX-C genes and functions to specify the identities of parasegments 5 (posterior T2 and anterior T3) and 6 (posterior T3 and anterior A1). The *Ubx* transcription unit is about 75 kb long and is differentially spliced

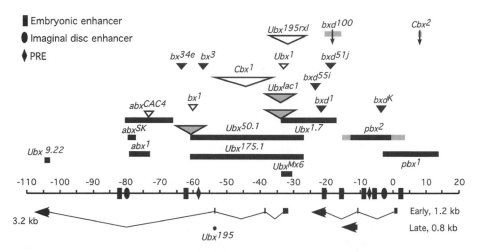

Figure 1   Map of the *Ubx* domain of the bithorax complex. Distances are marked in kb, using the standard zero position. The long transcription unit to the left encodes the *Ubx* proteins. Two transcription units from the *bxd* region (*early* and *late*) are also indicated. These transcripts probably do not encode proteins and are of unknown function. Rearrangement breakpoints are indicated by arrows, insertions by triangles, and deletions by bold lines. Breakpoint uncertainties are indicated in gray. Insertions of the *gypsy* element are indicated in black, whereas insertions of a *lacZ* enhancer trap element are indicated in gray. Ubx^{lac1}, $Ubx^{1.7}$, $Ubx^{50.1}$, and $Ubx^{175.1}$ are described in Reference (19). Ubx^{195rxl} and Ubx^{Mx6} are from (97). Ubx^{195} is a nonsense mutation in the microexon at -53kb (146a). For the remainder of the mutations, see (39) and references therein. The locations of enhancers and PRE elements are redrawn from (117, 119).

to produce mature RNAs encoding a family of closely related homeodomain proteins. A molecular map of *Ubx* showing the locations of a number of mutations important in studies of transvection is shown in Figure 1.

The differing morphologies of parasegment 5 (PS5) and PS6 are due to differences in the spatial and temporal control of *Ubx* expression within the two parasegments [(20); see also (40)]. Transcription of *Ubx* within PS5 is controlled by a large regulatory region located within the third intron. Two types of mutation have been identified within this region. These are *bithorax* (*bx*) mutations, which are located between about -65 and -55 kb on the standard map (see Figure 1), and *anterobithorax* (*abx*) mutations, which are located in the -70 to -80 kb interval. Both types of mutation cause transformations in the adult cuticle of PS5 toward PS4. They differ in that *abx* mutations transform posterior T2 and the very anterior portion of T3 more strongly than do *bx* mutations. Many of the *bx* alleles result from insertions of the *gypsy* transposable element, which contains a well-characterized enhancer blocking element (52). All of the *abx* mutations characterized are deletions. Several groups have searched for enhancers in the *abx* and *bx* regions by adding back fragments of BX-C DNA to a basal construct containing the *Ubx* promoter driving *lacZ*. These experiments led to the identification of two enhancers that are restricted in their activities along the body axis. A 1.7-kb fragment from the *abx* region drives expression in PS5, 7, 9, 11, and 13 (106, 137). A second enhancer is located within the *bx* region; a 500-bp *bx* fragment drives expression in PS6, 8, 10, and 12 (122). When combined, these enhancers drive expression from PS5 to PS13, the normal expression domain of *Ubx*. The 1.7-kb *abx* fragment contains an additional enhancer that drives expression in all imaginal discs (23, 30).

Patterns of *Ubx* expression that are specific to PS6 are controlled by a 40-kb upstream regulatory region. Rearrangements broken within this region cause a transformation of PS6 to PS5, and are called *bithoraxoid* (*bxd*) mutations. The severity of such rearrangements decreases with distance from the *Ubx* promoter. Several *gypsy* insertions are known within the *bxd* region; these cause milder transformations in the adult cuticle than rearrangements broken at similar locations. Two deletions in the *bxd* region have been described. These are called *postbithorax* (*pbx*) mutations, since their major effect is to transform posterior T3 to posterior T2. *pbx* hemizygotes also show a weak reduction of anterior A1. Molecular dissection of the *bxd* region has revealed a complex set of enhancers (119). Four of these enhancers are active early in embryogenesis and drive expression in PS6, 8, 10, and 12 (106, 119, 136, 137). The *bxd* region also contains a cluster of enhancers that are active after embryogenesis and drive expression in imaginal discs (23, 119) (see Figure 1). The *bxd* region is transcribed and produces a complex set of RNAs in the early embryo and late larval stages. These transcripts are probably not translated and are of unknown function.

In the late 1940s and early 1950s, E.B. Lewis mapped *bx*, *Ubx¹*, *bxd*, and *pbx* mutations by recombination. At the time, the dogma was that crossing over does not occur within genes. Therefore, despite the failure of complementation seen between *Ubx¹* and the other mutations, Lewis interpreted each mutation as being located in a separate gene [for Lewis's early work see (41, 42)]. Using the *cis-trans* test (which he developed), Lewis found that the bithorax group mutations show polarized position effects; *bx* mutations fail to complement *Ubx¹*, *Ubx¹* fails to complement *bxd* and *pbx*, and *bxd* mutations fail to complement *pbx*. To explain this polarity, Lewis suggested that the bithorax genes control localized reactions along the chromosome and that the reaction product of one gene serves as the substrate for the adjacent gene. Although incompatible with what we now know about gene expression, this model accounted very nicely for the *cis-trans* position effects observed, and explained why the different bithorax genes need to be adjacent for normal function. Lewis reasoned that since homologs are intimately synapsed in somatic cells of Drosophila, some reaction products could "leak" from one homolog to the other, resulting in partial complementation between mutations that block different steps in the reaction sequence. Lewis tested this prediction by introducing heterozygous rearrangements that disrupt somatic pairing of the bithorax locus. The prediction was confirmed: Lewis found pairing dependence for several heterozygotes involving BX-C mutations. Lewis coined the term transvection to describe such pairing dependence; the term conveyed well his idea that the basis of the pairing dependence was the leakage in *trans* of substances that are normally transported vectorially in *cis*.

Lewis's first paper on transvection (86) is extraordinary in its scope. In it he presented a thorough characterization of transvection involving a weak *bx* allele (bx^{34e}) and a *Ubx* null allele (Ubx^1). We now know that bx^{34e} is a *gypsy* insertion, whereas Ubx^1 is an insertion of a *Doc* transposable element into the 5′ untranslated region of *Ubx* (see Figure 1). Lewis found that bx^{34e} partially complements Ubx^1: bx^{34e}/Ubx^1 heterozygotes show essentially no transformation of the dorsal portion of T3 (the metanotum) to dorsal T2 (mesonotum), whereas this region is quite strongly transformed in bx^{34e} homozygotes. If a rearrangement that disrupts somatic pairing in the vicinity of the BX-C is introduced in heterozygous condition, this partial complementation is lost. Depending on the rearrangement used, the degree to which dorsal T3 is transformed when pairing is disrupted in bx^{34e}/Ubx^1 heterozygotes varies in extent from the presence of small tufts of bristled cuticle to the development of a broad region of well-formed mesonotum. The main experiment in Lewis's paper was to select for X ray–induced enhancers of the dorsal T3 transformation in bx^{34e}/Ubx^1 heterozygotes. The vast majority turned out to be chromosome rearrangements. When their breakpoints were determined by examination of salivary gland chromosomes, several rules became apparent. Most importantly, all had one breakpoint located between the centromere (polytene section 81) and the locus of the BX-C (the middle of 3R; polytene section 89E). Lewis called this the "critical region" and inferred that somatic pairing is initiated at the centromere and proceeds distally in each arm. Within the critical region, rearrangements broken in the proximal half (sections 81–85) tend to have their second breaks located distally in another chromosome arm, whereas rearrangements broken in the distal half can have their second breaks located anywhere in the genome. Although Lewis never quantitated the pairing disruptions caused by his rearrangements, the rules described above are generally consistent with the effects of rearrangements on pairing of the BX-C region in salivary gland nuclei. Three additional observations strongly supported pairing disruption as being the cause of the enhanced T3 transformation in rearrangement heterozygotes. First, Lewis found that it did not matter whether the rearrangement was in *cis* to bx^{34e} or Ubx^1: $R(bx^{34e})/Ubx^1$ had the same phenotype as $bx^{34e}/R(Ubx^1)$. Second, he showed that complementation between bx^{34e} and Ubx^1 is restored in rearrangement homozygotes [i.e., $R(bx^{34e})/R(Ubx^1)$ animals]. There was an exception to this rule: Rearrangements that placed the BX-C very distally showed a weak transformation of dorsal T3 when homozygous. Lewis suggested that somatic pairing initiated at the centromere frequently may not extend all the way to the tip of artificially long chromosomes. Third, Lewis showed that the transformation of dorsal T3 can be increased when two different rearrangements are present in the same animal.

In subsequent reports (87–90), Lewis described transvection in a number of other combinations of mutations in the *Ubx* domain, including *abx/Ubx*, *abx/bx³*, *abx/pbx²*, *bx³/Ubx¹*, *bx³/pbx¹*, *bx³/pbx²*, *Ubx¹/pbx²*, and *bxd/pbx²*. In each case, partial complementation of the alleles is suppressed when pairing of the BX-C is disrupted. Although not tested extensively, it would appear that all of these cases of transvection have the same or similar critical regions as *bx³⁴ᵉ/Ubx¹*, and have similar sensitivities to pairing reduction. A final case of transvection in the *Ubx* domain described by Lewis deserves special mention. It involves the gain-of-function mutation *Contrabithorax¹* (*Cbx¹*), which causes a transformation of posterior T2 to posterior T3. As shown in Figure 1, *Cbx* is an insertion of part of the *bxd* region into the second intron of *Ubx*. The inserted region is the exact complement of the region deleted in *pbx¹* (the two mutants were recovered together as a transposition) and contains two enhancers active in dorsal imaginal discs as well as one of the PS6, 8, 10, 12 enhancers from the *Ubx* upstream region. The relocation of these enhancers to the second intron apparently causes them to become active in PS5 (posterior T2 and anterior T3), where they are normally inactive. The result is that they drive high-level *Ubx* expression in the posterior wing (T2), causing it to transform to posterior haltere (a T3 structure) (17, 147). As expected, in the *Cbx¹ Ubx¹* double mutant, this transformation is largely suppressed. However, Lewis found that in *Cbx¹ Ubx¹/+ +* heterozygotes, the base of the wing continues to be weakly transformed, causing the wings to be held out from the body. Lewis showed that heterozygosity for rearrangements that disrupt pairing of the BX-C can completely suppress this residual transformation.

Two basic types of model have been put forth to explain transvection. First, it has been suggested that pairing-dependent interallelic complementation could be due to the production of joint RNAs either by *trans*-splicing (70) or by template switching by RNA polymerase during transcription (7). To test this type of model, Lewis asked whether transvection occurs between *Ubx¹* and *Ubx⁹·²²*, mutations located at opposite ends of the 75-kb *Ubx* primary transcript. His test was a sensitive one: He compared the phenotypes of *Ubx¹/Ubx¹*, *Ubx⁹·²²/Ubx⁹·²²*, and *Ubx¹/Ubx⁹·²²* animals that carried a duplication for the BX-C in 3L [*Dp(3;3)P47*] marked by *bx³⁴ᵉ*. The three genotypes were indistinguishable, indicating a complete failure of *Ubx¹* and *Ubx⁹·²²* to complement for the *bx⁺* function (E.B. Lewis, unpublished data). These observations argue strongly that joint RNAs are not produced within the *Ubx* transcription unit. Lewis also showed that *Ubx¹* and *Ubx⁹·²²* behave identically with respect to transvection opposite *abx¹*, *bx³⁴ᵉ*, *bx³*, and *pbx²* (E.B. Lewis, unpublished data). Both also show identical transvection effects when in *cis* to *Cbx¹* (22). Furthermore, *Ubx¹⁹⁵*, a nonsense mutation within the microexon at −53 kb, is indistinguishable from *Ubx¹* and *Ubx⁹·²²* when tested for transvection with *abx¹* and *bx³* (115) and *bx³⁴ᵉ* (97, 98). The identical behaviors of these widely spaced *Ubx* alleles provide additional evidence against the *trans*-splicing or template switching models for transvection, at least within *Ubx*. *Trans*-splicing has been demonstrated recently at another locus in Drosophila, *mod(mdg4)* (36a, 80a, 101b). Although it is not yet clear whether this *trans*-splicing is pairing or proximity dependent, this case suggests that the production of joint RNAs may well be involved in transvection at other genes in Drosophila.

The second type of model explains transvection as resulting from the action in *trans* of what are usually considered to be *cis*-regulatory regions. This model is supported by the finding that enhancers can activate promoters carried by different DNA molecules provided they are brought into close proximity, either by an artificial protein bridge [reviewed in (105)] or by catenation of plasmids (38, 128). For genes showing transvection, it has been suggested that regulatory regions could act in *trans* by recruiting a locally high concentration of regulatory factors (80), by diffusion of short-range regulatory RNAs from one homolog to the other (68, 90, 101), or by direct action of enhancer sequences in *trans* (121, 153). For the *Ubx* domain, *trans* action of enhancers has been tested only for the *Cbx¹* mutation. Goldsborough & Kornberg (58) examined *Ubx* transcription in *Cbx¹/+* heterozygotes. As expected, they found that *Ubx* is expressed at high levels in *Cbx¹/+* wing discs. Using a sequence polymorphism in the 3′ untranslated region of the *Ubx* mRNA to distinguish transcripts from the two alleles, they showed that about 14% of the transcripts induced by *Cbx¹* in the wing disc are transcribed from the + homolog. When chromosome rearrangements were introduced to disrupt pairing, this transactivation was reduced dramatically. Consistent with this analysis, *Ubx* protein expression is found in the wing discs of *Cbx¹ Ubx¹/+ +* animals when pairing is allowed, but not when pairing is disrupted (22). These observations indicate that transvection of *Cbx* and probably most or all other cases of transvection in the *Ubx* domain result from the action of regulatory regions in *trans*.

The transcript analysis described above shows that although the rearranged enhancers in *Cbx¹* can work in *trans*, they have a strong preference for action in *cis*. The same is likely true for all cases of transvection in the *Ubx* domain described by Lewis, as in each of these cases only rather weak, partial complementation is involved. However, in the past decade much more striking cases of transvection in the *Ubx* domain have been described in which it appears that *Ubx* enhancers are operating almost as well in *trans* as they are in *cis*. These cases involve mutations in which the *Ubx* promoter region is deleted or otherwise inactivated. Martínez-Laborda et al. (97) described two such mutations: One (*Ubx^{MX6}*) is a ~3.4-kb deletion centered on the transcription start site of *Ubx*, and the other (*Ubx^{195rxl}*) is an insertion of ~11 kb very close to this site. Like *Ubx¹*, these mutations show a null phenotype when homozygous in embryos and in imaginal clones, and lack *Ubx* protein expression in homozygotes. However, unlike *Ubx¹*, both almost fully complement all of the viable *Ubx* regulatory mutations tested, including *abx²*, *bx^{34e}*, *bx³*, *bxd¹*, *pbx¹*, and *pbx²*. In all cases, this complementation is lost if pairing is disrupted by chromosome rearrangements. In the absence of pairing, heterozygotes of the *Ubx* regulatory mutations and *Ubx^{MX6}* or *Ubx^{195rxl}* have phenotypes very similar or identical to those that Lewis described for unpaired heterozygotes with *Ubx¹*. The very strong complementing ability of *Ubx^{MX6}* and *Ubx^{195rxl}* as compared to *Ubx¹* suggests that the promoter region of *Ubx* somehow sequesters enhancers located in *cis*, largely precluding their action in *trans*. When the promoter is deleted or otherwise damaged, *Ubx* regulatory regions are liberated and free to act in *trans*. The sequestration of enhancers in *cis* by the promoter is also found at *yellow* (53), *Gpdh* (54), and *Sex combs reduced* (142), and may be a general feature of genes showing transvection.

In some ways, transvection of Ubx^{MX6} and Ubx^{195rx1} appears to be qualitatively different from transvection of Ubx^1. Ubx^{MX6} and Ubx^{195rx1} show transvection with all viable *Ubx* domain mutations tested, including weak *Ubx* alleles. Ubx^1, in contrast, does not show transvection with bxd^1 or pbx^1, nor does it show transvection with other *Ubx* alleles. Moreover, transvection of Ubx^{MX6} and Ubx^{195rx1} does not require function of the *zeste* gene (described in detail below), at least for transvection with bx^{34e}, whereas transvection of Ubx^1 does require *zeste*. Despite these differences and the much greater ability of the enhancers in Ubx^{MX6} and Ubx^{195rx1} to act in *trans*, the pairing rules for these cases are probably similar to those described for Ubx^1 by Lewis. Although only a few rearrangements have been tested, these inhibit transvection of Ubx^{MX6} and Ubx^{195rx1} to about the same extent as they do for Ubx^1, and they appear to lie in the same critical region.

Additional deletions ($Ubx^{1.7}$, $Ubx^{50.1}$, and $Ubx^{175.1}$; see Figure 1) that remove the *Ubx* promoter have been generated by mobilization of *lacZ* enhancer trap insertions (19). As expected, these show strongly enhanced (as compared to Ubx^1) complementation with recessive regulatory mutations. The $Ubx^{50.1}$ deletion is particularly informative. $Ubx^{50.1}$ retains the *lacZ* enhancer trap element, but deletes a region of 34 kb that includes the promoter and the first three exons of *Ubx* (see Figure 1). This deletion complements bx^3 and *pbx* much better than does Ubx^1, but shows weak transformations in heterozygotes with these mutations. However, when the P-element was removed from $Ubx^{50.1}$ by further P-mobilization, the *Ubx* allele produced ($Ubx^{175.1}$) almost fully complements bx^3 and *pbx*. These observations suggest that the P-element promoter can also sequester *Ubx* enhancers in *cis*, although much more weakly than the *Ubx* promoter itself. When both promoters are deleted, as in $Ubx^{175.1}$, the enhancers are free to act in *trans* with a paired homolog almost as well as they normally act in *cis*. An additional striking case of transvection within the *Ubx* domain is provided by a pairing-sensitive revertant of Cbx^1, called Cbx^{1RM} (101). In the absence of pairing, Cbx^{1RM} has no dominant effect. However, if paired with a normal homolog Cbx^{1RM} causes a strong transformation of posterior wing to haltere. This transformation almost certainly results from the ectopic activation of *Ubx* expression in *trans*, since the Cbx^{1RM} chromosome carries a null allele of *Ubx* (19). The strength of *trans* interaction suggests that Cbx^{1RM} might carry a deletion of the *Ubx* promoter. However, unlike the promoter deletions described above, Cbx^{1RM} requires $zeste^+$ for *trans* activation of Ubx^+ and does not show strong complementation ability with the recessive regulatory mutations within *Ubx* (101).

Two Polycomb Response Elements (PREs) have been identified within the *Ubx* gene (29, 117, 119, 136) (see Figure 1). These elements function to maintain the correct anatomical boundaries of *Ubx* expression throughout development. The initial patterns of BX-C gene expression are set up by the segmentation genes, which are expressed only transiently. Once the products of the segmentation genes decay, the correct limits of BX-C gene expression are maintained by the *Polycomb* group (PcG) of genes, which act as maintenance repressors. The PcG proteins act within large complexes that bind to PRE elements (117) of the BX-C and many other genes. Most PREs have been studied by introducing fragments using the P-element vector CaSpeR, which contains a mini-*white* gene. Remarkably, for almost all PREs, transformants carrying them show suppression of mini-*white* when homozygous. In one case, this suppression is relieved by chromosome rearrangements that disrupt pairing (55). In many other cases, suppression requires the proximity of two PRE elements. For these reasons, PREs showing silencing of mini-*white* are called "pairing sensitive" elements [see (74) for review]. The *Ubx* gene contains two PREs; one is within the *bxd* region, and the other is adjacent to the *bx* region (see Figure 1). Surprisingly, there is no evidence that either of these sequences plays an important role in transvection at *Ubx*. The *bx* region PRE is probably deleted in $Ubx^{50.1}$ and $Ubx^{175.1}$ (see Figure 1), both of which show strong pairing-dependent complementation with bx^3 and *pbx*. Likewise, the *bxd* region PRE is deleted in pbx^2, which shows transvection with bx^3, Ubx^1, and bxd^1 (90). Indeed, pbx^2 shows a greater ability to engage in transvection than pbx^1, which does not delete the *bxd* PRE. However, tests of the ability of chromosomes lacking both PREs to interact in *trans* have not yet been carried out.

Are there any specific sequence requirements for a chromosome to engage in transvection at *Ubx*? Based largely on the inability of pbx^1 to transvect with Ubx^1 and bxd^1, it has been suggested that the region deleted by pbx^1 but not pbx^2 might play an important role (39, 98). However, it was found subsequently that promoter deletions in *Ubx* transvect perfectly well with pbx^1 (97). With the possible exception of a small region between the breakpoints of $Tp(3;3)bxd^{100}$ or $Ubx^{1.7}$ and the pbx^2 deletion, no specific sequences to the left of +14 kb are required for interactions in *trans*. Moreover, there seems to be nothing special about the *Ubx* promoter in facilitating transvection, as the P-element promoter of the Ubx^{Jacl} enhancer trap can be activated in *trans* by Cbx^1 and Cbx^{1RM} (19).

Goldsborough & Kornberg [(58), see also (139)] have suggested that in addition to facilitating *trans*-action of enhancers, pairing causes a general enhancement of transcription. This conclusion was based on the finding that disruption of pairing by chromosome rearrangements in $Cbx^1/+$ wing imaginal discs causes a similar reduction of *Ubx* transcription from both the Cbx^1 and + homologs. For the strongest pairing disruption examined, this reduction was about 50%. Reduced transcription from the + homolog when pairing is disrupted is to be expected, since this transcription results from *trans*-activation by Cbx^1. However, reduced expression of the Cbx^1 homolog itself requires explanation. There are reasons to be skeptical that there is any significant effect of pairing on *Ubx* transcription in the wild type. First, the size of the haltere is a sensitive measure of the level of *Ubx* expression in T3: Animals carrying only one dose of Ubx^+ show an obvious swelling of the haltere. Therefore, if pairing disruption were to cause a reduction of *Ubx* transcription in T3 by as much as 50%, one would expect rearrangements that block pairing of the BX-C to cause some swelling of the haltere. In fact, they have no such effect. Second, if pairing were to promote transcription of the *Ubx* gene in both homologs, one would also expect the phenotypes of homozygotes for many *Ubx* domain partial loss-of-function mutations to be very sensitive to pairing. Only one such example has been found: Homozygotes for *bxd¹* show a stronger transformation of posterior haltere to posterior wing when paired than when pairing is disrupted [E.B. Lewis & I. Duncan, unpublished data; see also (21)]. This effect is the opposite of that predicted by the model; a stronger loss of function is seen in the paired than in the unpaired configuration. This effect is likely due to pairing-dependent associations of the *gypsy* element responsible for the *bxd¹* mutation. In other situations, interactions between *gypsy* elements have been well documented. For example, at both the *Ubx* and *yellow* loci, two *gypsy* elements located in *cis* can interact so as to block an intervening promoter from participating in transvection (9, 53). Interaction of *gypsies* in *trans* have also been described (9, 51). Finally, it seems unlikely that pairing promotes transcription broadly in the genome. If this were the case, heterozygosity for chromosome rearrangements would often cause phenotypes corresponding to the many haplo-abnormal genes of Drosophila, which they do not.

An alternative explanation for the reduced transcription of the *Cbx¹* homolog caused by pairing disruption is that the *Cbx¹* mutation itself is sensitive to pairing. There are ample precedents for such pairing dependence. The *cis* activities of the gain-of-function alleles *Cbx²*, *Scr^W*, and *ci¹* (*cubitus interruptus¹*) are all sensitive to pairing (93, 101, 114), although ectopic expression is suppressed by pairing in these cases (see below). Additional examples are known at the *yellow* locus, where pairing can activate (102) or suppress (51) the activity of enhancers in *cis*.

Transvection within the Bithorax Complex (BX-C): The Abdominal Domains

As summarized in Figure 2, the abdominal region of the BX-C comprises about 180 kb of DNA and contains two homeobox genes, *abdominal-A* (*abd-A*) and *Abdominal-B* (*Abd-B*) (39). *abd-A* specifies the identities of parasegments 7–9, while *Abd-B* controls parasegments 10–14. These genes are embedded within an array of parasegment-specific regulatory regions, named *infra-abdominal* (*iab*) regions, that are in the same order as the parasegments that each affects. The *iab* regions are numbered according to the location in the abdomen of the anterior compartment each controls. For example, *iab-2* specifies the identity of PS7, which consists of the anterior compartment of the second abdominal segment (aA2) and the posterior compartment of A1 (pA1), whereas *iab-3* controls PS8 (aA3 and pA2). Examination of *abd-A* and *Abd-B* expression in *iab* mutants has revealed that *iab-2*, *iab-3*, and *iab-4* control the level and pattern of expression of *abd-A* in parasegments 7–9 (73, 129), whereas *iab-5*–*iab-8* control *Abd-B* expression in parasegments 10–13 (24, 129). Using transgenes, distinct parasegment-specific enhancers have been identified in the *iab-2*, *iab-3*, *iab-5*, *iab-7*, and *iab-8* regions

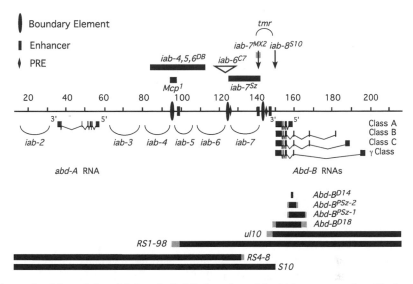

Figure 2   Map of the *abd-A* and *Abd-B* domains of the bithorax complex. [Redrawn from Figure 2 of (66).] The DNA scale and symbols are as in Figure 1. With the exception of *Abd-B^{PSz-1}* and *Abd-B^{PSz-2}*, which are described in (138), the references for all mutant breakpoints and the transcript exons are cited in (39, 66). The *iab-8^{S10}* mutation, previously known as *iab-7^{S10}*, is renamed here to reflect the position of its breakpoint. For *Abd-B*, only the class A transcript is expressed in A5–A7, the segments for which transvection of *Abd-B* has been described. The class B, class C, and the γ class transcripts are expressed in A8 or more posterior segments. The so-called *transvection mediating region* (*tmr*) is located between the breakpoints of *iab-7^{MX2}* and *Df(3R)ul10*(66). The *iab-8* region (not labeled) lies just to the right of *iab-7*. The right-hand boundary of *iab-8* has not been defined, but its left-hand boundary, which separates it from *iab-7*, is shown at about +144. This boundary element, as well as the *iab-7* and *iab-8* enhancers, and the *iab-8* PRE are from (10, 154). The boundary separating *iab-6* and *iab-7* and the *iab-7* PRE were mapped in (101a). The *iab-5* enhancer is described in (16). The *Mcp* region has long been interpreted as a boundary between *iab-4* and *iab-5*. Consistent with this idea, deletions of the *Mcp* region cause a transformation of A4 to A5, presumably due to the ectopic activation of *iab-5* in A4. However, careful analysis of the *Mcp* region (107) suggests that it may instead function primarily as a silencer.

(10, 11, 16, 137, 154) and are thought likely to be present in *iab-4* and *iab-6* as well. The anterior limit of activity for all of these enhancers corresponds to the parasegment for which each is named. All are active in more posterior parasegments as well, often with pair-rule modulation. For the best-studied *iab* regions (*iab-7* and *iab-8*), each contains an initiator enhancer element that responds to positional signals from the gap and pair-rule genes and a PRE that maintains repression of the region in anterior parasegments after the decay of the segmentation gene products. The *iab* regions are thought to be insulated from one another by distinct boundary elements. Such elements have been demonstrated to flank *iab-7* (10, 154) and appear also to be present between *iab-4* and *iab-5* (107) and between *bxd* and *iab-2* (11). The view now prevalent is that the *iab* regions are chromatin loop domains that are "opened out" in sequence as one proceeds from anterior to posterior in the animal (116).

Transvection within *abd-A* was first demonstrated by Lewis (90), who found that gonadal development in heterozygotes for a null allele (*abd-A^{D24}*) and a deletion for *iab-4* and *iab-5* (*iab-4,5,6^{DB}*) is sensitive to pairing. With structurally normal third chromosomes, almost all such heterozygotes show some gonadal development. However, if pairing is disrupted, most animals show no traces of gonadal tissue. A second case of transvection at *abd-A* has been reported by Jijakli & Ghysen (69a). These authors studied "lateral dots" (LD), paired structures in the third instar CNS revealed by staining with a monoclonal antibody. In wild type, these dots are present in A1 but not in more posterior segments. However, in animals heterozygous for a deficiency for all of *abd-A*, including the *iab-2*, *iab-3*, and *iab-4* regulatory regions, LD develop in segments A2–A6. Some null alleles of *abd-A* have a much weaker effect, causing strong LD development only in A2. For one such allele (*abd-A^{M1}*), it was found that strong LD development was extended to A6 when the wild-type copy of the BX-C was transposed to the X chromosome. These observations suggest that the *iab-3* and *iab-4* regions of *abd-A^{M1}* act on *abd-A^{+}* in *trans* in a pairing-dependent fashion to suppress LD development.

Transvection at *Abd-B* was also first reported by Lewis (90). In the abdominal epidermis of wild type, *Abd-B* is expressed at low level in PS10, intermediate level in PS11, and higher level in PS12 (78). These levels are driven by the *iab-5*, *iab-6*, and *iab-7* regions (24). In the male, *Abd-B* expression in PS10 and PS11 causes the dorsal cuticular plates (tergites) of A5 and A6 to become darkly pigmented. However, in A7, tergite development is completely repressed in males due to higher levels of *Abd-B* expression. It turns out that tergite development in A7 is very sensitive to the level of *Abd-B* expression: males having only one dose of *Abd-B* show a moderately large A7 tergite, indicating a partial transformation to A6. In his first report of transvection at *Abd-B*, Lewis showed that A7 tergite development in *iab-6^{C7}/Abd-B^{D14}* males is sensitive to pairing. With unrearranged chromosomes, such heterozygotes show weak A7 tergite development. However, when pairing is disrupted, the A7 tergite size is significantly increased in size. This transvection presumably results from the activation of *Abd-B^{+}* in *trans* by the *iab-6* or *iab-7* regions of the *Abd-B^{D14}* chromosome. Although not tested extensively, *iab-6^{C7}/Abd-B^{D14}* transvection appears to be sensitive to rearrangements that cause only moderate pairing disruption of the BX-C, as found for transvection in the *Ubx* domain.

A qualitatively different type of transvection has also been described for *Abd-B* (61, 66). This case involves *trans* interactions that are extraordinarily difficult to disrupt. The case was detected in experiments in which partial BX-C deletions entering the complex from the left were tested for complementation with deletions that enter the complex from the right. These experiments demonstrated that the *iab-5,6,7* region is able to specify A5 and A6 identities (recognized in the male by dark pigmentation and specific trichome patterns) in the absence of an *Abd-B* gene in *cis*. A key genotype is the heterozygote of *Df(3R)ul10*, which deletes the *Abd-B* transcription unit, but leaves *iab-5,6,7* intact, with *Df(3R)S10*, which removes essentially everything to the left of *Abd-B*, including *iab-5,6,7* (see Figure 2). Such heterozygotes show A5 and A6 identities in their posterior segments. Heterozygotes of *Df(3R)S10* with deficiencies that remove the *iab-5,6,7* region as well as *Abd-B* [e.g., *Df(3R)RS1-98*] lack these identities, demonstrating that they are dependent on *iab-5,6,7*. The obvious explanation was that *iab-5,6,7* could specify A5 and A6 identities by activating expression of *Abd-B*$^+$ in *trans*. However, when a number of rearrangements that block transvection in the *Ubx* domain were tested in appropriate deficiency heterozygotes, they had no effect on A5 or A6 identities. An alternate explanation, that in the absence of *Abd-B iab-5,6,7* can promote A5 and A6 identities by acting on *abd-A*$^+$ or *Ubx*$^+$ in *cis*, was tested by placing *Ubx*$^-$ and *abd-A*$^-$ mutations (as well as the *Abd-B*$^-$ deletion) in *cis* to *iab-5,6,7*. This also had no effect on the ability of *iab-5,6,7* to specify A5 and A6 identities. Initially, these results seemed to indicate that *iab-5,6,7* encodes *trans*-acting products of its own. Ultimately, however, a few rearrangements were recovered that reduce the ability of *iab-5,6,7* to act in *trans* (66). One is broken within the *bxd* region, whereas another transposes the BX-C, and little else, to a new location. Although these rearrangements probably disrupt pairing as much as is possible by rearrangement, they cause only a weak loss of *iab-5,6,7 trans* activity. These rearrangements show that *iab-5,6,7* can act by transvection to specify A5 and A6 identities, and that its *trans* interactions with *Abd-B* are extraordinarily avid. *iab-5,6,7* is able to control *Abd-B* in *trans* in the embryonic CNS as well as in the adult cuticle (61).

The *trans* activity of *iab-5,6,7* is the only case of transvection known to be independent of local synapsis. The deficiency chromosomes *Df(3R)ul10* and *Df(3R)S10* share no homology over the region from 89B through 90A, a distance of about a megabase. Despite this, the *iab-5,6,7* region (at 89E) from *Df(3R)ul10* interacts strongly in *trans* with the *Abd-B* gene of *Df(3R)S10*. This *trans* interaction can be disrupted significantly only by rearrangements broken very close to or within the BX-C and only in deficiency heterozygotes in which homologs share essentially no homology within or around the BX-C. It would appear that *iab-5,6,7* and *Abd-B* are able to locate one another and interact in *trans* without regard to their location in the genome or their ability to pair by synapsis. Unlike most cases of transvection in the *Ubx* domain, transvection of *iab-5,6,7* is independent of the *zeste* gene (61, 66).

That *iab-5,6,7* can act in *trans* in *Df(3R)ul10*, but not in *In(3LR)iab-7*^{MX2} (see Figure 2), suggests that the 9-kb region lying between the BX-C breakpoints of these rearrangements is required for *iab-5,6,7* transvection. This region lies between distal *iab-7* and *Abd-B* and must be in *cis* to *iab-5,6,7* for *trans* interactions to occur. Detailed characterization of this *"transvection mediating region"* (*tmr*)

(66) has revealed several elements, including initiator enhancer sequences for *iab-7* and *iab-8*, a PRE within *iab-8*, a boundary element between *iab-7* and *iab-8*, and a leftward directed promoter that lies just 3′ to *Abd-B* (10, 154). The latter is the most intriguing, as initiator elements, PREs, and boundary elements are probably present in all of the *iab* regions. Although transcription of the BX-C regulatory regions has been known many years, its significance has never been clarified. It will be of great interest to determine whether transcription facilitates the action of *iab-5,6,7* in *trans*. An additional intriguing element within the *tmr* has been found by Zhou & Levine (155). They call this novel element the promoter targeting sequence (PTS). The PTS lies adjacent to the boundary that separates *iab-7* and *iab-8* (*Fab-8*), and has anti-insulator activity. In transgene assays, the PTS can block both the *Fab-8* and the *gypsy* insulators. It can also confer promoter selectivity upon distal enhancers. Zhou & Levine suggest that the PTS functions to convert long-range insulators to strictly local elements that separate neighboring *iab* regions, thereby allowing the *iab-5,6,7* region to overcome the blocking effect of *Fab-8* and interact with the *Abd-B* promoter. However, the phenotypes of two mutations within the PTS indicate that this element is not essential for *iab-5,6,7* function in vivo (155). The role of the PTS in transvection remains to be tested.

Recently, Sipos et al. (138) have presented evidence that disruption of pairing actually facilitates long-range *trans* interactions of the *iab-5,6,7* region and that the critical region for pairing disruption is the *tmr*. These authors studied an inversion [$In(3R)Fab7iab-7^{R5}$] that separates the *iab-5,6,7* region from *Abd-B*. When homozygous, this inversion shows no interaction between the *iab-5,6,7* region and *Abd-B*. However, when this inversion is heterozygous with a deficiency for the entire BX-C, strong interactions are established, and *Abd-B* expression is activated in PS10-13 of the embryonic CNS. Elegant studies in which this inversion is combined with another having very similar breaks provide strong evidence that it is disruption of pairing in the vicinity of the *tmr* that strongly promotes long-range interactions with *Abd-B*.

Sipos et al. (138) have also addressed the question of whether there are specific elements at the *Abd-B* promoter that are required for interaction with the *iab-5,6,7* region. They found that although *Abd-B* point-mutation nulls do not appear to complement an *iab-7* region mutation (*iab-7^{Sz}*) at all, deletions that remove the *Abd-B* promoter region have complementing ability. By mobilizing a P-element inserted near the *Abd-B* promoter, they recovered two new deletions (*Abd-B^{PSz1}* and *Abd-B^{PSz2}*; see Figure 2) in the promoter region. Study of these as well as three other deletions recovered in other studies (*Abd-B^{D14}*, *Abd-B^{D18}*, and *Df(3R)ul10*) revealed a size effect: the larger the deletion, the greater its ability to complement *iab-7^{Sz}*. This complementing ability is dependent upon chromosome pairing and presumably results from action in *trans* of the *iab-7* region from the *Abd-B⁻* homolog upon the *Abd-B* gene of the *iab-7^{Sz}* homolog. To explain the size effect, Sipos et al. suggest that a large region (at least 7.6 kb) upstream of the *Abd-B* promoter contains multiple "tethering elements" that normally retain the *iab-5,6,7* region in *cis* and keep it from interacting in *trans*. When these elements are deleted, the *iab-7* region becomes free to act upon the homolog. Although this model is attractive, an alternate explanation not ruled out is that looping out of the unpaired *Abd-B* gene on the *iab-7^{Sz}* homolog renders it more available for *trans* interactions in the deletion heterozygotes studied.

Sipos et al. (138) also described a transvection effect involving *Fab-7*, a gain-of-function mutation in the BX-C that causes transformation of A6 to A7. *Fab-7* deletes a boundary element located between *iab-6* and *iab-7* (60). This causes the *iab-7* region to become active in A6, where it drives *Abd-B* expression at the level normally seen only in A7. Sipos et al. noted that the transformation of A6 to A7 caused by *Fab-7* is partially suppressed by the presence of an *Abd-B* null point mutation in *trans*. Deletions that remove the *Abd-B* gene do not have this effect. Almost certainly, this suppression results from the unproductive regulation of *Abd-B* in *trans* by *Fab-7*. Consistent with this model, suppression is sensitive to pairing. Surprisingly, the pairing requirements for this case of transvection are quite different from those of *iab-5,6,7* transvection described above. *Trans* suppression of *Fab-7* is reduced or eliminated by rearrangements broken within the same critical region Lewis defined for transvection at *Ubx*. How the same gene can show quite different pairing requirements for different transvection effects is discussed in the next section.

An additional case of very high affinity transvection at *Abd-B* is revealed by the *Mcp* mutation. *Mcp* is a small deletion thought to remove a boundary between *iab-4* and *iab-5*. The composite domain produced drives *Abd-B* expression in A4, transforming it to A5. Surprisingly, *Mcp* does not need to be in *cis* to *abd-A*⁺ or *Abd-B*⁺ to cause this transformation: Heterozygotes for the triple mutant *abd-A⁻ Mcp Abd-B⁻* and wild type show a transformation of A4 indistinguishable from that seen in *abd-A⁻ + Abd-B⁻/+ Mcp +* heterozygotes (I. Duncan, unpublished data). This suggests that the fused *iab-4/iab-5* domain interacts in *trans* as well as it does in *cis*. The interaction appears to be highly avid, as males homozygous for *abd-A⁻ Mcp Abd-B⁻* and carrying a small BX-C duplication on the X chromosome [*Dp(3;3)P68*] show an almost complete transformation of A4 to A5. Whether this case is mechanistically distinct from *iab-5,6,7* transvection is not clear. Recent work indicates that the DNA deleted in *Mcp* also has remarkable properties; the region contains a small (800-bp) element that can mediate very long distance pairing interactions (107).

Pairing Requirements for Transvection

The actual pairing disruptions caused by chromosome rearrangements have been determined in surprisingly few studies. In polytene chromosomes, a good correlation between the degree of pairing disruption and loss of transvection has been shown for *Ubx* (66) and *dpp* [cited in (144)]. For diploid cells, pairing disruptions have been monitored by fluorescent in situ hybridization (50) or inferred from the effects of rearrangements on mitotic recombination (47, 57). In his first paper on transvection (86), Lewis found that only rearrangements broken between the centromere and the BX-C (the "critical region") disrupt transvection within *Ubx*. The distal limit of the critical region is very sharp: Even rearrangements broken within the distal part of the *bxd* region or in the abdominal portion of the complex do not affect transvection in *Cbx^1 Ubx*/+ + heterozygotes (138; E.B. Lewis, personal communication). As Lewis pointed out, the critical region would seem to imply that pairing is initiated near the centromere and propagated more distally. However, cytological studies of pairing do not support this idea. Two groups (45, 50) have used fluorescent in situ hybridization to examine the establishment of pairing in the early embryo. They find that pairing is absent during the early cleavage divisions, which occur very rapidly. However, as the nuclear cycle lengthens in cycle 13, the association of homologous alleles begins. After gastrulation, pairing reaches a plateau at about 70% of nuclei showing pairing. Contrary to the Lewis "zippering" model, when probes from proximal, medial, and distal 3R (50) or 2L (45) were examined, no differences in the onset of pairing were found over most of these chromosome arms. Moreover, double labeling of single nuclei revealed that distal regions in 2L can be paired in the absence of pairing of proximal regions (45) in the same arm. Both groups conclude that pairing is initiated at numerous sites by random contacts, in a mechanism that can be compared to buttons on a cardigan. It is difficult to see how such a mechanism is consistent with Lewis's critical region. A further surprise was that quite small duplications for the BX-C inserted into other chromosomes [*Dp(3;1)P115* and *Dp(3;3)P47*] pair quite frequently (in about 20–30% of nuclei) with the BX-C in 3R (50), despite showing an almost complete failure of transvection with the *Ubx* gene [but see (19)]. Resolution of these apparent contradictions likely will require study of pairing in cells of the imaginal discs and abdominal histoblast nests, as these give rise to the structures in which transvection has been characterized genetically.

A striking feature of transvection is that different genes can have vastly different pairing requirements. This is well illustrated by the BX-C, where transvection within *Ubx* is relatively easy to disrupt by chromosome rearrangement, whereas *iab-5,6,7* transvection is extraordinarily difficult to disrupt. Other genes are of the *Ubx* type [e.g., *dpp* (48) and *eya* (84)], the *iab-5,6,7* type [e.g., the *zeste-white* interaction (144)] or perhaps somewhere in between [e.g., *y* (53)]. Smolik-Utlaut & Gelbart (144) tested two models to explain these differences in sensitivity to pairing disruption. First, genes showing very small critical regions could be very closely linked to or coincident with specific sites that promote pairing. Alternatively, all genes could have similar abilities to pair, but differ dramatically in their response to pairing disruption. To test these models, they examined transvection in strains carrying insertions of w^+ into the critical region for transvection at *Ubx* (proximal 3R) or *dpp* (distal 2L). As described below, w^+ has a very restricted critical region in its pairing-dependent interactions with *zeste^1* (144). They found that the critical regions for both *Ubx* and *dpp* were unaffected by the presence of the w^+ insertions, arguing that the w^+ region does not carry a strong pairing site. However, w^+ retains a very small critical region in these insertions. Therefore, rearrangements broken proximal to the w^+ insertion block transvection of *Ubx* or *dpp*, but have no effect on transvection at w. These observations rule out the first model and strongly support the second.

Why do different genes differ so dramatically in their response to pairing disruption? Golic & Golic (57) have suggested that these differences may be more apparent than real and may simply reflect the length of the cell cycle in the tissue in which the gene is active. That is, they suggest that in tissues that have very long cell cycle times, homologs will have sufficient time to reestablish pairing after each division even in the presence of rearrangements. In contrast, pairing in tissues having much shorter cycle times will appear to be much more strongly affected by rearrangement heterozygosity. In this view, transvection at w appears difficult to disrupt because w^+ is expressed in postmitotic cells in the eye, whereas transvection at Ubx and dpp appears much easier to disrupt because these genes are expressed within actively dividing imaginal disc cells. Based on their model, Golic & Golic (57) predicted that $Minute$ mutations, which are thought to lengthen the cell cycle, would enhance transvection. Results with two $Minutes$ confirmed this prediction for the transvection of $Cbx^1 Ubx^1/++$. Additional support comes from the study of tandem duplications. Since the two elements of such duplications are adjacent on the same chromosome, one would expect them to pair with great rapidity and therefore to show enhanced interactions with one another. Consistent with this expectation, E.B. Lewis (personal communication) found that transvection of $Cbx^1 Ubx^1$ and $++$ is very much enhanced when these are present in the two halves of a tandem duplication $[Dp(3;3)P5]$. Gubb et al. (59) showed that w^+ genes contained within an inverted repeat also interact very strongly.

However, several observations suggest that the Golic model is not the whole story. For example, transvection at Abd-B can be of a very avid type (for A5 and A6 identities) or of a much weaker type (for A7 identity) in exactly the same tissue, the abdominal histoblasts of A7. Similarly, in the eye the pairing-dependent suppression of w^+ in a transgene carrying part of the Scr regulatory region is easily disrupted by chromosome rearrangement (55), whereas the pairing-dependent suppression of w^+ in z^1 mutants is not (see below). These apparent contradictions to the Golic model may be explained by enhancers working at different times within the same tissue. However, they may also indicate differences in the underlying mechanisms by which genes interact in $trans$. What might such differences be? Perhaps at some genes transient regulatory interactions can have permanent effects. Transvection at such genes would presumably appear quite resistant to pairing disruption as compared to genes in which regulatory interactions must be sustained (84). Genes that interact by $trans$-splicing or by the diffusion of regulatory RNAs between homologs might also appear resistant to pairing disruption, since significant interactions likely could still occur when the genes are at some distance from one another.

Another possibility is that genes differ in their response to rearrangement heterozygosity as an indirect consequence of the mechanisms underlying pairing. Studies of salivary chromosome puffing suggest that pairing must be established within a restricted window of time. Ashburner (5, 5a) and Korge (79) have described salivary gland chromosome puff sites that show transvection. Puffing is variable at these sites: some strains show the presence of a puff, whereas other strains do not. In heterozygotes, both homologs are puffed when synapsed, but only the puffing-competent homolog is puffed in chromosomes that fail to synapse. These observations indicate that the pairing state of salivary gland chromosomes is stable; if pairing were dynamic one would not find strict autonomy of puffing states in asynapsed nuclei. This autonomy also shows that asynapsis is not induced by the squash preparation itself. The simplest explanation is that pairing is established only early in salivary gland development. The same may be true of other cell types as well.

How could a distinct establishment phase for pairing help explain the differing responses of genes to pairing disruption? Consider a region that remains stably asynapsed after the pairing-competent stage. For genes that require intimate synapsis for *trans* interactions, transvection will be completely blocked. However, for genes that contain sites that can interact over long distances, transvection will still be possible. For example, sites within homologous genes could be brought together in the absence of pairing by the action of DNA-binding proteins that interact with one another with high affinity. Such protein-protein interactions have been invoked to account for "homing" to the endogenous locus of P-element constructs containing enhancer regions (60a) and may also be responsible for similar targeting of plasmid sequences introduced into tissue culture cells (28a). If interactions of this type take place after the competent stage for synapsis has passed, they will not induce pairing of surrounding sequences, and interactions will remain restricted to the interacting elements themselves. Such a mechanism could account for the coexistence within the same gene (*Abd-B*) and tissue (the A7 tergite) of both extraordinarily avid and relatively weak *trans* interactions.

Transvection at *Yellow*

In a key paper published in 1990, Geyer et al. (53) demonstrated transvection at the *yellow* (y) locus. Because y is small (about 8 kb), and therefore easily manipulated, it has become an important model for studies of transvection. Transvection at y was first demonstrated for the y^2 allele, which had long been known to complement a number of other y alleles (92). y^2 is an insertion of a *gypsy* element at -700 bp from the transcription start site that separates wing and body enhancers from the y promoter (Figure 3). Because of the insulator activity of *gypsy*, y^2 flies have yellow cuticle in the wing and body. However, y^2 flies retain normal pigmentation in adult bristles, since the bristle enhancer is located within the y intron. Geyer et al. characterized a number of alleles that complement y^2 for both wing and body coloration. They found that all complementing alleles delete or damage the y promoter, but retain the wing and body enhancers. A good example is y^{59b}. This allele is a derivative of y^2 that deletes part of *gypsy*, the y promoter region, and the 5' part of the transcribed region. Although transcription of y in the pupal stage is absent in homozygotes for y^2 or y^{59b}, it is restored in y^2/y^{59b} heterozygotes. The transcripts produced are normal length, implying that they are initiated at the promoter of the y^2 allele. Complementation of y^2 and y^{59b} appears to be pairing

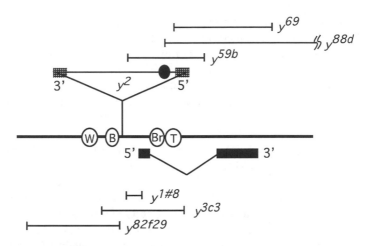

Figure 3   Map of the *yellow* locus. The central line shows the positions of the wing (W), body (B), bristle (Br), and tarsal claw (T) enhancers. Above this line are the *gypsy* insertion in the y^2 allele and the extents of three deletion derivatives of y^2. The insulator sequence of *gypsy*, which contains an array of 12 binding sites for the su(Hw) protein, is drawn as a *black circle*. Below the map of *y* enhancers are drawn the *y* transcript and the extents of three deletions discussed in the text. Drawing modified from (51).

dependent: y^2 transgenes inserted at ectopic locations do not complement y^{59b}. Moreover, a Y-borne duplication carrying y^2 does not complement y^{59b}, at least in homozygous females.

   Another derivative of y^2 (y^{88d}) removes all of the transcribed region of *y* as well as the promoter region and part of *gypsy*, but leaves the wing and body enhancers intact. This also complements y^2, indicating that it is the wing and body enhancers of y^{59b}, and not sequences from the 3' end of *y*, that act in *trans* to drive transcription of the y^2 promoter in the wing and body cuticle. A third y^2 derivative (y^{69}) also deletes part of *gypsy*, the *y* promoter, and the transcribed region of *y*. However, y^{69} does not complement y^2. Significantly, y^{69} does not delete the *gypsy* insulator region, whereas this is deleted in y^{59b}. Apparently, *gypsy* insulator can block activity of the wing and body enhancers in *trans* as well as *cis*. The y^{88d} allele removes part of the insulator, deleting 5 of the normal 12 su(Hw) protein binding sites from *gypsy*, and correspondingly shows only partial complementation with y^2. In subsequent work (102), an allele was identified (y^{82f29}) that deletes the wing and body enhancers of *y* but leaves the promoter and transcribed region intact. This allele causes a phenotype very similar to that of y^2, and has similar complementing abilities. This demonstrates that a *gypsy* need not be present for transvection to occur at *y*, and rules out the possibility that alleles like y^{59b} complement y^2 by allowing the wing and body enhancers of y^2 to overcome the *gypsy* insulator in *cis*.

   Characterization of additional *y* alleles points to the promoter region as key in controlling transvection. Alleles that can complement y^2 include a deletion of the promoter ($y^{1\#8}$) and four insertions into the promoter region. In contrast, three alleles that do not complement y^2 affect the transcribed region of the gene. To explore the promoter requirements for transvection in more detail, Morris et al. (104) used targeted gene replacement to alter the TATA and Inr elements of the y promoter. They found that both mutations allowed *trans* interactions and that the double mutant interacted more strongly than either single mutant. The picture that emerges is that in the absence of a functional promoter in *cis*, the wing and body enhancers of y are free to act in *trans*. Recently, a second mechanism by which transvection can occur at y has been discovered (102). By selection for derivatives of y^1 (a missense change in the initiating ATG of the y coding region) that can complement y^2, an allele (y^{3c3}) was recovered that deletes the body enhancer, promoter, and first exon of the gene, but not the wing enhancer (102, 103). Remarkably, y^{3c3} complements y^2 for both body and wing color, implying that this allele somehow allows the body enhancer of y^2 to bypass the *gypsy* insulator. An appealing model is that this bypass occurs because when y^{3c3} synapses with y^2, the *gypsy* element is looped out and the wing and body enhancers of the y^2 allele are brought into close proximity with the promoter. A similar mechanism may explain why enhancers are able to act over two copies of *gypsy* in *cis* (18, 108); by pairing, these *gypsies* may loop themselves out and bring the enhancer next to the promoter.

Transvection at *decapentaplegic* (*dpp*) and *eyes absent* (*eya*)

Two additional cases of transvection resulting from enhancers acting positively in *trans* have been well characterized in Drosophila. These are transvection at *dpp* and *eya*. Both genes are located in distal 2L and show similar pairing dependence. Transvection at *dpp* has been demonstrated for heterozygotes of the *held out* (*dpp*$^{d-ho}$) allele and *dpp*^{hr4} (48). *dpp*$^{d-ho}$ is a 3-kb deletion located some 23 kb 3′ of the *dpp* transcription unit in the *disk* regulatory region that when homozygous causes the wings to be held out at right angles from the body. *dpp*^{hr4} is a missense change in the *dpp* coding sequence that when homozygous causes a reduction of the wings to stumps. With structurally normal chromosomes, *dpp*$^{d-ho}$/*dpp*^{hr4} heterozygotes appear normal. However, introduction of rearrangements that disrupt pairing cause such heterozygotes to have a held-out phenotype. Transvection at *eya* has been characterized for heterozygotes of *eya*^{2}, an allele that deletes an eye-specific enhancer, and *eya*^{4}, an I-element insertion into the *eya* 5′ UTR (156). Homozygotes for *eya*^{2} lack eyes altogether, whereas homozygotes for *eya*^{4} have severely reduced eyes. With structurally normal homologs, heterozygotes have eyes that are only slightly smaller than normal. However, if rearrangements are introduced that disrupt pairing, eye size is reduced, sometimes dramatically. Rearrangements that disrupt complementation show a similar pattern of breakpoints for both *dpp* and *eya*. Consistent with Lewis's findings for *Ubx*, the critical region for both genes includes a large region proximal to each gene. For *dpp* (located in 22F), this region extends at least to 35E. In the presence of the $zeste^l$ allele, it extends all the way to the centromere (section 40) (144). For *eya*, the critical region extends from the locus (26EF) at least through section 33. For both genes, the second break of two simple break rearrangements tends strongly to be located proximally on one of the other chromosome arms. Such rearrangements strongly disrupt pairing of distal 2L in salivary chromosomes (84, 144). Complementation and pairing are restored when rearrangements with similar breakpoints are combined. Both genes are sensitive to the allelic status of *zeste* (49, 84).

Evidence for pairing-dependent enhancer action in *trans* has been described for several other genes in Drosophila. These include *w* (7), *Gpdh* (54), *pointed* (131), *wg* (15, 110), *hh* (83), *vg* (148), and *en* (31). As mentioned earlier, transvection has also been detected at three polytene puff sites (5, 5a,79), where non-puffing variants can be induced to puff by pairing with a normal homolog. One such puff is at the locus of a salivary gland secretion gene, *Sgs-4*, that has been characterized molecularly. Here a non-puffing variant that can be induced to puff by a normal homolog appears to be a deletion in an upstream regulatory region (109).

PAIRING-DEPENDENT REPRESSIVE INTERACTIONS

Transvection at *cubitus interruptus* (*ci*) and *Sex combs reduced* (*Scr*)

The behavior of the *ci* gene was for many years an enigma. The ci^l allele is normally recessive to ci^+, and ci^l homozygotes show disruptions of the fourth and fifth wing veins. However, the function of ci^+ is very sensitive to its position in the genome [see (85) for review of older literature]. Translocations that move ci^+ from its normal location on the small fourth chromosome [$R(ci^+)$] can cause ci^+ to lose its dominance over ci^l: $R(ci^+)/ci^l$ heterozygotes often show a variable *ci* wing phenotype. $R(ci^+)$ translocations selected on this basis tend to have breaks located distally in the other autosomal arms. What made this position effect so hard to understand was that $R(ci^+)$ rearrangements that fail to complement ci^l have no effect in the wing in $R(ci^+)/R(ci^+)$ homozygotes or in $R(ci^+)/Df$ hemizygotes. In addition, rearrangements induced upon ci^l can be dominant to wild type. That is, $R(ci^l)/+$ can show fourth and fifth wing vein gaps. Not until 1994 was a simple explanation forthcoming (93). When ci^+ was characterized at the molecular level (43, 111), its expression was found to be restricted to anterior compartments. Since the wing veins affected by ci^l are located in the posterior compartment, this suggested that ci^l causes ectopic expression in the posterior wing. This turns out to be the case (143). This finding suggested that the basis for the *ci* position effect was that ci^+ represses ectopic expression of ci^l in a pairing-dependent manner (93). Thus, the long enigmatic *ci* position effect was finally explained as a transvection effect in which a silencer element from ci^+ rather than an enhancer acts in *trans*. Molecular data support this interpretation: ci^l is a *gypsy* insertion that appears to insulate the *ci* promoter from a repressive element controlled by *engrailed* (93, 132).

A few additional examples are known in which repressive regulatory elements appear to act in *trans*. One intriguing case involves the *Sex combs reduced* (*Scr*) gene of the Antennapedia complex. Pattatucci & Kaufman (114) have characterized the transvection behavior of a gain-of-function allele, *Scr^W*, that is associated with a 50-kb inversion that has one break in the *Antp* gene and the other between the *Scr* imaginal leg enhancer and the *Scr* promoter. When heterozygous with wild type, *Scr^W* causes ectopic expression of *Scr* protein in T2 and T3, resulting in transformations of the second and third legs to first legs. When *Scr^W* is heterozygous with a point null allele (*Scr^4*), these transformations are totally suppressed, indicating that in *Scr^W*/+ heterozygotes transformations are caused exclusively by ectopic expression of the wild-type homolog. However, when pairing is disrupted in *Scr^W*/+ heterozygotes, ectopic expression of *Scr* protein in T2 and T3 is enhanced, not suppressed. Taken together, these observations indicate that *Scr^W* can act only in *trans* if paired, but also in *cis* if unpaired. The critical region for this case of transvection appears to be between *Scr* and the centromere. Southworth & Kennison (142) have shown that all inversions, translocations, and transpositions broken within a 60–70-kb region including *Scr* cause some level of ectopic expression of the homologous gene. They suggest that sites at each end of this region are important for maintaining repression of *Scr* in T2 and T3 and that breaking the continuity of one chromosome in this region hinders the functioning of these sites in *trans*. Other cases of negative interactions in *trans* include the repression of the *Ubx* gain-of-function mutation *Cbx^2* by pairing with a normal chromosome (101), and the repression of *y^+* in *trans* by the *gypsy* element in a *mod(mdg4)* mutant background (51).

The *zeste-white* Interaction

As described above, the product of the *zeste* gene is required for many cases of transvection in Drosophila. The first mutant allele of *zeste* was described by Gans (46). The properties of this allele (*z^1*) are most peculiar. The gene is X-linked, and Gans found that homozygous females have lemon yellow eye color (hence the name *zeste*), whereas hemizygous males appear to be wild type. Gans showed that this difference between the sexes is due to the dosage of the *w^+* gene, which is also X-linked: *z^1*/*z^1* females carrying only one dose of *w^+* have red eyes, whereas *z^1* males carrying two doses of *w^+* can have red eyes. The *zeste* eye color results from reduced transcription of *w^+*: The eyes of *z^1 w^+*/*z^1 w^+* females show a dramatic reduction in level of *w* transcripts as compared to *z^1 w^+*/Y males (153). Jack & Judd (68) made the key observation that *z^1* represses *w* transcription only when the two doses of *w^+* present are paired or in close proximity. They also constructed flies having three copies of *w*, only two of which could pair. When different alleles were used to distinguish the expression of these copies, the two paired alleles were repressed in *z^1* flies, whereas the third was not. The critical region for pairing of *w^+* alleles is very restricted. Indeed, only transportations of the *w* region to new locations disrupt the *z-w* interaction; no simple two-break rearrangements have been recovered that block this interaction by pairing disruption (144).

When the z gene was cloned (94, 120), it was found to encode a DNA-binding protein. Zeste binds strongly to two sites located within an eye-specific enhancer (121, 153) located between 1 and 2 kb upstream of the w promoter (12). These sites coincide with the location of deletion mutations (w^{sp} alleles) that render w insensitive to repression by z^l (12, 153). The Zeste protein has a very strong ability to aggregate. This aggregation depends upon the C-terminal 75 amino acids (26) and is likely caused by coiled-coil interactions of three hydrophobic heptad repeats in this region. z^l is a missense mutation causing the change of lys 425 to met. Although this lies outside of the C-terminal multimerization domain, z^l causes marked hyperaggregation of Zeste, apparently by creating a novel hydrophobic domain in the central part of the protein. The hyperaggregation of Zeste1 appears to be key in its repression of w. Chen et al. (27) created a large series of C-terminal and z^l-region mutations that enhance or reduce aggregation. When these mutations were tested in flies, a perfect correlation was found between hyperaggregation and the ability to repress paired copies of w^+. This suggests that aggregates of Zeste protein bound to the eye enhancer are responsible for w repression. Presumably a larger aggregate is assembled on two paired copies of w^+ than on a single copy, accounting for the repression of only paired copies of w^+ by z^l. Consistent with this model, overexpression of the z^l protein can cause a *zeste* eye color in males as well as in females (13). Moreover, a derivative of z^l (z^{op6}) that encodes a protein showing even more extreme hyperaggregation than z^l (27) causes repression of w^+ in both males and females (91).

Why does Zeste aggregation repress the w eye enhancer? Very likely repression is caused by recruitment of the PcG maintenance repressors. Wu et al. (151) showed that loss-of-function mutations in three different PcG genes [*Psc*, *Scm*, and *E(z)*] suppress the *zeste* eye color. Moreover, Zeste binds at many of the same sites as PcG proteins in polytene chromosomes (124) and has been shown to be associated with the PcG proteins in Polycomb repressive complex I (130). As described above, many PRE-containing transgenes carrying the mini-*white* gene show pairing-sensitive repression. Although *zeste* is not required for repression of these elements (75), the involvement of the Pc-group in the *zeste-white* interaction suggests that the repression of w in the two cases occurs by very similar mechanisms.

Surprisingly, the *zeste* gene is dispensable. Deletions for the gene are viable and fertile and have only slightly reduced eye pigmentation (56). Thus, z^+ would appear to have only a minor effect on the function of w^+ or other genes. However, loss-of-function alleles of z (z^a alleles) reduce pairing-dependent complementation at a number of loci. In his first paper on transvection, Lewis reported the recovery of a few X ray–induced changes in which transvection appeared to be blocked, but in which no rearrangement broken in the critical region was present. One of these cases was due to an X-linked mutation that Lewis called $e(bx)$. Kaufman et al. (76) showed that $e(bx)$ is a loss-of-function allele of z, and renamed the mutant $z^{ae(bx)}$. This and other z^a-type alleles block transvection of bx alleles with Ubx^l (8, 76), as well as the *trans* activation of Ubx^+ by Cbx^l in $Cbx^l\ Ubx^l/+ +$ heterozygotes (49). However, z^a alleles have no or only a slight effect on transvection of Ubx alleles that lack a functional promoter (97). z alleles affect transvection at several other loci, including dpp (49), y (53), and eya (84). z is also required for the ability of w^{sp} to complement other w alleles (7), which presumably occurs by *trans* activity of the w eye enhancer. How does z^+ facilitate transvection at these loci? At Ubx, Zeste has been shown to bind just upstream of the promoter, as well as at scattered sites in the bxd and bx regions (12). An appealing model is that transvection at Ubx is facilitated by self-association of Zeste protein bound at the promoter of one homolog with Zeste bound at the bx or bxd regions of the homolog (13). In support of this model, Zeste multimerization can crosslink two DNA molecules in vitro (12). The distribution of Zeste binding sites within w and dpp suggests that Zeste multimerization could facilitate enhancer-promoter interactions in *trans* for these genes as well.

The weak eye color phenotypes of z^a mutants and the fact that z^a mutants enhance the phenotypes of some homozygous bx mutations (76) suggest that z^+ has positive effects on transcription that are unrelated to transvection. Indeed, Zeste is an activator of Ubx reporters in vivo (81). Recent work indicates that Zeste also plays an important role in recruiting PcG maintenance repressors to the Ubx promoter region. In its role as an activator, Zeste is redundant with two other transcription factors, GAGA and NTF1. All three of these factors bind just upstream of the Ubx transcription start site (81). Substitution of this region by synthetic arrays of single sites for each of the factors indicates that any one of them is sufficient to activate transcription (67). However, only for Zeste and GAGA is the normal pattern of Ubx maintained during embryogenesis; if an NTF1 array is used, the expression pattern is normal initially, but then breaks down. Since pattern maintenance requires the PcG genes, these observations suggest that Zeste (and GAGA) play an important role in recruiting the PcG repressors as well as in activating transcription. Consistent with these two roles, Zeste has been shown to be a component of a PcG protein complex (130) and to interact with the Brahma Complex of *trithorax* group activators (72). Hur et al. (67) suggest that where early Ubx transcription is activated by the segmentation genes, Zeste and GAGA become complexed with activators, whereas in regions where Ubx is not activated, they are accessible for interaction with the PcG complexes.

brown-Dominant

When euchromatic genes are brought next to heterochromatin by chromosome re-arrangement, they frequently show variegated expression. In almost all such cases of position effect variegation (PEV), only genes within the rearranged chromosome are affected [(69a), reviewed in (87)]. An exception to this rule is the eye color gene *brown* (bw). Rearrangements that place bw^+ next to heterochromatin (bw^V rearrangements) are dominant (140, 141). Henikoff & Dreesen (65) showed that this dominance results from the transcriptional inactivation of bw^+ in both the re-arranged and normal homologs. The finding that bw^+ transgenes located at ectopic locations are not suppressed suggests that the suppression of bw^+ in *trans* depends on pairing (37). Consistent with this, when bw^V rearrangements broken next to bw^+ transgenes were recovered, these also caused *trans* suppression, but only of transgenes directly opposite their breakpoints. Surprisingly, *trans* suppression does not require the presence of bw^+ in *cis* to the rearrangement breakpoint: When the bw^+ element carried by a bw^V rearrangement was excised by P-element mobilization, suppression of the element in *trans* was still observed (37). The picture that emerges is that bw^+ is particularly sensitive to repression by heterochromatin, so that PEV takes place in *trans* as well as *cis*. The extent of *trans* suppression depends upon the proximity of bw to the coalesced heterochromatic regions of the genome (the chromocenter). Rearrangements that move the *bw-Dominant* (bw^D) allele (an insertion of 1–2 Mb of heterochromatin into the bw gene) further from the chromocenter suppress *trans*-inactivation, whereas rearrangements that bring bw^D closer to the chromocenter enhance *trans*-inactivation (145). The degree of *trans*-inactivation is correlated with the frequency that bw^D and its paired bw^+ allele become localized to the chromocenter, suggesting that *trans* suppression results from forced relocalization to the heterochromatic compartment (32, 33). There appear, therefore, to be two major factors affecting *trans*-suppression of bw: the extent of pairing at bw itself, and the proximity of the paired bw alleles to the chromocenter. This has made it difficult to define the pairing requirements for *trans*-suppression, since almost all rearrangements likely affect both factors. Nonetheless, the ability of bw^D to *trans*-inactivate bw^+ in some quite drastic rearrangements (65) indicates that the critical region is small.

Although bw is unusual in showing *trans*-suppression due to PEV, it is not unique. Indeed, Martin-Morris et al. (96) have shown that w is susceptible to *trans*-inactivation in the special case of mini-*white* transgenes located close to heterochromatin. *Trans*-suppression also appears to occur at the gene *Arista* (*Ata*) (92). Although heterozygotes for a deficiency for *Ata* appear normal, rearrangements that juxtapose the locus (47DE) and heterochromatin cuase dominant variegation for wing-hair development (I. Duncan, unpublished data). Other likely examples of *trans*-suppression due to PEV are known at *kar* (62) and *Pu* (77).

A phenomenon that may be the converse of bw^D has been described by Donaldson & Karpen (34). In characterizing a minichromosome [$Dp(1;f)1187$] that carries y^+, they found that PEV of y^+ is much enhanced by deficiencies that remove the telomere. However, this variegation is suppressed if another copy of the minichromosome is present, even if this copy lacks y^+. The authors show that suppression is not simply due to the addition of a block of heterochromatin by the second minichromosome. Rather, suppression depends on the overall structural similarity of the two minichromosomes, suggesting that pairing is important. To explain this *trans*-suppression, the authors suggest that for the full-length minichromosome, association of the telomere with the nuclear envelope pulls y^+ out of a heterochromatic domain (the chromocenter), largely preventing its suppression by PEV. When the telomere is removed, y^+ is engulfed by the heterochromatic domain and suppressed by PEV. However, pairing or association with another minichromosome carrying an intact telomere restores the force dragging y^+ out of the chromocenter, and suppresses y^+ variegation.

Homology-Dependent Silencing of Transgenes in Drosophila

There are numerous examples in diverse organisms in which multiple copies of a transgene cause gene silencing. The phenomenon is particularly common in plants [see (99) for review]. Although these silencing effects were initially thought to reflect DNA-DNA pairing interactions among the transgenes, in most cases evidence increasingly points to a mechanism based on RNA interference. Although dispersed multiple copies of transgenes in Drosophila do not generally show obvious silencing, Pal-Bhadra et al. (112) have described a case in which strong silencing does occur. The transgene they studied carried the promoter region and the eye enhancer of *w* driving the expression of the *Adh* structural gene. Three insertions of the transgene were studied. Surprisingly, when flies carrying two or more of these inserts were examined, expression was reduced rather than being proportional to dose. The expression of the endogenous *Adh* gene was also suppressed. Homozygotes for each insertion showed greater suppression than any of the pairwise combinations of elements at different sites. Remarkably, suppression of these transgenes is sensitive to mutations in the PcG genes, and PcG proteins were found to localize at the sites of suppressed, but not active, transgenes. Suppression of these transgenes has been shown to be at the transcriptional level (113).

Much weaker cosuppression was found when full-length *Adh* transgenes were examined. In this case, suppression clearly occurs by an RNA-based mechanism, and is posttranscriptional (113). Small interfering RNAs are detected and suppression is largely eliminated in *piwi* mutants, which block RNAi. Because homozygous insertions of the *w/Adh* transgenes are more strongly suppressed than combinations of two distant insertions, it has been argued that the establishment of transcriptional silencing must involve DNA-DNA pairing interactions (14). However, it turns out that transcriptional silencing is also sensitive to *piwi*, suggesting that it too may be controlled by RNAi. One possibility is that the pairing effect observed reflects the action of the PcG proteins in maintaining repression rather than the action of whatever RNAi mechanism is responsible for establishing this repression. Recent work has shown that RNAi can also trigger both transcriptional and posttranscriptional repression of transgenes in plants (135). There is some indication that the control of P-element cytotype in Drosophila may involve mechanisms similar to those of transgene cosuppression (125, 126).

Repression of tandemly arranged transgenes has also been documented in many organisms. An example in Drosophila has been described by Dorer & Henikoff (35, 36). These authors found that tandem arrays of transgenes carrying mini-*white* show variegated suppression. The degree of variegation depends upon copy number and orientation of the repeats, with inverted repeats showing much stronger suppression than direct tandem repeats. Suppression in these arrays is sensitive to known enhancers and suppressors of PEV, and the heterochromatin-specific protein

HP1 localizes to them (44), arguing that tandemly repeated copies of mini-*white* adopt a heterochromatic state. Consistent with this view, chromosome rearrangements that move these arrays closer to heterochromatin enhance the degree of suppression (36). Because of the tandem nature of these repeat arrays, suppression in this case very likely results from DNA-DNA pairing interactions. However, it cannot be excluded that some type of RNA-DNA interaction is also involved.

HOW GENERAL IS TRANSVECTION?

Transvection has usually been viewed as a bizarre and exceptional allelic interaction. Even in Drosophila, where homologs are intimately synapsed in somatic cells, transvection has been detected at only a small subset of loci. However, this subset is growing rapidly. Moreover, the work of Chen et al. (28) indicates that transvection is potentially very common. These authors recovered insertions into eight locations of a y^+ transgene in which the wing and body enhancers were flanked by target sites for cre recombinase (*loxP* sites) and the promoter region was flanked by FLP recombinase sites (FRT sites). For each site, separate derivatives deleted for the enhancers or the promoter were generated by the action of cre or FLP. For all eight sites, it was found that the promoterless and enhancerless elements complemented to produce a y^+ phenotype, indicating that the Drosophila genome is generally permissive for transvection. This suggests that transvection may be very common in flies. However, appropriate tests have yet to be carried out to determine what fraction of genes participate in pairing-dependent interactions.

Extensive somatic pairing does not occur in other organisms. However, the very avid *trans* interactions found at loci such as *w* or *Abd-B* in Drosophila suggest that transvection could occur generally. In plants, the allelic interactions seen in paramutation [reviewed in (25)] and transgene cosuppression [reviewed in (99)] have long been suspected to involve pairing interactions. Allelic interactions at the *nivea* locus of snapdragons have also been interpreted in terms of transvection (14a). However, in many or all of these cases RNA-mediated interactions are likely responsible. Although somatic pairing of homologs is not generally seen in plants, pairing has been detected in somatic floral tissue in wheat (1, 2). Matzke et al. (100) tested directly for transvection effects in tobacco by using the same cre-FLP strategy employed in Drosophila by Chen et al. (28). Transgenes at four sites were examined. Although *trans* action of an enhancer was not found at any of these sites, one undeleted element showed reduced expression in homozygotes as compared to hemizygotes, and elevated expression when heterozygous with the enhancer deletion, suggesting the possibility of some pairing dependence.

In fungi, evidence for pairing-dependent interactions is very strong. In Neurospora, interactions between duplicated regions in the sexual cycle cause these regions to become highly methylated and to undergo an extraordinary rate of G-C to A-T transitions [for reviews see (133, 134)]. Since this process, called "repeat induced point mutation" (RIP), occurs in a pairwise fashion (that is, a third copy can escape mutation), it seems almost certain that it involves DNA-DNA pairing. A similar process, involving only methylation, has been found in Ascobolus (127). Recently, evidence has been presented that sequences that remain unpaired in meiosis in *Neurospora* cause silencing of homologous sequences located elsewhere in the genome, even if these sequences are paired (3, 134a). Although this "meiotic silencing by unpaired DNA" appears to be controlled by DNA pairing, the silencing mechanism itself is probably RNA mediated, as it depends upon an RNA-directed RNA polymerase (134a).

Although somatic pairing is generally absent in mammals, examples of tissue- and stage-specific pairing of particular loci have been described (4). Homologous pairing of oppositely imprinted loci has also been demonstrated (82), suggesting that pairing may be important in imprinting. It has been speculated that the establishment of X inactivation may also depend on pairing (95). Ashe et al. (6) have described a most remarkable interaction in *trans* at the β-globin locus. They find that transcription of the intergenic regions of the β-globin cluster can be induced by transient transfection of a β-globin gene into nonerythroid cells in culture. In situ hybridization revealed that in 95% of transfected cells, the transfected plasmid colocalized with the endogenous β-globin locus. Recently, evidence for transvection during meiosis in the mouse has also been presented (123). When cre recombinase is expressed from a promoter specific for the pachytene stage of meiosis, it induces extensive methylation of *loxP* sites that spreads progressively in *cis*. Remarkably, when a normal homolog is made heterozygous with a methylated *loxP* homolog, it acquires the ability to cause extensive methylation of naive *loxP* homologs in subsequent generations. This ability does not depend upon the continued presence of cre. The fact that methylation is initiated in prophase I suggests that the transfer of methylation state from one homolog to another in this case involves pairing of homologous alleles.

ACKNOWLEDGMENTS

I am grateful to E.B. Lewis for discussions, permission to cite unpublished observations, and critical reading of the manuscript. I would also like to thank Jim Kennison for communicating results prior to publication, Dianne Duncan for preparation of the figures, and Dianne Duncan and Jennifer Brisson for review of the manuscript. This work was supported by a grant from the NIH.

The *Annual Review of Genetics* is online at http://genet.annualreviews.org

LITERATURE CITED

1. Abranches R, Santos AP, Wegel E, Williams S, Castilho A, et al. 2000. Widely separated multiple transgene integration sites in wheat chromosomes are brought together at interphase. *Plant J.* 24:713–23

2. Aragon-Alcaide L, Reader S, Beven A, Shaw P, Miller T, Moore G. 1997.

Association of homologous chromosomes during floral development. *Curr. Biol.* 7:905–8

3. Aramayo R, Metzenburg RL. 1996. Meiotic transvection in fungi. *Cell* 86:103–13

4. Arnoldus EPJ, Perters ACB, Bots GTAM, Raap AK, van der Ploeg M. 1989. Somatic pairing of chromosome 1 centromeres in interphase nuclei of human cerebellum. *Hum. Genet.* 83:231–34

5. Ashburner M. 1967. Gene activity dependent on chromosome synapsis in the polytene chromosomes of *Drosophila melanogaster. Nature* 214:1159–60

5a. Ashburner M. 1970. The genetic analysis of puffing in polytene chromosomes of *Drosophila. Proc. R. Soc. London Ser. B.* 176:319–27

6. Ashe HL, Monks J, Wijgerde M, Fraser P, Proudfoot NJ. 1997. Intergenic transcription and transinduction of the human β-globin locus. *Genes Dev.* 11:2494–509

7. Babu P, Bhat S. 1980. Effect of *zeste* on *white* complementation. In *Development and Neurobiology of Drosophila*, ed. O Siddiqi, P Babu, LM Hall, JC Hall, pp. 35–40. New York: Plenum

8. Babu P, Bhat SG. 1981. Role of *zeste* in transvection at the bithorax locus of *Drosophila. Mol. Gen. Genet.* 183:400–2

9. Babu P, Selvakumar KS, Bhosekar S. 1987. Studies on transvection at the bithorax complex in *Drosophila melanogaster. Mol. Gen. Genet.* 210:557–63

10. Barges S, Mihaly J, Galloni M, Hagstrom K, Müller M, et al. 2000. The *Fab-8* boundary defines the distal limit of the bithorax complex *iab-7* domain and insulates *iab-7* from initiation elements and a PRE in the adjacent *iab-8* domain. *Development* 127:779–90

11. Bender W, Hudson A. 2000. P element homing to the *Drosophila* bithorax complex. *Development* 127:3981–92

12. Benson M, Pirrotta V. 1988. The *Drosophila zeste* protein binds cooperatively to sites in many gene regulatory regions: implications for transvection and gene regulation. *EMBO J.* 7:3907–15

13. Bickel S, Pirrotta V. 1990. Self-association of the *Drosophila zeste* protein is responsible for transvection effects. *EMBO J.* 9:2959–67

14. Birchler JA, Bhadra MP, Bhadra U. 2000. Making noise about silence: repression of repeated genes in animals. *Curr. Opin. Genet. Dev.* 10:211–16

14a. Bollman J, Carpenter R, Coen ES. 1991. Allelic interactions at the *nivea* locus of *Antirrhinum. Plant Cell* 3:1327–36

15. Buratovitch MA, Phillips RG, Whittle JRS. 1997. Genetic relationships between the mutations *spade* and *Sternopleural* and the *wingless* gene in *Drosophila* development. *Dev. Biol.* 185:244–60

16. Busturia A, Bienz M. 1993. Silencers in *Abdominal-B*, a homeotic *Drosophila* gene. *EMBO J.* 12:1415–25

17. Cabrera CV, Botas J, Garcia-Bellido A. 1985. Distribution of *Ultrabithorax* proteins in mutants of *Drosophila* bithorax complex and its transregulatory genes. *Nature* 318:569–71

18. Cai HN, Shen P. 2001. Effects of *cis* arrangement of chromatin insulators on enhancer-blocking activity. *Science* 291:493–95

19. Casares F, Bender W, Merriam J, Sánchez-Herrero E. 1997. Interactions of Drosophila *Ultrabithorax* regulatory regions with native and foreign promoters. *Genetics* 145:123–37

20. Castelli-Gair J, Akam M. 1995. How the Hox gene *Ultrabithorax* specifies two different segments: the significance of spatial and temporal regulation within metameres. *Development* 121:2973–82

21. Castelli-Gair JE, Capdevila M-P, Micol J-L, García-Bellido A. 1992. Positive and negative *cis*-regulatory elements in the bithoraxoid region of the *Drosophila*

Ultrabithorax gene. *Mol. Gen. Genet.* 234:177–84

22. Castelli-Gair JE, Micol J-L, García-Bellido A. 1990. Transvection in the Drosophila *Ultrabithorax* gene: a *Cbx¹* mutant allele induces ectopic expression of a normal allele in *trans*. *Genetics* 126:177–84

23. Castelli-Gair J, Müller J, Bienz M. 1992. Function of an *Ultrabithorax* minigene in imaginal cells. *Development* 114:877–86

24. Celniker SE, Sharma S, Keelan D, Lewis EB. 1990. The molecular genetics of the bithorax complex of *Drosophila*: *cis*-regulation in the *Abdominal-B* domain. *EMBO J.* 9:4277–86

25. Chandler VL, Stam M, Sidorenko L. 2002. Long-distance *cis* and *trans* interactions mediate paramutation. *Adv. Genet.* 46:215–34

26. Chen JD, Chan CS, Pirrotta V. 1992. Conserved DNA binding and self-association domains of the *Drosophila* zeste protein. *Mol. Cell. Biol.* 12:598–608

27. Chen JD, Pirrotta V. 1993. Stepwise assembly of hyperaggregated forms of *Drosophila* Zeste mutant protein suppresses *white* gene expression in vivo. *EMBO J.* 12:2061–73

28. Chen J-L, Huisinga KL, Viering MM, Ou SA, Wu C-t, Geyer PK. 2002. Enhancer action in *trans* is permitted throughout the *Drosophila* genome. *Proc. Natl. Acad. Sci. USA* 99:3723–28

28a. Cherbas L, Cherbas P. 1997. "Parahomologous" gene targeting in Drosophila cells: an efficient, homology-dependent pathway of illegitimate recombination near a target site. *Genetics* 145:349–58

29. Chiang A, O'Connor MB, Paro R, Simon J, Bender W. 1995. Discrete Polycomb-binding sites in each parasegmental domain of the bithorax complex. *Development* 121:1681–89

30. Christen B, Bienz M. 1994. Imaginal disc silencers from *Ultrabithorax*: evidence for *Polycomb* response elements. *Mech. Dev.* 48:255–66

31. Condie JM, Brower DL. 1989. Allelic interactions at the *engrailed* locus of *Drosophila*: *engrailed* protein expression in imaginal discs. *Dev. Biol.* 135:31–42

32. Csink AK, Henikoff S. 1996. Genetic modification of heterochromatic association and nuclear organization in *Drosophila*. *Nature* 381:529–31

33. Dernburg AF, Borman KW, Fung JC, Marshall WF, Phillips J, et al. 1996. Perturbation of nuclear architecture by long-distance chromosome interactions. *Cell* 85:745–59

34. Donaldson KM, Karpen GH. 1997. *Trans*-suppression of terminal deficiency-associated position effect variegation in a Drosophila minichromosome. *Genetics* 145:325–37

35. Dorer DR, Henikoff S. 1994. Expansions of transgene repeats cause heterochromatin formation and gene silencing in Drosophila. *Cell* 77:993–1002

36. Dorer DR, Henikoff S. 1997. Transgene repeat arrays interact with distant heterochromatin and cause silencing in *cis* and *trans*. *Genetics* 147:1181–90

36a. Dorn R, Reuter G, Loewendorf A. 2001. Transgene analysis proves mRNA *trans*-splicing at the complex *mod(mdg4)* locus in *Drosophila*. *Proc. Natl. Acad. Sci. USA* 98:9724–29

37. Dreesen TD, Henikoff S, Loughney K. 1991. A pairing-sensitive element that mediates *trans*-inactivation is associated with the *Drosophila brown* gene. *Genes Dev.* 5:331–40

38. Dunaway M, Dröge P. 1989. Transactivation of the *Xenopus* rRNA gene promoter by its enhancer. *Nature* 341:657–59

39. Duncan I. 1987. The bithorax complex. *Annu. Rev. Genet.* 21:285–19

40. Duncan I. 1996. How do single homeotic genes control multiple segment identities? *BioEssays* 18:91–94

41. Duncan I, Montgomery G. 2002. E.B.

Lewis and the bithorax complex: Pt. I. *Genetics* 160:1265–72

42. Duncan I, Montgomery G. 2002. E.B. Lewis and the bithorax complex: Pt. II. From *cis-trans* test to the genetic control of development. *Genetics* 161:1–10

43. Eaton S, Kornberg TB. 1990. Repression of *ci-D* in posterior compartments of *Drosophila* by *engrailed*. *Genes Dev.* 4:1068–77

44. Fanti L, Dorer DR, Berloco M, Henikoff S, Pimpinelli S. 1998. Heterochromatin protein 1 binds transgene arrays. *Chromosoma* 107:286–92

45. Fung JC, Marshall WF, Dernburg A, Agard DA, Sedat JW. 1998. Homologous chromosome pairing in *Drosophila melanogaster* proceeds through multiple independent initiations. *J. Cell Biol.* 141:5–20

46. Gans M. 1953. Étude génétique et physiologique du mutant z de *Drosophila melanogaster. Bull. Biol. Fr. Belg. Suppl.* 38:1–90

47. García-Bellido A, Wandosell F. 1978. The effect of inversions on mitotic recombination in *Drosophila melanogaster. Mol. Gen. Genet.* 161:317–21

48. Gelbart WM. 1982. Synapsis-dependent allelic complementation at the decapentaplegic gene complex in *Drosophila melanogaster. Proc. Natl. Acad. Sci. USA* 79:2636–40

49. Gelbart WM, Wu C-t. 1982. Interactions of zeste mutations with loci exhibiting transvection effects in *Drosophila melanogaster. Genetics* 102:179–89

50. Gemkow MJ, Verveer PJ, Arndt-Jovin DJ. 1998. Homologous association of the Bithorax-Complex during embryogenesis: consequences for transvection in *Drosophila melanogaster. Development* 125:4541–52

51. Georgiev PG, Corces VG. 1995. The su(Hw) protein bound to *gypsy* sequences in one chromosome can repress enhancer-promoter interactions in the paired gene located in the other homolog.

Proc. Natl. Acad. Sci. USA 92:5184–88

52. Gerasimova TI, Corces VG. 2001. Chromatin insulators and boundaries: effects on transcription and nuclear organization. *Annu. Rev. Genet.* 35:193–208

53. Geyer PK, Green MM, Corces VG. 1990. Tissue-specific transcriptional enhancers may act in *trans* on the gene located in the homologous chromosome: the molecular basis of transvection in *Drosophila. EMBO J.* 9:2247–56

54. Gibson JB, Reed DS, Bartoszewski S, Wilks AV. 1999. Structural changes in the promoter region mediate transvection at the *sn*-glycerol-3-phosphate dehydrogenase gene of *Drosophila melanogaster. Biochem. Genet.* 37:301–15

55. Gindhart JG, Kaufman TC. 1995. Identification of *Polycomb* and *trithorax* group responsive elements in the regulatory region of the Drosophila homeotic gene *Sex combs reduced. Genetics* 139:797–814

56. Goldberg ML, Colvin RA, Mellin AF. 1989. The Drosophila *zeste* locus is nonessential. *Genetics* 123:145–55

57. Golic MM, Golic KG. 1996. A quantitative measure of the mitotic pairing of alleles in *Drosophila melanogaster* and the influence of structural heterozygosity. *Genetics* 143:385–400

58. Goldsborough AS, Kornberg TB. 1996. Reduction of transcription by homologue asynapsis in *Drosophila* imaginal discs. *Nature* 381:807–10

59. Gubb D, Ashburner M, Roote J, Davis T. 1990. A novel transvection phenomenon affecting the *white* gene of *Drosophila melanogaster. Genetics* 126:167–76

60. Gyurkovics H, Gausz J, Kummer J, Karch F. 1990. A new homeotic mutation in the *Drosophila* bithorax complex removes a boundary separating two domains of regulation. *EMBO J.* 9:2579–85

60a. Hama C, Ali Z, Kornberg TB. 1990. Region-specific recombination and expression are directed by portions of the

Drosophila engrailed promoter. *Genes Dev.* 4:1079–93

61. Hendrickson JE, Sakonju S. 1995. *Cis* and *trans* interactions between the *iab* regulatory regions and *abdominal-A* and *Abdominal-B* in *Drosophila melanogaster. Genetics* 139:835–48

62. Henikoff S. 1979. Position effects and variegation enhancers in an autosomal region of *Drosophila melanogaster. Genetics* 93:105–15

63. Henikoff S. 1997. Nuclear organization and gene expression: homologous pairing and long-range interactions. *Curr. Opin. Cell Biol.* 9:388–95

64. Henikoff S, Comai L. 1998. *Trans*-sensing effects: the ups and downs of being together. *Cell* 93:329–32

65. Henikoff S, Dreesen TD. 1989. *Trans*-inactivation of the Drosophila brown gene: evidence for transcriptional repression and somatic pairing dependence. *Proc. Natl. Acad. Sci. USA* 86:6704–8

66. Hopmann R, Duncan D, Duncan I. 1995. Transvection in the *iab-5,6,7* region of the bithorax complex of Drosophila: homology independent interactions in *trans. Genetics* 139:815–33

67. Hur M, Laney JD, Jeon S, Ali J, Biggin MD. 2002. Zeste maintains repression of *Ubx* transgenes: support for a new model of Polycomb repression. *Development* 129:1339–43

68. Jack JW, Judd BH. 1979. Allelic pairing and gene regulation: a model for the zeste-white interaction in *Drosophila melanogaster. Proc. Natl. Acad. Sci. USA* 76:1368–72

69. Jijakli H, Ghysen A. 1992. Segmental determination in *Drosophila* central nervous system: analysis of the *abdominal-A* region of the bithorax complex. *Int. J. Dev. Biol.* 36:93–99

69a. Judd BH. 1955. Direct proof of a variegated-type position effect at the *white* locus in *Drosophila melanogaster. Genetics* 40:739–44

70. Judd BH. 1979. Allelic complementation and transvection in *Drosophila melanogaster. ICN-UCLA Symp. Mol. Cell Biol.* 15:107–15

71. Judd BH. 1988. Transvection: allelic cross talk. *Cell* 53:841–43

72. Kal AJ, Mahmoudi T, Zak NB, Verrijzer CP. 2000. The *Drosophila* Brahma complex is an essential coactivator for the *trithorax* group protein Zeste. *Genes Dev.* 14:1058–71

73. Karch F, Bender W, Weiffenbach B. 1990. *abdA* expression in *Drosophila* embryos. *Genes Dev.* 4:1573–87

74. Kassis JA. 2002. Pairing-sensitive silencing, Polycomb group response elements, and transposon homing in *Drosophila. Adv. Genet.* 46:421–38

75. Kassis JA, VanSickle EP, Sensabaugh SM. 1991. A fragment of *engrailed* regulatory DNA can mediate transvection of the *white* gene in Drosophila. *Genetics* 128:751–61

76. Kaufman TC, Tasaka SE, Suzuki DT. 1973. The interaction of two complex loci, zeste and bithorax in *Drosophila melanogaster. Genetics* 75:299–321

77. Kennison JA, Southworth JW. 2002. Transvection in Drosophila. *Adv. Genet.* 46:399–420

78. Kopp A, Duncan I. 2002. Anteroposterior patterning in adult abdominal segments of *Drosophila. Dev. Biol.* 242:15–30

79. Korge G. 1977. Direct correlation between a chromosome puff and the synthesis of a larval saliva protein in *Drosophila melanogaster. Chromosoma* 62:155–74

80. Kornher JS, Brutlag D. 1986. Proximity-dependent enhancement of *Sgs-4* gene expression in *D. melanogaster. Cell* 44:879–83

80a. Labrador M, Mongelard F, Plata-Rengifo P, Baxter EM, Corces VG, Gerasimova TI. 2001. Protein encoding by both DNA strands. *Nature* 409:1000

81. Laney JD, Biggin MD. 1992. *zeste*, a nonessential gene, potently activates

Ultrabithorax transcription in the *Drosophila* embryo. *Genes Dev.* 6:1531–41

82. LaSalle JM, Lalande M. 1996. Homologous association of oppositely imprinted chromosomal domains. *Science* 272:725–28

83. Lee JJ, von Kessler DP, Parks S, Beachy PA. 1992. Secretion and localized transcription suggest a role in positional signaling for products of the segmentation gene hedgehog. *Cell* 71:33–50

84. Leiserson WM, Bonini NM, Benzer S. 1994. Transvection at the *eyes absent* gene of Drosophila. *Genetics* 138:1171–79

85. Lewis EB. 1950. The phenomenon of position effect. *Adv. Genet.* 3:73–115

86. Lewis EB. 1954. The theory and application of a new method of detecting chromosomal rearrangements in *Drosophila melanogaster*. *Am. Nat.* 88:225–39

87. Lewis EB. 1955. Some aspects of position pseudoallelism. *Am. Nat.* 89:73–89

88. Lewis EB. 1981. Developmental genetics of the bithorax complex of *Drosophila*. In *Develomental Biology Using Purified Genes. ICN-UCLA Symp. Mol. Cell. Biol.*, ed. DD Brown, CF Fox, pp. 189–208. New York: Academic

89. Lewis EB. 1982. Control of body segment differentiation in Drosophila by the bithorax gene complex. In *Embryonic Development: Genes and Cells, Proc. 9th Congr. Int. Soc. Dev. Biol.*, ed. M Burger, R Weber, pp. 269–89. New York: Liss

90. Lewis EB. 1985. Regulation of the genes of the bithorax complex in *Drosophila*. *Cold Spring Harbor Symp. Quant. Biol.* L:155–64

91. Lifschytz E, Green MM. 1984. The zeste white interaction: induction and genetic analysis of a novel class of zeste alleles. *EMBO J.* 3:999–1002

92. Lindsley DL, Grell EH. 1968. *Genetic Variations of* Drosophila melanogaster. Carnegie Inst. Washington Publ. No. 627

93. Locke J, Tartof KD. 1994. Molecular analysis of *cubitus interruptus* (*ci*) mutations suggests an explanation for the unusual *ci* position effects. *Mol. Gen. Genet.* 243:234–43

94. Mansukhani A, Crickmore A, Sherwood PW, Goldberg ML. 1988. DNA-binding properties of the *Drosophila melanogaster* zeste gene product. *Mol. Cell. Biol.* 8:615–23

95. Marahrens Y. 1999. X-inactivation by chromosomal pairing events. *Genes Dev.* 13:2624–32

96. Martin-Morris LE, Csink AK, Dorer DR, Talbert PB, Henikoff S. 1997. Heterochromatic *trans*-inactivation of Drosophila *white* transgenes. *Genetics* 147:671–77

97. Martínez-Laborda A, González-Reyes A, Morata G. 1992. *Trans* regulation in the *Ultrabithorax* gene of *Drosophila*: alterations in the promoter enhance transvection. *EMBO J.* 11:3645–52

98. Mathog D. 1990. Transvection in the *Ultrabithorax* domain of the bithorax complex of *Drosophila melanogaster*. *Genetics* 125:371–82

99. Matzke M, Aufsatz W, Kanno T, Mette MF, Jakowitsch J, Matzke AJM. 2002. Homology-dependent gene silencing and host defense in plants. *Adv. Genet.* 46:235–75

100. Matzke M, Mette MF, Jakowitsch J, Kanno T, Moscone EA, et al. 2001. A test for transvection in plants: DNA pairing may lead to *trans*-activation or silencing of complex heteroalleles in tobacco. *Genetics* 158:451–61

101. Micol JL, García-Bellido A. 1988. Genetic analysis of "transvection" effects involving Contrabithorax mutations in *Drosophila melanogaster*. *Proc. Natl. Acad. Sci. USA* 85:1146–50

101a. Mihaly J, Hogga I, Gausz J, Gyurkovics H, Karch F. 1997. In situ dissection of the *Fab-7* region of the bithorax complex into a chromatin domain boundary and a *Polycomb*-response element. *Development* 124:1809–20

101b. Mongelard F, Labrador M, Baxter EM, Gerasimova TI, Corces VG. 2002. *Trans*-splicing as a novel mechanism to explain interallelic complementation in Drosophila. *Genetics* 160:1481–87

102. Morris JR, Chen J, Geyer PK, Wu C. 1998. Two modes of transvection: enhancer action in *trans* and bypass of a chromatin insulator in *cis*. *Proc. Natl. Acad. Sci. USA* 95:10740–45

103. Morris JR, Chen J, Filandrinos ST, Dunn RC, Fisk R, Geyer PK, Wu C. 1999. An analysis of transvection at the *yellow* locus of *Drosophila melanogaster*. *Genetics* 151:633–51

104. Morris JR, Geyer PK, Wu C. 1999. Core promoter elements can regulate transcription on a separate chromosome in *trans*. *Genes Dev.* 13:253–58

105. Müller H-P, Schaffner W. 1990. Transcriptional enhancers can act in *trans*. *Trends Genet.* 6:300–4

106. Müller J, Bienz M. 1991. Long range repression conferring boundaries of *Ultrabithorax* expression in the *Drosophila* embryo. *EMBO J.* 10:3147–55

107. Muller M, Hagstrom K, Gyurkovics H, Pirrotta V, Schedl P. 1999. The *Mcp* element from the *Drosophila melanogaster* bithorax complex mediates long-distance regulatory interactions. *Genetics* 153:1333–56

108. Muravyova E, Golovnin A, Gracheva E, Parshikov A, Belenkaya T, et al. 2001. Loss of insulator activity by paired Su(Hw) chromatin insulators. *Science* 291:495–98

109. Muskavitch MAT, Hogness DS. 1980. Molecular analysis of a gene in a developmentally regulated puff of *Drosophila melanogaster*. *Proc. Natl. Acad. Sci. USA* 77:7362–66

110. Neumann CJ, Cohen SM. 1996. *Sternopleural* is a regulatory mutation of *wingless* with both dominant and recessive effects on larval development of *Drosophila melanogaster*. *Genetics* 142: 1147–55

111. Orenic T, Slusarski DC, Kroll KL, Holmgren R. 1990. Cloning and characterization of the segment polarity gene *cubitus interruptus Dominant* of Drosophila. *Genes Dev.* 4:1053–67

112. Pal-Bhadra M, Bhadra U, Birchler JA. 1997. Cosuppression in Drosophila: gene silencing of *Alcohol dehydrogenase* by *white-Adh* transgenes is *Polycomb* dependent. *Cell* 90:479–90

113. Pal-Bhadra M, Bhadra U, Birchler JA. 2002. RNAi related mechanisms affect both transcriptional and posttranscriptional transgene silencing in Drosophila. *Mol. Cell* 9:315–27

114. Pattatucci AM, Kaufman TC. 1991. The homeotic gene *Sex combs reduced* of *Drosophila melanogaster* is differentially regulated in the embryonic and imaginal stages of development. *Genetics* 129:443–61

115. Peifer M, Bender W. 1986. The anterobithorax and bithorax mutations of the bithorax complex. *EMBO J.* 5: 2293–303

116. Peifer M, Karch F, Bender W. 1987. The bithorax complex: control of segment identity. *Genes Dev.* 1:891–98

117. Pirrotta V. 1997. Chromatin-silencing mechanisms in *Drosophila* maintain patterns of gene expression. *Trends Genet.* 13:314–18

118. Pirrotta V. 1999. Transvection and chromosomal *trans*-interaction effects. *Biochim. Biophys. Acta* 1424:M1–M8

119. Pirrotta V, Chan CS, McCabe D, Qian S. 1995. Distinct parasegmental and imaginal enhancers and the establishment of the expression pattern of the *Ubx* gene. *Genetics* 141:1439–50

120. Pirrotta V, Manet E, Hardon E, Bickel SE, Benson M. 1987. Structure and sequence of the Drosophila *zeste* gene. *EMBO J.* 6:791–99

121. Pirrotta V, Steller H, Bozzetti MP. 1985. Multiple upstream regulatory elements control the expression of the *Drosophila white* gene. *EMBO J.* 4:3501–8

122. Qian S, Capovilla M, Pirrotta V. 1993.

Molecular mechanisms of pattern formation by the BRE enhancer of the *Ubx* gene. *EMBO J.* 12:3865–77

123. Rassoulzadegan M, Magliano M, Cuzin F. 2002. Transvection effects involving DNA methylation during meiosis in the mouse. *EMBO J.* 21:440–50

124. Rastelli L, Chan CS, Pirrotta V. 1993. Related chromosome binding sites for *zeste*, suppressors of *zeste* and *Polycomb* group proteins in *Drosophila* and their dependence on *Enhancer of zeste* function. *EMBO J.* 12:1513–22

125. Roche SE, Rio DC. 1998. *Trans*-silencing by P-elements inserted in subtelomeric heterochromatin involves the Drosophila Polycomb group gene, *Enhancer of zeste*. *Genetics* 149:1839–55

126. Ronsseray S, Marin L, Lehmann M, Anxolabéhère D. 1998. Repression of hybrid dysgenesis in *Drosophila melanogaster* by combinations of telomeric *P*-element reporters and naturally occurring *P* elements. *Genetics* 149:1857–66

127. Rossignol J-L, Faugeron G. 1994. Gene inactivation triggered by recognition between DNA repeats. *Experientia* 50: 307–17

128. Rothberg I, Hotaling E, Sofer W. 1991. A *Drosophila Adh* gene can be activated in *trans* by an enhancer. *Nucleic Acids. Res.* 19:5713–17

129. Sánchez-Herrero E. 1991. Control of the expression of the bithorax complex genes *abdominal-A* and *Abdominal-B* by *cis*-regulatory regions in *Drosophila* embryos. *Development* 111:437–49

130. Saurin AJ, Shao Z, Erdjument-Bromage H, Tempst P, Kingston RE. 2001. A *Drosophila* Polycomb group complex includes Zeste and dTAFII proteins. *Nature* 412:655–60

131. Scholz H, Deatrick J, Klaes A, Klämbt C. 1993. Genetic dissection of *pointed*, a Drosophila gene encoding two ETS-related proteins. *Genetics* 135:455–68

132. Schwartz C, Locke J, Nishida C, Kornberg TB. 1995. Analysis of *cubitus in-*

terruptus regulation in *Drosophila* embryos and imaginal discs. *Development* 121:1625–35

133. Selker EU. 1990. Premeiotic instability of repeated sequences in *Neurospora crassa*. *Annu. Rev. Genet.* 24:579–613

134. Selker EU. 2002. Repeat-induced gene silencing in fungi. *Adv. Genet.* 46:439–50

134a. Shiu PKT, Raju NB, Zickler D, Metzenberg RL. 2001. Meiotic silencing by unpaired DNA. *Cell* 107:905–16

135. Sijen T, Vijn I, Rebocho A, van Blokland R, Roelofs D, et al. 2000. Transcriptional and posttranscriptional gene silencing are mechanistically related. *Curr. Biol.* 11:436–40

136. Simon J, Chiang A, Bender W, Shimell MJ, O'Connor M. 1993. Elements of the *Drosophila* bithorax complex that mediate repression by *Polycomb* group products. *Dev. Biol.* 158:131–44

137. Simon J, Peifer M, Bender W, O'Connor M. 1990. Regulatory elements of the bithorax complex that control expression along the anterior-posterior axis. *EMBO J.* 9:3945–56

138. Sipos L, Mihaly J, Karch F, Schedl P, Gausz J, Gyurkovics H. 1998. *Trans*-vection in the Drosophila *Abd-B* domain: extensive upstream sequences are involved in anchoring distant *cis*-regulatory regions to the promoter. *Genetics* 149:1031–50

139. Sirén M, Portin P. 1988. Effect of transvection on the expression of the Notch locus in *Drosophila melanogaster*. *Heredity* 61:107–10

140. Slatis HM. 1955. Position effects at the *brown* locus in *Drosophila melanogaster*. *Genetics* 40:5–23

141. Slatis HM. 1955. A reconsideration of the *brown-Dominant* position effect. *Genetics* 40:246–51

142. Southworth JW, Kennison JA. 2002. Transvection and silencing of the *Scr* homeotic gene of *Drosophila melanogaster*. *Genetics*. In press

143. Slusarski DC, Motzny CK, Holmgren R. 1995. Mutations that alter the timing and pattern of *cubitus interruptus* gene expression in *Drosophila melanogaster*. *Genetics* 139:229–40

144. Smolik-Utlaut SM, Gelbart WM. 1987. The effects of chromosomal rearrangements on the *zeste-white* interaction in *Drosophila melanogaster*. *Genetics* 116: 285–98

145. Talbert PB, LeCiel CDS, Henikoff S. 1994. Modification of the Drosophila heterochromatic mutation *brown*^{Dominant} by linkage alterations. *Genetics* 136: 559–71

146. Tartof KD, Henikoff S. 1991. *Trans*-sensing effects from Drosophila to humans. *Cell* 65:201–3

146a. Weinzierl R, Axton JM, Ghysen A, Akam M. 1987. *Ultrabithorax* mutations in constant and variable regions of the protein coding sequence. *Genes Dev.* 1: 386–97

147. White RAH, Akam ME. 1985. *Contrabithorax* mutations cause inappropriate expression of *Ultrabithorax* products in *Drosophila*. *Nature* 318:567–69

148. Williams JA, Bell JB, Carroll SB. 1991. Control of *Drosophila* wing and haltere development by the nuclear *vestigial* gene product. *Genes Dev.* 5:2481–95

149. Wu C-t. 1993. Transvection, nuclear structure, and chromatin proteins. *J. Cell Biol.* 120:587–90

150. Wu C, Goldberg ML. 1989. The *Drosophila zeste* gene and transvection. *Trends Genet.* 5:189–94

151. Wu C, Jones RS, Lasko PF, Gelbart WM. 1989. Homeosis and the interaction of *zeste* and *white* in *Drosophila*. *Mol. Gen. Genet.* 218:559–64

152. Wu C, Morris JR. 1999. Transvection and other homology effects. *Curr. Opin. Genet. Dev.* 9:237–46

153. Zachar Z, Chapman CH, Bingham PM. 1985. On the molecular basis of transvection effects and the regulation of transcription. *Cold Spring Harbor Symp. Quant. Biol.* 50: 337–46

154. Zhou J, Ashe H, Burks C, Levine M. 1999. Characterization of the transvection mediating region of the *Abdominal-B* locus in *Drosophila*. *Development* 126:3057–65

155. Zhou J, Levine M. 1999. A novel *cis*-regulatory element, the PTS, mediates an anti-insulator activity in the *Drosophila* embryo. *Cell* 99:567–75

156. Zimmerman JE, Bui QT, Liu H, Bonini NM. 2000. Molecular genetic analysis of Drosophila *eyes absent* mutants reveals an eye enhancer element. *Genetics* 154:237–46

Annu. Rev. Genet. 2002. 36:557–615
doi: 10.1146/annurev.genet.36.052402.152652
Copyright © 2002 by Annual Reviews. All rights reserved

GENETICS OF *CRYPTOCOCCUS NEOFORMANS*

Christina M. Hull and Joseph Heitman

*Department of Molecular Genetics and Microbiology, Howard Hughes Medical Institute,
Duke University Medical Center, Durham, North Carolina 27710;
e-mail: chull@duke.edu; heitm001@duke.edu*

Key Words   Cryptococcus neoformans, mating, signaling, genetics, pathogenesis

■ **Abstract**   *Cryptococcus neoformans* is a pathogenic fungus that primarily afflicts immunocompromised patients, infecting the central nervous system to cause meningoencephalitis that is uniformly fatal if untreated. *C. neoformans* is a basidiomycetous fungus with a defined sexual cycle that has been linked to differentiation and virulence. Recent advances in classical and molecular genetic approaches have allowed molecular descriptions of the pathways that control cell type and virulence. An ongoing genome sequencing project promises to reveal much about the evolution of this human fungal pathogen into three distinct varieties or species. *C. neoformans* shares features with both model ascomycetous yeasts (*Saccharomyces cerevisiae*, *Schizosaccharomyces pombe*) and basidiomycetous pathogens and mushrooms (*Ustilago maydis*, *Coprinus cinereus*, *Schizophyllum commune*), yet ongoing studies reveal unique features associated with virulence and the arrangement of the mating type locus. These advances have catapulted *C. neoformans* to center stage as a model of both fungal pathogenesis and the interesting approaches to life that the kingdom of fungi has adopted.

CONTENTS

0066-4197/02/1215-0557$14.00

INTRODUCTION

Cryptococcus neoformans is a fungal pathogen that infects the central nervous system to cause meningoencephalitis that is uniformly fatal if untreated [reviewed in (18, 111, 126)]. The organism is largely an opportunistic pathogen but can also be a primary pathogen in a small cohort of patients with no apparent immune system defects. The majority of afflicted patients have compromised immune function as a consequence of HIV infection, steroid therapy, cancer, chemotherapy, or therapy to suppress rejection of transplanted solid organs. The existing arsenal of antifungal agents is limited, and toxic side effects and the emergence of drug resistance are significant impediments to effective therapy.

C. neoformans is a basidiomycete fungus, and is therefore quite distinct in evolution from ascomycete fungi such as the model yeast *Saccharomyces cerevisiae*, the fission yeast *Schizosaccharomyces pombe*, and many other common human fungal pathogens, including *Candida albicans*, *Aspergillus fumigatus*, *Histoplasma capsulatum*, *Coccidioides immitis*, and *Pneumocystis carinii*. Instead, *C. neoformans* is more similar to other basidiomycetes, including the corn and grain plant fungal pathogens *Ustilago maydis* and *Ustilago hordei*, the *Tremella* wood-rotting jelly fungi, and the model mushrooms *Coprinus cinereus* and *Schizophyllum commune*. *S. cerevisiae* and *C. neoformans* have been diverging for ~500 million years, and thus, studies have the potential to reveal molecular principles conserved between these two divergent fungi as well as those that make them unique.

C. neoformans has a defined life cycle that involves vegetative growth as a budding yeast combined with the ability to undergo filamentous dimorphic transitions [reviewed in (3)]. The organism exists in the environment as a haploid budding yeast with two mating types: α and **a**. Under appropriate nutrient-limiting conditions and in response to mating pheromones, the two mating partners produce conjugation tubes, and the cells fuse (46, 95, 96, 154). In contrast to model yeasts in which nuclear fusion occurs immediately (150), karyogamy is delayed in the basidiomycetes, and the resulting heterokaryon adopts a filamentous state. Ultimately, basidia are formed, and it is in these structures that nuclear fusion and meiosis occur. Spores are then produced on the surface of the basidium. Cells of the α mating type can also respond to nitrogen limitation, desiccation, and MF**a** pheromone signals and differentiate by a process known as haploid fruiting, which also involves filamentation and sporulation (183, 190).

C. neoformans is ubiquitous in the environment and is most commonly found in association with pigeon guano or certain tree species (18). The organism exists in two distinct groups, one found in temperate climates (serotypes A and D) that accounts for the vast majority of human infections, and one (serotypes B and C) that is restricted to tropical regions, associates with *Eucalyptus* trees, and infects immunocompetent individuals. Humans are exposed to the organism via inhalation, and the basidiospores that are produced by mating or haploid fruiting are thought to represent the infectious propagule because their size (1–2 μ) is optimal for deposition in the alveoli of the lung (Figure 1). Experimental studies suggest that basidiospores may be up to 100× more infectious than vegetative yeast form cells, which are larger (4–10 μ) and often encapsulated (165). Recent evidence suggests that infection usually occurs early in life, and can either be rapidly cleared or establish a dormant infection from which activation can occur later in life in response to waning immunity (52, 65, 70). The initial pulmonary infection spreads via the bloodstream to the central nervous system, the most common site of clinical presentation. Although *C. neoformans* grows in a filamentous form in vitro during mating, haploid fruiting, and diploid filamentation, the predominant form associated with clinical infection is the budding yeast. There is little evidence that the filamentous dimorphic transitions play a role in the infectious cycle beyond their potential to produce the highly infectious basidiospores. Instead, in response to host conditions *C. neoformans* produces two specialized virulence factors that are essential for infection: the polysaccharide capsule and the pigment melanin [reviewed in (15)].

One of the most interesting features of *C. neoformans* is that its prevalence in the environment, differentiation pattern, and virulence phenotype have each been linked to the α mating type. The majority of environmental isolates are of the α mating type, virtually all clinical isolates are α, and in studies of congenic strains in a murine model α strains were found to be more virulent than **a** strains (99, 101). One hypothesis for this α prevalence is that α cells gain an advantage in the environment via their ability to undergo haploid fruiting, and the resulting production of α basidiospores leads to more clinical infections by this mating type (190). However, in a murine tail vein injection model in which the lung stage of the infection is bypassed, an α strain was intrinsically more virulent than a congenic **a** strain (101), suggesting that the *MATα* allele of the mating-type locus makes additional contributions to virulence.

Several features and recent advances have led to the development of *C. neoformans* as an excellent genetic model of fungal pathogenesis. These include the following:

- The organism has a defined sexual cycle that can be manipulated for classic genetic studies.

- The organism is predominantly haploid, facilitating genetic analysis.

- Congenic α and **a** strains have been constructed.

- Gene disruption by biolistic transformation and homologous recombination is now routine, and nearly 100 genes have been analyzed by such approaches.

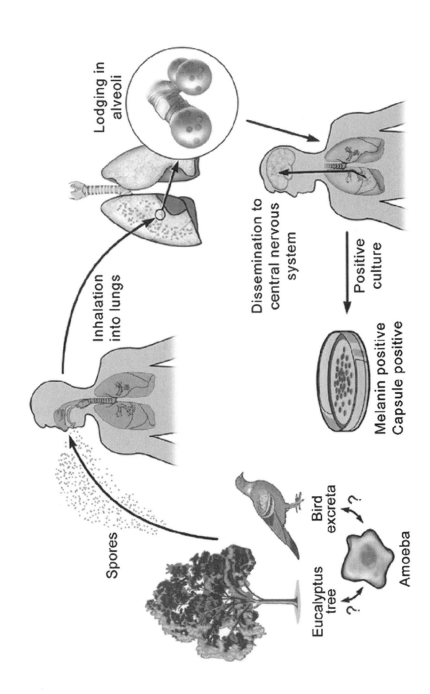

Lodging in alveoli

Inhalation into lungs

Dissemination to central nervous system

Positive culture

Melanin positive
Capsule positive

Spores

Eucalyptus tree

Bird excreta

Amoeba

- Both auxotrophic (*URA5*, *ADE2*) and dominant selectable markers (hygromycin, nourseothricin) are available. The *URA5* gene can be both positively and negatively selected using 5-fluoro-orotic acid.

- Stable, congenic diploid strains have been identified that enable studies of the life cycle and analysis of essential genes and regions of the genome.

- Several different and robust animal models have been implemented that allow the study of each phase of the infectious cycle and direct analyses of mutant strains and candidate therapies.

- Episomal linear telomeric plasmids have been developed that can be shuttled between bacterial and fungal cells. These plasmids allow genes to be identified by complementation or overexpression and can be readily rescued from fungal cells.

- A meiotic recombination map of the genome has been created for studies of genome structure and quantitative trait mapping.

- The *MATa* and *MATα* alleles of the mating type locus have been identified, enabling molecular studies of mating type that led to the discovery of sterile interspecies αADa and aADα hybrid strains and serotype A strains of the **a** mating type, which had been thought to be extinct.

- A genome sequencing project is ongoing for the serotype D laboratory-adapted reference strains JEC21 and B3501, and for the serotype A pathogenic isolate H99. Comparative genomic analysis is beginning for the divergent serotype B variety *gattii* strain WM276.

It is now possible in *C. neoformans* to identify a gene of interest, to disrupt this gene by transformation and homologous recombination, and to reintroduce the gene ectopically or at its normal chromosomal locus. It is then possible to test the effects of these genetic manipulations on the physiology of the organism both in vitro and in vivo with several different animal models. The application of these approaches has recently yielded insights into the genetic control of development and virulence and the role of the mating-type locus in the control of cell fate and identity. These studies also illustrate both conserved and organism-specific pathways that enable fungi to respond to challenges in their environment, including the ability to survive within and infect a mammalian host [reviewed in (109)].

Figure 1   The *C. neoformans* infectious cycle. *C. neoformans* resides in the environment and has been found associated primarily with pigeon droppings (serotypes A and D) and *Eucalyptus* trees (serotypes B and C). *C. neoformans* may also have interactions in the soil with amoebae. It is thought that infection of humans generally occurs when basidiospores produced by *C. neoformans* in nature are inhaled into the lungs. Inhaled spores are deposited into the alveoli and germinate to establish a dormant infection or disseminate to the central nervous system. Once dissemination has occurred, viable cells can be cultured from the cerebrospinal fluid of infected individuals and confirmed as *C. neoformans* by assaying for melanin and capsule production.

LIFE CYCLE AND SEXUAL CYCLE

C. neoformans is isolated from the environment and infected patients as a haploid, budding yeast. In growing cultures, the typical pattern of division by budding is observed, with larger mother cells and smaller daughter cells. Direct microscopic observations reveal that budding occurs predominantly in a bipolar fashion (J.A. Alspaugh, personal communication). As *C. neoformans* cells enter stationary phase, growth arrest occurs in both the G1 and the G2 phases of the cell cycle (167). This is in contrast to the budding yeast *S. cerevisiae*, in which nutrient limitation and stationary phase impose a G1 cell cycle arrest and is more similar to the fission yeast *S. pombe* in which a similar G2 arrest occurs when nutrients are exhausted. The rationale is that organisms that have evolved to exist predominantly as haploids, including *C. neoformans* and *S. pombe*, arrest in G2 rather than G1 to ensure that an extra copy of genomic information is available for repair.

C. neoformans can also adopt a filamentous growth form during several phases of its life cycle (Figure 2). These include the filamentous heterokaryon produced during mating, filamentation that occurs during haploid fruiting, and the temperature-controlled filament formation observed with diploid isolates (95, 96, 155, 190). Each of these filamentation pathways is described below.

Mating

The discovery of the *C. neoformans* sexual cycle was the key event in developing a workable genetic system for the organism and has resulted in an explosion of information about the virulence attributes, life cycle progression, and phylogenetic relationships among *C. neoformans* isolates. In 1975, Kwon-Chung identified the sexual state of *Cryptococcus neoformans* (for which the teleomorphic designation is *Filobasidiella neoformans*) by culturing different clinical isolates in pairwise combinations on various sporulation agars and scoring hyphae formation (95). After three weeks of incubation, a subset of strain combinations formed hyphae with clamp connections (associated with mating in other basidiomycetes). Careful observations revealed that the hyphae differentiate into basidia with spore chains similar to, but not identical to, members of the genus *Filobasidium*. This realization led Kwon-Chung to place the sexual state of *Cryptococcus neoformans* in the family Filobasidiaceae and name this sexual state *Filobasidiella neoformans*. Her subsequent cytologic work characterized the morphological events associated with basidiospore formation, and her pioneering genetic experiments revealed a bipolar mating system (two mating types, **a** and α) in which meiotic recombination occurs between strains.

Kwon-Chung also described differences in mating among the different serotypes of *C. neoformans*. She observed that mating was easily detected with strains of serotypes A and D but not with B and C. This prompted her to investigate serotype B and C matings and led to the finding that B and C strains can mate under certain conditions and form basidia and spores (97). These spores, however, were different

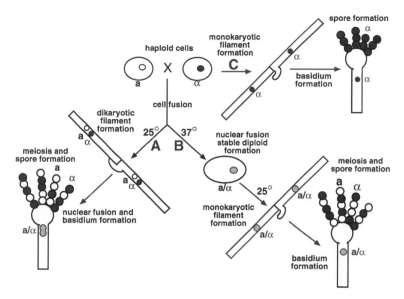

Figure 2 Phases of the *C. neoformans* life cycle. There are three developmental pathways for *C. neoformans* cells. The first pathway, known as mating (*A*), begins when two haploid cells of opposite mating types fuse and are maintained at 25°C. The haploid cells fuse and grow as filaments with distinct nuclei (dikaryons) and special clamp cells that are fused to the filaments. In response to unknown signals, the dikaryon produces a specialized sporulation structure called a basidium. It is in this basidium that nuclear fusion and meiosis take place. Many rounds of duplication and mitosis lead to the production of haploid **a** and α spore products that form long chains on the basidial head. The second developmental pathway, known as diploid filamentation (*B*), occurs when cells of the opposite mating type fuse and are maintained at 37°C. In this case, the haploid cells and their nuclei fuse to create a yeast form, diploid cell that grows as a yeast at 37°C. In response to a decrease in temperature (25°C), this diploid differentiates to form monokaryotic filaments with unfused clamp cells, basidia, and haploid **a** and α spores similar to those of mating filaments. The third possible fate occurs primarily in the α cell type. This pathway is known as haploid fruiting (*C*) and occurs when α cells are grown under nutrient limited, desiccating conditions. In this pathway, haploid α cells form monokaryotic filaments with unfused clamp cells, basidia, and haploid α spores. In each case the resulting spores are competent to germinate and grow as haploid cells or initiate one of the three developmental pathways again.

from those formed in A and D matings, and Kwon-Chung designated the sexual state of serotypes B and C as *Filobasidiella bacillispora* to reflect the elongated bacillus form spores produced by this species.

The identification of different *C. neoformans* mating types allowed Edman et al. to develop congenic serotype D strains through a series of ten backcrosses (79, 101).

The two parental strains for the crosses were NIH12, a *MATα* clinical isolate from a patient who presented with osteomyelitis caused by *C. neoformans* and NIH433, a *MATa* strain that was isolated from the environment (pigeon guano) in Denmark. Importantly, the ultimate congenic strains are the products of backcrossing to the F2 generation and thus are not congenic with either NIH12 or NIH433 and share, on average, half their genomes with each parent. The resulting congenic strain pair (JEC20/B4476 and JEC21/B4500) was then used in controlled experiments both in vivo and in vitro. The availability of easily manipulated strains has proven invaluable; however, one serious drawback to these laboratory-adapted serotype D strains is that they are significantly attenuated in animal models, and this attenuation hampers analysis of virulence attributes. On the other hand, JEC20 and JEC21 represent an exceptional resource for conducting genetic analyses in this system. A variety of mutant and auxotrophic derivatives are currently available, and the use of these strains has led directly to a better understanding of the life-cycle processes (like mating) that are crucial to *C. neoformans* growth.

An important step in creating useful strains was refining the conditions under which mating can occur. Efficient mating requires strains of opposite mating types and a nitrogen-limited growth medium or a medium containing a plant extract. A common and efficient mating medium is solid agar containing V8 juice (100). V8 juice (and carrot juice) media are limiting for several amino acids and nucleotides, but also contain a water-soluble, heat-stable, small-molecular-weight component that stimulates mating (J.A. Alspaugh & J. Heitman, unpublished observations). Mating occurs most efficiently on V8 medium at room temperature in the dark. When strains of the opposite mating type confront each other under these conditions, α cells make conjugation tubes, and **a** cells expand to form round, enlarged cells. The **a** and α cells fuse with one another to form elongated cells that contain two distinct nuclei. These dikaryons grow as filaments until they differentiate to form a specialized cell called a basidium. It is in this basidium that nuclear fusion and meiosis occur. The meiotic products are repeatedly duplicated mitotically, packaged into spores, and moved to the surface of the basidium to form long chains of randomly ordered basidiospores extending from the basidial head (95, 98).

The signals that govern the stages of this differentiation process are suspected to mirror those of other basidiomycetes such as *C. cinereus* and *S. commune*, but in fact very little is known about the gene products that control these events in *C. neoformans*. What is clear is that *C. neoformans* develops quite differently from other model fungi. For example, in contrast to its ascomycete relatives (like *S. cerevisiae*), the spores of this basidiomycete are not packaged in an ascus. In the case of *C. neoformans*, they are loosely attached to the basidium and free to dislodge and be carried away to another venue (such as the lung of a human host). There are several other major differences between the mating systems in *S. cerevisiae* and *C. neoformans*. Unlike in *C. neoformans*, mating in the ascomycete *S. cerevisiae* takes place under nutrient-replete conditions. The diploid state is stable, and meiosis and sporulation take place under low nutrient conditions (81). In this system the term "mating" has traditionally been meant to include only the initial steps of cell and nuclear fusion to form a stable diploid. In *C. neoformans*, however, the predilection of heterokaryons and diploids to grow filamentously has led the term "mating" to include the processes of cell fusion and filament formation. Mating assays to date assess the ability of a strain to form filaments and do not speak directly to the issue of partner cell fusion. As a result, strains that fuse with a mating partner but fail to make filaments have generally been scored as sterile. More recent experiments have addressed the issue of cell fusion using assays to score the production of prototrophic dikaryons following fusion of two auxotrophic strains (26, 154, 198).

These mating and filamentation assays are very similar to those carried out in a relative of *C. neoformans*, the plant pathogen *U. maydis* (94). In *U. maydis* dikaryotic filaments are formed when compatible mating partners come into contact and fuse with one another. Mating in this system is also generally scored as the ability to form dikaryotic filaments. The dikaryon is infectious to corn plants and can differentiate only in planta to form the black spores that are characteristic of this smut fungus. Although the spores of *C. neoformans* are thought to be the infectious propagule in cryptococcosis, there is no evidence that the diploid or filamentous state is required for any part of pathogenesis within the human host.

Haploid Fruiting

The second phase of the *C. neoformans* life cycle that can lead to filamentation is known as haploid or monokaryotic fruiting (Figure 2). This phase involves filamentous differentiation that occurs when cells of the α mating type are grown under conditions of desiccation and severe nitrogen limitation (190). Haploid α cells produce filaments that are monokaryotic and septated, but the filament cells have unfused clamp connections. Basidia are produced that often yield four short chains of basidiospores. Robust blastospores (yeast-form cells) are produced along the filaments via budding from the clamp cell, and proliferation of these cells gives rise to chains of microcolonies that have a beads-on-a-string appearance. Haploid fruiting has been proposed as a mechanism by which the organism could forage for nutrients, in a manner analogous to the pseudohyphal filamentous differentiation that occurs in *S. cerevisiae* in response to nitrogen limitation (67). Fruiting might also promote survival by producing spores that could survive nutrient-poor conditions. As discussed further below, haploid fruiting may play a role in generating the infectious propagule, the basidiospore. Recent observations indicate that haploid fruiting is dramatically stimulated by mating partner cells, and thus likely functions in the early steps of mating, allowing α cells to detect **a** cells at a distance (183). The mating partners might then come into contact via filamentation, or the dispersal of spores from aerial basidia. Haploid fruiting of α cells driven by **a** cells might also provide a selective pressure to maintain both mating types in the environment without a strict requirement for mating.

The process of haploid fruiting is thought to be asexual and to involve mitotic production of spores. However, it has not been examined experimentally whether meiosis occurs during haploid fruiting. In *S. cerevisiae*, meiosis is normally restricted to diploid cells that are heterozygous at the *MAT* locus (**a**/α)

(124). However, **a**/**a** and α/α yeast strains that lack the Rme1 repressor of meiosis (*rme1/rme1*) can undergo meiosis and sporulate (125). Thus, by analogy *C. neoformans* α haploid cells might undergo fusion to produce α/α homozygous strains that then undergo meiosis.

Two other conditions have recently been discovered that dramatically stimulate haploid fruiting: darkness and pheromone secreted by **a** cells (154, 183). Limited haploid fruiting of the standard haploid α lab strain JEC21 typically occurs in 2 to 4 weeks when incubated on SLAD (nitrogen-limiting) medium at room temperature. Under the same conditions, cultures incubated in the dark undergo much more robust haploid fruiting in less than 1 week. This finding came as a surprise, given that *S. cerevisiae* has no apparent light-sensing mechanisms. However, many other fungi, including the light-entrained clock-containing *Neurospora crassa*, clearly sense light [reviewed in (113)]. The mechanisms by which *C. neoformans* senses and responds to light, or even the wavelength detected remain to be explored.

A second signal that robustly stimulates haploid fruiting of α cells is the presence of factors secreted by **a** cells (183). This was discovered during efforts to develop confrontation assays to examine the morphological actions of pheromones during mating. When a line of α cells is streaked in close proximity to but not touching a line of **a** cells, both cell types respond to factors secreted by the opposite cell type. Initially, both cell types produce short conjugation tubes. The filaments produced by the **a** cells fail to elongate and instead enlarged, round cells are observed. In contrast, the α cells produce conjugation tubes that dramatically extend and then haploid fruit, producing basidia and basidiospores. Haploid fruiting of α cells is stimulated by **a** cells at a distance with no requirement for contact between opposite cell types. Stimulation of α haploid fruiting by **a** cells can also occur through a dialysis membrane. Finally, recent studies have revealed that haploid fruiting is intimately controlled by pheromones (154). When the gene encoding the MF**a** pheromone that is normally produced by **a** cells is transformed into α cells, haploid fruiting is dramatically stimulated in an autocrine signaling fashion. In addition, overexpression of the MFα pheromone in α cells also enhances haploid fruiting, and deletion of the *MFα1,2,3* genes results in a pheromoneless triple mutant strain that exhibits reduced haploid fruiting. These observations provide evidence that the pheromones of *C. neoformans* function in both paracrine signaling between cells of the opposite mating type and autocrine signaling pathways that act on the pheromone producing cells themselves.

Recent observations suggest that **a** strains can also undergo haploid fruiting under certain conditions. The original studies that demonstrated that α strains undergo haploid fruiting while **a** strains do not included a genetic analysis showing that the ability to haploid fruit and the α mating type cosegregated in genetic crosses (190). In more recent studies involving stimulated haploid fruiting (darkness, confrontation with **a** cells), haploid fruiting remains restricted to cells of the α mating type. However, an unusual isolate of the serotype D *MATa* strain B3502, one of the ancestral strains of the JEC20/JEC21 congenic strain series undergoes robust haploid fruiting on SLAD or V8 medium (K. Lengeler & J. Heitman, unpublished observations). These findings suggest that mutations might enable **a** strains to haploid fruit under certain conditions. Similarly, a recently developed improved medium (sucrose-proline agar) supports haploid fruiting of the serotype D *MATa* strain JEC20 (K.J. Kwon-Chung, personal communication). These findings suggest that there is a relative but not absolute difference between α and a strains in their potential to undergo haploid fruiting. This also suggests that genes within the *MAT* region likely work in concert with nonmating type-specific genes, such as elements of the MAP kinase cascade, to promote fruiting in both α and a cells. The conditions under which fruiting of a cells is physiologically relevant remain to be defined.

The realization that both α and a cells of *C. neoformans* can undergo haploid fruiting under different conditions is analogous to the recent discovery that haploid cells of *S. cerevisiae* can undergo pseudohyphal differentiation [reviewed in (109)]. Previously, pseudohyphal growth was thought to be restricted to only diploid cells, to function in foraging for nutrients, and to not involve the pheromones, pheromone receptors, or the coupled heterotrimeric G-protein (67, 115). Yet recent studies reveal that haploid yeast cells respond to low doses of mating pheromone by switching from axial to unipolar budding, elongating, and invading the growth substrate, all features of pseudohyphal differentiation (58, 146). The revised view is that filamentous growth allows diploid yeast cells to forage for nutrients and allows haploid yeast cells to search for mating partners. Pseudohyphal growth of haploid yeast cells in response to pheromone requires the mating pheromones, their receptors, and the coupled G-protein. By analogy, *C. neoformans* α cells may haploid fruit in response to MFa mating pheromone to search for mating partners. In contrast, a cells have not yet been observed to filament in response to a mating partner but rather in response to different environmental conditions or in mutant strain backgrounds. Thus, in a manner similar to *S. cerevisiae*, the functions of filamentous growth and sporulation may differ between the two cell types of *C. neoformans*.

Some serotype A strains have been reported to haploid fruit (190) but this has not been widely observed for most serotype A strains under a variety of conditions. These include altered growth mediums, incubation in the dark, and in confrontation with serotype A or D strains. It is likely that serotype A strains will fruit more readily under certain conditions that remain to be defined, but currently, this poses a dilemma for the theory that haploid fruiting represents the mechanism by which the infectious basidiospores are produced, since most clinical infections are produced by serotype A strains.

Diploid Filamentation

The third and final phase of the *C. neoformans* life cycle in which a filamentous transition occurs is during the growth of diploid strains (Figure 2). *C. neoformans* is thought to exist predominantly as a haploid budding yeast in which the diploid stage of the life cycle is transient and unstable. However, the dissection of basidiospores during standard genetic crosses results in a small number of unusual self-filamentous segregants (187, 188). Sia et al. characterized these isolates and found that they grow as budding yeasts with smooth, uniform colonies when cultured at 37°C, yet when grown at 24°C these isolates spontaneously filament and sporulate on a variety of media (155). Upon further examination, these strains were found to be uninucleate but with nuclei larger than the nuclei of the haploid parental strains. FACS analysis revealed that these strains were diploid, and PCR analysis showed that these strains were heterozygous at the mating-type locus (**a**/α) (155). When diploid strains that were heterozygous for several genetic markers were sporulated and dissected, the resulting progeny were haploid and exhibited meiotic segregation of genetic markers. These congenic diploid strains can be transformed and used to study key features of cell identity and to establish that genes are essential (43, 83). The filaments produced by these congenic serotype D diploid strains share some features with haploid fruiting filaments and other features with the filaments produced during mating (155). For example, the diploid filaments are monokaryotic and have unfused clamp connections, similar to haploid fruiting. But like mating, the diploid filaments produce abundant basidia with long chains of basidiospores.

Serotype AD Diploids

The relevance of the diploid state in nature has also come to the fore in recent years with the discovery of naturally occurring diploid strains in the patient population. These strains are known as AD hybrids and appear to be the result of a fusion event between a serotype A strain and a serotype D strain. Serotype A strains are the predominant form of *C. neoformans* infection. ~95% of all *C. neoformans* infections are caused by serotype A isolates, and >99% of infections in AIDS patients are serotype A (18). In some regions of the world, such as Europe, serotype D strains are more common and have been thought to represent up to 30% of infections (51). Even in New York City up to 12% of infections may be caused by serotype D (159). Recent studies, however, indicate that many of these strains in fact represent unusual serotype AD hybrid strains (11).

Early speculation was that AD strains might represent unusual mutants of the enzymes that produce the capsular polysaccharide antigens. This speculation may have been fueled by the finding that strains that initially typed as AD often reverted to only serotype A or D with passage. However, a variety of studies suggested that the serotype AD strains might represent unusual hybrid strains (10, 12, 179). Following the rediscovery of congenic serotype D diploid strains by Sia et al., Lengeler et al. applied molecular methods to analyze a collection of serotype AD strains (108). These strains were found to be aneuploid or diploid by FACS analysis, to be uninucleate, to often contain both serotype A and D alleles of several genes, and to be heterozygous at the mating-type locus (**a**/α). A few serotype AD strains were found to be self-filamentous, but the spores produced germinated poorly, and those spores that did germinate gave rise to isolates that were usually still diploid and serotype AD. Thus, meiosis fails to occur properly in AD hybrid strains, likely as a result of genome differences between the original serotype A and D parental strains.

The AD hybrid strains fall into two classes (108). In one type, the *MAT*α allele was inherited from the serotype A parent, and the *MAT***a** allele was inherited from the serotype D parent. In the second type, the *MATa* allele was inherited from the serotype A parent, which, until recently, had been thought to be extinct, and the *MATα* allele was inherited from a serotype D parent. We call these two classes of AD hybrid strains the αADa and the aADα strains to reflect their distinct origins. Somewhat surprisingly, from the analysis of 10 serotype AD strains, four were found to be of the atypical aADα origin. This suggests that serotype A *MATa* strains might still be common in some unique environmental niche that remains to be discovered. Alternatively, the serotype A *MATa* strains might have begun to disappear after the time of origin of the AD hybrids, which has been estimated from population genetic studies to be recent but could still be as long as several million years ago (194).

Several complementary studies come to similar conclusions but by different approaches. Boekhout et al. have used AFLP (amplified fragment length polymorphism) analysis to derive evolutionary trees for a vast collection of environmental and clinical isolates of *C. neoformans* serotype A, D, B/C, and AD strains (11). First, they find evidence that serotype A and D strains are quite distinct from serotype B/C strains, leading to suggestions that these represent different species, as discussed further below. Second, they find that serotype AD strains are hybrids containing AFLP banding patterns that are a superimposition of the patterns found in haploid serotype A and D strains. Third, they find that many European strains reported to be serotype D are in fact serotype AD hybrid strains. These findings support the hypothesis that serotype AD strains are intervarietal or interspecies hybrids between parental A and D haploid strains, and they suggest that serotype A strains and the intervarietal AD diploid strains are more virulent than haploid serotype D strains. The virulence of serotype D strains may be enhanced by fusion with more pathogenic serotype A strains, and the molecular analysis of AD hybrid strains might represent one avenue to analyze the molecular determinants of virulence on a genome-wide level.

Cogliati et al. (29) recently analyzed a large series of serotype AD strains by molecular approaches similar to those employed by Lengeler et al. (108). Additional strains were analyzed, and the analysis included both the *STE20α/a* genes and the mating pheromone genes. The conclusions are strikingly similar: Namely, the serotype AD strains represent hybrids between parental A and D haploid strains and occur in the two general classes of the αADa and aADα types. That two independent approaches with two different strain collections reached concordant conclusions provides strong support for the hypothesis that the serotype AD strains originated via hybridization. In addition, the studies of Lengeler et al. and Cogliati et al. in which the serotype A *MATa* locus was found to be present in the unusual class of aADα hybrid strains both served as the impetus to search for and independently discover the first two haploid serotype A *MATa* strains (described further below). Finally, by population genetic approaches using AFLP analysis, Xu and colleagues provide evidence that the serotype AD hybrid strains arose via several independent origins, fully consistent with the previous finding that there are two distinct classes of AD hybrids (αADa and aADα) (194). In a recent study,

Yan et al. examined the distribution of the mating-type locus by PCR analysis and mating assays with a collection of 358 strains (197). Their analysis revealed no additional serotype A *MATa* strains from 324 different serotype A strains, and provided independent confirmation that two classes of serotype AD hybrids exist.

In addition, Lengeler et al. found that AD hybrid strains could be generated readily in the laboratory with defined serotype A and D parental strains (108). In these studies, genetically marked auxotrophic serotype A H99–derived and serotype D JEC20–derived strains were co-cultured under mating conditions, and prototrophic strains resulting from cell-cell fusion were isolated. These strains are diploid by FACS analysis, and by molecular analysis using PCR, these strains are diploid and heterozygous for both the mating-type locus and the serotype A and D alleles of several genes. Curiously, most of these AD hybrid strains do not serotype as AD strains; instead the majority type as serotype D. Perhaps the serotype D phenotype is dominant [as is the case in progeny of serotype B by D crosses (100)] and as these diploid AD hybrids become aneuploid upon passage, they become serotype AD by standard antibody tests. As the resulting serotype AD strains become further aneuploid, they have frequently been observed to type as only serotype A or D. Thus, careful analysis of environmental and clinical isolates is necessary to ascertain whether strains are haploid or hybrid.

Xu and colleagues conducted a similar series of laboratory serotype A by D crosses and observed that mitochondria are inherited in a uniparental fashion in the resulting AD hybrid progeny (193). They employed RFLP markers in the mitochondrial genome and found that in 570 progeny of six A by D crosses, all progeny inherited the mitochondrial pattern from the serotype D parent. This led to an initial suggestion that the mitochondria from serotype D might be dominantly inherited. To address this hypothesis, Xu and colleagues recently analyzed the pattern of mitochondrial inheritance in serotype AD hybrid strains (192). To their surprise, they found that some serotype AD hybrid strains inherited the mitochondria from the serotype A parent, whereas others inherited the mitochondria from the serotype D parent. These findings again reflect a uniparental inheritance of mitochondria, but underscore that it is not determined by the serotype from which they originate.

The finding that the behavior of α and **a** cells differs during mating provides one possible explanation. In response to mating pheromones, the α cells produce filaments that extend and search out **a** mating partner cells. In turn, the **a** cells dramatically enlarge to form large swollen cells that serve as ready targets for the filaments of α cells. Perhaps following fusion of the tip of an α filament with a swollen **a** cell, the nucleus from the α partner migrates through the filament into the **a** cell, the two nuclei pair, and dikaryotic filaments are then launched from the swollen **a** cell as the base. In this scenario, the mitochondria that migrate into the dikaryotic filament cell might be preferentially derived from the **a** cell parent, rather than from the α cell. Additional mechanisms could also function to destroy the mitochondria from one parent in the cross, analogous to the nucleolytic mechanisms that destroy chloroplasts derived from one parent in the green alga, *Chlamydomonas reinhardtii* (173).

Serotype A *MATa* Strains

In addition to information about mitochondrial inheritance, the studies of AD hybrids provided the impetus to explore the issue of the elusive serotype A *MATa* strains. As mentioned previously, until recently, no serotype A *MATa* strain had been identified and it was thought that this mating type might have become extinct, possibly reflecting an evolution of serotype A strains toward an asexual life cycle. Two unusual serotype A *MATa* strains have been discovered recently, again forcing a reevaluation of the life cycle and the role of the sexual cycle in this organism.

During an examination of clinical isolates from Tanzania, Lengeler et al. identified an unusual isolate, strain 125.91, that typed as serotype A but lacked several genes encoded by the *MATα* allele of the mating type locus (112). Two hypotheses were proposed: either this strain contained a large deletion of the *MATα* allele, or the strain represented a serotype A *MATa* isolate. Using primers to the *STE20a* gene encoded by the serotype D *MATa* mating-type allele, they amplified a serotype A-specific *STE20a* allele by PCR, proving that strain 125.91 represents an elusive serotype A *MATa* strain. Lengeler et al. showed that this strain is haploid by FACS analysis, and the strain contains serotype A-specific genes and lacks serotype D-specific genes by PCR analysis. Moreover, using two novel transposable elements (T1 and T2) that are common in serotype D but rare or absent in serotype A (M.C. Cruz & J. Heitman, in preparation) as probes, strain 125.91 exhibited the serotype A transposon fingerprint (112). Strain 125.91 was sterile when crossed with the standard serotype A *MATα* strain H99 under a large variety of conditions. Conditions have recently been discovered that support mating of this unusual serotype A *MATa* strain (K. Nielsen & J. Heitman, unpublished results), and the entire *MATa* allele for this strain has been cloned and sequenced (110), opening the door to further genetic analysis in the predominant serotype causing infection.

The serotype A *MATa* strain 125.91 is a clinical isolate from a Tanzanian AIDS patient with cryptococcal meningitis, and is therefore pathogenic. In the murine tail vein injection model, the serotype A *MATa* strain 125.91 is less virulent than the serotype A *MATα* strain H99, suggesting that the *MATα* allele might also be linked to virulence potential in the serotype A clinical isolates. A second serotype A *MATa* strain has recently been reported, IUM96-2828, an environmental isolate from the soil in Italy (178). The IUM96-2828 strain has been reported to be fertile in crosses with serotype A *MATα* strains. 125.91 and IUM96-2828 represent the only two serotype A *MATa* strains that have been identified from a search of ~1500 strains in several labs worldwide (197; T. Boekhout, personal communication; W. Meyer, personal communication; K.J. Kwon-Chung, personal communication; K. Lengeler & J. Heitman, unpublished results). Thus, serotype A *MATa* strains are rare but not extinct, and exist as about 0.1% of the serotype

A population compared to the serotype D population in which **a** strains occur at about 2% of the population (99).

The discovery of two serotype A *MATa* strains is an exciting breakthrough because these strains provide the platform to construct congenic serotype A strains. Once congenic strains become available, much of the genetic work in this organism should shift to studies in the more pathogenic serotype A isolates. Reasons for switching to serotype A backgrounds include that serotype A strains represent the overwhelming majority of clinical isolates, and that most serotype A strains are more virulent than most serotype D strains, facilitating genetic analysis of virulence properties. Finally, an increasing number of mutants exhibit differences in phenotypes between the serotype A H99 strain background and the serotype D JEC20/JEC21 congenic strain series (26, 40, 182, 198).

IDENTIFICATION AND CHARACTERIZATION OF THE *MAT* LOCUS

Mating type in fungi is controlled by the information encoded by the mating-type or *MAT* locus. This region of the genome is special because homologous chromosomes contain nonhomologous information that specifies the genetic differences between cell types. In ascomycete fungi, this region has been studied extensively, particularly in *S. cerevisiae* [reviewed in (81, 90)]. In *S. cerevisiae* information at *MAT* is different between **a** and α cells. *MATa* encodes the transcriptional regulator **a**1, and *MATα* encodes the transcriptional regulators α1 and α2. The region of difference between the two chromosomes containing *MAT* is about 700 base pairs. The size and composition of this locus is common for ascomycetes, which generally encode the key transcriptional regulators of cell type at a single *MAT* locus of relatively small size. In basidiomycetes, there are usually two, unlinked *MAT* loci; one encodes pheromones and pheromone receptors, while the other encodes homeodomain transcriptional regulators. Both *MAT* loci are necessary to specify cell type. The basidiomycete *MAT* loci are typically much larger than the ascomycete *MAT* loci, encode more than one gene product, and exist in multiple alleles, giving rise to thousands of different mating types [reviewed in (20, 94)].

Given the intriguing relationship in *C. neoformans* between the α mating type and prevalence in patients and the environment (99), an early effort was made to clone the *MAT* region from α cells. Moore & Edman used a difference cloning technique to identify regions of the α genome that were not present in the **a** genome (129). They identified an approximately 35-kb region that was present in only α cells. They also defined a 2-kb fragment of this region that when transformed into **a** cells conferred the ability to form conjugation tubes (mating structures) in response to conditions that support mating (V8 medium). The sequence of this fragment revealed an open reading frame with strong similarity to the mating pheromones of other fungi. This gene has subsequently been shown to encode one of three α pheromones used for signaling to a mating partner in *C. neoformans* (46, 154).

To understand the basis of filament production in *C. neoformans*, Wickes et al. carried out a screen to identify genes that when overexpressed enhance the production of filaments by α cells in the absence of an **a** mating partner (189). A genomic library of α DNA fragments in a telomeric vector was transformed into α cells, which were then tested for filament formation on filament agar. One consistently hyperfilamentous strain was selected for characterization, and the plasmid insert was recovered. The insert contained a gene with extensive similarity to the *STE12* gene found in other fungi. Ste12 in other fungi is a transcription factor that activates the expression of many different genes in response to signals from conserved MAP kinase cascades regulating mating and filament formation.

Preliminary experiments in *C. neoformans* suggested roles consistent with those in other fungi; namely, Ste12 induces the expression of at least one mating-type gene, the pheromone gene *MFα1*, and can control gene expression in both **a** and α cells. Expression of the *CNLAC1* gene encoding the melanin biosynthetic enzyme laccase was found to be induced upon overexpression of *STE12* in **a** and α cells. This established the first molecular connection between a mating pathway component and a virulence factor in *C. neoformans* (even though *ste12* mutants have no melanin synthesis defect). These expression experiments also revealed perhaps the most interesting feature of the *STE12* gene: This allele of *STE12* in *C. neoformans* is specific to the α cell type. That is, Southern and northern blot analyses showed hybridization in only the α mating type in several different backgrounds, and no signal in **a** cells of the same background was detected. This intriguing finding suggested that unlike *STE12* in other organisms, the *C. neoformans STE12α* was carrying out a function specific to the α cell type, like traditionally recognized components of the *MAT* locus. However, *STE12α* did not reside in the 35-kb region of the *MAT* locus originally identified by Moore & Edman, and probes to the gene hybridized to a previously unidentified cosmid of α-specific DNA. Identification and localization of another α-specific gene, *STE20α* revealed that it also did not map to the original 35-kb region of *MATα* (112, 182). These findings suggested that the *MAT* locus was larger than previously realized.

To understand the organization of *MATα* further, characterization of the *MATα* region was carried out by Karos et al. in which they identified cosmid clones containing α-specific DNA (91). Endpoints were defined using probes to both **a** and α cells to identify 5–7-kb boundaries of an approximately 50-kb MATα region. This analysis confirmed the suspicion that the locus was larger than that identified by Moore & Edman, placed several previously identified α-specific genes in the locus (*STE11α*, *STE20α*, *STE12α*), and identified several unexpected genes in the locus including a myosin homolog (*MYO2α*), an apparent translation initiation factor (*PRT1α*), two additional pheromone genes (*MFα2* and *MFα3*), and a pheromone receptor gene (*STE3α* a.k.a. *CPRα*). This expanded locus was unusual in its size and architecture, but additional experiments revealed that it was still not the complete *C. neoformans MAT* locus.

Recent work by Hull et al. in which the phenotype of a deletion of this 50-kb *MATα* region was analyzed revealed the presence of additional α-specific components (83). In this study, a complete deletion of the 50-kb *MATα* region was generated in **a**/α diploid cells (no deletions were recovered in α haploids due to the presence of one or more essential genes in the locus). Because the process of diploid filamentation requires information from both **a** and α cells, it was predicted that diploid filamentation would be abolished in an **a**/Δ strain. However, diploid filamentation was still intact, suggesting that an additional factor or mechanism was functioning to maintain the **a**/α cell identity. Further experiments showing that a 2n-1 diploid strain missing the *MATα*-containing chromosome was deficient in diploid filamentation (and mated like an **a** cell) suggested that the required component(s) is located on the *MATα*-containing chromosome.

Two models for the regulation of diploid filamentation were proposed and evaluated. In the first model, the ploidy of the *MAT* chromosome could be regulating filamentation analogous to the regulatory system found in other organisms in which ploidy and/or dosage establish mating type (9, 141) or to *S. cerevisiae* in which recent studies have shown that expression of a set of genes is ploidy dependent (64). This ploidy-dependent model was ruled out by experiments in which **a**/**a** and α/α diploid cells were created and tested for diploid filamentation. Homozygous diploid strains were created using a novel assisted mating or "ménage à trois" reaction in which strains of the same mating type were induced to fuse by the addition of an opposite mating partner to the mix (83). The resulting homozygous diploid strains were unable to undergo diploid filamentation (although α/α diploids can still form filaments in a fruiting assay). This result showed that ploidy alone does not control diploid filamentation and supported a second model in which an additional α-specific factor(s) on the *MAT* chromosome acts to induce filamentation. The ploidy result suggested that a second *MATα* region resided on the same chromosome or that the locus was larger than previously thought. Two approaches were taken to explore these possibilities.

In the first approach, Hull et al. hypothesized that a second *MATα* locus might contain homeodomain DNA-binding proteins. Basidiomycete *MAT* loci generally consist of two separate regions, one that contains pheromones and pheromone receptors, and one that contains homeodomain proteins (20). Because the previously defined locus contained pheromones and a pheromone receptor (91, 129, 154), Hull et al. hypothesized that a second locus could contain DNA-binding proteins. Using a bioinformatics approach, they identified putative homeodomain protein sequences from the *C. neoformans* Genome Project at Stanford University. One of the identified sequences hybridized in a Southern blot to DNA from only α cells and was located on the *MAT*-containing chromosome. Deletion of the gene (*SXI1α* for Sex Inducer 1α) resulted in a severe reduction in diploid filamentation and mating. In addition a double deletion strain in which the 50-kb *MATα* region and the *SXI1α* gene were both deleted (**a**/Δ *sxi1α*Δ) was no longer capable of diploid filamentation and mated like an **a** cell. This result indicates that Sxi1α is a key α cell-identity factor required for specifying cell fate, analogous to the cell-fate determinants encoded by the *MAT* loci of other fungi (e.g., **a**1 and α2 of *S. cerevisiae*) (90).

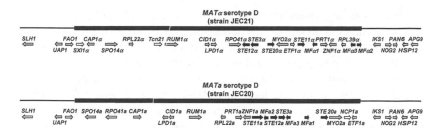

Figure 3   The **a** and α alleles of the mating-type (*MAT*) locus. The *MAT*α and *MAT***a** mating-type alleles and the adjacent genomic regions are depicted for serotype D. The mating-type specific regions are shown as bold lines and flanking regions as thinner lines. Sequences were analyzed using the BLASTX algorithm, and identified genes are shown as arrows in the direction of transcription. Genes encoding pheromone response pathway elements are shown as black arrows, and the remaining genes in the fragment are shown as gray arrows with the exception of the mating-type–specific genes *SXI1*α, *Tcn21*, and *NCP1***a**, which are represented by open arrows.

In the second approach to understand the nature of the *MAT* locus, Lengeler et al. set out to elucidate the nucleotide sequence of the locus (110). Genomic BAC libraries were created for both the **a** and α mating types from the A and D serotypes (strains JEC20 and JEC21, H99 and 125.91). BAC clones containing *MAT* sequences were identified by hybridization, and overlapping clones were isolated and mapped in conjunction with BAC end sequences from Schein and colleagues (152). The BACs were sequenced and assembled to generate a *MAT* region for each strain (Figure 3). The sequences were aligned so that sequences identical between mating types flanked sequences that were different between mating types, thus defining a region of nonhomologous DNA sequence. The complete serotype D *MAT*α locus was found to contain the original 50-kb locus defined by Karos et al. as well as another 55 kb of *MAT* sequence. The sizes of all of the *MAT* alleles are in the range of 105 to 130 kb. The additional sequence contains apparent homologs for a mitochondrial RNA polymerase (*RPO41*α), a dihydrolipoamide dehydrogenase (*LPD1*α), a caffeine-induced death protein (*CID1*α), an Rb binding protein 2 homolog similar to a regulatory protein in *U. maydis* that controls the expression of some sexual cycle genes (*RUM1*α), a mariner-type transposon (*Tcn21*), a ribosomal protein (*RPL22*α), a phospholipase D homolog (*SPO14*α), a capsule-associated protein (*CAP1*α), and the cell identity factor *SXI1*α. Also identified within the original 50-kb locus was another ribosomal protein homolog (*RPL39*α) and a putative zinc finger protein (*ZNF1*α).

For almost every gene, there are similar, but not identical alleles in both *MAT*α and *MAT***a**. Exceptions include the α-specific gene *SXI1*α, the α-only transposon *Tcn21*, and the **a**-specific gene *NCP1***a**. This is the only **a**-specific gene, and it shows sequence similarity to a single gene, an open reading frame identified by the *N. crassa* sequencing project that shows no sequence similarity to any protein motifs and has no known function. Although *MATa* is clearly allelic to *MATα*, the organization of *MATa* reveals extensive rearrangement of the genes compared to *MATα* and may explain the absence of recombination in this *MAT* locus. Overall, minor differences occur between the two serotypes, but the final conclusion is that the *C. neoformans MAT* locus is >100-kb in length and encodes approximately 20 genes of diverse function (Figure 3). This substantial effort has resulted in a molecular definition of *MAT* in *C. neoformans*.

The *MAT* locus in *C. neoformans* is significantly larger than any other known fungal *MAT* locus, and its architecture differs from that of its basidiomycete relatives. Instead of containing two distinct loci, *C. neoformans* has one large locus encoding all of the traditional *MAT* components (pheromones, pheromone receptor, and DNA binding regulator) as well as a number of genes never seen in a *MAT* locus before. A few examples of *MAT* loci containing unusual genes exist (84, 175), but they are rare and involve only a few genes. *C. neoformans* is apparently the founding member of a new class of *MAT* loci in which genes not traditionally present in the *MAT* locus (such as signal transduction components) have mating type–specific alleles. A region of the genome of *Pneumocystis carinii* contains apparent signal transduction components as well, suggesting a strong similarity with the *C. neoformans MAT* locus (156). Perhaps these loci represent evolutionary intermediates between *MAT* loci in fungi and the sex chromosomes of multicellular eukaryotes. Because the regulation of the sexual cycle in *S. cerevisiae* has been studied extensively, and the molecular details of how cell fate is established and how sexual cycle progression occurs have been mapped in detail, we can look to this system for clues as to how the process might work in *C. neoformans*. In *S. cerevisiae*, **a** cells produce the homeodomain protein **a**1, and α cells produce the homeodomain protein α2. When **a** and α cells mate to form a diploid **a**/α cell, **a**1 and α2 interact to form a novel transcriptional regulatory activity that represses the expression of haploid-specific genes, thus establishing the diploid state (Figure 4) [reviewed in (90)]. The Sxi1α protein might interact with an **a**-specific partner protein to create a novel transcriptional regulatory complex in *C. neoformans* that would signify the dikaryotic state and promote progression through sporulation (Figure 4). This heteromeric regulatory complex model has been characterized in other basidiomycetes, including *U. maydis* where heterodimers composed of different forms of the bE and bW homeodomain proteins interact to induce sexual cycle progression [reviewed in (20)]. Although such a scheme for *C. neoformans* is appealing, no obvious partner proteins are apparent in the *MATa* allele. The single **a**-specific gene, *NCP1a*, shows no sequence similarity to any known protein motif, making it a poor candidate for a DNA-binding protein.

An alternative model is one in which one of the **a** alleles in *MATa* has a disparate function from its partner allele in *MATα*. For example, the *ZNF1α/a* genes may encode proteins (predicted zinc-finger DNA-binding proteins) that interact with Sxi1α with different affinities to create complexes with different regulatory properties. Such complexes would be similar to the regulatory complex in *S. cerevisiae*

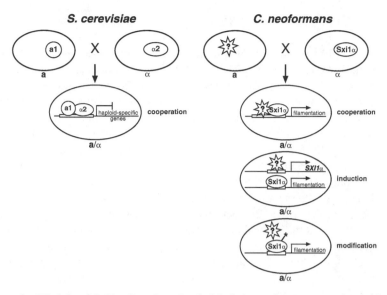

Figure 4 Models of Sxi1α function. Ovals labeled **a** and α represent haploid cells. Larger ovals labeled **a**/α represent diploid cells formed after mating. Objects within the ovals represent proteins. *Left panel*: In *S. cerevisiae*, the homeodomain proteins **a**1 and α2 are produced in **a** and α cells, respectively. Cell fusion produces an **a**/α diploid cell, and these proteins interact in a cooperative manner to create a novel transcriptional regulator. Through the repression of haploid-specific genes, this site-specific DNA binding activity specifies the diploid state. *Right panel*: Three models are proposed for how Sxi1α controls cell fate in *C. neoformans*. (i) In the Cooperation Model a factor from **a** cells (*star*) and a factor from α cells (Sxi1α) interact to form a new activity that controls expression of genes important for development, similar to that seen in *S. cerevisiae* and other fungi. (ii) In the Induction Model *SXI1α* is controlled at the transcriptional level by an **a**-specific regulator that induces the expression (or stabilizes the transcript) of *SXI1α* upon cell fusion. Once Sxi1α is produced, it acts as a transcriptional regulator to control sexual development genes. (iii) In the Modification Model an **a**-specific modifier protein acts directly on Sxi1α, modulates its activity, and leads to altered gene expression.

composed of the homeodomain protein Pho2 (a.k.a. Grf10) and the zinc-finger protein Swi5 (14). In α cells, a Sxi1α-Znf1α complex could confer α cell identity, while after cell fusion, a Sxi1α-Znf1**a** complex could form and specify the dikaryotic state. Alternatively, an **a**-specific allele could alter *SXI1α* transcript levels by inducing expression in the dikaryon, by regulating splicing of the pre-mRNA product, or by stabilizing the transcript (Figure 4). The regulation could also be a posttranslational event in which an **a**-specific kinase phosphorylates and activates the Sxi1α protein to allow it to function as a regulator of the dikaryotic state.

MOLECULAR APPROACHES

The application of Koch's postulates revolutionized the study of microbial pathogenesis by providing a sound scientific basis to link pathogenic organisms to the diseases they cause. The four postulates are (a) all individuals with the disease harbor the presumptive causative agent; (b) the organism can be isolated from the infected host and propagated in the laboratory; (c) reintroduction of the organism results in the original disease in susceptible individuals; and (d) the organism can be reisolated from experimentally or accidentally infected individuals. Falkow and colleagues modified these postulates to formulate a molecular version of Koch's postulates: (a) The property studied should be associated with pathogenicity or infectivity; (b) mutation of genes involved in this property should lead to a loss of virulence; and (c) reintroduction of the gene should restore virulence of the organism. Thus, fulfilling Falkow's molecular postulates requires the identification of genes involved in virulence attributes, specific approaches to mutate these genes such as transformation and homologous recombination, methods to reintroduce the wild-type gene, and animal models to assess the impact of these molecular manipulations on virulence.

Recent genetic and molecular advances have allowed Falkow's molecular postulates to be fulfilled for a variety of different phenotypic traits and signaling cascades in *C. neoformans*. Early studies linking the polysaccharide capsule and production of the pigment melanin to virulence of *C. neoformans* utilized mutants isolated following chemical mutagenesis. Acapsular and albino strains were avirulent in animal models, and these mutant phenotypes cosegregated in genetic crosses, providing evidence that both properties are linked to virulence (87, 103).

Gene Disruption

In the early 1990s, two different transformation systems were developed for *C. neoformans*. In the first, DNA is introduced by electroporation (54). In the second, DNA is delivered by biolistic delivery of gold microprojectiles decorated with DNA (170). This second approach was developed to circumvent difficulties in delivering exogenous DNA through the polysaccharide capsule. Both transformation systems relied on the use of genes involved in nucleotide biosynthesis, *URA5* or *ADE2*, as selectable markers (54, 164). Recipient strains for transformation were obtained either by the selection of *ura5* auxotrophic strains on medium containing 5-fluoro-orotic acid (which is converted to a toxin by the Ura5 enzyme) (104), or by the isolation of *ade2* mutants (which form pink or red colonies) following mutagenesis with γ-rays or UV light (143). DNA introduced into the cell by electroporation was found to frequently undergo the spontaneous addition of telomeres and to replicate in an extrachromosomal linear fashion (53). DNA delivered by biolistics was more frequently found to integrate into the chromosomal DNA, either ectopically by nonhomologous recombination or sequence specifically via homologous recombination. These studies set the stage for an analysis of gene function by transformation and gene disruption by homologous integration.

An initial set of gene disruptions was conducted by biolistic transformation in the serotype A *ade2* recipient strains M001 and M049 (derived from strain H99) and by electroporation in *ura5* serotype D recipient strains of the JEC21/B3501 strain series (4, 21, 116, 136, 151). Two general principles emerged. First, biolistic transformation yields higher rates of integration and homologous recombination than electroporation in both serotype A and D strains (45). Second, it is essential that gene disruption alleles be constructed with DNA sequences that are isogenic to the recipient strain, because DNA polymorphisms result in abortion of homologous recombination events by the DNA mismatch repair system (45). In several recent studies, the frequency of homologous recombination that can be achieved with ~1000 bp of flanking sequence on each side of the selectable marker is in the range of 2% to 25%, with some exceptional cases at higher efficiency (42, 43, 154, 182, 183, 198). Most recent studies employ either the *URA5* marker and a spontaneous 5-FOA-resistant *ura5* mutant of the serotype A strain H99, or a dominant selectable marker that confers resistance to the aminoglycoside nourseothricin (123). The wild-type gene can be reintroduced ectopically by cotransformation with a linked selectable marker, either *URA5* or the hygromycin resistance gene (36, 82). Alternatively, the wild-type genetic locus can be reconstituted by starting with the *xyz1::URA5* disruption strain, transforming with the wild-type gene, and selecting for loss of the integrated *URA5* marker on 5-FOA medium to result in a wild-type *ura5* auxotrophic strain in which the wild-type gene is reconstituted at the endogenous locus (182). For virulence studies, the *URA5* gene is then reintroduced ectopically in a third transformation event. Although the resulting strains have been subjected to three independent transformation events, which some have argued might be deleterious if transformation itself is mutagenic, in several recent examples this approach has resulted in the full reconstitution of virulence to the wild-type level (182).

Essential Genes

Three genes have been shown to be essential in *C. neoformans*. In the first example, the N-myristoyl transferase gene, *NMT1*, a temperature-sensitive allele was isolated based on a known, temperature-sensitive *nmt1* mutation in *S. cerevisiae* (116). The corresponding allele of the *C. neoformans NMT1* gene was found to be temperature sensitive in an *S. cerevisiae nmt1* mutant host. Next, the *nmt1-ts* allele was integrated into the *C. neoformans* genome linked to the *ADE2* selectable marker, and transformants in which the temperature-sensitive allele had replaced the wild-type locus were identified as temperature-sensitive transformants and confirmed by molecular analysis (116).

In the second example, the *TOP1* gene encoding topoisomerase I was shown to be essential in *C. neoformans* (47). In this case, no disruption mutants were identified from a screen of 8000 transformants obtained with a *top1::ADE2* deletion allele, suggesting the gene was essential. When a second copy of the *TOP1* gene was introduced into the genome ectopically, the endogenous *TOP1* gene was readily disrupted by the same *top1::ADE2* disruption allele at 7% efficiency, demonstrating that the *TOP1* gene is essential (47).

In the third example, the *FKS1* gene encoding the unique β-1,3 glucan synthase gene was shown to be essential using a novel targeting strategy (169). In this case, the *ADE2* selectable marker was cloned between the 5′ flanking region of the *FKS1* gene and a 5′ truncated portion of the gene. Integration of the circular form of this disruption allele can occur in two ways. Integration into the upstream region on one side of the *ADE2* marker leaves the *FKS1* gene intact, whereas integration into the internal portion of the *FKS1* gene disrupts the gene. If the *FKS1* gene were nonessential, both types of integration events would be observed. In this case, homologous integration occurred at a frequency of 7.9% and 0/26 transformants analyzed contained a targeting event into the *FKS1* locus, indicating that the gene is essential. This finding is of particular interest given that the Fks1 glucan synthase is the target of the antifungal drug caspofungin, which exhibits poor activity against *C. neoformans* even though the *FKS1* gene is essential. Other factors involved in the action of this novel antifungal agent likely remain to be defined.

Given that only three genes have been documented to be essential in *C. neoformans* and that three different methods were used, more facile approaches would clearly be welcome. Stable congenic **a**/α diploid strains of *C. neoformans* were recently described (155). These strains grow as a budding yeast at 37°C, yet when grown at 24°C spontaneously filament, form basidia, and undergo meiosis and sporulation. These strains can be used to study an essential region of the *MATα* allele of the mating-type locus (83). Deletion of an ∼50-kb region of the *MATα* allele resulted in a *MATa*/Δ diploid strain that sporulates to produce only *MATa* segregants that are all *ura5* and lack the *URA5* marker integrated at the site of deletion in the *MAT* locus. These studies reveal that one or more of the 12 genes contained in this interval is essential. Thus, these congenic diploid strains can be used to demonstrate that genes are essential, analogous to tetrad dissection in the model ascomycetous yeast *S. cerevisiae*.

Two recent reports suggest alternative approaches might also be applicable to study gene functions and essential proteins in *C. neoformans*. First, expression of two different antisense mRNAs was found to inhibit function of the *ADE2* gene or the gene encoding the calcineurin A catalytic subunit Cna1, which is required for both growth at 37°C and virulence (73). In a related study, expression of dsRNA was found to repress expression of the corresponding gene (114). In this case, the actin gene promoter was utilized to drive expression of a tandem gene in an inverted repeat orientation, to result in expression of a hairpin double-stranded RNA molecule. Hairpin structures were sufficient to repress expression of the *ADE2* gene, resulting in pink/red auxotrophic colonies. In a second example, dsRNA repression of the *CAP59* gene resulted in acapsular cells. Both approaches might find utility in further studies, but have thus far been applied to only three genes (*ADE2*, *CNA1*, *CAP59*) whose mutant phenotypes have been extensively characterized by conventional mutational or gene disruption approaches. In addition, these novel approaches do not provide stable mutants for further genetic or phenotypic studies, such as virulence tests.

Plasmids, Multicopy Libraries, and the Two-Hybrid System

Several different types of plasmids are available for molecular studies in *C. neoformans*, including both integrating and non-integrating vectors (Figure 5). First, a series of plasmids containing the *URA5* or HYG^r selectable markers are available for integration of foreign DNA into the *C. neoformans* genome (36, 53, 54, 82). These plasmids have frequently been used to introduce dominant alleles for epistasis analysis, or to reintroduce wild-type genes into mutant strains to ensure that phenotypes are complemented and satisfy the third of Falkow's molecular postulates of virulence.

A second series of plasmids for *C. neoformans* are non-integrating plasmids that can be shuttled between *E. coli* and fungal cells. Two commonly used non-integrating plasmids are pCnTel1 and pPM8 (Figure 5) (53, 127). Both plasmids are maintained as circular plasmids in *E. coli* and contain both an origin of replication derived from pBR322 and an ampicillin resistance marker for selection in bacteria. These plasmids also contain two other special features. First, both contain the *C. neoformans URA5* gene as the marker for selection in *C. neoformans ura5* mutant cells. Importantly, the *URA5* gene can also be counterselected on medium containing 5-FOA, allowing tests of plasmid dependence to be readily performed (104).

A second important feature of both of these plasmids is that two telomeric sequences are present in an inverted orientation flanking the kanamycin resistance gene for selection in bacteria. Circular plasmid DNA from *E. coli* can be cleaved with a rare cutting restriction enzyme (Meganuclease I-SceI), which releases the Kan resistance gene cassette and reveals telomeric sequences that are converted into functional telomeres when introduced into *C. neoformans* (Figure 5). The plasmids are thus maintained as circular DNA molecules in bacteria and as linear, telomeric plasmids in fungal cells. The copy number of these plasmids has been determined to be ∼5 to 10 copies per cell, and studies are in progress to determine how these plasmids replicate in fungal cells (C. Newlon, personal communication). Both plasmids can be rescued from fungal cells by isolating DNA, cleaving with NotI (which removes the telomeric repeats), circularizing under dilute ligation conditions, and rescuing in *E. coli* by transformation and selection for ampicillin resistance (Figure 5). These approaches allow the introduction of genes into fungal cells, the identification of genes by complementation approaches, and the rescue of these genes in bacterial cells for molecular analysis.

The pPM8 plasmid has an additional feature that makes it exceptionally well suited for certain cloning procedures and the construction of genomic libraries (127). The pCnTel1 plasmid is somewhat unstable in *E. coli*, possibly due to the inverted telomeric repeats, and it has been difficult to construct stable genomic

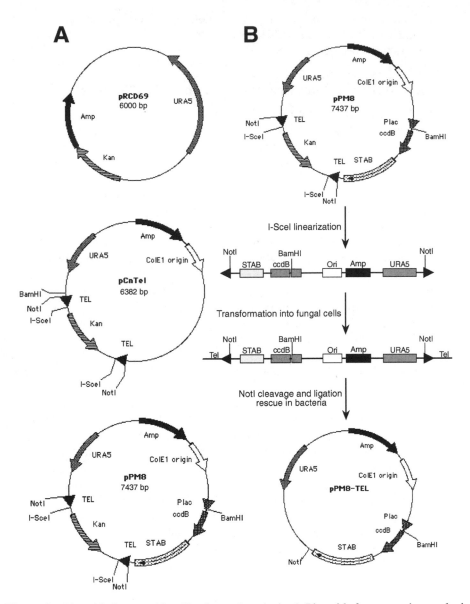

Figure 5 Plasmids for gene identification and analysis. *A.* Plasmids for expression analysis in *C. neoformans.* pRCD69 (*top*) is an integrating vector. pCnTel (*middle*) is a telomeric vector that can be linearized prior to transformation and maintained as a stable linear fragment. pPM8 (*bottom*) is a telomeric plasmid derived from pCnTel that is optimized for use with genomic library fragments. All contain the *URA5* gene as a selectable marker. *B.* Schematic shows how pPM8 vector can be used to transform *C. neoformans* and then be recovered in *E. coli,* facilitating library screens.

libraries in this plasmid. To circumvent this issue, Mondon et al. constructed the pCnTel1 derivative pPM8 that allows plasmids containing inserts to be positively selected in bacterial host cells (127). The *ccdB* gene encoding a gyrase inhibitor was cloned under the control of the lac promoter in the pCnTel1 plasmid to create pPM8. Introduction of this plasmid into *E. coli* laci^q cells is tolerated because *ccdB* expression is repressed, whereas addition of IPTG induces *ccdB* and kills the cells. Introduction of DNA inserts into the BamHI cloning site in the middle of the *ccdB* gene inactivates the toxin, allowing the positive selection of clones bearing inserts on medium containing IPTG.

Using this plasmid system, we have constructed a *C. neoformans* serotype D genomic library using the congenic α strain JEC21. Sau3A partial fragments in the size range of 6 to 11 kb were cloned in the vector. The library contains >90% inserts and represents ∼6.7X coverage of the genome. This library has been successfully used in two independent screens to identify multicopy suppressors of two different mutants. In the first, a gene that restores growth at elevated temperature in a calcineurin B (*cnb1*) mutant strain was identified as a candidate component of the calcineurin signaling pathway required for growth at elevated temperature and virulence (39; D. Fox & J. Heitman, in preparation). In the second screen, clones that restore melanin production in a mutant strain lacking the Gpa1 Gα protein involved in nutrient sensing were identified and may include novel components involved in melanin biosynthesis or its regulation (B. Allen & J.A. Alspaugh, personal communication).

The pPM8 plasmid also contains an additional unique sequence element known as the STAB sequence. The STAB region was thought to have been originally derived from a *C. neoformans* minichromosome and to confer increased stability to episomal plasmids (176). More recent studies have revealed that the STAB sequence is actually a region of the *E. coli* chromosome that has no effect on plasmid stability in *C. neoformans* (C. Newlon, personal communication).

A two-hybrid library has been constructed from cDNA grown under a variety of conditions and used to identify the novel calcineurin binding protein Cbp1, which is homologous to the calcineurin inhibitor Rcn1 in *S. cerevisiae* and DSCR1/calcipressin in humans (72).

GENOME SEQUENCING PROJECTS AND GENETIC AND PHYSICAL MAPS

One of the most exciting recent developments in microbial genetics and pathogenesis is the application of genomic technologies and their integration with genetic and molecular biology approaches. In this regard, *C. neoformans* is no exception. Starting in 1999, an international consortium was formed to focus on the *C. neoformans* genome sequencing project whose goal is to provide the complete and annotated genome sequences, including bioinformatics tools and reagents, for at least two and possibly as many as four strains representing the major serotypes/varieties/species for this organism (80). Several overlapping approaches will complement the direct sequencing. First, physical and genetic maps will ensure that the genome sequences faithfully represent the structure and architecture of the genome and provide tools to analyze the genome experimentally. Second, extensive EST analysis will be used to identify genes and assist in the annotation of this intron-rich genome. Finally, the genome information will be used to develop genome arrays for expression profiling and genotyping. The genome of this organism spans some 18 to 24 Mb and is divided into ~12 chromosomes. Thus, determining the genomic sequences of three related but diverged strains will require the acquisition of ~75 Mb of genomic sequence information.

The *C. neoformans* genome project has made considerable progress since its inception. BAC libraries were constructed for representative serotype A (H99/125.91), serotype D (JEC20/JEC21), and serotype B (WM276/E566) strains of both mating types. The serotype A, D, and B BAC libraries have been fingerprinted and end-sequenced at the Vancouver Genome Centre by Kronstad and colleagues, providing BAC maps for the genomes, which are serving as scaffolds for the genome wide assembly (152) (VGC website: http://rcweb.bcgsc.bc.ca/cgi-bin/cryptococcus/cn.pl). Shotgun sequencing for the serotype D strains JEC21 and the ancestral precursor strain B3501 have reached ~6X coverage for each strain at Stanford University and The Institute for Genome Research, and chromosome size assemblies have been achieved in large part based on paired plasmid reads and BAC end reads to scaffold the assembly (Stanford website: http://www-sequence.stanford.edu/group/C.neoformans/index.html, TIGR website: http://www.tigr.org/tdb/e2k1/cna1/index.shtml). It is anticipated that closure and annotation of the JEC21 reference serotype D genome will be achieved by the end of 2003. A comparative genomics project to determine the genome sequence for the serotype A strain H99 has begun at the Duke Center for Genome Technology (Duke CGT website: http://cneo.genetics.duke.edu). A BAC library has been fingerprinted and end-sequenced to serve as a scaffold for the sequencing project (152). ~61,000 sequence traces or 1.5X coverage has been determined by shotgun sequencing of paired plasmid reads. Initial comparisons reveal that global genome architecture may be conserved between the divergent serotype A and D strains (F. Dietrich, personal communication). Finally, studies on a representative serotype B variety *gattii* strain, WM276, have begun to extend the comparative aspects of the genome project to three divergent varieties or sibling species of this human fungal pathogen.

A challenging aspect of annotating the *C. neoformans* genome is the presence of multiple small introns in virtually every gene. This challenge is being met by two complementary approaches. First, a large-scale EST sequencing project at Oklahoma Health Science Center has provided cDNA sequence information for several thousand genes from both serotypes A and D, and the information revealed about intron-exon borders is being used to train gene finding software for annotation (OHSC website: http://www.genome.ou.edu/cneo.html). Second, comparative genomics approaches between serotype A and D genomic sequences provide a novel means to identify the most highly conserved regions of the genome, which include exon sequences (but not intron sequences). Thus, the determination of two different but related genome sequences has the power to make the analysis of the genome much more tractable than analysis of a single reference genome.

In addition to these physical and sequence-based maps of the genome, a meiotic-based recombination map of the genome has recently been generated (60). This map was based on an analysis of meiotic segregants from the serotype D *MATα* strain B3501 and the *MATa* strain B3502. 280 meiotic progeny were analyzed by AFLP analysis of polymorphic markers, and a map consisting of 14 major linkage groups was established. This map is currently being extended and refined using microsatellite loci. These studies provide a starting point for the definition of higher-resolution maps to be used as tools in the analysis of quantitative trait loci and their contribution to physiology and virulence of the organism. These types of analysis will also provide mapping information and resources that can be used to identify centromeres using an array of microsatellite markers and basidiospores analyzed from individual basidia in which a single meiotic event gives rise to parental ditype, nonparental ditype, or tetratype segregation patterns (98).

ANIMAL MODELS OF VIRULENCE

One of the most attractive features of *C. neoformans* as an experimental system for studies of pathogenesis is that several robust animal model systems have been developed and widely implemented. These include models in several different animal species and which probe unique aspects of the infectious cycle. As described above, *C. neoformans* most commonly occurs in immunocompromised hosts but can also be a primary pathogen in individuals with no apparent immune system dysfunction. Second, *C. neoformans* is acquired from the environment by inhalation of spores or desiccated yeast cells, and thus initially infects the lung (Figure 1). While virtually any organ can be infected by *C. neoformans*, the most common clinical manifestation is hematogenous dissemination to the central nervous system where meningoencephalitis ensues. An important aspect is that latent infections can be established in the lung, which can then lead to dissemination years or decades later. The animal models in current use seek to emulate these features of the infectious cycle.

The animal models most commonly used are the murine tail vein injection model, the murine inhalation model, the rat inhalation model, and the rabbit intracerebral model. Each has strengths, and together these models provide a robust platform to examine virtually every aspect of *C. neoformans* pathogenesis. The murine models have distinct advantages in terms of the size and cost of animals and the large number of inbred and genetically altered lines that are available. The murine inhalation model recapitulates virtually all aspects of the natural course of infection in humans. Animals are anesthetized and infected with drops of fungal cells deposited on the nares, which are inhaled (35, 42, 182). The infection begins in the lung and spreads to the CNS. Virtually all infected animals die from cryptococcal meningitis within 30 days of infection with the serotype A strain H99 delivered at an infecting inoculum of 5×10^4 cells. A/Jcr mice are commonly used in this model. These animals are immunocompetent but can be readily infected with pathogenic serotype A strains. For the congenic serotype D lab strains, which are significantly attenuated compared to serotype D or A clinical isolates, a tail vein injection model has been commonly employed. These studies are generally carried out using DBA mice lacking the C5 component of complement, although studies using BALB/c mice have also been reported (21, 23, 40, 154, 182). These two models allow study of the lung-blood-CNS virulence cycle following infection by inhalation, or to bypass the lung and study only the blood-CNS cycle following tail vein injection. Models employing direct intrathecal inoculation in either rabbits or mice allow one to bypass both the lung and the bloodstream to analyze only the CNS part of the infectious cycle.

The rabbit model has advantages in that the animals are immunosuppressed with steroids (mimicking a common human host state), and that larger quantities of CSF and fungal cells can be obtained for analysis (142). Disadvantages of the rabbit model are the large size and high cost of animals, that expertise is required to deliver the infecting inoculum and recover CSF fluid, and that the normal body temperature of the rabbit is $39°C$ (higher than normal human body temperature). Nevertheless, the rabbit model has been enormously useful for studies of antifungal drug action and mutants and will continue to play a valuable role in allowing access to large numbers of fungal cells and RNA for gene expression profiling.

Finally, the rat inhalation and intratracheal models provide a valuable animal model system in which to study latency and dissemination of infection (68, 69, 71). In contrast to the murine and rabbit models, in which infection progresses rapidly to symptomatic disease, infection in the rat leads to a latent infection that can persist for as long as 18 months. This provides a window of opportunity to study the latent phase of the infection in the lung and to analyze molecular determinants that could contribute to understanding how this important aspect of the natural history of infection might be controlled in the human host. In practice, for new investigators in this area, establishing the murine inhalation and tail vein injection models would provide a robust and flexible platform to conduct a variety of virulence studies.

ROLE OF MAPK AND cAMP SIGNALING IN CELL DIFFERENTIATION

Two conserved signaling pathways, the pheromone-activated MAP kinase pathway and a nutrient-sensing cAMP-protein kinase A pathway, control differentiation of *C. neoformans* (Figure 6). Of particular importance is that several components of the MAP kinase pathway (Ste20, Ste11, and Ste12) are encoded by the mating-type locus, and thus α and **a** cells express distinct alleles of these signaling elements. While the role of the MAP kinase pathway in controlling mating and haploid fruiting has been definitively established, the role of the MAP kinase pathway in

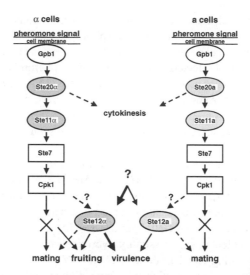

Figure 6 MAP kinase signaling cascades. Some components of the signaling cascades leading to mating, fruiting, and virulence in *C. neoformans* are mating-type specific. In α cells, pheromone activates a MAP kinase cascade through a G-protein–coupled receptor and a heterotrimeric G-protein. The Gβ subunit Gpb1 transduces the signal to the mating-type–specific PAK kinase Ste20α, which relays the signal to the MAP kinase cassette consisting of the mating-type–specific MEK, Ste11α and the nonmating-type–specific kinases Ste7 and Cpk1. Signals from this cassette likely impinge on the α-specific transcription factor Ste12α to control fruiting and virulence and an as-yet unknown factor to control mating. This signaling pathway is mirrored in **a** cells utilizing the cell type-specific factors Ste20**a**, Ste11**a**, and Ste12**a**.

controlling virulence remains less clear. Instead, the protein kinase A pathway plays a central role in controlling virulence, and mutations that disable signaling in this pathway attenuate virulence, whereas those that activate signaling enhance virulence.

A MAP kinase signaling cascade has been defined that functions in sensing pheromones during mating [reviewed in (109, 181)]. Three genes encoding the MFα pheromone have been identified in *MATα* (46, 129, 154). These genes encode a precursor peptide that is modified by proteolysis and C-terminal farnesylation at a CAAX motif. The MFα1 pheromone has been produced synthetically, and induced conjugation tube formation in **a** cells. By both reporter gene assays and northern blots, the *MFα1, 2*, and *3* genes are induced in response to nutritional limitation and co-culture with **a** cells. Deletion of the three genes results in an *mfα1,2,3* pheromoneless triple mutant strain that is mating impaired but still capable of producing mating filaments and recombinant basidiospores at about 1% the efficiency of wild-type cells. This is in stark contrast to *S. cerevisiae*, in which mating pheromones are essential for mating, but similar to other basidiomycetes in which cell-cell fusion events are more promiscuous. The MFα pheromone is required for α cells to induce the morphological changes associated with mating in confronting **a** cells. The *MFα1* pheromone gene is induced in vivo in the central nervous system of infected animals (48), but the pheromoneless triple mutant strain exhibits only a very modest defect in virulence (154). Three related genes encoding the MF**a** pheromone produced by **a** cells have also recently been identified (121). The three genes are encoded by *MAT***a** and encode lipid modified peptide pheromones that share some limited amino acid sequence similarity with the MFα pheromones.

Two G-protein–coupled receptor homologs that likely function in pheromone sensing are encoded by the *MATα* and *MAT***a** alleles of the mating-type locus, and both share sequence identity with the *S. cerevisiae* **a**-factor receptor Ste3 (27a, 110). These receptors are coupled to a heterotrimeric G-protein whose Gβ subunit is encoded by the *GPB1* gene (183). Mutants lacking Gpb1 are sterile, and this mating defect is suppressed by overexpression of the MAP kinase homolog Cpk1, implicating Gpb1 in control of the MAP kinase cascade. Overexpression of Gpb1 promotes conjugation tube formation in both α and **a** cells, supporting a model in which the $\beta\gamma$ subunit plays an active signaling role analogous to the $\beta\gamma$ heterodimer Ste4-Ste18 in *S. cerevisiae*. The α subunit coupled to Gpb1 has not been definitively identified but may be encoded by the *GPA3* gene. The γ subunit remains to be identified.

Recent studies have implicated a second GTP binding protein, the Ras1 protein, in cellular responses to pheromone (2, 185). *ras1* mutant cells are sterile, defective in responses to pheromone in confrontation assays, and fail to express mating pheromones or pheromone-activated genes in co-culture. Overexpression of Gpb1, or of the MFα1 pheromone, restores mating of *ras1* mutant cells. Finally, when the MFα1 mating pheromone is expressed from a heterologous promoter in α *ras1* mutant cells, the ability to promote pheromone-induced morphology changes in confronting **a** cells is restored, but the α *ras1* mutant cells still fail to respond to MF**a** pheromone. These findings illustrate that Ras1 is required both for pheromone production and pheromone response and suggest that Ras1 functions early in the pheromone response pathway in *C. neoformans*. An analogous situation occurs in *S. pombe* in which the ras1 homolog functions with the Gα subunit gpa1 to promote activation of the MAP kinase cascade by mating pheromones (132). In *S. cerevisiae*, the Ras2 protein plays a dual signaling role and activates both cAMP production by adenylyl cyclase and MAP kinase signaling during filamentous and invasive growth (130). The functions of Ras appear to be more restricted to MAP kinase signaling in fission yeast and in the basidiomycete *C. neoformans*.

The genes encoding several kinases that function in pheromone sensing and mating have been identified, including *STE20α/**a***, *STE11α/**a***, *STE7*, and *CPK1* (28, 44, 182). Three genes encoding homologs of the PAK kinases have been identified and studied in detail, including the mating type–specific Ste20α and Ste20**a** kinases encoded by the *MAT* locus and the nonmating type–specific Pak1 kinase (182). Mutants lacking the Ste20 kinases exhibit a bilateral mating defect, and *ste20α* mutants fail to respond to MF**a** pheromone in confrontation assays. Mating is restored in *ste20* mutant strains by overexpression or mutational activation of the Ste11α or Cpk1 kinase, providing epistasis evidence that the PAK kinases function upstream in activating the MAP kinase signaling pathway.

Importantly, the genes encoding the principle components of the MAP kinase pathway, *STE11α/a*, *STE7*, and *CPK1*, are both mating-type specific and non-specific. Genetic evidence links these three kinases in a common pathway that senses pheromones and controls mating (44). Mutants lacking any of the three kinases exhibit a very similar severe sterile phenotype and fail to respond to mating pheromones in confrontation assays. The function of all three kinases is required for induction of the MFα mating pheromone genes in response to MF**a** pheromone secreted by **a** cells. Epistasis evidence supports a model in which components of the Ste11-Ste7-Cpk1 module function downstream of both the pheromone-activated heterotrimeric G-protein and the Ste20α PAK kinase. This pathway is thus composed of both mating-type–specific components (Ste20α/**a**, Ste11α/**a**) and nonmating-type–specific components (Ste7, Cpk1). The Ste7 and Cpk1 kinases therefore must have evolved to function with two different types of partner subunits depending on the cell type in which they are expressed. As a consequence of this unique specialization, α and **a** cells use two somewhat different signaling cascades to respond to pheromone during mating. This may in part give rise to some of the unique features exhibited by the two cell types during mating: namely, that α cells respond to **a** cells by haploid fruiting, while **a** cells respond to α cells by initially making short conjugation tubes and then dramatically enlarging, possibly to provide an easier target for cell fusion with an extending filament tip or basidiospore.

One candidate target of the MAP kinase pathway is Ste12, which is encoded by the mating-type locus and therefore exists as both Ste12α and Ste12**a** (25, 91, 189, 198). However, *ste12α* mutant cells exhibit only a modest defect in mating, respond normally to pheromones in cell-cell confrontation assays, and pheromone gene expression is inducible by co-culture with cells of the opposite mating type (26, 44, 198). Moreover, overexpression of Ste12α suppresses the mating defects of mutants lacking the Ste11α, Ste7, or Cpk1 kinases, indicating that simply increasing Ste12 levels in the cell can bypass the normal requirement for MAP kinase signaling (44). These findings could be accommodated by models in which Ste12 is one of two redundant downstream targets in which overexpression of Ste12 bypasses a requirement for MAPK action. Alternatively, Ste12 might be the target of a different signaling pathway, such as the PKA cascade (42). The *C. neoformans* Ste12 homologs contain an N-terminal homeodomain presumed to be involved in specific DNA binding, and a C-terminal zinc finger region of unknown function. The *C. neoformans* Ste12 homologs lack the peptide binding sites for the Dig1/2 repressor proteins found in *S. cerevisiae*, and thus are likely to be regulated in a distinct fashion from their ascomycete counterparts.

In *S. cerevisiae*, one component of the MAP kinase pathway has been functionally specialized into two divergent forms. The Fus3 MAP kinase promotes mating, whereas the Kss1 MAP kinase promotes pseudohyphal growth in the diploid (33, 120). While Kss1 can in part support mating in haploid yeast cells, Fus3 is not expressed in diploids and plays no role in diploid pseudohyphal growth. This theme is revisited in *C. neoformans* but now three different signaling components, Ste20 and Ste11 (kinases), and Ste12 (a transcription factor) are all expressed from the mating-type locus in α and **a** cell type–specific forms.

Studies on the pheromones and pheromone-activated G-protein reveal an autocrine signaling loop that promotes haploid fruiting of α cells in the absence of **a** cells (154). Mutants lacking the MFα mating pheromones or the G-protein β subunit, Gpb1, exhibit defects in haploid fruiting, providing evidence that an autocrine signaling loop involving MFα pheromone acting on α cells contributes to cellular differentiation (183). One analogy is to self-filamentous mutants of *S. commune* and *C. cinereus*. In these organisms, single amino acid changes allow a mutant pheromone to act on a normally nonresponsive receptor, or broaden the ligand-binding specificity of a pheromone receptor such that a cell responds to its own pheromone (61, 137). Similar mutations may have arisen in *C. neoformans* and could explain the specialized ability of α cells to differentiate in response to nutrient limitation. Haploid fruiting conditions result in the induction of *MFα* expression, which could then act as a partial agonist on the MF**a** pheromone receptor or another as-yet undefined receptor.

The MAP kinase cascade that plays a role in pheromone sensing during mating also functions during haploid fruiting in response to nitrogen limitation and desiccation. Mutants lacking the Ste20α, Ste11α, Ste7, or Cpk1 kinases all exhibit profound defects in haploid fruiting (28, 44, 182). Although the Ste12α kinase plays only a very minor role in mating, *ste12α* mutants exhibit a profound defect in haploid fruiting in response to nitrogen limitation and desiccation (26, 198). In epistasis experiments, the haploid fruiting defect of *ste12α* mutant cells is suppressed by overexpression of the MFα pheromone, a dominant activated Ste11α-1 kinase, or overexpression of Cpk1 (44). These findings do not support a model in which Ste12α functions downstream of the MAP kinase pathway in a strictly linear fashion, since activation of the upstream components bypasses the requirement for Ste12α. In addition *ste12α* mutant cells can haploid fruit in response to MF**a** pheromone produced by confronting **a** cells, whereas *ste20α*, *ste11α*, *ste7*, and *cpk1* mutants do not. These findings reveal that the MAP kinase cascade and Ste12α both control haploid fruiting, but again reveal that the functions of the upstream elements of the pathway are distinguished from the transcription factor that has been thought to be a possible target of the cascade. These findings underscore a molecular link between the signaling pathways that allow the cell to differentiate during haploid fruiting and mating. This may reflect a role for haploid fruiting in response to pheromone in the earliest steps of mating to assist cells in locating a mating partner. In turn, the ability to haploid fruit to a more limited extent in the absence of a mating partner may reflect evolution of an asexual pathway from a sexual one to promote survival under adverse conditions.

The finding that *MATα* is linked to virulence fueled early speculation that the pheromone-activated MAP kinase pathway might play a central role in virulence. Yet a different picture has emerged from recent studies of the pathogenic potential of mutant strains lacking defined elements of this signaling pathway. First, the Ste11α, Ste7, and Cpk1 kinases are completely dispensable for virulence of the serotype D strain JEC21, and a serotype A *cpk1* mutant strain is also fully virulent (44). Independent studies by Wickes and colleagues reveal that *ste11α* mutants are nearly fully virulent (28). These findings clearly indicate that the MAP kinase pathway does not play an important role in virulence.

Two other components have been linked to virulence but appear to play serotype-specific roles. When the Ste20α PAK kinase is mutated in the serotype A pathogenic isolate H99, virulence is attenuated in both rabbits and mice (182). In contrast, serotype D mutants lacking either Ste20α or Ste20a exhibit no defect in virulence compared to wild-type cells. The virulence defect of serotype A strains lacking the Ste20α kinase is likely attributable to several findings. First, these mutants exhibit a cytokinesis defect that becomes more severe at elevated temperature and results in a temperature-sensitive growth defect. Second, the *ste20α* mutants adopt a filamentous growth form in vivo in which elongated buds are produced and many of the mutant cells exhibit a defect in capsule production. In contrast, serotype D strains lacking Ste20α or Ste20a exhibit a cytokinesis defect but no temperature-sensitive growth defect, which may explain why they are dispensable for virulence in serotype D.

A similar scenario has emerged with respect to the Ste12 transcription factor homolog. In this case, serotype D mutants lacking Ste12α or Ste12a are attenuated for virulence, whereas serotype A *ste12α* mutants are fully virulent (25, 26, 198). *ste12α* mutant cells exhibit a modest defect in capsule production in vivo. These findings link a second component of the *MAT* locus to virulence and also underscore the divergence that has occurred between these two varieties or species.

During mating, the MAP kinase pathway functions in parallel with a nutrient-sensing Gα protein-cAMP-protein kinase A signaling cascade that also plays a central role in controlling virulence of this organism (4, 5, 42). The Gα protein Gpa1 was first identified by Courchesne and colleagues who hypothesized that it might play a role in mating, possibly as an element of the pheromone response pathway (171). Studies in *S. cerevisiae* on the homologous Gα protein Gpa2 revealed a distinct role in nutrient sensing during diploid pseudohyphal growth (117). Alspaugh et al. established that *C. neoformans gpa1* mutants are viable and exhibit a sterile phenotype in the serotype A pathogenic strain H99 (4). This initial finding could be explained by a role in nutrient sensing, pheromone sensing, or both, since both signals are required for mating. Alspaugh et al. went on to show that *gpa1* mutant cells exhibit defects in producing two inducible virulence factors: capsule in response to iron limitation and melanin in response to carbohydrate limitation. Both capsule and melanin had previously been linked to virulence in classic studies, and mutants lacking either capsule or melanin are avirulent or severely attenuated for virulence. Thus, the *gpa1* mutant cells have two strikes against them, and in two different animal models (the rabbit CNS model and the murine inhalation model), *gpa1* mutant strains are dramatically attenuated (4, 42).

The *S. cerevisiae* Gpa2 Gα protein has been implicated in activation of cAMP production and a protein kinase A signaling pathway that promotes filamentous growth (30, 117, 138, 139, 140, 148). Analogously, provision of exogenous cAMP restores mating, capsule production, and melanin production in *C. neoformans gpa1* mutant cells. Direct measurement of intracellular cAMP concentrations reveals that Gpa1 is required for cAMP production in response to glucose in *C. neoformans* (42). Mutants lacking the enzyme adenylyl cyclase (Cac1) exhibit phenotypes strikingly similar to mutants lacking the heterotrimeric Gα protein Gpa1 (5). The *CAC1* gene encoding adenylyl cyclase was identified by low-stringency PCR with primers targeted to conserved regions of homologous fungal enzymes, and the gene was disrupted by transformation and homologous recombination. The resulting *cac1* mutant cells are viable, sterile, and defective in producing melanin and capsule and are avirulent in the murine inhalation model (5). These cells lack any detectable cAMP, illustrating that adenylyl cyclase is not essential for viability in this pathogenic basidiomycete, in contrast to the essential function of the homologous enzyme and pathway in *S. cerevisiae*. Provision of exogenous cAMP restores mating, capsule, and melanin production in cells lacking adenylyl cyclase, indicating that the functions of this large peripheral membrane protein are catalytic and not structural.

The target of cAMP, the cAMP-dependent protein kinase PKA, plays a central role in controlling the virulence of *C. neoformans* (42). PKA is highly conserved and exists as a tetramer of two regulatory subunits bound to two catalytic subunits. In cells with low cAMP concentrations, the enzyme is in an inactive complex; however, when cAMP concentrations increase, the binding of cAMP to the regulatory subunits induces conformational changes that release and activate the catalytic subunits. D'Souza et al. identified and characterized the genes encoding the catalytic and regulatory subunits of protein kinase A from *C. neoformans* (42). Two genes encode the catalytic subunits, *PKA1* and *PKA2*, whereas the regulatory subunit is encoded by a single unique gene, *PKR1* (42; C. D'Souza, J. Hicks, G.M. Cox & J. Heitman, in preparation). In the pathogenic serotype A strain H99, *pka1* mutants exhibited phenotypes strikingly similar to *gpa1* and *cac1* mutant cells lacking the Gα protein Gpa1 or adenylyl cyclase (42). *pka1* mutant strains are sterile, fail to produce capsule or melanin in response to inducing conditions, and are avirulent in the murine inhalation model. *pka1* mutant strains generated in the congenic lab-adapted serotype D JEC20/JEC21 background exhibit no phenotype, are fertile, produce melanin and capsule, and are fully virulent (C. D'Souza, J. Hicks, G.M. Cox & J. Heitman, in preparation). Instead, serotype D *pka2* mutant cells exhibit phenotypes in common with serotype A *pka1* mutants and are sterile in bilateral crosses and fail to produce melanin or capsule in response to inducing conditions.

The two PKA catalytic subunits share limited sequence identity, and their functions have clearly diverged. This specialization of PKA catalytic subunitsunderscores the molecular divergence that has occurred in the ~20 million years that separate the serotype A var. *grubii* and the serotype D var. *neoformans* from a common ancestor (195) and is reminiscent of similar functional specialization that has occurred in *S. cerevisiae*, in which the three PKA catalytic subunits Tpk1, Tpk2, and Tpk3 play a redundant role in growth but unique roles to promote (Tpk2) or inhibit (Tpk1/3) filamentous growth (138, 148). Mutations altering the PKA regulatory subunit gene *PKR1* constitutively activate the PKA signaling cascade and result in increased capsule production and a hypervirulent phenotype in animal models. The *PKR1* gene encoding the protein kinase A regulatory subunit was identified by PCR with primer pools under low-stringency conditions (42). The gene is unique in *C. neoformans* and was disrupted by transformation and homologous recombination. In contrast to mutations in the catalytic subunit genes, *PKA1* and *PKA2*, which prevent signaling via the pathway, mutations in the regulatory subunit have two different physiological outcomes. First, *pkr1* mutations release the Pka1 and Pka2 catalytic subunits in an active form, and second, the pathway is severed from upstream regulatory signals that control cAMP levels.

pkr1 mutant strains lacking the Pkr1 regulatory subunit are viable and produce enlarged capsules under normally inducing conditions, similar to wild-type cells exposed to exogenous cAMP (42). These observations provide further evidence that the PKA pathway controls capsule production, that signaling via the pathway is limiting for capsule production and can be increased, and that signaling via the cAMP-PKA pathway functions independently of the sensor for the iron-limitation signal that induces capsule production. The consequences of increasing capsule production on virulence were examined in animal models. In both the murine inhalation and tail vein injection models, *pkr1* mutant strains were hypervirulent in comparison with the congenic wild-type strain (42). The fungal burden in the central nervous systems of infected animals was increased four- to sixfold with the *pkr1* mutant compared to the wild-type strain. Importantly, when the fungal cells were examined directly in the CNS of infected animals, the *pkr1* mutant cells produced dramatically enlarged capsules whose volumes were increased 12-fold compared to those produced in response to host signals by the wild-type strain. The capsule of wild-type and *pkr1* mutant cells in the inoculum (grown in vitro in YPD-rich medium) did not differ, indicating that this dramatic induction is in response to host signals following infection. These findings suggest that the PKA pathway modulates the ability of another sensing mechanism to effect changes in virulence factor production. The finding that *pkr1* mutant strains are hypervirulent may be relevant in two clinical settings. First, although *C. neoformans* is classified as an opportunistic pathogen, infections can occur in hosts with no known immune dysfunction (1, 18, 135). These individuals may have subtle immune defects or, alternatively, may be infected with hypervirulent strains. In fact, two clinical case reports identified patients who were infected with atypical forms of

C. neoformans that produce dramatically enlarged capsules (37, 119). Whether these represent naturally occurring lesions in the *PKR1* gene or of other elements of the PKA-controlled pathway regulating capsule remains to be explored experimentally. A second potentially relevant clinical finding is that whereas serotype A and D strains infect predominantly immunocompromised hosts, strains of the serotype B and C *gattii* variety infect immunocompetent hosts and are rarely isolated from AIDS patients [reviewed in (157)]. Thus, the serotype B and C strains may be hypervirulent with respect to host immune function. Alternatively, it has been proposed recently that serotype B and C strains do not establish latent infections and thus give rise only to primary infections in immunocompetent hosts rather than to reactivation of latent infections in response to immunosuppression which occurs with serotype A and D strains (T. Sorrell, personal communication).

The hypervirulent *pkr1* mutant strain is one of the few mutants to be molecularly defined in any pathogenic system. Two other mutations that enhance virulence of *C. neoformans* have been identified recently. Janbon and colleagues identified the *CAS1* gene, which encodes an integral membrane protein involved in capsule acetylation that generates specific epitopes detectable by anticapsular monoclonal antibodies (89). *cas1* mutant serotype D strains have lost these capsular epitopes, and were hypervirulent in the murine tail vein injection model, leading to 100% lethal infections by day 20 postinfection compared to ~60 days with the congenic wild-type strain (89). A third hypervirulent mutant strain was recently identified in a signature-tagged mutagenesis approach (131). Following random insertion of the hygromycin resistance gene into the *C. neoformans* genome, mutants were identified that were altered in ability to survive in vivo in mice compared to the parental wild-type strain (H99, serotype A). One such mutant, the 2A2 strain, resulted in a tenfold greater burden of fungal cells in the CNS and animals infected with this strain exhibited frank signs of infection in as short as 10 days following infection, a time at which animals infected with the wild-type remained largely healthy. Identification of the insertional mutation in the 2A2 mutant strain, and demonstration that it causes the mutant phenotype, should provide significant insight into the nature of this intriguing hypervirulent isolate.

What remains to be learned about the Gα-cAMP-PKA pathway and its role in virulence? First, the molecular identity of the receptor that activates this pathway has not been established. One hypothesis is that the receptor will be a member of the G-protein–coupled receptor family, and in fact GPCRs that are linked to the homologous Gα proteins in both *S. cerevisiae* and *S. pombe* have been identified. In these model yeasts, the Gpr1 and git3 receptors are GPCRs that span the membrane seven times, share homology with each other, and are involved in sensing glucose and activating cAMP production (93, 118, 186, 196). In contrast, although a homologous Gα protein exists in the basidiomycetes *C. neoformans* and *U. maydis*, no receptor homolog has as yet been identified. A second interesting feature of the signaling pathway is that no Gβγ subunits linked to the Gα protein Gpa1 have been identified. Gpa1 shares marked sequence identity with known heterotrimeric Gα protein subunits, but the only Gβ subunit identified, Gpb1, is linked to pheromone sensing and plays no role in virulence factor production or virulence and does not function as a subunit with Gpa1 (183). A similar situation exists in *S. cerevisiae* in which the Gα protein Gpa2 does not interact with the only known G$\beta\gamma$ subunits (Ste4-Ste18) that function in mating but not filamentous growth (115, 147). Very recent studies have identified several novel proteins that function as Gβ subunit structural mimics in the Gpa2 signaling pathway in budding yeast (76). These proteins, Gpb1 and Gpb2, physically interact with Gpa2 and play an inhibitory signaling role. Remarkably, these proteins lack the signature seven WD-40 repeats of all known Gβ subunits but contain instead seven copies of a different sequence, the kelch repeat, previously implicated in mediating protein-protein interactions. The x-ray structure of one enzyme (galactose oxidase) with seven kelch repeats is known, and the protein folds into a seven bladed β–propeller that is essentially superimposable on the known structure of the G$\beta\gamma$ complex (86). Finally, the targets of the PKA pathway that function to promote mating, capsule and melanin production, and virulence also remain to be identified. By analogy with previous studies on PKA signaling in *S. cerevisiae*, *S. pombe*, and *U. maydis*, it is anticipated that one or more transcription factors that control target gene expression may lie downstream of PKA [reviewed in (41, 109)]. For example, in *S. cerevisiae* the Tpk2 catalytic subunit of PKA is nuclear localized and functions to promote expression of the cell surface flocculin Flo11 that is necessary for filamentous and invasive growth (140). Tpk2 governs differentiation by a dual mechanism in which phosphorylation of the Sfl1 repressor promotes dissociation of the dimeric repressor, causing it to disengage from the *FLO11* promoter (31, 140, 148). Tpk2 also phosphorylates the activator Flo8, enabling it to bind to the *FLO11* gene promoter and activate transcription (140). In *U. maydis*, a key transcription factor target of the PKA pathway is the Prf1 pheromone response factor, a member of the HMG box transcription factor family (77, 78). PKA controls Prf1 by promoting expression of the *PRF1* gene, and it likely phosphorylates and activates Prf1 as well. A homolog of the Prf1 factor has recently been identified in *C. neoformans*, and studies are in progress to establish whether this is the target of the Pka1 catalytic subunit that controls virulence factor production and pathogenesis (C. D'Souza, J. Hicks & J. Heitman, unpublished results). Additional studies are in progress using genome microarray approaches to define the transcriptional profiles of cells with engaged or defective PKA signaling routes.

GENETICS OF VIRULENCE

C. neoformans produces two specialized virulence factors that play a central role in its ability to survive in the harsh host environment and cause symptomatic infection. These specialized virulence factors are the polysaccharide capsule and the antioxidant pigment melanin (18). The production of a darkly pigmented cell ensheathed in a thick capsule makes a stark contrast between this pathogenic budding yeast and the model yeast *S. cerevisiae*. Growth of *C. neoformans* cells under certain conditions results in the dramatic induction of a complex polysaccharide capsule. These inducing conditions include growth in limiting iron concentrations or in physiological concentrations of CO_2, both conditions found in the infected host (74, 177). The capsule can be readily detected microscopically by its ability to exclude India ink particles, resulting in the appearance of a halo surrounding the cell. The capsule can also be detected with polyclonal or monoclonal sera raised against carbohydrate epitopes, and this forms the basis of the common serotyping assay that detects differences in the capsule structure that exist in strains of different serotypes. In addition to cell-associated capsule, the growing fungal cells shed capsular material into the growth medium, both in vitro and in vivo in infected animals [reviewed in (50)]. The capsule plays myriad roles in virulence, including inhibiting phagocytosis in some settings and promoting intracellular survival in macrophages following phagocytosis (15). In addition, the shed capsular antigen wreaks havoc on the host immune system by multiple mechanisms, further enfeebling the already immunocompromised host. Of all the virulence traits associated with *C. neoformans*, the capsule most closely fulfills the criteria often associated with virulence factors of, for example, bacterial pathogens. All clinical isolates make capsule, mutants lacking capsule are avirulent, and restoration of capsule production restores virulence.

A panoply of genes involved in capsular biosynthesis have already been identified, and many more likely remain to be discovered. In early studies, both Bulmer et al. and Kozel & Cazin identified a limited number of acapsular mutant strains (16, 92). Subsequently, Jacobson and colleagues identified a series of acapsular mutant strains (*cap43*, *cap44*, *cap48*, *cap53*, *cap54*, *cap55*, *cap59*, and *cap64*) following mutagenesis with N-nitrosoguanidine or UV irradiation in a serotype D strain background (87). Acapsular and hypocapsular mutants were identified initially as nonglistening colonies and then screened by microscopic examination with India ink. These mutants were isolated in the serotype D background in the B3501 mating-type α and B3502 mating-type **a** strains. Importantly, these strains are the f1 progeny of the ancestral precursor strains NIH12 and NIH433 that gave rise to the congenic JEC20/JEC21 strain pair [reviewed in (79)]. B3501 and B3502 are not congenic with each other, with NIH12 or NIH433, or with JEC20 or JEC21. They do, however, share ~50% of their genomes with their parents and ~75% with their offspring, and the B3501 genome is one of the two serotype D strains being subjected to sequence analysis.

Genetic outcrosses were conducted with the isolated hypo- and acapsular mutant strains and the four spore chains were either isolated as a spore suspension, or else directly micromanipulated and dissected from individual basidia (162). The mutant phenotypes segregated in a ratio of 1 wild-type:1 mutant in meiotic progeny. Subsequent studies involving genetic crosses between two different acapsular mutant strains yielded evidence for independent segregation of alleles in a limited number of cases (~3:1 ratio of mutant to wild-type), but in most cases the mutations appeared to be linked (all mutant progeny) (162). Subsequent molecular studies revealed in some cases that *CAP* genes thought to be linked by genetic analysis were in fact located on different chromosomes (24). These studies were among the first to indicate that standard yeast genetic approaches could be applied to this pathogenic basidiomycete. They also suggest caution in isolating recombinants by the wire loop en masse approach, which is not a suitable substitute for careful basidiospore dissection.

Subsequently, these acapsular mutant strains served as a valuable resource to clone the corresponding genes. In an elegant series of molecular biological studies, Chang & Kwon-Chung cloned and characterized the *CAP10*, *CAP59*, *CAP60*, and *CAP64* genes (21–24). These genes were isolated by introducing a genomic library into the respective acapsular mutant strains, and complementing clones were isolated by the alteration in density and charge of capsular vs. acapsular mutant strains in a polyethylene glycol partitioning gradient. The cloned genes were sequenced and then used to produce gene disruption alleles to mutate the corresponding genes. Each gene was disrupted with the *ADE2* selectable marker for positive selection, and the transforming DNA was introduced by electroporation. Homologous recombination was achieved with a selection method in which nonhomologous recombination events were counterselected using a flanking *URA5* marker distal to the region of homology and selection of the desired *ura5* transformants on 5-FOA medium. These approaches are similar to those originally developed for gene disruption in murine embryonic stem (ES) cells. Although homologous recombination was achieved, the frequency was extremely low in several cases (~1/10,000). Two factors likely contributed to these low frequencies. First, electroporation is less efficient than biolistic methods for transformation and gene disruption in this organism (45). Second, in several cases the gene disruption alleles were likely constructed from DNA sequences that were not strictly isogenic with the transformation recipient strain, and DNA polymorphisms inhibit targeted integration of homologous sequences in bacteria, yeasts, and murine ES cells.

In these examples, targeted disruption of the *CAP10*, *CAP54*, *CAP59*, and *CAP64* genes each resulted in acapsular mutant strains that were avirulent in the murine tail-vein injection model, and reintroduction of the wild-type gene restored virulence. Several of these genes that were thought to be linked based on earlier genetic studies (162) were found to be encoded by different chromosomes (24). Finally, the sequences of this series of *CAP* genes provide little or no clue as to their functions in capsule biosynthesis.

More recent studies have begun to attack the construction of the capsule by biochemical, immunological, and cell biological approaches wedded to molecular biology and genetics. For example, Janbon et al. identified the *CAS1* gene, which encodes a membrane protein that is essential for O-acetylation of sugar hydroxyl groups on the GXM polysaccharide (89). The *cas1* mutant strain was isolated in the serotype D JEC21 strain background and was identified following UV irradiation as nonreactive in ELISA assays with a monoclonal antibody raised against capsular antigen. The gene was cloned by complementation with a genomic library constructed in the pCnTel1 telomeric shuttle vector by colony hybridization for restoration of antibody cross-reactivity. The gene was cloned and sequenced, and shown to be the correct gene by identification of a mutation in the *CAS1* gene in the original *cas1* mutant strain. The gene was disrupted by electroporation and positive-negative selection. Importantly, the *cas1* disruption mutant failed to react with the original monoclonal, and most interestingly, this mutant was hypervirulent in mice. These studies illustrate the power of combining immunological reagents against the capsule with genetic and molecular biology approaches to dissect the characteristics of this virulence trait.

Doering and colleagues have pioneered the use of cell biological and biochemical approaches to analyze capsule structure and construction (50). In an elegant series of studies, novel imaging techniques were employed to examine how and where new capsule material is formed (144). Newly synthesized capsular material is deposited near the cell wall, displacing outward older capsular antigen. In addition, as the outer capsule is displaced and therefore occupies a greater volume, the original capsular material becomes diluted and appears to acquire additional newly synthesized material. Finally, the capsule surrounding the bud is newly synthesized and originally has a looser configuration that can give rise to an unusual India ink staining pattern in which grains of the ink penetrate the capsule of the bud but not the mother cell. This finding may be analogous to the recent finding in budding yeast that the septins form a diffusion barrier between the plasma membranes of mother and daughter cells (168).

In addition to these cell biological studies, Doering and colleagues have begun to apply biochemical approaches to identify enzymes involved in capsule biosynthesis. These studies have identified a unique α-1,3 mannosyltransferase that transfers mannose from a donor molecule to a dimannoside receptor to establish the critical α-1,3 linkages that buttress the capsule (49). Doering and colleagues also recently identified the enzyme that produces the UDP-xylose donor for capsule synthesis and the *USX1* gene encoding it (8).

The second specialized virulence factor elaborated by *C. neoformans* in response to nutritional and host signals is the pigment melanin [reviewed in (19)]. This dark polymer is produced from precursors such as dopa or caffeic acid following their oxidation by the enzyme laccase, and expression of the laccase enzyme is induced by carbohydrate limitation. The production of melanin serves to identify *C. neoformans*, which produces darkly pigmented colonies on dopa or bird seed agar (which contains diphenolic precursors) (158, 163). Melanin is deposited in the cell wall, and protects the cell from oxidative and nitrosative challenge by the host immune system and likely also plays a role in defense against other stresses. Melanized cells are extremely durable, and following an overnight exposure to boiling acid, the shells of these cells (so called melanin ghosts) remain (133, 149). Melanin ghosts can be recovered by this procedure from the organs of infected mice and rats, and from the brain tissue of human patients with cryptococcal meningitis examined at autopsy, proving that melanin is produced during infection (133, 134). Virtually all clinical isolates of *C. neoformans* make both capsule and melanin, although a few rare nonpigmented variants that retain virulence in animal models have been reported (103, 174).

Mutants that fail to produce melanin are significantly attenuated in animal models. In early studies, Mel⁻ mutant strains were isolated following mutagenesis and found to have defects in the enzyme required for melanin production, and an apparent defect in active transport of precursors required for melanin biosynthesis (102). These Mel⁻ mutant strains were significantly attenuated in animal models, and those fungal cells that could be isolated from the CNS had reverted to a Mel⁺ phenotype (145). In other cases, strains that exhibit a temperature-sensitive phenotype and produce melanin at 24°C or 30°C but not at 37°C have been reported (103). Even wild-type strains produce significantly less melanin at 37°C than at lower growth temperatures in vitro, which introduces a paradox with respect to this virulence factor unless other conditions mitigate this effect in vivo (88). Importantly, mutations in the Gpa1-cAMP-PKA pathway confer a defect in melanin production that is most apparent at 37°C and less so at 30°C (4), suggesting that earlier temperature-sensitive Mel⁻ mutants might have altered elements of this signaling cascade that controls virulence factor production.

The *LAC1* gene encoding the melanin biosynthetic enzyme laccase has been disrupted by transformation and homologous recombination (151, 191). Following a series of backcrosses, the *lac1* mutant strain was found to be attenuated compared to the congenic wild-type in a murine tail vein injection model. However, in contrast to acapsular mutants, which are completely avirulent, the *lac1* mutant strains were significantly attenuated but not avirulent and some lethal infections were observed. Thus, in rare apigmented clinical isolates, other mutations may mitigate the loss of melanin and thereby restore virulence.

Genetic approaches analogous to those applied to capsule have also been applied to melanin biosynthesis. Torres-Guererro & Edman subjected strains of the congenic serotype D JEC21/JEC20 strain series to mutagenesis with the alkylating agent EMS (172). A series of mutants that produce no melanin or reduced levels of melanin was isolated as those that produced white or light tan colonies on birdseed or dopamine medium. Crosses between wild-type and Mel⁻ mutant strains yielded wild-type and mutant progeny in a ratio of 1:1, indicating single nuclear mutations confer the observed phenotypes. Genes were identified as allelic by segregation tests in which crosses between Mel⁻ strains yielded either 100% mutant progeny (allelic) or 75% mutant/25% wild-type (nonallelic). This approach assigned 18 mutants to seven different genes defined as segregation groups (by analogy to complementation groups). Importantly, the *mel2* mutant strain was found to be complemented by the cloned *LAC1* gene encoding laccase and shown to harbor a mutation in a conserved histidine residue (H164Y) in a putative copper-binding site of the laccase enzyme (151).

Several of the other Mel⁻ mutants exhibited additional interesting phenotypes (172). First, melanin production was restored by the addition of exogenous copper ions to the *mel1*, *mel3*, *mel5*, and *mel7* mutant strains. Many laccase enzymes are copper-zinc enzymes, and thus mutations affecting proteins involved in metal ion transport and delivery to charge the active site of the cryptococcal laccase enzyme might result in melanin defects. The second intriguing phenotype is that the *mel1*, *mel5*, and *mel7* mutants also exhibited a recessive sterile phenotype. These mutants mate with wild-type strains, but α *mel1* strains fail to mate with **a** *mel1* strains and no filaments, basidia, or spores were produced. These mutants are therefore bilaterally sterile. This sterile phenotype cosegregated in genetic crosses to wild-type with the Mel⁻ phenotype, indicating a single mutation confers both phenotypes. Copper not only suppresses the melanin defect of *mel1* mutant cells but also restores mating in a *mel1* by *mel1* cross. Finally, two other mutants, *mel3* and *mel6*, exhibited an unusual bilateral mating defect and produced filaments but no basidiospores. These studies reveal an unexpected link between mating and laccase production. Thus far, only the *mel2* mutant has been characterized at the molecular level, and the remaining Mel⁻ mutants represent a unique vantage point to begin a molecular dissection of melanogenesis and its contribution to virulence.

All pathogenic organisms must survive within the infected host, and thus an important property essential for virulence is the ability to grow at 37°C. Of the many thousands of fungal species, only a few dozen cause human infection. The only common trait that links those fungi capable of causing systemic disease is their shared ability to proliferate at 37°C. Other fungi that colonize and infect humans but are not capable of growth above ~35°C are relegated to causing dermatophytic infections of the skin, hair, and nails. Although the model yeast *S. cerevisiae* is an uncommon human pathogen, more than 50 clinical isolates have been reported, and their one common feature is an ability to grow at temperatures up to 42°C, whereas laboratory strains of yeast do not grow above 39°C (122). Novel quantitative trait-mapping approaches have begun to pinpoint the molecular basis of this virulence trait in yeast (161).

Early studies in *C. neoformans* defined temperature-sensitive, avirulent mutants (102). More recent molecular studies have revealed important roles for calcineurin, Ras1, and the Ste20α kinase in growth at elevated temperature and virulence (2, 40, 62, 136, 182, 184). Calcineurin is a serine-threonine specific protein phosphatase that consists of two subunits, the calcineurin A catalytic subunit and the calcineurin B regulatory subunit. In addition the calcineurin AB heterodimer is bound and activated by calmodulin in response to intracellular calcium increases. Calcineurin is the molecular target of the immunosuppressive drugs cyclosporin A (CsA) and FK506, and both drugs were found to inhibit growth of *C. neoformans* at 37°C but not at 24°C (136). The isolation of FK506-resistant mutants that retained sensitivity to CsA identified two genes involved in FK506 antifungal action (136). Three mutations (*frr1-1*, *frr1-2*, *frr1-3*) conferred resistance to both FK506 and another related drug, rapamycin, and were recessive in genetic crosses. All three harbor mutations within the *FRR1* gene encoding the FK506/rapamycin binding protein FKBP12 (38). The remaining mutation, *FKR1-1*, conferred resistance to FK506 but not to rapamycin, was dominant in heterokaryon analysis, and segregated independently from the *FRR1* gene encoding FKBP12. Subsequent genetic and molecular analysis revealed the *FKR1-1* mutation results from a tandem duplication of 6 bp that inserts two amino acids into the calcineurin B subunit that prevents FKBP12-FK506 binding (62).

The finding that calcineurin inhibitors prevented growth at 37°C but not at 24°C suggested that calcineurin might be essential for growth at elevated temperature and for virulence. The genes encoding the calcineurin A and B subunits were cloned and disrupted by transformation and homologous recombination (40, 62, 136). The resulting *cna1* and *cnb1* mutant strains are viable at 24°C, but inviable at 37°C and completely avirulent in animal models. These studies were the first to define a molecular determinant of growth at elevated temperature necessary for infection. Importantly, this phenotype could not have been predicted from studies in model yeasts or even other pathogenic yeasts. Calcineurin is not essential for growth at elevated temperature in either lab strains or pathogenic isolates of *S. cerevisiae*. Calcineurin is also dispensable for growth of *C. albicans* at 37°C or 39°C, but interestingly does play a role in virulence (J. Blankenship & J. Heitman, in preparation). Thus, the functions of calcineurin are linked to virulence in two different fungal pathogens but via distinct mechanisms.

Several other genes have recently been linked to growth at elevated temperature and virulence, including those encoding the small GTP binding protein Ras1, the protein kinase Ste20α encoded by the mating-type locus (discussed above), the cyclophilin A protein Cpa1 (180), and a component of the vacuolar proton ATPase Vph1 (59). Mutations in all of these genes impair growth at elevated temperature and compromise the ability of the organism to establish infection and cause disease.

Finally, recent studies have revealed that two secreted enzyme activities, phospholipase B and urease encoded by the *PLB1* and *URE1* genes, respectively, also contribute to the full virulence composite but are not strictly essential for virulence. Serotype A *plb1* mutant strains were found to be significantly attenuated for virulence in both the murine inhalation model and the rabbit meningitis model but remained capable of causing 100% lethal infections in mice (34). Serotype A *ure1* mutant strains lacking urease showed no defect in fungal cell survival in the rabbit CNS, but in the murine inhalation and tail vein injection model the mutant was attenuated compared to wild type (35). Mice infected with the wild-type or the *ure1* reconstituted strain in which the wild-type gene was introduced exhibited signs of pulmonary distress and hydrocephalus at late stages of infection, whereas animals infected with the urease mutant strain showed only hydrocephalus. Thus, the ability of urease to produce ammonia and regulate local pH might play a role in promoting fungal cell survival in the lung.

EVOLUTION OF A FUNGAL PATHOGEN

Serotype A variety *grubii* and serotype D variety *neoformans* strains are both found in association with pigeon droppings worldwide. Virtually all humans are exposed to *C. neoformans*, and in most individuals the immune system rapidly clears an asymptomatic infection. There is recent evidence, however, that the majority of children in the Bronx, NY, are infected early in life, and some of these infections are symptomatic (70). In other individuals, the initial infection in the lung enters a pulmonary lymph node complex and establishes a dormant phase in which the organism may replicate and survive inside macrophages (6, 65).

This macrophage survival is thought to enable *C. neoformans* to establish latent infections. Studies in France reveal that Asians and Africans who emigrated to France many years previously present with *C. neoformans* infections with strains endemic from Asia or Africa, rather than those present in France (52, 65). A series of autopsy studies in the 1950s (7), and studies repeated recently in Japan (85), reveal that 0.5–1% of normal individuals exhibit evidence of cryptococcal infection in hilar lymph nodes. These findings suggest that in the 2–3% of transplant recipients in which *C. neoformans* develops many of these infections may represent reactivation of latent infections acquired much earlier in life.

Although many bacterial and fungal pathogens have adopted strategies to survive inside macrophages, it is curious when these pathogens are organisms generally found in the soil. How have saprophytes developed mechanisms to survive in response to the mammalian immune system? One possibility is that the defense mechanisms developed by organisms like *C. neoformans* to survive in the presence of other microorganisms can be put to use in the presence of the host immune response. It has been proposed that the interaction of *C. neoformans* with soil amoebae as competitors in the environment has allowed *C. neoformans* to develop strategies to survive within macrophages (160). Macrophages and amoebae are similar in many respects, including the abilities to phagocytose particles into vacuoles and digest them with secreted enzymes. Steenbergen et al. found that *C. neoformans* is in fact phagocytosed by the amoeba *Acanthamoeba castellanii*, and that once inside the amoeba, divides, and leads to amoeba killing (160). This apparent survival strategy is very similar to that seen in macrophages where *C. neoformans* can be phagocytosed and then persist for years (71). Using the amoeba as a comparative model for macrophage survival may provide valuable insights into the success of *C. neoformans* as a human pathogen, and allow a better understanding of the latency phase of infection.

Fully understanding the epidemiology of infection and latent disease depends on knowing the environmental reservoirs for *C. neoformans*. *C. neoformans* var. *neoformans* (serotype D) and var. *grubii* (serotype A) are clearly associated with birds, particularly pigeons (57, 66, 166), but the environmental reservoir for var. *gattii* (serotype B) has been much more elusive. The *gattii* variety is found primarily in subtropical and tropical regions. In Australia, Ellis & Pfeiffer found this variety to be associated with *Eucalyptus* trees (primarily the red gum *E. camaldulensis*) (55). They conducted large-scale sampling of air, soil, and vegetation on a weekly basis over an 8-month period and found var. *gattii* in vegetation only from *E. camaldulensis* (56). Var. *gattii* can be cultured out of tree hollows and the koalas that inhabit these trees (32), and it appears to be cultured most easily from flowering trees (56). The authors suggest that this discovery explains the unexpected epidemiology of var. *gattii*: It is not common among AIDS patients in Australia.

Because the majority of the AIDS population in Australia lives where there are no *E. camaldulensis* trees, the opportunity for exposure is low. As a result, most of the cases of cryptococcosis in Australia are found in individuals who are immunocompetent and likely to be outdoors in areas of the country where eucalypts grow.

Finding var. *gattii* in eucalypts may have revealed the environmental reservoir in Australia, but it does not explain the distribution of var. *gattii* worldwide. For example, almost all cases of cryptococcosis in Papua New Guinea (PNG) are caused by var. *gattii*, yet there are almost no *Eucalyptus* trees in PNG. This finding suggests that other natural habitats must exist for var. *gattii*, and thus far no one has succeeded in isolating var. *gattii* from the environment in PNG (105). On the other hand, efforts to culture *Cryptococcus* from the environment in South America have resulted in the discovery of both var. *gattii* and var. *neoformans* on trees other than eucalypts. (106, 107, 128). Clearly, there is great variation in environmental habitats among the *C. neoformans* varieties around the world.

Another difference between var. *gattii* and the other varieties is that most *gattii* infections appear to represent primary infections, and susceptibility is not limited to immunocompromised individuals (157). In fact most infections are in apparently healthy people. Interestingly, an outbreak of cryptococcosis caused by var. *gattii* in Vancouver, B.C. beginning in 1999 has affected more than 40 individuals, 27 of them exhibiting no apparent immune deficiencies (63). Investigations are under way to identify an environmental source.

Given the striking differences in epidemiology, distribution, environmental habitats, and molecular and biochemical characteristics between var. *gattii* and vars. *grubii* and *neoformans*, Boekhout et al. have proposed that the *gattii* variety be reclassified as a different *Cryptococcus* species (11). This idea is supported by their thorough investigation of the AFLPs of 153 var. *neoformans* and *grubii* isolates and 54 var. *gattii* isolates from AIDS and non-AIDS patients, the environment and animals from all over the world. Boekhout et al. showed that the AFLP patterns for the two varieties cluster into two main branches, which seem to correspond with reproductively isolated populations. Based on their data and the accumulated evidence of differences between the varieties, they recommend designating var. *neoformans* and var. *grubii* (serotypes A, D, and AD) as *Cryptococcus neoformans* and var. *gattii* (serotypes B and C) as *Cryptococcus bacillisporus*. The *bacillisporus* designation was first used by Kwon-Chung in 1976 to distinguish the sexual states of serotypes A/D (*Filobasidiella neoformans*) from serotypes B/C (*Filobasidiella bacillisporus*) on the basis of observations of the spores resulting from intraserotype crosses (96). Subsequent mating experiments suggesting successful interserotype mating led to the designation of varieties instead of separate species for the two classes of serotypes (100, 153). Now, however, the overwhelming phenotypic and molecular data indicate separating the serotypic groups into species. These conclusions support the idea that the species concept, both biologically and phylogenetically, is a continuum, and as such, speciation determinations need to be made on the basis of many criteria.

CONCLUSION

Genetic analysis of *C. neoformans* over the past two decades has revolutionized what we know about its life cycle and virulence properties. As we proceed, the value of having defined the sexual cycle will continue to be high. Although scientists have found the *C. neoformans* sexual cycle to be invaluable, it is much less clear what its value is to *C. neoformans*. Even though it can be seduced to mate in the laboratory, there is limited evidence that *C. neoformans* undergoes sexual reproduction in nature.

The absence of active sexual reproduction in *C. neoformans* is curious given that the sexual cycle is an important part of the life cycles of most eukaryotes. The purpose of sexual reproduction is a matter of debate, but what is known is that sexual reproduction affords a fitness advantage to most organisms, and those that lose or never develop this ability are likely to face extinction more rapidly than those that do [reviewed in (17)]. The simple explanation for the advantage that sexual reproduction affords is that it allows recombination between different genetic backgrounds to occur, facilitating the propagation of beneficial mutations into a population. Alternatively, recombination could afford the opportunity to purge deleterious mutations from the population (199). Either way, it appears that recombination is important enough to drive many organisms to invest energy in the processes of mating and meiosis.

Even with the strong selection for sexuality, it appears that *C. neoformans* is not actively engaging in sexual reproduction. Sporulation structures have not been observed in patients or the environment, and the populations that have been studied show linkage disequilibrium consistent with clonal growth (10, 13, 27). Clonal growth could, in part, be explained by the finding that almost all clinical and environmental isolates are of a single mating type (α) providing little opportunity for sex. Of ~1500 serotype A strains screened, only two were of the **a** mating type, suggesting that very little opportunity exists for natural crossing to occur (112, 178). On the other hand, there is very little evidence of recombination even when strains of different mating types are found in the same environmental location and, presumably, have every opportunity to mate. Serotype B *MATa* and *MATα* isolates from *Eucalyptus* trees in Australia have been found on the same tree, but show no crossing, and all attempts to induce mating between these strains in the laboratory have failed (75; W. Meyer, personal communication).

It seems that a piece of the lifestyle puzzle for *C. neoformans* is still missing. Perhaps extremely infrequent mating is all that is required for healthy *C. neoformans* populations, so the sexual cycle is maintained, but used very rarely in an as-yet unidentified environmental niche. Maybe **a** and α strains rarely engage in sexual reproduction but still enhance fruiting to promote spore formation and thus increase survival in response to harsh conditions. Alternatively, perhaps *C. neoformans* is evolving to be an asexual organism. Maybe the ability of α strains to form spores in the absence of **a** cells has led to their dispersal and prevalence in the environment, while **a** strains are disappearing. Whatever the intricacies of the *C. neoformans* life cycle, there is still much to be learned about how this unusual pathogen survives in both the environment and human beings. Understanding the complexities of when, where, and how *C. neoformans* reproduces will provide insights not only into how *C. neoformans* infects humans but also into how all fungal pathogens live, reproduce, and evolve strategies to survive.

ACKNOWLEDGMENTS

The authors thank J.A. Alspaugh, R. Brazas, J. Fraser, A. Goldstein, and J. Perfect for comments on the manuscript; K. Lengeler for providing Figure 3; D. Fox for assistance with Figure 5; and R. Casey for assistance with manuscript preparation. Some of the studies reviewed here were supported by RO1 grants AI39115, AI42159, AI50438, AI50113, and PO1 grant AI44975 from the NIAID/NIH. C. M. H. is supported by Damon Runyon Cancer Research Fund Fellowship #DRG-1694. J. H. is a Burroughs-Wellcome Scholar in Molecular Pathogenic Mycology and an Associate Investigator of the Howard Hughes Medical Institute.

The *Annual Review of Genetics* is online at http://genet.annualreviews.org

LITERATURE CITED

1. Aberg JA, Mundy LM, Powderly WG. 1999. Pulmonary cryptococcosis in patients without HIV infection. *Chest* 115:734–40

2. Alspaugh JA, Cavallo LM, Perfect JR, Heitman J. 2000. *RAS1* regulates filamentation, mating and growth at high temperature of *Cryptococcus neoformans*. *Mol. Microbiol.* 36:352–65

3. Alspaugh JA, Davidson RC, Heitman J. 2000. Morphogenesis of *Cryptococcus neoformans*. In *Dimorphism in Human Pathogenic and Apathogenic Yeasts*, ed. JF Ernst, A Schmidt, pp. 217–38. Basel: Karger Contrib. Microbiol.

4. Alspaugh JA, Perfect JR, Heitman J. 1997. *Cryptococcus neoformans* mating and virulence are regulated by the G-protein α subunit Gpa1 and cAMP. *Genes Dev.* 11:3206–17

5. Alspaugh JA, Pukkila-Worley R, Harashima T, Cavallo LM, Funnell D, et al. 2002. Adenylyl cyclase functions downstream of the Gα protein Gpa1 and con-

trols mating and pathogenicity of *Cryptococcus neoformans*. *Eukaryot. Cell* 1:75–84

6. Baker RD. 1976. The primary pulmonary lymph node complex of cryptococcosis. *Am. J. Clin. Pathol.* 65:83–92

7. Baker RD, Haugen RK. 1955. Tissue changes and tissue diagnosis in cryptococcosis: a study of twenty-six cases. *Am. J. Clin. Pathol.* 25:14–24

8. Bar-Peled M, Griffith CL, Doering TL. 2001. Functional cloning and characterization of a UDP-glucuronic acid decarboxylase: the pathogenic fungus *Cryptococcus neoformans* elucidates UDP-xylose synthesis. *Proc. Natl. Acad. Sci. USA* 98:12003–8

9. Beukeboom LW. 1995. Sex determination in *Hymenoptera*: a need for genetic and molecular studies. *BioEssays* 17:813–17

10. Boekhout T, Belkum AV. 1997. Variability of karyotypes and RAPD types in genetically related strains of *Cryptococcus neoformans*. *Curr. Genet.* 32:203–8

11. Boekhout T, Theelen B, Diaz M, Fell JW, Hop WC, et al. 2001. Hybrid genotypes in the pathogenic yeast *Cryptococcus neoformans. Microbiology* 147:891–907

12. Brandt ME, Bragg SL, Pinner RW. 1993. Multilocus enzyme typing of *Cryptococcus neoformans. J. Clin. Microbiol.* 31:2819–23

13. Brandt ME, Hutwagner LC, Klug LA, Baughman WS, Rimland D, et al. 1996. Molecular subtype distribution of *Cryptococcus neoformans* in four areas of the United States. *J. Clin. Microbiol.* 34:912–17

14. Brazas RM, Stillman DJ. 1993. The Swi5 zinc-finger and Grf10 homeodomain proteins bind DNA cooperatively at the yeast HO promoter. *Proc. Natl. Acad. Sci. USA* 90:11237–41

15. Buchanan KL, Murphy JW. 1998. What makes *Cryptococcus neoformans* a pathogen? *Emerg. Infect. Dis.* 4:71–83

16. Bulmer GS, Sans MD, Gunn CM. 1967. *Cryptococcus neoformans.* I. Nonencapsulated mutants. *J. Bacteriol.* 94:1475–79

17. Burt A. 2000. Perspective: sex, recombination, and the efficacy of selection—Was Weismann right? *Evolution Int. J. Org. Evolution* 54:337–51

18. Casadevall A, Perfect JR. 1998. *Cryptococcus neoformans.* Washington, DC: ASM Press

19. Casadevall A, Rosas AL, Nosanchuk JD. 2000. Melanin and virulence in *Cryptococcus neoformans. Curr. Opin. Microbiol.* 3:354–58

20. Casselton LA, Olesnicky NS. 1998. Molecular genetics of mating recognition in basidiomycete fungi. *Microbiol. Mol. Biol. Rev.* 62:55–70

21. Chang YC, Kwon-Chung KJ. 1994. Complementation of a capsule-deficient mutation of *Cryptococcus neoformans* restores its virulence. *Mol. Cell. Biol.* 14:4912–19

22. Chang YC, Kwon-Chung KJ. 1998. Isolation of the third capsule-associated gene, *CAP60*, required for virulence in *Cryptococcus neoformans. Infect. Immun.* 66:2230–36

23. Chang YC, Kwon-Chung KJ. 1999. Isolation, characterization, and localization of a capsule-associated gene, *CAP10*, of *Cryptococcus neoformans. J. Bacteriol.* 181:5636–43

24. Chang YC, Penoyer LA, Kwon-Chung KJ. 1996. The second capsule gene of *Cryptococcus neoformans, CAP64*, is essential for virulence. *Infect. Immun.* 64:1977–83

25. Chang YC, Penoyer LA, Kwon-Chung KJ. 2001. The second *STE12* homologue of *Cryptococcus neoformans* is *MATa*-specific and plays an important role in virulence. *Proc. Natl. Acad. Sci. USA* 98:3258–63

26. Chang YC, Wickes BL, Miller GF, Penoyer LA, Kwon-Chung KJ. 2000. *Cryptococcus neoformans STE12α* regulates virulence but is not essential for mating. *J. Exp. Med.* 191:871–82

27. Chen F, Currie BP, Chen L-C, Spitzer SG, Spitzer ED, Casadevall A. 1995. Genetic relatedness of *Cryptococcus neoformans* clinical isolates grouped with the repetitive DNA probe CNRE-1. *J. Clin. Microbiol.* 33:2818–22

27a. Chung S, Karos M, Chang YC, Lukszo J, Wickes BL, Kwon-Chung KJ. 2002. Molecular analysis of *CPRα,* a *MATα*-specific pheromone receptor gene of *Cryptococcus neoformans. Eukaryot. Cell* 1:432–39

28. Clarke DL, Woodlee GL, McClelland CM, Seymour TS, Wickes BL. 2001. The *Cryptococcus neoformans STE11α* gene is similar to other fungal mitogen-activated protein kinase kinase kinase (MAPKKK) genes but is mating type specific. *Mol. Microbiol.* 40:200–13

29. Cogliati M, Esposto MC, Clarke DL, Wickes BL, Viviani MA. 2001. Origin of *Cryptococcus neoformans* var. *neoformans* diploid strains. *J. Clin. Microbiol.* 39:3889–94

30. Colombo S, Ma P, Cauwenberg L, Winderickx J, Crauwels M, et al. 1998. Involvement of distinct G-proteins, Gpa2 and Ras, in glucose- and intracellular acidification-induced cAMP signalling in the yeast *Saccharomyces cerevisiae*. *EMBO J.* 17:3326–41

31. Conlan RS, Tzamarias D. 2001. Sfl1 functions via the co-repressor Ssn6-Tup1 and the cAMP-dependent protein kinase Tpk2. *J. Mol. Biol.* 309:1007–15

32. Connolly JH, Krockenberger MB, Malik R, Canfield PJ, Wigney DI, Muir DB. 1999. Asymptomatic carriage of *Cryptococcus neoformans* in the nasal cavity of the koala (*Phascolarctos cinereus*). *Med. Mycol.* 37:331–38

33. Cook JG, Bardwell L, Thorner J. 1997. Inhibitory and activating functions for MAPK Kss1 in the *S. cerevisiae* filamentous-growth signalling pathway. *Nature* 390:85–88

34. Cox GM, McDade HC, Chen SC, Tucker SC, Gottfredsson M, et al. 2001. Extracellular phospholipase activity is a virulence factor for *Cryptococcus neoformans*. *Mol. Microbiol.* 39:166–75

35. Cox GM, Mukherjee J, Cole GT, Casadevall A, Perfect JR. 2000. Urease as a virulence factor in experimental Cryptococcosis. *Infect. Immun.* 68:443–48

36. Cox GM, Toffaletti DL, Perfect JR. 1996. Dominant selection system for use in *Cryptococcus neoformans*. *J. Med. Vet. Mycol.* 34:385–91

37. Cruickshank JG, Cavill R, Jelbert M. 1973. *Cryptococcus neoformans* of unusual morphology. *Appl. Microbiol.* 25:309–12

38. Cruz MC, Cavallo LM, Görlach JM, Cox G, Perfect JR, et al. 1999. Rapamycin antifungal action is mediated via conserved complexes with FKBP12 and TOR kinase homologs in *Cryptococcus neoformans*. *Mol. Cell. Biol.* 19:4101–12

39. Cruz MC, Fox DS, Heitman J. 2001. Calcineurin is required for hyphal elongation during mating and haploid fruiting

in *Cryptococcus neoformans*. *EMBO J.* 20:1020–32

40. Cruz MC, Sia RAL, Olson M, Cox GM, Heitman J. 2000. Comparison of the roles of calcineurin in physiology and virulence in serotype D and serotype A strains of *Cryptococcus neoformans*. *Infect. Immun.* 68:982–85

41. D'Souza C, Heitman J. 2001. Conserved cAMP signaling cascades regulate fungal development and virulence. *FEMS Microbiol. Rev.* 25:349–64

42. D'Souza CA, Alspaugh JA, Yue C, Harashima T, Cox GM, et al. 2001. Cyclic AMP-dependent protein kinase controls virulence of the fungal pathogen *Cryptococcus neoformans*. *Mol. Cell. Biol.* 21:3179–91

43. Davidson RC, Blankenship JR, Kraus PR, De Jesus Berrios M, Hull CM, et al. 2002. A PCR-based strategy to generate integrative targeting alleles with large regions of homology. *Microbiology.* 148:2607–15

44. Davidson RC, Cox GM, Perfect JR, Heitman J. 2002. Divergence of a MAP kinase cascade into cell type specific and nonspecific elements imparts signaling specificity in fungal mating and differentiation. *Mol. Microbiol.* Submitted

45. Davidson RC, Cruz MC, Sia RAL, Allen BM, Alspaugh JA, Heitman J. 2000. Gene disruption by biolistic transformation in serotype D strains of *Cryptococcus neoformans*. *Fungal Genet. Biol.* 29:38–48

46. Davidson RC, Moore TDE, Odom AR, Heitman J. 2000. Characterization of the MFα pheromone of the human fungal pathogen *Cryptococcus neoformans*. *Mol. Microbiol.* 39:1–12

47. Del Poeta M, Toffaletti DL, Rude TH, Dykstra CC, Heitman J, Perfect JR. 1999. Topoisomerase I is essential in *Cryptococcus neoformans*: role in pathobiology and as an antifungal target. *Genetics* 152:167–78

48. Del Poeta M, Toffaletti DL, Rude TH, Sparks SD, Heitman J, Perfect JR.

1999. *Cryptococcus neoformans* differential gene expression detected in vitro and in vivo with green fluorescent protein. *Infect. Immun.* 67:1812–20

49. Doering TL. 1999. A unique α-1,3 mannosyltransferase of the pathogenic fungus *Cryptococcus neoformans*. *J. Bacteriol.* 181:5482–88

50. Doering TL. 2000. How does *Cryptococcus* get its coat? *Trends Microbiol.* 8:547–53

51. Dromer F, Mathoulin S, Dupont B, Laporte A. 1996. Epidemiology of *Cryptococcosis* in France: a 9-year survey (1985–1993). *Clin. Infect. Dis.* 23:82–90

52. Dromer F, Ronin O, Dupont B. 1992. Isolation of *Cryptococcus neoformans* var. *gattii* from an Asian patient in France: evidence for dormant infection in healthy subjects. *J. Med. Vet. Mycol.* 30:395–97

53. Edman JC. 1992. Isolation of telomere-like sequences from *Cryptococcus neoformans* and their use in high-efficiency transformation. *Mol. Cell. Biol.* 12:2777–83

54. Edman JC, Kwon-Chung KJ. 1990. Isolation of the *URA5* gene from *Cryptococcus neoformans* var. *neoformans* and its use as a selective marker for transformation. *Mol. Cell. Biol.* 10:4538–44

55. Ellis DH, Pfeiffer TJ. 1990. Ecology, life cycle, and infectious propagule of *Cryptococcus neoformans*. *Lancet* 336:923–25

56. Ellis DH, Pfeiffer TJ. 1990. Natural habitat of *Cryptococcus neoformans* var. *gattii*. *J. Clin. Microbiol.* 28:1642–44

57. Emmons CW. 1955. Saprophytic sources of *Cryptococcus neoformans* associated with the pigeon (*Columbia livia*). *Am. J. Hyg.* 62:227–52

58. Erdman S, Snyder M. 2001. A filamentous growth response mediated by the yeast mating pathway. *Genetics* 159:919–28

59. Erickson T, Liu L, Gueyikian A, Zhu X, Gibbons J, Williamson PR. 2001. Multiple virulence factors of *Cryptococcus neoformans* are dependent on *VPH1*. *Mol. Microbiol.* 42:1121–31

60. Forche A, Xu J, Vilgalys R, Mitchell TG. 2000. Development and characterization of a genetic linkage map of *Cryptococcus neoformans* var. *neoformans* using amplified fragment length polymorphisms and other markers. *Fungal Genet. Biol.* 31:189–203

61. Fowler TJ, Mitton MF, Vaillancourt LJ, Raper CA. 2001. Changes in mate recognition through alterations of pheromones and receptors in the multisexual mushroom fungus *Schizophyllum commune*. *Genetics* 158:1491–503

62. Fox DS, Cruz MC, Sia RAL, Ke H, Cox GM, et al. 2001. Calcineurin regulatory subunit is essential for virulence and mediates interactions with FKBP12-FK506 in *Cryptococcus neoformans*. *Mol. Microbiol.* 39:835–49

63. Fyfe M, Black W, Romney M, Kibsey P, MacDougall L, et al. 2002. *Unprecedented outbreak of* Cryptococcus neoformans var. gattii *infections in British Columbia, Canada*. Presented at Int. Conf. *Cryptococcus* and Cryptococcosis, 5th, Adelaide, Aust.

64. Galitski T, Saldanha AJ, Styles CA, Lander ES, Fink GR. 1999. Ploidy regulation of gene expression. *Science* 285:251–54

65. Garcia-Hermoso D, Janbon G, Dromer F. 1999. Epidemiological evidence for dormant *Cryptococcus neoformans* infection. *J. Clin. Microbiol.* 37:3204–9

66. Garcia-Hermoso D, Mathoulin-Pelissier S, Couprie B, Ronin O, Dupont B, Dromer F. 1997. DNA typing suggests pigeon droppings as a source of pathogenic *Cryptococcus neoformans* serotype D. *J. Clin. Microbiol.* 35:2683–85

67. Gimeno CJ, Ljungdahl PO, Styles CA, Fink GR. 1992. Unipolar cell divisions in the yeast *S. cerevisiae* lead to filamentous growth: regulation by starvation and *RAS*. *Cell* 68:1077–90

68. Goldman D, Lee SC, Casadevall A. 1994. Pathogenesis of pulmonary *Cryptococcus neoformans* infection in the rat. *Infect. Immun.* 62:4755–61

69. Goldman DL, Casadevall A, Cho Y, Lee SC. 1996. *Cryptococcus neoformans* meningitis in the rat. *Lab. Invest.* 75:759–70

70. Goldman DL, Khine H, Abadi J, Lindenberg DJ, Pirofski L, et al. 2001. Serologic evidence for *Cryptococcus neoformans* infection in early childhood. *Pediatrics* 107:1–6

71. Goldman DL, Lee SC, Mednick AJ, Montella L, Casadevall A. 2000. Persistent *Cryptococcus neoformans* pulmonary infection in the rat is associated with intracellular parasitism, decreased inducible nitric oxide synthase expression, and altered antibody responsiveness to cryptococcal polysaccharide. *Infect. Immun.* 68:832–38

72. Görlach J, Fox DS, Cutler NS, Cox GM, Perfect JR, Heitman J. 2000. Identification and characterization of a highly conserved calcineurin binding protein, Cbp1/calcipressin, in *Cryptococcus neoformans*. *EMBO J.* 19:3618–29

73. Görlach JM, McDade HC, Perfect JR, Cox GM. 2002. Antisense repression in *Cryptococcus neoformans* as a laboratory tool and potential antifungal strategy. *Microbiology* 148:213–19

74. Granger DL, Perfect JR, Durack DT. 1985. Virulence of *Cryptococcus neoformans*: regulation of capsule synthesis by carbon dioxide. *J. Clin. Invest.* 76:508–16

75. Halliday CL, Bui T, Krockenberger M, Malik R, Ellis DH, Carter DA. 1999. Presence of α and **a** mating types in environmental and clinical collections of *Cryptococcus neoformans* var. *gattii* strains from Australia. *J. Clin. Microbiol.* 37:2920–26

76. Harashima T, Heitman J. 2002. The Gα protein Gpa2 controls yeast differentiation by interacting with kelch repeat proteins that mimic Gβ subunits. *Mol. Cell* 10:163–73

77. Hartmann HA, Kahmann R, Bölker M. 1996. The pheromone response factor coordinates filamentous growth and pathogenicity in *Ustilago maydis*. *EMBO J.* 15:1632–41

78. Hartmann HA, Krüger J, Lottspeich F, Kahmann R. 1999. Environmental signals controlling sexual development of the corn smut fungus *Ustilago maydis* through the transcriptional regulator Prf1. *Plant Cell* 11:1293–305

79. Heitman J, Allen B, Alspaugh JA, Kwon-Chung KJ. 1999. On the origins of the congenic *MAT*α and *MAT***a** strains of the pathogenic yeast *Cryptococcus neoformans*. *Fungal Genet. Biol.* 28:1–5

80. Heitman J, Casadevall A, Lodge JK, Perfect JR. 1999. The *Cryptococcus neoformans* genome sequencing project. *Mycopathologia* 148:1–7

81. Herskowitz I, Rine J, Strathern J. 1992. Mating-type determination and mating-type interconversion in *Saccharomyces cerevisiae*. In *The Molecular and Cellular Biology of the Yeast Saccharomyces*. Vol. 2. *Gene Expression*, ed. EW Jones, JR Pringle, JR Broach, pp. 583–656. Cold Spring Harbor Lab., NY: Cold Spring Harbor Press

82. Hua J, Meyer JD, Lodge JK. 2000. Development of positive selectable markers for the fungal pathogen *Cryptococcus neoformans*. *Clin. Diagn. Lab. Immunol.* 7:125–28

83. Hull CM, Davidson RC, Heitman J. 2002. A novel region of the *Cryptococcus neoformans* mating type locus encodes Sxi1α, a homeodomain protein that controls cell fate in this human pathogen. Submitted

84. Hull CM, Johnson AD. 1999. Identification of a mating type-like locus in the asexual pathogenic yeast *Candida albicans*. *Science* 285:1271–5

85. Ito M, Kobayashi M. 2002. *Clinical and histopathologic analysis of cryptococcal infection at autopsy: a review of twenty-eight cases*. Presented at Int. Conf. *Cryptococcus* and Cryptococcosis, 5th, Adelaide, Aust.

86. Ito N, Phillips SE, Stevens C, Ogel ZB, McPherson MJ, et al. 1991. Novel

thioether bond revealed by a 1.7 Å crystal structure of galactose oxidase. *Nature* 350:87–90

87. Jacobson ES, Ayers DJ, Harrell AC, Nicholas CC. 1982. Genetic and phenotypic characterization of capsule mutants of *Cryptococcus neoformans. J. Bacteriol.* 150:1292–96

88. Jacobson ES, Jenkins ND, Todd JM. 1994. Relationship between superoxide dismutase and melanin in a pathogenic fungus. *Infect. Immun.* 62:4085–86

89. Janbon G, Himmelreich U, Moyrand F, Improvisi L, Dromer F. 2001. Cas1p is a membrane protein necessary for the *O*-acetylation of the *Cryptococcus neoformans* capsular polysaccharide. *Mol. Microbiol.* 42:453–67

90. Johnson AD. 1995. Molecular mechanisms of cell-type determination in budding yeast. *Curr. Opin. Genet. Dev.* 5:552–58

91. Karos M, Chang YC, McClelland CM, Clarke DL, Fu J, et al. 2000. Mapping of the *Cryptococcus neoformans MATα* locus: presence of mating type-specific mitogen-activated protein kinase cascade homologs. *J. Bacteriol.* 182:6222–27

92. Kozel TR, Cazin JJ. 1971. Nonencapsulated variant of *Cryptococcus neoformans*. I. Virulence studies and characterization of soluble polysaccharide. *Infect. Immun.* 3:287–94

93. Kraakman L, Lemaire K, Ma P, Teunissen AWRH, Donaton MCV, et al. 1999. A *Saccharomyces cerevisiae* G-protein coupled receptor, Gpr1, is specifically required for glucose activation of the cAMP pathway during the transition to growth on glucose. *Mol. Microbiol.* 32:1002–12

94. Kronstad JW, Staben C. 1997. Mating type in filamentous fungi. *Annu. Rev. Genet.* 31:245–76

95. Kwon-Chung KJ. 1975. A new genus, *Filobasidiella*, the perfect state of *Cryptococcus neoformans. Mycologia* 67:1197–200

96. Kwon-Chung KJ. 1976. Morphogenesis of *Filobasidiella neoformans*, the sexual state of *Cryptococcus neoformans. Mycologia* 68:821–33

97. Kwon-Chung KJ. 1976. A new species of *Filobasidiella*, the sexual state of *Cryptococcus neoformans* B and C serotypes. *Mycologia* 68:943–46

98. Kwon-Chung KJ. 1980. Nuclear genotypes of spore chains in *Filobasidiella neoformans* (*Cryptococcus neoformans*). *Mycologia* 72:418–22

99. Kwon-Chung KJ, Bennett JE. 1978. Distribution of α and a mating types of *Cryptococcus neoformans* among natural and clinical isolates. *Am. J. Epidemiol.* 108:337–40

100. Kwon-Chung KJ, Bennett JE, Rhodes JC. 1982. Taxonomic studies on *Filobasidiella* species and their anamorphs. *Antonie van Leeuwenhoek J. Microbiol. Serol.* 48:25–38

101. Kwon-Chung KJ, Edman JC, Wickes BL. 1992. Genetic association of mating types and virulence in *Cryptococcus neoformans. Infect. Immun.* 60:602–5

102. Kwon-Chung KJ, Polacheck I, Popkin TJ. 1982. Melanin-lacking mutants of *Cryptococcus neoformans* and their virulence for mice. *J. Bacteriol.* 150:1414–21

103. Kwon-Chung KJ, Rhodes JC. 1986. Encapsulation and melanin formation as indicators of virulence in *Cryptococcus neoformans. Infect. Immun.* 51:218–23

104. Kwon-Chung KJ, Varma A, Edman JC, Bennett JE. 1992. Selection of *ura5* and *ura3* mutants from the two varieties of *Cryptococcus neoformans* on 5-fluoroorotic acid medium. *J. Med. Vet. Mycol.* 30:61–69

105. Laurenson IF, Lalloo DG, Naraqi S, Seaton RA, Trevett AJ, et al. 1997. *Cryptococcus neoformans* in Papua New Guinea: a common pathogen but an elusive source. *J. Med. Vet. Mycol.* 35:437–40

106. Lazera MS, Cavalcanti MAS, Londero AT, Trilles L, Nishikawa MM, Wanke B. 2000. Possible primary ecological niche

of *Cryptococcus neoformans. Med. Mycol.* 38:379–83

107. Lazera MS, Pires FD, Camillo-Coura L, Nishikawa MM, Bezerra CC, et al. 1996. Natural habitat of *Cryptococcus neoformans* var. *neoformans* in decaying wood forming hollows in living trees. *J. Med. Vet. Mycol.* 34:127–31

108. Lengeler KB, Cox GM, Heitman J. 2001. Serotype AD strains of *Cryptococcus neoformans* are diploid or aneuploid and are heterozygous at the mating-type locus. *Infect. Immun.* 69:115–22

109. Lengeler KB, Davidson RC, D'Souza C, Harashima T, Shen W-C, et al. 2000. Signal transduction cascades regulating fungal development and virulence. *Microbiol. Mol. Biol. Rev.* 64:746–85

110. Lengeler KB, Fox DS, Fraser JA, Allen A, Forrester K, et al. 2002. The mating type locus of *Cryptococcus neoformans*: a step in the evolution of sex chromosomes. *Eukaryot. Cell.* In press

111. Lengeler KB, Heitman J. 2002. *Cryptococcus neoformans* as a model fungal pathogen. In *Molecular Biology of Fungal Development*, ed. HD Osiewacz, pp. 513–57. New York: Dekker

112. Lengeler KB, Wang P, Cox GM, Perfect JR, Heitman J. 2000. Identification of the *MATa* mating-type locus of *Cryptococcus neoformans* reveals a serotype A *MATa* strain thought to have been extinct. *Proc. Natl. Acad. Sci. USA* 97:14455–60

113. Linden H, Ballario P, Macino G. 1997. Blue light regulation in *Neurospora crassa. Fungal Genet. Biol.* 22:141–50

114. Liu H, Cottrell TR, Pierini LM, Goldman WE, Doering TL. 2002. RNA interference in the pathogenic fungus *Cryptococcus neoformans. Genetics* 160:463–70

115. Liu H, Styles CA, Fink GR. 1993. Elements of the yeast pheromone response pathway required for filamentous growth of diploids. *Science* 262:1741–44

116. Lodge JK, Jackson-Machelski E, Toffaletti DL, Perfect JR, Gordon JI. 1994. Targeted gene replacement demonstrates that myristoyl-CoA: protein *N*-myristoyltransferase is essential for viability of *Cryptococcus neoformans. Proc. Natl. Acad. Sci. USA* 91:12008–12

117. Lorenz MC, Heitman J. 1997. Yeast pseudohyphal growth is regulated by Gpa2, a G-protein α homolog. *EMBO J.* 16:7008–18

118. Lorenz MC, Pan X, Harashima T, Cardenas ME, Xue Y, et al. 2000. The G protein-coupled receptor Gpr1 is a nutrient sensor that regulates pseudohyphal differentiation in *Saccharomyces cerevisiae. Genetics* 154:609–22

119. Love GL, Boyd GD, Greer DL. 1985. Large *Cryptococcus neoformans* isolated from brain abscess. *J. Clin. Microbiol.* 22:1068–70

120. Madhani HD, Styles CA, Fink GR. 1997. MAP kinases with distinct inhibitory functions impart signaling specificity during yeast differentiation. *Cell* 91:673–84

121. McClelland CM, Fu J, Woodlee GL, Seymour TS, Wickes BL. 2002. Isolation and characterization of the *Cryptococcus neoformans MATa* pheromone gene. *Genetics* 160:935–47

122. McCusker JH, Clemons KV, Stevens DA, Davis RW. 1994. Genetic characterization of pathogenic *Saccharomyces cerevisiae* isolates. *Genetics* 136:1261–69

123. McDade HC, Cox GM. 2001. A new dominant selectable marker for use in *Cryptococcus neoformans. Med. Mycol.* 39:151–54

124. Mitchell AP. 1994. Control of meiotic gene expression in *Saccharomyces cerevisiae. Microbiol. Rev.* 58:56–70

125. Mitchell AP, Herskowitz I. 1986. Activation of meiosis and sporulation by repression of the *RME1* product in yeast. *Nature* 319:738–42

126. Mitchell TG, Perfect JR. 1995. Cryptococcosis in the era of AIDS—100 years after the discovery of *Cryptococcus neoformans. Clin. Microbiol. Rev.* 8:515–48

127. Mondon P, Chang YC, Varma A, Kwon-Chung KJ. 2000. A novel episomal shuttle

vector for transformation of *Cryptococcus neoformans* with the *ccdB* gene as a positive selection marker in bacteria. *FEMS Microbiol. Lett.* 187:41–45

128. Montenegro H, Paula CR. 2000. Environmental isolation of *Cryptococcus neoformans* var. *gattii* and *C. neoformans* var. *neoformans* in the city of Sao Paulo, Brazil. *Med. Mycol.* 38:385–90

129. Moore TDE, Edman JC. 1993. The α mating-type locus of *Cryptococcus neoformans* contains a peptide pheromone gene. *Mol. Cell. Biol.* 13:1962–70

130. Mösch HU, Kubler E, Krappmann S, Fink GR, Braus GH. 1999. Crosstalk between the Ras2p-controlled mitogen-activated protein kinase and cAMP pathways during invasive growth of *Saccharomyces cerevisiae*. *Mol. Biol. Cell* 10:1325–35

131. Nelson RT, Hua J, Pryor B, Lodge JK. 2001. Identification of virulence mutants of the fungal pathogen *Cryptococcus neoformans* using signature-tagged mutagenesis. *Genetics* 157:935–47

132. Nielsen O, Davey J, Egel R. 1992. The *ras1* function of *Schizosaccharomyces pombe* mediates pheromone-induced transcription. *EMBO J.* 11:1391–95

133. Nosanchuk JD, Rosas AL, Lee SC, Casadevall A. 2000. Melanisation of *Cryptococcus neoformans* in human brain tissue. *Lancet* 355:2049–50

134. Nosanchuk JD, Valadon P, Feldmesser M, Casadevall A. 1999. Melanization of *Cryptococcus neoformans* in murine infection. *Mol. Cell. Biol.* 19:745–50

135. Nunez M, Peacock JE, Chin RJ. 2000. Pulmonary cryptococcosis in the immunocompetent host. Therapy with oral fluconazole: a report of four cases and a review of the literature. *Chest* 118:527–34

136. Odom A, Muir S, Lim E, Toffaletti DL, Perfect J, Heitman J. 1997. Calcineurin is required for virulence of *Cryptococcus neoformans*. *EMBO J.* 16:2576–89

137. Olesnicky NS, Brown AJ, Dowell SJ, Casselton LA. 1999. A constitutively active G-protein-coupled receptor causes mating self-compatibility in the mushroom *Coprinus*. *EMBO J.* 18:2756–63

138. Pan X, Heitman J. 1999. Cyclic AMP-dependent protein kinase regulates pseudohyphal differentiation in *Saccharomyces cerevisiae*. *Mol. Cell. Biol.* 19:4874–87

139. Pan X, Heitman J. 2000. Sok2 regulates yeast pseudohyphal differentiation via a transcription factor cascade that regulates cell-cell adhesion. *Mol. Cell. Biol.* 20:8364–72

140. Pan X, Heitman J. 2002. Protein kinase A operates a molecular switch that governs yeast pseudohyphal differentiation. *Mol. Cell. Biol.* 22:3981–93

141. Parkhurst SM, Meneely PM. 1994. Sex determination and dosage compensation: lessons from flies and worms. *Science* 264:924–32

142. Perfect JR, Lang SDR, Durack DT. 1980. Chronic cryptococcal meningitis: a new experimental model in rabbits. *Am. J. Pathol.* 101:177–94

143. Perfect JR, Toffaletti DL, Rude TH. 1993. The gene encoding phosphoribosylaminoimidazole carboxylase (*ADE2*) is essential for growth of *Cryptococcus neoformans* in cerebrospinal fluid. *Infect. Immun.* 61:4446–51

144. Pierini LM, Doering TL. 2001. Spatial and temporal sequence of capsule construction in *Cryptococcus neoformans*. *Mol. Microbiol.* 41:105–15

145. Rhodes JC, Polacheck I, Kwon-Chung KJ. 1982. Phenoloxidase activity and virulence in isogenic strains of *Cryptococcus neoformans*. *Infect. Immun.* 36:1175–84

146. Roberts CJ, Nelson B, Marton MJ, Stoughton R, Meyer MR, et al. 2000. Signaling and circuitry of multiple MAPK pathways revealed by a matrix of global gene expression profiles. *Science* 287:873–80

147. Roberts RL, Fink GR. 1994. Elements of a single MAP kinase cascade in *Saccharomyces cerevisiae* mediate two developmental programs in the same cell type:

mating and invasive growth. *Genes Dev.* 8:2974–85

148. Robertson LS, Fink GR. 1998. The three yeast A kinases have specific signaling functions in pseudohyphal growth. *Proc. Natl. Acad. Sci. USA* 95:13783–87

149. Rosas AL, Nosanchuk JD, Feldmesser M, Cox GM, McDade HC, Casadevall A. 2000. Synthesis of polymerized melanin by *Cryptococcus neoformans* in infected rodents. *Infect. Immun.* 68:2845–53

150. Rose MD, Price BR, Fink GR. 1986. *Saccharomyces cerevisiae* nuclear fusion requires prior activation by α factor. *Mol. Cell. Biol.* 6:3490–97

151. Salas SD, Bennett JE, Kwon-Chung KJ, Perfect JR, Williamson PR. 1996. Effect of the laccase gene, *CNLAC1*, on virulence of *Cryptococcus neoformans*. *J. Exp. Med.* 184:377–86

152. Schein J, Tangen K, Chiu R, Shin H, Lengeler KB, et al. 2002. Physical maps for genome analysis of serotype A and D strains of the fungal pathogen *Cryptococcus neoformans*. *Genome Res.* In press

153. Schmeding KA, Jong SC, Hugh R. 1981. Sexual compatibility between serotypes of *Filobasidiella neoformans* (*Cryptococcus neoformans*). *Curr. Microbiol.* 5:133–38

154. Shen W-C, Davidson RC, Cox GM, Heitman J. 2002. Pheromones control mating and differentiation via paracrine and autocrine signaling in *Cryptococcus neoformans*. *Eukaryot. Cell* 1:366–77

155. Sia RA, Lengeler KB, Heitman J. 2000. Diploid strains of the pathogenic basidiomycete *Cryptococcus neoformans* are thermally dimorphic. *Fungal Genet. Biol.* 29:153–63

156. Smulian AG, Sesterhenn T, Tanaka R, Cushion MT. 2001. The *STE3* pheromone receptor gene of *Pneumocystis carinii* is surrounded by a cluster of signal transduction genes. *Genetics* 157:991–1002

157. Sorrell TC. 2001. *Cryptococcus neoformans* variety *gattii*. *Med. Mycol.* 39:155–68

158. Staib F. 1962. *Cryptococcus neoformans* und *Guizotia abyssinica* (syn. G. *oleifera*) Farbreaktion für *Cr. neoformans*. *Z. Hyg.* 148:466–75

159. Steenbergen JN, Casadevall A. 2000. Prevalence of *Cryptococcus neoformans* var. *neoformans* (serotype D) and *Cryptococcus neoformans* var. *grubii* (serotype A) isolates in New York City. *J. Clin. Microbiol.* 38:1974–76

160. Steenbergen JN, Shuman HA, Casadevall A. 2001. *Cryptococcus neoformans* interactions with amoebae suggest an explanation for its virulence and intracellular pathogenic strategy in macrophages. *Proc. Natl. Acad. Sci. USA* 98:15245–50

161. Steinmetz LM, Sinha H, Richards DR, Spiegelman JI, Oefner PJ, et al. 2002. Dissecting the architecture of a quantitative trait locus in yeast. *Nature* 416:326–30

162. Still CN, Jacobson ES. 1983. Recombinational mapping of capsule mutations in *Cryptococcus neoformans*. *J. Bacteriol.* 156:460–62

163. Strachan AA, Yu RJ, Blank F. 1971. Pigment production of *Cryptococcus neoformans* grown with extracts of *Guizotia abyssinica*. *Appl. Microbiol.* 22:478–79

164. Sudarshan S, Davidson RC, Heitman J, Alspaugh JA. 1999. Molecular analysis of the *Cryptococcus neoformans ADE2* gene, a selectable marker for transformation and gene disruption. *Fungal Genet. Biol.* 27:36–48

165. Sukroongreung S, Kitiniyom K, Nilakul C, Tantimavanich S. 1998. Pathogenicity of basidiospores of *Filobasidiella neoformans* var. *neoformans*. *Med. Mycol.* 36:419–24

166. Swinne-Desgain D. 1974. The pigeon as reservoir for *Cryptococcus neoformans*. *Lancet* 2:842–43

167. Takeo K, Tanaka R, Miyaji M, Nishimura K. 1995. Unbudded G2 as well as G1 arrest in the stationary phase of the basidiomycetous yeast *Cryptococcus neoformans*. *FEMS Microbiol. Lett.* 129:231–35

168. Takizawa PA, DeRisi JL, Wilhelm JE, Vale RD. 2000. Plasma membrane compartmentalization in yeast by messenger RNA transport and a septin diffusion barrier. *Science* 290:341–44

169. Thompson JR, Douglas CM, Li W, Jue CK, Pramanik B, et al. 1999. A glucan synthase *FKS1* homolog in *Cryptococcus neoformans* is single copy and encodes an essential function. *J. Bacteriol.* 181:444–53

170. Toffaletti DL, Rude TH, Johnston SA, Durack DT, Perfect JR. 1993. Gene transfer in *Cryptococcus neoformans* by use of biolistic delivery of DNA. *J. Bacteriol.* 175:1405–11

171. Tolkacheva T, McNamara P, Piekarz E, Courchesne W. 1994. Cloning of a *Cryptococcus neoformans* gene, *GPA1*, encoding a G-protein α-subunit homolog. *Infect. Immun.* 62:2849–56

172. Torres-Guererro H, Edman JC. 1994. Melanin-deficient mutants of *Cryptococcus neoformans*. *J. Med. Vet. Mycol.* 32:303–13

173. Umen JG, Goodenough UW. 2001. Chloroplast DNA methylation and inheritance in *Chlamydomonas*. *Genes Dev.* 15:2585–97

174. Uno J, Tanaka R, Branchini ML, Aoki FH, Yarita K, et al. 2001. Atypical *Cryptococcus neoformans* isolate from an HIV-infected patient in Brazil. *Nippon Ishinkin Gakkai Zasshi* 42:127–32

175. Urban M, Kahmann R, Bölker M. 1996. The biallelic *a* mating type locus of *Ustilago maydis*: remnants of an additional pheromone gene indicate evolution from a multiallelic ancestor. *Mol. Gen. Genet.* 250:414–20

176. Varma A, Kwon-Chung KJ. 1998. Construction of stable episomes in *Cryptococcus neoformans*. *Curr. Genet.* 34:60–66

177. Vartivarian SE, Anaissie EJ, Cowart RE, Sprigg HA, Tingler MJ, Jacobson ES. 1993. Regulation of cryptococcal capsular polysaccharide by iron. *J. Infect. Dis.* 167:186–90

178. Viviani MA, Esposto MC, Cogliati M, Montagna MT, Wickes BL. 2001. Isolation of a *Cryptococcus neoformans* serotype A *MATα* strain from the Italian environment. *Med. Mycol.* 39:383–86

179. Viviani MA, Wen H, Roverselli A, Caldarelli-Stefano R, Cogliati M, et al. 1997. Identification by polymerase chain reaction fingerprinting of *Cryptococcus neoformans* serotype AD. *J. Med. Vet. Mycol.* 35:355–60

180. Wang P, Cardenas ME, Cox GM, Perfect J, Heitman J. 2001. Two cyclophilin A homologs with shared and distinct functions important for growth and virulence of *Cryptococcus neoformans*. *EMBO Rep.* 2:511–18

181. Wang P, Heitman J. 1999. Signal transduction cascades regulating mating, filamentation, and virulence in *Cryptococcus neoformans*. *Curr. Opin. Microbiol.* 2:358–62

182. Wang P, Nichols CB, Lengeler KB, Cardenas ME, Cox GM, et al. 2002. Mating-type-specific and nonspecific PAK kinases play shared and divergent roles in *Cryptococcus neoformans*. *Eukaryot. Cell* 1:257–72

183. Wang P, Perfect JR, Heitman J. 2000. The G-protein β subunit Gpb1 is required for mating and haploid fruiting in *Cryptococcus neoformans*. *Mol. Cell. Biol.* 20:352–62

184. Waugh MS, Nichols CB, DeCesare CM, Cox GM, Heitman J, Alspaugh JA. 2002. Ras1 and Ras2 contribute shared and unique roles in physiology and virulence of *Cryptococcus neoformans*. *Microbiology* 148:191–201

185. Waugh MS, Vallim MA, Heitman J, Alspaugh JA. 2002. Ras1 controls pheromone expression and response during mating in *Cryptococcus neoformans*. *Fungal Genet. Biol.* In press

186. Welton RM, Hoffman CS. 2000. Glucose monitoring in fission yeast via the gpa2 Gα, the git5 Gβ and the git3 putative glucose receptor. *Genetics* 156:513–21

187. Whelan WL, Kwon-Chung KJ. 1986.

Genetic complementation in *Cryptococcus neoformans. J. Bacteriol.* 166:924–29

188. White CW, Jacobson ES. 1985. Occurrence of diploid strains of *Cryptococcus neoformans. J. Bacteriol.* 161:1231–32

189. Wickes BL, Edman U, Edman JC. 1997. The *Cryptococcus neoformans STE12α* gene: a putative *Saccharomyces cerevisiae STE12* homologue that is mating type specific. *Mol. Microbiol.* 26:951–60

190. Wickes BL, Mayorga ME, Edman U, Edman JC. 1996. Dimorphism and haploid fruiting in *Cryptococcus neoformans*: association with the α-mating type. *Proc. Natl. Acad. Sci. USA* 93:7327–31

191. Williamson PR. 1994. Biochemical and molecular characterization of the diphenol oxidase of *Cryptococcus neoformans*: identification as a laccase. *J. Bacteriol.* 176:656–64

192. Xu J. 2002. Mitochondrial DNA polymorphisms in the human pathogenic fungus *Cryptococcus neoformans. Curr. Genet.* 41:43–47

193. Xu J, Ali RY, Gregory DA, Amick D, Lambert SE, et al. 2000. Uniparental mitochondrial transmission in sexual crosses in *Cryptococcus neoformans. Curr. Microbiol.* 40:269–73

194. Xu J, Luo G, Vilgalys RJ, Brandt ME, Mitchell TG. 2002. Multiple origins of hybrid strains of *Cryptococcus neoformans* with serotype AD. *Microbiology* 148: 203–12

195. Xu J, Vilgalys R, Mitchell TG. 2000. Multiple gene genealogies reveal recent dispersion and hybridization in the human pathogenic fungus *Cryptococcus neoformans. Mol. Ecol.* 9:1471–81

196. Xue Y, Batlle M, Hirsch JP. 1998. *GPR1* encodes a putative G protein-coupled receptor that associates with the Gpa2p Gα subunit and functions in a Ras-independent pathway. *EMBO J.* 17:1996–2007

197. Yan Z, Li X, Xu J. 2002. Geographic distribution of mating type alleles of *Cryptococcus neoformans* in four areas of the United States. *J. Clin. Microbiol.* 40:965–72

198. Yue C, Cavallo LM, Alspaugh JA, Wang P, Cox GM, et al. 1999. The *STE12α* homolog is required for haploid filamentation but largely dispensable for mating and virulence in *Cryptococcus neoformans. Genetics* 153:1601–15

199. Zeyl C, Bell G. 1997. The advantage of sex in evolving yeast populations. *Nature* 388:465–68

Annu. Rev. Genet. 2002. 36:617–56
doi: 10.1146/annurev.genet.36.060402.113540

TOWARD MAINTAINING THE GENOME:
DNA Damage and Replication Checkpoints

Kara A. Nyberg, Rhett J. Michelson, Charles W. Putnam, and Ted A. Weinert

Molecular and Cellular Biology Department, University of Arizona, Tucson, Arizona, 85721; e-mail: knyberg@email.arizona.edu, rhettm@email.arizona.edu, cwp@email.arizona.edu, tweinert@email.arizona.edu

Key Words genomic instability, repair, ATM, BRCA1, p53

■ **Abstract** DNA checkpoints play a significant role in cancer pathology, perhaps most notably in maintaining genome stability. This review summarizes the genetic and molecular mechanisms of checkpoint activation in response to DNA damage. The major checkpoint proteins common to all eukaryotes are identified and discussed, together with how the checkpoint proteins interact to induce arrest within each cell cycle phase. Also discussed are the molecular signals that activate checkpoint responses, including single-strand DNA, double-strand breaks, and aberrant replication forks. We address the connection between checkpoint proteins and damage repair mechanisms, how cells recover from an arrest response, and additional roles that checkpoint proteins play in DNA metabolism. Finally, the connection between checkpoint gene mutation and genomic instability is considered.

CONTENTS

INTRODUCTION

Growth and division of a single cell to yield two daughter cells requires the coordination of numerous events, in particular the faithful replication and partitioning of the cell's genetic material to each daughter cell. At the extreme, errors in this process could mean death for a unicellular organism, and for multicellular organisms, faulty cell division may ultimately culminate in developmental defects or oncogenesis.

To ensure the fidelity of division, cells have evolved general mechanisms called checkpoints that monitor the successful completion of cell cycle events. Checkpoints are typically not essential for cell cycle events per se, but rather they make certain that events are completed correctly and in the proper order. When one cell cycle event has not been successfully completed, checkpoints will delay progression until the step is correctly accomplished, and only then will they relieve the arrest to allow the cell to move to the next phase.

Given that cells are constantly under the assault of endogenous and exogenous forms of damage, maintaining a complete, undamaged genome is a continual challenge and of vital importance to the cell and future cellular generations. All eukaryotic cells, except certain embryonic cells, possess checkpoints to monitor virtually every cell cycle event involving DNA metabolism (86). Cells not only supervise the process of DNA replication during S-phase to ensure correct completion of this event, but they also monitor the state of DNA throughout the entire cell cycle to minimize the accruement of damage. In addition to arresting cells with compromised DNA, these checkpoints also mediate repair of the damage, a role still being elucidated. Therefore, these DNA checkpoints—or, perhaps more specifically, DNA maintenance checkpoints—include what may have traditionally been thought of as two classes: (a) the DNA damage checkpoints that recognize and respond to DNA damage, and (b) the replication checkpoints that monitor the fidelity of copying DNA.

The importance of DNA checkpoints in human pathology, especially cancer, is now well established. The p53 tumor suppressor, the most infamous of checkpoint genes, is the only one routinely mutated in sporadic human cancers [reviewed in (11)]. Although p53's role in G1 arrest was first described 11 years ago (98), our understanding of its function is incomplete. As for many checkpoint genes, p53 is not essential for cell viability yet plays key roles in at least two response pathways after DNA damage: G1 arrest and apoptosis [reviewed in (186)]. Early hypotheses that arrest provides time for DNA repair whereas apoptosis eliminates cells with irreparable DNA damage from cycling proves insufficient. In addition, how each p53 function—G1 arrest, G2 arrest, and apoptosis—relates to human cancer awaits the description of molecular mechanisms in specific cellular contexts.

Other human disease genes have links to cancer pathology, although again those links are unresolved. ATM (ataxia telangiectasia mutated) is a protein kinase central to all DNA maintenance responses that, when absent, results in the disease ataxia telangiectasia (AT). AT is characterized at the cellular level by gross chromosomal rearrangements and radioresistant DNA synthesis, and at the organismal level by immune deficiency, cerebellar degeneration, and an increased predisposition to cancer (107, 164). Not known is whether the disease is due to defects in DNA damage responses or to defects in some other cellular activity regulated by ATM. Humans harboring mutations in proteins that have recently been shown to interact with ATM in DNA damage recognition, Mre11 and Nbs1, also present with AT-like symptoms, thereby suggesting a connection to checkpoint and DNA repair defects (26, 180). Additional links include mutation of either Chk2 or p53, both of which act downstream of ATM, resulting in a genetic disease called Li Fraumeni syndrome that dramatically predisposes patients to cancer development (14, 125). Mutations in the checkpoint protein BRCA1, another substrate of ATM, result in defects in S-phase and G2 arrest (220) and also cause defects in DNA repair (1, 15, 135). Notably, an estimated 50% of inherited cases of breast cancer are likely due to mutations in the *BRCA1* gene, and *BRCA1* mutation is implicated in almost all families with histories of both ovarian and breast cancer (73). Such results clearly demonstrate that DNA checkpoints play a central role in human pathology, and we now must face the challenge of understanding the molecular intricacies as to how and why.

In the past few years, studies in organisms from budding and fission yeasts, the fungus Aspergillus, nematodes, Drosophila, mouse, to mammalian systems have combined to enhance our understanding of these DNA maintenance checkpoints in cell cycle biology. The recurrent theme is the importance of these checkpoints in maintaining genome stability and the correlation of loss of this function with the human pathology of cancer. Cells lacking functional checkpoints display genomic instability due to a failure to properly respond to DNA damage, faulty DNA replication, or aberrant chromosome segregation, resulting in an accelerated mutator phenotype (116). Recent understanding of molecular mechanisms summarized in this review may provide some insight into the causes of genomic instability and suggest strategies for the development of therapeutics. Of promise, researchers hope to exploit checkpoint defects in cancer cells to selectively kill them.

This review describes our current understanding of the molecular and genetic pathways of the DNA damage and replication checkpoints. We present working models based on available data from studies of various organisms, most notably those performed in mammalian cells and in budding and fission yeast, and point out that many molecular and biochemical details needed to support some of the proposed models are lacking. Checkpoint controls in different organisms indicate a high degree of conservation, although occasional divergences do occur.

THE CHECKPOINT PROTEINS

The checkpoint pathways involve three major groups of proteins that act in concert to translate the signal of damaged DNA into the response of cell cycle arrest and repair. These groups include (*a*) sensor proteins that recognize damaged DNA directly or indirectly and function to signal the presence of abnormalities, initiating a biochemical cascade of activity; (*b*) transducer proteins, typically protein kinases, that relay and amplify the damage signal from the sensors by phosphorylating other kinases or downstream target proteins; and (*c*) effector proteins, which include the most downstream targets of the transducer protein kinases, and are thus regulated, usually by phosphorylation, to prevent cell cycle progression. Table 1 summarizes the principal orthologous checkpoint proteins identified thus far in mammals and budding and fission yeasts. Within the text, the mammalian nomenclature is used

TABLE 1   Orthologous checkpoint proteins

Protein function	Mammals	*S. pombe*	*S. cerevisiae*
Sensors			
RFC1-like	Rad17	Rad17	Rad24
PCNA-like	Rad9	Rad9	Ddc1
	Rad1	Rad1	Rad17
	Hus1	Hus1	Mec3
BRCT-containing	BRCA1	Crb2/Rph9	Rad9
	TopBP1	Cut5	Dpb11
DSB recognition/repair	Mre11	Rad32	Mre11
	Rad50	Rad50	Rad50
	Nbs1		Xrs2
Replication proteins			
recruits polymerases	TopBP1	Cut5	Dpb11
needed for replication		Drc1	Drc1
DNA polymerase	Pol2	Cdc20	Pol2
DNA helicase	BLM, WRN*	Rhq1/Rad12	Sgs1
Topoisomerase	Top3	Top3	Top3
clamp loader	Rfc2–5	Rfc2–5	Rfc2–5
binds ssDNA	Rpa2		Rfa2
Transducers			
PI3-kinases (PIKK)	ATR	Rad3	Mec1
	ATM	Tel1	Tel1
PIKK binding partner	ATRIP	Rad26	Ddc2/Lcd1
Effector Kinases	Chk1	Chk1	Chk1
	Chk2	Cds1	Rad53
Replication fork	—	—	Tof1
stabilizers	Claspin	Mrc1	Mrc1

*WRN—mutated in Werner syndrome.

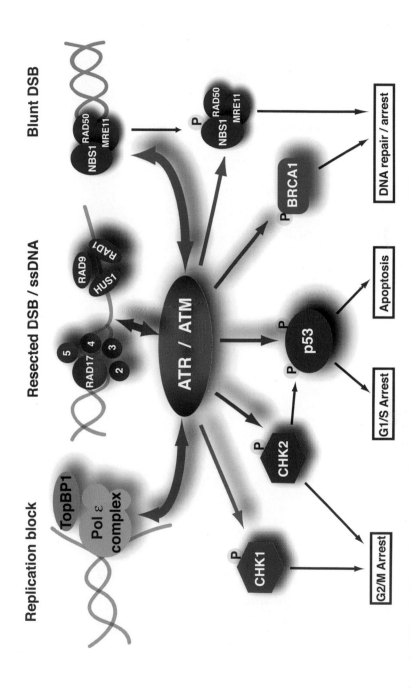

Figure 1 ATR and ATM are the central proteins regulating checkpoint responses to various forms of DNA damage. Activation of ATR and its downstream effects are indicated in red, whereas activation and downstream effects of ATM are indicated in blue (not absolute; some exceptions of overlapping function do occur). Signals and downstream effects common to both kinases are designated in purple.

unless otherwise stated. Below, we name and briefly discuss the primary proteins involved in DNA checkpoint responses. More detailed discussion of their roles follows in subsequent sections.

Sensors

Proteins have been placed in the sensor class based on genetic and biochemical findings or inference, as little direct biochemical evidence yet exists. Some sensors are thought to directly associate with damaged DNA; others are most likely indirectly associated by interactions with the aforementioned sensors. The types of sensor proteins that associate with a particular DNA lesion may serve as recognition complexes to modulate, recruit, and localize specific target proteins for that lesion type.

Rad17-RFC AND 9-1-1   These two complexes of proteins are presumed to act in concert with one another and to have functional analogs in DNA replication. As first shown in budding and fission yeasts, Rad17 interacts with four replication factor C subunits (Rfc2, Rfc3, Rfc4, Rfc5) to form a pentameric structure, referred to as the Rad17- RFC complex (79, 81, 114). Another complex is a heterotrimeric ring composed of Rad9, Hus1, and Rad1, termed the 9-1-1 complex (23, 28, 84, 179, 198, 201). The RFC subunits found in Rad17-RFC are more commonly recognized for their association with Rfc1 during replication to form what is termed the "clamp loader." The replication RFC clamp loader complex recognizes single-strand/double-strand DNA junctions and loads the homotrimeric PCNA "sliding clamp" complex, which encircles the DNA to act as a general scaffold upon which DNA polymerases and other DNA replication proteins are assembled (193). One infers from sequence and structural similarities that during a checkpoint response the Rad17-RFC complex may recognize damage and load the 9-1-1 sliding clamp onto DNA. Indeed, association of Rad17-RFC and the 9-1-1 heterotrimer has been identified in vitro, and each is inferred to bind to sites of DNA damage in vivo as the Rad17 -RFC complex binds preferentially to DNA with primer-template-like structures (114). The 9-1-1 complex may have an additional role of processing DNA damage to generate more single-stranded DNA, perhaps required to enhance damage signaling to other checkpoint components (122, 146). Although the two human Rad17-RFC and 9-1-1 complexes have been reconstituted individually in vitro from recombinant proteins, loading of the 9-1-1 complex onto DNA by Rad17-RFC has not yet been shown. A recent report, however, indicates that mammalian Rad17 recruits the 9-1-1 complex to sites of damage in vivo as in yeast, although the consequences of this association have yet to be elucidated (235).

BRCA1   BRCA1 appears to function as an adaptor of checkpoint initiation by localizing additional substrates for transducer kinase phosphorylation and by perhaps linking checkpoint arrest to DNA damage repair. BRCA1 colocalizes with a number of proteins involved in DNA repair and/or replication, including Rad51, PCNA, Mre11-Rad50-Nbs1, histone deacetylases, the DNA helicase BLM (mutated in Bloom's Syndrome), and mismatch repair proteins (167, 207, 225, 232). Additionally, BRCA1 directly binds specific DNA structures that are sequence independent in vitro (147). Consistent with these associations, BRCA1 affects homologous recombination, mismatch repair, and transcription-coupled repair (1, 77, 135, 178, 206, 207).

Based on functional and sequence similarities, we suggest that the fission yeast Crb2 and the budding yeast Rad9 proteins should be considered BRCA1 orthologs. All three proteins share limited sequence similarity in that each contains two BRCT (Brca1 Carboxy Terminal) domains, which are motifs most likely involved in dimerization [reviewed in (92)]. In addition, their regulation is similar: BRCA1, Crb2, and Rad9 are all phosphorylated by ATR and ATM orthologs after damage (39, 56, 92, 158, 190, 200), and each is required to activate downstream protein kinases (76, 80, 109, 158, 159, 226). Furthermore, each acts to induce a G2/M cell cycle arrest to allow time for DNA repair (158, 211, 215, 220). Disparities in other protein functions are expected, but the functional core similarities listed here warrant their being termed orthologs.

Mre11-Rad50-Nbs1   The Mre11-Rad50-Nbs1 (MRN) complex localizes to sites of double-strand breaks (DSBs) in vivo and plays vital roles in DNA metabolism, including DSB repair, meiotic recombination, and telomere maintenance (32, 50, 128, 234). Cells deficient for Mre11 or Nbs1 continue DNA replication after X-ray damage, known as radioresistant DNA synthesis (RDS), indicating defective checkpoint signaling during S-phase (170). Additionally, such mutations affect checkpoints at all phases of the cell cycle (24, 223).

REPLICATION PROTEINS   We have grouped these proteins because they are all localized at sites of replication forks, play roles in DNA replication, and have checkpoint defects within S-phase. This group includes budding yeast topoisomerase III [Top3 (31)], the Pol2 subunit of the DNA polymerase from budding yeast (138), the budding and fission yeast Drc1 protein required for DNA replication (138a, 203), the mammalian DNA helicase BLM and orthologous proteins (45, 52, 65, 66), the mammalian polymerase recruiter TopBP1 and orthologous proteins (8, 124, 199, 222), eukaryotic members of the RFC clamp loader complex (102, 137, 139, 172), and the Rpa heterotrimer that binds single-strand DNA in mammals and budding yeast (163, 205).

Transducers

Transducers include the protein kinases that, when activated by the presence of DNA damage, initiate a signal transduction cascade that propagates and amplifies the damage signal to ultimately cause cell cycle arrest.

ATM AND ATR   Both ATM and ATR (A̲T̲M and R̲ad3-related) are members of the phosphoinositide 3-kinase related kinases (PIKKs), which are large proteins ranging between 275-500 kDa that possess a unique protein kinase domain at their C termini with little sequence conservation outside this region (2).

Following observations in yeast, ATR has recently been shown to form a heterodimer with the associated protein ATRIP, which is required for the checkpoint signaling pathway (38, 54, 142, 157, 202, 216). The mechanism of ATR-ATRIP activation remains elusive. As ATR is a member of the PIKK family that includes DNA-PK, it is interesting to note that in vitro DNA-PK kinase activity is mostly activated upon association with the Ku heterodimer that binds double-stranded DNA ends (176). Whether ATRIP possesses detectable DNA binding activity or is required for protein kinase activity of ATR is not yet known, although ATR and ATRIP colocalize to intranuclear foci after DNA damage or inhibition of replication (38). In a recent budding yeast study, Rouse & Jackson show that the ATRIP ortholog, Ddc2 (also known as Lcd1/Pie1), can bind DNA independently of Mec1, the ATR ortholog, and that Ddc2 is required for the recruitment of Mec1 to sites of DNA damage (157a).

ATR in mammalian cells and Mec1 in budding yeast are essential for cell viability, whereas Rad3 in fission yeast is not. In *Saccharomyces cerevisiae*, this essentiality is due to Mec1's role in regulating dNTP levels (229, 231). As ATR's essential function has yet to be clearly defined, regulation of dNTP pools in mammalian cells is an obvious place to look. Alternately, because mammalian genomes are so large, the DNA damage sustained on a continuous basis may require the constant monitoring via ATR to ensure viability. For instance, the Rad51 recombination protein is not essential in yeast but it is in mammalian cells (194); perhaps mammalian and yeast genomes incur the same number of breaks per unit of DNA, but the mere fact that the mammalian genome is 250 times larger than the yeast genome makes repair proteins essential.

Chk1 AND Chk2   Chk1 and Chk2 are classic serine-threonine kinases that are required for cell cycle arrest in response to DNA damage. As downstream kinases, they are phosphorylated by ATM/ATR-dependent processes, may additionally undergo autophosphorylation [demonstrated for *S. cerevisiae* Rad53 (76) and Chk1 (202a)], and potentiate phosphorylation of downstream targets.

Effectors

Many targets of transducer phosphorylation have yet to be identified, and known targets are typically not well conserved across species. We discuss the effects of effector phosphorylation throughout the remaining sections only where appropriate.

THE MOLECULAR SIGNALS THAT
ACTIVATE CHECKPOINT RESPONSES

The nature of DNA structures that are recognized by the sensor proteins to activate checkpoint responses remains obscure. Even in the well-defined bacterial SOS response system that leads to activation of the RecA protein, defining the DNA damage signal in vivo has been a formidable task, though it is likely single-strand DNA (53). In eukaryotic cells, even the simplest hypothesis becomes complex when accounting for the type of damage, the quantity of lesions, and the various protein kinases and substrates involved. Results from in vivo studies that utilize irradiation or alkylating agents to induce DNA damage—both of which cause a vigorous checkpoint response—are ambiguous owing to the myriad of DNA lesions generated (209). Furthermore, there are no defined biochemical systems that can be used to determine mechanisms of activation in vitro.

Despite these difficulties, a prevailing view for eukaryotic cells is that different kinds of lesions are converted to single-strand DNA (ssDNA) and double-strand DNA breaks (DSBs), two common structures that then signal checkpoints. Moreover, the weight of evidence suggests that the two signals activate different arms of the checkpoint pathways: DSBs activate a pathway containing the ATM kinase family, whereas ssDNA activates a checkpoint pathway containing the ATR kinase family (Figure 1, see color insert).

Activation of the ATR checkpoint pathway by ssDNA rests on extrapolation from three key observations in budding yeast. In yeast, a single defined DSB generated by the HO endonuclease is rapidly converted to long tracts (>5 kb) of ssDNA by degradation of the 5′ strand (181). This ssDNA causes an acute arrest of chromosome segregation that requires the ATR homolog, Mec1, and presumably not the ATM homolog, Tel1 (161). In the second study, inactivation of Cdc13, a protein normally associated with telomeres that aids in their protection, leads to extensive ssDNA near telomere ends. This defect also elicits a profound Mec1-dependent arrest that is Tel1 independent (67, 69). Third, unrepaired DSBs (due to mutation of the RecA homolog Dmc1) generated during meiosis undergo 5′ end resection to create ssDNA and require Mec1 for meiotic arrest (121). The length or quantity of ssDNA required to activate arrest is not known, nor are the proteins needed to associate with ssDNA, though RPA is a plausible candidate (117, 163).

In budding yeast, evidence regarding the activation of Tel1 by DSBs is less extensive, as these breaks are typically converted rapidly to ssDNA. However, a recent study shows activation of Tel1 using cells where DSB resection is inferred to be blocked in meiosis by *rad50S* or *sae2* mutation (195). Meiotic cells with a resected DSB require Mec1 for arrest (121), whereas meiotic cells with an unresected DSB require Tel1 and, enigmatically, Mec1 for arrest (195). This study also shows that mitotic cells with a *mec1* mutation show extreme sensitivity to the damaging agent methylmethane sulfonate (MMS) (195). *mec1 sae2* or *mec1 rad50S* mutants, however, show partial suppression of sensitivity, re-establishment of downstream Rad53 phosphorylation, and restoration of cell cycle arrest, and this suppression requires an intact TEL1 gene. These data support a model whereby suppression of sensitivity in *mec1 sae2* or *mec1 rad50S* mutants is achieved because of enhanced activation of the parallel checkpoint pathway mediated by Tel1, which may specifically respond to DSBs. Elimination of both Mec1- and Tel1-dependent pathways results in enhanced lethality when DNA damage is incurred (195), presumably because both branches of the DNA damage response have been abrogated.

Many other indirect observations support the view that ssDNA activates ATR homologs and that DSBs activate ATM-like kinases. γ-irradiation of mammalian cells, which is thought to induce DSBs, causes rapid accumulation of ATM kinase activity, whereas ATR activity increases only at a later time, perhaps in response to DSB processing (217, 221). Moreover, immunoprecipitated ATM is more enzymatically active after extraction from cells treated with IR or radiomimetic agents (10, 25). In contrast, disruption of DNA replication in budding and fission yeasts and mammalian cells, thought to generate extensive ssDNA regions, results in responses that require activated ATR but not ATM (4, 83, 88, 149).

One structure critical for activating checkpoint controls is hypothesized to arise from aberrant replication forks. Of principal importance may be the structure and quantity of stalled forks required to signal; whether ssDNA, DSBs, or other structures specific to faulty forks are needed to signal is unknown. The structures of replication forks from wild-type and checkpoint-mutated cells have recently been reported in an electron micrograph study of replication forks isolated from budding yeast (178a). These results suggest that stalled forks contain more ssDNA than replicating forks, and that this is part of the signal for arrest. Just as one DSB is sufficient to cause arrest in yeast cells (161), we envision that one stalled fork is likewise sufficient to signal. Whether a single unperturbed DNA replication fork may also be sufficient to form a signal is unclear [see (196)].

THE MOLECULAR PATHWAYS ASSOCIATED WITH DNA DAMAGE

Numerous genes and proteins that act in checkpoint pathways have now been identified. We present the general mechanisms for pathways responding to ssDNA and DSBs (Figure 1), recognizing that most molecular aspects rely on inference and are based, in part, on the expectation of functional conservation of orthologous proteins.

ATR and Single-Strand DNA Response Complexes

The ATR-ssDNA checkpoint pathway likely involves four sets of protein complexes that assemble directly or indirectly on DNA to initiate and propagate a signal for arrest and to recruit damage repair proteins. How these four protein complexes interact in concert to mediate a DNA damage response is largely speculative. Like the replicative RFC complex with PCNA, the Rad17-RFC pentamer may bind ssDNA and facilitate the binding of the 9-1-1 heterotrimer to damage; the ATR-ATRIP heterodimer may bind damage independently, as suggested by yeast studies (106, 132, 157). The Rad17-RFC complex together with the 9-1-1 complex may then recruit substrates, including BRCA1 and its associated proteins, for phosphorylation by ATR. Indeed, in fission and budding yeast, the ATR-like kinase requires components of both the yeast Rad17-RFC complex and the yeast 9-1-1 complex to phosphorylate Crb2 and Rad9 (56, 158). Once phosphorylated, BRCA1 may activate downstream protein kinase cascades, aiding in amplification of the damage signal by recruiting additional phosphorylation substrates. ATR-like protein kinases can phosphorylate some substrates, like the H2AX component of nucleosomes, independent of sensor proteins (208). Phosphorylation of other substrates by ATR is dependent on Rad17-RFC, 9-1-1, and BRCA1 complexes, which suggests that these complexes serve to localize some but not all substrates.

How ATR becomes activated during a checkpoint response is uncertain, although *Xenopus* studies may provide some indication. Guo et al. (83) purified ATR from egg extracts using DNA-cellulose and found enhanced ATR activity, 10- to 20-fold above that of ATR immunoprecipitated directly from egg extracts. This suggests that binding to DNA activates ATR activity or selectively purifies an activated form of ATR (2).

ATM and Double-Strand Break Response Complexes

The ATM protein kinase, orthologous to Tel1 in both budding and fission yeast, has been studied more extensively than ATR for two reasons: (*a*) in 1995, ATM was identified as the gene defective in the syndrome AT, whereas no association between ATR deficiency and a disease yet exists, and (*b*) unlike ATR, ATM is not essential for cell viability, (18, 46, 164). Cells lacking ATM display chromosomal instability, RDS, and extreme sensitivity to ionizing radiation (IR) and radiomimetic drugs. AT cells are defective for cell cycle checkpoints at G1/S, S, and G2/M phase transitions in response to radiation-induced damage [reviewed in (2)].

In accord with the view that ATM recognizes dsDNA ends, purified ATM associates with nonspecific DNA fragments in vitro, demonstrating a modest affinity for linear DNA over supercoiled molecules, and often localizing at DNA ends (40, 175). At DSBs, ATM may associate with the MRN complex of proteins, which could then serve to recruit ATM substrates for phosphorylation, much like the 9-1-1 and Rad17-RFC complexes do for ATR. That ATM and MRN act together at DSBs is inferred from the fact that cells from patients with mutations in *NBS1*, presenting as Nijmegan breakage syndrome, also exhibit radiosensitivity and chromosomal fragility like AT cells (17), and patients with mutations in *MRE11* present with clinical symptoms and cellular defects much like those seen in AT, thus giving rise to the name "AT-like disorder" (180). Budding yeast studies additionally support the model that ATM and MRN act coordinately at DSBs. Usui et al. (195) determined that unresected DSBs created during meiosis activate Tel1 in vivo through the MRX complex, where checkpoint activation is dependent on Mre11 acting as a damage sensor. The MRX complex consists of Mre11, Rad50, and the budding yeast protein, Xrs2, which is replaced by Nbs1 in mammals.

Identifying the targets of ATM phosphorylation will likely provide insight into the mechanism of checkpoint activation in response to DSBs [reviewed in (97)]. What can be gleaned at this point, however, is that, at least for two cases, ATM regulates several proteins within the same pathway to ensure the control of a key target. In regulating p53, ATM employs three different means to stabilize the protein. ATM, or a closely associated kinase, phosphorylates p53 on Ser[15] in vitro, and most likely also in vivo, as AT cells show much reduced levels of p53 phosphorylated at the Ser[15] residue in response to IR (25). This phosphorylation by ATM likely causes p53 transcriptional activation (51), while phosphorylation of p53 at Ser[20] by Chk2 acting downstream of ATM, reduces the ability of ubiquitin ligase Mdm2 to bind p53, thus promoting its stabilization (34, 87, 90, 169). ATM further ensures p53 stabilization by phosphorylating Mdm2, thereby preventing export of p53 to the cytoplasm where it is degraded (101). Another "tri-strategy" approach involves regulation of BRCA1. ATM directly phosphorylates BRCA1 after activation by DSBs caused by ionizing radiation, although the consequence of this phosphorylation is unknown (39, 71). Moreover, activated Chk2, downstream of ATM, additionally phosphorylates BRCA1 on another site that is required for the dissociation of BRCA1 and Chk2 (109). ATM also phosphorylates CtIP, forcing dissociation of this inhibitor from BRCA1 (111).

THE MOLECULAR PATHWAYS ASSOCIATED WITH REPLICATION FORKS

The regulation of DNA replication by checkpoint controls may be most important in both genome stability and potentially in cancer therapy. Indeed, a major role of checkpoint proteins may be to stabilize stalled replication forks, since fork collapse can lead to chromosome rearrangements and, thus, genome instability or cell death. Genome instability can activate oncogenes leading to unregulated, abnormal cell growth. Further destabilizing replication forks in cancer cells might provide an avenue for therapy. Much of the current interest in replication fork biology has been motivated by findings in bacteria where replication fork collapse is thought to be a common feature (occurring about once in every two cell divisions), with RecA acting to restore failed forks by homologous recombination (41).

Replication forks slow or stall when they encounter DNA adducts, creating a physical impediment, or when dNTP pools are limiting. Replication disruption elicits four cellular responses: a block to initiation of replication (origin firing), slowing of elongation, maintenance of slowed or stalled replication forks, and a block to mitosis (Figure 2). In budding yeast, checkpoint proteins regulate three of these four responses. After DNA damage, if the key Mec1 and Rad53 regulators are mutated, late origins of replication continue to fire (162), stalled replication forks are not stabilized (48, 119, 188), and cells enter mitosis (213).

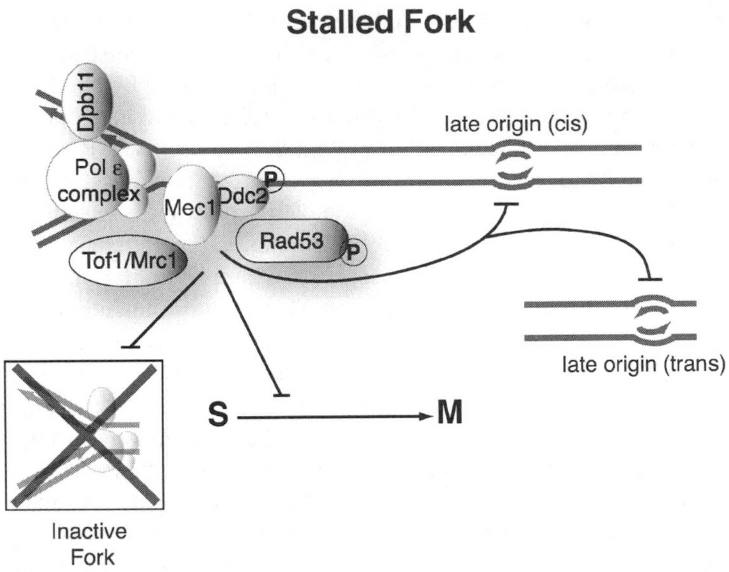

Stalled Fork

Figure 2   A stalled replication fork activates three common checkpoint mechanisms: (*a*) stabilization of the fork by blocking fork collapse, (*b*) a block to mitosis, and (*c*) blockage of late origin firing both in *cis* and in *trans*.

Many of the mechanisms controlling these responses are at least partially understood. Surprisingly, slowing of elongation does not require the yeast checkpoint genes; the replication apparatus inherently slows upon encountering damage, likely due to a physical block in fork progression. The Mec1 and Rad53 checkpoint proteins in *S. cerevisiae* are, however, required for the other three responses. Blocking of late origin firing in budding yeast likely occurs though Rad53 phosphorylation of Dbf4-Cdc7, a protein kinase required for the firing of all origins (214), and in higher cells through inhibition of Cdc25A and Orc function (59). Inhibition of mitosis in budding yeast involves regulation of Esp1 and Pds1, which in turn regulate cleavage of cohesins between sister chromatids required for chromosome segregation (191). Inhibition of mitosis in higher eukaryotes involves inhibitory phosphorylation of Cdc2 by blocking Cdc25C activity (160).

The roles of Mec1 and Rad53 proteins in preventing catastrophic termination of forks are likely crucial (48, 119, 188). That checkpoint proteins might regulate stability of replication forks was first suggested by Enoch et al. (57), who found two mutants, *rad3* and *cdc2-3w*, defective in blocking mitosis of S-phase-arrested cells in *Schizosaccharomyces pombe*. The *rad3* mutant dies in S-phase, whereas the *cdc2-3w* mutant dies only upon entry into mitosis. The *S. pombe* Rad3 (ATR-like) protein kinase was thereby considered to be required for recovery from replication stress. This recovery function probably entails stabilizing slowed or stalled replication forks, now shown most convincingly in budding yeast studies.

The Diffley and Foiani groups independently showed in *S. cerevisiae* that wild-type cells slow the rate of replication when treated either with MMS to induce damage or with hydroxyurea (HU) to deplete dNTP pools. In wild-type cells fork stability is maintained such that replication by the stalled or slowed fork can resume when cellular conditions permit (119, 188). In Mec1- and Rad53-deficient cells, however, replication forks collapse and fail to resume replication. One can now infer that stability of forks is a primary determinant for damage-sensitivity in checkpoint mutants from the following observations. Hypomorphic *mec1* mutants and *mrc1* null mutants are relatively resistant to MMS, yet fail to block late origin firing or entry into mitosis (5, 143). Hence, it is inferred that *mec1* and *mrc1* mutants are damage sensitive owing to their inability to stabilize replication forks and not owing to defects in blocking replication from late origins or preventing mitotic entry. An additional implication is that checkpoint proteins may act efficiently in *cis* to retain a fork, but require more activity to act in *trans* to block late origin firing and mitotic progression. For these latter purposes, Mrc1 (and perhaps Tof1 and Rad9) may be a potentiator or amplifier of the damage signal in S-phase.

Interestingly, one can make a similar inference for mammalian cells in that fork stability, and not cell cycle progression per se, may be a key determinant of damage sensitivity. Mutants for ATM are defective for inhibition of origin firing or RDS. Suppressors of ATM$^{-/-}$ mutations have been identified that restore inhibition of late replication origins, yet the strains remain damage sensitive (95).

The mechanism of fork stabilization is not known [although it may involve regulation of DNA primase (126)], nor is the exact fate of stalled, intact replication forks known. Recently, an interaction was reported between a Holliday resolving protein and the Cds1 checkpoint protein kinase in fission yeast (16). Checkpoint proteins may well stabilize forks by preventing cleavage of key intermediates. Thus, in a larger scheme, checkpoint proteins may prevent repair of stalled forks by recombination pathways.

THE G1 CHECKPOINT

Most eukaryotic cells damaged in G1 exhibit a pronounced delay prior to S-phase. This arrest in G1 allows vital time for repair and prevents replication of a damaged template. Otherwise, uninhibited DNA replication would convert one gapped chromosome into two sister chromosomes, one of which contains a DSB. Although its role in cell physiology seems clear, the role of the G1 checkpoint is often the subject of debate because it is very different in higher compared to in lower eukaryotes.

In budding yeast, the G1 checkpoint exists but is very weak; damage induces a delay that lasts far less than an hour, and most damage remains unrepaired (61, 174). For example, a double-strand break induces no detectable delay in G1 but leads to a prolonged arrest in G2 (213). This may reflect the ability of budding yeast to efficiently repair DSBs by homologous recombination using the intact sister homolog. What little delay is detected in G1 involves Rad53-dependent

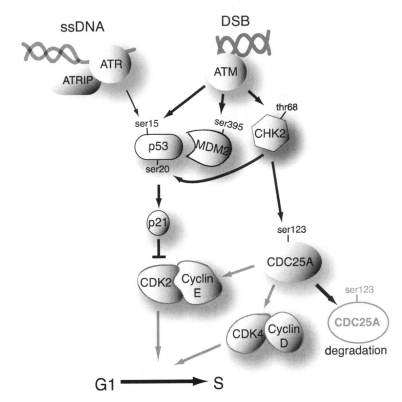

Figure 3   The G1 checkpoint in mammalian cells primarily functions to block Cdk2-cyclin E activity. This is achieved by stabilizing p53 and degrading Cdc25A to maintain Cdk2 inhibitory phosphorylation. Gray arrows denote functions that are lost upon activation of the checkpoint cascade, and labeled amino acids on proteins indicate sites of phosphorylation.

phosphorylation of the Swi4/6 transcription factors. This phosphorylation inhibits transcription of G1 cyclins, thereby slowing entry into S-phase (173).

In contrast to the weak G1 arrest in yeast, DNA damage induces a very robust G1 arrest via action of p53 in higher eukaryotes from *Xenopus* to mammals. This G1 arrest now appears to have two waves of action (Figure 3). The first immediate delay provides time for repair but lasts only several hours, whereas a slower and sometimes irreversible delay may function to remove cells from the cell cycle (12, 123).

The first delay occurs within minutes after damage, employs posttranslational modification of proteins, and is p53-independent. ATM-dependent activation of Chk2 leads to phosphorylation of Cdc25A, thereby priming it for ubiquitination and proteasome destruction (123). Loss of Cdc25A consequently regulates Cdk2 by maintaining its inhibitory phosphorylation on Tyr[15] and inhibiting its association with cyclin E, a necessary step for progression into S-phase. An elegant set of experiments using a cell-free system derived from *Xenopus* eggs showed that Cdk2-cyclin E inhibition also prevents complete formation of prereplicative complexes on DNA to inhibit the start of replication (40). All prereplication complex components are present—ORC, Cdc6, CDC7, and MCM proteins—except Cdc45, which requires Cdk2-cyclin E activity for assembly at origins where it then attracts DNA polymerases. Cdc25A also regulates Cdk4 activity, and, as indicated by a much earlier study performed with UV treatment, destruction of Cdc25A results in maintenance of the Tyr^{17} inhibitory phosphorylation on Cdk4 that also prevents S-phase entry (187).

A second G1 delay mechanism also involves ATM/ATR and Chk2, but takes several hours to initiate because regulation for this block to S-phase entry occurs at the transcriptional level. Here, ATM/ATR and Chk1/Chk2 activate and stabilize p53, as discussed earlier, causing transcriptional induction of p21, which then inhibits the Cdk2-cyclin E complex (55, 168). The p53-dependent block in G1 maintains the arrest initiated by the Cdc25A pathway to allow sufficient time for repair of DNA damage. How recovery from arrest is achieved or whether, in fact, the p53-dependent delay is reversible in all cell types is poorly understood (27, 49). Also undetermined is whether these two checkpoint cascades can occur independently; that is, can limited damage that requires little time for repair induce only the posttranslational cascade for a brief delay, or are both cascades always activated together, regardless of the extent of damage (12)?

Whether G1 arrest or apoptosis is more important in mammalian cells in preventing cancer is as yet unresolved. If both mechanisms function to remove damaged cells from cycling, then both may play prominent roles in cancer evasion.

THE S-PHASE CHECKPOINT

Description of S-phase checkpoint pathways is under way, driven predominantly by studies of budding and fission yeast. Based on details inferred from genetic observations, we present models of each organism in an attempt at deriving unifying mechanisms. Details of the fates of forks in the two yeasts are similar, but not identical, and the possibility that mammalian cells exhibit similar features is addressed.

A traditional view held by many is that cells in S-phase activate one of two checkpoint pathways to either signal DNA damage through the intra-S checkpoint or aberrant replication forks through the replication checkpoint. Due to key features shared by these two pathways, we believe that these two responses can be integrated into one pathway, termed simply the S-phase checkpoint. Much of this model is based on key observations of alternative pathways in S-phase control both in budding yeast (Figure 4) and fission yeast (Figure 5) (5, 64, 184).

The first step in S-phase checkpoint activation is stalling of replication forks owing either to the depletion of dNTPs or to an encounter with DNA damage. We consider these two forms of damage similar because they both slow replication fork

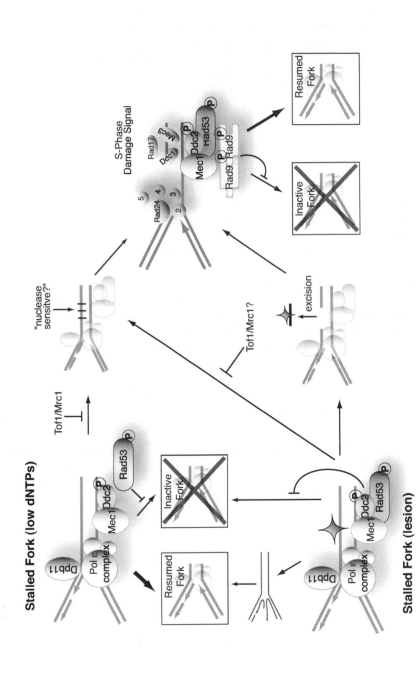

Figure 4 In budding yeast, replication forks stall upon encountering DNA lesions or low dNTP pools, signaled by proteins localized at sites of replication. Tof1 and Mrc1 block the conversion of stalled forks prone to form alternative structures. Potential attack by nucleases or lesion processing induces activation of the checkpoint sensors that specifically recognize DNA damage to further maintain fork stability.

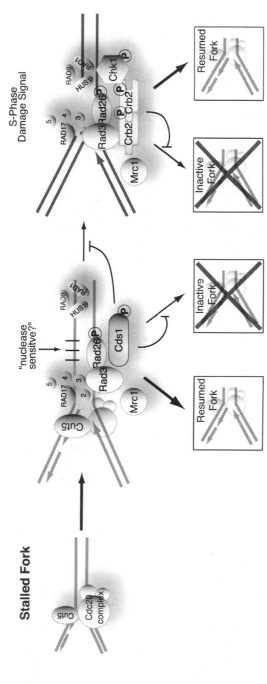

Figure 5   In fission yeast, stalled replication forks are immediately converted to alternative structures that are blocked from possible nuclease attack and collapse. Mrc1 additionally functions to block possible nuclease attack. Those forks that undergo nuclease processing activate the checkpoint sensors that respond to DNA damage to maintain fork stability.

progression and either insult can culminate in fork collapse in checkpoint mutants. Both mechanisms of stalling replication forks activate the S-phase checkpoint pathway by proteins that are localized at the fork sites. The proteins associated with replication forks that display checkpoint activity include the DNA helicases Sgs1 and Top3 (31, 66); Pol2, a subunit of DNA pol [essential for DNA replication and implicated in DNA repair (138)]; Dpb11, a protein that interacts with and recruits Pol2 to replication origins (204); Drc1, an essential protein for DNA replication that interacts with Dpb11 (203); and RFC subunits 2, 3, 4, and 5 (102, 137, 139, 172). These proteins, along with contributing functions from Mrc1 and Tof1, somehow activate Mec1 and Rad53 to carry out the common S-phase responses detailed in Figure 4 (5, 64).

To explain the role of budding yeast Rad9 within S-phase, we propose that a stalled fork can be converted to a different fork form that activates other checkpoint proteins, like Rad9, to somehow maintain or restore fork stability (Figure 4). Based on genetic analysis, we suggest that the primary stalled fork is prevented from conversion to this other form by Tof1 and Mrc1, because in the absence of Tof1 or Mrc1, S-phase checkpoints become dependent on Rad9 (5, 64). These "converted" forks can resume replication provided that they activate Mec1 and Rad53 for fork stabilization.

We suggest that a stalled fork at a lesion is subject to an additional fate. Lesion processing followed by replication generates a fork with a DSB. In this situation, the DNA damage created is sufficient to signal the damage checkpoint proteins, such as Rad9, that then prevent late origin firing and spindle elongation via Mec1 and Rad53. Therefore, the damage checkpoint serves as a backup to the replication checkpoint. This model explains why damage induces a weak Rad9-dependent S-phase response that becomes greatly enhanced in cells lacking either Tof1 or Mrc1 (5, 64).

We propose that in fission yeast, and therefore perhaps in mammalian cells, a similar, though not identical, progression of replication fork conversions may occur (Figure 5). First, replication forks blocked in fission yeast always activate Rad3 using the damage sensor proteins and Mrc1 (184). We speculate that a stalled replication fork may constitutively convert to an altered form that then signals using damage sensor proteins. Next, this fork intermediate may convert to another form, a process actively prevented by Cds1/Chk2 (113). This is supported by the key observation that *cds1* mutants are not appreciably sensitive to hydroxyurea, and when HU-treated require Chk1 activity for viability (113, 127). We suggest that Cds1 prevents the conversion of the first broken fork structure, while Chk1 acts as a failsafe mechanism, like budding yeast Rad9, to rescue any forks that have converted to the second altered fork form. Tests of these models linking DNA fork structures to protein functions are close at hand. Foiani and colleagues have shown that replication forks formed in yeast can be isolated and their structures determined by electron microscopy (178a).

In mammalian cells, the genetic basis of S-phase checkpoints has also been well documented (144), though the underlying mechanisms remain obscure. BRCA1,

ATM, and MRN complexes are involved after γ-irradiation, and ATR is likely involved after UV damage (59, 112, 208, 220, 227, 228). Cdc25A is also implicated as a downstream effector since it appears to regulate ORC complexes at initiation (59). It is unknown whether mammalian replication forks experience similar options of conversion as we propose in yeast. Differences between lower and higher eukaryotes include much longer tracts of DNA between origins in higher eukaryotes, suggesting that processivity might be more critical in mammals and may require other or additional forms of regulation by checkpoints. Whatever mechanisms operate in eukaryotic cells, fork stability mediated by checkpoint proteins is likely to be critical for genome stability. Furthermore, if replication defects are ongoing and persistent in cancer cells, then the potential for developing therapeutic strategies based on defects in replication fork regulation are attractive.

THE G2/M CHECKPOINT

Though well characterized in both yeast systems, the details of G2/M arrest in higher eukaryotic cells are still being determined. The picture thus far is that fission yeast and higher eukaryotes, both target Cdc2 to maintain its inhibitory phosphorylations as a principal means to block the G2/M transition. This is achieved by targeting various parallel pathways, including the phosphatases that promote mitosis, the kinases that block Cdc2 function, and other proteins that regulate these regulators, which all converge to regulate Cdc2 activity (Figure 6).

DNA damage in mammalian cells causes the activation of four different downstream kinases that independently block the activity of Cdc25C, the phosphatase that promotes mitosis by removing phosphates at Cdc2 inhibitory sites. First, activation of ATR by damage leads to Chk1 phosphorylation (83). Chk1 then negatively regulates Cdc25C by phosphorylating it at Ser^{216} (151, 160). This phosphorylation of Cdc25C creates a binding site for 14-3-3 proteins, and in the bound state renders Cdc25C either catalytically less inactive, sequestered in the cytoplasm, or both (78, 120). In addition to activation by ATR, Chk1 may be activated by BRCA1 to induce a G2/M arrest, although this has yet to be confirmed by in vitro analysis (226). By a second mechanism, ATM activation by DNA damage in G2 leads to Chk2 activation that also phosphorylates Cdc25C at Ser^{216} to block its function (129, 233). And now two new regulators of Cdc25C are emerging, both members of the Polo-like kinase (Plk) family. Plk3 resembles Chk1 and Chk2, as it is activated in an ATM-dependent manner, physically interacts with Cdc25C, and phosphorylates this protein at Ser^{216} to inhibit its activity (177, 218). Unlike the kinases that inhibit Cdc25C activity by Ser^{216} phosphorylation, Plk1 is recognized for its role in promoting mitotic entry by phosphorylating Cdc25C to activate it. Both ATR and ATM are thought to phosphorylate Plk1, perhaps via Chk1 and Chk2 activation, to block Plk1 activation of Cdc25C (177, 197). Plk1 inhibition correlates with a decrease in Cdc2-cyclin B kinase activity, and this inhibition is completely

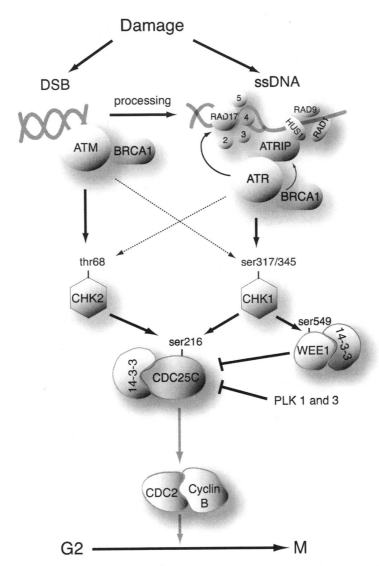

Figure 6   The G2 checkpoint in mammalian cells primarily functions to block Cdc2-cyclin B activity. The common means of maintaining Cdc2 inhibitory phosphorylation is by blocking Cdc25C phosphatase activity, namely by promoting its association with 14-3-3-proteins. Gray arrows denote functions that are lost upon activation of the checkpoint cascade, and labeled amino acids on proteins indicate sites of phosphorylation.

blocked by caffeine, an inhibitor of ATM/ATR. Additionally, expression of an un-phosphorylatable Plk1 cannot be inhibited in the presence of DNA damage and does not initiate a G2/M arrest, further implicating Plk1 as another regulator of Cdc25C activity (177). Adding an additional level of G2/M control, Plk1 protein stability now appears to be regulated by the checkpoint protein Chfr, which delays entry into mitosis when cells are under mitotic stress (96, 166). Using *Xenopus* cell-free extracts, Kang et al. showed that Chfr ubiquitinates the human Plk1 to target it for degradation, thereby inhibiting Cdc25C activation, resulting in a delay in Cdc2 activation (96). The mechanism of Chfr activation has yet to be determined.

While the majority of 14-3-3-bound Cdc25C localizes to the cytoplasm, some Cdc25C remains in the nucleus (233), implying that other mechanisms are required for inhibition of Cdc25C phosphatase activity. In mammalian cells, inhibition may be accomplished by the Cdk inhibitor p21 by the following mechanism (7). p21, Cdc25C, and Cdc2-cyclinB can all bind PCNA, but not all at the same time (85, 99, 219). PCNA may act as a platform to mediate protein interactions, and binding of p21 and Cdc25C to a PCNA-Cdc2-cyclin B complex may be a mutually exclusive event (7). Therefore, p21 interaction with PCNA-Cdc2-cyclin B may exclude Cdc25C from interacting with Cdc2 to dephosphorylate it for mitotic progression. By an additional method, p21 may act directly on Cdc2. Although p21 does not affect the inhibitory phosphorylation sites (Thr^{14} and Tyr^{15}) on Cdc2, p21 does block the activating phosphorylation of Cdc2 on Thr^{161}, mediated by the Cdk-activating kinase, CAK (177).

Another major player in the G2/M arrest is the 14-3-3 protein. 14-3-3 not only regulates Cdc25C as discussed previously, but it also regulates Wee1. Experiments using *Xenopus* extracts demonstrated that 14-3-3 binds to Wee1 when phospho-rylated at Ser^{549} and that this site on Wee1 is phosphorylated by Chk1 (108). By both recombinant protein and immunodepletion analysis, Chk1 phosphorylation of Wee1 promotes 14-3-3 binding, which significantly enhances the inhibitory kinase activity of Wee1 toward Cdc2. By another mechanism identified in colorectal cell lines, p53 mediates 14-3-3σ induction after damage; σ expression is needed for G2/M arrest and to prevent "mitotic catastrophe" (33, 89). Based on Western analysis of immunoprecipitated complexes, 14-3-3σ appears to bind to the Cdc2-cyclin B complex to sequester it in the cytoplasm to maintain G2 arrest (33). Whether the same mechanism functions in non-colorectal cells is unknown.

Myt1 is yet another possible target of checkpoint mechanisms in mammalian cells. The Myt1 fission yeast ortholog, Mik1, is regulated by checkpoint components. After DNA damage or replication defects Mik1 protein accumulates in a Chk1-dependent manner that is required for both initiation and maintenance of a G2/M arrest (9, 154). Checkpoint regulation of Myt1 in mammalian cells has not yet been demonstrated, however overexpression of Myt1 induces a G2 delay by retaining Cdc2-cyclin B in the cytoplasm (115). Additionally, Myt1 is down-regulated in a cell-cycle-dependent manner through phosphorylation by Akt, leading to the meiotic G2/M transition (141).

Regulation of cyclin B may also contribute to the G2/M arrest (21). Studies from two laboratories demonstrated that induction of DNA damage leads to retention of cyclin B in the cytoplasm and that forcing nuclear localization of cyclin B partially abrogated the G2/M damage-induced arrest (94, 192). Moreover, cyclin B and Cdc2 are both transcriptionally down-regulated upon checkpoint induction, correlating with a reduction in protein, which further reinforces the G2 arrest (43, 185).

In response to UV damage only the mitogen-activated protein kinase p38 is rapidly activated to induce a G2/M delay. Active p38 phosphorylates Cdc25B at Ser^{309}, demonstrated both in vitro and in vivo, to induce 14-3-3 binding, which then leads to the arrest response (22). 14-3-3 binding to Cdc25B blocks access of substrates to the catalytic site to inhibit Cdc25B activity (63). Notably, p38 apparently plays no role in affecting 14-3-3 binding to Cdc25C in vivo (22).

The budding yeast G2/M checkpoint arrests cells principally at the metaphase-anaphase transition, and the effectors targeted to cause arrest differ from those in fission yeast and higher eukaryotes. This checkpoint primarily halts arrest by two mechanisms that converge on the regulation of degradation of cohesins [(191, 224; reviewed in (6)]. The anaphase promoting complex (APC) plays a major role in regulating G2/M in budding yeast. If future work demonstrates that the APC in higher eukaryotes is subject to checkpoint regulation in response to DNA damage, with the polo-like kinases likely targets (159, 177), then insights from budding yeasts mechanisms can be drawn to direct experimentation in mammalian cells or *Xenopus* egg extracts.

DAMAGE REPAIR AND RECOVERY

DNA Checkpoints and Their Link to Repair

The checkpoint proteins arrest the cell cycle in the presence of DNA damage, presumably to allow the cell time for repair. The checkpoints are proposed to act not simply as a switch for arrest, but also to directly recruit DNA repair machinery or perhaps even play direct roles in repairing DNA. A role in repair is borne out by the fact that checkpoint-deficient cells temporarily arrested with drugs after infliction of DNA damage are not as viable as similarly treated wild-type cells [reviewed in (156)]. Thus, checkpoint proteins clearly have roles presumably in DNA repair above and beyond that of inducing cell cycle arrest and fork stability. Here we present some of the indirect links and the few direct links between checkpoint machinery and DNA repair mechanisms uncovered thus far.

Yeast and mammalian cells indirectly contribute to repair responses by inducing transcription of a number of genes encoding proteins involved in repair (12, 62, 70, 111, 153, 165). Checkpoint proteins also regulate dNTP pools, perhaps a common feature involved in DNA repair, since limiting dNTP pools enhances damage sensitivity (183, 210, 231).

Hard evidence supporting the notion that checkpoint and repair proteins interact to enhance cell survival is scarce, whereas soft evidence abounds. ATM is linked to the repair of DNA damage by homologous recombination (HR) based on the fact that AT cells defective in HR components (*rad54* mutation) are no more sick than AT cells, whereas AT cells defective for non-homologous end-joining (NHEJ; *ku70* mutation) show drastic increases in chromosomal instability (134). The inference is that ATM$^{-/-}$cells are already defective for HR, resulting in no increased defect upon *rad54* mutation, while they are proficient for NHEJ. BRCA1, also implicated in HR repair (135) as well as transcription-coupled repair (1, 77), is thought to be involved in various repair processes. BRCA1 transiently associates with a number of repair proteins, including mismatch repair proteins, Nbs1, and the BLM helicase, in the BRCA1-associated genome surveillance complex (BASC) after DNA damage treatment (207). Colocalization after damage induction has also been demonstrated for BRCA1 and Rad51 (15), BRCA1 and PCNA (167), and BRCA1 and the MRN complex (232). Based on the DNA damage-sensitivity of Mre11-Rad50-Nbs1 mutants, their ability to process DSBs, and their genetic grouping with the Rad52 epistasis group, it was assumed that the MRN complex also plays a role in HR. However, numerous studies suggest that the roles of the MRN complex in arrest and in repair (either by HR or by NHEJ) are distinct [reviewed in (44)].

One protein likely to be involved in both signaling and repairing damage is RPA, the heterotrimeric ssDNA binding protein. RPA contributes essential functions to DNA replication, repair, and recombination [reviewed in (93)]. Studies in both mammals and yeast indicate that the middle RPA subunit, Rpa2, is phosphorylated by ATM (DNA-PK phosphorylation has also been demonstrated) following fork stalling and DNA damage induction, and this phosphorylation abrogates DNA replication as measured by in vitro SV40 replication (20, 72, 140, 205). Since Rpa2 phosphorylation is likely ATM-dependent and, at least in budding yeast, downstream factors (Rad9 and Rad53) are not required for this event, Rpa2 phosphorylation may be necessary to stabilize stalled replication forks or to halt replication until DNA damage is repaired. The involvement of Rpa2 phosphorylation in mediating DNA repair is currently unclear. Interestingly, in budding yeast the large RPA subunit (Rpa1) is additionally phosphorylated by Mec1, and this event requires Rad9 and the downstream kinase Rad53, placing the event farther down the checkpoint cascade (19). Whereas Rpa2 is also normally phosphorylated during the cell cycle, as well as in response to damage, Rpa1 is only phosphorylated upon genotoxic and HU insult, and the extent of phosphorylation correlates with the amount of damage induced. Given the late timing and modulation by levels of DNA damage within the cell, Rpa1 phosphorylation mediated by Mec1 may be involved in some aspect of DNA repair (117).

The only conclusive evidence tying the checkpoint response directly to DNA repair is that of Rad55 activation in budding yeast and Crb2 regulation in fission yeast. In budding yeast, Rad55, a recombinational repair protein, undergoes specific phosphorylation mediated by Mec1 following treatment by numerous damaging agents (13). Mutant *rad55* cells are proficient for arrest and induced gene expression in response to DNA damage, but a *mec1* mutant that cannot phosphorylate Rad55 shows an 88-fold reduction in the amount of damage-induced recombination. That Rad55 is a terminal substrate of the checkpoint cascade and that a major repair defect is evident when Mec1 fails to phosphorylate Rad55 suggests a link between checkpoints and repair processes. In fission yeast, the checkpoint protein Crb2 has a role in regulating topoisomerase III (Top3; 29). Crb2 hyperphosphorylation in response to DNA damage not only initiates a checkpoint arrest response, but Crb2's hyperphosphorylated state is needed to regulate Top3 activity to prevent aberrant hyperrecombination mediated by the RecQ-like helicase, additionally linking checkpoints and repair mechanisms.

Recovery, Adaptation, and Apoptosis

Upon repair of damage, cells resume cycling. But is the completion of repair per se sufficient to turn off the checkpoint arrest? Indeed, data indicate that the checkpoint cascade must be actively down-regulated before the cell cycle can resume. In budding yeast, overexpression of Ddc2 (the ATRIP ortholog) results in an irreversible G2/M arrest after DNA damage (36). One interpretation of this result is that the ability to shut off the checkpoint response is abrogated in Ddc2-overexpressing cells. Additional evidence that recovery from damage is an active process is supported by the fact that yeast undergo adaptation, a process by which cells resume cell cycle progression despite the presence of unrepaired DNA (161). The activity of Rad53 and Chk1 must be down-regulated to permit adaptation, indicating that the kinase cascade must be shut off (150). Adaptation involves Cdc5, which may operate in a feedback loop to inactivate the checkpoint cascade (159). Additionally, the inactivation event may be a regulatory phosphorylation on Rad9, a necessary protein for Rad53 activity. Fission yeast employ such a mechanism of terminating arrest whereby they shut off the activity of Crb2, the Rad9 ortholog, via inhibitory phosphorylation carried out by Cdc2 (58). The clearest example of checkpoint inactivation in mammalian cells involves the p53-inducible protein Wip1. Genotoxic stress leads to p53 activation by p38 phosphorylation, consequently inducing transcription of a number of genes, including Wip1 (182). Accumulation of the Wip1 phosphatase then inactivates p38, demonstrated both in vivo and in vitro, leading to p53 transcriptional down-regulation and a reduced apoptotic response. Attenuation of cell cycle arrest was not specifically examined, but it, too, is predicted to be affected.

In many higher eukaryotes, extensive DNA damage that cannot be repaired during a checkpoint-controlled cell cycle arrest is thought to initiate cell death via the apoptotic pathway. That p53 is a key effector of ATM/ATR protein kinases during checkpoint responses strongly suggests that checkpoint controls and apoptotic controls are coordinated processes. For example, mice harboring mutations in p53 or ATM are defective in response to DNA damage for both checkpoint activation and apoptosis [reviewed in (100)]. Also, studies in *Caenorhabditis elegans* suggest a prominent role in arrest and apoptosis for an orthologous Rad1 sensor protein in germ cells (68). (Curiously, somatic cells in *C. elegans* do not undergo either fate, only their germ cells do.) Whether checkpoint protein interaction with damage is somehow central to the switch between arrest and apoptosis or whether the decision lies elsewhere in signal transduction is unknown. Association between checkpoint pathways and apoptosis is suggested by the interaction between the human Rad9 damage sensor protein and BCL apoptotic proteins (105).

ADDITIONAL ROLES OF CHECKPOINT PROTEINS IN DNA METABOLISM

The checkpoint proteins initially identified as controlling the cell cycle have many additional roles in DNA metabolism. These include roles in transcription of damage-responsive genes and roles in DNA repair (discussed in the "Repair" section), telomere biology, meiotic functions different from those operating in mitotic cells, and control of dNTP pools. Furthermore, proteins that act together in one pathway or response may not act together in a second response, indicating that "complexes" do not necessarily function constitutively.

Prominent among the additional roles of checkpoint proteins is their part in telomere biosynthesis. In mammals, ATM mutants have shorter telomeres, as do budding yeast *tel1* mutants. These initial observations led the way to the remarkable finding that the fission yeast Rad3 and Tel1 proteins are required for telomere addition (130), followed by similar findings for budding yeast Mec1 and Tel1 proteins (155). Tel1 appears to recruit telomerase to sites of novel telomere addition, although exactly how Tel1 does so and whether Mec1 does so as well is unknown. In budding yeast, Mec1 and Dun1 also affect gene expression of telomere-proximal genes, suggesting a role in chromatin structure (42, 47). An effect on telomere chromatin is also indicated by the finding that in budding yeast some repair and chromatin proteins are relocalized from the telomeres to sites of DNA damage— recruitment that requires checkpoint protein functions [Mec1, Rad9, and Rad53 (118, 131, 133)]. An additional and opposite effect on telomeres is mediated by budding yeast Mec3 and Rad17 proteins; inexplicably, *mec3* mutants have shorter telomeres and *rad17* mutants have longer telomeres (37, 118). Recall that Mec3 and Rad17 both act coordinately in cell cycle arrest in what is the equivalent of the 9-1-1 mammalian complex. *rad1* mutants in *C. elegans* also have shorter telomeres (3), implying a role for this class of proteins in telomere biosynthesis in many organisms. That checkpoint proteins might interact with telomeres is not unexpected, given that telomeres, by their nature, might resemble a DSB. Hence, checkpoint proteins might be expected to interact with telomeres because they may resemble damage and/or to perhaps aid in their protection

Study of meiosis has provided yet another fertile ground for discovery of checkpoint protein functions. Checkpoint proteins play many roles in meiosis, including imparting two cell cycle delays and regulating partner choice in recombination (82, 121, 189). Checkpoint protein functions in meiosis are sometimes divergent from their roles in mitosis. For example, Rad9 is not required for early meiotic arrest (121), and members of the 9-1-1 complex act at independent steps during meiotic arrest (91, 189). Meiotic studies also reveal the diversity of arrest mechanisms involving Cdc2 in different organisms. The budding yeast meiotic arrest (called the pachytene arrest) and the fission yeast mitotic G2/M arrest after DNA damage involves inhibitory phosphorylation of Cdc2, yet the budding yeast mitotic G2/M arrest and the fission yeast meiotic arrest enigmatically do not. The mechanisms of meiotic arrest in fission yeast remain unclear.

Finally, budding yeast checkpoint proteins have other important roles in the regulation of DNA replication independent of cell cycle progression and replication fork stability. Mec1 and Rad53 regulate the synthesis of dNTPs by blocking activity of Sml1, an inhibitor of ribonucleotide reductase (30). Mec1 and Rad53 inhibit Sml1, illustrated by the fact that *mec1* and *rad53* mutants are inviable due to Sml1 inhibition of the ribonucleotide reductase protein (Rnr1) and that *mec1* and *rad53* mutants become viable when *sml1* is deleted (230). Mec1 also appears to regulate replication fork movement in specific "replication slow zones" within the yeast chromosome, and slow replication is again relieved by mutation of *sml1* or upregulation of *RNR1* (48; R. Cha & N. Kleckner, manuscript submitted). To this end, Mec1 may somehow regulate replication fork progression and/or more simply regulate the level of dNTPs.

CONCLUDING REMARKS REGARDING GENOME STABILITY

The importance of checkpoints in biology, and especially in human cancer, is likely due to their preeminent roles in maintaining genomic stability. Multiple significant links between checkpoint proteins and genome stability have been identified in model organisms, as well as in mammalian cells. In mammals, ATM, ATR, BRCA1, Chk2, and p53 mutants all show instability. For ATM, some of this instability may be due to shortening of telomeres, leading to telomere fusions and breakage-fusion-bridge events [reviewed in (145)]. However, like most other checkpoints proteins, ATM is likely involved in preserving stability by several mechanisms independent of telomere biology. For example, ATM$^{-/-}$ mutants exhibit elevated mitotic recombination (104). In another example, BRCA1 mutants have partial defects in homologous recombination and transcription-coupled repair (1, 135). As better systems to study instability in mammalian cells are developed, our understanding of these processes should improve.

In budding yeast, the roles of checkpoint proteins in genome stability are being examined. General mechanisms linking checkpoint protein function and genome stability are illustrated by four examples discussed below. In the first example, Paulovich et al. (148) showed that after UV irradiation, checkpoint proteins promote error-prone repair carried out by Rev3/Rev7, which form a polymerase complex that replicates over DNA damage by inserting noncognate nucleotides. Whether a cell cycle delay is required for Rev3/7 function or whether checkpoint proteins act more directly, perhaps by recruiting Rev3/7 to UV lesions, is unknown. Second, checkpoint proteins seem to promote allelic recombination (13). Defects in this process lead to an increased frequency of ectopic recombination in meiotic (82) and mitotic cells (60), as well as defects in intergenic mitotic recombination (13). Third, *rad9* and other checkpoint mutants spontaneously lose chromosomes (103, 212). One mechanism of loss was, until recently, thought to involve mitosis of a cell with a chromosome that has an unrepaired DSB. Surprisingly, a recent study indicates that a cell completely defective for the G2/M checkpoint (a *dun1 chk1* double mutant) undergoes no detectable spontaneous chromosome loss (103). Therefore, chromosome loss in *rad9* mutants is likely due to a defect in some other cellular activity.

Finally, and in our view most significantly, checkpoint proteins may play key roles in maintaining stalled replication forks, because when defective the collapsed forks may lead either to cell death or to genomic rearrangements. Myung & Kolodner (136) analyzed the effects of checkpoint mutations on a spectrum of chromosomal arrangements that they call gross chromosome rearrangements, or GCR. They suggest that GCRs in yeast structurally resemble those rearrangements seen in cancer cells. By surveying checkpoint mutants for GCR they found that the proteins regulating S-phase responses generally had significant roles in maintaining stability. Defects in both Mec1 and Tel1 lead to a 10,000-fold increase in instability. But which of their many functions contributes to stability remains uncertain; they mediate essentially all responses to DNA damage, are required for telomere synthesis, and play roles in DNA replication. Defects in Pds1 also strongly affect stability, probably due to a role in sister chromatin cohesion (rather than to a defect in its role in G2/M arrest, since *chk1* mutants having a similar arrest defect demonstrate no instability). Linking complex replication fork biology, checkpoint protein function, and genome stability remains a major challenge, but one we anticipate is fundamental to understanding cancer etiology and treatment.

The *Annual Review of Genetics* is online at http://genet.annualreviews.org

LITERATURE CITED

1. Abbott DW, Thompson ME, Robinson-Benion C, Tomlinson G, et al. 1999. BRCA1 expression restores radiation resistance in BRCA1-defective cancer cells through enhancement of transcription-coupled DNA repair. *J. Biol. Chem.* 274:18808–12
2. Abraham RT. 2001. Cell cycle check-point signaling through the ATM and ATR kinases. *Genes Dev.* 15:2177–96
3. Ahmed S, Hodgkin J. 2000. MRT-2 checkpoint protein is required for germline immortality and telomere replication in *C. elegans. Nature* 403:159–64
4. al-Khodairy F, Fotou E, Sheldrick KS, Griffiths DJ, Lehmann AR, Carr AM.

1994. Identification and characterization of new elements involved in checkpoint and feedback controls in fission yeast. *Mol. Biol. Cell* 5:147–60

5. Alcasabas AA, Osborn AJ, Bachant J, Hu F, Werler PJ, et al. 2001. Mrc1 transduces signals of DNA replication stress to activate Rad53. *Nat. Cell Biol.* 3:958–65

6. Amon A. 2001. Together until separin do us part. *Nat. Cell Biol.* 3:E12–14

7. Ando T, Kawabe T, Ohara H, Ducommun B, Itoh M, Okamoto T. 2001. Involvement of the interaction between p21 and proliferating cell nuclear antigen for the maintenance of G2/M arrest after DNA damage. *J. Biol. Chem.* 276:42971–77

8. Araki H, Leem SH, Phongdara A, Sugino A. 1995. Dpb11, which interacts with DNA polymerase II(epsilon) in *Saccharomyces cerevisiae*, has a dual role in S-phase progression and at a cell cycle checkpoint. *Proc. Natl. Acad. Sci. USA* 92:11791–95

9. Baber-Furnari BA, Rhind N, Boddy MN, Shanahan P, Lopez-Girona A, Russell P. 2000. Regulation of mitotic inhibitor Mik1 helps to enforce the DNA damage checkpoint. *Mol. Biol. Cell* 11:1–11

10. Banin S, Moyal L, Shieh S, Taya Y, Anderson CW, et al. 1998. Enhanced phosphorylation of p53 by ATM in response to DNA damage. *Science* 281:1674–77

11. Bargonetti J, Manfredi JJ. 2002. Multiple roles of the tumor suppressor p53. *Curr. Opin. Oncol.* 14:86–91

12. Bartek J, Lukas J. 2001. Mammalian G1- and S-phase checkpoints in response to DNA damage. *Curr. Opin. Cell Biol.* 13:738–47

13. Bashkirov VI, King JS, Bashkirova EV, Schmuckli-Maurer J, Heyer WD. 2000. DNA repair protein Rad55 is a terminal substrate of the DNA damage checkpoints. *Mol. Cell. Biol.* 20:4393–404

14. Bell DW, Varley JM, Szydlo TE, Kang DH, Wahrer DC, et al. 1999. Heterozygous germ line hCHK2 mutations in Li-Fraumeni syndrome. *Science* 286:2528–31

15. Bhattacharyya A, Ear US, Koller BH, Weichselbaum RR, Bishop DK. 2000. The breast cancer susceptibility gene BRCA1 is required for subnuclear assembly of Rad51 and survival following treatment with the DNA cross-linking agent cisplatin. *J. Biol. Chem.* 275: 23899–903

16. Boddy MN, Lopez-Girona A, Shanahan P, Interthal H, Heyer WD, Russell P. 2000. Damage tolerance protein Mus81 associates with the FHA1 domain of checkpoint kinase Cds1. *Mol. Cell. Biol.* 20:8758–66

17. Bressan DA, Baxter BK, Petrini JH. 1999. The Mre11-Rad50-Xrs2 protein complex facilitates homologous recombination-based double-strand break repair in *Saccharomyces cerevisiae. Mol. Cell. Biol.* 19:7681–87

18. Brown EJ, Baltimore D. 2000. ATR disruption leads to chromosomal fragmentation and early embryonic lethality. *Genes Dev.* 14:397–402

19. Brush GS, Kelly TJ. 2000. Phosphorylation of the replication protein A large subunit in the *Saccharomyces cerevisiae* checkpoint response. *Nucleic Acids Res.* 28:3725–32

20. Brush GS, Morrow DM, Hieter P, Kelly TJ. 1996. The ATM homologue MEC1 is required for phosphorylation of replication protein A in yeast. *Proc. Natl. Acad. Sci. USA* 93:15075–80

21. Bulavin DV, Amundson SA, Fornace AJ. 2002. p38 and Chk1 kinases: different conductors for the G(2)/M checkpoint symphony. *Curr. Opin. Genet. Dev.* 12:92–97

22. Bulavin DV, Higashimoto Y, Popoff IJ, Gaarde WA, Basrur V, et al. 2001. Initiation of a G2/M checkpoint after ultraviolet radiation requires p38 kinase. *Nature* 411:102–7

23. Burtelow MA, Roos-Mattjus PM, Rauen M, Babendure JR, Karnitz LM. 2001.

Reconstitution and molecular analysis of the hRad9-hHus1-hRad1 (9-1-1) DNA damage responsive checkpoint complex. *J. Biol. Chem.* 276:25903–9

24. Buscemi G, Savio C, Zannini L, Micciche F, Masnada D, et al. 2001. Chk2 activation dependence on Nbs1 after DNA damage. *Mol. Cell. Biol.* 21:5214–22

25. Canman CE, Lim DS, Cimprich KA, Taya Y, Tamai K, et al. 1998. Activation of the ATM kinase by ionizing radiation and phosphorylation of p53. *Science* 281:1677–79

26. Carney JP, Maser RS, Olivares H, Davis EM, Le Beau M, et al. 1998. The hMre11/hRad50 protein complex and Nijmegen breakage syndrome: linkage of double-strand break repair to the cellular DNA damage response. *Cell* 93:477–86

27. Carr AM. 2000. Cell cycle. Piecing together the p53 puzzle. *Science* 287:1765–66

28. Caspari T, Dahlen M, Kanter-Smoler G, Lindsay HD, Hofmann K, et al. 2000. Characterization of *Schizosaccharomyces pombe* Hus1: a PCNA-related protein that associates with Rad1 and Rad9. *Mol. Cell. Biol.* 20:1254–62

29. Caspari T, Murray JM, Carr AM. 2002. Cdc2-cyclin B kinase activity links Crb2 and Rqh1-topoisomerase III. *Genes Dev.* 16:1195–208

30. Chabes A, Domkin V, Thelander L. 1999. Yeast Sml1, a protein inhibitor of ribonucleotide reductase. *J. Biol. Chem.* 274:36679–83

31. Chakraverty RK, Kearsey JM, Oakley TJ, Grenon M, de La Torre Ruiz MA, et al. 2001. Topoisomerase III acts upstream of Rad53p in the S-phase DNA damage checkpoint. *Mol. Cell. Biol.* 21:7150–62

32. Chamankhah M, Xiao W. 1999. Formation of the yeast Mre11-Rad50-Xrs2 complex is correlated with DNA repair and telomere maintenance. *Nucleic Acids Res.* 27:2072–79

33. Chan TA, Hermeking H, Lengauer C, Kinzler KW, Vogelstein B. 1999. 14-3-3 Sigma is required to prevent mitotic catastrophe after DNA damage. *Nature* 401:616–20

34. Chehab NH, Malikzay A, Appel M, Halazonetis TD. 2000. Chk2/hCds1 functions as a DNA damage checkpoint in G(1) by stabilizing p53. *Genes Dev.* 14:278–88

35. Cheng L, Hunke L, Hardy CF. 1998. Cell cycle regulation of the *Saccharomyces cerevisiae* polo-like kinase cdc5p. *Mol. Cell. Biol.* 18:7360–70

36. Clerici M, Paciotti V, Baldo V, Romano M, Lucchini G, Longhese MP. 2001. Hyperactivation of the yeast DNA damage checkpoint by TEL1 and DDC2 overexpression. *EMBO J.* 20:6485–98

37. Corda Y, Schramke V, Longhese MP, Smokvina T, Paciotti V, et al. 1999. Interaction between Set1p and checkpoint protein Mec3p in DNA repair and telomere functions. *Nat. Genet.* 21:204–8

38. Cortez D, Guntuku S, Qin J, Elledge SJ. 2001. ATR and ATRIP: partners in checkpoint signaling. *Science* 294:1713–16

39. Cortez D, Wang Y, Qin J, Elledge SJ. 1999. Requirement of ATM-dependent phosphorylation of brca1 in the DNA damage response to double-strand breaks. *Science* 286:1162–66

40. Costanzo V, Robertson K, Ying CY, Kim E, Avvedimento E, et al. 2000. Reconstitution of an ATM-dependent checkpoint that inhibits chromosomal DNA replication following DNA damage. *Mol. Cell* 6:649–59

41. Cox MM, Goodman MF, Kreuzer KN, Sherratt DJ, Sandler SJ, Marians KJ. 2000. The importance of repairing stalled replication forks. *Nature* 404:37–41

42. Craven RJ, Petes TD. 2000. Involvement of the checkpoint protein Mec1p in silencing of gene expression at telomeres in *Saccharomyces cerevisiae*. *Mol. Cell. Biol.* 20:2378–84

43. Crawford DF, Piwnica-Worms H. 2001. The G(2) DNA damage checkpoint delays expression of genes encoding mitotic regulators. *J. Biol. Chem.* 276: 37166–77

44. D'Amours D, Jackson SP. 2002. The mre11 complex: at the crossroads of DNA repair and checkpoint signalling. *Nat. Rev. Mol. Cell. Biol.* 3:317–27

45. Davey S, Han CS, Ramer SA, Klassen JC, Jacobson A, et al. 1998. Fission yeast rad12$^+$ regulates cell cycle checkpoint control and is homologous to the Bloom's syndrome disease gene. *Mol. Cell. Biol.* 18:2721–28

46. de Klein A, Muijtjens M, van Os R, Verhoeven Y, Smit B, et al. 2000. Targeted disruption of the cell-cycle checkpoint gene ATR leads to early embryonic lethality in mice. *Curr. Biol.* 10:479–82

47. de la Torre Ruiz MA, Lowndes NF. 2000. DUN1 defines one branch downstream of RAD53 for transcription and DNA damage repair in *Saccharomyces cerevisiae. FEBS Lett.* 485:205–6

48. Desany BA, Alcasabas AA, Bachant JB, Elledge SJ. 1998. Recovery from DNA replicational stress is the essential function of the S-phase checkpoint pathway. *Genes Dev.* 12:2956–70

49. Di Leonardo A, Linke SP, Clarkin K, Wahl GM. 1994. DNA damage triggers a prolonged p53-dependent G1 arrest and long-term induction of Cip1 in normal human fibroblasts. *Genes Dev.* 8:2540–51

50. Dolganov GM, Maser RS, Novikov A, Tosto L, Chong S, et al. 1996. Human Rad50 is physically associated with human Mre11: identification of a conserved multiprotein complex implicated in recombinational DNA repair. *Mol. Cell. Biol.* 16:4832–41

51. Dumaz N, Meek DW. 1999. Serine15 phosphorylation stimulates p53 transactivation but does not directly influence interaction with HDM2. *EMBO J.* 18:7002–10

52. Dutertre S, Sekhri R, Tintignac LA, Onclercq-Delic R, Chatton B, et al. 2002. Dephosphorylation and subcellular compartment change of the mitotic Bloom's syndrome DNA helicase in response to ionizing radiation. *J. Biol. Chem.* 277: 6280–86

53. Eder E, Kutt W, Deininger C. 2001. On the role of alkylating mechanisms, O-alkylation and DNA-repair in genotoxicity and mutagenicity of alkylating methanesulfonates of widely varying structures in bacterial systems. *Chem. Biol. Interact.* 137:89–99

54. Edwards RJ, Bentley NJ, Carr AM. 1999. A Rad3-Rad26 complex responds to DNA damage independently of other checkpoint proteins. *Nat. Cell Biol.* 1: 393–98

55. Ekholm SV, Reed SI. 2000. Regulation of G(1) cyclin-dependent kinases in the mammalian cell cycle. *Curr. Opin. Cell Biol.* 12:676–84

56. Emili A. 1998. MEC1-dependent phosphorylation of Rad9p in response to DNA damage. *Mol. Cell* 2:183–89

57. Enoch T, Carr AM, Nurse P. 1992. Fission yeast genes involved in coupling mitosis to completion of DNA replication. *Genes Dev.* 6:2035–46

58. Esashi F, Yanagida M. 1999. Cdc2 phosphorylation of Crb2 is required for reestablishing cell cycle progression after the damage checkpoint. *Mol. Cell* 4: 167–74

59. Falck J, Mailand N, Syljuasen RG, Bartek J, Lukas J. 2001. The ATM-Chk2-Cdc25A checkpoint pathway guards against radioresistant DNA synthesis. *Nature* 410:842–47

60. Fasullo M, Bennett T, AhChing P, Koudelik J. 1998. The *Saccharomyces cerevisiae* RAD9 checkpoint reduces the DNA damage-associated stimulation of directed translocations. *Mol. Cell. Biol.* 18:1190–200

61. Fitz Gerald JN, Benjamin JM, Kron SJ. 2002. Robust G1 checkpoint arrest

in budding yeast: dependence on DNA damage signaling and repair. *J. Cell Sci.* 115:1749–57

62. Foiani M, Pellicioli A, Lopes M, Lucca C, Ferrari M, et al. 2000. DNA damage checkpoints and DNA replication controls in *Saccharomyces cerevisiae*. *Mutat. Res.* 451:187–96

63. Forrest A, Gabrielli B. 2001. Cdc25B activity is regulated by 14-3-3. *Oncogene* 20:4393–401

64. Foss EJ. 2001. Tof1p regulates DNA damage responses during S phase in *Saccharomyces cerevisiae*. *Genetics* 157:567–77

65. Frei C, Gasser SM. 2000. RecQ-like helicases: the DNA replication checkpoint connection. *J. Cell Sci.* 113 (Pt 15):2641–46

66. Frei C, Gasser SM. 2000. The yeast Sgs1p helicase acts upstream of Rad53p in the DNA replication checkpoint and colocalizes with Rad53p in S-phase-specific foci. *Genes Dev.* 14:81–96

67. Gardner R, Putnam CW, Weinert T. 1999. RAD53, DUN1 and PDS1 define two parallel G2/M checkpoint pathways in budding yeast. *EMBO J.* 18:3173–85

68. Gartner A, Milstein S, Ahmed S, Hodgkin J, Hengartner MO. 2000. A conserved checkpoint pathway mediates DNA damage-induced apoptosis and cell cycle arrest in *C. elegans*. *Mol. Cell* 5:435–43

69. Garvik B, Carson M, Hartwell L. 1995. Single-stranded DNA arising at telomeres in cdc13 mutants may constitute a specific signal for the RAD9 checkpoint. *Mol. Cell. Biol.* 15:6128–38

70. Gasch AP, Huang M, Metzner S, Botstein D, Elledge SJ, Brown PO. 2001. Genomic expression responses to DNA-damaging agents and the regulatory role of the yeast ATR homolog Mec1p. *Mol. Biol. Cell* 12:2987–3003

71. Gatei M, Scott SP, Filippovitch I, Soronika N, Lavin MF, et al. 2000.

Role for ATM in DNA damage-induced phosphorylation of BRCA1. *Cancer Res.* 60:3299–304

72. Gately DP, Hittle JC, Chan GK, Yen TJ. 1998. Characterization of ATM expression, localization, and associated DNA-dependent protein kinase activity. *Mol. Biol. Cell* 9:2361–74

73. Gayther SA, Pharoah PD, Ponder BA. 1998. The genetics of inherited breast cancer. *J. Mammary Gland Biol. Neoplasia* 3:365–76

74. Geske FJ, Nelson AC, Lieberman R, Strange R, Sun T, Gerschenson LE. 2000. DNA repair is activated in early stages of p53-induced apoptosis. *Cell Death Differ.* 7:393–401

75. Geyer RK, Nagasawa H, Little JB, Maki CG. 2000. Role and regulation of p53 during an ultraviolet radiation-induced G1 cell cycle arrest. *Cell Growth Differ.* 11:149–56

76. Gilbert CS, Green CM, Lowndes NF. 2001. Budding yeast Rad9 is an ATP-dependent Rad53 activating machine. *Mol. Cell* 8:129–36

77. Gowen LC, Avrutskaya AV, Latour AM, Koller BH, Leadon SA. 1998. BRCA1 required for transcription-coupled repair of oxidative DNA damage. *Science* 281:1009–12

78. Graves PR, Lovly CM, Uy GL, Piwnica-Worms H. 2001. Localization of human Cdc25C is regulated both by nuclear export and 14-3-3 protein binding. *Oncogene* 20:1839–51

79. Green CM, Erdjument-Bromage H, Tempst P, Lowndes NF. 2000. A novel Rad24 checkpoint protein complex closely related to replication factor C. *Curr. Biol.* 10:39–42

80. Grenon M, Tillit J, Piard K, Baldacci G, Francesconi S. 1999. The S/M checkpoint at 37 degrees C and the recovery of viability of the mutant poldeltats3 require the crb2+/rhp9+ gene in fission yeast. *Mol. Gen. Genet.* 260:522–34

81. Griffiths DJ, Barbet NC, McCready S, Lehmann AR, Carr AM. 1995. Fission yeast rad17: a homologue of budding yeast RAD24 that shares regions of sequence similarity with DNA polymerase accessory proteins. *EMBO J.* 14:5812–23

82. Grushcow JM, Holzen TM, Park KJ, Weinert T, Lichten M, Bishop DK. 1999. *Saccharomyces cerevisiae* checkpoint genes MEC1, RAD17 and RAD24 are required for normal meiotic recombination partner choice. *Genetics* 153:607–20

83. Guo Z, Kumagai A, Wang SX, Dunphy WG. 2000. Requirement for Atr in phosphorylation of Chk1 and cell cycle regulation in response to DNA replication blocks and UV-damaged DNA in *Xenopus* egg extracts. *Genes Dev.* 14:2745–56

84. Hang H, Lieberman HB. 2000. Physical interactions among human checkpoint control proteins HUS1p, RAD1p, and RAD9p, and implications for the regulation of cell cycle progression. *Genomics* 65:24–33

85. Harper JW, Adami GR, Wei N, Keyomarsi K, Elledge SJ. 1993. The p21 Cdk-interacting protein Cip1 is a potent inhibitor of G1 cyclin-dependent kinases. *Cell* 75:805–16

86. Hartwell LH, Weinert TA. 1989. Checkpoints: controls that ensure the order of cell cycle events. *Science* 246:629–34

87. Haupt Y, Maya R, Kazaz A, Oren M. 1997. Mdm2 promotes the rapid degradation of p53. *Nature* 387:296–99

88. Hekmat-Nejad M, You Z, Yee MC, Newport JW, Cimprich KA. 2000. *Xenopus* ATR is a replication-dependent chromatin-binding protein required for the DNA replication checkpoint. *Curr. Biol.* 10:1565–73

89. Hermeking H, Lengauer C, Polyak K, He TC, Zhang L, et al. 1997. 14-3-3 sigma is a p53-regulated inhibitor of G2/M progression. *Mol. Cell* 1:3–11

90. Hirao A, Kong YY, Matsuoka S, Wakeham A, Ruland J, et al. 2000. DNA damage-induced activation of p53 by the checkpoint kinase Chk2. *Science* 287:1824–27

91. Hong EJ, Roeder GS. 2002. A role for Ddc1 in signaling meiotic double-strand breaks at the pachytene checkpoint. *Genes Dev.* 16:363–76

92. Huyton T, Bates PA, Zhang X, Sternberg MJ, Freemont PS. 2000. The BRCA1 C-terminal domain: structure and function. *Mutat. Res.* 460:319–32

93. Iftode C, Daniely Y, Borowiec JA. 1999. Replication protein A (RPA): the eukaryotic SSB. *Crit. Rev. Biochem. Mol.Biol.* 34:141–80

94. Jin P, Hardy S, Morgan DO. 1998. Nuclear localization of cyclin B1 controls mitotic entry after DNA damage. *J. Cell Biol.* 141:875–85

95. Kamijo T, van de Kamp E, Chong MJ, Zindy F, Diehl JA, et al. 1999. Loss of the ARF tumor suppressor reverses premature replicative arrest but not radiation hypersensitivity arising from disabled ATM function. *Cancer Res.* 59:2464–69

96. Kang D, Chen J, Wong J, Fang G. 2002. The checkpoint protein Chfr is a ligase that ubiquitinates Plk1 and inhibits Cdc2 at the G2 to M transition. *J. Cell Biol.* 156:249–59

97. Kastan MB, Lim DS. 2000. The many substrates and functions of ATM. *Nat. Rev. Mol. Cell. Biol.* 1:179–86

98. Kastan MB, Onyekwere O, Sidransky D, Vogelstein B, Craig RW. 1991. Participation of p53 protein in the cellular response to DNA damage. *Cancer Res.* 51:6304–11

99. Kawabe T, Suganuma M, Ando T, Kimura M, Hori H, Okamoto T. 2002. Cdc25C interacts with PCNA at G2/M transition. *Oncogene* 21:1717–26

100. Khanna KK, Jackson SP. 2001. DNA double-strand breaks: signaling, repair and the cancer connection. *Nat. Genet.* 27:247–54

101. Khosravi R, Maya R, Gottlieb T, Oren M,

Shiloh Y, Shkedy D. 1999. Rapid ATM-dependent phosphorylation of MDM2 precedes p53 accumulation in response to DNA damage. *Proc. Natl. Acad. Sci. USA* 96:14973–77

102. Kim HS, Brill SJ. 2001. Rfc4 interacts with Rpa1 and is required for both DNA replication and DNA damage checkpoints in *Saccharomyces cerevisiae*. *Mol. Cell. Biol.* 21:3725–37

103. Klein HL. 2001. Spontaneous chromosome loss in *Saccharomyces cerevisiae* is suppressed by DNA damage checkpoint functions. *Genetics* 159:1501–9

104. Kohn PH, Whang-Peng J, Levis WR. 1982. Chromosomal instability in ataxia telangiectasia. *Cancer Genet. Cytogenet.* 6:289–302

105. Komatsu K, Miyashita T, Hang H, Hopkins KM, Zheng W, et al. 2000. Human homologue of *S. pombe* Rad9 interacts with BCL-2/BCL-xL and promotes apoptosis. *Nat. Cell Biol.* 2:1–6

106. Kondo T, Wakayama T, Naiki T, Matsumoto K, Sugimoto K. 2001. Recruitment of Mec1 and Ddc1 checkpoint proteins to double-strand breaks through distinct mechanisms. *Science* 294:867–70

107. Lavin MF, Shiloh Y. 1997. The genetic defect in ataxia-telangiectasia. *Annu. Rev. Immunol.* 15:177–202

108. Lee J, Kumagai A, Dunphy WG. 2001. Positive regulation of Wee1 by Chk1 and 14-3-3 proteins. *Mol. Biol. Cell* 12:551–63

109. Lee JS, Collins KM, Brown AL, Lee CH, Chung JH. 2000. hCds1-mediated phosphorylation of BRCA1 regulates the DNA damage response. *Nature* 404:201–4

110. Leu JY, Roeder GS. 1999. The pachytene checkpoint in *S. cerevisiae* depends on Swe1-mediated phosphorylation of the cyclin-dependent kinase Cdc28. *Mol. Cell* 4:805–14

111. Li S, Ting NS, Zheng L, Chen PL, Ziv Y, et al. 2000. Functional link of BRCA1 and ataxia telangiectasia gene product in DNA damage response. *Nature* 406:210–15

112. Lim DS, Kim ST, Xu B, Maser RS, Lin J, et al. 2000. ATM phosphorylates p95/nbs1 in an S-phase checkpoint pathway. *Nature* 404:613–17

113. Lindsay HD, Griffiths DJ, Edwards RJ, Christensen PU, Murray JM, et al. 1998. S-phase-specific activation of Cds1 kinase defines a subpathway of the checkpoint response in *Schizosaccharomyces pombe*. *Genes Dev.* 12:382–95

114. Lindsey-Boltz LA, Bermudez VP, Hurwitz J, Sancar A. 2001. Purification and characterization of human DNA damage checkpoint Rad complexes. *Proc. Natl. Acad. Sci. USA* 98:11236–41

115. Liu F, Rothblum-Oviatt C, Ryan CE, Piwnica-Worms H. 1999. Overproduction of human Myt1 kinase induces a G2 cell cycle delay by interfering with the intracellular trafficking of Cdc2-cyclin B1 complexes. *Mol. Cell. Biol.* 19:5113–23

116. Loeb LA. 1991. Mutator phenotype may be required for multistage carcinogenesis. *Cancer Res.* 51:3075–79

117. Longhese MP, Neecke H, Paciotti V, Lucchini G, Plevani P. 1996. The 70 kDa subunit of replication protein A is required for the G1/S and intra-S DNA damage checkpoints in budding yeast. *Nucleic Acids Res.* 24:3533–37

118. Longhese MP, Paciotti V, Neecke H, Lucchini G. 2000. Checkpoint proteins influence telomeric silencing and length maintenance in budding yeast. *Genetics* 155:1577–91

119. Lopes M, Cotta-Ramusino C, Pellicioli A, Liberi G, Plevani P, et al. 2001. The DNA replication checkpoint response stabilizes stalled replication forks. *Nature* 412:557–61

120. Lopez-Girona A, Kanoh J, Russell P. 2001. Nuclear exclusion of Cdc25 is not required for the DNA damage checkpoint in fission yeast. *Curr. Biol.* 11:50–54

121. Lydall D, Nikolsky Y, Bishop DK, Weinert T. 1996. A meiotic recombination checkpoint controlled by mitotic checkpoint genes. *Nature* 383:840–43

122. Lydall D, Weinert T. 1995. Yeast checkpoint genes in DNA damage processing: implications for repair and arrest. *Science* 270:1488–91

123. Mailand N, Falck J, Lukas C, Syljuasen RG, Welcker M, et al. 2000. Rapid destruction of human Cdc25A in response to DNA damage. *Science* 288:1425–29

124. Makiniemi M, Hillukkala T, Tuusa J, Reini K, Vaara M, et al. 2001. BRCT domain-containing protein TopBP1 functions in DNA replication and damage response. *J. Biol. Chem.* 276: 30399–406

125. Malkin D, Li FP, Strong LC, Fraumeni JF Jr, Nelson CE, et al. 1990. Germline p53 mutations in a familial syndrome of breast cancer, sarcomas, and other neoplasms. *Science* 250:1233–38

126. Marini F, Pellicioli A, Paciotti V, Lucchini G, Plevani P, et al. 1997. A role for DNA primase in coupling DNA replication to DNA damage response. *EMBO J.* 16:639–50

127. Martinho RG, Lindsay HD, Flaggs G, DeMaggio AJ, Hoekstra MF, et al. 1998. Analysis of Rad3 and Chk1 protein kinases defines different checkpoint responses. *EMBO J.* 17:7239–49

128. Maser RS, Monsen KJ, Nelms BE, Petrini JH. 1997. hMre11 and hRad50 nuclear foci are induced during the normal cellular response to DNA double-strand breaks. *Mol. Cell. Biol.* 17:6087–96

129. Matsuoka S, Huang M, Elledge SJ. 1998. Linkage of ATM to cell cycle regulation by the Chk2 protein kinase. *Science* 282:1893–97

130. Matsuura A, Naito T, Ishikawa F. 1999. Genetic control of telomere integrity in *Schizosaccharomyces pombe*: rad3(+) and tel1(+) are parts of two regulatory

networks independent of the downstream protein kinases chk1(+) and cds1(+). *Genetics* 152:1501–12

131. McAinsh AD, Scott-Drew S, Murray JA, Jackson SP. 1999. DNA damage triggers disruption of telomeric silencing and Mec1p-dependent relocation of Sir3p. *Curr. Biol.* 9:963–66

132. Melo JA, Cohen J, Toczyski DP. 2001. Two checkpoint complexes are independently recruited to sites of DNA damage in vivo. *Genes Dev.* 15:2809–21

133. Mills KD, Sinclair DA, Guarente L. 1999. MEC1-dependent redistribution of the Sir3 silencing protein from telomeres to DNA double-strand breaks. *Cell* 97:609–20

134. Morrison C, Sonoda E, Takao N, Shinohara A, Yamamoto K, Takeda S. 2000. The controlling role of ATM in homologous recombinational repair of DNA damage. *EMBO J.* 19:463–71

135. Moynahan ME, Chiu JW, Koller BH, Jasin M. 1999. Brca1 controls homology-directed DNA repair. *Mol. Cell* 4: 511–18

136. Myung K, Kolodner RD. 2002. Suppression of genome instability by redundant S-phase checkpoint pathways in *Saccharomyces cerevisiae*. *Proc. Natl. Acad. Sci. USA* 99:4500–7

137. Naiki T, Shimomura T, Kondo T, Matsumoto K, Sugimoto K. 2000. Rfc5, in cooperation with rad24, controls DNA damage checkpoints throughout the cell cycle in *Saccharomyces cerevisiae*. *Mol. Cell. Biol.* 20:5888–96

138. Navas TA, Sanchez Y, Elledge SJ. 1996. RAD9 and DNA polymerase epsilon form parallel sensory branches for transducing the DNA damage checkpoint signal in *Saccharomyces cerevisiae*. *Genes Dev.* 10:2632–43

138a. Noguchi E, Shanahan P, Noguchi C, Russell P. 2002. CDK phosphorylation of Drc1 ewgulates DNA replication in fission yeast. *Curr. Biol.* 12:599–605

139. Noskov VN, Araki H, Sugino A. 1998.

The RFC2 gene, encoding the third-largest subunit of the replication factor C complex, is required for an S-phase checkpoint in *Saccharomyces cerevisiae*. *Mol. Cell. Biol.* 18:4914–23

140. Oakley GG, Loberg LI, Yao J, Risinger MA, Yunker RL, et al. 2001. UV-induced hyperphosphorylation of replication protein A depends on DNA replication and expression of ATM protein. *Mol. Biol. Cell* 12:1199–213

141. Okumura E, Fukuhara T, Yoshida H, Hanada Si S, Kozutsumi R, et al. 2002. Akt inhibits Myt1 in the signalling pathway that leads to meiotic G2/M-phase transition. *Nat. Cell Biol.* 4:111–16

142. Paciotti V, Clerici M, Lucchini G, Longhese MP. 2000. The checkpoint protein Ddc2, functionally related to *S. pombe* Rad26, interacts with Mec1 and is regulated by Mec1–dependent phosphorylation in budding yeast. *Genes Dev.* 14:2046–59

143. Paciotti V, Clerici M, Scotti M, Lucchini G, Longhese MP. 2001. Characterization of mec1 kinase-deficient mutants and of new hypomorphic mec1 alleles impairing subsets of the DNA damage response pathway. *Mol. Cell. Biol.* 21:3913–25

144. Painter RB, Young BR. 1980. Radiosensitivity in ataxia-telangiectasia: a new explanation. *Proc. Natl. Acad. Sci. USA* 77:7315–17

145. Pandita TK. 2002. ATM function and telomere stability. *Oncogene* 21:611–18

146. Parker AE, Van de Weyer I, Laus MC, Oostveen I, Yon J, et al. 1998. A human homologue of the *Schizosaccharomyces pombe* rad1 + checkpoint gene encodes an exonuclease. *J. Biol. Chem.* 273:18332–39

147. Paull TT, Cortez D, Bowers B, Elledge SJ, Gellert M. 2001. From the cover: direct DNA binding by Brca1. *Proc. Natl. Acad. Sci. USA* 98:6086–91

148. Paulovich AG, Armour CD, Hartwell LH. 1998. The *Saccharomyces cerevisiae* RAD9, RAD17, RAD24 and MEC3 genes are required for tolerating irreparable, ultraviolet-induced DNA damage. *Genetics* 150:75–93

149. Paulovich AG, Hartwell LH. 1995. A checkpoint regulates the rate of progression through S phase in *S. cerevisiae* in response to DNA damage. *Cell* 82:841–47

150. Pellicioli A, Lee SE, Lucca C, Foiani M, Haber JE. 2001. Regulation of *Saccharomyces* Rad53 checkpoint kinase during adaptation from DNA damage-induced G2/M arrest. *Mol. Cell* 7:293–300

151. Peng CY, Graves PR, Thoma RS, Wu Z, Shaw AS, Piwnica-Worms H. 1997. Mitotic and G2 checkpoint control: regulation of 14-3-3 protein binding by phosphorylation of Cdc25C on serine-216. *Science* 277:1501–5

152. Raghuraman MK, Winzeler EA, Collingwood D, Hunt S, Wodicka L, et al. 2001. Replication dynamics of the yeast genome. *Science* 294:115–21

153. Ren B, Cam H, Takahashi Y, Volkert T, Terragni J, et al. 2002. E2F integrates cell cycle progression with DNA repair, replication, and G(2)/M checkpoints. *Genes Dev.* 16:245–56

154. Rhind N, Russell P. 2001. Roles of the mitotic inhibitors Wee1 and Mik1 in the G(2) DNA damage and replication checkpoints. *Mol. Cell. Biol.* 21:1499–508

155. Ritchie KB, Mallory JC, Petes TD. 1999. Interactions of TLC1 (which encodes the RNA subunit of telomerase), TEL1, and MEC1 in regulating telomere length in the yeast *Saccharomyces cerevisiae*. *Mol. Cell. Biol.* 19:6065–75

156. Rotman G, Shiloh Y. 1999. ATM: a mediator of multiple responses to genotoxic stress. *Oncogene* 18:6135–44

157. Rouse J, Jackson SP. 2000. LCD1: an essential gene involved in checkpoint control and regulation of the MEC1 signalling pathway in *Saccharomyces cerevisiae*. *EMBO J.* 19:5801–12

157a. Rouse, Jackson SP. 2002. Lcd1p recruits

Mec1p to DNA lesion in vitro and in vivo. *Mol. Cell.* 9:857–69

158. Saka Y, Esashi F, Matsusaka T, Mochida S, Yanagida M. 1997. Damage and replication checkpoint control in fission yeast is ensured by interactions of Crb2, a protein with BRCT motif, with Cut5 and Chk1. *Genes Dev.* 11:3387–400

159. Sanchez Y, Bachant J, Wang H, Hu F, Liu D, et al. 1999. Control of the DNA damage checkpoint by chk1 and rad53 protein kinases through distinct mechanisms. *Science* 286:1166–71

160. Sanchez Y, Wong C, Thoma RS, Richman R, Wu Z, et al. 1997. Conservation of the Chk1 checkpoint pathway in mammals: linkage of DNA damage to Cdk regulation through Cdc25. *Science* 277:1497–501

161. Sandell LL, Zakian VA. 1993. Loss of a yeast telomere: arrest, recovery, and chromosome loss. *Cell* 75:729–39

162. Santocanale C, Diffley JF. 1998. A Mec1- and Rad53-dependent checkpoint controls late-firing origins of DNA replication. *Nature* 395:615–18

163. Santocanale C, Neecke H, Longhese MP, Lucchini G, Plevani P. 1995. Mutations in the gene encoding the 34 kDa subunit of yeast replication protein A cause defective S phase progression. *J. Mol. Biol.* 254:595–607

164. Savitsky K, Bar-Shira A, Gilad S, Rotman G, Ziv Y, et al. 1995. A single ataxia telangiectasia gene with a product similar to PI-3 kinase. *Science* 268:1749–53

165. Schramke V, Neecke H, Brevet V, Corda Y, Lucchini G, et al. 2001. The set1Delta mutation unveils a novel signaling pathway relayed by the Rad53-dependent hyperphosphorylation of replication protein A that leads to transcriptional activation of repair genes. *Genes Dev.* 15:1845–58

166. Scolnick DM, Halazonetis TD. 2000. Chfr defines a mitotic stress checkpoint that delays entry into metaphase. *Nature* 406:430–35

167. Scully R, Chen J, Plug A, Xiao Y, Weaver D, et al. 1997. Association of BRCA1 with Rad51 in mitotic and meiotic cells. *Cell* 88:265–75

168. Sherr CJ, Roberts JM. 1999. CDK inhibitors: positive and negative regulators of G1-phase progression. *Genes Dev.* 13:1501–12

169. Shieh SY, Ahn J, Tamai K, Taya Y, Prives C. 2000. The human homologs of checkpoint kinases Chk1 and Cds1 (Chk2) phosphorylate p53 at multiple DNA damage-inducible sites. *Genes Dev.* 14:289–300

170. Shiloh Y. 1997. Ataxia-telangiectasia and the Nijmegen breakage syndrome: related disorders but genes apart. *Annu. Rev. Genet.* 31:635–62

171. Shiloh Y. 2001. ATM and ATR: networking cellular responses to DNA damage. *Curr. Opin. Genet. Dev.* 11:71–77

172. Shimada M, Okuzaki D, Tanaka S, Tougan T, Tamai KK, et al. 1999. Replication factor C3 of *Schizosaccharomyces pombe*, a small subunit of replication factor C complex, plays a role in both replication and damage checkpoints. *Mol. Biol. Cell* 10:3991–4003

173. Sidorova JM, Breeden LL. 1997. Rad53-dependent phosphorylation of Swi6 and down-regulation of CLN1 and CLN2 transcription occur in response to DNA damage in *Saccharomyces cerevisiae*. *Genes Dev.* 11:3032–45

174. Siede W, Friedberg AS, Dianova I, Friedberg EC. 1994. Characterization of G1 checkpoint control in the yeast *Saccharomyces cerevisiae* following exposure to DNA-damaging agents. *Genetics* 138:271–81

175. Smith GC, Cary RB, Lakin ND, Hann BC, Teo SH, et al. 1999. Purification and DNA binding properties of the ataxia-telangiectasia gene product ATM. *Proc. Natl. Acad. Sci. USA* 96:11134–39

176. Smith GC, Jackson SP. 1999. The DNA-dependent protein kinase. *Genes Dev.* 13:916–34

177. Smits VA, Klompmaker R, Arnaud L, Rijksen G, Nigg EA, Medema RH. 2000. Polo-like kinase-1 is a target of the DNA damage checkpoint. *Nat. Cell Biol.* 2:672–76

178. Snouwaert JN, Gowen LC, Latour AM, Mohn AR, Xiao A, et al. 1999. BRCA1 deficient embryonic stem cells display a decreased homologous recombination frequency and an increased frequency of non-homologous recombination that is corrected by expression of a brca1 transgene. *Oncogene* 18:7900–7

178a. Sogo JM, Lopes M, Foiani M. 2002. Fork reversal and ssDNA accumulation of stalled replication forks owing to checkpoint defects. *Science* 297:599–602

179. St Onge RP, Udell CM, Casselman R, Davey S. 1999. The human G2 checkpoint control protein hRAD9 is a nuclear phosphoprotein that forms complexes with hRAD1 and hHUS1. *Mol. Biol. Cell* 10:1985–95

180. Stewart GS, Maser RS, Stankovic T, Bressan DA, Kaplan MI, et al. 1999. The DNA double-strand break repair gene hMRE11 is mutated in individuals with an ataxia-telangiectasia-like disorder. *Cell* 99:577–87

181. Sugawara N, Haber JE. 1992. Characterization of double-strand break-induced recombination: homology requirements and single-stranded DNA formation. *Mol. Cell. Biol.* 12:563–75

182. Takekawa M, Adachi M, Nakahata A, Nakayama I, Itoh F, et al. 2000. p53-inducible wip1 phosphatase mediates a negative feedback regulation of p38 MAPK-p53 signaling in response to UV radiation. *EMBO J.* 19:6517–26

183. Tanaka H, Arakawa H, Yamaguchi T, Shiraishi K, Fukuda S, et al. 2000. A ribonucleotide reductase gene involved in a p53-dependent cell-cycle checkpoint for DNA damage. *Nature* 404:42–49

184. Tanaka K, Russell P. 2001. Mrc1 channels the DNA replication arrest signal to checkpoint kinase Cds1. *Nat. Cell Biol.* 3:966–72

185. Taylor WR, DePrimo SE, Agarwal A, Agarwal ML, Schonthal AH, et al. 1999. Mechanisms of G2 arrest in response to overexpression of p53. *Mol. Biol. Cell* 10:3607–22

186. Taylor WR, Stark GR. 2001. Regulation of the G2/M transition by p53. *Oncogene* 20:1803–15

187. Terada Y, Tatsuka M, Jinno S, Okayama H. 1995. Requirement for tyrosine phosphorylation of Cdk4 in G1 arrest induced by ultraviolet irradiation. *Nature* 376:358–62

188. Tercero JA, Diffley JF. 2001. Regulation of DNA replication fork progression through damaged DNA by the Mec1/Rad53 checkpoint. *Nature* 412:553–57

189. Thompson DA, Stahl FW. 1999. Genetic control of recombination partner preference in yeast meiosis. Isolation and characterization of mutants elevated for meiotic unequal sister-chromatid recombination. *Genetics* 153:621–41

190. Tibbetts RS, Cortez D, Brumbaugh KM, Scully R, Livingston D, et al. 2000. Functional interactions between BRCA1 and the checkpoint kinase ATR during genotoxic stress. *Genes Dev.* 14:2989–3002

191. Tinker-Kulberg RL, Morgan DO. 1999. Pds1 and Esp1 control both anaphase and mitotic exit in normal cells and after DNA damage. *Genes Dev.* 13:1936–49

192. Toyoshima F, Moriguchi T, Wada A, Fukuda M, Nishida E. 1998. Nuclear export of cyclin B1 and its possible role in the DNA damage-induced G2 checkpoint. *EMBO J.* 17:2728–35

193. Tsurimoto T, Stillman B. 1991. Replication factors required for SV40 DNA replication in vitro. II. Switching of DNA polymerase alpha and delta during initiation of leading and lagging strand synthesis. *J. Biol. Chem.* 266:1961–68

194. Tsuzuki T, Fujii Y, Sakumi K, Tominaga Y, Nakao K, et al. 1996. Targeted

disruption of the Rad51 gene leads to lethality in embryonic mice. *Proc. Natl. Acad. Sci. USA* 93:6236–40

195. Usui T, Ogawa H, Petrini JH. 2001. A DNA damage response pathway controlled by Tel1 and the Mre11 complex. *Mol. Cell* 7:1255–66

196. van Brabant AJ, Buchanan CD, Charboneau E, Fangman WL, Brewer BJ. 2001. An origin-deficient yeast artificial chromosome triggers a cell cycle checkpoint. *Mol. Cell* 7:705–13

197. van Vugt MA, Smits VA, Klompmaker R, Medema RH. 2001. Inhibition of Polo-like kinase-1 by DNA damage occurs in an ATM- or ATR-dependent fashion. *J. Biol. Chem.* 276:41656–60

198. Venclovas C, Thelen MP. 2000. Structure-based predictions of Rad1, Rad9, Hus1 and Rad17 participation in sliding clamp and clamp-loading complexes. *Nucleic Acids Res.* 28:2481–93

199. Verkade HM, O'Connell MJ. 1998. Cut5 is a component of the UV-responsive DNA damage checkpoint in fission yeast. *Mol. Gen. Genet.* 260:426–33

200. Vialard JE, Gilbert CS, Green CM, Lowndes NF. 1998. The budding yeast Rad9 checkpoint protein is subjected to Mec1/Tel1-dependent hyperphosphorylation and interacts with Rad53 after DNA damage. *EMBO J.* 17:5679–88

201. Volkmer E, Karnitz LM. 1999. Human homologs of *Schizosaccharomyces pombe* rad1, hus1, and rad9 form a DNA damage-responsive protein complex. *J. Biol. Chem.* 274:567–70

202. Wakayama T, Kondo T, Ando S, Matsumoto K, Sugimoto K. 2001. Pie1, a protein interacting with Mec1, controls cell growth and checkpoint responses in *Saccharomyces cerevisiae*. *Mol. Cell. Biol.* 21:755–64

202a. Walworth NC, Bernards R. 1996. rad-dependent response of the Chk1-encoded protein kinase at the DNA damage checkpoint. *Science* 271:353–56

203. Wang H, Elledge SJ. 1999. DRC1, DNA replication and checkpoint protein 1, functions with DPB11 to control DNA replication and the S-phase checkpoint in *Saccharomyces cerevisiae*. *Proc. Natl. Acad. Sci. USA* 96:3824–29

204. Wang H, Elledge SJ. 2002. Genetic and Physical Interactions Between DPB11 and DDC1 in the Yeast DNA Damage Response Pathway. *Genetics* 160:1295–304

205. Wang H, Guan J, Perrault AR, Wang Y, Iliakis G. 2001. Replication protein A2 phosphorylation after DNA damage by the coordinated action of ataxia telangiectasia-mutated and DNA-dependent protein kinase. *Cancer Res.* 61:8554–63

206. Wang H, Zeng ZC, Bui TA, DiBiase SJ, Qin W, et al. 2001. Nonhomologous end-joining of ionizing radiation-induced DNA double-stranded breaks in human tumor cells deficient in BRCA1 or BRCA2. *Cancer Res.* 61:270–77

207. Wang Y, Cortez D, Yazdi P, Neff N, Elledge SJ, Qin J. 2000. BASC, a super complex of BRCA1-associated proteins involved in the recognition and repair of aberrant DNA structures. *Genes Dev.* 14:927–39

208. Ward IM, Chen J. 2001. Histone H2AX is phosphorylated in an ATR-dependent manner in response to replicational stress. *J. Biol. Chem.* 276:47759–62

209. Ward JF. 1985. Biochemistry of DNA lesions. *Radiat. Res. Suppl.* 8:S103–11

210. Weinert T. 1998. DNA damage checkpoints update: getting molecular. *Curr. Opin. Genet. Dev.* 8:185–93

211. Weinert TA, Hartwell LH. 1988. The RAD9 gene controls the cell cycle response to DNA damage in *Saccharomyces cerevisiae*. *Science* 241:317–22

212. Weinert TA, Hartwell LH. 1990. Characterization of RAD9 of *Saccharomyces cerevisiae* and evidence that its function acts posttranslationally in cell cycle arrest after DNA damage. *Mol. Cell. Biol.* 10:6554–64

213. Weinert TA, Kiser GL, Hartwell LH. 1994. Mitotic checkpoint genes in budding yeast and the dependence of mitosis on DNA replication and repair. *Genes Dev.* 8:652–65

214. Weinreich M, Stillman B. 1999. Cdc7p-Dbf4p kinase binds to chromatin during S phase and is regulated by both the APC and the RAD53 checkpoint pathway. *EMBO J.* 18:5334–46

215. Willson J, Wilson S, Warr N, Watts FZ. 1997. Isolation and characterization of the *Schizosaccharomyces pombe* rhp9 gene: a gene required for the DNA damage checkpoint but not the replication checkpoint. *Nucleic Acids Res.* 25:2138–46

216. Wolkow TD, Enoch T. 2002. Fission yeast rad26 is a regulatory subunit of the rad3 checkpoint kinase. *Mol. Biol. Cell* 13:480–92

217. Wright JA, Keegan KS, Herendeen DR, Bentley NJ, Carr AM, et al. 1998. Protein kinase mutants of human ATR increase sensitivity to UV and ionizing radiation and abrogate cell cycle checkpoint control. *Proc. Natl. Acad. Sci. USA* 95:7445–50

218. Xie S, Wu H, Wang Q, Cogswell JP, Husain I, et al. 2001. Plk3 functionally links DNA damage to cell cycle arrest and apoptosis at least in part via the p53 pathway. *J. Biol. Chem.* 276:43305–12

219. Xiong Y, Zhang H, Beach D. 1992. D type cyclins associate with multiple protein kinases and the DNA replication and repair factor PCNA. *Cell* 71:505–14

220. Xu B, Kim S, Kastan MB. 2001. Involvement of Brca1 in S-phase and G(2)-phase checkpoints after ionizing irradiation. *Mol. Cell. Biol.* 21:3445–50

221. Xu Y, Baltimore D. 1996. Dual roles of ATM in the cellular response to radiation and in cell growth control. *Genes Dev.* 10:2401–10

222. Yamane K, Wu X, Chen J. 2002. A DNA damage-regulated   BRCT-containing protein, TopBP1, is required for cell survival. *Mol. Cell. Biol.* 22:555–66

223. Yamazaki V, Wegner RD, Kirchgessner CU. 1998. Characterization of cell cycle checkpoint responses after ionizing radiation in Nijmegen breakage syndrome cells. *Cancer Res.* 58:2316–22

224. Yang SS, Yeh E, Salmon ED, Bloom K. 1997. Identification of a mid-anaphase checkpoint in budding yeast. *J. Cell Biol.* 136:345–54

225. Yarden RI, Brody LC. 1999. BRCA1 interacts with components of the histone deacetylase complex. *Proc. Natl. Acad. Sci. USA* 96:4983–88

226. Yarden RI, Pardo-Reoyo S, Sgagias M, Cowan KH, Brody LC. 2002. BRCA1 regulates the G2/M checkpoint by activating Chk1 kinase upon DNA damage. *Nat. Genet.* 30:285–89

227. You Z, Kong L, Newport J. 2002. The role of single-stranded DNA and Pol a in establishing the ATR, Hus1 DNA replication checkpoint. *J. Biol. Chem.* In press

228. Zhao S, Weng YC, Yuan SS, Lin YT, Hsu HC, et al. 2000. Functional link between ataxia-telangiectasia and Nijmegen breakage syndrome gene products. *Nature* 405:473–77

229. Zhao X, Chabes A, Domkin V, Thelander L, Rothstein R. 2001. The ribonucleotide reductase inhibitor Sml1 is a new target of the Mec1/Rad53 kinase cascade during growth and in response to DNA damage. *EMBO J.* 20:3544–53

230. Zhao X, Muller EG, Rothstein R. 1998. A suppressor of two essential checkpoint genes identifies a novel protein that negatively affects dNTP pools. *Mol. Cell* 2: 329–40

231. Zhao X, Rothstein R. 2002. The Dun1 checkpoint kinase phosphorylates and regulates the ribonucleotide reductase inhibitor Sml1. *Proc. Natl. Acad. Sci. USA* 99:3746–51

232. Zhong Q, Chen CF, Li S, Chen Y, Wang CC, et al. 1999. Association of BRCA1 with the hRad50-hMre11-p95 complex

and the DNA damage response. *Science* 285:747–50

233. Zhou BB, Chaturvedi P, Spring K, Scott SP, Johanson RA, et al. 2000. Caffeine abolishes the mammalian G(2)/M DNA damage checkpoint by inhibiting ataxia-telangiectasia-mutated kinase activity. *J. Biol. Chem.* 275:10342–48

234. Zhu XD, Kuster B, Mann M, Petrini JH, Lange T. 2000. Cell-cycle-regulated association of RAD50/MRE11/NBS1 with TRF2 and human telomeres. *Nat. Genet.* 25:347–52

235. Zou L, Cortez D, Elledge SJ. 2002. Regulation of ATR substrate selection by Rad17-dependent loading of Rad9 complexes onto chromatin. *Genes Dev.* 16:198–20843

Annu. Rev. Genet. 2002. 36:657–86
doi: 10.1146/annurev.genet.36.060602.145553
First published online as a Review in Advance on September 12, 2002

THE FELINE GENOME PROJECT[1]

Stephen J. O'Brien, Marilyn Menotti-Raymond,
William J. Murphy, and Naoya Yuhki
Laboratory of Genomic Diversity, National Cancer Institute–Frederick,
Frederick, Maryland 21702-1201; e-mail:OBRIEN@ncifcrf.gov

Key Words   feline, comparative, genome, genetic map, cat

■ **Abstract**   The compilation of a dense gene map and eventually a whole genome sequence (WGS) of the domestic cat holds considerable value for human genome annotation, for veterinary medicine, and for insight into the evolution of genome organization among mammals. Human association and veterinary studies of the cat, its domestic breeds, and its charismatic wild relatives of the family Felidae have rendered the species a powerful model for human hereditary diseases, for infectious disease agents, for adaptive evolutionary divergence, for conservation genetics, and for forensic applications. Here we review the advantages, rationale, and present strategy of a feline genome project, and we describe the disease models, comparative genomics, and biological applications posed by the full resolution of the cat's genome.

CONTENTS

INTRODUCTION

In the century since the rediscovery and replication of Mendel's seminal tenets of hereditary transmission by Hugo DeVries, Carl Correns, and Erick von Tschermak in 1900, the field of genetics has blossomed into a defining discipline of biology. A half-century of deductive genetic experimentation followed by a generation of advances in molecular biology laid the conceptual groundwork for the modern

---

[1]The U.S. Government has the right to retain a nonexclusive, royalty-free license in and to any copyright covering this paper.

genomics era. Genetic maps of microorganisms, plants, and animals were detailed, leading to whole genome sequencing (WGS) of traditional model organisms including *Escherichia coli*, yeast, *Drosophila*, *Arabidopsis*, *Caenorhabditis elegans*, rice, mouse, and human. The prospect of WGS analysis of the genomes of selected species is providing an unprecedented view of the genetic instructions that specify development, distinctiveness, adaptation, and reproduction. The combination of automated DNA sequencing, gene annotation algorithms, and other computational routines hold exceptional promise for explaining gene action, gene retention, and gene interactions in virtually any species we choose to study.

Among mammals, genomic advances have been driven by the completion of the human and mouse full genome sequences (30, 36, 117). The next mammal whose WGS is under way is the rat, a powerful biomedical model for hypertension, diabetes, and other complex polygenic human diseases (http://rgd.mcw.edu) (170). Close behind in gene map and WGS development are the close primate relatives of man, chimpanzee and rhesus macaque, as well as agriculturally important species, cow, pig, and chicken, which will surely follow (136). In parallel, the companion animal species, domestic dog and cat, have stimulated increasing enthusiasm for genome study because of important advantages that these species bring to biomedical genomic research (16, 137, 140). In this review, we describe the rationale, progress, applications, and potential of the genome project targeting the domestic cat, *Felis catus*.

The Feline Genome Project began two decades ago when the first gene map of the species was published and compared to the developing human gene map of that time (138). We selected the cat because it offered a particularly good model for feline transmissible cancer, caused by Feline Leukemia virus (FeLV) (60, 61). We were hoping to develop a new tool to study genetic and infectious disease and also to add an additional mammalian order to the long-term project of studying the mammalian radiations by tracking the changes and adaptations in the genomes of modern mammals (134, 137). The genomic advances and promise for comparative genetics offered by the cat genome project offer a cogent reason for emphasis and eventual WGS for the cat.

THE CASE FOR A FELINE GENOME PROJECT

The cost of WGS is high, estimated at 50 million U.S. dollars to complete a mammalian genome sequence of 2.7–3.2 billion bases at 5X coverage, comparable to the size of the murine and human genomes, respectively (59, 76, 117, 136). Thus choosing a species for WGS requires considerable biological utility, particularly for human biomedical inference, since the largest funding agencies support health-related research. Once the principal mouse and rat WGS are achieved, there are a number of advantages for WGS of additional mammalian species (136). The most compelling features for targeting the domestic cat are described below.

There are approximately 65 million cats in the United States and several times that number worldwide, so many that overpopulation of feral cats is considered a serious nuisance in many countries (3, 6, 11, 147). The contributing reasons for the population increase include mankind's fascination and domestication of the species combined with a relatively high fecundity. Ease of breeding plus our historic adulation and domestication of cats increase their potential as a genomic model for medical and biological application. Our personal affinity for companion animals, notably cats and dogs, provides a medical surveillance matched only by human biology. The world's veterinary schools produce hundreds of practitioners each year, most of whom carefully document genetic and chronic diseases of our pets. The result is a comprehensive veterinary literature that has described over 258 feline genetic diseases and 437 canine genetic diseases (http://www.angis.org.au/Databases/BIRX/omia/). Approximately half of these diseases have established homology with human genetic defects. A list of feline homologs of single-gene defects found in humans is presented in Table 1. The clinical and physiological study of these feline hereditary diseases provides a strong comparative medicine opportunity for prevention, diagnostic, and treatment studies in a laboratory setting.

The cat also has provided several invaluable models for infectious disease. These include endemic feline leukemia virus and feline sarcoma virus, Type C retroviruses that interact with cellular oncogenes to induce leukemia, lymphoma, and sarcoma (60, 61). Historically, many of the human oncogenes that define signal transduction pathways were originally discovered in the context of FeLV interaction in cat models. The cat provides the only naturally occurring model for human AIDS pathogenesis in its endemic fatal transmissible feline immunodeficiency virus (FIV) (97a, 148, 184). Similar to its close phylogenetic relative HIV, FIV induces CD4-T lymphocyte depletion in affected cats, immune system collapse, and susceptibility to adventitious microbial agents as a prelude to wasting disease and death (148). Interestingly, over 20 wildcat species (including lions, leopards, cheetahs, ocelots, pumas, and other big cats) are epidemic with their own species-specific strain of FIV (18, 24, 25, 142). In contrast to domestic cats, the endemic FIV strains do not appear to cause acute immunodeficiency in the wildcat species, perhaps a consequence of historic natural selection of host genetic resistance to the fatal virus (25, 134).

The feline panleukopenia (distemper) virus has revealed a natural history parable in its abrupt transformation of the cat virus to an epidemic, fatal canine parvovirus that emerged in the world's puppy population in 1978 (145). In another chilling episode, the canine distemper virus, which is normally restricted to canid species, precipitously adapted to and decimated a large African lion population in 1994, killing one third of the lions in the Serengeti ecosystem within a nine-month period (156). A clear involvement of host defense mechanisms in these and other infectious disease outbreaks renders the cats and their pathogens an excellent candidate species for characterizing the interaction of microbial adaptation and host disease gene defenses. Given the critical importance of infectious disease in scores of chronic and acute human disease, there are powerful research opportunities in the cat family (45, 134).

TABLE 1 Hereditary human diseases with feline models

Phenotype	Human candidate loci	Human locus	Feline chromosomal position	Inheritance pattern in cats	Ref.
ALBINISM, OCULOCUTANEOUS TYPE I	TYR	11q14-q21	D1 pcen	AR	97
CARDIOMYOPATHY, HYPERTROPHIC	SEVERAL			AD	91
CEREBELLAR DEGENERATION	SEVERAL			AR	75
CEROID LIPOFUSCINOSIS	CLN1-6	1q42.1-q42.2			180
CHEDIAK-HIGASHI SYNDROME	CHS1		D2	AR	150
DEAFNESS	SEVERAL				70
DIABETES MELLITUS, TYPE I	IDDM1	6p21.3	B2 cen		89
DIABETES MELLITUS, TYPE II	IDDM2	11p15.5	D1q		102
DWARFISM	ACH	4p16.3	B1q	AD	69
EHLERS-DANLOS SYNDROME, TYPE VII	ADAMTS2	5q23	A1q	AR	94
FACTOR X DEFICIENCY	F10	13q34	A1p		57
FACTOR XII DEFICIENCY	F12	5q33-qter	A1q	AR	88
G6PD DEFICIENCY	G6PD	Xq28	Xq	XR	167
GANGLIOSIDOSIS, GM1	*GLB1	3p21.3	C2q	AR	38
GANGLIOSIDOSIS, GM2	*HEXB	5q13	A1q	AR	116***
GLYCOGEN STORAGE DISEASE II	GAA	17q25.2-q25.3	E1q		153
GLYCOGEN STORAGE DISEASE IV	*GBE1	3p12	C2q	AR	50, 51
HAEMOPHILIA A	F8	Xq28	Xq	X	98
HAEMOPHILIA B	F9	Xq27.1-q27.2	Xq	X	104
HYPERLIPOPROTEINAEMIA	*LPL	8p22	B1 pcen	AR	55, 99***
LUPUS ERYTHEMATOSUS	SEVERAL			UNK	174

MANNOSIDOSIS, ALPHA	*MAN2B1	19cen-q12	E2p	AR	10, 21, 165, 176***
MONO-CRYPTORCHIDISM	NC			UNK	114
MUCOLIPIDOSIS II	GNPTA	4q21-q23	B1q	AR	14
MUCOPOLYSACCHARIDOSIS I	*IDUA	4p16.3	B1q**	AR	64, 68, 62a**
MUCOPOLYSACCHARIDOSIS VI	*ARSB	5q11-q13	A1q	AR	32, 37, 46, 54, 66, 67, 81, 192, 107***
MUCOPOLYSACCHARIDOSIS VII	*GUSB	7q21.11	E3cen	AR	56, 52***
MUSCULAR DYSTROPHY, DMD, BECKER	*DMD	Xp21.2	Xp	X	23, 53, 185
NEUROAXONAL DYSTROPHY	NC			UNK	177
NIEMANN-PICK DISEASE, TYPE C	NPC1	18q11-q12	D3q	AR	17, 101***
PELGER-HUET ANOMALY	NC			UNK	179
POLYCYSTIC KIDNEY DISEASE	PKD1,2,3	16p13.31-p13, 4q21-q23	E3q, B1q	AD	12***
POLYDACTYLY	SEVERAL			AD	35
PROGRESSIVE RETINAL ATROPHY	SEVERAL			AR	157
ROD-CONE DYSPLASIA	SEVERAL			AD	96
SPINA BIFIDA	Mouse T/t locus			AD	90
SPINAL MUSCULAR ATROPHY, TYPE III	SMN1	5q12.2-q13.3	A1q	AR	49***
TESTICULAR FEMINIZATION	DHTR	Xq11-q12	Xq	X	95
URTICARIA PIGMENTOSA	NC				175
VITAMIN-K-DEPENDENT COAGULATION DEFECT	GGCX	2p12	A3q	AD	163

UNK: Unreported pattern of inheritance, but heritable.

NC: No gene identified in human.

*Mutational mechanism identified in the cat.

**Tentative assignment.

***Reported breeding colony.

The cat also possesses several advantages from a comparative genomics perspective. Gene mapping and chromosome painting experiments have shown that the feline genome, which is composed of 19 chromosome pairs, is extensively conserved in gene content (conserved synteny) and G-banded chromosome appearance among other Felidae species, among other carnivore species, and indeed across many placental mammals (115, 124, 137, 138, 152, 182). The extent of chromosome segment conservation between the cat and human genomes is among the highest observed between mammalian orders (120, 121, 137, 152, 182). For example, the feline genome assembly is 3 to 4 times less rearranged relative to the human genome than are the genomes of murid rodent species (mouse and rats) (120, 121, 137). Overall, there seems to have been an extremely slow rate of chromosome translocation exchange between cats and primates. The remarkable colinear parallel of the cat and human genome provides an opportunity to inspect rather long stretches of conserved synteny between the two species, as well as the patterns and details of global reshuffling that are apparent in other lineages.

The domestic cat is one of 37 species of the Felidae family, itself one of 11 Carnivore families. The Felidae family dates back to around 15 mya (84), leading to the adaptive occupation of ecological niches throughout the world. The relative success of these majestic predators combined with humankind's fascination with the great cats for thousands of years has produced an extensive literature on human-cat interactions. Several species have been the object of long-term field ecology projects, and most can be observed and sampled in zoological collections. Thus, biological specimens are accessible in zoos, from field projects, and museums. In the past two decades, scientists at NCI's Laboratory of Genomic Diversity, in cooperation with scientists from the Smithsonian's NOAHS Center, have assembled over 40,000 tissue specimens from cats and their wild relatives collected across the world. This collection is unprecedented in its scope and utility for population-based research inquiries of free-ranging species (132, 133).

Further, the genomes of the Felidae family species are nearly identical to the domestic cat, with 15 of 19 domestic cat chromosomes invariant among all the other Felidae species (115, 190, 191). The genetic tools and resources developed for the domestic cat (e.g., microsatellites, coding gene PCR primers, libraries, etc.; see below) are readily applied to the study of wildcat species (34, 41, 42, 113, 172). Application of these molecular genetic tools and resulting evolutionary inferences has provided considerable insight into the history and peril of endangered Felidae species (cheetahs, pumas, lions, tigers, and others), laying the groundwork for the important new discipline of conservation genetics (33, 41, 42, 48, 83–86, 112, 132, 133, 172).

There are a number of practical advantages to a domestic cat model as well. The cats breed well in a captive setting and domestication dating back between 6000 to 8000 years ago has produced nearly 40 recognized breeds that have experienced moderate degrees of inbreeding and artificial selection across their recent ancestry (47, 71). The breeds provide recent phylogenetic lineages that capture different combinations of coat color, coat length, patterning, appearance, and behavioral traits suitable for genetic analysis (155). Modern breeds reflect different combinations at around 12 monogenic coat color trait loci, most with homologous counterparts in coat color genes of mouse and other domestic species (154). The same gene homologs of pigmentation loci in other mammalian species have been implicated in anemia, sterility, and neurological and metabolic disorders (7, 8, 77). Further, the history of modest inbreeding in cat breeds supplies important populations ideal for linkage disequilibrium mapping of complex quantitative characters as have also been recognized in dog breeds (143). Dense genetic maps combined with existing cat pedigrees offer a rare opportunity to interpret a large body of hereditary trait inference.

The cat's reproductive apparatus and physiology have been extensively studied, leading to a fascinating comparative database that describes hormonal, behavioral, and reproductive distinctions among Felidae species (19, 73, 183). The strikingly different reproductive strategies seen among different cat species (e.g., induced or spontaneous ovulation, hormone ratios, mating systems, variable sperm quality, etc.) illustrate the adaptive coevolution of reproductive physiology, sexual selection, and behavioral ecology in graphic, well-studied situations (22, 27, 144). The experimental knowledge of cat reproduction has allowed considerable advances in assisted reproduction in cat species, notably artificial insemination, sperm, and embryo cryo-presentation, and in vitro fertilization (19, 40, 74, 80, 187, 189). Embryo transfer technology has led in December 2001 to the birth of CC, the first cloned domestic cat (160). The development of nuclear cloning for cats nearly ensures the likelihood of stem cell development and therapy, as well as the prospect of gene-specific knockout technology in the species, so powerful in rodent models (78, 87, 171).

Finally, the cat has evolved within the mammal superorder Laurasiatheria, one of four mammal clades that predated the radiation of modern mammals (118, 136). The three mammalian species already scheduled for WGS, human, rat, and mouse, are all members of a different clade, Euarchontoglires. For this reason, the cat genome with its conserved syntenic organization relative to human would represent a significant genomic expansion of the evolutionary diversity present among modern mammals. This divergence offers considerable breadth in sampling available genetic diversity among the living species of mammals.

PROGRESS IN ASSEMBLING THE FELINE GENE MAP

The domestic cat carries 18 autosomal pairs, X and Y chromosomes, in a genome containing around 3×10^9 nucleotides, comparable to the human and mouse genomes. Early feline gene maps employed somatic cell hybrid panels derived from fusion of cat lymphocytes and genetically selectable rodent cell lines (135, 138). A combination of somatic cell hybrid mapping coupled with FISH mapping of heterologous molecular clones to cat metaphase chromosomes led to a skeleton map of 105 coding genes (135). Comparison of the linkage arrangement of cat genes to their human counterparts revealed a high degree of conserved synteny, strings of homologous genes on a single chromosome in both species (135, 138, 140, 141). The extended genome conservation revealed by the gene map comparisons of cat and man was affirmed, virtually by direct observation using chromosome painting methods, or Zoo-FISH (152, 182). By hybridizing fluorescent-labeled chromosomes isolated from human chromosome libraries to cat metaphase chromosomes, the precise regions or segments of gene sequence homology in the cat genomes for each human chromosome were identified (140, 152, 182). The reciprocal human conserved syntenic segments were demarcated by painting human metaphase chromosome spreads with individual flow-sorted cat chromosomes (182).

The chromosome painting procedures identified 32 contiguous cat chromosome segments that were homologous to single human chromosomes and 30 conserved human chromosome segments that were painted by a single cat chromosome probe. These initial comparisons suggested that as few as 13 scissor-cuts could rearrange the cat genome into the human genome or vice versa. This value is lower than similar Zoo-FISH comparisons between other mammal species and the human genome. For example, the cattle genome would require 27 scissor cuts to reassemble it to the human genome arrangement. Horse would require 34 cuts, pig 28, dog 45, and mouse 160 cuts (28, 137). If we postulate a date of approximately 90 million years as the age of the common ancestor of humans and cats (43, 118), then it takes an average of 14 million years for a single translocation exchange to occur. This is among the slowest rate observed among all the mammalian genome comparisons with humans reported and emphasizes the highly conservative "default" mode of genome evaluation documented in many primate and carnivore species (137).

The Zoo-FISH methodology is limited by its inability to resolve homologous chromosome segments less than 5 Mb and also by its failure to reveal intrachromosomal inversion rearrangements within studied mammalian genomes. To increase the comparative genomic resolution of such inversions as well as to achieve higher power in the phenotype mapping, the cat map was expanded in several ways to provide a dense representation of both Type I coding genes and Type II hypervariable microsatellite markers (108, 121, 166). Mapping the coding genes is critical for establishing homologs to the full-length whole genome sequence (WGS) maps of human and mouse. Type II microsatellite locus markers are required to map feline phenotypes to specific genomic locations. Three mapping resources (Table 2), each with particular advantages were developed: (*a*) an interspecies backcross (ISB) between domestic cats and a closely related species Asian leopard cat (*Prionailarus bengalensis*) (Figure 1, see color insert); (*b*) a domestic cat pedigree established from outbred cats by Nestlé-Purina PetCare (St. Louis, MO), and segregating six coat color loci plus other tractable phenotypes; (*c*) a 5000-rad radiation hybrid (RH) panel of the domestic cat that allows for physical mapping to a resolution of less than a megabase (119, 121). First- and second-generation linkage and physical maps using each of these tools have been developed (108, 109, 121, 166) to produce a feline genetic map integrating (*a*) comparative anchor–Type I loci for alignment with human and mouse genomes (103, 121), (*b*) microsatellite loci

TABLE 2 Progress in feline gene mapping, December 2002

	Type I coding genes	Type II microsatellite loci	Other	Total markers	Average marker density
I. Linkage map—ISB*	81	248	0	329	10 cM
II. Linkage map—Nestlé Purina	0	705	0	705	4.7 cM
III. Radiation hybrid map-5000 rad	775	954	11	1740	1.9 cM
IV. Integrated map	784	1086	11	1881	1.8 cM

*ISB—Interspecies backcross, see Figure 1.

placed on average 11 cM apart (108), and (*c*) selected genes with important phenotypes. A summary of the most recent maps, including markers as of December 2002, is presented in Table 2.

The ISB pedigree, composed of 108 individuals and 66 informative meioses, was constructed to maximize the chance of obtaining genetic variants between Type I loci, as was demonstrated in building the mouse gene map (29, 108, 109). To date we have placed 81 Type I markers on the linkage map. In addition, 248 microsatellites were mapped on the linkage map from a larger group of 600 characterized microsatellites to provide highly informative loci for phenotype mapping. The sex-averaged length of the feline genome was estimated from the ISB at 3300 cM, with an average density of 10 cM between each marker. Forty-seven linkage groups were physically mapped to cat chromosomes by using the cat-rodent somatic cell hybrid panel (135), the radiation hybrid mapping data, or by conserved synteny with the human genome (121). A separate intraspecific pedigree, developed in collaboration with Nestlé-Purina, utilizes 256 cats with 483 meiosis derived from 27 founders. A total of 705 microsatellite loci are mapped on this pedigree, which provides an average density of one marker per 4.7 cM.

The 5000 rad RH map consists of a panel of 93 hybrid lines analyzed for concordant retention of markers to develop a higher density ordered physical map of Type I and Type II markers. The RH map contains 775 Type I markers (density = 4.3 cM/Type I marker) and 954 Type II markers (density = 3.5 cM/marker). The Type I markers are derived from a variety of sources including those detected by the Comparative Anchor Tagged Sequence (CATS) primer design method (82, 103), feline mRNAs deposited in GenBank, human expressed sequence tags (ESTs), and feline ESTs matched to human or mouse Type 1 coding genes (109, 121).

By comparing the linkage order derived from the ISB and the RH map, an alignment of the markers mapped in two or three approaches can be used to develop an integrated map (Figure 2). The derived marker order derived from the ISB and RH map are remarkably concordant, providing a high confidence in integration strategy. The current integrated RH-linkage map contains a total of 1881 markers,

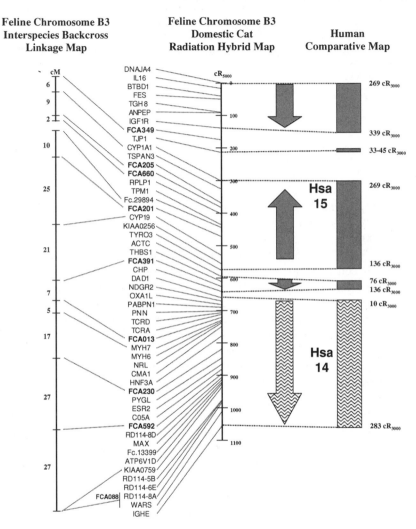

Figure 2 Comparison of the feline chromosome B3 maps derived from the feline interspecies backcross linkage panel (*left*), feline radiation hybrid (RH) panel (*center*) compared with the human genome (*right*). Feline microsatellite markers are shown in bold, while Type I coding markers are shown in regular type. The patterned and shaded blocks for the human genome denote regions of conserved order with the adjoining feline chromosome (demarcated by *dashed connectors*), based on human radiation hybrid maps. Shaded blocks are homologous to human chromosome 15, while patterned blocks are homologous to human chromosome 14. Human Genebridge4 RH centiray positions are shown to the right of each block. The orientation of the collinear homology segments are depicted by arrows.

an average interval of 1.8 cM or 16.8 cR between each marker. Updated genetic maps of the integrated marker map can be viewed at (www.lgd.nci.nih.gov). The integrated gene map provides a powerful tool both for tracking cat phenotypes and comparing the processes that mold genome organizations that determined the evolution of mammals (120, 137, 140).

THE CAT AS AN INDEX FOR COMPARATIVE GENOMICS

The rapid development of comparative genomic data is now beginning to reveal some important evolutionary features around the organization of modern genomes. Gene maps and chromosome painting clearly indicate a bimodal pattern of genome conservation among placental mammals. The ancestral or default pattern of genome rearrangement is very slow, roughly one translocation exchange every 14 million years. The slow rate is evident in Felidae species, humans, many primates, and also in other mammals such as mink, ferret, dolphin, and shrews (120, 137). Nested within and among these conserved genomic lineages, however, are species whose genomes have been reshuffled moderately (e.g., pigs and cows) and other mammalian lineages where the genome reshuffling is even more extensive (mouse, rat, gibbons, New World monkeys, dogs, and bears) (79, 120, 122, 123, 125, 181).

Once the unequal rates of chromosome exchange among different lineages were appreciated it became possible to use evolutionary principles to reconstruct the ancestral genome organization and to interpret the genomic changes that have occurred on each mammal lineage. By identifying certain common human homologue combinations (fusions) or separations (fissions), it has been possible to postulate the disposition of the primitive common ancestor of primates (139), of carnivores (120), and of all placental mammals (120), based on maximum parsimony (Figure 3, see color insert). The ancestral placental mammal genome consisted of 24 autosomes plus X and Y, and included 32 ancestral conserved syntenies or segments, compared to the human genome (Figure 3a). Similar imputations from comparative mapping and painting have allowed postulation of the common ancestor of all carnivore species (Figure 3b). That genome has 21 chromosome pairs and is composed of 26 conserved syntenies as compared with the ancestral mammal, 23 conserved syntenies with the modern cat and 34 conserved syntenies with the human genome. Since human and cat both display the conserved "slow" pattern of genome evolution, apparent chromosome exchanges are remarkably few in number. By contrast, the extensive genomic exchanges that occur in the "fast" lineage species (dogs, bears, gibbons, New World monkeys) are attributed to more recent reshuffling events that occurred abruptly (in evolutionary terms), subsequent to divergence from the common ancestor but before divergence into the modern genomically divergent bears, dogs, and other species (123, 125, 137). This point is illustrated by Table 3 where the number of conserved segments (unordered), revealed by painting human chromosome probes onto karyotypes of 15 representative mammal species from seven placental orders, were used to

TABLE 3   Variation in genomic evolutionary rate across placental mammals

Species	Zoo-FISH conserved segments with placental mammal ancestor[a]	Haploid number	# of rearrangements relative to ancestral placental mammal genome[b]	Whole genomic rate[c]
Dolphin	25	22	3	0.034
Cat	27	20	6	0.068
Human	29	23	6	0.068
Macaque	28	21	7	0.080
Mink	26	15	10	0.114
Lemur	35	30	10	0.114
Tree shrew	35	31	10	0.114
Horse	40	32	15	0.170
Cow	41	30	16	0.182
Bat	35	16	19	0.216
Shrew	30	10	20	0.227
Pig	41	19	22	0.250
Gibbon	51	22	26	0.295
Spider monkey	48	17	30	0.341
Dog	64	39	39	0.443
Rat[d]	85	20	65	0.739
Mouse[d]	103	20	83	0.943

[a]References (140, 181).

[b]Equal to the number of conserved unordered segments minus the lower haploid number of the compared species (137). Estimated based on chromosome painting data using human probes.

[c]Equal to the number of rearrangements relative to ancestral placental mammal genome divided by the estimated time of divergence from the ancestral Boreoeutherian mammal [~88 mya (43, 118)].

[d]Estimated from gene mapping data, excluding smaller segments defined by one or two genes to be more comparable to estimates derived from Zoo-FISH alone.

compute the genomic rate of translocation exchange based upon recent phylo-genetic relationships and dating (43, 118). The disparate rates of chromosome translocation are evident by a 12-fold difference between the slowest species (dolphin) and the fastest (dog) (Table 3).

As mentioned above, genome comparisons derived from chromosome painting and banding homologies can reveal neither small (<5 cm) conserved segments nor intra-chromosomal inversions. Painting provides a considerable underestimate as shown by the analysis of 353 feline Type I markers with the human map (121). In that analysis we discerned 100 conserved segment orders—CSOs—between human and cat (Figure 4, see color insert), nearly three times as many as the 30–32 conserved syntenies (140, 152, 182) revealed by chromosome painting (Table 4). A similar result was also obtained with comparison of ordered Type I gene maps of

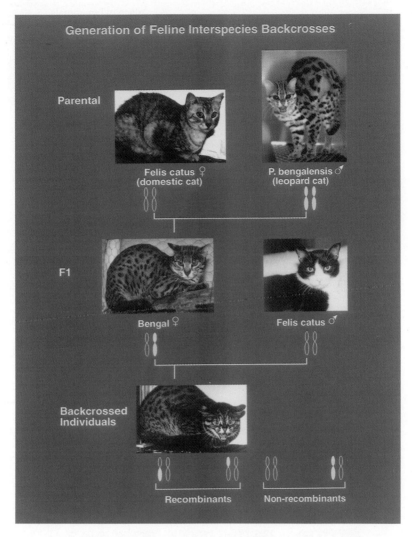

Figure 1 Domestic and Asian leopard cat Interspecific Backcross Pedigree. Parental species used to generate F_1 individuals included domestic cat females and leopard cat males. Bengal: F_1 hybrid females; backcrossed individuals: progeny of F_1 hybrid females backcrossed to domestic cat males.

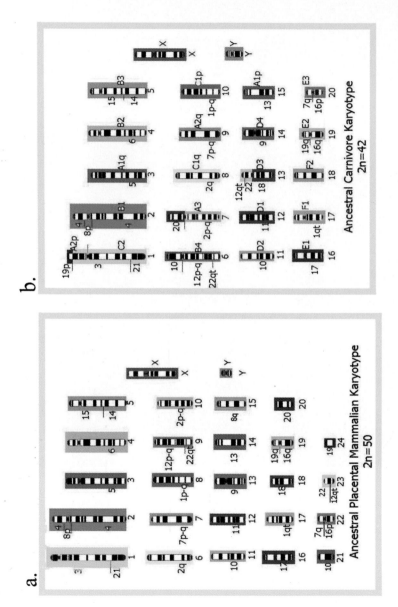

Figure 3 Hypotheses for the chromosome content of the (*a*) ancestral placental mammal karyotype and (*b*) ancestral carnivore karyotype. Numbers to the left and right of each chromosome represent homologous regions of the human and cat genomes, respectively. Numbers below each chromosome are identifiers for the chromosome in each ancestral genome. The colors represent the chromosome in the ancestral mammalian genome from which each region originated. The G-banded karyotype depicted for the ancestral placental genome is imputed.

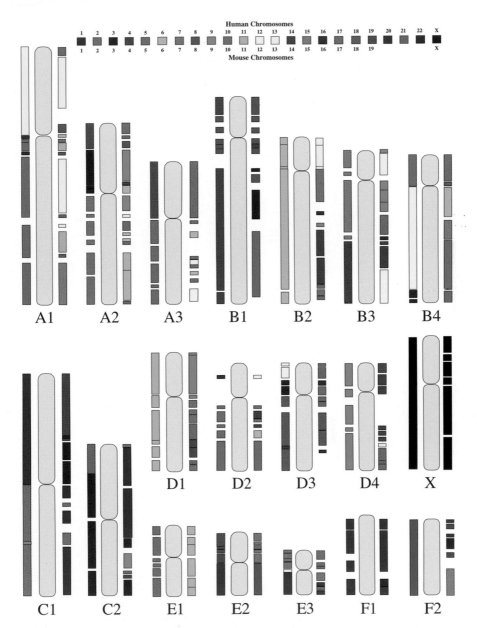

Figure 4 Cat-human (*left*) and cat-mouse (*right*) homology maps showing conserved ordered segments between each genome, using the feline karyotype, drawn to scale, as an index genome. The Y chromosome is not included in these comparisons. The homology of human and mouse chromosome segments is indicated by the color key shown at the top of the figure. The human and mouse conserved ordered segments are drawn to their approximate position relative to the feline genome, assuming a physical scale proportional to the feline radiation hybrid maps. Gaps would indicate regions of homology not covered by Type I markers (microsatellites only) or inferred spans of ordered conserved segments.

TABLE 4   Conserved segments estimated using different mapping technologies

Species comparison	Conserved segments: Zoo-FISH	Conserved segments: Unordered via mapping	Conserved segments: Ordered via mapping
Human-cat	30–32	100	
Human-cow	50	82	149
Human-rat	NA	152	190
Human-mouse	NA	180	200

cow to human (Table 4). This increasing focus of CSOs over paint-conserved syntenies suggests there are twice as many interstitial inversions produced as translocations in these lineages (Figure 4). These higher-level resolution comparisons increase the precision of conserved segment identification to approach the number of conserved segments predicted by the theoretical calculations (121, 122). They also would suggest that about 500 ordered Type I markers would be sufficient to reveal >90% of the conserved syntenic segments between any two mammalian species with slow to moderate rates of genomic exchange.

The most precise comparisons of mammalian genomes will derive from aligned homologs of DNA sequences, eventually across the entire genome. As has been recently demonstrated, mouse and human WGS comparisons reveal the full spectrum of gene inclusion, gene birth and death, transpositions, repetitive element expansions, and conserved syntenies (30, 36, 59, 117). A provocative glimpse of a three species genome sequence comparisons has been achieved with the recent full sequence comparison of the major histocompatibility complex (MHC) class II sequence of human *HLA*, mouse *H2*, and cat *FLA* (Figure 5) (193). The human *HLA* region consists of 224 genes of which 128 are expressed while 96 are pseudogenes (113a). Nearly half of the *HLA* genes play a role in immune defenses and about 50% of the HLA sequences consist of repetitive elements (LINES, SINES, LTRs, and STRs). Sequence alignment of human mouse and cat MHC class II region homologs (Figure 5) revealed several fascinating evolutionary features (1, 193). First, the three species differ considerably in the gene cluster length with *HLA*-998kb, *FLA*-758kb, and *H2*-495kb. The three species each contained around 35 functional genes but differed markedly in gene disposition and pseudogene numbers: *HLA* has 23 pseudogenes, *FLA* has 7, and H2 has 5 within the same region. In addition, cats have appreciable differences in their class II gene families. The *DQ* family is absent and *DP* genes are vestigial, represented by two pseudogenes. The loss of *DP* and *DQ* genes is compensated for in cats by an expansion of the *DR* region to seven modern *DR* genes derived from gene duplication and inversion events in the history of this family (193). The extinction of *DP* and *DQ* gene function in the cat is a likely explanation for the rather inefficient humoral response to maternal antigens (149) or to graft rejection seen in domestic cats (186). Additional differences in pseudogene transposition, retro element density,

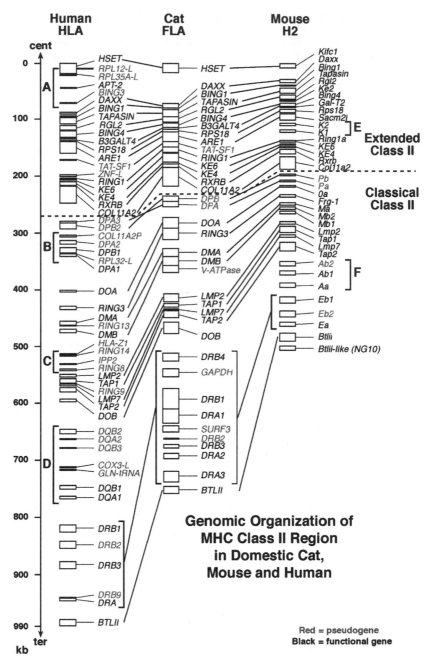

Figure 5 Comparative genomic organization at the MHC Class II region in mouse, domestic cat, and human (193). Brackets (*A–C*) indicate gene segments in human, but absent in mouse and cat. *D* is a human gene segment absent in cat. *E* represents a mouse gene segment absent in human and cat. *F* is a mouse segment absent in cat.

length expansion, and conserved sequence blocks provide a tantalizing portrait of the adaptive events that have shaped this important genomic region in the cats, mice, and men.

PROGRESS IN FELINE GENETIC DISEASE MODELS

Some 258 hereditary pathologies have been reported in the domestic cat (http://www.angis.org.au/Databases/BIRX/omia/), largely due to intensive medical surveillance of cats in the veterinary profession. To date, disease-causing mutations have been characterized in nine cat genes, although several other pathologies have been well characterized on a biochemical or protein level, including Niemann-Pick Type C, spinal muscular atrophy, Chediak-Higashi syndrome, dwarfism, hypertrophic cardiomyopathy, and mucolipidosis (Table 1). The largest representation comes from lysosomal storage enzyme disorders that arise from defects in genes that play a role in degradation of macromolecules such as mucopolysaccharides by lysosomes. As most lysosomal enzymes are secreted and can be taken up by neighboring cells (128, 168), an array of corrective therapeutic strategies has been proposed and many of these have been examined in the cat, including enzyme replacement, bone marrow transplantation, and gene therapy (54, 62, 99, 165, 178, 188). Feline models have been important in elucidating molecular pathogenesis as well as in playing a critical role in evaluating and optimizing therapeutic strategies prior to clinical trials in humans.

The cat is a model for mucopolysaccharidosis Types I, VI, and VII disorders, which result from lysosomal enzyme deficiencies involved in mucopolysaccharide degradation. Mucopolysaccharidosis Type I (MPS I), which results from deficient activity of the enzyme alpha-L-iduronidase (IDUA) (130), can lead to mental retardation, growth abnormalities, and shortened lifespan in humans (129). Naturally occurring models have been characterized in the cat (64, 65) and dog (164). A 3-base pair (bp) in-frame mutation characterized in the feline *IDUA* gene of affected individuals (68) results in deletion of an aspartic acid residue highly conserved in human, dog, cat, and mouse. The cat model provides an ideal system to study mechanisms of brain neurodegeneration and neural-directed strategies, especially given a large body of pre-existing literature on cat neurology.

MPS VI or Marteaux-Lamy disease, a deficiency for activity of arylsulfatase B (ARSB), is associated with growth retardation, coarse facial features, and skeletal deformities in humans and cat (31, 66, 67, 81). A missense mutation in *MPS VI* in affected cats results in a nonsynonymous substitution (L476P) (192) in a residue conserved across six other sulfatases, suggesting its critical enzymatic role. Two other independent mutations have been identified (32, 37). Affected cats respond to allogeneic bone marrow transplantation (54), while in vivo studies have demonstrated retroviral-mediated correction of MPS VI-deficient fibroblasts, chondrocytes, and bone marrow cells in both humans and cat (46).

MPS VII results from deficiency of beta glucuronidase (GUSB) (130), which in humans manifests as cartilaginous and bony malformations, growth and mental retardation, abdominal organ enlargement, and corneal clouding (130). Naturally occurring animal models have been described in mice (13), dogs (63), and cat (56). The molecular basis for feline MPS VII (52) results from a nonsynonymous substitution (E351K) in a highly conserved amino acid, likely involved in maintaining protein conformation. Enzymatic activity has been restored in fibroblasts and restored by retroviral gene transfer of rat beta-glucuronidase cDNA. As GUSB is an essential housekeeping enzyme, this feline model is important for examination of exogenous genes and gene product delivery to a variety of tissue types, and could prove especially valuable in light of the extensive research conducted on the anatomy and physiology of the cat central nervous and visual systems.

Deficiency of lysosomal alpha mannosidase leads to accumulation of mannose-rich oligosaccharides (169), leading to mental retardation, recurrent infections, skeletal changes, and hearing impairment (21). A 4-bp deletion leads to a frameshift mutation and premature termination codon in affected Persian cats (10). Mutational molecular heterogeneity has been demonstrated in a domestic long-haired cat. The feline model has served as a powerful model for bone marrow transplantation (BMT) of lysosomal storage diseases, exhibiting dramatic improvement of α-mannosidase activity in brain tissue of affected Persian cats (62, 178). These results have provided direct evidence of the efficacy of BMT as corrective strategy for neuronal storage diseases of the CNS and the potential of hematopoietic stem cells as corrective strategy for lysosomal storage disorders. Retroviral constructs of a human cDNA have also been demonstrated to correct enzymatic activity in deficient human and feline fibroblasts (165).

Glycogen storage disease Type IV is a rare disorder of glycogen metabolism caused by deficiency in glycogen branching enzyme (151). Glycogen deposits, found in numerous tissues, result in a failure to thrive and death from cirrhosis (151). The Norwegian Forest cat is the only animal model reported for this pathology (50). The feline mutation mechanism has been identified as a complex rearrangement resulting in the deletion of a 172-bp exon (51).

Lipoprotein lipase (LPL) is a crucial enzyme involved in the regulation of lipoprotein and lipid metabolism (20). Cats with LPL deficiency share remarkably similar phenotype to humans, including severe pancreatitis, chylomicronemia, and failure to thrive (55). A nonsynonymous substitution in feline LPL (Gly412 Arg) results in LPL deficiency in affected cats (55). Cats could prove to be a most valuable animal model of LPL deficiency as, of the numerous animal model systems examined including the mouse, the cat most closely resembles the lipoprotein pattern and lipid transport system of humans. This feline model offers great potential as an in vivo system to examine increased triglyceride levels associated with LPL deficicency on atherosclerosis (55).

A separate class of lysosomal storage disorder characterized in the cat are the gangliosidoses, GM1 and GM2, heritable neurodegenerative diseases. Excessive neuronal accumulation of the gangliosides GM1 and GM2 results from deficiency of lysosomal beta-galactosidase (BGAL) and hexosaminidase (HEX) A and B activity, respectively (131), leading to neuronal distortion and degeneration.

Feline models have been especially important in characterizing the pathobiology and molecular biology of these diseases. The mutational mechanism of GM1 in the cat results from a nonsynonymous substitution (Arg to Prol at base 1486) with subsequent loss of hydrolytic activity (4). GM2-gangliosidosis has been characterized in two independent cat models (58, 116) exhibiting remarkably similar pathology to human Sandhoff's disease (131). In affected cats, deletion of a cytosine residue results in a frameshift and premature stop codon (116), while a 25-bp inversion within the reading frame was characterized in a non-bred domestic cat (4). Limited reduction in GM2 neuronal storage has been reported following bone marrow therapy (178). Feline models will be critical in the development of therapeutic strategies for these disorders. Whereas acid beta-galactosidase deficiency has been corrected in human fibroblasts by retroviral mediated gene transfer (158), gene therapy of the CNS presents a challenging front as corrective retroviral constructs require mitotically dividing cells for integration and expression (26). On this note, targeted delivery of hexosaminidase A, covalently bound to the nontoxic fragment C of tetanus toxin, increased in vitro enzyme binding and uptake by cultured brain cells from *GM2*-affected cats (39).

X-linked muscular dystrophy in man is characterized by progressive degeneration of skeletal and cardiac muscle. Mutations, which have been exhaustively characterized in this disorder in man, lead to either absence or abnormality in the protein product dystrophin (72, 92). Models for X-linked muscular dystrophy have been characterized in mouse (161), cat (53), and dog (159). A deletion in the dystrophin muscle promoter characterized in the cat eliminates expression of muscle and Purkinje neuronal dystrophin isoforms (185). The marked clinical heterogeneity observed in these models, from severe disability exhibited in man and dog, to little muscle fibrosis and an actual regenerative process leading to muscle hypertrophy in mouse and cat (2, 23, 53), could be important in characterizing immediate and secondary consequences of the lack of dystrophin (146) and points out the importance of multiple animal models.

A ROLE FOR THE CAT GENOME IN FORENSICS

The use of DNA markers to identify sources of biological traces left at crime scenes is heralded as the most important advance in the forensic sciences since fingerprinting. The report of microsatellite loci as a source of polymorphism in human DNA has revolutionized the forensic community in the past several years (44a, 127). Forensic DNA typing with human microsatellite loci has become widespread and is now routinely used in hundreds of public and private crime laboratories in the United States and throughout the world. The feasibility of genotyping multiple microsatellite loci in PCR-based multiplex analysis with as little as a single nanogram of genomic DNA allows forensic accession to biological materials previously considered inappropriate because of the age of the sample, quality, or quantity of DNA yield. These technological developments have also made realistic the genetic screening of trace biological specimens of animal tissues, particularly hairs and blood, from individual animals including pets inadvertently left at crime scenes (110). The definition of species-specific microsatellite maps, such as those of cats and dogs, have made such forensic assessments an important new tool in forensic laboratories (5, 15, 16, 106, 108).

Utilizing microsatellite genotypic characterization of forensic hairs from a pet cat, we contributed to the establishment of a legal precedent for employing genetic individualization of animal tissue in homicide cases (110, 111). With support from the National Institutes of Justice, we have expanded this application by developing a forensic typing system for the cat and a genetic database of cat breeds for the genetic individualization of domestic cat tissue specimen.

A set of 11 tetranucleotide microsatellite loci were selected for a forensic panel based on distribution in the cat genome (121), heterozygosity observed across a reference panel of 29 cat breeds (5–10 animals/breed, n = 230), and Mendelian inheritance testing. The number of alleles observed for the panel in the 230 animals genotyped ranged from 9 to 33, with an average of 17 alleles/locus. Average locus heterozygosities across the 29 breeds for the independent loci ranged from 0.78 to 0.95, with an average locus heterozygosity for the forensic panel of 0.90. Breed-specific locus heterozygosities for the 11 loci ranged from 0.61 (Birman) to 0.86 (Norwegian Forest cat). A multiplex genotyping is under validation in a panel of approximately 1300 cats representing 37 of the major recognized cat breeds in the United States to generate a genetic database of cat breeds.

High discriminating potential for genetic individualization was observed in the sample data set for the forensic panel. The average probabilities of finding a matching genotype was estimated as 5.5×10^{-7} (British short hair) to 3.25×10^{-13} (Norwegian Forest cat), given observed allele frequencies in the 29 breeds. The allele frequency database for specific cat breeds provides a powerful analytical resource for assessing the statistical strength of genetic individualizations of cat specimens discovered at crime scenes.

CONCLUSIONS AND FUTURE PROSPECTS

Advances in comparative genomics have transformed this discipline from a cottage industry to the framework for annotation of the human and mouse WGS and the basis for future species genome exploration. The feline genome project, now entering its third decade and armed with a broad array of advanced genomic resources, is positioning the domestic cat and its wild relative species to make substantive contributions to a number of scientific fields.

Over 258 hereditary pathologies have been reported in the domestic cat, largely due to intensive medical surveillance of cats in the veterinary profession (Table 1). To date, mutations have been characterized in nine hereditary disease genes. Though few in number, these feline models have been important in elucidating molecular pathogenesis and are playing a critical role in evaluating and optimizing therapeutic strategies prior to clinical trials in humans. With continued development of a high-resolution integrated map of the cat (Table 2), mapping and

TABLE 5   Developed feline genome project resources (May 2002)

Resource	Citation
I. Somatic cell hybrid panel framework physical map >100 Type I genes	(135, 138)
II. Interspecies Backcross (ISB) Genetic Linkage Map	(108)
III. Intra species Nestlé-Purina pedigree Genetic Linkage Map	(44)
IV. 5000-rad radiation hybrid panel and map	(119, 121)
V. Arrayed BAC and PAC libraries	(9)
VI. Flow sorted feline chromosome libraries: reciprocal chromosome paint map	(140, 152, 182)
VII. Tissue/cell line DNA repository of >10,000 exotic and domestic feline specimens	(84, 162)
VIII. Domestic cat breed forensic database 40 breeds, 11multiplexed, optimized STRs, ISTs	Unpublished
IX. Domestic cat Y chromosome cosmid library	Unpublished
X. Complete sequence	
a. mtDNA genome	(100)
b. Major histocompatibility complex	(193)

characterization of many hereditary pathologies in the domestic cat are anticipated in the future.

The feline model shows continued promise for resolution, diagnostics, vaccine and treatment of human infectious disease. The identification of FIV in domestic cats offers a viable model for HIV pathogenesis as it provides the only known naturally occurring model for human AIDS pathogenesis. The revelation that strains of FIV in exotic felid species, such as lion and puma, show little immune depletion implicates naturally evolved adaptation to FIV in protecting wild cat species in the face of constant exposure to the virus (25). Future applications of available feline project resources (Table 5) to studying FIV in domestic and exotic felids hold the potential to unlock mechanisms behind this resilience.

The feline genome project has also contributed to the field of criminal justice. A Canadian homicide case involving the defendant's pet cat provided a strong legal precedent for the introduction of animal DNA as evidence in a criminal case (111). As a result of this precedent, the National Institute of Justice has endorsed the creation of a microsatellite-based forensic typing system and genetic database for the domestic cat that will contribute to the analysis of physical evidence of criminal investigations.

The domestic cat was the first nonhuman, nonrodent gene map developed that illustrated the strong degree of conserved synteny among mammalian orders (138). Further investigations using chromosome painting and ultimately, linkage and radiation hybrid mapping have begun to reveal a more dynamic view of comparative genomic organization, where intrachromosomal change plays a fundamental role in reshaping modern genomes. Nonetheless, the overall slow rate of genomic change in the cat lineage provides an important opportunity for understanding the persistence of very large stretches of conserved gene order since the earliest divergence events in mammalian evolution. Through these studies, it is becoming increasingly clear that the reshaping of genomes is not random. Comparisons of the cat map with other species maps are revealing deserts and hotspots of genomic instability across mammalian orders and may ultimately contribute to characterizing fundamental processes involved with chromosomal rearrangement, and their potential contributions to speciation and to disease. At the sequence level, multispecies megabase sequence comparisons of the cat, human, and mouse MHC have begun to reveal the nature of gene loss and genomic adaptation (59, 193). As the Human Genomic Sequencing Consortium winds down on the drafts of the human, mouse, and rat genomes, other species are jumping in line to reap the benefits of WGS comparisons (136). The discussions laid forth here present a compelling case for considering the domestic cat as one of the next candidates for whole genome sequencing of mammalian species.

The *Annual Review of Genetics* is online at http://genet.annualreviews.org

LITERATURE CITED

1. Allcock RJN, Martin AM, Price P. 2000. The mouse as a model for the effects of MHC genes on human disease. *Immunol. Today* 21:328–32

2. Anderson JE, Bressler BH, Ovalle WK. 1988. Functional regeneration in the hindlimb skeletal muscle of the mdx mouse. *J. Muscle Res. Cell Motil.* 9:499–515

3. Ashmole NP, Ashmole MJ, Simmons KEL. 1994. Seabird conservation and feral cats on Ascension Island, South Atlantic. In *Seabirds on Islands: Threats, Case Studies and Action Plans*, ed. DN Nettleship, J Burger, M Gochfeld, pp. 94–121. Cambridge, UK: BirdLife Int.

4. Baker HJ, Smith BF, Martin DR, Foureman P, Castagnaro M, et al. 1998. *The molecular bases of feline GM1 and GM2 gangliosidoses.* Presented at Int. Feline Genet. Dis. Conf., 1st, Philadelphia

5. Barendse W, Vaiman D, Kemp SJ, Sugimoto Y, Armitage SM, et al. 1997. A medium-density genetic linkage map of the bovine genome. *Mamm. Genome* 8:21–28

6. Barratt DG. 1998. Predation by house cats, *Felis catus* (L.), in Canberra, Australia. II. Factors affecting amount of prey caught and estimates of the impact on wildlife. *Wildl. Res.* 25:475–87

7. Barsh GS. 1995. Pigmentation, pleiotropy, and genetic pathways in humans and mice. *Am. J. Hum. Genet.* 57:743–47

8. Barsh GS. 1996. The genetics of pigmentation: from fancy genes to complex traits. *Trends Genet.* 12:299–305

9. Beck TW, Menninger J, Voight G, Newmann K, Nishigaki Y, et al. 2001. Comparative feline genomics: BAC/PAC contig map of the major histocompatibility complex class II region. *Genomics* 71:282–95

10. Berg T, Tollersrud OK, Walkley SU, Siegel D, Nilssen O. 1997. Purification of feline lysosomal alpha-mannosidase, determination of its cDNA sequence and identification of a mutation

causing alpha-mannosidosis in Persian cats. *Biochem. J.* 328:863–70

11. Berruti A. 1986. The predatory impact of feral cats *Felis catus* and their control on Dassen Island. *S. Afr. J. Antarct. Res.* 16: 123–27

12. Biller DS, DiBartola SP, Eaton KA, Pflueger S, Wellman ML, Radin MJ. 1996. The Inheritance of polycystic kidney disease in Persian cats. *J. Hered.* 87: 1–5

13. Birkenmeier EH, Davisson MT, Beamer WG, Ganschow RE, Vogler CA, et al. 1989. Murine mucopolysaccharidosis type VII. Characterization of a mouse with beta-glucuronidase deficiency. *J. Clin. Invest.* 83:1258–6

14. Bosshard NU, Hubler M, Arnold S, Briner J, Spycher MA, et al. 1996. Spontaneous mucolipidosis in a cat: an animal model of human I-cell disease. *Vet. Pathol.* 33:1–13

15. Breen M, Jouquand S, Renier C, Mellersh CS, Hitte C, et al. 2001. Chromosome-specific single-locus FISH probes allow anchorage of an 1800–marker integrated radiation-hybrid/linkage map of the domestic dog genome to all chromosomes. *Genome Res.* 11:1784–95

16. Breen M, Lindgren G, Binns MM, Norman J, Irvin Z, et al. 1997. Genetical and physical assignments of equine microsatellites—first integration of anchored markers in horse genome mapping. *Mamm. Genome* 8:267–73

17. Brown DE, Thrall MA, Walkley SU, Wengler DA, Mitchell TW, et al. 1994. Feline Niemann-Pick disease type C. *Am. J. Pathol.* 144:1–6

18. Brown EW, Yuhki N, Packer C, O'Brien SJ. 1994. A lion lentivirus related to feline immunodeficiency virus: epidemiologic and phylogenetic aspects. *J. Virol.* 68:5953–68

19. Brown JL, Graham LH, Wielebnowski N, Swanson WF, Wildt DE, Howard JG. 2001. Understanding the basic reproductive biology of wild felids by monitoring of faecal steroids. *J. Reprod. Fertil. Suppl.* 57:71–82

20. Brunzell JD. 1995. Familial lipoprotein lipase deficiency and other causes of the chylomicronemia syndrome. See Ref. 156a, pp. 1913–32

21. Burditt LJ, Chotai K, Hirani S, Nugent PG, Winchester BG, Blakemore WF. 1980. Biochemical studies on a case of feline mannosidosis. *Biochem. J.* 189: 467–73

22. Caro TM. 1994. Cheetahs of the Serengeti Plains. In *Wildlife Behavior and Ecology*, ed. GB Schaller, p. 478. Chicago: Univ. Chicago Press

23. Carpenter JL, Hoffman EP, Romanul FC, Kunkel LM, Rosales RK, et al. 1989. Feline muscular dystrophy with dystrophin deficiency. *Am. J. Pathol.* 135:909–19

24. Carpenter MA, Brown EW, Culver M, Johnson WE, Pecon-Slattery J, et al. 1996. Genetic and phylogenetic divergence of feline immunodeficiency virus in the puma (*Puma concolor*). *J. Virol.* 70:6682–93

25. Carpenter MA, O'Brien SJ. 1995. Coadaptation and immunodeficiency virus: lessons from the Felidae. *Curr. Opin. Genet. Devel.* 5:739–45

26. Chavany C, Jendoubi M. 1998. Biology and potential strategies for the treatment of GM2 gangliosidoses. *Mol. Med. Today* 4:158–65

27. Christen Y. 2000. *Le Peuple Leopard.* Paris: Ed. Michalon

28. Chowdhary BP, Fronicke L, Gustavsson I, Scherthan H. 1996. Comparative analysis of the cattle and human genomes: detection of ZOO-FISH and gene mapping-based chromosomal homologies. *Mamm. Genome* 7:297–302

29. Copeland NG, Jenkins NA, Gilbert DJ, Eppig JT, Maltais LJ, et al. 1993. A genetic linkage map of the mouse: current applications and future prospects. *Science* 262:57–66

30. Copeland NG, Jenkins NA, O'Brien SJ. 2002. Mmu 16—comparative genomic highlights. *Science* 296:1617–18

31. Cowell KR, Jezyk PF, Haskins ME, Patterson DF. 1976. Mucopolysaccharidosis in a cat. *J. Am. Vet. Med. Assoc.* 169:334–39

32. Crawley AC, Yogalingam G, Muller VJ, Hopwood JJ. 1998. Two mutations within a feline mucopolysaccharidosis type VI colony cause three different clinical phenotypes. *J. Clin. Invest.* 101:109–19

33. Culver M, Johnson WE, Pecon-Slattery J, O'Brien SJ. 2000. Genomic ancestry of the American puma (*Puma concolor*). *J. Hered.* 91:186–97

34. Culver M, Menotti-Raymond MA, O'Brien SJ. 2001. Patterns of size homoplasy at 10 microsatellite loci in pumas (*Puma concolor*). *Mol. Biol. Evol.* 18:1151–56

35. Danforth CH. 1947. Heredity of polydactyly in the cat. *J. Hered.* 38:107–12

36. Dehal P, Predki P, Olsen AS, Kobayashi A, Folta P, et al. 2001. Human chromosome 19 and related regions in mouse: conservative and lineage-specific evolution. *Science* 293:104–11

37. De Luca T, Minichiello L, Leone A, Di Natale P. 1993. Preliminary molecular analysis of a case of feline mucopolysaccharidosis VI. *Biochem. Biophys. Res. Commun.* 196:1177–82

38. De Maria R, Divari S, Bo S, Sonnio S, Lotti D, et al. 1998. Beta-galactosidase deficiency in a Korat cat: a new form of feline GM1–gangliosidosis. *Acta Neuropathol.* 96:307–14

39. Dobrenis K, Joseph A, Rattazzi MC. 1992. Neuronal lysosomal enzyme replacement using fragment C of tetanus toxin. *Proc. Natl. Acad. Sci. USA* 89: 2297–301

40. Donoghue AM, Byers AP, Johnston LA, Armstrong DL, Wildt DE. 1996. Timing of ovulation after gonadotrophin induction and its importance to successful intrauterine insemination in the tiger (*Panthera tigris*). *J. Reprod. Fertil.* 107:53–58

41. Driscoll CA, Menotti-Raymond M, Nelson G, Goldstein D, O'Brien SJ. 2002. Genomic microsatellites as evolutionary chronometers: a test in wild cats. *Genome Res.* 12:414–23

42. Eizirik E, Kim JH, Menotti-Raymond M, Crawshaw PG Jr, O'Brien SJ, Johnson WE. 2001. Phylogeography, population history and conservation genetics of jaguars (*Panthera onca*, Mammalia, Felidae). *Mol. Ecol.* 10:65–79

43. Eizirik E, Murphy WJ, O'Brien SJ. 2001. Molecular dating and biogeography of the early placental mammal radiation. *J. Hered.* 92:212–19

44. Eizirik E, Yuhki N, Johnson WE, Menotti-Raymond M, Hannah S, O'Brien SJ. 2002. Multiple origins of melanism in Felidae. Submitted

44a. Evett IW, Weir BS. 1998. *Interpreting DNA Evidence, Statistical Genetics for Forensic Scientists.* Sunderland, MA: Sinauer

45. Ewald PW. 2000. *Plague Time: How Stealth Infections Cause Cancer, Heart Disease, and Other Deadly Ailments.* New York: Free Press

46. Fillat C, Simonaro CM, Yeyati PL, Abkowitz JL, Haskins ME, Schuchman EH. 1996. Arylsulfatase B activities and glycosaminoglycan levels in retrovirally transduced mucopolysaccharidosis type VI cells. Prospects for gene therapy. *J. Clin. Invest.* 98:497–502

47. Fogle B. 2001. *The New Encyclopedia of the Cat.* New York: DK Publ. 288 pp.

48. Frankham R, Ballou JD, Briscoe DA. 2002. *Introduction to Conservation Genetics.* London: Cambridge Univ. Press

49. Fyfe J, Lowrie C, Bell TG, Shelton GD. 2001. *Spinal muscle atrophy in cats.* Presented at Proc. An. For. Am. Coll. Vet. Intern. Med. Denver, 19th, pp. 411–15

50. Fyfe JC, Giger U, Winkle TJV, Haskins ME, Steinberg SA, et al. 1992. Glycogen

storage disease type IV: inherited deficiency of branching enzyme activity in cats. *Pediatr. Res.* 32:719–25

51. Fyfe JC, Kurzhals RL. 1998. *Glycogen storage disease Type IV in Norwegian forest cats: molecular detection of carriers.* Presented at Int. Feline Genet. Dis. Conf., Univ. Penn., 1st

52. Fyfe JC, Kurzhals RL, Lassaline ME, Henthorn PS, Alur PR, et al. 1999. Molecular basis of feline beta-glucuronidase deficiency: an animal model of mucopolysaccharidosis VII. *Genomics* 58: 121–28

53. Gaschen FP, Hoffman EP, Gorospe JR, Uhl EW, Senior DF, et al. 1992. Dystrophin deficiency causes lethal muscle hypertrophy in cats. *J. Neurol. Sci.* 110: 149–59

54. Gasper PW, Thrall MA, Wenger DA, Macy DW, Ham L, et al. 1984. Correction of feline arylsulphatase B deficiency (mucopolysaccharidosis VI) by bone marrow transplantation. *Nature* 312: 467–69

55. Ginzinger DG, Lewis MES, Ma YH, Jones BR, Liu GQ, Jones SD. 1996. A mutation in the lipoprotein lipase gene is the molecular basis of chylomicronemia in a colony of domestic cats. *J. Clin. Invest.* 97:1257–66

56. Gitzelmann R, Bosshard NU, Superti-Furga A, Spycher MA, Briner J, et al. 1994. Feline mucopolysaccharidosis VII due to beta-glucuronidase deficiency. *Vet. Pathol.* 31:435–43

57. Gookin JL, Brooks MB, Catalfamo JL, Bunch SE, Munana KR. 1997. Factor X deficiency in a cat. *J. Am. Vet. Med. Assoc.* 211:576–79

58. Gravel RA. 1995. The GM2 gangliosidoses. See Ref. 156a, pp. 2839–79

59. Green ED. 2001. Strategies for the systematic sequencing of complex genomes. *Nat. Rev. Genet.* 2:573–83

60. Hardy WD, Essex M, McClelland AJ, eds. 1980. *Feline Leukemia Virus.* New York: Elsevier

61. Hardy WD Jr. 1993. Feline oncoretroviruses. See Ref. 97a, 2:109–180

62. Haskins M, Abkowitz J, Aguirre GD, Evans S, Hasson C, et al. 1997. Bone marrow transplantation in animal models of lysosomal storage diseases. In *Correction of Genetic Diseases by Transplantation,* ed. O Ringden, J Hobbs, C Stewart, pp. 1–11. London: Cogent

62a. Haskins ME, Aguirre GD, Jezyk PF, Desnick RJ, Patterson DF. 1983. The pathology of the feline model of mucopolysaccharidosis. *Am. J. Pathol.* 112: 27–36

63. Haskins ME, Desnick RJ, DiFerrante N, Jezyk PF, Patterson DF. 1984. Beta-glucuronidase deficiency in a dog: a model of human mucopolysaccharidosis VII. *Pediatr. Res.* 18:980–84

64. Haskins ME, Jezyk PF, Desnick RJ, McDonough SK, Patterson DF. 1979. Alpha-L-iduronidase deficiency in a cat: a model of mucopolysaccharidosis I. *Pediatr. Res.* 13:1294–97

65. Haskins ME, Jezyk PF, Desnick RJ, McDonough SK, Patterson DF. 1979. Mucopolysaccharidosis in a domestic short-haired cat—a disease distinct from that seen in the Siamese cat. *J. Am. Vet. Med. Assoc.* 175:384–87

66. Haskins ME, Jezyk PF, Desnick RJ, Patterson DF. 1981. Animal model of human disease: mucopolysaccharidosis VI Maroteaux-Lamy syndrome, arylsulfatase B-deficient mucopolysaccharidosis in the Siamese cat. *Am. J. Pathol.* 105:191–93

67. Haskins ME, Jezyk PF, Patterson DF. 1979. Mucopolysaccharide storage disease in three families of cats with arylsulfatase B deficiency: leukocyte studies and carrier identification. *Pediatr. Res.* 13:1203–10

68. He X, Li CM, Simonaro CM, Wan Q, Haskins ME, et al. 1999. Identification and characterization of the molecular lesion causing mucopolysaccharidosis

type I in cats. *Mol. Genet. Metab.* 67: 106–12

69. Hegreberg GA, Norby DE, Hamilton MJ. 1974. Lysosomal enzyzme changes in an inherited dwarfism of cats. *Fed. Proc.* 33:598

70. Heid S, Hartmann R, Klinke R. 1998. A model for prelingual deafness, the congenitally deaf white cat—population statistics and degenerative changes. *Hear. Res.* 115:101–12

71. Helgren JA. 1997. *Barron's Encyclopedia of Cat Breeds: A Complete Guide to the Domestic Cats of North America.* Hauppauge, NY: Barron's Educ. Ser. 312 pp.

72. Hoffman EP, Brown RH Jr, Kunkel LM. 1987. Dystrophin: the protein product of the Duchenne muscular dystrophy locus. *Cell* 51:919–28

73. Howard JG. 1993. Semen collection and analysis in carnivores. In *Zoo and Wild Animal Medicine, Current Therapy*, ed. M Fowler, pp. 390–99. Philadelphia: Saunders. 3rd ed.

74. Howard JG, Roth TL, Byers AP, Swanson WF, Wildt DE. 1997. Sensitivity to exogenous gonadotropins for ovulation induction and laparoscopic artificial insemination in the cheetah and clouded leopard. *Biol. Reprod.* 56:1059–68

75. Inada S, Mochizuki M, Izumo S, Kuriyama M, Sakamoto H, et al. 1996. Study of hereditary cerebellar degeneration in cats. *Am. J. Vet. Res.* 57:296–301

76. Int. Hum. Genome Seq. Consort. 2001. Initial sequencing and analysis of the human genome. *Nature* 409:860–921

77. Jackson IJ. 1994. Molecular and developmental genetics of mouse coat color. *Annu. Rev. Genet.* 28:189–217

78. Jackson IJ, Abbott CM. 2000. *Mouse Genetics and Transgenics, A Practical Approach.* New York: Oxford Univ. Press

79. Jauch A, Wienberg J, Stanyon R, Arnold N, Tofanelli S, et al. 1992. Reconstruction of genomic rearrangements in great apes and gibbons by chromosome painting. *Proc. Natl. Acad. Sci. USA* 89:8611–15

80. Jewgenow K, Meyer HH. 1998. Comparative binding affinity study of progestins to the cytosol progestin receptor of endometrium in different mammals. *Gen. Comp. Endocrinol.* 110:118–24

81. Jezyk PF, Haskins ME, Patterson DF, Mellman WJ, Greenstein M. 1977. Mucopolysaccharidosis in a cat with arylsulfatase B deficiency: a model of Maroteaux-Lamy syndrome. *Science* 198:834–36

82. Jiang Z, Priat C, Galibert F. 1998. Traced orthologous amplified sequence tags (TOASTs) and mammalian comparative maps. *Mamm. Genome* 9:577–87

83. Johnson WE, Culver M, Iriarte JA, Eizirik E, Seymour KL, O'Brien SJ. 1998. Tracking the evolution of the elusive Andean mountain cat (*Oreailurus jacobita*) from mitochondrial DNA. *J. Hered.* 89:227–32

84. Johnson WE, O'Brien SJ. 1997. Phylogenetic reconstruction of the Felidae using 16S rRNA and NADH-5 mitochondrial genes. *J. Mol. Evol.* 44:S98–S116

85. Johnson WE, Shinyashiku F, Menotti-Raymond M, Driscoll C, Leh C, et al. 1999. Molecular genetic characterization of two insular Asian cat species, Bornean Bay Cat and Iriomote Cat. In *Evolutionary Theory and Processess: Modern Perspectives*, ed. SP Wasser, pp. 223–48. Netherlands: Kluwer

86. Johnson WE, Slattery JP, Eizirik E, Kim JH, Raymond MM, et al. 1999. Disparate phylogeographic patterns of molecular genetic variation in four closely related South American small cat species. *Mol. Ecol.* 8:S79–94

87. Joyner AL. 2000. *Gene Targeting, A Practical Approach.* New York: Oxford Univ. Press

88. Kier AB, Bresnahan JF, White FJ, Wagner JE. 1980. The inheritance pattern of factor XII (Hageman) deficiency in

domestic cats. *Can. J. Comp. Med.* 44:309–14

89. Kirk CA, Feldman EC, Nelson RW. 1993. Diagnosis of naturally acquired type-I and type-II diabetes mellitus in cats. *Am. J. Vet. Res.* 54:463–67

90. Kitchen H, Murray RE, Cockrell BY. 1972. Animal model for human disease. Spina bifida, sacral dysgenesis and myelocele. Animal model: Manx cats. *Am. J. Pathol.* 68:203–6

91. Kittleson MD, Meurs KM, Munro MJ, Kittleson JA, Liu SK, et al. 1999. Familial hypertrophic cardiomyopathy in Maine coon cats: an animal model of human disease. *Circulation* 99:3172–80

92. Koenig M, Monaco AP, Kunkel LM. 1988. The complete sequence of dystrophin predicts a rod-shaped cytoskeletal protein. *Cell* 53:219–26

93. Deleted in proof

94. Lapiere CM, Nusgens BV. 1993. Ehlers-Danlos type VII-C, or human dermatosparaxis. The offspring of a union between basic and clinical research. *Arch. Dermatol.* 129:1316–19

95. Lawhorn B. 1989. Testicular feminization in a cat. *J. Am. Vet. Med. Assoc.* 195:1456–58

96. Leon A, Hussain AA, Curtis R. 1991. Autosomal dominant rod-cone dysplasia in the Rdy cat. 2. Electrophysiological findings. *Exp. Eye Res.* 53:489–502

97. Leventhal AG, Vitek DJ, Creel DJ. 1985. Abnormal visual pathways in normally pigmented cats that are heterozygous for albinism. *Science* 229:1395–97

97a. Levy JA, ed. 1993. *Viruses: The Retroviridae.* New York: Plenum. Vol. 2

98. Littlewood JD, Evans RJ. 1990. A combined deficiency of factor VIII and contact activation defect in a family of cats. *Br. Vet. J.* 146:30–35

99. Liu G, Ashbourne Excoffon KJ, Wilson JE, McManus BM, Rogers QR, et al. 2000. Phenotypic correction of feline lipoprotein lipase deficiency by aden-oviral gene transfer. *Hum. Gene Ther.* 11:21–32

100. Lopez JV, Cevario S, O'Brien SJ. 1996. Complete nucleotide sequence of the domestic cat (*Felis catus*) mitochondrial genome and a transposed mtDNA tandem repeat (*Numt*) in the nuclear genome. *Genomics* 33:229–46

101. Lowenthal AC, Cummings JF, Wenger DA, Thrall MA, Wood PA, de Lahunta A. 1990. Feline sphingolipidosis resembling Niemann-Pick disease type C. *Acta Neuropathol.* 81:189–97

102. Lutz TA, Rand JS. 1993. A review of new developments in type 2 diabetes in human beings and cats. *Br. Vet. J.* 149:527–36

103. Lyons LA, Laughlin TF, Copeland NG, Jenkins NA, Womack JE, O'Brien SJ. 1997. Comparative anchor tagged sequences (CATS) for integrative mapping of mammalian genomes. *Nat. Genet.* 15:47–56

104. Maggio-Price L, Dodds WJ. 1993. Factor IX deficiency (hemophilia B) in a family of British shorthair cats. *J. Am. Vet. Med. Assoc.* 203:1702–4

105. Deleted in proof

106. Marklund L, Johansson Moller M, Hoyheim B, Davies W, Fredholm M, et al. 1996. A comprehensive linkage map of the pig based on a wild pig-Large White intercross. *Anim. Genet.* 27:255–69

107. McGovern MM, Mandell N, Haskins M, Desnick RJ. 1985. Animal model studies of allelism: characterization of arylsulfatase B mutations in homoallelic and heteroallelic (genetic compound) homozygotes with feline mucopolysaccharidosis VI. *Genetics* 110:733–49

108. Menotti-Raymond M, David VA, Lyons LA, Schäffer AA, Tomlin JF, et al. 1999. A genetic linkage map of microsatellites in the domestic cat (*Felis catus*). *Genomics* 57:9–23

109. Menotti-Raymond M, David VA, Chen ZQ, Menotti KA, et al. 2002. Second generation integrated genetic linkage

radiation hybrid maps in the domestic cat (*Felis catus*). *J. Hered.* In press

110. Menotti-Raymond M, David VA, Stephens JC, Lyons LA, O'Brien SJ. 1997. Genetic individualization of domestic cats using feline STR loci for forensic applications. *J. Forensic Sci.* 42:1039–51

111. Menotti-Raymond MA, David VA, O'Brien SJ. 1997. Pet cat hair implicates murder suspect. *Nature* 386:774

112. Menotti-Raymond M, O'Brien SJ. 1993. Dating the genetic bottleneck of the African cheetah. *Proc. Natl. Acad. Sci. USA* 90:3172–76

113. Menotti-Raymond MA, O'Brien SJ. 1995. Evolutionary conservation of ten microsatellite loci in four species of Felidae. *J. Hered.* 86:319–22

113a. MHC Seq. Consort. 1999. Complete sequence and gene map of a human major histocompatibility complex. *Nature* 401:921–23

114. Millis DL, Hauptman JG, Johnson CA. 1992. Cryptorchidism and monorchism in cats: 25 cases (1980–1989). *J. Am. Vet. Med. Assoc.* 200:1128–30

115. Modi WS, O'Brien SJ. 1998. Quantitative cladistic analyses of chromosomal banding data among species in three orders of mammals: Hominoid primates, felids and arvicolid rodents. In *Chromosome Structure and Function*, ed. JP Gustafson, R Appels, pp. 215–42. New York: Plenum

116. Muldoon LL, Neuwelt EA, Pagel MA, Weiss DL. 1994. Characterization of the molecular defect in a feline model for type II GM2–gangliosidosis (Sandhoff disease). *Am. J. Pathol.* 144:1109–18

117. Mural RJ, Adams MD, Myers EW, Smith HO, Miklos GLG, et al. 2002. A comparison of whole-genome shotgun-derived mouse chromosome 16 and the human genome. *Science* 296:1661–67

118. Murphy WJ, Eizirik E, O'Brien SJ, Madsen O, Scally M, et al. 2001. Resolution of the early placental mammal radiation using Bayesian phylogenetics. *Science* 294:2348–51

119. Murphy WJ, Menotti-Raymond M, Lyons LA, Thompson ME, O'Brien SJ. 1999. Development of a feline whole-genome radiation hybrid panel and comparative mapping of human chromosome 12 and 22 loci. *Genomics* 57:1–8

120. Murphy WJ, Stanyon R, O'Brien SJ. 2001. Evolution of mammalian genome organization inferred from comparative gene mapping. *Genome Biol.* 2:0005.1–5.8

121. Murphy WJ, Sun S, Chen Z, Yuhki N, Hirschmann D, et al. 2000. A radiation hybrid map of the cat genome: implications for comparative mapping. *Genome Res.* 10:691–702

122. Nadeau JH, Taylor BA. 1984. Lengths of chromosomal segments conserved since divergence of man and mouse. *Proc. Natl. Acad. Sci. USA* 81:814–18

123. Nash WG, Menninger JC, Wienberg J, O'Brien SJ. 2001. Resolving the sequence of dynamic genomic exchange in the evolution of Canidae. *Cytogenet. Cell Genet.* 95:210–24

124. Nash WG, O'Brien SJ. 1982. Conserved regions of homologous G-banded chromosomes between orders in mammalian evolution: carnivores and primates. *Proc. Natl. Acad. Sci. USA* 79:6631–35

125. Nash WG, Wienberg J, Ferguson-Smith M, Menninger J, O'Brien SJ. 1998. Comparative genomics: tracking chromosome evolution in the family Ursidae using reciprocal chromosome painting. *Cytogenet. Cell Genet.* 83:182–92

126. Deleted in proof

127. Natl. Res. Counc. 1996. *The Evaluation of Forensic DNA Evidence.* Washington, DC: Natl. Acad. Press

128. Neufeld EF, Fratantoni JC. 1970. Inborn errors of mucopolysaccharide metabolism. *Science* 169:141–46

129. Neufeld EF, Lim TW, Shapiro LJ. 1975. Inherited disorders of lysosomal metabolism. *Annu. Rev. Biochem.* 44:357–76

130. Neufeld EF, Muenzer J. 1995. The mucopolysaccharidoses. See Ref. 156a, pp. 2465–95

131. Neuwelt EA, Johnson WG, Blank NK, Pagel MA, Maslen-McClure C, et al. 1985. Characterization of a new model of GM2–gangliosidosis (Sandhoff's disease) in Korat cats. *J. Clin. Invest.* 76: 482–90

132. O'Brien SJ. 1994. A role for molecular genetics in biological conservation. *Proc. Natl. Acad. Sci. USA* 91:5748–55

133. O'Brien SJ. 1994. Genetic and phylogenetic analyses of endangered species. *Annu. Rev. Genet.* 28:467–89

134. O'Brien SJ. 1995. Genomic prospecting. *Nat. Med.* 1:742–44

135. O'Brien SJ, Cevario SJ, Martenson JS, Thompson MA, Nash WG, et al. 1997. Comparative gene mapping in the domestic cat *(Felis catus). J. Hered.* 88: 408–14

136. O'Brien SJ, Eizirik E, Murphy WJ. 2001. Genomics. On choosing mammalian genomes for sequencing. *Science* 292:2264–66

137. O'Brien SJ, Menotti-Raymond M, Murphy WJ, Nash WG, Wienberg J, et al. 1999. The promise of comparative genomics in mammals. *Science* 286:458–62, 79–81

138. O'Brien SJ, Nash WG. 1982. Genetic mapping in mammals: chromosome map of domestic cat. *Science* 216:257–65

139. O'Brien SJ, Stanyon, R. 1999. Phylogenomics: ancestral primate viewed. *Nature* 402:365–66

140. O'Brien SJ, Wienberg J, Lyons LA. 1997. Comparative genomics: lessons from cats. *Trends Genet.* 13:393–99

141. O'Brien SJ, Haskins ME, Winkler CA, Nash WG, Patterson DF. 1986. Chromosomal mapping of beta-*globin* and *albino* loci in the domestic cat: a conserved mammalian chromosome group. *J. Hered.* 77:374–78

142. Olmsted RA, Langley R, Roelke ME, Goeken RM, Adger-Johnson D, et al. 1992. Worldwide prevalence of lentivirus infection in wild feline species: epidemiologic and phylogenetic aspects. *J. Virol.* 66:6008–18

143. Ostrander EA, Kruglyak L. 2000. Unleashing the canine genome. *Genome Res.* 10:1271–74

144. Packer C, Gilbert DA, Pusey AE, O'Brien SJ. 1991. Kinship, cooperation and inbreeding in African lions: a molecular genetic analysis. *Nature* 351:562–65

145. Parrish CR. 1994. The emergence and evolution of canine parvovirus—an example of recent host range mutation. *Semin. Virol.* 5:121–32

146. Partridge T. 1991. Animal models of muscular dystrophy—what can they teach us? *Neuropathol. Appl. Neurobiol.* 17:353–63

147. Patronek GJ. 1998. Free-roaming and feral cats—their impact on wildlife and human beings. *J. Am. Vet. Med. Assoc.* 212:218–26

148. Pedersen NC. 1993. The feline immunodeficiency virus. See Ref. 97a, pp. 181–228

149. Pollack MS, Mastrota F, Chin-Louie J, Monney S, Hayes A. 1982. Preliminary studies of the feline histocompatibility system. *Immunogenetics* 16:339–47

150. Prieur DJ, Collier LL. 1981. Inheritance of the Chediak-Higashi syndrome in cats. *J. Hered.* 72:175–77

151. Reed GB Jr, Dixon JF, Neustein JB, Donnell GN, Landing BH. 1968. Type IV glycogenosis. Patient with absence of a branching enzyme alpha-1,4-glucan:alpha-1,4-glucan 6-glycosyl transferase. *Lab. Invest.* 19:546–57

152. Rettenberger G, Klett C, Zechner U, Bruch J, Just W, et al. 1995. ZOO-FISH analysis: cat and human karyotypes closely resemble the putative ancestral mammalian karyotype. *Chromosome Res.* 3:479–86

153. Reuser AJJ. 1993. Molecular biology, therapeutic trials and animal models of lysosomal diseases-type II glycogenosis

as an example. *Ann. Biol. Clin.* 51:218–19

154. Robinson R. 1976. Homologous genetic variation in the Felidae. *Genetica* 46:1–31

155. Robinson R. 1991. *Genetics for Cat Breeders.* Oxford: Pergamon. 234 pp.

156. Roelke-Parker ME, Munson L, Packer C, Kock R, Cleaveland S, et al. 1996. A canine distemper virus epidemic in Serengeti lions (*Panthera leo*). *Nature* 379:441–45

156a. Scriver CR, Beaudet AL, Sly WS, Valle D, eds. 1995. *The Metabolic and Molecular Bases of Inherited Diseases.* New York: McGraw-Hill

157. Seeliger MW, Narfstrom K. 2000. Functional assessment of the regional distribution of disease in a cat model of hereditary retinal degeneration. *Invest. Ophthalmol. Vis. Sci.* 41:1998–2005

158. Sena-Esteves M, Camp SM, Alroy J, Breakefield XO, Kaye EM. 2000. Correction of acid beta-galactosidase deficiency in GM1 gangliosidosis human fibroblasts by retrovirus vector-mediated gene transfer: higher efficiency of release and cross-correction by the murine enzyme. *Hum. Gene Ther.* 11:715–27

159. Sharp NJ, Kornegay JN, Van Camp SD, Herbstreith MH, Secore SL, et al. 1992. An error in dystrophin mRNA processing in golden retriever muscular dystrophy, an animal homologue of Duchenne muscular dystrophy. *Genomics* 13:115–21

160. Shin T, Kraemer D, Pryor J, Liu L, Rugila J, et al. 2002. A cat cloned by nuclear transplantation. *Nature* 415:859

161. Sicinski P, Geng Y, Ryder-Cook AS, Barnard EA, Darlison MG, Barnard PJ. 1989. The molecular basis of muscular dystrophy in the mdx mouse: a point mutation. *Science* 244:1578–80

162. Slattery J, O'Brien SJ. 1998. Patterns of Y and X chromosome DNA sequence divergence during the Felidae radiation. *Genetics* 148:1245–55

163. Soute BA, Ulrich MM, Watson AD, Maddison JE, Ebberink RH, Vermeer C. 1992. Congenital deficiency of all vitamin K-dependent blood coagulation factors due to a defective vitamin K-dependent carboxylase in Devon Rex cats. *Thromb. Haemost.* 68:521–25

164. Spellacy E, Shull RM, Constantopoulos G, Neufeld EF. 1983. A canine model of human alpha-L-iduronidase deficiency. *Proc. Natl. Acad. Sci. USA* 80:6091–95

165. Sun H, Yang M, Haskins ME, Patterson DF, Wolfe JH. 1999. Retrovirus vector-mediated correction and cross-correction of lysosomal alpha-mannosidase deficiency in human and feline fibroblasts. *Hum. Gene Ther.* 10:1311–19

166. Sun S, Murphy WJ, Menotti-Raymond M, O'Brien SJ. 2001. Integration of the feline radiation hybrid and linkage maps. *Mamm. Genome* 12:436–41

167. Tanaka KR. 1971. Introduction to discussion of glucose-6-phosphate dehydrogenase deficiency. *Exp. Eye Res.* 11:396–401

168. Taylor RM, Wolfe JH. 1994. Cross-correction of beta-glucuronidase deficiency by retroviral vector-mediated gene transfer. *Exp. Cell Res.* 214:606–13

169. Thomas GH, Beaudet AL. 1995. Disorders of glycoprotein degradation: α-mannosidosis, β-mannosidosis, sialidosis, asportylglucosaminuria, and carbohydrate-deficient glycoprotein syndrome. See Ref. 156a, pp. 2529–61

170. Twigger S, Lu J, Shimoyama M, Chen D, Pasko D, et al. 2002. Rat Genome Database (RGD): mapping disease onto the genome. *Nucleic Acids Res.* 30:125–28

171. Tymms MJ, Kola I. 2001. *Gene Knockout Protocols. Methods in Molecular Biology.* Totowa, NJ: Humana

172. Uphyrkina O, Johnson WE, Quigley H, Miquelle D, Marker L, et al. 2001. Phylogenetics, genome diversity and origin of modern leopard, *Panthera pardus. Mol. Ecol.* 10:2617–33

173. Deleted in proof
174. Vitale CB, Ihrke PJ, Gross TL, Werner LL. 1997. Systemic lupus erythematosus in a cat-fulfillment of the American rheumatism association criteria with supportive skill histopathology. *Vet. Dermatol.* 8:133–38
175. Vitale CB, Ihrke PJ, Olivry T, Stannard AA. 1996. Feline urticaria pigmentosa in three related sphinx cats. *Vet. Dermatol.* 7:227–33
176. Vite CH, McGowan JC, Braund KG, Drobatz KJ, Glickson JD, et al. 2001. Histopathology, electrodiagnostic testing, and magnetic resonance imaging show significant peripheral and central nervous system myelin abnormalities in the cat model of alpha-mannosidosis. *J. Neuropathol. Exp. Neurol* 60:817–28
177. Walkley SU, Baker HJ, Rattazzi MC, Haskins ME, Wu JY. 1991. Neuroaxonal dystrophy in neuronal storage disorders: evidence for major GABAergic neuron involvement. *J. Neurol. Sci.* 104:1–8
178. Walkley SU, Thrall MA, Dobrenis K, Huang M, March PA, et al. 1994. Bone marrow transplantation corrects the enzyme defect in neurons of the central nervous system in a lysosomal storage disease. *Proc. Natl. Acad. Sci. USA* 91:2970–74
179. Weber SE, Feldman BF, Evans DA. 1981. Pelger-Huet anomaly of granulocytic leukocytes in two feline littermates. *Feline Pract.* 11:44–47
180. Weissenbock H, Rossel C. 1997. Neuronal ceroid-lipofuscinosis in a domestic cat: clinical, morphological and immunohistochemical findings. *J. Comp. Pathol.* 117:17–24
181. Wienberg J, Stanyon R. 1997. Comparative painting of mammalian chromosomes. *Curr. Opin. Genet. Dev.* 7:784–91
182. Wienberg J, Stanyon R, Nash WG, O'Brien PC, Yang F, et al. 1997. Conservation of human vs. feline genome organization revealed by reciprocal chromosome painting. *Cytogenet. Cell Genet.* 77:211–17
183. Wildt DE, Brown JL, Swanson WF. 1998. Reproduction in cats. In *Encyclopedia of Reproduction*, ed. E Knobil, J Neill, pp. 497–510. New York: Academic
184. Willett BJ, Flynn JN, Hosie MJ. 1997. FIV infection of the domestic cat: an animal model for AIDS. *Immunol. Today* 182–89
185. Winand NJ, Edwards M, Pradhan D, Berian CA, Cooper BJ. 1994. Deletion of the dystrophin muscle promoter in feline muscular dystrophy. *Neuromuscular Disord.* 4:433–45
186. Winkler C, Schultz A, Cevario S, O'Brien SJ. 1989. Genetic characterization of FLA, the cat major histocompatibility complex. *Proc. Natl. Acad. Sci. USA* 86:943–47
187. Wolfe BA, Wildt DE. 1996. Development to blastocysts of domestic cat oocytes matured and fertilized in vitro after prolonged cold storage. *J. Reprod. Fertil.* 106:135–41
188. Wolfe JH, Sands MS. 1996. Murine mucopolysaccharidosis type VII: a model system for somatic gene therapy of the central nervous system. In *Gene Protocols for Gene Transfer in Neuroscience: Towards Gene Therapy of Neurologic Disorders*, ed. PR Lowenstein, LW Enquist, pp. 263–74. Essex, England: Wiley
189. Wood TC, Wildt DE. 1997. Effect of the quality of the cumulus-oocyte complex in the domestic cat on the ability of oocytes to mature, fertilize and develop into blastocysts in vitro. *J. Reprod. Fertil.* 110:355–60
190. Wurster-Hill DH, Centerwall WR. 1982. The interrelationships of chromosome banding patterns in canids, mustelids, hyena, and felids. *Cytogenet. Cell Genet.* 34:178–92
191. Wurster-Hill DH, Gray CW. 1975. The interrelationships of chromosome banding patterns in procyonids, viverrids,

and felids. *Cytogenet. Cell Genet.* 15: 306–31

192. Yogalingam G, Litjens T, Bielicki J, Crawley AC, Muller V, et al. 1996. Feline mucopolysaccharidosis type VI. Characterization of recombinant N-acetyl-galactosamine 4-sulfatase and identifica-tion of a mutation causing the disease. *J. Biol. Chem.* 271:27259–65

193. Yuhki N, Beck T, Stephens RM, Nishigaki Y, Newmann K, O'Brien SJ. Comparative genome organization of human, murine and feline MHC class II region. *Genome Res.* Submitted

Annu. Rev. Genet. 2002. 36:687–720
doi: 10.1146/annurev.genet.36.062802.091007
Copyright © 2002 by Annual Reviews. All rights reserved
First published online as a Review in Advance on September 12, 2002

GENETIC APPROACHES TO MOLECULAR AND CELLULAR COGNITION: A Focus on LTP and Learning and Memory

Anna Matynia, Steven A. Kushner, and Alcino J. Silva

*Departments of Neurobiology, Psychiatry, and Psychology, Brain Research Institute,
University of California, Los Angeles, Los Angeles, California 90095;
e-mail: amatynia@mednet.ucla.edu, skushner@ucla.edu, silvaa@mednet.ucla.edu*

Key Words   plasticity, hippocampus, synapse, gene targeting, transgenic

■ **Abstract**   Long-term potentiation (LTP) is the predominant experimental model for the synaptic plasticity mechanisms thought to underlie learning and memory. This review is focused on the contributions of genetics to the understanding of the role of LTP in learning and memory. These studies have used a combination of genetics, molecular biology, neurophysiology, and psychology to demonstrate that molecular mechanisms of synaptic plasticity are critical for learning and memory. Because of the large scope of this literature, we focus primarily on genetic studies of hippocampal-dependent learning. Altogether, these findings not only demonstrate a role for plasticity in learning, they also lay down the foundations for the new field of molecular and cellular cognition.

CONTENTS

INTRODUCTION

Synaptic Mechanisms of Learning and Memory

Memory allows organisms to make adaptive changes that improve their fitness. For example, the ability to remember the territories of rival males allows a mouse to minimize possible life-threatening confrontations. There is now an overwhelming amount of data that suggest that synaptic changes play a key role in learning and memory. Thirty years before the discovery of long-term potentiation, Donald Hebb proposed that memory is stored in patterns of synaptic connections established during learning. The idea is that neurons that are repeatedly coactivated during learning, strengthen their synaptic connections, and that these changes in neuronal connectivity can be used to recreate the information acquired during learning. This simple idea is the basis of the most influential theory of learning and memory and accounts for a very large body of molecular, neurophysiological, and behavioral data in multiple systems and organisms.

While studying the basic electrophysiological properties of excitatory neurons in the dentate gyrus of the hippocampus, Lomo & Bliss discovered a phenomenon in which excitatory postsynaptic potentials (EPSPs) were stably enhanced after specific types of high-frequency stimulation [reviewed in (15)]. This electrophysiological finding was the first demonstration of long-term synaptic plasticity and a confirmatory hint of Hebbian ideas about learning and memory. The high-frequency stimulation protocols used by Lomo & Bliss were subsequently adapted to studying LTP in hippocampal slices. Additional studies have also discovered long-term depression (LTD) in which a stable decrease in baseline EPSPs occurs after low-frequency activation of the postsynaptic cell. Theoretical studies had predicted the existence of LTD, since without some mechanism to balance LTP, synapses in neuronal circuits would progressively become saturated and thus unable to store information.

Synaptic plasticity in the dentate gyrus, CA1, and CA3 neurons of the hippocampus is an attractive candidate mechanism for learning and memory, including declarative memory, a form of hippocampal-dependent memory for facts and events [reviewed in (34)]. Studies of HM, a patient who received medial temporal lobe lesions to alleviate his intractable epilepsy, revealed that the hippocampus plays a pivotal role in this form of memory. The studies with HM had an enormous impact because they demonstrated that memory is a phenomenon separate from other cognitive abilities, such as attention and general intelligence. The hippocampus also has a key role in place learning in mammalian species as diverse as mice and men. Therefore, spatial and contextual learning have had a prominent role in studies of hippocampal-dependent learning and memory. Pharmacological and genetic manipulations that interfere with LTP are often found to affect place learning and memory, although there are some interesting exceptions to this correlation. Although mechanisms that result in stable and long-lasting changes in synaptic function are ensconced as learning and memory mechanisms, many questions remain unanswered. Among these are how synaptic inputs are modulated during learning, how cells integrate their inputs, and how that code is transferred, processed, and stored over multiple networks of neurons in the brain.

The Genetic Basis of Learning and Memory

The idea that there is a genetic basis for learning and memory originated nearly a century ago, even before the physical mode of inheritance was known. Since then, many studies have shown that learning and other cognitive abilities have a genetic basis [reviewed in (117)]. Current genetic approaches to learning and memory came from three main sources: forward genetic studies in fruit flies heralded the advent of the genetic dissection of behavior in model systems (32), mouse reverse genetic approaches to behavior started by asking how genes affect plasticity and learning and memory in mice (101, 103), and more recent studies ascertained the feasibility of performing mutagenic screens in mice (60, 115). The original forward genetic studies of learning and memory in Drosophila not only showed that there are discrete loci for learning and memory, but also identified key signaling pathways that continue to be intensively investigated [reviewed in (116)].

The genetic study of molecular and cellular cognition in mice benefited tremendously from both behavioral neuroscience and neurophysiology. A key contribution of reverse genetic studies of molecular and cellular cognition has been to bring psychology and neurophysiology closer together. The availability of mutants has allowed behavioral neuroscientists and neurophysiologists to study and analyze the same genetically modified mice. This has led to a more coherent integration of findings in both fields, to the development of a common language between these areas, and to the emergence of a new field, molecular and cellular cognition, where genetics, physiology, and behavior are used side by side in integrative explanations of cognition.

Important developments have also been made in forward genetic approaches. Quantitative trait loci (QTL) that account for small differences in behavior have been identified in many inbred mice strains (86), including those associated with Pavlovian fear conditioning (114, 119). These and other findings indicated that genetic loci can account for the complex behavioral variance among inbred mouse strains. These studies have also highlighted the critical importance of controlling for genetic background when carrying out genetic studies of behavior (26, 102).

Transgenic Approaches

Manipulations that alter gene function include functional deletions of genes, overexpression of wild-type or mutated genes, and gene replacements with mutant alleles. Collectively, these manipulations are referred to as transgenetic manipulations, to distinguish them from transgenic approaches that refer specifically to studies done with mice derived by injections of constructs into the pro-nuclei of oocytes. Germline deletions of genes typically include the insertion of a neomycin cassette near the beginning of the open reading frame of a gene, causing truncation and instability of resulting transcripts. The neomycin cassette can sometimes disrupt nearby loci and thus confound the interpretation of the results. Therefore, targeted mutations in which the neomycin cassette is removed prior to blastocyst injection are being generated. Although this approach is the mainstay of reverse genetic studies of learning and memory, considerable effort has been placed into developing techniques that improve the temporal, regional, and biochemical specificity of the mutations studied. Proteins can have multiple biochemical roles, in different tissues and over time. Therefore, it is critical to develop methods that unravel the multiple roles of single genes in specific cells and brain regions. The development of the Cre/loxP system represents a significant step toward this goal. Cre recombinase deletes sequences flanked by its 34-bp recognition sites, loxP. Therefore, specific exons flanked by loxP (or floxed) can be deleted in tissues expressing this recombinase. The expression of the recombinase can be restricted with tissue-specific promoters. For example, an 8.5-kb segment of the αCaMKII promoter is sufficient to direct expression of a reporter gene to postnatal excitatory neurons of the forebrain (74). Furthermore, transcriptional regulatory elements in the genomic integration site can further modify the expression profile of the transgenic Cre recombinases. Thus, gene expression can be limited to specific regions of the brain, such as the CA1 pyramidal field of the hippocampus (112). Localized transgenic expression can also be obtained by region-specific injection of virally driven genes [e.g., (58)].

Temporal control over gene expression in the brain has also been achieved with both transcriptional and posttranscriptional methods. The tetracycline transactivator (tTA) system uses the bacterial tetracycline-sensitive activator to direct expression from the Tet operon promoter (tetO). In this system, administration of a tetracycline analog (doxycycline) turns off gene expression driven from the tetO promoter. In a mutated version of this system (the reverse tTA or rtTA system), administration of the tetracycline analog induces expression from the tetO promoter. In the LBD posttranscriptional system, a mutant ligand-binding domain (LBD) of the estrogen or progesterone receptor (12, 118) is fused with a protein of interest. The LBD fusion will be transcribed and translated, but will be inactive until the LBD binds its ligand. This system allows activation of proteins of interest within hours rather than days (62), as is the case with the tTA systems. Improvements in these inducible and tissue-restricted systems will continue to benefit studies of learning and memory.

Searching for Learning and Memory Genes in the Mouse

Taking a cue from learning and memory mutant screens in fruit flies, recent forward genetic experiments with traditional ENU mutagenesis are attempting to identify new mouse learning and memory mutations. Similarly, our laboratory has also developed phenotypic screens of previously derived knock-outs (KO). Although most of these KO mutant mice were derived to study other biological problems, such as development, cancer, or immunology, many of the mutated genes are expressed in the brain. Therefore, those mutations could also have behavioral phenotypes. Indeed, a limited screen of randomly selected KO mice has already identified several new learning and memory genes (A. Matynia & A.J. Silva, unpublished data).

Analysis of genes expressed after behavioral training has also been used to identify potential new learning and memory genes. Numerous microarray and SAGE (serial analysis of gene expression) experiments have identified genes normally expressed in structures with a known role in certain forms of learning, such as the amygdala (127) and hippocampus (29, 82) as well as genes that are up- or down-regulated during associative learning (21; Y. Robles & S. Pena-de-Ortiz, personal communication). Data from these types of experiments show the diverse array of proteins required for learning and memory (21; Y. Robles & S. Pena-de-Ortiz, personal communication). Attempts have also been made to develop methods to visualize gene expression in the brains of behaving animals. Gene expression profiles may very likely represent functional signatures of different memory processes. In combination with genetics, these genomic approaches will be an important tool to dissect and uncover novel learning and memory mechanisms. Thus, transgenetic approaches should continue to play an important role in uncovering the mechanisms responsible for the acquisition, processing, storage, and recall of information.

Below, we highlight key findings and insights gained from the analysis of mutant mice in the field of molecular and cellular cognition. Because of the large scope of this literature, we focus our review on transgenetic studies of hippocampal-dependent learning and memory.

MUTATIONS AND INSIGHTS

NMDAR: A Molecular Coincidence Detector

The NMDA receptor is a glutamate-gated receptor composed of four to five subunits with at least one ζ and one ε subunit. The $\zeta 1$, also called the NR1 subunit, is ubiquitously expressed during development. Deletion of this subunit causes neonatal lethality (65). NR2B ($\varepsilon 2$) and NR2D ($\varepsilon 4$) are expressed embryonically, whereas all four subunits are expressed after birth. Postnatally, NR2A is expressed throughout the brain, NR2B is restricted to the forebrain, NR2C is expressed mainly in the cerebellum, and NR2D has limited expression in the brain stem. Activation of the NMDARs requires both binding of glutamate by any of the four NR2 subunits and release from a voltage-dependent magnesium block. Thus, sodium or calcium influx (e.g., through local depolarization or back-propagating spikes) alters the membrane potential, thereby alleviating the block and allowing activation. Since these two events are required simultaneously for activation, NMDARs have been called molecular coincidence detectors. After activation, calcium then enters through this channel and induces LTP. Both NR2A and NR2B bind the alpha isoform of calcium/calmodulin-dependent kinase II (αCaMKII) as well as PostSynaptic Density-95 (PSD-95) via their carboxy-terminal tails, presumably linking channel activation with intracellular signaling pathways.

The role of NMDA receptors (NMDAR) in LTP and memory has been long established. Treatment with NMDAR blockers severely impaired performance in the spatial version of the Morris water maze. In this task, mice have to learn to find a submerged platform in a large pool of water using distal guiding cues (see Figure 1A, see color insert) (83). In addition to their clear impact on learning in the Morris water maze, a task very sensitive to hippocampal lesions, NMDAR blockers also disrupted LTP in the hippocampus, suggesting that NMDAR-dependent induction of LTP in the hippocampus is critical for spatial learning. This study was one of the first to provide compelling evidence for a link between LTP and learning. However, follow-up studies showed that NMDAR blockers have many behavioral effects, other than blocking spatial learning (8, 93). With genetic approaches it has been possible to manipulate specific subunits and properties of these receptors. Individual subunits can not only be deleted, but also deleted in specific regions of the brain, thereby obviating the need to consider functional effects from regions other than the one of interest. This is particularly important when studying the NMDA receptor since deletion of the NR1, NR2A, or NR2B subunits results in neonatal lethality (38, 65, 66).

NR1   The properties of the NR1 subunit determine not only affinities for the channel modulators glycine and glutamine, but also Ca^{++} permeability, gating properties, potentiation, and the voltage-dependent Mg^{++} block. A key amino acid for function of this subunit is N598 that lies in the pore-forming region of the protein. To determine the role of the NR1 subunit in both LTP and learning, mice with a substitution of N598 for glutamine (conserving the charge) or arginine (altering the charge) were studied. Surviving only as heterozygotes, the N598Q mice exhibited normal LTP but increased mortality, whereas the N598R mice had decreased NMDA currents, decreased calcium permeability and died within 2 days (104). Given the essential nature of this gene, very careful manipulations must be made in the NR1 subunit to molecularly dissect the specific contributions of this subunit to LTP and learning and memory.

To investigate the role of glycine binding to the NR1 subunit and its phenotypic consequences, Kew et al. used the knock-in approach to make point mutations that disrupt glycine binding. Mice with a D481N mutation, which moderately decreases the affinity of the subunit for glycine, were viable. In acutely dissociated neurons, glutamate binding to NMDA receptors was normal and the decreased binding of glycine was confirmed. Hippocampal LTP could not be induced using theta-burst stimulation. Consistent with a relationship between LTP and spatial learning, these mice were also deficient in the hidden-platform version of the water maze (61).

To circumvent the lethality of NR1 deletions, this subunit was deleted specifically in the CA1 pyramidal field of the hippocampus, a region with a well-demonstrated role in learning and memory. This was done with the Cre/LoxP system under the regulation of the αCaMKII promoter (75). This promoter normally directs gene expression to postnatal excitatory neurons of the forebrain. However, interactions with putative transcriptional elements in the genomic site of transgene insertion result in specific expression patterns. For example, a specific insertion site restricted deletions of NR1 to the CA1 region (112). The loss of NR1 from CA1 cells resulted in a selective loss of NMDAR function without affecting other glutamate receptors (e.g., AMPA receptors). Neither NMDA-dependent LTP nor LTD could be induced in the CA1 region, although LTP and LTD was not affected in other regions studied, confirming the neuroanatomical specificity of this genetic manipulation. This CA1-specific deletion of NR1 also resulted in spatial learning deficits in the water maze. The CA1 NR1$^{-/-}$ mice were unable to learn to locate the hidden platform, even though they could learn to escape the water by swimming towards and climbing to a visible-platform in the water maze, suggesting that their spatial learning deficit was not due to abnormalities in motor coordination, motivation to escape the water or vision required for spatial navigation (113).

Consistent with the idea that abnormal CA1 function accounts for the spatial learning deficits of the CA1 NR1$^{-/-}$ mutant mice, single unit in vivo electrophysiological recordings of CA1 cells revealed interesting spatial deficits. The activity of these cells reflects the position of the animal in its environment. The firing pattern of different CA1 cells, as the animal travels through a given environment, can be used to reconstruct its trajectory. This and other results suggested that place cell maps have a role in spatial learning [reviewed in (96)]. Multi-electrode recordings in the CA1 region of CA1 NR1$^{-/-}$ mice showed that place-related firing is not affected, but that the average size of the place where CA1 cells fire (place fields) was larger in the mutant mice. Additionally, the coordinated firing of pairs of neurons in similar spatial locations was decreased in the mutants (78), suggesting that NMDAR-dependent LTP in the CA1 region is required for the orchestration and coordination of place maps during learning. This was one of the first attempts to use circuit phenomena to bridge synaptic physiology with behavior. This illustrates an important feature of molecular and cellular cognition studies: the integration of findings at multiple levels of biological complexity, including biochemical, subcellular, cellular, network, and systems.

NR2A   CA1 cells from mice carrying a deletion of NR2A have smaller NMDA currents and, not surprisingly, show attenuated LTP (91). Stronger stimulation, however, can trigger LTP in the mutants that is comparable to WT mice, suggesting that this mutation did not disrupt the core mechanisms of LTP induction (63). Instead, it may have increased the threshold for LTP induction, a finding consistent with the idea that NR2A modulates NMDA currents during LTP induction. The increased threshold for LTP induction observed in these mutants may account for the higher thresholds for learning revealed in the NR2A mutant mice. In contextual fear conditioning, a hippocampal-dependent test, mice learn to fear a context (conditioned stimulus or CS) in which they receive a footshock (unconditioned stimulus or US) (see Figure 1B, see color insert). When placed back in the context in which they received the shock, mice show freezing, a cessation of all movement except for respiration. Freezing is a normal fear response used to evade natural predators since predators often use movement to detect their prey. While these NR2A$^{-/-}$ mice were able to perform contextual fear conditioning normally, they were unable to process contextual information as readily as wild-type mice (63). The NR2A mutant mice can learn to associate the context with the shock, but they require longer exposures to it. Thus, just as more intensive LTP protocols revealed normal levels of potentiation, longer intervals to process the context allowed normal levels of learning in these mutants. These data suggest that the NR2A subunit's ability to modulate NMDAR currents may be involved in the modulation of thresholds for LTP and learning.

NR2B   Deletion of the NR2B subunit or its carboxy terminus results in neonatal lethality. Through intervention, the NR2B$^{-/-}$ mice can be kept alive for a few days, at which time LTD was assessed in hippocampal slices. Although in wild-type mice LTD could be induced, it was absent from the NR2B$^{-/-}$ mice (65). This subunit is normally expressed at high levels during hippocampal development, before the hippocampus is thought to be functional, and is down-regulated in the adult. Nevertheless, Tang et al. tested whether overexpression of this subunit, which should increase Ca^{++} flow through the NMDA receptor, affected plasticity and learning. Two-fold overexpression of the wild-type NR2B subunit with the αCaMKII promoter, resulted in an increase in LTP and in enhanced performance on multiple learning tests. This is an important result in evaluating the role of LTP in learning because it demonstrates a positive correlation between these two phenomena. However, not all transgenetically modified animals with increased LTP result in enhanced learning due to deleterious effects of the mutation on other cellular processes.

NMDAR INTERACTING PROTEINS   A number of proteins are known to interact with NMDA receptor subunits and therefore are also likely to be involved in LTP, place cells, and memory formation. One such protein is PSD-95 (PostSynaptic Density 95), which binds NR2A and NR2B via PDZ domains. A dominant-negative mutation of PSD-95, which leaves the first two PDZ domains intact, results in enhanced paired pulse facilitation (PPF) and dramatically enhanced LTP. Inappropriate potentiation of synapses during learning may obscure the pertinent information, resulting in the apparent learning and memory deficits in these knockout mice (80).

Receptor tyrosine kinases (RTKs) are a large family of membrane-associated proteins that undergo autophosphorylation in response to activation by their respective ligands. Ephrin receptors comprise the largest RTK family. EphA receptors are membrane associated via GPI anchors, whereas EphB receptors have transmembrane domains; both contain PDZ protein interaction domains. Important during development as guidance cues, Ephs also function in the adult brain in learning and memory. EphA5 is expressed in the hippocampus, cortex, and olfactory bulb.

Sequestration of the EphA5 ligand, ephrin-A5, resulted in decreased LTP and impaired contextual fear conditioning (41, 43). Alternately, activation of this receptor enhanced paired pulse facilitation, LTP, and contextual fear conditioning (41, 43). The mechanism for these effects is unknown, but it may involve cytoskeletal regulation.

EphB2, another ephrin receptor, is also expressed in the hippocampus, amygdala, and neocortex. It physically interacts via its extracellular domain with the NR1, NR2A, and NR2B subunits of the NMDA receptor (27). In fact, activation of EphB2 results in clustering of NMDA receptors with other synaptic proteins, including αCaMKII (27). EphB2 activation enhances Ca^{++} influx through the NMDAR, and is dependent on the phosphorylation of specific tyrosines in NR2B (108). Furthermore, EphB2 activation results in phosphorylation of ERK and CREB, two other proteins involved in LTP and memory (see below). In two different transgenic lines, EphB2 deletion caused deficits in the late phases of LTP, LTD, and in depotentiation (47, 50). A corresponding deficit in the spatial version of the Morris water maze was also observed. Interestingly, these effects were not dependent on the kinase activity of EphB2 since mice carrying a kinase-deleted form of EphB performed similarly to wild-type mice (47). Instead, the lower levels of NR1 in the postsynaptic densities of EphB2$^{-/-}$ mutants may account for their LTP and learning deficits (50).

Another RTK, TrkB, is activated by the growth factors NT-4/5 and BDNF (Brain-Derived Neurotrophic Factor). TrkB$^{-/-}$ mice have increased apoptosis in the hippocampus and cortex. Regional postnatal deletion, using a floxed TrkB and Cre recombinase expressed from the αCaMKII promoter, resulted in no gross abnormalities, including normal levels of apoptosis. However, these mice displayed decreased LTP as well as impaired memory in a variety of hippocampal-dependent tasks (81).

TrkB requires processing by Presenilin 1 (PS1). This protease is involved in a multitude of functions, including intracellular trafficking, intracellular Ca^{++} handling, processing APP and Notch, and interaction with β-catenin in mediating apoptosis. In addition, PS1 mutations are commonly found in early onset familial Alzheimer's disease. In the PS1$^{+/-}$ heterozygous mouse, hippocampal LTP induction was normal but its maintenance was impaired (84). In two mouse lines with PS1 deleted specifically in regions of the forebrain, γ-secretase activity was impaired in the cortex and hippocampus, and no increase in apoptosis was observed. Deletion of PS1 in these two lines also did not appear to affect hippocampal LTP. However, one of the mouse lines exhibited mild spatial water maze deficits (123). The other mouse displayed enhanced contextual fear conditioning but only in circumstances in which neurogenesis was altered, suggesting a role for PS1 in coordinating these phenomena (37).

αCaMKII also interacts with NMDARs, and has a profound effect on LTP, place cells, and learning and memory. Furthermore, αCaMKII, the protein kinase A (PKA) pathway, and the Ras/MAPK pathway can all be activated by Ca^{++} influx through the NMDA channel (see Figures 2–5). These pathways culminate in new

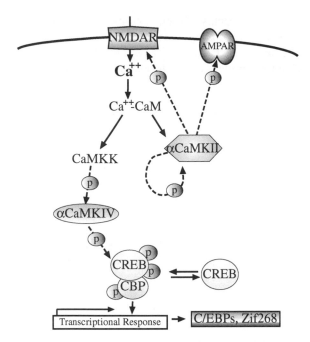

Figure 2   Calcium plays a key role in synaptic plasticity and memory formation. Ca^{++} influx via the NMDA receptor binds calmodulin, activating Ca^{++}/calmodulin-dependent kinases. The role of the boxed proteins in LTP and in learning and memory is described.

gene transcription, frequently mediated by the CREB transcription factor. The role these pathways play in the processes underlying learning and memory is discussed below.

αCaMKII: Triggering the Molecular Processes of LTP and Learning

Once NMDA receptors are activated (e.g., during the induction of LTP and during learning), calcium enters dendritic spines and activates a number of kinases, including αCaMKII. αCaMKII is a Ca^{++}/calmodulin-dependent kinase that is heavily expressed in postnatal forebrain structures such as the hippocampus and cortex. Autophosphorylation of αCaMKII at threonine-286 (T286) converts the enzyme into an active state that is Ca^{++}/calmodulin independent, and stabilizes its association with proteins in the postsynaptic density (PSD), such as the NMDAR. Subsequent autophosphorylation at TT305/306 blocks activation by calcium/calmodulin and drives the kinase from the PSD. This kinase is thought to have a key role in synaptic transmission because its phosphorylation of AMPA receptors is thought to affect their conductance and/or stability [reviewed in (67)].

Evidence for an interaction between NMDARs and αCaMKII during LTP and learning was obtained using a novel method with considerable potential. A heterozygous αCaMKII null mutation was combined with a subthreshold dose of an NMDAR antagonist. The heterozygous αCaMKII null mouse does not exhibit deficits in acquisition of contextual fear conditioning. Similarly, subthreshold doses of CPP or MK801, two different NMDA receptor antagonists, do not affect contextual fear conditioning in wild-type mice. When the subthreshold doses of drugs are combined with the heterozygous mutation, however, a deficit in contextual fear conditioning is revealed. These data suggest that NMDAR-dependent activation of αCaMKII is required for learning and memory (87). A similar experiment also suggested that NMDAR-dependent activation of αCaMKII is required for LTP (87). These studies are applicable to other heterozygous recessive mutations, and provide a novel strategy to test gene function in a temporal and tissue-specific manner (regulated by where and when the drug is administered).

The NMDA receptor and αCaMKII were the first two molecules implicated in both LTP and spatial learning and memory. The first study using knockout mice to investigate the connection between LTP and learning examined the effect of deleting αCaMKII. In these mice, NMDA currents were unaffected, as would be predicted if the action of αCaMKII were downstream of calcium influx through this receptor. Even LTP induced by high-frequency stimulation (3×100 Hz) was severely decreased in these mice (103). The mutant mice also showed profound spatial learning deficits: they were unable to learn to find the hidden platform although they were able to master a version of the water maze (visible platform) that is insensitive to hippocampal lesions.

Similarly, deregulation of αCaMKII function with a transgenically expressed constitutively active mutant kinase also resulted in deficient LTP (6, 75). Under conditions where the controls showed LTP, these transgenics showed LTD, suggesting that this kinase modulates how stimulation frequency affects synaptic plasticity. Place cell studies revealed profoundly unstable place cells in these mutants, suggesting that this kinase is also required for place field stability. This study was one of the first to address the genetic basis of place fields. The effects on LTP and learning were not due to changes during development since this phenotype was reproduced with the same αCaMKII transgene under the control of the tTA system. With this inducible system, it was also possible to show that deregulating αCaMKII function affects not only learning, but also memory and/or recall (75).

In contrast to homozygotes, heterozygous αCaMKII$^{+/-}$ mutants can acquire spatial and contextual information. Accordingly, they also show normal hippocampal LTP tested with a variety of protocols (39). LTP in various areas of the neocortex, however, was profoundly abnormal in these mutants. The lower levels of this kinase in the neocortex, compared to the hippocampus, may explain why the neocortex is more sensitive than the hippocampus to the loss of half of the levels of the kinase in the heterozygous mutant mice. Lesion studies have suggested that the hippocampus is initially involved in memory, but that the neocortex is the final site for memory storage. Since these mice had normal hippocampal but abnormal neocortical LTP, it was predicted that memory would be stable as long as it was supported by hippocampal circuits, but that memory would weaken as it became more dependent on neocortical sites. Consistent with this idea, the heterozygous αCaMKII$^{+/-}$ mice showed normal spatial memory when tested 3 days after training, but revealed profound amnesia when tested at later time points (39). Imaging studies with 2-deoxyglucose had shown predominant hippocampal activation during spatial learning, but neocortical activation during spatial recall at approximately the same time point that the heterozygous αCaMKII$^{+/-}$ mice showed spatial memory deficits (17). These studies suggested that αCaMKII-dependent LTP is required for memory storage in the neocortex. Other studies had already involved this molecule in neocortical plasticity [reviewed in (67)].

Calcium influx leads to autophosphorylation of αCaMKII at T286, converting it to a Ca^{++}/calmodulin-independent and active kinase. Induction of LTP results in the activation and concomitant autophosphorylation at this site (11). The importance of this site has been tested using knock-in mice with a T286A mutation that prevents phosphorylation. The T286A mice showed unaltered presynaptic plasticity, NMDA currents, GABA$_A$ inhibition, and synaptic transmission. However, this mutation resulted in a clear disruption of CA1 LTP tested with a number of different induction protocols. This LTP deficit, which very likely affects other hippocampal and neocortical regions, also disrupted the stability of hippocampal place cells. The mutants showed place fields, but they tended to change their mapping locations every time the animal was removed from the experimental room (23). Not surprisingly, with unstable LTP and place maps, these animals revealed profound spatial learning deficits (44). These findings demonstrate the critical role that this kinase plays on LTP, place cell stability and learning.

GLUR1, AN αCaMKII SUBSTRATE   A known substrate of αCaMKII is the GluR1 subunit of the AMPA receptor. Hippocampal pyramidal cells express mainly the GluR1 and GluR2 subunits. The GluR1 subunit is phosphorylated by αCaMKII and PKA after LTP induction (7, 11, 49), which is thought to result in a concomitant increase in AMPA receptor function. Deletion of the GluR1 subunit demonstrated that this subunit is required for LTP induced by high-frequency tetanization in the hippocampus (124). This deficit in LTP could be rescued by exogenous expression (<20% of endogenous GluR1) of GluR1 both during development and postnatally (69). Further studies of LTP in these mice revealed that a more physiologically relevant stimulus (theta burst stimulation) induced nearly normal LTP (52). This physiologically relevant LTP probably accounts for the apparent lack of spatial learning deficits in this mutant mouse (124). This study is a good example of the complexities of testing connections between physiology and behavior.

The PKA Pathway

An influx of Ca^{++} through the NMDA receptor can also activate the Protein Kinase A (PKA) signaling pathway. When high cAMP levels are produced (e.g.,

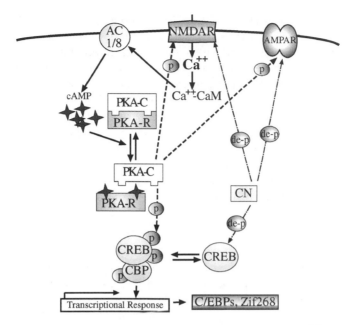

Figure 3 The PKA pathway is required for LTP and memory. Ca^{++} influx via the NMDA receptor activates Ca^{++} sensitive adenylyl cyclases 1 and 8, PKA and calcineurin (CN). The role of the boxed proteins in LTP and in learning and memory is described.

through Ca^{++} sensitive adenlyl cyclases), cAMP binds to the regulatory subunit of PKA. This binding alters the conformation of the regulatory subunit, thereby releasing the catalytic subunit. PKA can then phosphorylate numerous substrates, including the NMDA receptor, Inhibitor-1 (I-1) and the cAMP Response Element Binding protein (CREB) (see Figure 3). Opposing the actions of PKA are the protein phosphatases PP1A and calcineurin (PP2B), since they dephosphorylate PKA substrates. Calcineurin (CN), a calcium-sensitive serine/threonine phosphatase, is highly expressed in the hippocampus and is enriched at synapses. The CN regulatory subunit (B) is required for the phosphatase activity of the catalytic subunit (Aα or Aβ) that contains an autoinhibitory domain. In brain, there is only one regulatory subunit, B1. Among its substrates are the NMDA receptor, I-1, and CREB, some of the same proteins that are targets of PKA phosphorylation.

Adenylyl Cyclases (AC) catalyze the production of cAMP, a classic second-messenger signaling molecule that activates PKA. Ten different ACs have been identified and most are expressed in all cell types, including nine different ones in the hippocampus. Of these nine, AC1 and AC8 are stimulated by Ca^{++}/CaM. Deletion of either AC1 or AC8 does not alter LTP; however, the double mutant, which retains essentially no Ca^{++}/CaM-stimulated activity, exhibits profound deficits in the late phase of LTP (L-LTP) (122). Forskolin, a chemical activator of all ACs, rescues the L-LTP of these mutants, thereby suggesting that downstream components of L-LTP are intact in these mutants. In agreement with the hypothesis that L-LTP is required for long-term memory formation, the memory deficit of the ΔAC1/ΔAC8 double mutant mice is more profound when tested eight days after training. Finally, just as forskolin rescued the L-LTP deficits, it also reversed the memory impairments of these mice, demonstrating that both deficits were due to a decrease in cAMP production. The infusion of forskolin also increased expression of a CRE-lacZ reporter, showing that the increase in cAMP via ACs probably involves CREB activation (see below). These data show that Ca^{++} activation of adenylyl cyclases, and therefore cAMP signaling, is required for L-LTP and LTM formation, a finding first suggested in pioneering studies in flies and in the sea slug Aplysia (100).

DECREASING PHOSPHORYLATION OF PKA/CALCINEURIN SUBSTRATES   The opposing effects of PKA and calcineurin on learning and memory have been studied in multiple transgenic mice using overexpression of a dominant negative mutant of the regulatory domain of PKA, R(AB), or overexpression of the calcineurin Aα subunit lacking its autoinhibitory domain, ΔCaM-AI. Both of these manipulations were expected to result in an overall decrease in the phosphorylation of PKA substrates.

Increased activity of calcineurin in the ΔCaM-AI mice affected LTP and memory. Pharmacological activation of PKA rescued these LTP deficits. This misregulation of phosphorylation also affected spatial memory tested in two separate hippocampal-dependent tasks. The mice displayed intact short-term memory but severely impaired long-term memory (24 h after training) (72).

Evidence for a role of PKA on LTP was also obtained from studies of mice over-expressing R(AB), an inhibitory form of the regulatory subunit of PKA. These transgenics show a decrease in both the basal and cAMP-stimulated activity of PKA with a concomitant decrease in L-LTP 1 to 3 h after high-frequency stimulation. Early stages of LTP, however, are not affected by this transgene. Similarly, although hippocampal-dependent memory was disrupted when tested 24 h after training, short-term memory was not (1). These results provide a compelling parallel between the time course of LTP and learning deficits.

Besides a requirement for PKA activity during training, pharmacological studies with Rp-cAMP, a PKA inhibitor, showed that PKA activity is also needed for memory 4 h after training. Blocking PKA activity at other times does not seem to have an effect on contextual memory. The requirement for PKA at 4 h was only observed in mice that received weak training. Strong training seemed to circumvent this requirement. Similar results were also obtained with protein synthesis inhibitors. Weak training revealed that protein synthesis is also required during training and 4 h later, but not at other times (18).

Electrophysiological studies showed that PKA is required for an intermediate phase of LTP that is insensitive to protein synthesis inhibitors. Mild synaptic

Figure 1 Behavioral tasks used to test learning and memory. (*A*) The hidden version of the Morris water maze is used to test hippocampal-dependent spatial learning. Picture courtesy of Ype Elgersma. (*B*) Contextual fear conditioning is used to test associative learning in which a footshock is associated with the context in which the footshock occurred. Close-up picture courtesy of Paul Frankland, train and test images reproduced with permission from Anagnostaras et al. 2000 Learning and Memory vol 7:58.

stimulation in wild-type mice results in an intermediate level of LTP that is insensitive to protein synthesis inhibitors, but sensitive to PKA inhibitors. In the ΔCaM-AI mutant, this intermediate level of synaptic stimulation elicited decreased LTP compared to wild-type mice and was insensitive to PKA inhibitors (121). These data show a dissociation between the early phase of LTP (E-LTP), which is independent of protein synthesis and PKA/calcineurin regulation, intermediate LTP (I-LTP), which is dependent on PKA/calcineurin regulation, and L-LTP, which is dependent on both.

The I-LTP deficits in the R(AB) also resulted in unstable place fields (90), confirming the importance of stable synaptic changes for the stability of representations in hippocampal circuits. Note that like the LTP deficits in the αCaMKII mutants, the LTP deficit in the R(AB) almost certainly affects brain regions other than CA1. Altogether, the experiments just reviewed defined a time-dependent relationship between long-term memory (LTM) formation and PKA activity, and suggest a systems-level mechanism (place cell stability) upon which this relationship may rely. These elegant data further define an intermediate memory phase with behavioral and electrophysiological correlates.

Using the reverse tetracycline regulated system, rtTA, in which doxycycline turns on gene expression, thereby obviating concerns regarding developmental effects of doxycycline, Mansuy et al. confirmed that ΔCaM-AI overexpression does indeed affect I-LTP and spatial memory in the Morris water maze (72). Taking full advantage of the rtTA system, the role of calcineurin in acquisition, retrieval, and consolidation was studied. Transgene expression during training in the water maze prevented acquisition of spatial information. Further testing showed that it also seemed to interfere with retrieval of stored information: expression of the transgene specifically during testing disrupted performance. This deficit seemed to be due to a retrieval failure, because a later test done with the transgene repressed revealed normal memory, suggesting that the expression of the transgene does not disrupt memory. This study provides both the first molecular dissociation between memory storage and memory retrieval, and the first mechanistic insight into the processes underlying memory retrieval.

INCREASING PHOSPHORYLATION OF PKA/CALCINEURIN SUBSTRATES   The data discussed above support the idea that calcineurin and PKA regulate key substrates for LTP and learning and memory, using manipulations that decrease overall phosphorylation. Alternately, phosphorylation can generally be increased by inhibition of calcineurin via overexpression of the inhibitory domain of Aα, deletion of Aα or deletion of the brain-specific regulatory domain, B1.

The Aα subunit is predominantly expressed in the hippocampus, cerebral cortex, cerebellum, and striatum. The rtTA system was used to overexpress the carboxy terminus of the Aα subunit containing the autoinhibitory domain (AI). Biochemical studies showed that this resulted in decreased phosphatase activity in the hippocampus and cortex. These mutants displayed an increase in E-LTP that was blocked by PKA inhibition (71). Thus, overexpression of the catalytic domain in ΔCaM-AI mice (decreasing overall phosphorylation) impaired LTP, whereas overexpression of the inhibitory domain in AI mice (increasing overall phosphorylation) enhanced LTP. In the AI mice, enhanced LTP was also observed in vivo in anaesthetized and awake mice.

Consistent with enhanced LTP, the AI mutant mice also showed enhanced memory in hippocampal-dependent tasks. For example, in the Morris water maze, the mutant mice searched selectively for the hidden platform after only five days of training, whereas wild-type mice required ten days of training to attain the same level of performance (71). These data suggest that calcineurin normally acts as a negative regulator of LTP and memory. This may serve as a filter so that only strong and important memories are stored. There are several psychological studies that suggest that the inability to filter information is as debilitating as the inability to store it.

The overexpression of the auto-inhibitory domain of Aα reduced phosphatase activity, but did not eliminate it. To completely eliminate calcineurin activity, Zeng et al. deleted the B1 subunit (CNB1$^{-/-}$ mice), the only regulatory subunit in the brain. In the CNB1$^{-/-}$ mice, LTD was diminished, whereas LTP induced by one train of 100 Hz was intact. Depotentiation was also normal in the mutant mice. This is in contrast to the CNA$\alpha^{-/-}$ mice that had normal LTD and LTP but a diminished ability to depotentiate LTP (125). Since phosphatases and kinases affect the thresholds of LTP and LTD induction, Zeng et al. investigated these thresholds in the CNB1$^{-/-}$ mice. Consistent with the idea that increased kinase activity favors LTP, while increased phosphatase activity favors LTD, the CNB1$^{-/-}$ mutation lowered the threshold for LTP while increasing the threshold for LTD. These mice showed normal performance in cued and contextual conditioning as well as in spatial learning tested in the Morris water maze. However, CNB1$^{-/-}$ showed severe deficits in both a working memory version of the water maze and the radial arm maze, suggesting that CNB1 is required for working memory in which fast, one-trial learning is required. These one-trial-learning tasks require that the animal disregard previous trials and focus its attention on information relevant to the current trial. The inability to reverse or erase information pertaining to prior trials could conceivably interfere with performance in future trials. Thus, the deficits in LTD in the CNB1$^{-/-}$ mice may account for their disrupted performance in the working memory tasks used (125). This is a good illustration of the complex relation between LTP/LTD and learning and memory. To successfully interpret studies testing the connection between LTP and learning, it is critical to consider the details of the experiments performed, including the possible presence of other electrophysiological changes and the specific requirements of the tasks used. Note also that changes in synaptic function may represent only one class of physiological changes involved in learning and memory, and that interaction with other relevant mechanisms may further complicate the interpretation of experiments testing the role of LTP in learning.

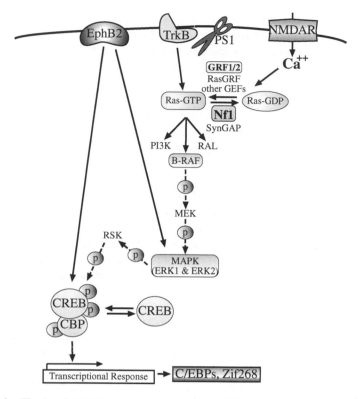

Figure 4   The Ras/MAPK pathway plays a role in LTP and memory. Ca^{++} influx via the NMDA receptor activates the Ras pathway. The receptor tyrosine kinase TrkB and the Ephrin receptor also activate the Ras pathway. The role of the boxed proteins in LTP and in learning and memory is described.

The Ras/Raf/MAPK Pathway is Involved in Learning and Memory

The Ras/Raf/MAPK signaling pathway is another pathway that figures prominently in LTP and learning and memory. Ras is the head of the superfamily of small GTPases that bind GTP. Ras signaling is stimulated by tyrosine kinase receptors, G-protein–coupled receptors, NMDA receptors, and some voltage-dependent Ca^{++} channels. The activity of Ras is further regulated by GTPase Activating Proteins (GAP) and Guanine Exchange Factors (GEF or GRF) (see Figure 4). Although Ras is primarily thought of as an oncogene with a key role in cellular proliferation, it also plays a role in key biological processes in post-mitotic cells, such learning and memory. Pioneering pharmacological work from Sweatt and colleagues implicated MEK/MAPK signaling in LTP and learning [reviewed in (106)].

Nf1, a RasGAP, is predominantly expressed in adult neurons. Mutations in this gene lead to neurofibromatosis I, the commonest single-gene genetic disorder resulting in learning disabilities (see below). A homozygous deletion of this gene is lethal whereas heterozygotes are viable. These $Nf1^{+/-}$ mice display learning deficits in the water maze; they perform poorly but with additional training are able to reach the same performance levels as their littermate controls (102). The Nf1 gene has multiple splice forms and this diversity has been explored in learning and memory studies. For example, an isoform containing exon23a (Nf1 type II) has increased affinity for Ras, but decreased GAP activity, thereby competing with the type I isoform and thus decreasing the conversion of Ras into its GDP-bound state. Analysis of the Nf1 $23a^{-/-}$ mice showed normal life expectancy, normal levels of Nf1 type I, seemingly normal development and no hints of increased tumor predisposition similar to the $Nf1^{+/-}$ heterozygotes. They do display, however, clear hippocampal-dependent learning deficits. These mice showed impairments in both the water maze and contextual discrimination, which can be partially rescued with extended training (24). These findings show that the learning deficits associated with Nf1 mutations are not due to tumors or to deficits in development.

Neurofibromin, the protein encoded by the Nf1 gene, is a large protein with multiple biochemical functions, including GAP activity, adenylyl cyclase modulation, and microtubule binding. However, rescue experiments based on manipulations that decrease Ras function showed that the learning deficits of the $Nf1^{+/-}$ mice (and by implication NF1 patients) are due to increased Ras signaling. Deleting N-ras, for example, rescued the spatial learning deficits of the $Nf1^{+/-}$ mutant mice. A K-ras heterozygous null deletion, which on its own causes spatial learning deficits, was also able to rescue the learning deficits of the $Nf1^{+/-}$ mutant mice (25). Furthermore, a farnesyl transferase inhibitor, which affects the farnesylation and hence the activity of Ras, was also able to rescue the learning deficits of the $Nf1^{+/-}$ mutant mice, suggesting that these deficits are not due to irreversible developmental problems. Importantly, the K-ras$^{+/-}$ mutation, which rescued the spatial learning impairments of the $Nf1^{+/-}$ mutants, also rescued their LTP impairments, a result that suggests a causal connection between the LTP deficits of the $Nf1^{+/-}$ mutants and their learning impairments. Further electrophysiological studies indicated that the LTP deficits of the $Nf1^{+/-}$ mutants are caused by increased GABA-mediated inhibition, and that the K-Ras$^{+/-}$ mutation could also rescue these deficits. These data suggest that Ras/Nf1 signaling modulates inhibition and that inhibition may have a role in regulating LTP during learning.

While RasGAPs inactivate Ras-GTP by catalyzing the hydrolysis of GTP to GDP, GRFs (Guanine nucleotide Releasing Factors) stimulate the release of GDP thereby promoting the active, GTP-bound state of Ras. Ras GRF1 and GRF2 both couple Ras activation to Ca^{2+}/CaM, G protein, cAMP, and receptor tyrosine kinase signaling. Ras-GRF1 is expressed almost exclusively in brain (97). It is found primarily in the PSD, suggesting a role in plasticity and learning. Two separate deletions of this gene were generated, one disrupting the nucleotide exchange activity domain (ex) and the other, a more N-terminal insertion of neo, disrupting the Dbl domain (dbl) that activates Ras (20, 45). Both, however, show a complete loss of GRF1 protein in the mutant mice. Although both manipulations affected learning and memory, they differ in important ways. The GRF1ex$^{-/-}$ mutant strain had normal hippocampal, but impaired amygdala-dependent memory (20, 45), whereas the dbl-mutant showed normal amygdala, but impaired hippocampal-dependent learning (20, 45). It is difficult to reconcile these data because the mice were studied in seemingly identical genetic backgrounds and with very similar procedures. However, the GRF1 locus is imprinted and differential disruption of imprinting by these two different mutations could account for the disparities in their behavioral phenotypes. Nevertheless, these discrepancies could be due to a number of other factors, such as the differential disruption of nearby genes by the neomycin selectable marker inserted into this locus during gene targeting. These studies underscore the importance of studying multiple alleles of the same mutation. Allelic series of point mutations, complete deletions, and overexpression of wild-type or mutant alleles will give a richer understanding of the role of GRF1 in learning and memory. Although the history of biology has clearly demonstrated the enormous power of genetics, the most compelling evidence usually comes from convergent data obtained with different approaches (e.g., KO, transgenics, viral vectors, pharmacology). Taken together, these studies show that regulation of Ras activity through perturbation of either its GAP or GRF affect learning and memory. Why one GRF allele affects learning and the other memory remains unclear, but further genetic and genomic analysis should shed light on this interesting difference.

Members of the MAPK pathway, ERK1 and ERK2, are highly expressed in postmitotic neurons. Their activation via phosphorylation is required for theta-frequency induced LTP and for hippocampal-dependent learning. For example, phosphorylated ERK is seen after electrical corticostriatal stimulation, high-frequency stimulation in living rats, after amygdala-dependent CTA training and after hippocampal-dependent contextual fear conditioning and water maze training. Furthermore, these effects are blocked by ERK kinase (MEK) inhibitors (5, 13, 31, 36, 95). Increased ERK1/2 phosphorylation and nuclear translocation were observed after learning in the water maze, but not after a single training trial that results in no obvious learning (16). This phosphorylation was observed only in the CA1 and CA3 regions of the dorsal hippocampus. As inhibition of ERK phosphorylation after training blocked recall, it was suggested that MAPK activation in the CA1/3 regions is required for memory consolidation.

The role of ERK1 in learning and memory was further analyzed in two different ERK1$^{-/-}$ mice (77, 94). Behavioral analyses in these mice gave somewhat contradictory results. One of the ERK1$^{-/-}$ mouse strains showed no alterations in ERK2 protein levels or phosphorylation, LTP, general activity, motor skills, or learning and memory as assessed by fear conditioning and passive avoidance tasks (94). These results were later challenged in a second ERK1$^{-/-}$ mouse line in which alterations in the phosphorylation levels of ERK2 were enhanced, and LTP in the nucleus accumbens, but not the hippocampus or amygdala, was enhanced (77).

Similarly, these mice showed enhanced learning or memory observed in active and passive avoidance tests.

Dopaminergic stimulation resulted in enhanced ERK2 phosphorylation in these ERK1$^{-/-}$ mutant mice (77). Consistent with the idea that this mutation led to a potentiation of the reward system, the mice displayed enhanced place preference for a context associated with a drug stimulus. These results support the hypothesis that MAPK activity, and specifically ERK1, is required for synaptic plasticity and learning and memory, as well as in drug reward behavior. Unfortunately, ERK2 is an essential gene (2) so further studies of its role in LTP and learning and memory have to wait for the generation of conditional alleles.

These studies have shown that Ras and its modulators, the GAP (Nf1) and the GEFs (GRF1 and GRF2), as well as downstream signaling molecules (ERK1 and potentially ERK2) have functional roles in both synaptic plasticity and learning and memory. How this signaling pathway interacts with either NMDAR/CaM Kinase or PKA/calcineurin signaling may lead to a better understanding of the molecular events underlying memory formation. Another important question concerns the behavioral relevance of the transcriptional events known to be downstream to these signaling pathways.

Transcriptional Regulation and Memory: CREB and Long-Term Memory

The signaling pathways required for learning are thought to ultimately drive the transcription of genes required for memory consolidation, a process that filters and stabilizes information in the brain. Early work using pharmacological inhibitors showed the requirement for protein synthesis in long-term memory. In contrast, protein synthesis inhibitors did not affect short-term memory (30). One of the transcription factors activated during learning and required for memory consolidation is the transcription factor cAMP Response Element Binding protein (CREB). In response to elevated levels of cAMP and Ca^{++}, this transcription factor is phosphorylated at S133 (as well as S142 and S143), and then directs transcription of genes containing cAMP Response Elements (CRE) in their promoters. CREB is phosphorylated by multiple kinases including PKA, RSK-2, CaMKII, and CaMKIV (see Figure 5). Note that CREB has roles in multiple cellular processes from metabolism to apoptosis. Thus, CRE sites are found in the promoters of many genes (76), not just learning and memory genes. The CREB gene encodes several isoforms, derived from alternate splicing, that act as either activators or inhibitors. The CREB family of transcription factors forms functional heterodimers. Previous studies in numerous species and brain systems have demonstrated that CREB has a clear role in the formation of long-term memory (LTM): behavioral and electrophysiological protocols that induce LTP and LTM also activate CREB, lesions of CREB function block long-term plasticity and long-term memory, and overexpression of CREB can facilitate plasticity and learning [reviewed in (99)]. Thus, CREB plays a pivotal role in the transition between short- and long-term memory.

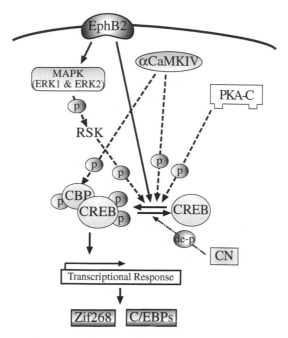

Figure 5   The CREB transcription factor is required for L-LTP and long-term memory. The Ca^{++}/calmodulin-dependent kinases, the PKA pathway and the Ras pathway in part regulate the phosphorylation state of CREB. Phosphorylated CREB is required for long-term memory formation.

Early studies showed that injection of CRE-containing oligomers in cultured Aplysia neurons specifically blocked long-term facilitation, a cellular model of synaptic changes thought to underlie behavioral sensitization (28). Later studies in Drosophila and mice suggested that CREB was universally required for memory. The deletion of the CREB α and δ isoforms (CREB$^{\alpha\Delta-}$) resulted in deficits in the stability of hippocampal LTP, place cells, and place memory (18, 23). Follow-up studies showed that CREB affected long-term memory for a large number of other tasks, irrespective of the brain system involved. In contrast, short-term memory was unaffected in these mice. The loss of the CREB α and δ isoforms led to up-regulation of the β isoform, as well as to changes in the expression of specific CREM isoforms, which may explain why the severity of the phenotype of the CREB$^{\alpha\Delta-}$ mutants appears to be so sensitive to genetic background (42, 46). Variations in the compensatory expression of other CREB-family genes may account for the variable penetrance of the CREB$^{\alpha\Delta-}$ phenotype in different genetic backgrounds. Importantly, transgenic expression of a dominant-negative form of CREB that heterodimerizes with CREB, CREM, and ATF1 but inhibits DNA binding, also impaired hippocampal-dependent long-term memory. This transgenic manipulation also attenuated forms of LTP that were thought to be sensitive to CREB manipulations, such as forskolin-induced and dopamine-regulated LTP (89).

The analysis of the CREB$^{\alpha\Delta-}$ mice also revealed a role for this transcription factor in regulating the training required for LTM formation. Pioneering studies by Ebbinghaus in the nineteenth century established that the schedule of training affects long-term memory formation: Longer intervals between training sessions (spaced training) are more effective than shorter intervals (massed training) (33). Strikingly, the long-term memory deficits of the CREB$^{\alpha\Delta-}$ mice (i.e., fear conditioning and spatial learning) could be rescued by spaced training (58). Accordingly, increases in CREB levels with viral delivery systems eased training requirements for the formation of long-term memory (58). With increased CREB levels, animals were able to show LTM formation under training conditions (massed training) that are ineffective in controls, suggesting that CREB regulates how training schedules affect memory formation. Similarly, overexpressing CREB in mice enhances LTP, suggesting that the level of this transcription factor correlates with the ease of induction of synaptic changes underlying memory formation (10). Similar results were also obtained with Drosophila and Aplysia [reviewed in (99)]. These results suggest that complex behavioral phenomena (i.e., the effect of training schedules on memory) can be directly modulated by single molecular mechanisms (i.e., CREB activity), a finding that supports the idea that integration of some aspects of behavioral experience can occur at the molecular level.

The studies with the CREB$^{\alpha\Delta-}$ mice indicated a role for CREB in LTM. However, it was difficult to pinpoint exactly when CREB function was required for memory formation. To circumvent this limitation, a new CREB mutant was derived with the inducible LBD system (88). A mutant ligand-binding domain of the estrogen receptor (LBD), which binds tamoxifen instead of estrogen (68), was fused to a dominant-negative CREB gene (S133A substitution). This fusion gene (LBD-CREB) was placed under the regulation of the αCaMKII promoter, which restricted its expression to postnatal excitatory neurons. With this system, the fusion protein is inactive until tamoxifen binds the receptor. Injections of tamoxifen activated the dominant negative CREB-fusion protein and repressed the transcription of 14-3-3η, a CREB-dependent gene whose transcription is activated during learning (62). With this inducible system, it was possible to show that CREB is crucial for the consolidation of long-term memory, but not for the storage or retrieval of memory.

The studies with the LBD-CREB mice also showed that CREB is required for the stability of memory after retrieval (62). Retrieved memories can be fine-tuned and changed, and there is extensive evidence that some of the same mechanisms involved in memory formation may also be required for the stabilization of retrieved memories, a process referred to as reconsolidation (92). For example, blocking protein synthesis during re-exposure to the conditioning context can attenuate fear conditioning (85). These studies showed that memory not only depends upon CREB function during training, but also requires it during retrieval. Interestingly, C/EBPβ, another transcription factor required for memory consolidation is not required for reconsolidation after retrieval (110).

In addition to experiments demonstrating a requirement for CREB function in memory, there is also evidence indicating that CREB-dependent transcription is activated during learning. For example, experiments with antibodies specific to the phosphorylated version of CREB showed that behavioral training as well as memory retrieval can induce the phosphorylation (activation) of CREB (48). Elegant transgenic experiments with a transgenic β-galactosidase gene under the regulation of a CREB-dependent promoter showed that both LTP and learning can activate CREB-dependent transcription (55).

CREB AND CaMKIV   CaMKIV counts among its substrates many transcription factors including CREB and its cofactor, CBP, as well as proteins that interact with actin and microtubules. This kinase is found in adult cortical, cerebellar, hippocampus and amygdala neurons and is both cytoplasmic and nuclear. Because of its nuclear localization, this kinase is hypothesized to phosphorylate and activate CREB. Furthermore, CREB-dependent transcription requires phosphorylation of CBP at S301 by CaMKIV and this phosphorylation is dependent on NMDA activation (54).

Sustained low-frequency stimulation of hippocampal neurons results in increased expression of immediate early genes (IEGs) and a long-lasting increase in phosphorylated CREB. Phosphorylation of CREB in this system is specifically blocked by decreasing α/β CaMKIV levels (14). In addition to its role in CREB phosphorylation, CaMKIV is required for hippocampal, amygdala, and cortical L-LTP (51, 120). Deletion of CaMKIV α and β isoforms, as well as overexpression of a dominant negative form of CaMKIV, both result in decreased L-LTP (51, 59). The CaMKIV$^{-/-}$ mouse displays deficits in both context and cued fear conditioning, whereas mice overexpressing the dominant negative protein show deficits in both water maze and contextual fear conditioning (59, 120). These studies suggest that CaMKIV is required for L-LTP and for long-term spatial memory formation.

DOWNSTREAM OF CREB   Although CREB clearly plays a role in memory consolidation, the downstream effectors that mediate this role are less well understood. C/EBP and Zif268 are two transcription factors with a role in learning and memory and whose transcription is regulated by CREB.

The C/EBP bZIP transcription factor family consists of 5 isoforms that bind to CCAAT promoter elements. C/EBPs can heterodimerize with C/ATF (CREB family member) and bind specific CRE sites, suggesting that it could competitively inhibit CREB function. Furthermore, C/EBP itself is regulated at least in part by CRE sites in its promoter. C/EBPs in Aplysia are regulated by cAMP and are required for long-term facilitation (3). Consistent with the idea that C/EBP isoforms play a role in memory consolidation, β and δ mRNA levels are upregulated after inhibitory avoidance training in rats (111). Deletion of the α isoform results in perinatal lethality whereas the health of mice with a deletion of the β isoform declines in the weeks following birth. Deletion of the δ isoform, however, produces viable, healthy animals with no gross histological neuronal abnormalities (105). Consistent with a possible CREB inhibitory role, C/EBP$\delta^{-/-}$ mice display a specific enhancement in context conditioning 24 h, but not 30 min after training. Similarly, the mutant mice seemed to acquire spatial information faster than controls. Thus, these data suggest that C/EBPδ has an inhibitory role in LTM formation. Perhaps, this molecule is part of a feedback mechanism that down-regulates CREB-mediated transcription.

While CREB is required for memory consolidation and reconsolidation, the role of C/EBPs in memory consolidation does not parallel that of CREB. Disruption of C/EBPβ via injection of antisense oligomers into the hippocampus blocks memory consolidation. However, the same injection into the hippocampus does not affect memory reconsolidation (110). These data suggest that reconsolidation is independent of hippocampal expression of C/EBPβ. They also suggested that the role of CREB in reconsolidation is not mediated by C/EBPβ.

A second protein up-regulated after LTP is Zif268, a zinc finger transcription factor. In monkeys, Zif268 is up-regulated after associative learning. In rats, novel stimuli activate the transcription of this gene in the hippocampus. Mice with a deletion of Zif268 show no obvious histological differences, but exhibit deficits in L-LTP and hippocampal-dependent memory; short-term memory is normal in these mice. In awake, freely moving mice, stimulation of the perforant path induced normal LTP in the dentate gyrus for at least one hour post-stimulation. However, LTP in these mutants was dramatically reduced one day and three days later (56). Performance in a number of behavioral tasks confirmed that these mice exhibited normal short-term memory but decreased long-term memory, consistent with the idea that Zif268 is acting downstream of CREB during memory formation. As with the CREB$^{\alpha\Delta-}$ mice, spaced training was able to rescue the memory deficits of the Zif268$^{-/-}$ mice. Taken together, these data support the hypothesis that Zif268 acts as part of a cascade of transcription factors required for protein-synthesis-dependent L-LTP and long-term memory formation.

Both C/EBPδ and Zif268 are transcription factors that in turn regulate expression of other genes. The identity of these genes that relate to LTP and learning and memory remains unknown. Careful phenotypic characterization combined with expression profiling and promoter element analyses may yield the identity of these genes, and thus further our understanding of the molecular mechanisms underlying long-term memory.

Restructuring Synapses

So far, we have only discussed intracellular mechanisms of learning and memory. The synaptic cleft, however, is also a target for modification during short-term and long-term memory formation. Two mediators of extracellular remodeling are urokinase-plasminogen activator (uPA) and tissue type-plasminogen activator (tPA). These serine proteases convert zymogens into their active, proteolytic state (e.g., conversion of plasminogen into plasmin, which in turn activates metalloproteases). These proteases are thought to be involved in structural alterations triggered by synaptic plasticity. Both uPA and tPA are expressed in the CNS although tPA is the more abundant of the two.

These proteases also play a role in synaptic plasticity and learning and memory. Mice overexpressing tPA show enhanced PPF as well as enhanced early and late LTP. These mice show a corresponding enhancement in spatial memory tested with the Morris water maze (70). If enhanced function of tPA results in enhanced synaptic plasticity and memory, presumably increasing the malleability of the synaptic milieu, then disruption of tPA would be expected to conversely affect these two processes. Abrogation of tPA activity via genetic or pharmacological intervention blocks L-LTP (9, 40, 53). Accordingly, tPA null mice show memory deficits. However, they show a complex behavioral phenotype that includes a deficit in active avoidance but normal performance in the water maze as well as a female-specific enhancement in contextual fear conditioning (53). Barring the sex differences, regulation of uPA may provide some functional redundancy in the $tPA^{-/-}$ mice, thereby confounding interpretation of the direct role of tPA in learning and memory. Mice over-expressing uPA, however, show impairments in spatial learning in the water maze (79). Thus, these studies indicate that while uPA and tPA may share some degree of functional redundancy, they do not have identical roles in learning and memory. Other extracellular proteases, such as metalloprotease, are also induced upon neuronal activation, suggesting that multiple proteases remodel synaptic connections and thus affect synaptic plasticity and learning (107).

MOUSE MODELS OF COGNITIVE DISORDERS

An important facet of the new field of molecular and cellular cognition is transgenic mouse models of cognitive disorders. The study of these models will be key in understanding molecular and cellular mechanisms of human cognition, since they will enable us to apply ideas and findings derived from animal studies to human cognitive disorders. These studies may also bring us closer to designing rational therapies for these problems. For example, some mouse models of hemostatic diseases have effectively recapitulated hemophilic diseases and have been used to design effective clinical therapies [reviewed in (65a)].

However, a model is a simplified representation amenable to in-depth studies. Therefore, to be useful, mouse models do not have to recapitulate every feature of the human disease. Actually, simpler models often point to natural subdivisions of a more complex cluster of phenotypes associated with the human disorder. For example, mouse models of Alzheimer's disease do not show the neurodegeneration characteristic of the human disorder. Nevertheless, they show accelerated age-related cognitive decline, a finding that supports the idea that cognitive decline cannot be solely attributed to neurodegeneration [reviewed in (4)]. Some mouse models of NF1 do not have any tumors, but show clear learning deficits, confirming that the learning disabilities of the patients are probably unrelated to their tumors (24).

There are also cases where mouse models do not exhibit any signs of the disorder, suggesting that functional redundancy or alternate pathways exist. For example, mice deficient for HPRT, the gene mutated in Lesch-Nyhan syndrome, have no overt phenotype. Since purine metabolism differs in mice and men, other purine monophosphate synthesizing enzymes may play a more important role in disease formation. The HPRT-APRT double mutant, however, showed no self-injurious behavior (35). Regardless of the results, modeling disorders in mice always leads to new insights into the biology associated with the gene or pathways studied. It is possible that careful study of the HPRT mouse, for example, will lead to strategies to circumvent the problems associated with the human condition.

Despite these caveats, mouse models for human learning and memory disorders have been very successful. Insights from studies of animals models of Neurofibromatosis I, the most common hereditary form of learning disabilities, Fragile X mental retardation, the most common hereditary form of mental retardation, and Alzheimer's disease, the most common cause of age-related dementia, have already furthered our understanding of these disorders and figure prominently in our efforts to develop cures. An in-depth treatment of this topic is outside of the scope of this review.

CONCLUSIONS

Memory is a multilevel problem that spans the whole organizational complexity of biology from molecular mechanisms to the properties of interactions in complex societies. So far, the new field of molecular and cellular cognition has focused its studies on the interface between molecular, cellular, circuit, and behavioral processes. Although the majority of the studies reviewed here involved knockouts and transgenic mice, a growing number of other gene-based approaches also show considerable promise, including antisense and viral approaches. Their potential clinical value makes these approaches quite appealing. Additionally, emerging genomic approaches, such as expression studies with microarrays, promise to complement genetic studies, since it is problematic to imply function solely from studies of dysfunction (i.e., genetics experiments).

Molecular and cellular cognitive studies came of age with the advent of gene knock-out technology. This technology allowed molecular biologists, neurophysiologists, and psychologists to work side by side with the same animals. This led to an unprecedented explosion of multidisciplinary studies that tested the role of key physiological mechanisms in cognitive phenomena, such as learning and memory. Although the current review focused on hippocampal-dependent learning and memory, studies in this emerging field have encompassed many other memory systems (cerebellum, amygdala, striatum, cortex) and other cognitive functions (fear, attention). Interestingly, as transgenetic analysis of mechanisms of learning and memory developed, the field tapped into classic, tried and true genetic methods such as determining epistatic interactions. The field is also beginning the process of archiving all genes involved in learning and memory. This process is in its infancy and will involve the collection and characterization of all learning and memory genes. This will entail the complementary and systematic characterization of transgenic mice, as well as forward genetic searches for novel allelic variants derived from mutagenic screens. We estimate that more than 5000 genes have already been knocked out, representing a vastly underused resource. Cataloging the learning and memory phenotypes of these mice will undoubtedly identify novel signaling pathways and cellular processes affecting learning and memory. Mutagenic screens, on the other hand, will be used to scan for mutations with robust effects on learning and memory. The challenge will not be in identifying these genes, as this will assuredly happen. Rather, the difficulty will be to determine when, where and how these novel genes affect memory formation. This is a challenge that will require the full range of expertise of this emerging field, since it will involve extensive genetic, molecular, biochemical, electrophysiological, and behavioral studies. It is rewarding to see a growing number of young scientists trained in molecular and cellular cognition labs who have developed considerable expertise in all of these areas. These scientists and their students may be able to synthesize the complex body of molecular, cellular, circuit, and psychological data resulting from these studies.

Besides educating a new generation of interdisciplinary neuroscientists, what has been the major contribution of the first 10 years of this young field? The single most important contribution of this young field to neuroscience is the demonstration that changes in synaptic function are critical to learning and memory. This is not to say that the field accomplished this alone. Instead, the overwhelming amount of data provided in favor of this hypothesis by this young field synergized with and complemented existing electrophysiological and behavioral studies carried out in a number of other species, ranging from the sea slug Aplysia to monkeys and people. In the past ten years, the sheer volume and surprising internal consistency of the transgenetic experiments testing the idea that synaptic changes are required for learning and memory, established the validity of this hypothesis beyond a reasonable doubt. This is not to say that there are no results that question this hypothesis. The complexity of both the biological phenomena and the experimental approaches involved are such that it is surprising that there are not more apparent contradictions of this simple hypothesis. Recent studies have already identified other mechanisms that work in parallel and synergize with synaptic plasticity in memory formation (for example, see (25).

Although, molecular and cellular cognition studies have provided compelling evidence that synaptic plasticity is required for learning, it is still unclear how this happens, what other cellular mechanisms are involved, and where and how they act. The field is full of major questions that are even difficult to frame, such as the nature of the molecular and cellular mechanisms that encode, retrieve, edit, orchestrate, and use stored information. However, the last ten years have prepared us for the exciting journey ahead. We know that mechanisms of plasticity are central to cognitive function and thus, we have a starting point. We have an experimental and theoretical framework to guide and instruct our searches, and perhaps most important, we are armed with the invincible optimism of a young field.

The *Annual Review of Genetics* is online at http://genet.annualreviews.org

LITERATURE CITED

1. Abel T, Nguyen PV, Barad M, Deuel TA, Kandel ER, Bourtchouladze R. 1997. Genetic demonstration of a role for PKA in the late phase of LTP and in hippocampus-based long-term memory. *Cell* 88:615–26

2. Adams JP, Sweatt JD. 2002. Molecular psychology: roles for the ERK MAP kinase cascade in memory. *Annu. Rev. Pharmacol. Toxicol.* 42:135–63

3. Alberini CM, Ghirardi M, Metz R, Kandel ER. 1994. C/EBP is an immediate-early gene required for the consolidation of long-term facilitation in Aplysia. *Cell* 76:1099–114

4. Ashe KH. 2001. Learning and memory in transgenic mice modeling Alzheimer's disease. *Learn. Mem.* 8:301–8

5. Atkins CM, Selcher JC, Petraitis J, Trzaskos J, Sweatt J. 1998. The MAPK cascade is required for mammalian associative learning. *Nat. Neurosci.* 1:602–9

6. Bach ME, Hawkins RD, Osman M, Kandel ER, Mayford M. 1995. Impairment of spatial but not contextual memory in CaMKII mutant mice with a selective loss of hippocampal LTP in the range of the theta frequency. *Cell* 81:905–15

7. Banke TG, Bowie D, Lee H, Huganir RL, Schousboe A, Traynelis SF. 2000. Control of GluR1 AMPA receptor function by cAMP-dependent protein kinase. *J. Neurosci.* 20:89–102

8. Bannerman DM, Good MA, Butcher SP, Ramsay M, Morris RG. 1995. Distinct components of spatial learning revealed by prior training and NMDA receptor blockade. *Nature* 378:182–86

9. Baranes D, Lederfein D, Huang YY, Chen M, Bailey CH, Kandel ER. 1998. Tissue plasminogen activator contributes to the late phase of LTP and to synaptic growth in the hippocampal mossy fiber pathway. *Neuron* 21:813–25

10. Barco A, Alarcon JM, Kandel ER. 2002. Expression of constitutively active CREB protein facilitates the late phase of long-term potentiation by enhancing synaptic capture. *Cell* 108:689–703

11. Barria A, Muller D, Derkach V, Griffith LC, Soderling TR. 1997. Regulatory phosphorylation of AMPA-type glutamate receptors by CaM-KII during long-term potentiation. *Science* 276:2042–45

12. Beekman JM, Allan GF, Tsai SY, Tsai MJ, O'Malley BW. 1993. Transcriptional activation by the estrogen receptor requires a conformational change in the ligand binding domain. *Mol. Endocrinol.* 7:1266–74

13. Berman DE, Hazvi S, Rosenblum K, Seger R, Dudai Y. 1998. Specific and differential activation of mitogen-activated protein kinase cascades by unfamiliar taste in the insular cortex of the behaving rat. *J. Neurosci.* 18:10037–44

14. Bito H, Deisseroth K, Tsien RW. 1996. CREB phosphorylation and dephosphorylation: a Ca^{2+}- and stimulus duration–dependent switch for hippocampal gene expression. *Cell* 87:1203–14

15. Bliss TVP, Lomo T. 1995. Long-lasting potentiation of synaptic transmission in the dentate area of the anaesthetized rabbit following stimulation of the perforant path. *J. NIH Res.* 7:59–67

16. Blum S, Moore AN, Adams F, Dash PK. 1999. A mitogen-activated protein kinase cascade in the CA1/CA2 subfield of the dorsal hippocampus is essential for long-term spatial memory. *J. Neurosci.* 19:3535–44

17. Bontempi B, Laurent-Demir C, Destrade C, Jaffard R. 1999. Time-dependent reorganization of brain circuitry underlying long-term memory storage. *Nature* 400:671–75

18. Bourtchouladze R, Abel T, Berman N, Gordon R, Lapidus K, Kandel ER. 1998. Different training procedures recruit either one or two critical periods for contextual memory consolidation, each of which requires protein synthesis and PKA. *Learn. Mem.* 5:365–74

19. Bourtchouladze R, Frenguelli B, Blendy J, Cioffi D, Schutz G, Silva AJ. 1994. Deficient long-term memory in mice with a targeted mutation of the cAMP-responsive element-binding protein. *Cell* 79:59–68

20. Brambilla R, Gnesutta N, Minichiello L, White G, Roylance AJ, et al. 1997. A role for the Ras signalling pathway in synaptic transmission and long-term memory. *Nature* 390:281–86

21. Cavallaro S, Schreurs BG, Zhao W, D'Agata V, Alkon DL. 2001. Gene expression profiles during long-term memory consolidation. *Eur. J. Neurosci.* 13:1809–15

22. Deleted in proof

23. Cho YH, Giese KP, Tanila H, Silva AJ, Eichenbaum H. 1998. Abnormal hippocampal spatial representations in alpha-CaMKIIT286A and CREBalphaDelta-mice. *Science* 279:867–69

24. Costa R, Yang T, Huynh D, Pulst S, Viskochil D, et al. 2001. Learning deficits but normal development and tumor predisposition in mice lacking exon 23a of the Neurofibromatosis type 1 gene. *Nat. Genet.* 27:399–405

25. Costa RM, Federov NB, Kogan JH, Murphy GG, Stern J, et al. 2002. Mechanism for the learning deficits in a mouse model of neurofibromatosis type 1. *Nature* 415:526–30

26. Crawley JN, Belknap JK, Collins A, Crabbe JC, Frankel W, et al. 1997. Behavioral phenotypes of inbred mouse strains: implications and recommendations for molecular studies. *Psychopharmacology* 132:107–24

27. Dalva MB, Takasu MA, Lin MZ, Shamah SM, Hu L, et al. 2000. EphB receptors interact with NMDA receptors and regulate excitatory synapse formation. *Cell* 103:945–56

28. Dash PK, Hochner B, Kandel ER. 1990. Injection of the cAMP-responsive element into the nucleus of *Aplysia* sensory neurons blocks long-term facilitation. *Nature* 345:718–21

29. Datson NA, van der Perk J, de Kloet ER, Vreugdenhil E. 2001. Expression profile of 30,000 genes in rat hippocampus using SAGE. *Hippocampus* 11:430–44

30. Davis HP, Squire LR. 1984. Protein synthesis and memory. *Psychol. Bull.* 96:518–59

31. Davis S, Vanhoutte P, Pages C, Caboche J, Laroche S. 2000. The MAPK/ERK cascade targets both Elk-1 and cAMP response element-binding protein to control long-term potentiation-dependent gene expression in the dentate gyrus in vivo. *J. Neurosci.* 20:4563–72

32. Dudai Y, Jan YN, Byers D, Quinn WG, Benzer S. 1976. dunce, a mutant of Drosophila deficient in learning. *Proc. Natl. Acad. Sci. USA* 73:1684–88

33. Ebbinghaus H. 1885. *Uber das Gedachtnis*. New York: Dover

34. Eichenbaum H. 2001. The hippocampus and declarative memory: cognitive mechanisms and neural codes. *Behav. Brain Res.* 127:199–207

35. Engle SJ, Womer DE, Davies PM, Boivin G, Sahota A, et al. 1996. HPRT-APRT-deficient mice are not a model for lesch-nyhan syndrome. *Hum. Mol. Genet.* 5:1607–10

36. English JD, Sweatt JD. 1997. A requirement for the mitogen-activated protein kinase cascade in hippocampal long term potentiation. *J. Biol. Chem.* 272:19103–6

37. Feng R, Rampon C, Tang YP, Shrom D, Jin J, et al. 2001. Deficient neurogenesis

in forebrain-specific presenilin-1 knockout mice is associated with reduced clearance of hippocampal memory traces. *Neuron* 32:911–26

38. Forrest D, Yuzaki M, Soares H, Ng L, Luk D, et al. 1994. Targeted disruption of NMDA receptor 1 gene abolishes NMDA response and results in neonatal death. *Neuron* 13:325–38

39. Frankland PW, O'Brien C, Ohno M, Kirkwood A, Silva AJ. 2001. Alpha-CaMKII-dependent plasticity in the cortex is required for permanent memory. *Nature* 411:309–13

40. Frey U, Muller M, Kuhl D. 1996. A different form of long-lasting potentiation revealed in tissue plasminogen activator mutant mice. *J. Neurosci.* 16:2057–63

41. Gao WQ, Shinsky N, Armanini MP, Moran P, Zheng JL, et al. 1998. Regulation of hippocampal synaptic plasticity by the tyrosine kinase receptor, REK7/EphA5, and its ligand, AL-1/Ephrin-A5. *Mol. Cell. Neurosci.* 11:247–59

42. Gass P, Wolfer DP, Balschun D, Rudolph D, Frey U, et al. 1998. Deficits in memory tasks of mice with CREB mutations depend on gene dosage. *Learn. Mem.* 5:274–88

43. Gerlai R, Shinsky N, Shih A, Williams P, Winer J, et al. 1999. Regulation of learning by EphA receptors: a protein targeting study. *J. Neurosci.* 19:9538–49

44. Giese KP, Fedorov NB, Filipkowski RK, Silva AJ. 1998. Autophosphorylation at Thr286 of the alpha calcium-calmodulin kinase II in LTP and learning. *Science* 279:870–73

45. Giese KP, Friedman E, Telliez JB, Fedorov NB, Wines M, et al. 2001. Hippocampus-dependent learning and memory is impaired in mice lacking the Ras-guanine-nucleotide releasing factor 1 (Ras-GRF1). *Neuropharmacology* 41:791–800

46. Graves L, Dalvi A, Lucki I, Blendy JA, Abel T. 2002. Behavioral analysis of CREB alphadelta mutation on a B6/129 F1 hybrid background. *Hippocampus* 12:18–26

47. Grunwald IC, Korte M, Wolfer D, Wilkinson GA, Unsicker K, et al. 2001. Kinase-independent requirement of EphB2 receptors in hippocampal synaptic plasticity. *Neuron* 32:1027–40

48. Hall J, Thomas KL, Everitt BJ. 2001. Fear memory retrieval induces CREB phosphorylation and Fos expression within the amygdala. *Eur. J. Neurosci.* 13:1453–58

49. Hayashi Y, Shi SH, Esteban JA, Piccini A, Poncer JC, Malinow R. 2000. Driving AMPA receptors into synapses by LTP and CaMKII: requirement for GluR1 and PDZ domain interaction. *Science* 287:2262–67

50. Henderson JT, Georgiou J, Jia Z, Robertson J, Elowe S, et al. 2001. The receptor tyrosine kinase EphB2 regulates NMDA-dependent synaptic function. *Neuron* 32:1041–56

51. Ho N, Liauw JA, Blaeser F, Wei F, Hanissian S, et al. 2000. Impaired synaptic plasticity and cAMP response element-binding protein activation in Ca^{2+}/calmodulin-dependent protein kinase type IV/Gr-deficient mice. *J. Neurosci.* 20:6459–72

52. Hoffman DA, Sprengel R, Sakmann B. 2002. Molecular dissection of hippocampal theta-burst pairing potentiation. *Proc. Natl. Acad. Sci. USA* 99:7740–45

53. Huang YY, Bach ME, Lipp HP, Zhuo M, Wolfer DP, et al. 1996. Mice lacking the gene encoding tissue-type plasminogen activator show a selective interference with late-phase long-term potentiation in both Schaffer collateral and mossy fiber pathways. *Proc. Natl. Acad. Sci. USA* 93:8699–704

54. Impey S, Fong AL, Wang Y, Cardinaux JR, Fass DM, et al. 2002. Phosphorylation of CBP mediates transcriptional activation by neural activity and CaM kinase IV. *Neuron* 34:235–44

55. Impey S, Mark M, Villacres EC, Poser S, Chavkin C, Storm DR. 1996. Induction of

CRE-mediated gene expression by stimuli that generate long-lasting LTP in area CA1 of the hippocampus. *Neuron* 16:973–82

56. Jones MW, Errington ML, French PJ, Fine A, Bliss TV, et al. 2001. A requirement for the immediate early gene Zif268 in the expression of late LTP and long-term memories. *Nat. Neurosci.* 4:289–96

57. Deleted in proof

58. Josselyn SA, Shi C, Carlezon WA Jr, Neve RL, Nestler EJ, Davis M. 2001. Long-term memory is facilitated by cAMP response element-binding protein overexpression in the amygdala. *J. Neurosci.* 21: 2404–12

59. Kang H, Sun LD, Atkins CM, Soderling TR, Wilson MA, Tonegawa S. 2001. An important role of neural activity-dependent CaMKIV signaling in the consolidation of long-term memory. *Cell* 106: 771–83

60. Kasarskis A, Manova K, Anderson KV. 1998. A phenotype-based screen for embryonic lethal mutations in the mouse. *Proc. Natl. Acad. Sci. USA* 95:7485–90

61. Kew JN, Koester A, Moreau JL, Jenck F, Ouagazzal AM, et al. 2000. Functional consequences of reduction in NMDA receptor glycine affinity in mice carrying targeted point mutations in the glycine binding site. *J. Neurosci.* 20:4037–49

62. Kida S, Josselyn SA, de Ortiz SP, Kogan JH, Chevere I, et al. 2002. CREB required for the stability of new and reactivated fear memories. *Nat. Neurosci.* 5:348–55

63. Kiyama Y, Manabe T, Sakimura K, Kawakami F, Mori H, Mishina M. 1998. Increased thresholds for long-term potentiation and contextual learning in mice lacking the NMDA-type glutamate receptor epsilon1 subunit. *J. Neurosci.* 18:6704–12

64. Kogan JH, Frankland PW, Blendy JA, Coblentz J, Marowitz Z, et al. 1997. Spaced training induces normal long-term memory in CREB mutant mice. *Curr. Biol.* 7:1–11

65. Kutsuwada T, Sakimura K, Manabe T, Takayama C, Katakura N, et al. 1996. Impairment of suckling response, trigeminal neuronal pattern formation, and hippocampal LTD in NMDA receptor epsilon 2 subunit mutant mice. *Neuron* 16:333–44

65a. Leadley RJ, Chi L, Rebello SS, Gagnon A. 2000. Contributions of in vivo models of thrombosis to the discovery and development of novel antithrombotic agents. *J. Pharmacol. Toxicol. Methods* 43:101–16

66. Li Y, Erzurumlu RS, Chen C, Jhaveri S, Tonegawa S. 1994. Whisker-related neuronal patterns fail to develop in the trigeminal brainstem nuclei of NMDAR1 knockout mice. *Cell* 76:427–37

67. Lisman J, Schulman H, Cline H. 2002. The molecular basis of CaMKII function in synaptic and behavioural memory. *Curr. Biol.* 3:175–90

68. Logie C, Stewart AF. 1995. Ligand-regulated site-specific recombination. *Proc. Natl. Acad. Sci. USA* 92:5940–44

69. Mack V, Burnashev N, Kaiser KM, Rozov A, Jensen V, et al. 2001. Conditional restoration of hippocampal synaptic potentiation in Glur-A-deficient mice. *Science* 292:2501–4

70. Madani R, Hulo S, Toni N, Madani H, Steimer T, et al. 1999. Enhanced hippocampal long-term potentiation and learning by increased neuronal expression of tissue-type plasminogen activator in transgenic mice. *EMBO J.* 18: 3007–12

71. Malleret G, Haditsch U, Genoux D, Jones MW, Bliss TV, et al. 2001. Inducible and reversible enhancement of learning, memory, and long-term potentiation by genetic inhibition of calcineurin. *Cell* 104:675–86

72. Mansuy IM, Winder DG, Moallem TM, Osman M, Mayford M, et al. 1998. Inducible and reversible gene expression with the rtTA system for the study of memory. *Neuron* 21:257–65

73. Mayford M, Bach ME, Huang YY, Wang L, Hawkins RD, Kandel ER. 1996. Control of memory formation through

regulated expression of a CaMKII transgene. *Science* 274:1678–83

74. Mayford M, Baranes D, Podsypanina K, Kandel ER. 1996. The 3′-untranslated region of CaMKII alpha is a *cis*-acting signal for the localization and translation of mRNA in dendrites. *Proc. Natl. Acad. Sci. USA* 93:13250–55

75. Mayford M, Wang J, Kandel ER, O'Dell TJ. 1995. CaMKII regulates the frequency-response function of hippocampal synapses for the production of both LTD and LTP. *Cell* 81:891–904

76. Mayr B, Montminy M. 2001. Transcriptional regulation by the phosphorylation-dependent factor CREB. *Nat. Rev. Mol. Cell Biol.* 2:599–609

77. Mazzucchelli C, Vantaggiato C, Ciamei A, Fasano S, Pakhotin P, et al. 2002. Knockout of ERK1 MAP kinase enhances synaptic plasticity in the striatum and facilitates striatal-mediated learning and memory. *Neuron* 34:807–20

78. McHugh TJ, Blum KI, Tsien JZ, Tonegawa S, Wilson MA. 1996. Impaired hippocampal representation of space in CA1–specific NMDAR1 knockout mice. *Cell* 87:1339–49

79. Meiri N, Masos T, Rosenblum K, Miskin R, Dudai Y. 1994. Overexpression of urokinase-type plasminogen activator in transgenic mice is correlated with impaired learning. *Proc. Natl. Acad. Sci. USA* 91:3196–200

80. Migaud M, Charlesworth P, Dempster M, Webster LC, Watabe AM, et al. 1998. Enhanced long-term potentiation and impaired learning in mice with mutant postsynaptic density-95 protein [see comments]. *Nature* 396:433–39

81. Minichiello L, Korte M, Wolfer D, Kuhn R, Unsicker K, et al. 1999. Essential role for TrkB receptors in hippocampus-mediated learning. *Neuron* 24:401–14

82. Mody M, Cao Y, Cui Z, Tay KY, Shyong A, et al. 2001. Genome-wide gene expression profiles of the developing mouse

hippocampus. *Proc. Natl. Acad. Sci. USA* 98:8862–67

83. Morris RG, Anderson E, Lynch GS, Baudry M. 1986. Selective impairment of learning and blockade of long-term potentiation by an N-methyl-D-aspartate receptor antagonist, AP5. *Nature* 319:774–76

84. Morton RA, Kuenzi FM, Fitzjohn SM, Rosahl TW, Smith D, et al. 2002. Impairment in hippocampal long-term potentiation in mice under-expressing the Alzheimer's disease related gene presenilin-1. *Neurosci. Lett.* 319:37–40

85. Nader K, Schafe GE, Le Doux JE. 2000. Fear memories require protein synthesis in the amygdala for reconsolidation after retrieval. *Nature* 406:722–26

86. Neiderhiser JM, Plomin R, McClearn GE. 1992. The use of CXB recombinant inbred mice to detect quantitative trait loci in behavior. *Physiol. Behav.* 52:429–39

87. Ohno M, Frankland PW, Silva AJ. 2002. A pharmacogenetic inducible approach to the study of NMDA/alphaCaMKII signaling in synaptic plasticity. *Curr. Biol.* 12:654–56

88. Picard D. 1993. Steroid-binding domains for regulating functions of heterologous proteins in cis. *Trends Cell Biol.* 3:278–80

89. Pittenger C, Huang YY, Paletzki RF, Bourtchouladze R, Scanlin H, et al. 2002. Reversible inhibition of CREB/ATF transcription factors in region CA1 of the dorsal hippocampus disrupts hippocampus-dependent spatial memory. *Neuron* 34:447–62

90. Rotenberg A, Abel T, Hawkins RD, Kandel ER, Muller RU. 2000. Parallel instabilities of long-term potentiation, place cells, and learning caused by decreased protein kinase A activity. *J. Neurosci.* 20:8096–102

91. Sakimura K, Kutsuwada T, Ito I, Manabe T, Takayama C, et al. 1995. Reduced hippocampal LTP and spatial learning in mice lacking NMDA receptor epsilon 1 subunit. *Nature* 373:151–55

92. Sara SJ. 2000. Retrieval and reconsolidation: toward a neurobiology of remembering. *Learn. Mem.* 7:73–84

93. Saucier D, Cain DP. 1995. Spatial learning without NMDA receptor-dependent long-term potentiation. *Nature* 378:186–89

94. Selcher JC, Nekrasova T, Paylor R, Landreth GE, Sweatt JD. 2001. Mice lacking the ERK1 isoform of MAP kinase are unimpaired in emotional learning. *Learn. Mem.* 8:11–19

95. Sgambato V, Pages C, Rogard M, Besson MJ, Caboche J. 1998. Extracellular signal-regulated kinase (ERK) controls immediate early gene induction on corticostriatal stimulation. *J. Neurosci.* 18:8814–25

96. Shapiro ML, Eichenbaum H. 1999. Hippocampus as a memory map: synaptic plasticity and memory encoding by hippocampal neurons. *Hippocampus* 9:365–84

97. Shou C, Farnsworth CL, Neel BG, Feig LA. 1992. Molecular cloning of cDNAs encoding a guanine-nucleotide-releasing factor for Ras p21 [see comments]. *Nature* 358:351–54

98. Silva AJ, Frankland PW, Marowitz Z, Friedman E, Lazlo G, et al. 1997. A mouse model for the learning and memory deficits associated with neurofibromatosis type I. *Nat. Genet.* 15:281–84

99. Silva AJ, Kogan JH, Frankland PW, Kida S. 1998. CREB and memory. *Annu. Rev. Neurosci.* 21:127–48

100. Silva AJ, Murphy GG. 1999. cAMP and memory: a seminal lesson from *Drosophila* and *Aplysia*. Special issue: Highlights in twentieth century neuroscience. Brain Res. Bull.

101. Silva AJ, Paylor R, Wehner JM, Tonegawa S. 1992. Impaired spatial learning in alpha-calcium-calmodulin kinase II mutant mice. *Science* 257:206–11

102. Silva AJ, Simpson EM, Takahashi JS, Lipp HP, Nakanishi S, et al. 1997. Mutant mice and neuroscience: recommendations concerning genetic background. *Neuron* 19:755–59

103. Silva AJ, Stevens CF, Tonegawa S, Wang Y. 1992. Deficient hippocampal long-term potentiation in alpha-calcium-calmodulin kinase II mutant mice. *Science* 257:201–6

104. Single FN, Rozov A, Burnashev N, Zimmermann F, Hanley DF, et al. 2000. Dysfunctions in mice by NMDA receptor point mutations NR1(N598Q) and NR1(N598R). *J. Neurosci.* 20:2558–66

105. Sterneck E, Paylor R, Jackson-Lewis V, Libbey M, Przedborski S, Tessarollo L, et al. 1998. Selectively enhanced contextual fear conditioning in mice lacking the transcriptional regulator CCAAT/enhancer binding protein delta. *Proc. Natl. Acad. Sci. USA* 95:10908–13

106. Sweatt JD. 2001. The neuronal MAP kinase cascade: a biochemical signal integration system subserving synaptic plasticity and memory. *J. Neurochem.* 76:1–10

107. Szklarczyk A, Lapinska J, Rylski M, McKay RD, Kaczmarek L. 2002. Matrix metalloproteinase-9 undergoes expression and activation during dendritic remodeling in adult hippocampus. *J. Neurosci.* 22:920–30

108. Takasu MA, Dalva MB, Zigmond RE, Greenberg ME. 2002. Modulation of NMDA receptor-dependent calcium influx and gene expression through EphB receptors. *Science* 295:491–95

109. Tang YP, Shimizu E, Dube GR, Rampon C, Kerchner GA, et al. 1999. Genetic enhancement of learning and memory in mice. *Nature* 401:63–69

110. Taubenfeld SM, Milekic MH, Monti B, Alberini CM. 2001. The consolidation of new but not reactivated memory requires hippocampal C/EBPbeta. *Nat. Neurosci.* 4:813–18

111. Taubenfeld SM, Wiig KA, Monti B, Dolan B, Pollonini G, Alberini CM. 2001. Fornix-dependent induction of hippocampal CCAAT enhancer-binding protein [beta] and [delta] co-localizes with phosphorylated cAMP response

element-binding protein and accompanies long-term memory consolidation. *J. Neurosci.* 21:84–91

112. Tsien JZ, Chen DF, Gerber D, Tom C, Mercer EH, et al. 1996. Subregion- and cell type-restricted gene knockout in mouse brain. *Cell* 87:1317–26

113. Tsien JZ, Huerta PT, Tonegawa S. 1996. The essential role of hippocampal CA1 NMDA receptor-dependent synaptic plasticity in spatial memory. *Cell* 87:1327–38

114. Valentinuzzi VS, Kolker DE, Vitaterna MH, Shimomura K, Whiteley A, et al. 1998. Automated measurement of mouse freezing behavior and its use for quantitative trait locus analysis of contextual fear conditioning in (BALB/cJ × C57BL/6J)F2 mice. *Learn. Mem.* 5:391–403

115. Vitaterna MH, King DP, Chang AM, Kornhauser JM, Lowrey PL, et al. 1994. Mutagenesis and mapping of a mouse gene, Clock, essential for circadian behavior. *Science* 264:719–25

116. Waddell S, Quinn WG. 2001. Flies, genes, and learning. *Annu. Rev. Neurosci.* 24:1283–309

117. Wahlsten D. 1972. Genetic experiments with animal learning: a critical review. *Behav. Biol.* 7:143–82

118. Wang Y, Xu J, Pierson T, O'Malley BW, Tsai SY. 1997. Positive and negative regulation of gene expression in eukaryotic cells with an inducible transcriptional regulator. *Gene Ther.* 4:432–41

119. Wehner JM, Radcliffe RA, Rosmann ST, Christensen SC, Rasmussen DL, et al. 1997. Quantitative trait locus analysis of contextual fear conditioning in mice [In Process Citation]. *Nat. Genet.* 17: 331–34

120. Wei F, Qiu CS, Liauw J, Robinson DA, Ho N, Chatila T, Zhuo M. 2002. Calcium

calmodulin-dependent protein kinase IV is required for fear memory. *Nat. Neurosci.* 5:573–579

121. Winder DG, Mansuy IM, Osman M, Moallem TM, Kandel ER. 1998. Genetic and pharmacological evidence for a novel, intermediate phase of long-term potentiation suppressed by calcineurin. *Cell* 92:25–37

122. Wong ST, Athos J, Figueroa XA, Pineda VV, Schaefer ML, et al. 1999. Calcium-stimulated adenylyl cyclase activity is critical for hippocampus-dependent long-term memory and late phase LTP. *Neuron* 23:787–98

123. Yu H, Saura CA, Choi SY, Sun LD, Yang X, et al. 2001. APP processing and synaptic plasticity in presenilin-1 conditional knockout mice. *Neuron* 31:713–26

124. Zamanillo D, Sprengel R, Hvalby O, Jensen V, Burnashev N, et al. 1999. Importance of AMPA receptors for hippocampal synaptic plasticity but not for spatial learning [see comments]. *Science* 284:1805–11

125. Zeng H, Chattarji S, Barbarosie M, Rondi-Reig L, Philpot BD, et al. 2001. Forebrain-specific calcineurin knockout selectively impairs bidirectional synaptic plasticity and working/episodic-like memory. *Cell* 107:617–29

126. Zhuo M, Zhang W, Son H, Mansuy I, Sobel RA, et al. 1999. A selective role of calcineurin alpha in synaptic depotentiation in hippocampus. *Proc. Natl. Acad. Sci. USA* 96:4650–55

127. Zirlinger M, Kreiman G, Anderson DJ. 2001. Amygdala-enriched genes identified by microarray technology are restricted to specific amygdaloid subnuclei. *Proc. Natl. Acad. Sci. USA* 98: 5270–75

Annu. Rev. Genet. 2002. 36:721–50
doi: 10.1146/annurev.genet.36.050802.093940
Copyright © 2002 by Annual Reviews. All rights reserved
First published online as a Review in Advance on September 12, 2002

ESTIMATING F-STATISTICS

B. S. Weir

Program in Statistical Genetics, Department of Statistics, North Carolina State University, Raleigh, North Carolina 27695-7566

W. G. Hill

Institute for Cell, Animal and Population Biology, University of Edinburgh, Edinburgh EH9 3JT, United Kingdom

Key Words population structure, forensic profiles, inbreeding, relatedness

■ **Abstract** A moment estimator of θ, the coancestry coefficient for alleles within a population, was described by Weir & Cockerham in 1984 (100) and is still widely cited. The estimate is used by population geneticists to characterize population structure, by ecologists to estimate migration rates, by animal breeders to describe genetic variation, and by forensic scientists to quantify the strength of matching DNA profiles. This review extends the work of Weir & Cockerham by allowing different levels of coancestry for different populations, and by allowing non-zero coancestries between pairs of populations. All estimates are relative to the average value of θ between pairs of populations. Moment estimates for within- and between-population θ values are likely to have large sampling variances, although these may be reduced by combining information over loci. Variances also decrease with the numbers of alleles at a locus, and with the numbers of populations sampled. This review also extends the work of Weir & Cockerham by employing maximum likelihood methods under the assumption that allele frequencies follow the normal distribution over populations. For the case of equal θ values within populations and zero θ values between populations, the maximum likelihood estimate is the same as that given by Robertson & Hill in 1984 (70). The review concludes by relating functions of θ values to times of population divergence under a pure drift model.

CONTENTS

0066-4197/02/1215-0721$14.00

INTRODUCTION

In 1984, Weir & Cockerham (100) published a set of equations for estimating the parameter F_{ST} or θ that describes the genetic structure of populations. The paper is still widely cited; in the first three months of 2002 the methods it described were applied to data on ash trees (59), Barbus (86), barley (42), barnacle (22), butterfly (18), cherry (54), cod (44), cord grass (85), Drosophila (32), eelgrass (64), frog (84), housefly (23), insects (58, 103), ladybird beetle (92), mackerel (11), moose (41), mountain lion (24), pig (45), pine (66, 68), quelea (19), red drum (33), redfish (72), river otter (9), rodent (14), salmon (37), scallops (67), sea trout (94), seaweed (88), shrimp (28), snail (13), stonefly (76), sugar beet (89), trout (38, 48), tsetse fly (47), wombat (7), zooplankton (34), and humans (1, 36, 53) among other species. Population biologists, ecologists and human geneticists have a substantial interest in being able to quantify the genetic relationships among their populations; it is therefore timely to re-visit the 1984 paper they cite. It may be especially useful to allow for different values of θ in different populations.

This discussion regards population structure, or the genetic differentiation of populations within the same species, as allelic frequency variation over populations. The restriction to allele frequencies, as opposed to genotypic frequencies, carries an implicit assumption of Hardy-Weinberg equilibrium at the loci under consideration. Even if two populations are maintained under the same evolutionary conditions they will have different allele frequencies because of the stochastic nature of these forces. Different evolutionary conditions for a set of populations will increase the differentiation among them, and θ can be defined in terms of variances and covariances of allele frequencies. The magnitude of these coefficients therefore reflects the evolutionary history of the populations being studied, although the observed allele frequencies also reflect the sampling processes within each population. The various approaches to estimating θ can differ according to whether they use only expected variances and covariances of allele frequencies or the entire frequency distributions. Use of the whole distribution may appear to be better, but there is an implicit constraint on the class of evolutionary scenarios if second-moment parameters are assumed to completely characterize a distribution.

The emphasis on within-species variation, and the usual use of unlinked loci means that coalescent approaches for non-recombining DNA sequences and deep evolutionary divergences [e.g., (61, 93)] are not considered.

LITERATURE REVIEW

Estimation Strategies

The search for the best estimators of θ, and the evaluation of existing estimators, continues. One way of distinguishing estimators is to consider how much of the distribution of allele frequencies across populations is used. It is shown below that the variances and covariances of allele frequencies across populations depend on θ as well as on the mean frequencies. This suggests that θ can be estimated from just the first and second moments of the allele frequency distribution, and this is the essence of the method of moments used by Weir & Cockerham (100). No particular evolutionary model leading to specific values for θ is assumed. Other methods assume the form of the whole distribution, which constrains applicability to certain evolutionary scenarios. The Dirichlet distribution used by Balding & Nichols (4) and Lange (49) assumes an evolutionary equilibrium, and is appropriate under the infinite alleles mutation model. Strictly, it is the Multinomial-Dirichlet distribution that is needed. The Dirichlet distribution is not appropriate for the stepwise mutation model (35). It is not clear that there is an evolutionary model for which the normal distribution used by Smouse & Williams (81), Long (51), and Nicholson et al. (60) and employed below in this review is appropriate, but it is justified by convenience and an appeal to large sample theory.

More statistical issues were addressed by Weicker et al. (95). The estimator of θ described by Weir & Cockerham (100) used the actual sample sizes in each sample in order to reduce bias, and Weicker et al. showed that good approximations to that estimator can be found that use the average sample size. These authors also presented confidence intervals found by bootstrapping over loci, with an implicit assumption that the number of loci is not small. Questions of both bias and variance were covered by Raufaste & Bonhomme (62) for loci with multiple alleles. The simplest models assume that allele frequency distributions have the same variances and covariances for all alleles, so that θ could be estimated separately for each allele. Raufaste & Bonhomme confirmed the prediction of Weir & Cockerham (100) that their weighting was satisfactory for larger values of θ, whereas an alternative weighting of Robertson & Hill (70) was better for small θ. The Robertson & Hill approach is equivalent to the multivariate approaches (51) described below.

This review is concerned with the relationships of pairs of alleles within and between populations, but a further hierarchy of relationships when there are sub-populations nested within populations, sub-subpopulations nested within subpopulations, and so on (97, 105). The nested analysis of variance structure is a natural framework for the analysis of that situation, and a generic definition of population-structure parameters for a hierarchy of populations was given by Rousset (75).

The growing use of Bayesian methods to population genetics is reflected by several papers that use such methods to characterize population structure (30, 39, 40, 71). Allele frequencies are assumed to follow a Dirichlet distribution across populations, or a beta distribution in the case of loci with two alleles.

Non-Frequency Measures

Although θ is defined in terms of variances of allele frequencies, there are parallel measures that use other parameters. The fact that mutation at microsatellite markers is generally between pairs of alleles with similar numbers of repeat units suggests that allele size (i.e., number of repeats) can be used in place of allele frequency (79). Balloux & Goudet (5) and Balloux & Lugin-Moulin (6) were concerned with the case where the stepwise mutation model holds for microsatellite markers. They compared two estimators of the form $\sum_{\text{loci}} V_a / \sum_{\text{loci}} V_t$ where the variance components (V_a among populations and V_t total) were for allele frequencies (100) or allele sizes (57). They compared the estimators for data simulated under a finite island model and concluded that neither estimator was best overall, although the Weir-Cockerham estimator was better for higher levels of gene flow. Weir & Cockerham (100) pointed out that the performance of their estimator reflects the method they used for combining information over multiple alleles at a locus, and they predicted better behavior for higher values of θ. It is the magnitudes of the parameter, rather than the forces leading to those values, that should affect the quality of the estimator in the multiple-alleles case.

Merilä & Crnoka (56) compared estimates of θ from various genetic markers with an analogous quantity, Q_{ST}, defined for quantitative traits (83). The estimate is based on the genetic variances of an additive quantitative trait, V_a among populations and V_w within populations, and is given by $V_a/(V_a + 2V_w)$. If allele frequencies are available for the same loci that affect the quantitative trait, values of θ and Q_{ST} should be equal.

Estimation of Migration Rates

Molecular ecologists, in particular, have been interested in inferring migration rates from estimates of θ, usually by employing the equilibrium result for the infinite-island migration model: $\theta = 1/(1 + 4Nm)$. Here N is the effective population size of each island and m is the migration rate between each pair of islands. Because this is a monotonic transformation of θ, it is not clear that much is gained over simply presenting θ estimates, especially as real populations are unlikely to conform to the many assumptions that lead to this result (101). Cockerham & Weir (15, 16) discussed more general relationships between θ and m. Kinnison et al. (46) fitted Nm to estimated θ values without assuming equilibrium. A recent review is given by Rousset (74), and a multivariate normal approach was adopted by Tufto et al. (87). Analogous work uses estimates of θ to estimate effective population size (8, 90).

Allocation of Individuals to Populations

Even though the genetic variation within human populations tends to be much greater than that among populations, there is often sufficient genetic differentiation among populations, as described by θ, to allow individuals to be allocated to populations. The problem was discussed for blood-type markers by Spielman & Smouse (82) and Smouse & Spielman (80). More recent studies, primarily by forensic scientists, have used microsatellite markers (10, 25, 52, 77, 78). Cornuet et al. (17) evaluated several methods for allocating individuals by assessing their behavior as functions of θ. Dawson & Belkhir (20) assessed the quality of their Bayesian method for assigning individuals to groups within a population by estimating θ from the resulting grouped data.

Forensic Applications

Genetic profiles are now widely used for human identification in a forensic setting, and also for inferring relationships in cases of disputed parentage or the identification of remains. The key question generally involves determining the probability of a set of profiles under alternative hypotheses about the sources of those profiles. In the simplest forensic situation where the profile of a suspect matches that of a stain found at the scene of a crime, this reduces to determining the probability that an unknown person in a population has the profile given that a suspect is known to have the profile (26). When allele frequencies are assumed to have a Dirichlet distribution over populations, this probability is a function of θ (3, 4), and forensic scientists routinely estimate θ for the populations with which they work (4, 30, 102).

ESTIMATION OF θ

The parameter θ provides a description of the relationship between pairs of alleles in a population. It could be defined as the probability that the two alleles are identical by descent, but this is restrictive in that its values are then constrained to lie in the range [0,1]. A more general definition is in terms of correlation coefficients, and can be expressed in terms of indicator variables x_{ju} for the jth allele in a sample:

$$x_{ju} = \begin{cases} 1 & \text{allele is of type } A_u \\ 0 & \text{otherwise.} \end{cases}$$

Then θ is the correlation between x_{ju} and $x_{j'u}$ for different alleles ($j \neq j'$), where the underlying expectation process is over replicates of the population. This correlation should be written as θ_u to allow for selection or mutation differences for different allelic types, but these differences generally are assumed not to exist. Although θ is designed to capture evolutionary variation, values of its estimates also reflect the sampling process leading to the data employed. Weir (97) made the distinction between genetic and statistical sampling for these two sources of variation. Another way of expressing this concept is to say that θ measures relatedness of pairs of alleles within a population relative to the total (i.e., the expected) population and this is why Wright (104) used the notation F_{ST}, where S denotes subpopulation and T denotes the total population.

Under the random mating assumption, expectations of the indicator variables do not depend on the particular values of j, and

$$\mathcal{E}(x_{ju}) = p_u$$

$$\mathcal{E}\left(x_{ju}^2\right) = p_u$$

$$\mathcal{E}(x_{ju}x_{j'u}) = p_u^2 + p_u(1 - p_u)\theta, \quad j \neq j',$$

where p_u is the population frequency of allele A_u, an expected value over replicates of the population. The expression for $\mathcal{E}(x_{ju}x_{j'u})$ can be taken as a definition of θ, and clearly $\mathrm{Var}(x_{ju}) = p_u(1 - p_u)$, $\mathrm{Cov}(x_{ju}, x_{j'u}) = p_u(1 - p_u)\theta$ so that θ is indeed a correlation coefficient over replicate populations.

It may be convenient to write the expected value of $x_{ju}x_{j'u}$ as $P_{u,u}$, the probability with which the two alleles are both of type A_u. However, for a population mating by random union of gametes, this quantity is the same as the homozygote frequency P_{uu}. For nonrandom mating populations, it is necessary to distinguish the cases where the alleles are in the same or different individuals and the indicator variables need to be defined as x_{jku} for the kth allele in the jth individual. Expectations are then

$$\mathcal{E}(x_{jku}) = p_u$$

$$\mathcal{E}\left(x_{jku}^2\right) = p_u$$

$$\mathcal{E}(x_{jku}x_{j'k'u}) = \begin{cases} p_u^2 + p_u(1 - p_u)F, & j = j', k \neq k' \\ p_u^2 + p_u(1 - p_u)\theta & j \neq j' \end{cases},$$

where F is the total inbreeding coefficient (sometimes written as F_{IT}). Then $P_{uu} = p_u^2 + p_u(1 - p_u)F$ differs from $P_{u,u} = p_u^2 + p_u(1 - p_u)\theta$.

Because θ refers to variation over the evolutionary process, it cannot be estimated from a sample from a single population. Inferences made from a single sample are for within-population parameters such as the within-population inbreeding coefficient f, or F_{IS}. This quantity satisfies $f = (F - \theta)/(1 - \theta)$, and it describes the relationship of pairs of alleles within individuals relative to that between individuals within the same population. There is generally little interest in within-population analogs of θ, as the point of estimating θ is to make inferences about evolutionary processes.

MOMENT ESTIMATES

With the assumption of no local inbreeding, $F_{IS} = 0$, $F_{IT} = F_{ST} = \theta$, estimation of θ makes use only of sample allele frequencies, although these need to be inferred from sample genotype frequencies. Second moments of allele frequencies can be expressed in terms of θ, suggesting that estimators can be constructed from sample second moments.

Overall Estimates

The variation described by θ is estimated in practice from allele frequency variation among different populations, and it has been customary to regard extant populations as providing the replicates inherent in its definition. This carries the assumption that each sampled population has the same θ value, and this will now be relaxed. To distinguish the populations sampled, an index i is added to the indicator variables for the ith sample. A general set of expectations for the jth allele in the ith sample are

$$\mathcal{E}(x_{iju}) = p_u$$

$$\mathcal{E}(x_{iju}^2) = p_u$$

$$\mathcal{E}(x_{iju}x_{i'j'u}) = \begin{cases} p_u^2 + p_u(1 - p_u)\theta_i & i = i', j \neq j' \\ p_u^2 + p_u(1 - p_u)\theta_{ii'} & i \neq i' \end{cases}.$$

Each population is assumed to have the same (expected) allele frequency. Weir & Cockerham (100) assumed that $\theta_{ii'} = 0$ for all $i' \neq i$. Later they relaxed those assumptions (15, 98).

Sample allele frequencies are denoted by tildes, and the average frequency over samples is denoted by a bar. If there are n_i alleles sampled from the ith of r populations:

$$\tilde{p}_{iu} = \frac{1}{n_i} \sum_{j=1}^{n_i} x_{iju}$$

$$\bar{p}_u = \frac{1}{\sum_i n_i} \sum_{i=1}^{r} n_i \tilde{p}_{iu},$$

so that

$$\mathcal{E}(\tilde{p}_{iu}) = p_u$$

$$\mathcal{E}(\bar{p}_u) = p_u$$

$$\text{Var}(\tilde{p}_{iu}) = \frac{1}{n_i} p_u(1 - p_u)[1 + (n_i - 1)\theta_i] \qquad\qquad 1.$$

$$\text{Cov}(\tilde{p}_{iu}, \tilde{p}_{i'u}) = p_u(1 - p_u)\theta_{ii'}. \qquad\qquad 2.$$

Subsequent developments are simplified with additional notation:

$$\pi_u = p_u(1 - p_u)$$

$$\phi_i = \frac{1}{n_i}[1 + (n_i - 1)]\theta_i]$$

Equations 1, 2 can be taken as defining the θ parameters and therefore can serve as a starting point. They could be derived by considering two sets of expectations, one within (W) and one among (A) populations. If p_{iu} is the frequency of allele A_u in the ith population, the usual multinomial distribution gives:

$$\left.\begin{aligned} \mathcal{E}_W(\tilde{p}_{iu}) &= p_{iu} \\ \mathrm{Var}_W(\tilde{p}_{iu}) &= \frac{1}{n_i} p_{iu}(1 - p_{iu}). \end{aligned}\right\} \qquad 3.$$

Among populations, the moments are

$$\left.\begin{aligned} \mathcal{E}_A(p_{iu}) &= p_u \\ \mathrm{Var}_A(p_{iu}) &= p_u(1 - p_u)\theta_i \end{aligned}\right\} \qquad 4.$$

to introduce the θ's. The method of moments for estimating θ makes no more statements concerning the distribution of the p_{iu}'s about p_u. Balding & Nichols (3, 4) assumed a Dirichlet distribution with parameters $(1 - \theta_i)p_u/\theta_i$ for A_u which also gives Equations 4, as does the normal distribution $N(p_u, \pi_u\theta_i)$ assumed by Nicholson et al. (60). Combining Equations 3 and 4 leads to Equations 1 and 2, emphasizing that expectations in such equations are total (within and among populations). Foulley & Hill (31) contrasted the use of the normal and Dirichlet distributions.

When it is assumed that $\theta_i = \theta$ for all i and $\theta_{ii'} = 0$ for all $i \neq i'$, Weir & Cockerham (100) note that there are two unknown quantities, π_u and θ, and define two mean squares. In the notation of Weir (97):

$$\mathrm{MSP}_u = \frac{1}{r - 1} \sum_{i=1}^{r} n_i(\tilde{p}_{iu} - \bar{p}_u)^2$$

$$\mathrm{MSG}_u = \frac{1}{\sum_{i=1}^{r}(n_i - 1)} \sum_{i=1}^{r} n_i \tilde{p}_{iu}(1 - \tilde{p}_{iu}).$$

The average allele frequency $\bar{p}_u$ includes sample size weights. An alternative is to use an unweighted average $\bar{p}_u^* = \sum_{i=1}^{r} \tilde{p}_{iu}/r$. Estimates based on $\bar{p}_u$ or $\bar{p}_u^*$ will be better when θ or $(1 - \theta)/n_i$, respectively, are larger. Following Robertson (69), a weighted estimate could be obtained from the two.

Under the general model, the mean squares have expected values

$$\mathcal{E}(\mathrm{MSP}_u) = \frac{\pi_u}{r - 1}\left[\sum_{i=1}^{r} n_{ic}\phi_i - \frac{1}{\sum_{i=1}^{r} n_i} \sum_{\substack{i,i'=1 \\ i \neq i'}}^{r} n_i n_{i'}\theta_{ii'}\right]$$

$$\mathcal{E}(\mathrm{MSG}_u) = \frac{\pi_u}{\sum_{i=1}^{r}(n_i = 1)}\left(\sum_{i=1}^{r} n_i - \sum_{i=1}^{r} n_i\phi_i\right),$$

where $n_{ic} = n_i - n_i^2/\sum_{i=1}^{r} n_i$. There are two special cases that lead to simplification.

In the special case that $\theta_i = \theta$ for all i and $\theta_{ii'} = 0$ for all $i \neq i'$,

$$\mathcal{E}(\text{MSP}_u) = \pi_u[(1 - \theta) + n_c\theta]$$

$$\mathcal{E}(\text{MSG}_u) = \pi_u(1 - \theta),$$

where

$$n_c = \frac{1}{r-1}\left(\sum_{i=1}^{r} n_i - \frac{\sum_{i=1}^{r} n_i^2}{\sum_{i=1}^{r} n_i}\right) = \frac{1}{r-1}\sum_{i=1}^{r} n_{ic}.$$

This led Weir & Cockerham (100) to their moment estimator of θ:

$$\hat{\theta}_{Mu} = \frac{\text{MSP}_u - \text{MSG}_u}{\text{MSP}_u + (n_c - 1)\text{MSG}_u}.$$

To the extent that the expected value of this quantity is the ratio of expectations of its numerator and denominator, it is unbiased for θ.

In the special case of balanced data, $n_i = n$ for all i,

$$\mathcal{E}(\text{MSP}_u) = \pi_u[(1 - \theta_w) + n(\theta_w - \theta_a)]$$

$$\mathcal{E}(\text{MSG}_u) = \pi_u(1 - \theta_w),$$

where

$$\theta_w = \frac{1}{r}\sum_{i=1}^{r} \theta_i$$

$$\theta_a = \frac{1}{r(r-1)}\sum_{\substack{i,i'=1 \\ i \neq i'}}^{r} \theta_{ii'},$$

so that the moment estimate, now written as $\hat{\beta}$, is providing an estimate of $(\theta_w - \theta_a)/(1 - \theta_a)$. This result should also hold if all of the sample sizes are large and approximately equal. In general, however, the usual moment estimate is of a complex function of the θ_i's and $\theta_{ii'}$'s. Alternative statistics lead to estimates of weighted averages of θ_i's and $\theta_{ii'}$'s, as shown below.

Under the assumption that the same value of θ applies to each allele at a locus, Weir & Cockerham (100) combined information over alleles by summing numerator and denominator separately

$$\hat{\theta}_M = \frac{\sum_{u=1}^{m}(\text{MSP}_u - \text{MSG}_u)}{\sum_{u=1}^{m}[\text{MSP}_u + (n_c - 1)\text{MSG}_u]}, \qquad 5.$$

and they found by simulation that this method of weighting over alleles generally provides low bias and variance. No explicit account is taken of the correlation among frequencies of different alleles. If data are collected from a series of L loci, and if θ is assumed to apply equally to each locus, then an obvious extension is to add mean squares over loci:

$$\hat{\theta}_M = \frac{\sum_{l=1}^{L}\sum_{u=1}^{m_l}(\text{MSP}_{lu} - \text{MSG}_{lu})}{\sum_{l=1}^{L}\sum_{u=1}^{m_l}[\text{MSP}_{lu} + (n_c - 1)\text{MSG}_{lu}]}.$$

Properties of Moment Estimate

Because of the difficulty in describing the properties of ratio estimates, Dodds (21) and Weir (97) suggested numerical resampling for obtaining the sampling distribution of $\hat{\theta}_M$. Resampling over populations would change the structure of the data, but resampling over loci would exploit the assumption that (unlinked) loci provide independent replicates of the evolutionary process. Resampling was also used by Raymond & Rousset (63). Jiang (43) used a Taylor series expansion and approximate higher-order moments of sample allele frequencies to obtain the mean and variance of $\hat{\theta}_M$. Li (50) appealed to asymptotic theory to show that the mean square MSP_u has a chi-square distribution in the two-allele case,

$$\text{MSP}_u \sim \pi_u[1 + (n_c - 1)\theta]\chi^2_{(r-1)},$$

and that the mean square MSG_u tends to a constant value of $\pi_u(1 - \theta)$. This assumes that the θ_i's are equal and that the $\theta_{ii'}$'s are zero. These results allowed her to derive expressions for the mean and variance of $\hat{\theta}$:

$$\mathcal{E}(\hat{\theta}_M) = \theta - \frac{2(1-\theta)}{r-1}\left(\frac{1 + (n_c - 1)\theta}{n_c}\right)^2$$

$$\text{Var}(\hat{\theta}_M) = \frac{2(1-\theta)^2}{r-1}\left(\frac{1 + (n_c - 1)\theta}{n_c}\right)^2.$$

The variance formula differs slightly from the variance of the intraclass correlation given by Fisher (29), but is equal to that result for large sample sizes.

Population-Specific Estimates

If independent populations have different values of θ_i, maybe reflecting the differences in population size or differences in environmental influences, there is the danger of having an over-parameterized model. There are r independent sample allele frequencies $\tilde{p}_{iu}$ for allele A_u. In the two-allele case, this means r observations but $(r + 1)$ parameters: the frequency p_u and the r values of θ_i. It is possible to construct estimates, but they will not be unique. For $m > 2$ alleles at a locus, however, there are more $[r(m - 1)]$ independent sample allele frequencies than there are parameters: $(m-1)$ parameters p_u plus r parameters θ_i. Similarly, for $L > 1$ diallelic loci, there are more observations (rL allele frequencies) than there are parameters (L allele frequencies and r θ's). The following discussion assumes that there are at least as many allele frequencies in the data as there are parameters to be estimated.

If the terms in the mean square within populations are weighted by n_{ic} instead of n_i, the sums of squares corresponding to MSP and MSG have expectations

$$\mathcal{E}\left[\sum_{i=1}^{r} n_i(\tilde{p}_{iu} - \bar{p}_u)^2\right] = \pi_u\left[\sum_{i=1}^{r} n_{ic}\phi_i - \frac{1}{\sum_{i=1}^{r} n_i}\sum_{\substack{i,i'=1\\i\neq i'}}^{r} n_i n_{i'}\theta_{ii'}\right]$$

$$\mathcal{E}\left[\sum_{i=1}^{r} n_{ic}\tilde{p}_{iu}(1 - \tilde{p}_{iu})\right] = \pi_u\left[\sum_{i=1}^{r} n_{ic} - \sum_{i=1}^{r} n_{ic}\phi_i\right],$$

suggesting that, for independent populations ($\theta_{ii'} = 0$), π_u can be estimated as

$$\hat{\pi}_u = \frac{\sum\limits_{i=1}^{r} n_i(\tilde{p}_{iu} - \bar{p}_u)^2 + \sum\limits_{i=1}^{r} n_{ic}\tilde{p}_{iu}(1 - \tilde{p}_{iu})}{\sum\limits_{i=1}^{r} n_{ic}}.$$

Therefore, from the relationship

$$\mathcal{E}\left[\sum_{u=1}^{m} \tilde{p}_{iu}(1 - \tilde{p}_{iu})\right] = \left(\sum_{u=1}^{m} \pi_u\right)(1 - \phi_i),$$

a moment estimate of ϕ_i for independent populations is

$$\hat{\phi}_i = 1 - \frac{\left(\sum\limits_{i=1}^{r} n_{ic}\right)\sum\limits_{u=1}^{m} \tilde{p}_{iu}(1 - \tilde{p}_{iu})}{\sum\limits_{u=1}^{m}\sum\limits_{i=1}^{r}\left[n_i(\tilde{p}_{iu} - \bar{p}_u)^2 + n_{ic}\tilde{p}_{iu}(1 - \tilde{p}_{iu})\right]}.$$ 6.

The estimate of the mean of the ϕ_i's is

$$\hat{\bar{\phi}} = 1 - \frac{\left(\sum\limits_{i=1}^{r} n_{ic}\right)\sum\limits_{u=1}^{m}\sum\limits_{i=1}^{r} \tilde{p}_{iu}(1 - \tilde{p}_{iu})}{r\sum\limits_{u=1}^{m}\sum\limits_{i=1}^{r}\left[n_i(\tilde{p}_{iu} - \bar{p}_u)^2 + n_{ic}\tilde{p}_{iu}(1 - \tilde{p}_{iu})\right]}.$$

When the sample sizes are equal, $n_i = n$ for all i,

$$\hat{\phi}_i = 1 - \frac{\displaystyle\sum_{u=1}^{m} \tilde{p}_{iu}(1 - \tilde{p}_{iu})}{\displaystyle\sum_{u=1}^{m} \left[\frac{1}{r-1} \sum_{i=1}^{r} (\tilde{p}_{iu} - \bar{p}_u)^2 + \frac{1}{r} \sum_{i=1}^{r} \tilde{p}_{iu}(1 - \tilde{p}_{iu}) \right]}.$$

Further, when the number r of samples is large

$$\hat{\phi}_i \approx 1 - \frac{\displaystyle\sum_{u=1}^{m} \tilde{p}_{iu}(1 - \tilde{p}_{iu})}{\displaystyle\sum_{u=1}^{m} \bar{p}_u(1 - \bar{p}_u)}$$

$$\hat{\phi} \approx \frac{\displaystyle\sum_{u=1}^{m}\sum_{i=1}^{r} (\tilde{p}_{iu} - \bar{p}_u)^2}{r \displaystyle\sum_{u=1}^{m} \bar{p}_u(1 - \bar{p}_u)}.$$

For each independent locus indexed by $l = 1, 2, \ldots L$, the estimate of ϕ_i may be written as $1 - x_{li}/y_l$ where

$$x_{li} = \sum_{u=1}^{m} \tilde{p}_{liu}(1 - \tilde{p}_{liu})$$

$$y_l = \frac{1}{\displaystyle\sum_{i=1}^{r} n_{lic}} \sum_{u=1}^{m}\sum_{i=1}^{r} \left[n_{li}(\tilde{p}_{liu} - \bar{p}_{lu})^2 + n_{lic}\tilde{p}_{liu}(1 - \tilde{p}_{liu}) \right],$$

showing the addition of locus subscripts on sample sizes and allele frequencies. These terms have expectations

$$\mathcal{E}(x_{li}) = (1 - \phi_i) \sum_{u=1}^{m_l} \pi_{lu}$$

$$\mathcal{E}(y_l) = \sum_{u=1}^{m_l} \pi_{lu}.$$

Information from loci with the same values of ϕ_i can be combined as for the earlier Weir & Cockerham estimator (100): $\hat{\phi}_i = 1 - (\sum_l x_{li})/(\sum_l y_l)$. The sampling distribution of this combined estimate may be found by bootstrapping over loci if L is not small.

Nicholson et al. (60) were especially interested in SNP loci, which generally have only two alleles. In that case, the two summands in the sums over alleles u

are the same and only one needs to be used. If $\tilde{p}_i$ is the frequency of one of the alleles at a locus, the equal sample size estimate is

$$\hat{\phi}_i = 1 - \frac{\tilde{p}_i(1 - \tilde{p}_i)}{\frac{1}{r-1}\sum_{i=1}^{r}(\tilde{p}_i - \bar{p})^2 + \frac{1}{r}\sum_{i=1}^{r}\tilde{p}_i(1 - \tilde{p}_i)},$$

and, for a large number of samples,

$$\hat{\phi}_i \approx 1 - \frac{\tilde{p}_i(1 - \tilde{p}_i)}{\bar{p}(1 - \bar{p})}.$$

Averaging over samples recovers the "classical" estimate (27)

$$\hat{\phi} \approx \frac{\sum_{i=1}^{r}(\tilde{p}_i - \bar{p})^2}{r\,\bar{p}(1 - \bar{p})}.$$

Care is needed in interpreting the values of the estimates $\hat{\phi}_i$, as differences may reflect differences among the sample sizes n_i or among the coefficients θ_i, or both.

When the populations are not independent, $\theta_{ii'} \neq 0$, the estimate of ϕ_i shown in Equation 6 is actually estimating $(\phi_i - \theta_A)/(1 - \theta_A)$, where

$$\theta_A = \frac{\sum_{\substack{i,i'=1 \\ i \neq i'}}^{r} n_i n_{i'} \theta_{ii'}}{\sum_{\substack{i,i'=1, \\ i \neq i'}}^{r} n_i n_{i'}}.$$

The weighted average θ_A reduces to the simple arithmetic mean, θ_a, of the $\theta_{ii'}$'s when the sample sizes are equal. An estimate of $\beta_{ii'} = (\theta_{ii'} - \theta_A)/(1 - \theta_A)$ is given by

$$\beta_{ii'} = \frac{\theta_{ii'} - \theta_A}{1 - \theta_A} \triangleq 1 - \frac{\left(\sum_{i=1}^{r} n_{ic}\right)\sum_{u=1}^{m}\left[\tilde{p}_{iu}(1 - \tilde{p}_{i'u}) + \tilde{p}_{i'u}(1 - \tilde{p}_{iu})\right]}{2\sum_{u=1}^{m}\sum_{i=1}^{r}\left[n_i(\tilde{p}_{iu} - \bar{p}_u)^2 + n_{ic}\tilde{p}_{iu}(1 - \tilde{p}_{iu})\right]}. \quad 7.$$

where $\triangleq$ denotes "is estimated by." These estimates sum to zero. In the case of only two samples, this estimate is zero as required. The corresponding single-population equation is

$$\beta_i = \frac{\theta_i - \theta_A}{1 - \theta_A} \triangleq 1 - \frac{\left(\sum_{i=1}^{r} n_{ic}\right)\sum_{u=1}^{m}\frac{n_i}{n_i - 1}\tilde{p}_{iu}(1 - \tilde{p}_{iu})}{\sum_{u=1}^{m}\sum_{i=1}^{r}\left[n_i(\tilde{p}_{iu} - \bar{p}_u)^2 + n_{ic}\tilde{p}_{iu}(1 - \tilde{p}_{iu})\right]}. \quad 8.$$

This is to replace Equation 6, although the difference between them is trivial for large sample sizes.

By analogy to θ_A, the weighted average θ_W can be defined as

$$\theta_W = \frac{\displaystyle\sum_{i=1}^{r} n_i \theta_i}{\displaystyle\sum_{i=1}^{r} n_i},$$

which reduces to the simple arithmetic average, θ_w, when the sample sizes are equal. The quantity $\beta_W = (\theta_W - \theta_A)/(1 - \theta_A)$ can be estimated as

$$\hat{\beta}_W = 1 - \frac{\left(\displaystyle\sum_{i=1}^{r} n_{ic}\right) \displaystyle\sum_{u=1}^{m} \frac{n_i^2}{n_i - 1} \tilde{p}_{iu}(1 - \tilde{p}_{iu})}{\left(\displaystyle\sum_{i=1}^{r} n_i\right) \displaystyle\sum_{u=1}^{m} \sum_{i=1}^{r} \left[n_i(\tilde{p}_{iu} - \bar{p}_u)^2 + n_{ic}\tilde{p}_{iu}(1 - \tilde{p}_{iu})\right]}. \qquad 9.$$

For equal sample sizes this reduces to the estimator in Equation 5 given by Weir & Cockerham (100). Because it serves as an estimator in the case of unequal sample sizes, however, it may be preferred to the Weir & Cockerham estimator.

There are two unsatisfactory aspects of this development. In the first place, it is seen that the quantities being estimated depend on the sample sizes, unless those sizes are equal. A more serious problem is the involvement of the average between-population relatedness quantity θ_A. Unless there are grounds for assuming this quantity is zero, all estimates are relative to that value. This does not prevent a comparison among the values of θ_i or $\theta_{ii'}$, but it does prevent their absolute value being estimated. There is the same need for a reference population when inbreeding coefficients F_{IT} are to be estimated. The issue is similar to that faced in the reconstruction of phylogenetic trees. Trees cannot be rooted unless there is information from an outgroup.

Finally, for large numbers of large samples,

$$\frac{\theta_i - \theta_A}{1 - \theta_A} \doteq 1 - \frac{\displaystyle\sum_{u=1}^{m} \tilde{p}_{iu}(1 - \tilde{p}_{iu})}{\displaystyle\sum_{u=1}^{m} \bar{p}_u(1 - \bar{p}_u)} \qquad 10.$$

$$\frac{\theta_{ii'} - \theta_A}{1 - \theta_A} \doteq 1 - \frac{\displaystyle\sum_{u=1}^{m} [\tilde{p}_{iu}(1 - \tilde{p}_{i'u}) + \tilde{p}_{i'u}(1 - \tilde{p}_{iu})]}{2\displaystyle\sum_{u=1}^{m} \bar{p}_u(1 - \bar{p}_u)}. \qquad 11.$$

NORMAL THEORY APPROACH

Moment estimators have the property of being unbiased but little else is known about their sampling properties. If the sampling distribution for the data is known, then likelihood methods can be employed. If individuals, and hence genotypes, are sampled randomly from a single population their counts follow a multinomial distribution among samples from the same population. When there is random union of gametes in the population, allele counts are also multinomially distributed over samples from the population. For large samples, the multinomial distribution can be approximated by the multivariate normal distribution, and it will now be assumed that the normal distribution applies also across populations. Normality has also been assumed by previous authors (51, 60, 81, 87). If $\tilde{\mathbf{P}}$ is the vector of sample allele frequencies:

$$\tilde{\mathbf{P}} \sim \text{MVN}(\mathbf{P}, \mathbf{V}),$$

where

$$\tilde{\mathbf{P}} = \begin{bmatrix} \tilde{\mathbf{p}}_1 \\ \tilde{\mathbf{p}}_2 \\ \cdots \\ \tilde{\mathbf{p}}_r \end{bmatrix}, \quad \mathbf{P} = \begin{bmatrix} \mathbf{p} \\ \mathbf{p} \\ \cdots \\ \mathbf{p} \end{bmatrix}, \quad \mathbf{V} = \begin{bmatrix} \mathbf{V}_{11} & \mathbf{V}_{12} & \cdots & \mathbf{V}_{1r} \\ \mathbf{V}_{21} & \mathbf{V}_{22} & \cdots & \mathbf{V}_{2r} \\ \cdots & \cdots & \cdots & \cdots \\ \mathbf{V}_{r1} & \mathbf{V}_{r2} & \cdots & \mathbf{V}_{rr} \end{bmatrix}.$$

The vectors $\tilde{\mathbf{p}}_i$ and $\mathbf{p}$ have $(m-1)$ components $\tilde{p}_{iu}$ and p_u, one for each of $(m-1)$ of the alleles at the locus. The $(m-1) \times (m-1)$ matrices $\mathbf{V}_{ii'}$ have elements $V_{ii'uu'}$. When $i = i'$ and $u = u'$ these elements are the variances of $\tilde{p}_{iu}$, otherwise they are the covariances of $\tilde{p}_{iu}$ and $\tilde{p}_{i'u'}$. Their values are:

$$V_{ii'uu'} = \begin{cases} p_u(1 - p_u)\phi_i & i = i', u = u' \\ -p_u p_{u'}\phi_i & i = i', u \neq u' \\ p_u(1 - p_u)\theta_{ii'} & i \neq i', u = u' \\ -p_u p_{u'}\theta_{ii'} & i \neq i', u \neq u'. \end{cases}$$

Overall Estimate

If there is no relationship among alleles from different populations, $\theta_{ii'} = 0$, then the vectors $\tilde{\mathbf{p}}_i$ are independent. These vectors also have the same expected value, but they have the same variances only if the ϕ_i values are the same. Unless the sample sizes are very large, this requires not only equal θ_i values, but also equal sample sizes n_i. Suppose now that $\phi_i = \phi$, because $\theta_i = \theta$ and because the n_i's are either equal or so large that they are approximately equal. The sample allele frequency vectors $\tilde{\mathbf{p}}_i$ are then independently and identically distributed and, from standard theory, the quadratic form

$$Q = \sum_{i=1}^{r} (\tilde{\mathbf{p}}_i - \bar{\mathbf{p}})' \mathbf{V}_{ii}^{-1} (\tilde{\mathbf{p}}_i - \bar{\mathbf{p}})$$

$$= \frac{1}{\phi} \sum_{i=1}^{r} \sum_{u=1}^{m} \frac{(\tilde{p}_{iu} - \bar{p}_u)^2}{\bar{p}_u}$$

has a chi-square distribution

$$Q \sim \phi \chi^2_{(r-1)(m-1)}.$$

The mean allele frequencies are $\bar{p}_u = \sum_{i=1}^{r} n_i \tilde{p}_{iu} / \sum_{i=1}^{r} n_i$ as before, and the estimate of the common value θ is

$$\hat{\theta}_N = \frac{1}{n-1} \left(\frac{n}{(r-1)(m-1)} \sum_{i=1}^{r} \sum_{u=1}^{m} \frac{(\tilde{p}_{iu} - \bar{p}_u)^2}{\bar{p}_u} - 1 \right) \qquad 12.$$

when the sample sizes are equal, or

$$\hat{\theta}_N = \frac{1}{(r-1)(m-1)} \sum_{i=1}^{r} \sum_{u=1}^{m} \frac{(\tilde{p}_{iu} - \bar{p}_u)^2}{\bar{p}_u} \qquad 13.$$

when the sample sizes are large (70). If data are available from L independent loci, the lth of which has m_l alleles, the sum over loci of the quadratic forms has a chi-square distribution with $d = (r-1) \sum_{l=1}^{L} (m_l - 1)$ df, and the estimates are simply averaged over loci.

From the properties of the chi-square distribution

$$\mathcal{E}(\hat{\theta}_N) = \theta$$

$$\mathrm{Var}(\hat{\theta}_N) = \frac{2[1 + (n-1)\theta]^2}{(n-1)^2 d} \approx \frac{2\theta^2}{d}.$$

Similar expressions were given by Foulley & Hill (31).

The chi-square distribution also provides confidence intervals. For example, if $X_{0.025}$ and $X_{0.975}$ are the 2.5th and 97.5th percentiles of the χ^2_d distribution, a 95% confidence interval is

$$\left(\frac{d}{X_{0.975}} \left[\hat{\theta}_N + \frac{1}{n-1} \right] - \frac{1}{n-1}, \frac{d}{X_{0.025}} \left[\hat{\theta}_N + \frac{1}{n-1} \right] - \frac{1}{n-1} \right)$$

for equal sample sizes, and

$$\left(\frac{d\hat{\theta}_N}{X_{0.975}}, \frac{d\hat{\theta}_N}{X_{0.025}} \right)$$

for large sample sizes.

Population-Specific Estimates

When the populations are independent, $\theta_{ii'} = 0$ for all $i \neq i'$, but with different values of θ_i, the variance matrix $\mathbf{V}$ can be written as a Kronecker product:

$$\mathbf{V} = \boldsymbol{\Pi} \otimes \boldsymbol{\Phi},$$

where

$$\boldsymbol{\Pi} = \begin{bmatrix} p_1(1 - p_1) & -p_1 p_2 & \cdots \\ -p_1 p_2 & p_2(1 - p_2) & \cdots \\ \cdots & \cdots & \cdots \end{bmatrix}$$

$$\boldsymbol{\Phi} = \begin{bmatrix} \phi_1 & 0 & \cdots \\ 0 & \phi_2 & \cdots \\ \cdots & \cdots & \cdots \end{bmatrix}.$$

If there are r samples and m alleles at the locus, $\mathbf{V}$ has determinant

$$|\mathbf{V}| = \left(\prod_{i=1}^{r} \phi_i \right)^m \left(\prod_{u=1}^{m} p_i \right)^r$$

and inverse

$$\mathbf{V}^{-1} = \boldsymbol{\Phi}^{-1} \otimes \boldsymbol{\Pi}^{-1},$$

where

$$\boldsymbol{\Pi}^{-1} = \begin{bmatrix} \frac{1}{p_1} + \frac{1}{p_m} & \frac{1}{p_m} & \cdots \\ \frac{1}{p_m} & \frac{1}{p_2} + \frac{1}{p_m} & \cdots \\ \cdots & \cdots & \cdots \end{bmatrix}$$

$$\boldsymbol{\Phi}^{-1} = \begin{bmatrix} \frac{1}{\phi_1} & 0 & \cdots \\ 0 & \frac{1}{\phi_2} & \cdots \\ \cdots & \cdots & \cdots \end{bmatrix}.$$

Ignoring terms that do not include the parameters of interest in likelihood expressions, the log-likelihood function is

$$\ln L = -\frac{1}{2} \ln(|\mathbf{V}|) - \frac{1}{2}(\tilde{\mathbf{P}} - \mathbf{P})' \mathbf{V}^{-1}(\tilde{\mathbf{P}} - \mathbf{P})$$

$$= -\frac{m}{2} \sum_{i=1}^{r} \ln(\phi_i) - \frac{r}{2} \sum_{u=1}^{m} \ln(p_u) - \frac{1}{2} \sum_{i=1}^{r} \sum_{u=1}^{m} \frac{(\tilde{p}_{iu} - p_u)^2}{\phi_i p_u}.$$

Because the p_u's sum to one, it is necessary to add a Lagrangian term before maximizing this function in order to find the maximum likelihood estimates of the

p_u's and ϕ_i's. The modified function and its derivatives are

$$\ln L = -\frac{m}{2} \sum_{i=1}^{r} \ln(\phi_i) - \frac{r}{2} \sum_{u=1}^{m} \ln(p_u) - \frac{1}{2} \sum_{i=1}^{r} \sum_{u=1}^{m} \frac{\tilde{p}_{iu}^2}{\phi_i p_u}$$

$$+ \frac{1}{2} \sum_{i=1}^{r} \frac{1}{\phi_i} + \lambda \left(\sum_{u=1}^{m} p_u - 1 \right)$$

$$\frac{\partial \ln L}{\partial \phi_i} = -\frac{m}{2\phi_i} - \frac{1}{2} \sum_{u=1}^{m} \frac{\tilde{p}_{iu}^2}{\phi_i^2 p_u} - \frac{1}{2\phi_i^2}$$

$$\frac{\partial \ln L}{\partial p_u} = -\frac{r}{2p_u} + \frac{1}{2} \sum_{i=1}^{r} \frac{\tilde{p}_{iu}^2}{\phi_i p_u^2} + \lambda$$

$$\frac{\partial \ln L}{\partial \lambda} = \sum_{u=1}^{m} p_u - 1.$$

Setting the derivatives to zero provides equations that need to be solved numerically. One approach would be to iterate

$$\phi_i = \frac{1}{m} \sum_{u=1}^{m} \frac{(\tilde{p}_{iu} - p_u)^2}{p_u} \qquad\qquad 14.$$

$$p_u = \frac{\sum_{i=1}^{r} \left(1 - \dfrac{\tilde{p}_{iu}^2}{\phi_i p_u} \right)}{\sum_{u=1}^{m} \sum_{i=1}^{r} \left(1 - \dfrac{\tilde{p}_{iu}^2}{\phi_i p_u} \right)}.$$

The θ_i's are then recovered from the ϕ_i's.

In the special case of equal ϕ_i's (which implies equal sample sizes as well as equal θ_i's), the log-likelihood becomes

$$\ln L = -\frac{rm}{2} \ln(\phi) - \frac{r}{2} \sum_{u=1}^{m} \ln(p_u) - \frac{1}{2\phi} \sum_{i=1}^{r} \sum_{u=1}^{m} \frac{\tilde{p}_{iu}^2}{p_u} + \frac{r}{2\phi} + \lambda \left(\sum_{u=1}^{m} p_u - 1 \right)$$

This leads to the iterative equations

$$\phi = \frac{1}{rm} \sum_{i=1}^{r} \sum_{u=1}^{m} \frac{(\tilde{p}_{iu} - p_u)^2}{p_u}$$

$$p_u = \frac{\sum_{i=1}^{r} \left(1 - \dfrac{\tilde{p}_{iu}^2}{\phi p_u} \right)}{\sum_{u=1}^{m} \sum_{i=1}^{r} \left(1 - \dfrac{\tilde{p}_{iu}^2}{\phi p_u} \right)}.$$

A comparison with the estimate of θ in Equations 12 and 13 emphasizes that the maximum likelihood estimates of allele frequencies are not the sample allele frequencies (see Appendix), although the two will be equal for large m and r. It appears to be satisfactory in practice (simulation results not shown) to replace p_u in the estimates of ϕ_i and ϕ by the sample average values $\bar{p}_u$ and change the m divisor to $(m-1)$:

$$\hat{\theta}_{iN} = \frac{1}{n-1} \left(\frac{rn}{(r-1)(m-1)} \sum_{u=1}^{m} \frac{(\tilde{p}_{iu} - \bar{p}_u)^2}{\bar{p}_u} - 1 \right). \qquad 15.$$

Averaging the estimates from Equation 15 over samples gives the estimate in Equation 12 and there is a corresponding simplification for large sample sizes n. This approximation requires independent populations.

The advantage of the likelihood approach is that hypotheses about the ϕ_i's can be tested. The hypothesis H_0: $\phi_i = \phi$ can be tested by comparing the likelihoods maximized under no constraint and under the constraint of the hypothesis.

NUMERICAL RESULTS

The moment estimators discussed here were applied to the simple case of three populations having the tree structure shown in Figure 1. Data were simulated assuming a pure drift model, and means and standard deviations of estimates from 1000 replicates are shown in Table 1. The simulation was for a single locus with $m = 5$ alleles, all equally frequent initially. Population $i = 0$, of size 500 alleles, resulted from 5 generations of random mating. Population $i = 3$ was of size 300 alleles, and $t_1 + t_2$ was 20 generations. Population $i = 4$, of size 500 alleles, resulted from $t_2 = 10$ generations of random mating from population $i = 0$.

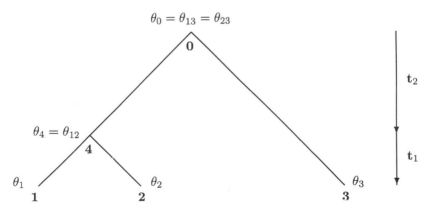

Figure 1   Three-population tree.

TABLE 1   Moment estimates, using Equations 7 and 8, for populations in Figure 1. (Parameter values given in text.)

	Populations						
	1	**2**	**3**	**1&2**	**1&3**	**2&3**	β_W
θ parameter	.210	.053	.076	.032	.010	.010	
β parameter*	.196	.036	.060	.015	−.007	−.007	.097
β estimate	.195	.033	.060	.017	−.008	−.008	.096
SD of estimate	.130	.047	.066	.049	.046	.037	.052

$^{*}\beta = (\theta - \theta_A)/(1 - \theta_A)$

Populations $i = 1$ and $i = 2$, of 50 and 500 alleles, respectively, resulted from $t_1 = 10$ generations of random mating after population $i = 4$. All sample sizes were $n_i = 100$, $i = 1, 2, 3$.

The moment methods were then applied to data made publicly available by the FBI (12). Three samples, each of about 200 people, were collected from the United States and typed at 13 microsatellite markers, the "CODIS" set of loci. Sample properties for these loci are shown in Table 2: the locus name, the number of alleles m_l and the adjusted sample size terms n_{lc} for the lth locus. Estimates of the within-population coancestries θ_i are shown in Table 3, and of the between-population coancestries $\theta_{ii'}$ in Table 4.

TABLE 2   Sample properties of FBI data (12)

			Heterozygosity		
Locus	**No. Alleles**	**Sample size**	**AA**	**CA**	**HI**
D3S135	10	414.6	.763	.795	.719
vWA	10	385.5	.809	.811	.769
FGA	22	385.5	.863	.860	.878
D8S117	13	385.5	.778	.797	.792
D21S11	21	384.8	.861	.853	.811
D18S51	17	385.5	.873	.876	.875
D5S818	10	384.9	.739	.682	.718
D13S31	9	384.8	.688	.771	.827
D7S820	10	414.6	.782	.806	.772
CSF1PO	11	414.6	.781	.734	.707
TPOX	11	414.0	.763	.621	.607
THO1	8	414.6	.727	.783	.757
D16S53	8	412.6	.798	.767	.771

AA: African American, CA: Caucasian, HI: Hispanic.

TABLE 3   Single-population estimates, from Equation 8, for FBI data (12)

Locus	β_i AA	β_i CA	β_i HI	β_i Average	$\hat{\beta}_W$
D3S135	.010	−.030	.069	.017	.019
vWA	.000	−.002	.050	.017	.019
FGA	.007	.012	−.008	.003	.006
D8S117	.026	.003	.009	.012	.015
D21S11	−.012	−.003	.047	.012	.014
D18S51	.011	.008	.010	.010	.012
D5S818	−.018	.061	.012	.019	.021
D13S31	.132	.028	−.042	.036	.040
D7S820	.014	−.016	.026	.008	.011
CSF1PO	−.048	.015	.051	.006	.008
TPOX	−.118	.090	.112	.027	.030
THO1	.078	.008	.041	.043	.045
D16S53	−.011	.028	.024	.014	.016
All loci	.010	.017	.032	.020	.020

AA: African American, CA: Caucasian, HI: Hispanic.

The development based on normal theory shown above suggests that sample variances decrease with the number of alleles per locus, the number of loci, and the number of samples. The simulation results shown in Table 1 show rather large standard deviations for the case of only three samples, and this may account for the very large variation among loci for the results in Tables 3 and 4. Of course it may also be that the different loci are not providing replicates of the same evolutionary history. Loci may have been subjected to different selection pressures, for example, and variation among θ values has been suggested as a means of detecting selection, as recently reviewed by Vitalis et al. (91) and applied by Marshall & Ritland (55). If loci can be regarded as providing replication of the same process, however, then averaging over loci is appropriate. The variation among loci is much reduced when the three population-specific estimates are averaged, or when only a common value is estimated.

DISCUSSION

This review has extended Weir & Cockerham (100) in two directions. Most significantly, it has allowed the separate estimation of population- and population-pair specific values of θ. Previously it was assumed that populations were independent

TABLE 4 Two-population estimates, from Equation 7, for FBI data (12)

Locus	$\hat{\beta}_{ii'}$			$\hat{\beta}_W$		
	AA&CA	AA&HI	CA&HI	AA&CA	AA&HI	CA&HI
D3S135	−.018	.026	−.009	.010	.016	.030
vWA	−.018	.006	.010	.019	.021	.017
FGA	.006	−.002	−.004	.006	.004	.008
D8S117	−.012	.002	.009	.029	.018	.000
D21S11	−.008	−.008	.015	.003	.029	.010
D18S51	−.004	−.008	.011	.016	.021	.001
D5S818	.003	−.039	.033	.021	.037	.006
D13S31	.058	−.021	−.032	.026	.067	.026
D7S820	.001	.004	−.006	.000	.019	.013
CSF1PO	−.024	−.009	.034	.010	.012	.002
TPOX	−.043	−.053	.097	.030	.049	.007
THO1	−.034	.028	.006	.077	.035	.021
D16S53	−.009	−.009	.018	.020	.018	.011
Total	−.008	−.006	.014	.021	.023	.020

AA: African American, CA: Caucasian, HI: Hispanic.

and that either each population had the same value of θ or a population-average value was being estimated. The other extension has been the adoption of multivariate normal methods as an alternative to the method of moments. There may be an increase in computational burden and increase in bias with these methods, but there is the gain of a distributional form for the estimates.

Natural populations of the same species are unlikely to have the same value of θ, if only because they have different sizes. Although the reconstruction of intra-specific trees can proceed satisfactorily on the basis of the usual estimates of average θ values (65, 98), there are occasions when population-specific values are needed. There is the immediate issue of degrees of freedom. For r populations, there are r within-population values and $r(r − 1)/2$ between-population values to be estimated. As there are $m − 1$ independent allele frequencies for a locus with m alleles, there are only $r(m − 1)$ independent observations in all, so only loci with large numbers of alleles can be used. With L loci, there is an increase in the number of observed allele frequencies to $Lr(m − 1)$ and an increase to $r(r + 1)/2 + L(m − 1)$ parameters, so that even diallelic SNPs can be used. The constraints are less severe if the between-population coefficients $\theta_{ii'}$ are ignored, but it needs to be recognized that the estimates are then actually for a combination of within- and between-population values.

Under a pure drift model, values of θ are simple functions of population size and time. For a pair of populations, the values of θ within each can be expressed

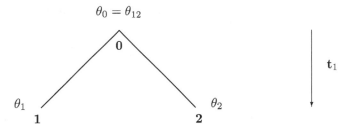

Figure 2   Two populations.

in terms of θ for their most recent common ancestral population. For the situation in Figure 2:

$$\theta_i = 1 - (1 - \theta_{12})X_i^{t_1}, \quad i = 1, 2,$$

where $X_i = (2N_i - 1)/2N_i$ and N_i is the constant population size for populations $i = 1, 2$. Therefore,

$$\beta_i = \frac{\theta_i - \theta_{12}}{1 - \theta_{12}} = 1 - X_i^{t_1} \approx \frac{t_1}{2N_i}.$$

The β parameters estimated by Equation 8 for a pair of populations are therefore furnishing estimates of the time since those populations diverged from an ancestral population. Although the two times must be the same, the pure drift model shows that the estimates will be different when the two population sizes are different. The estimate of Weir & Cockerham (100) is for

$$\beta_W = \frac{\theta_W - \theta_{12}}{1 - \theta_{12}} = 1 - \frac{X_1^{t_1} + X_2^{t_1}}{2}$$

$$\approx \frac{1}{2}\left(\frac{1}{2N_1} + \frac{1}{2N_2}\right)t_1 = \frac{t_1}{2N_h},$$

where N_h is the harmonic mean of the two population sizes. The quantity β_W is proportional to the divergence time t_1 (65).

If populations $i = 1, 2, 3, 4$ in Figure 1 have sizes N_i, and if $X_i = (2N_i - 1)/2N_i$:

$$\theta_{12} = 1 - (1 - \theta_0)X_4^{t_2}$$

$$\theta_i = 1 - (1 - \theta_{12})X_i^{t_1} = 1 - (1 - \theta_0)X_i^{t_1}X_4^{t_2}, \quad i = 1, 2$$

$$\theta_3 = 1 - (1 - \theta_0)X_3^{t_1 + t_2}$$

$$\theta_{13} = \theta_{23} = \theta_0.$$

The β parameters being estimated from the three extant populations 1, 2 and 3 involve the average between-population quantity $\theta_A = (\theta_{12} + 2\theta_0)/3$ although this cancels out of the expressions needed to estimate the times:

$$\frac{\beta_i - \beta_{12}}{1 - \beta_{12}} = 1 - X_i^{t_1} \approx \frac{t_1}{2N_i}, \quad i = 1, 2$$

$$\frac{\beta_i - \beta_{i3}}{1 - \beta_{i3}} = 1 - X_3^{t_1+t_2} \approx \frac{t_1 + t_2}{2N}, \quad i = 1, 2.$$

The θ's of interest can be expressed in terms of the estimable β's:

$$\frac{\theta_i - \theta_{12}}{1 - \theta_{12}} = \frac{\beta_i - \beta_{12}}{1 - \beta_{12}}, \quad i = 1, 2.$$

If θ_0 is assumed to be zero, the outgroup population 3 allows estimation of all three measures θ_1, θ_2 and θ_{12} for populations 1 and 2 since then $\beta_{12} = 2\theta_{12}/(3 - \theta_{12})$ and $\beta_i = (3\theta_i - \theta_{12})/(3 - \theta_{12}), i = 1, 2$.

Moment estimates of the θ's involve only the second moments of sample allele frequencies, whereas likelihood or Bayesian methods use the whole distribution. Higher-order moments can be expressed in terms of analogs of θ (96). Ignoring sample-size terms

$$\mathcal{E}(\tilde{p}_{iu} - p_u)^2 = p_u(1 - p_u)\theta$$

$$\mathcal{E}(\tilde{p}_{iu} - p_u)^3 = p_u(1 - p_u)(1 - 2p_u)\gamma$$

$$\mathcal{E}(\tilde{p}_{iu} - p_u)^4 = p_u(1 - p_u)(1 - 2p_u)(1 - 3p_u)\delta + 3p_u^2(1 - p_u)^2\Delta.$$

The normal distribution assumption implies that $\gamma = \delta = 0$, $\Delta = \theta^2$, or that there are no dependencies among a set of four alleles in addition to those between any pair of them. Assuming that allele frequencies have a Dirichlet distribution over populations, or that p_{iu} has a Beta distribution with parameters $(1 - \theta)p_u/\theta$ and $(1-\theta)(1-p_u)/\theta$ (4) implies that $\gamma = 2\theta^2/(1+\theta)$, $\delta = 6\theta^3/[(1+\theta)(1+2\theta)]$, $\Delta =$ (99). These relations hold for the infinite-allele mutation model, but not for the stepwise mutation model (35).

ACKNOWLEDGMENTS

This work was supported in part by NIH Grant GM 45344. Very helpful discussions were held with Peter Donnelly and George Nicholson, and the review was completed while the first author enjoyed the hospitality of both the Mathematical Genetics group in the Department of Statistics and the Wellcome Trust Centre for Human Genetics at the University of Oxford.

APPENDIX

The failure of the maximum likelihood estimate of mean allele frequencies to equal their observed values reflects, in part, the approximation of a multinomial distribution by a multivariate normal. In the general setting of a population with proportions Q_u in the uth of m categories, the probability of category counts n_u in a sample of size $n = \sum_{u=1}^{m} n_u$ is

$$\Pr(\{n_u\}) = \frac{n!}{\prod_{u=1}^{m} n_u!} \prod_{u=1}^{m} (Q_u)^{n_u},$$

and the means, variances, and covariances of the counts are

$$\mathcal{E}(n_u) = nQ_u$$

$$\mathrm{Var}(n_u) = nQ_u(1 - Q_u)$$

$$\mathrm{Cov}(n_u, n_{u'}) = -nQ_uQ_{u'}, \quad u \neq u'.$$

The log-likelihood for the category probabilities is

$$\ln(L(\{Q_u\})) = \sum_{u=1}^{m} n_u \ln(Q_u).$$

To accommodate the dependency caused by $\sum_{u=1}^{m} Q_u = 1$, the Lagrange multiplier term $\lambda(1 - \sum_{u=1}^{m} Q_u)$ is added to the log-likelihood. Differentiating with respect to Q_u gives

$$\frac{\partial \ln(L)}{\partial Q_u} = \frac{n_u}{Q_u} - \lambda,$$

which leads to the maximum likelihood estimates (MLEs) $\hat{Q}_u = \tilde{Q}_u$ where $\tilde{Q}_u = n_u/n$.

For large sample sizes, the multivariate normal distribution provides a good approximation to the multinomial. The appropriate normal distribution for category counts will have variance matrix $n\mathbf{V}$ where $\mathbf{V}$ has uth diagonal element $Q_u(1 - Q_u)$ and off-diagonal elements $- Q_uQ_{u'}, u \neq u'$. Omitting the mth row and column removes the singularity of this matrix. The mean vector is then $n\mathbf{Q} = n[Q_1, Q_2, \ldots, Q_{m-1}]'$. The determinant of the reduced matrix is $\prod_{u=1}^{m} Q_u$ and its inverse has uth diagonal element $[1/(Q_u) + 1/(Q_m)]$ and all off-diagonal elements equal to $1/(Q_m)$. These results lead to the log-likelihood

$$\ln(L) = -\frac{1}{2} \ln\left(\prod_{u=1}^{m} Q_u \right) - \frac{1}{2} \sum_{u=1}^{m} \frac{(n_u - nQ_u)^2}{nQ_u} - \lambda\left(1 - \sum_{u=1}^{m} Q_u \right),$$

where the Lagrange multiplier λ allows all m unknowns Q_u to be included. Setting the derivative with respect to each Q_u equal to zero gives

$$\frac{1}{n}\left(\lambda - \frac{1}{2Q_u} \right) + \left(\frac{\tilde{Q}_u - Q_u}{Q_u} + \frac{(\tilde{Q}_u - Q_u)^2}{2Q_u^2} \right) = 0.$$

Only for large n will these equations be satisfied by $Q_u = \tilde{Q}_u$, so that $\hat{Q}_u = \tilde{Q}_u$ are approximations to the MLEs in the normal approximation formulation. In general, however, the MLEs are not simply the observed values.

The *Annual Review of Genetics* is online at http://genet.annualreviews.org

LITERATURE CITED

1. Anaya JM, Correa PA, Mantilla RD. 2002. Rheumatoid arthritis association in Colombian population is restricted to HLA-DRB1*04 QRRAA alleles. *Genes Immun.* 3:56–58

2. Balding DJ, Bishop M, Cannings C, eds. 2001. *Handbook of Statistical Genetics.* New York: Wiley. 890 pp.

3. Balding DJ, Nichols RA. 1994. DNA profile match probability calculations: how to allow for population stratification, relatedness, database selection and single bands. *Forensic Sci. Int.* 64:125–40

4. Balding DJ, Nichols RA. 1995. A method for characterizing differentiation between populations at multi-allelic loci and its implications for establishing identity and paternity. *Genetica* 96:3–12

5. Balloux F, Goudet J. 2002. Statistical properties of population differentiation estimators under stepwise mutation in a finite island model. *Mol. Ecol.* 11:771–83

6. Balloux F, Lugon-Moulin N. 2002. The estimation of population differentiation with microsatellite markers. *Mol. Ecol.* 11:155–65

7. Banks SC, Skerratt LF, Taylor AC. 2002. Female dispersal and relatedness structure in common wombats (*Vombatus ursinis*). *J. Zool.* 256:389–99

8. Basset P, Balloux F, Perrin N. 2001. Testing demographic models of effective population size. *Proc. R. Soc. London Ser. B* 268:311–17

9. Blundell GM, Ben-David M, Groves P, Bowyers RT, Geffen E. 2002. Characteristics of sex-biased dispersal and gene flow in coastal river otters: implications for natural recolonization of extirpated populations. *Mol. Ecol.* 11:289–303

10. Brenner CH. 1998. Difficulties in the estimation of ethnic affiliation. *Am. J. Hum. Genet.* 62:1559–60

11. Broughton RE, Stewart LB, Gold JR. 2002. Microsatellite variation suggests substantial gene flow between king mackerel (*Scomberomorus cavalla*) in the western Atlantic Ocean and Gulf of Mexico. *Fish. Res.* 54:305–16

12. Budowle B, Moretti T. 1999. Genotype profiles for six population groups at the 13 CODIS short random repeat core loci and other PCR-based loci. *Forensic Sci. Comm.* http://www.fbi.gov/hq/lab/fsc/backissu/july1999/budowle.htm

13. Charbonnel N, Angers B, Rastavonjizay R, Bremond P, Jarne P. 2002. Evolutionary aspects of the metapopulation dynamics of *Biomphataria pfeifferi*, the intermediate host of *Schistosoma mansoni*. *J. Evol. Biol.* 15:248–61

14. Chiappero MB, Sabattini MS, Blanco A, Calderon GE, Gardenal CN. 2002. Gene flow among *Calomys musculinis* (Rodentia, Muridae) populations in Argentina. *Genetica* 114:63–72

15. Cockerham CC, Weir BS. 1987. Correlations, descent measures: drift with migration and mutation. *Proc. Natl. Acad. Sci. USA* 84:8512–14

16. Cockerham CC, Weir BS. 1993. Estimation of gene flow from *F*-statistics. *Evolution* 47:855–63

17. Cornuet J-M, Piry S, Luikart G, Estoup A, Solignac M. 1999. New methods employing multilocus genotypes to select or exclude populations as origins of individuals. *Genetics* 153:1989–2000

18. Davies N, Bermingham E. 2002. The historical biogeography of two Caribbean

butterflies (*Lepidoptera: Heliconiidae*) as inferred from genetic variation at multiple loci. *Evolution* 56:573–89

19. Dallimer M, Blackburn C, Jones PJ, Pemberton JM. 2002. Genetic evidence for male biased dispersal in the red-billed quelea *Quelea quelea. Mol. Ecol.* 3:529–33

20. Dawson KJ, Belkhir K. 2001. A Bayesian approach to the identification of panmictic populations and the assignment of individuals. *Genet. Res.* 78:59–77

21. Dodds KG. 1986. *Resampling methods in genetics and the effect of family structure in genetic data.* PhD thesis, North Carolina State Univ. 110 pp.

22. Dufresne F, Bourget E, Bernatchez L. 2002. Differential patterns of spatial divergence in microsatellite and allozyme alleles: further evidence for locus-specific selection in the acorn barnacle, *Semibalanus balanoides? Mol. Ecol.* 11:113–23

23. Endsley MA, Baker MD, Krafsur ES. 2002. Microsatellite loci in the house fly *Musca domestica* L (Diptera: Muscidae). *Mol. Ecol. Notes* 2:72–74

24. Ernest HB, Rubin ES, Boyce WM. 2002. Fecal DNA analysis and risk assessment of mountain lion predation of bighorn sheep. *J. Wildlife Manage.* 66:75–85

25. Evett IW, Pinchin R, Buffery C. 1992. An investigation of the feasibility of inferring ethnic origin from DNA profiles. *J. Forensic Sci.* 32:301–6

26. Evett IW, Weir BS. 1998. *Interpreting DNA Evidence.* Sunderland, MA: Sinauer. 285 pp.

27. Excoffier L. 2001. Analysis of population subdivision. See Ref. 2, pp. 271–307

28. Fievet E, Eppe R. 2002. Genetic differentiation among populations of the amphidromous shrimp *Atya innocous* (HERBST) and obstacles to their upstream migration. *Arch. Hydrobiol.* 153:287–300

29. Fisher RA. 1921. On the "probable error" of a coefficient of correlation deduced from a small sample. *Metron* 1:3

30. Foreman LA, Lambert JA. 2000. Genetic differentiation within and between for UK ethnic groups. *Forensic Sci. Int.* 114:7–20

31. Foulley JL, Hill WG. 1999. On the precision of estimation of genetic distance. *Genet. Sel. Evol.* 31:457–64

32. Frydenberg J, Pertoldi C, Dahlgaard J. 2002. Genetic variation in original and colonizing *Drosophila buzzatii* populations analyzed by microsatellite loci isolated with a new PCR screening method. *Mol. Ecol.* 11:181–90

33. Gold JR, Turner TF. 2002. Population structure of red drum (*Sciaenops ocellatus*) in the northern Gulf of Mexico, as inferred from variation in nuclear encoded microsatellites. *Mar. Biol.* 140:249–65

34. Gomez A, Adcock GJ, Lunt DH, Carvalho GR. 2002. The interplay between colonization history and gene flow in passively dispersing zooplankton: microsatellite analysis of rotifer resting egg banks. *J. Evol. Biol.* 15:158–71

35. Graham J, Curran J, Weir BS. 2000. Conditional genotypic probabilities for microsatellite loci. *Genetics* 155:1973–80

36. Grimaldi MC, Crouau, Roy B, Contu L, Amoros JP. 2002. Molecular variation of HLA class I genes in the Corsican population: approach to its origin. *Eur. J. Immun.* 29:101–7

37. Hawkins SL, Varnavskya NV, Matzak EA, Efremov VV, Guthrie CM III, et al. 2002. Population structure of odd-broodline Asian pink salmon and its contrast to even–broodline structure. *J. Fish. Biol.* 60:370–88

38. Heath DD, Busch C, Kelly J, Atagi DY. 2002. Temporal change in genetic structure and effective population size in steelhead trout (*Oncorhynchus mykiss*). *Mol. Ecol.* 11:197–214

39. Holsinger KE. 1999. Analysis of genetic diversity in geographically structured populations: a Bayesian perspective. *Hereditas* 130:245–55

40. Holsinger KE, Lewis PO, Dey DK. 2002. A Bayesian approach to inferring

population structure from dominant markers. *Mol. Ecol.* 11:1157–64

41. Hundertmark KJ, Shields GF, Udina IG. 2002. Mitochondrial phylogeography of moose (*Alces alces*): Late Pleistocene divergence and population expansion. *Mol. Phylogenet. Evol.* 22:375–87

42. Ivandic V, Hackett CA, Nevo E, Keith R, Thomas WTB, Forster BP. 2002. Analysis of simple sequence repeats (SSRs) in wild barley from the Fertile Crescent: associations with ecology, geography and flowering time. *Plant Mol. Biol.* 48:511–27

43. Jiang C. 1987. *Estimation of F-statistics in subdivided populations.* PhD thesis, North Carolina State Univ. 95 pp.

44. Jonsdottir ODB, Imsland AK, Danielsdottir AK, Marteinsdttir G. 2002. Genetic heterogeneity and growth properties of different genotypes of Atlantic cod (*Gadus morhua L.*) at two spawning sites off south Iceland. *Fish. Res.* 55:37–47

45. Kim KS, Choi CB. 2002. Genetic structure of Korean native pig using microsatellite markers. *Korean J. Genet.* 24:1–7

46. Kinnison MT, Bentzen B, Unwin MJ, Quinn TP. 2002. Reconstructing recent divergence: evaluating nonequilibrium population structure in New Zealand chinook salmon. *Mol. Ecol.* 11:739–54

47. Krafsur ES. 2002. Population structure of the tsetse fly *Gossina pallidipes* estimated by allozyme, microsatellite and mitochondrial gene diversities. *Insect Mol. Biol.* 11:37–45

48. Laikre L, Jarvi T, Johansson L, Palm S, Rubin JF, et al. 2002. Spatial and temporal population structure of sea trout at the Island of Gotland, Sweden, delineated from mitochondrial DNA. *J. Fish. Biol.* 60:49–71

49. Lange K. 1995. Applications of the Dirichlet distribution to forensic match probabilities. *Genetica* 96:107–17

50. Li Y-J. 1996. *Characterizing the structure of genetics populations.* PhD thesis, North Carolina State Univ. 106 pp.

51. Long J. 1986. The allelic correlation structure of Gainj- and Kalam-speaking people I. The estimation and interpretation of Wright's F-statistics. *Genetics* 112:629–47

52. Lowe AL, Urquhart A, Foreman LA, Evett IW. 2001. Inferring ethnic origin by means of an STR profile. *Forensic Sci. Int.* 119:17–22

53. Manzano C, de la Rua C, Iriondo M, Mazn LI, Vicario A, Aguirre A. 2002. Structuring the genetic heterogeneity of the Basque population: a view from classical polymorphisms. *Hum. Biol.* 74:51–74

54. Margis R, Felix D, Caldas JF, Salgueiro F, de Araujo DSD, et al. 2002. Genetic differentiation among three neighboring Brazil-cherry (*Eugenia uniflora L.*) populations within the Brazilian Atlantic rain forest. *Biodivers. Conserv.* 11:149–63

55. Marshall HD, Ritland K. 2002. Genetic diversity and differentiation of Kermode bear populations. *Mol. Ecol.* 11:685–97

56. Merilä J, Crnokrak P. 2001. Comparison of genetic differentiation at marker loci and quantitative traits. *J. Evol. Biol.* 14:892–903

57. Michalakis Y, Excoffier L. 1996. A generic estimation of population subdivision using distances between alleles with special reference for microsatellite loci. *Genetics* 142:1061–64

58. Monaghan MT, Spaak P, Robinson CT, Ward JV. 2002. Population genetic structure of 3 alpine stream insects: influences of gene flow, demographics, and habitat fragmentation. *J. N. Am. Benthol. Soc.* 21:114–31

59. Morand ME, Brachet S, Rossignol P, Dufour J, Frascaria-Lacoste N. 2002. A generalized heterozygote deficiency assessed with microsatellites in French common ash populations. *Mol. Ecol.* 11:377–85

60. Nicholson G, Smith AV, Jónsson F, Gústafsson Ó, Stefánsson K, Donnelly P.

2002. Assessing population differentiation and isolation from single nucleotide polymorphism data. *Proc. R. Stat. Soc.* In press

61. Nielsen R, Wakeley J. 2001. Distinguishing migration from isolation: a Markov chain Monte Carlo approach. *Genetics* 158:885–96

62. Raufaste N, Bonhomme F. 2000. Properties of bias and variance of two multiallelic estimators of F_{ST}. *Theoret. Pop. Biol.* 57:285–96

63. Raymond M, Rousset F. 1995. An exact test for population differentiation. *Evolution* 49:1280–83

64. Reusch TBH. 2002. Microsatellites reveal high population connectivity in eelgrass (*Zostera marina*) in two contrasting coastal areas. *Limnol. Oceanogr.* 47:78–85

65. Reynolds J, Weir BS, Cockerham CC. 1983. Estimation of the coancestry coefficient: basis for a short-term genetic distance. *Genetics* 105:767–79

66. Ribeiro MM, LeProvost G, Gerber S. 2002. Origin identification of maritime pine stands in France using chloroplast simple-sequence repeats. *Ann. For. Sci.* 59:53–62

67. Rios C, Sanz S, Saavedra C, Pea JB. 2002. Allozyme variation in populations of scallops, *Pecten jacobaeus* (L.) and *P. maximus* (L.) (Bivalvia: Pectinidae), across the Almeria-Oran front. *J. Exp. Mar. Biol. Ecol.* 267:223–44

68. Richardson BA, Brunsfeld J, Klopfenstein NB. 2002. DNA from bird-dispersed seed and wind-disseminated pollen provides insights into postglacial colonization and population genetic structure of whitebark pine (*Pinus albicaulis*). *Mol. Ecol.* 11:215–27

69. Robertson AR. 1962. Weighting in the estimation of variance components in the unbalanced single classification. *Biometrics* 18:413–17

70. Robertson A, Hill WG. 1984. Deviations from Hardy-Weinberg proportions: sampling variances and use in estimation of inbreeding coefficients. *Genetics* 107:703–18

71. Roeder K, Escobar M, Kadane J, Balasz I. 1998. Measuring heterogeneity in forensic databases using hiererachical Bayes models. *Biometika* 85:269–87

72. Roques S, Sevigny JM, Bernatchez L. 2002. Genetic structure of deep-water redfish, *Sebastes mentella*, populations across the North Atlantic. *Mar. Biol.* 140:297–307

73. Rothman ED, Sing CF, Templeton AR. 1974. A model for analysis of population structure. *Genetics* 76:943–60

74. Rousset F. 2001. Inferences from spatial population genetics. See Ref. 2, pp. 239–69

75. Rousset F. 2002. Inbreeding and relatedness coefficients: What do they measure? *Heredity* 88:371–80

76. Schultheis AS, Hendricks AC, Weigt LA. 2002. Genetic evidence for 'leaky' cohorts in the semivoltine stonefly *Peltoperla tarteri* (Plecoptera: Peltoperlidae). *Freshwater Biol.* 47:367–76

77. Shriver MD, Smith MW, Li Jin. 1998. Reply to Brenner. *Am. J. Hum. Genet.* 62:1560–61

78. Shriver MD, Smith MW, Li Jin, Marcini A, Akey JM, et al. 1997. Ethnic-affiliation estimation by use of population-specific DNA markers. *Am. J. Hum. Genet.* 60:957–64

79. Slatkin M. 1995. A measure of population subdivision based on microsatellite frequencies. *Genetics* 139:457–62

80. Smouse PE, Spielman RS. 1977. How allocation of individuals depends on genetic differences among populations. In *Human Genetics*, ed. S Armendares, R. Lisker, pp. 255–60. Amsterdam: Excerpta Medica

81. Smouse PE, Williams RC. 1982. Multivariate analysis of HLA-disease associations. *Biometrics* 38:757–68

82. Spielman RS, Smouse PE. 1975. Multivariate classification of human populations. I. Allocation of Yanomama Indians to villages. *Am. J. Hum. Genet.* 28:317–31

83. Spitze K. 1993. Population structure in *Daphnia abtusa*: quantitative genetic and allozymic variation. *Genetics* 135:367–74

84. Squire T, Newman RA. 2002. Fine-scale population structure in the wood frog (*Rana sylvatica*) in a northern woodland. *Herpetologica* 58:119–30

85. Travis SE, Proffitt CE, Lowenfield RC, Mitchell TW. 2002. A comparative assessment of genetic diversity among differently-aged populations of *Spartina alterniflora* on restored versus natural wetlands. *Restor. Ecol.* 10:37–42

86. Tsigenopoulos CS, Kotlik P, Berrebi P. 2002. Biogeography and pattern of gene flow among Barbus species (*Teleostei: Cyprinidae*) inhabiting the Italian Peninsula and neighbouring Adriatic drainages as revealed by allozyme and mitochondrial sequence data. *Biol. J. Linn. Soc.* 75:83–99

87. Tufto J, Engen S, Hindar K. 1996. Inferring patterns of migration from gene frequencies under equilibrium conditions. *Genetics* 144:1911–21

88. Van der Strate HJ, Van de Zande L, Stam WT. 2002. The contribution of haploids, diploids and clones to fine-scale population structure in the seaweed *Cladophoropsis membranacea* (Chlorophyta). *Mol. Ecol.* 11:329–45

89. Viard F, Bernard J, Despalnque B. 2002. Crop-weed interactions in the *Beta vulgaris* complex at a local scale: allelic diversity and gene flow within sugar beet fields. *Theor. Appl. Genet.* 104:688–97

90. Vitalis R, Couvet D. 2001. Estimation of effective population size and migration rate from one- and two-locus identity measures. *Genetics* 157:911–25

91. Vitalis R, Dawson K, Boursot P. 2001. Interpretation of variation across marker loci as evidence of selection. *Genetics* 158:1811–23

92. von der Schulenburg JHJ, Hurst GDD, Tetzlaff D, Booth GE, Zakharov IA, Majerus MEN. 2002. History of infection with different male-killing bacteria in the two-spot ladybird beetle *Atalia bipunctata* revealed through mitochondrial DNA sequence analysis. *Genetics* 160:1075–86

93. Wakeley J. 2001. The coalescent in an island model of population subdivision with variation among demes. *Theor. Pop. Biol.* 59:133–44

94. Was A, Wenne R. 2002. Genetic differentiation in hatchery and wild sea trout (*Salmo trutta*) in the Southern Baltic at microsatellite loci. *Aquaculture* 204:493–506

95. Weicker JJ, Brumfield RT, Winker K. 2001. Estimating the unbiased estimator θ for population genetic survey data. *Evolution* 55:2601–5

96. Weir BS. 1994. The effects of inbreeding on forensic calculations. *Annu. Rev. Genet.* 28:597–621

97. Weir BS. 1996. *Genetic Data Analysis II.* Sunderland, MA: Sinauer 376 pp.

98. Weir BS. 2000. What is the structure of human populations? *Evol. Biol.* 32:195–202

99. Weir BS. 2001. Forensics. See Ref. 2, pp. 721–39

100. Weir BS, Cockerham CC. 1984. Estimating F-statistics for the analysis of population structure. *Evolution* 38:1358–70

101. Whitlock MC, McCauley DE. 1999. Indirect measures of gene flow and migration: $F_{ST} \neq 1/(4Nm + 1)$. *Heredity* 82:1385–70

102. Wolańska-Nowak P. 2000. Application of subpopulation theory to evaluation of DNA evidence. *Forensic Sci. Int.* 113:63–69

103. Wondji C, Simard F, Fontenille D. Evidence for genetic differentiation between the molecular forms M and S within the Forest chromosomal form of *Anopheles gambiae* in an area of sympatry. *Insect Mol. Biol.* 11:11–19

104. Wright S. 1951. The genetical structure of populations. *Ann. Eugen.* 15:323–54

105. Yang R-C. 1998. Estimating hierarchical F-statistics. *Evolution* 52:950–56

Subject Index

A

abdominal genes
transvection effects in
Drosophila and, 530–35
Acanthamoeba spp.
Cryptococcus neoformans
genetics and, 602
Drosophila actin ovarian
cytoskeleton and, 460
Acentrosomal spindle
meiotic recombination and
chromosome regulation in
Drosophila females,
221–23
achaete-scute genes
sensory
mechanotransduction and,
426
Achiasmate segregation
meiotic recombination and
chromosome regulation in
Drosophila females, 210
Acquired immuno deficiency
syndrome (AIDS)
Cryptococcus neoformans
genetics and, 602–3
β-Actin
conditional gene activation
in eukaryotes and, 161
Actin-binding proteins
Drosophila oogenesis and,
455–80
Activation
allosteric cascade of
spliceosome activation,
333–44
conditional gene activation
in eukaryotes and, 153–68
DNA checkpoints and
genome stability, 624–25
primordial genetics and

phenotype of ribocyte,
135–36
Ac transposon
chromosome
rearrangements and
transposable elements,
399
ada mutant
angiosperm female
gametophyte development
and function, 106, 109
Adaptation
DNA checkpoints and
genome stability, 640–41
ADE2 gene
Cryptococcus neoformans
genetics and, 561,
578–80, 597
Adenylyl cyclases
learning, memory, and
synaptic plasticity, 699
AGL genes
angiosperm female
gametophyte development
and function, 117
ald mutant
meiotic recombination and
chromosome regulation in
Drosophila females, 209
Aligned sequences
recombination in
evolutionary genomics
and, 75–90
Allelic frequency variation
F-statistics and, 721–46
alligator mutant
sensory
mechanotransduction and,
432–33
Allium spp.
meiotic recombination and

chromosome regulation in
Drosophila females, 211
Allocation of individuals to
populations
F-statistics and, 724
Alternative reproductive
tactics
genetic mating systems and
reproductive natural
histories of fishes, 19, 21,
23, 26–27
Alvinella pompejana
spirochete chemotaxis and
motility, 50
Amiloride
sensory
mechanotransduction and,
417
Amino acid activation
primordial genetics and
phenotype of ribocyte,
135–36
Aminoacyl-tRNA synthesis
primordial genetics and
phenotype of ribocyte,
136–38
Ancestral recombination
graph
recombination in
evolutionary genomics
and, 82
Anemone nemorosa
chromosome
rearrangements and
transposable elements,
395
Angiosperm female
gametophyte
development and function
Arabidopsis, 105–9
expression-based

751

CUMULATIVE INDEXES

CONTRIBUTING AUTHORS, VOLUMES 32–36

CHAPTER TITLES, VOLUMES 32–36

Viral Genetics